Make: Rockets

Down-to-Earth Rocket Science

Mike Westerfield

Make: Rockets

Down-to-Earth Rocket Science

by Mike Westerfield

Printed in Canada.

Published by Maker Media, Inc., 1005 Gravenstein Highway North, Sebastopol, CA 95472.

Maker Media books may be purchased for educational, business, or sales promotional use. Online editions are also available for most titles (*http://safaribooksonline.com*). For more information, contact O'Reilly Media's corporate/institutional sales department: 800-998-9938 or *corporate@oreilly.com*.

Editor: Patrick Di Justo
Production Editor: Kristen Brown
Copyeditor: Rachel Head
Proofreader: Charles Roumeliotis

Indexer: Judith McConville
Cover Designers: Juliann Brown and Brian Jepson
Interior Designer: Nellie McKesson
Illustrator: Mike Westerfield

September 2014: First Edition

Revision History for the First Edition:

2014-08-19: First release

See *http://oreilly.com/catalog/errata.csp?isbn=9781457182921* for release details.

ISBN: 978-1-457-18292-1

[TI]

Table of Contents

Preface

Model rocketry has had a profound impact on my life, and I hope it does on yours, too. If you are associated with education, you've probably heard the term *STEM* tossed around. It stands for Science, Technology, Engineering, and Math—all those skills we as a nation are trying to promote. Model rocketry is the perfect STEM hobby. Let me share how it changed me.

While I have always been bright, some might argue precocious, I was a more or less average student through most of my early education. I did well enough in science and math to be placed in the advanced classes, but generally didn't work hard enough to excel. The classes simply didn't interest me much. Who cared that a screw was an example of a wedge, one of the basic tools? Seriously? That was science? Why would I bother?

Around the time I turned 14, I started building and flying model rockets. I read everything I could find on how they flew and how to make them fly higher and faster. As a sophomore in high school, I was cruising along with a 2.3 or 2.4 grade point average. I got interested in what I thought was a simple problem: how high would my rockets go?

That summer I took an algebra book home. I came back the next fall and tested out of second-year algebra. I enrolled in physics, devouring the book, working every problem, not just the ones that were assigned. My GPA that year was 3.7. It was 4.0 my senior year. I won an appointment to the United States Air Force Academy, where I majored in physics. Upon graduation, my first assignment was as the resident physicist for a classified satellite program. I was, quite literally, a rocket scientist. Four years later I earned an MS in physics, and finally felt I had answered that problem I started on when I was a sophomore in high school. I could determine how high my rockets would go to my satisfaction.

Now I'm not saying model rockets will turn every underachieving sophomore into a rocket scientist. Even if it would, we don't need that many rocket scientists! I introduced both of my daughters to model rocketry, along with about a dozen fifth-grade classes over the years. Neither of my daughters is a rocket scientist. One is a special education teacher; the other is a research scientist working on cancer vaccines. My granddaughter is seven; she has been flying rockets on and off since she was two, and could push the launch button. She may not end up being a rocket scientist, either, but she will still see a little more about what all those math and science classes can be used for. At the very least, she still likes science, and comes over to play with the toys regularly. Her current specialty is polymers—she loves learning how they mix and react, especially the ones that change state or swell with water.

Taking a look at my life and those of many of my friends, we got different things from rocketry at different times in our lives. This book is written to let you get different things out of it, too, whether you are reading it for yourself, as a teacher or parent helping out an interested child, or as an adult taking up a new hobby.

For younger children, say ages 8 to 15, most of the math and a lot of the science in this book will be too advanced. Skip it. You can build and fly all of the rockets and launchers without digging into any of the math, and with little or no understanding of the science. After all, rockets are fun to build and fly. This book will show you how to build solid propellant rockets, water rockets, and air rockets. You'll see how to build launchers for each type of rocket and how to track them to see how high they go. You'll learn to build high-performance rockets that can hit 500 mph and fly a half mile into the sky, and experience the thrill of seeing a rocket glider roar into the air and glide safely back to earth.

As children become young adults, say ages 14 to 18, they rapidly pick up new skills in math and science. While some of the math may still be a bit too advanced, most isn't. Very simple trigonometry shows how to track the altitude of a rocket. Basic principles of physics and aeronautical engineering explain how rockets fly and what makes them stable—or unstable! Simple computer programs or free rocket simulators let budding scientists safely design and build their own rockets, showing which designs will fly well. They also do a great job of telling you how high a rocket will go. Most of the math, most of the science, and all of the computer programs are well within reach for high school students. This is the stage when browsing through the book will either answer your questions or give you a good start on finding the answers.

Model rocketry is a hobby for a lot of adults, too. As you make your way into college and take ever more advanced classes in calculus, physics, electrical engineering, and aeronautical engineering, you will find fun projects in this book. That's when you'll dive into the more advanced parts of the book. Basic electrical engineering explains why launchers are designed as they are. Computer programs and simulations combined with aeronautical engineering help you coax the most from a particular model rocket design. You will push beyond some of the information in this book. Perhaps you will design your own GPS device, use high-powered rockets to explore transsonic flight, build sensor packages to track the speed and altitude of a rocket, or design control mechanisms to keep a rocket from twisting so it is a better platform for photography—all projects that have been worked on by members of the local rocketry club I belong to. By now, you can handle anything in this book. It will give you a basic understanding, although you'll almost certainly jump off to more advanced material in physics, math, electrical engineering, mechanical engineering, computer science, and aeronautical engineering to satisfy the particular projects you find interesting.

So whether you are just getting started with model rocketry or are an experienced hobbyist looking for ways to sharpen your skills for a contest, this book will help you reach your goals. Whether you end up with a career in space exploration or not, you're about to become a rocket scientist!

Conventions Used in This Book

The following typographical conventions are used in this book:

Italic
: Indicates new terms, URLs, email addresses, filenames, and file extensions.

`Constant width`
: Used for program listings, as well as within paragraphs to refer to program elements such as variable or function names.

This icon signifies a tip, warning, or general note.

Safari® Books Online

Safari Books Online is an on-demand digital library that delivers expert content in both book and video form from the world's leading authors in technology and business.

Technology professionals, software developers, web designers, and business and creative professionals use Safari Books Online as their primary resource for research, problem solving, learning, and certification training.

Safari Books Online offers a range of plans and pricing for enterprise, government, education, and individuals.

Members have access to thousands of books, training videos, and prepublication manuscripts in one fully searchable database from publishers like O'Reilly Media, Prentice Hall Professional, Addison-Wesley Professional, Microsoft Press, Sams, Que, Peachpit Press, Focal Press, Cisco Press, John Wiley & Sons, Syngress, Morgan Kaufmann, IBM Redbooks, Packt, Adobe Press, FT Press, Apress, Manning, New Riders, McGraw-Hill, Jones & Bartlett, Course Technology, and hundreds more. For more information about Safari Books Online, please visit us online.

How to Contact Us

Please address comments and questions concerning this book to the publisher:

Make:
1005 Gravenstein Highway North
Sebastopol, CA 95472
800-998-9938 (in the United States or Canada)
707-829-0515 (international or local)
707-829-0104 (fax)

Make: unites, inspires, informs, and entertains a growing community of resourceful people who undertake amazing projects in their backyards, basements, and garages. Make: celebrates your right to tweak, hack, and bend any technology to your will. The Make: audience continues to be a growing culture and community that believes in bettering ourselves, our environment, our educational system—our entire world. This is much more than an audience, it's a worldwide movement that Make: is leading—we call it the Maker Movement.

For more information about Make:, visit us online:

Make: magazine: *http://makezine.com/magazine/*
Maker Faire: *http://makerfaire.com*
Makezine.com: *http://makezine.com*
Maker Shed: *http://makershed.com/*

We have a web page for this book, where we list errata, examples, and any additional information. You can access this page at:

http://bit.ly/make-rockets

To comment or ask technical questions about this book, send email to:

bookquestions@oreilly.com

Acknowledgments

There are a lot more people involved in writing a book than you might expect. The author does the writing—most of it, anyway—and gets almost all of the credit, but lots of other people contribute to the finished book. This is my chance to thank those people for helping me look like a literate professional who knows what he's talking about.

Brian Jepson started off as the editor for this book, and was most responsible for making it happen. As with my last book, though, he moved on. I'd check my deodorant, but we've never met in person. Still, he kept up with the book despite his many other obligations, helping me with several issues. When his time was stretched too thin by the new Maker books label, he also found a great new editor to take over. Patrick Di Justo patiently went through the entire book, offering sage advice, encouragement, and more than a few helpful edits.

I also got to work with Rachel Head again, a great copyeditor who turns my prose into consistent, grammatically correct English. She lives in France, world famous for great cooking, and also contributed in another way, with a slight tweak to the recipe for rocket eggs in Chapter 9. You can thank her when your eggs are just a tad creamier.

I got to work with an amazing set of reviewers for this book. Several were from the rocket club I belong to. Todd Kerns helped check my math and made great suggestions throughout the book, particularly in the launcher chapter, which went through a major revision because of his excellent suggestions. He also built the beautiful minimum-diameter rocket that appears in Figure 15-9. Bob Finch is one of the old hands at the club. He reviewed a number of sections and made suggestions in several long conversations at launches, at club meetings, and in phone calls. Tony Lazzaro helped with several issues, especially in the launcher chapter. Steve Lubliner, the chair of the NAR safety committee, helped improve the book in a number of places, especially Chapters 2, 4, and 5.

The Ryan family holds an annual rocket day where they get together to fly all sorts of rockets, including water rockets and air rockets. I've been very fortunate to be invited to the recent events. The extended Ryan family of Kenny Moreland, Kevin Ryan, and Jess Ryan Finley offered suggestions and clarifications throughout the book. In particular, Kenny helped get a nasty terminology error fixed, and Kevin made some great suggestions that simplified the presentation of the electronics theory in the launcher chapter. Both went through the entire book to find numerous other small glitches. It's a tedious task, but they did it well!

I was certainly no expert on water rockets or air rockets when I started this book. Fortunately, I could lean on the expertise of several people who were. Mike Sinclair from US Water Rockets did a careful review of the water rocket chapters. I suppose that's only fitting, since both the water rocket launcher and water rocket parachute deployment system were based on designs from the US Water Rockets website. The team of Keith Violette and Rick Schertle developed the air rocket launcher and air rocket glider you see in Chapter 6 and Chapter 21, and reviewed my version to offer advice from their vast experience.

My wife Patty was amazing. She let me take over the living room with launchers and rockets as I developed the projects for the book; went with me to fly, photograph, and recover the rockets; and was the first person to review each chapter —sometimes multiple times. I know a lot of you have very happy marriages, but sorry—I have the best wife there is.

Photo Credits

Most of the photographs of space vehicles and real airplanes are from Wikimedia Commons's vast photo library (*http://commons.wikimedia.org/*).

Roger Smith of JonRocket.com kindly gave permission to use two photos from his collection, Figures 20-28 and 23-48.

Marc Bonem built the flying Tardis shown in Figure 23-47, and kindly supplied the photo of his creation.

Jim Jewell is the unofficial photographer for the Albuquerque Rocket Society. He kindly gave me permission to use any of his photos, and I did so on several occasions, including Figures 15-9 and 22-3..

1 Let's Fly Some Rockets!

There are three basic kinds of rockets in hobby rocketry, and we're going to start right off by building small examples of each. Part of the reason is for you to learn some basic concepts before moving on to bigger rockets. Part is just to have some fun with matches, balloons, and projectiles.

The first kind of hobby rocket is the one you may have seen in hobby stores or the local park. Solid propellant rockets burn some sort of chemical, usually black powder in model rocketry. We'll take a look at solid propellant rockets using a match head as rocket fuel.

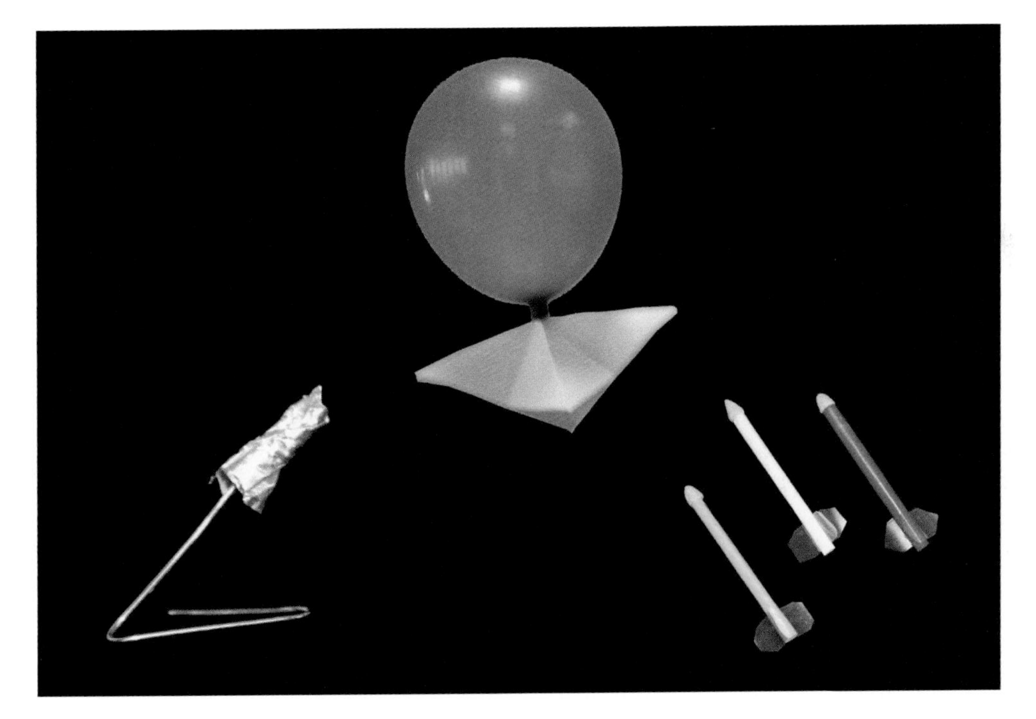

Figure 1-1. *Match head, balloon, and straw rockets are simple examples of the three kinds of hobby rocketry.*

Next most common is the water rocket. Water rockets usually use soda bottles under pressure to force the water out at high speed. We'll use a balloon to get started, forcing air out under pressure instead of water. It's the same idea, though.

Another fun form of rocketry is air rockets. Air rockets use a blast of pressurized air to propel a rocket from the launch pad. Our first air rocket is made from paper, and uses a straw for the launcher.

Shopping List

Table 1-1 lists all of the parts you will need to build the rockets in this chapter. All can be found at your local grocery store or variety store. Some are almost certainly lying around the house, so collect those first.

Tools are pretty basic for this chapter. You will need a pair of scissors or some other way to cut paper, such as a hobby knife. A pair of needle-nose pliers is handy but not absolutely necessary.

Table 1-1. Parts list

Part	Description
Paper matches	We will use a variety of matches. In some cases, the kind doesn't really matter, but this one is important. Get the cheap paper matches that come in a book where the match tears out of the matchbook. Get at least two books of matches, preferably three or four.
Wooden matches	Here you have some choice. You can get by with paper matches, but you will be holding them for a while. Wooden matches work better. I like the long kitchen matches. You could also use a fireplace lighter, one of those long gadgets that create a small flame at the end of a metal rod. They are great if you have one, but the wooden matches are a lot cheaper.
Paper clips	You will need one or two small metal paper clips. Smaller ones are better for our purpose, for reasons we'll see in a moment. Plastic paper clips or paper clips coated in plastic will not work. Thin, stiff metal wire can be used if you have wire cutters and don't mind a little artful bending.
Paper-backed foil	Here's one item you can't find sitting on the store shelf—or can you? It turns out you can, but this one needs some careful repurposing. A good choice is the foil from chewing gum, but be careful: most companies are moving away from foil-wrapped gum. One I found that works well is Wrigley's Extra. I tried lots of foils, and you can, too. One of my favorite failures was the foil used to wrap Hershey's Kisses. The foil was a complete flop, but disposing of the unneeded material wrapped in the foil was quite nice. Remember, experimentation and lab time can be fun. What a great excuse to buy a bag of candy!
Sandpaper	A small piece of fine grit sandpaper. Any old scrap will do.
Straws	Plastic drinking straws are best, but paper will do. Make sure they are fairly large.
Several sheets of paper	Notebook paper, printer paper, or any other thin paper will do. Origami paper is perfect. Construction paper will be too heavy and porous, though—stick with a paper designed for writing or printing that is thin enough to fold.
Transparent tape	You will just need a few inches, so grab a roll from last year's holiday wrapping paper box.
Modeling clay	You won't need much—about enough to form two balls the diameter of the drinking straw.
Balloons	A small variety pack with 25 or so balloons is ideal. If you get individual balloons, try to get both round and long balloons, and get several of each. Avoid the balloons intended for twisting into shapes. They are fun for other reasons, but are not thick enough for our purpose. You're looking for a nice, strong balloon.

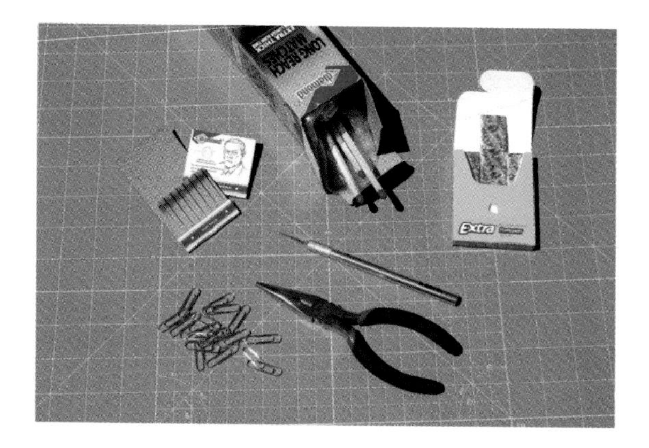

Figure 1-2. *Parts and tools for match head rockets include paper matches, wooden matches, paper clips, foil from gum wrappers, sandpaper, a hobby knife, and needle-nose pliers.*

Figure 1-3. *Parts and tools for balloon rockets include balloons, paper, and scissors.*

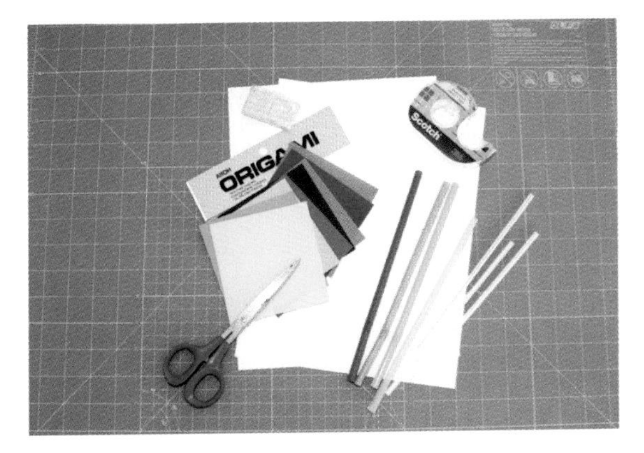

Figure 1-4. *Parts and tools for straw rockets include straws, tape, perhaps a small amount of clay, paper, and scissors.*

The Match Head Rocket

You'll need these items for your first rocket:

- Paper matches
- Wooden matches or lighter
- Paper clip
- Paper-backed foil
- Sandpaper
- Scissors or hobby knife

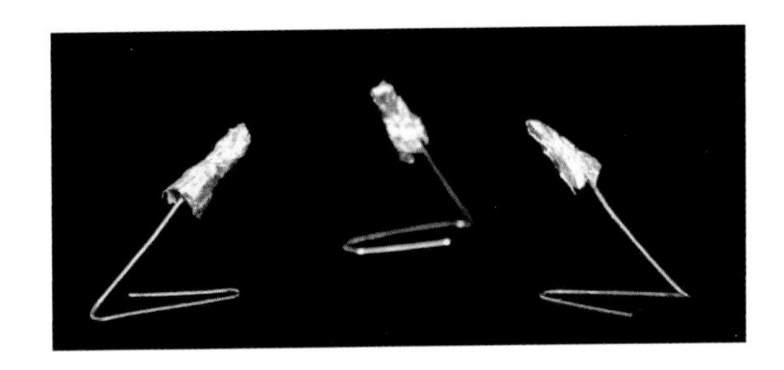

Figure 1-5. *An arsenal of match head rockets, waiting for flight.*

Playing with Fire

You are literally playing with fire. As with almost everything else in hobby rocketry, that means you need to exercise a bit of common sense.

This rocket should be built and flown outdoors or in a large room devoid of flammable materials, such as an aircraft hanger or gymnasium. Take a look around. Is there anything within five feet or so that would burn if you held a match under it? If the answer is yes, move it out of the way or pick another location. It's also a good idea to have a hose or fire extinguisher handy.

Our first rocket is a solid propellant rocket capable of flight up to a few feet, but usually traveling only a foot or two. It's made with a single match head from a paper match and a small piece of paper-backed aluminum foil —a gum wrapper or candy wrapper. The launcher is a bent paper clip, and the igniter is another match or fire starter, preferably a wooden match with a long splinter of wood.

Start by bending the paper clip as shown in Figures 1-6 through 1-8. The long, straight part of the wire is the launch rod for the rocket, and the remaining part of the paper clip forms the stand. The angle between the launch rod and the ground should be about 45 degrees (°). Be sure the paper clip is very straight; take out small kinks by very carefully straightening the wire. Needle-nose pliers work well for this.

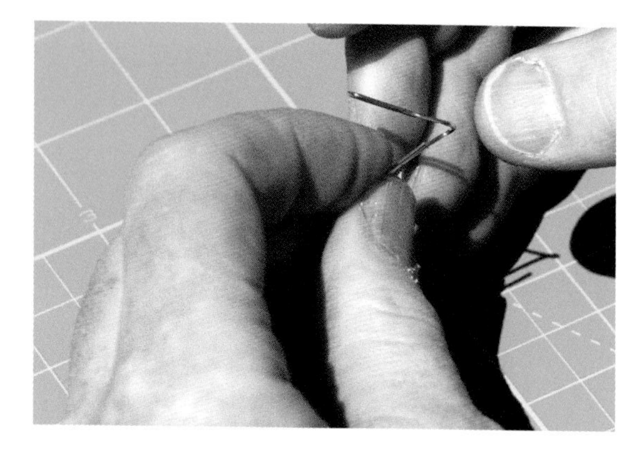

Figure 1-6. *Bend the paper clip up at about a 45-degree angle.*

Figure 1-7. *Straighten the end of the wire that sticks up (the launch rail).*

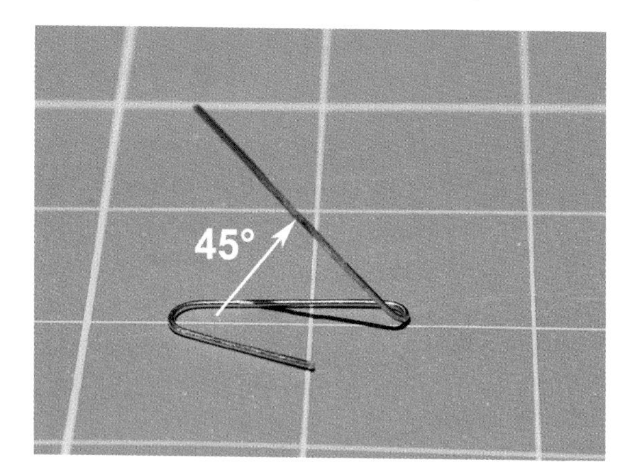

Figure 1-8. *The finished launcher.*

The rocket is made from foil and a match head. Cut one match head from a paper match, keeping as much of the match head and as little of the paper stem as possible. Place this on the paper side of the gum wrapper and fold the foil around the launch rail. Pull it off and use a pair of scissors or a hobby knife to cut all but about a half-inch of the material away. Slide the rocket back onto the launch rail.

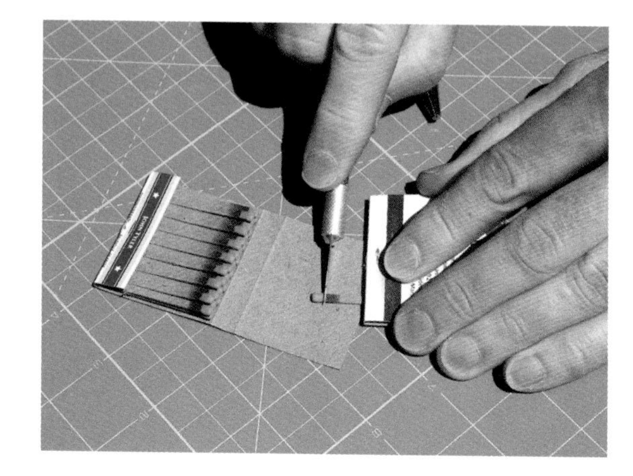

Figure 1-9. *Cut off the match head using a hobby knife or scissors.*

Figure 1-10. *Wrap the foil around the match head.*

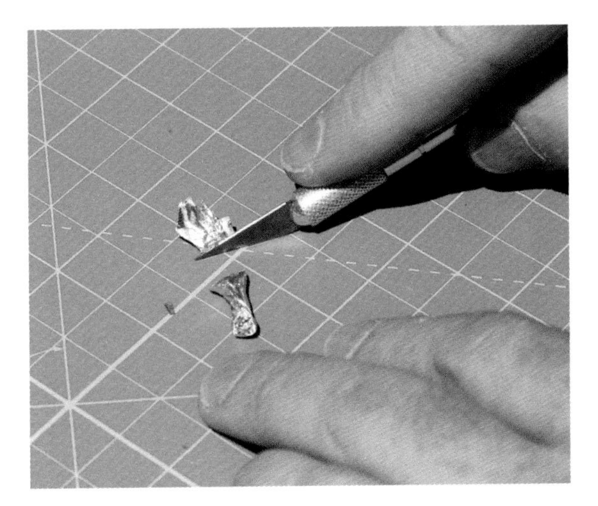

Figure 1-11. *Trim excess foil.*

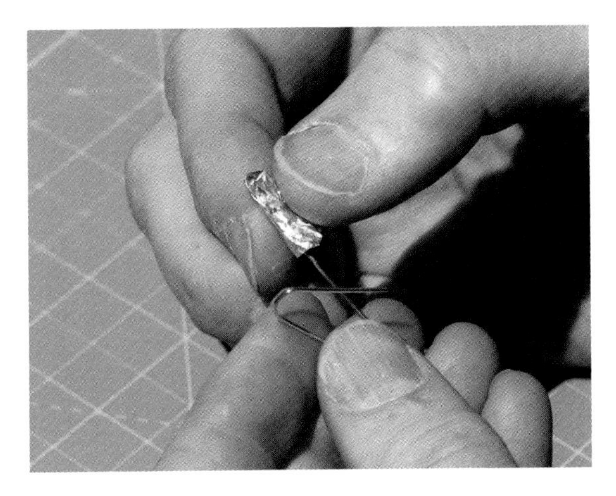

Figure 1-12. *Slip the rocket onto the launch rail, making sure the fit is fairly loose.*

The most critical—and difficult—part of building this rocket is getting the foil just tight enough, but not too tight. The rocket needs to slip off of the launch rail effortlessly. If there is any resistance, especially any snagging, the rocket will sit on the pad, trapped on the launch rail, instead of taking flight. Slip it up and down the launch rail to be sure it moves easily. Sand rough edges on the paper clip if necessary.

When you are satisfied with the fit, hold a long wooden match or lighter directly under the bulge in the rocket where the match head is wrapped in foil. In a moment, the match will ignite and…

Figure 1-13. *Light the rocket by holding the starter under the fuel. Not every try is a success! Here the fuel burned through the side of the rocket.*

What, you got a dud? It happens. (Even NASA has a few mistakes on its record.) If the rocket started to slide but stopped, you likely had a snag on the launcher. Clean the launch rail of any carbon using sandpaper, and check it for rough spots or bends where the rocket might get caught. If the rocket just sat there, it might not have been tight enough on the launch rod. The foil needs to be crimped fairly tightly around the paper clip so there is not a large opening. It can't be too tight, of course—make sure the rocket slides freely—but if it is too loose, the match doesn't provide any thrust. Set up another rocket and try again. Don't expect a high percentage of successes, though. On some days, I do well to get one successful flight out of ten.

Figure 1-14. *A successful launch! The rocket flew about 15,000 mils!*

Newton's Third Law of Motion

Rockets work based on a very simple but often misunderstood principle known as *Newton's Third Law of Motion*. You will see it written many different and equivalent ways. A common way to state it is:

> For every action, there is an equal and opposite reaction.

What does that mean?

Imagine for a moment that you are sitting in a boat on a calm lake. You put a shotgun to your shoulder and fire parallel to the water. Either from experience or from movies, you know the shotgun will push back on your shoulder when it fires. The force will also move the boat. It's a simple example of a rocket-propelled boat. By flinging the shotgun pellets and hot gas in one direction, you move your shoulder and the boat in the other.

Keep in mind that the shotgun boat didn't need air or the water to push against. In fact, it's better off *without* the friction of the air and water. The whole contraption will move even better in the vacuum of space, where there is no friction from air or water to slow it down.

Rockets take this idea to an extreme. They push something—usually, hot gases—very, very fast in one direction. The rocket itself moves in the other direction, but because it is so much heavier, it moves much more slowly than the gases. There is a lot of engineering involved in making all of this work efficiently.

Figure 1-15. *Pellets and gas shot in one direction cause the boat to move in the other.*

One piece of engineering you saw with the match head rocket was nozzle design. The *nozzle* is the small opening that allows the gas, water, or other material to escape out of the back of the rocket. If it is too big, the rocket fuel simply burns, just like a match that isn't enclosed at all. If the opening is too small, the gas cannot escape fast enough, and too much pressure builds up inside the rocket motor, causing the motor to rupture. In extreme cases, we call this an explosion. If the nozzle is just right, the rocket propels itself as efficiently as possible with the available rocket fuel.

Newton's Third Law is a basic principle of physics taught in the first few weeks of pretty much any introductory physics course. Such courses are the perfect place to start if you would like to know more about the science behind rocket propulsion.

Did you try two match heads? In my tests, two match heads were simply too much fuel for a rocket whose combustion chamber was thin foil. The fuel burned through the foil every time.

Keep All Fingers Intact

You might be tempted to take this to the next level. Please don't. Match head rockets are safe and fun, but too many match heads in too strong of a container can create a dangerous explosive.

In particular, do not try to stuff match heads into a CO_2 cylinder like the kind used for pellet guns. It is disturbingly easy to accidentally ignite a match as it is squeezed through the small metal opening, and the result can be an explosion rather than a rocket motor. According to reports, several people lost fingers or eyes attempting to do just this before prepackaged model rocket motors became available.

Stick to one or, at most, two match heads. Use thin foil. Leave the design and construction of larger rocket motors to people with the proper training and facilities.

Balloon Rockets

Figure 1-16. *Up, up, and away!*

Here's what you will need for your second rocket:

- Balloons
- Writing or copying paper
- Scissors or hobby knife

At some point, anyone who plays with balloons blows one up and lets it go. It's a perfect example of a simple rocket. Air is pushed out one end of the balloon, and the balloon itself is pushed in the opposite direction. Go ahead—give it a try.

As you saw, the balloon moves, but there is very little direction to the flight; the balloon twists and twirls around at random. We'd say that the balloon rocket has very little *stability*.

Rocket Stability

Balance a pencil on your finger. The place where it balances is called the *center of gravity*. If the pencil's density is the same along the entire length, the center of gravity is the location on the pencil where there is an equal amount of weight on either side. Finding the center of gravity is a bit more complicated for objects that have some spots that are heavier than others, a topic we'll return to in Chapter 7.

The center of gravity is important because it's also the point the pencil will spin around if you toss it into the air. The same idea applies to the balloon: any spinning object always spins around its center of gravity, unless there is some force throwing it off balance.

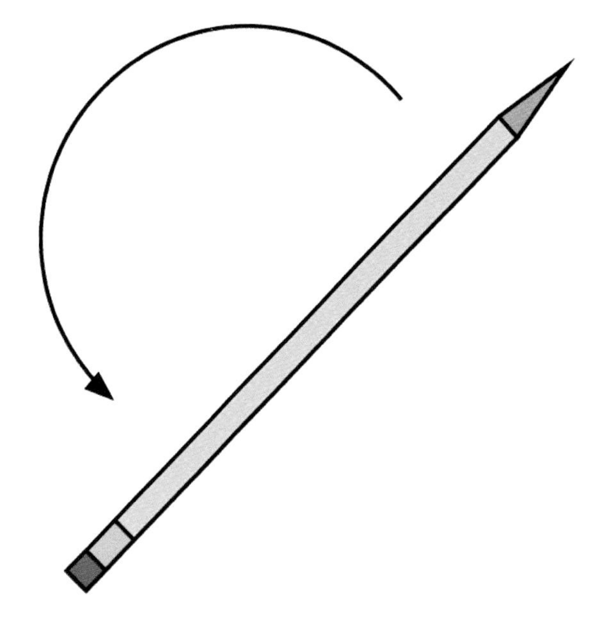

Figure 1-17. *Objects spin around their center of gravity unless there is an outside force.*

Now imagine a weather vane for a moment. You can make one from a pencil by attaching a piece of paper to one end and poking a pin through the center of gravity. When the wind blows from the side, there is more force on the side with the paper vane than the side that just has the pencil, so the pencil spins around until the pencil point is pointing into the wind. You could, of course, move the pin back to a point where the force from the wind was the same on either side of the pin, and the pencil would no longer spin in the wind. That point is called the *center of pressure*.

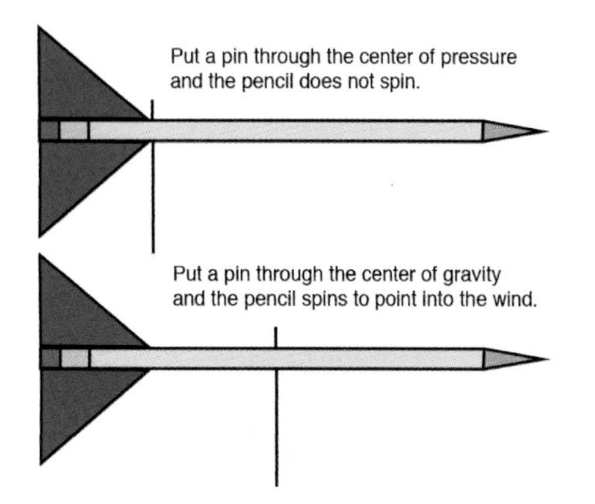

Figure 1-18. *Objects spin in the wind unless they are pinned at the center of pressure.*

Putting these two ideas together gives the classic design for a rocket. The center of gravity is about halfway down the body of the rocket, but the fins near the base of the rocket put the center of pressure well behind the center of gravity. If the rocket starts to spin for some reason—a gust of wind or a slightly off-center thrust from the rocket motor—the fins act to keep it flying straight.

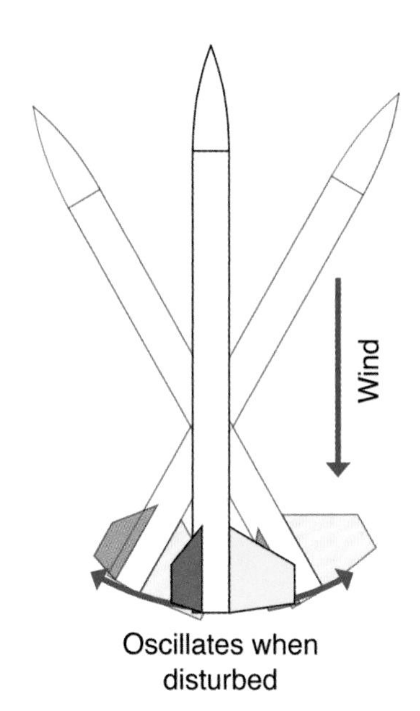

Figure 1-19. *Objects spin in the wind unless the point they turn around is at the center of pressure.*

The skeptics in the crowd may note that NASA's rockets don't all have fins, and they would be right. Some, like the Mercury-Redstone that carried the first American astronaut (Alan Shepherd) into space, did have fins for guidance, but others, like the Mercury-Atlas that carried John Glenn into orbit, did not. Still others, like the Vostok that carried the first human (the Soviet Union's Yuri Gagarin) into space, have tiny fins.

Figure 1-20. *Some rockets don't have fins. They use active guidance instead. From left to right, the Mercury-Redstone has fins, the Mercury-Atlas does not, and the Vostok (Mikhail Olykainen/Shutterstock) has very small fins.*

All of these rockets, though, used active guidance, where computers and gyroscopes monitor the flight of the rocket, redirecting the thrust of the rocket motor to keep the rocket flying straight. Our simpler rockets use fins instead of active guidance.

We'll see later that all of our rockets must have the center of pressure behind the center of gravity by at least the diameter of the body tube, and we'll look at ways to change our rocket designs to make this happen.

Building the Balloon Rocket

We'll need to add something to our balloon to stabilize it. Fins are a little impractical, but a sheet of paper will do nicely. Take a look at Figure 1-21, which shows our balloon rocket with a stabilization system attached.

Figure 1-21. *Adding crude fins to a balloon rocket straightens its flight.*

Cut a square piece of paper about the diameter of a long balloon, or 2/3 the diameter of a round balloon. Fold it corner to corner, unfold it, then fold it corner to corner across the other corners. Fold the edges down to form a fin-like shape. Trim out a small opening in the center where the folds intersect. Make the hole just big enough for the neck of the balloon to fit without pinching off the air flow.

Fold the paper as much as possible to form a more fin-like shape. You probably won't be able to make it perfect because you need room to blow up the balloon.

Figure 1-22. *Create a guidance system for the balloon using a piece of paper.*

Finally, blow up the balloon and launch your rocket!

Figure 1-23. *The balloon rocket takes flight.*

If all goes well, the rocket will power across the room flying a lot straighter than before, although definitely not perfectly straight. These are fairly crude fins, after all.

If the rocket is still very unstable, increase the size of the paper. If it is very slow or can't lift itself at all, reduce the size of the paper. The balloon can wear out, too. If the balloon doesn't shrink back to roughly its original size, or if it gets very easy to blow up, it won't have as much power—the balloon's thrust comes from the springiness of the rubber trying to get back into its smallest shape. Switch to a new balloon and try again.

The idea of using a pressurized fluid to power a rocket is something you will see again later in the book. A plastic soda bottle will replace the balloon, a bicycle pump or compressor will replace blowing up the balloon, and we will use water rather than air as a propellant, but the idea is exactly the same.

Exploration

1. Try long and round balloons. Which works better?
2. Try large and small balloons. Again, which works better?
3. How many times can the balloon be used before it loses too much thrust?
4. Can you come up with a design that has better fins?
5. Build a boat from a small piece of wood or Styrofoam and design a rocket propulsion system for it using a balloon for power.
6. Why not use water? Add a small amount of water to the balloon, and then launch it vertically so the water is at the bottom and gets pushed out by the air. Does it work? Why?

Air Rockets

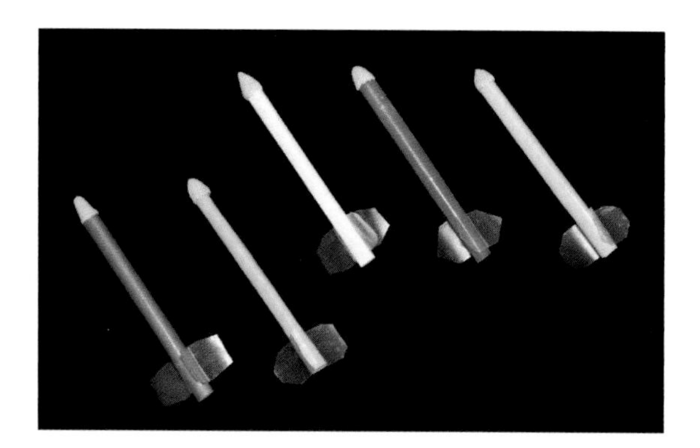

Figure 1-24. *Colorful air rockets.*

You will need these items for an air rocket:

- A straw
- Paper
- Tape
- Clay (optional, but nice)
- Scissors or hobby knife

Air rockets still use Newton's Third Law, but unlike the match head rocket or the balloon rocket, an air rocket doesn't carry the propellant along with it. In this case, the rocket is a projectile, and it is expelled from the launcher with a blast of air.

Theory: Is an Air Rocket Really a Rocket?

A purist might object that an air rocket really isn't a rocket at all, since it doesn't carry along its own fuel or expel something for propulsion.

I tend to get caught in those ivory towers, too. Come down for a moment, and consider that this is a book about rocket science using hobby rocketry. Many of the principles of rocket flight, like stability, apply just as much to air rockets as to chemical or water rockets. Also, not all serious rocket science uses classical rocket propulsion. Consider the *mass driver* as just one example. This NASA concept is a rail gun, generally envisioned as a way to launch material from the Moon for pickup by an orbiting platform. It uses electricity rather than air, but like the air rocket, the energy is provided at launch, not during flight.

Figure 1-25. *NASA's concept for a mass driver to launch material from the surface of the Moon.*

Besides, air rockets are fun!

The straw is the launcher, and doesn't need any preparation.

Create a *body tube* for the rocket from a strip of paper about four inches long and about as wide as four straw diameters. A sheet of plain paper works well, but origami paper makes a fun and colorful substitute. The straw I used is about 3/16" in diameter, so the width of the paper strip was 12/16", or 3/4". (See, fractions really are useful!)

Wrap the paper around the straw, making it as snug as you can while still letting it slide easily along the straw. Use two small pieces of tape about one inch from each end to secure the paper while you add a final piece of tape along the entire length of the paper tube.

Figure 1-26. *Construct the rocket tube from the paper strip. Tape the ends to hold the paper in place.*

Figure 1-27. *Finish the body of the rocket with a piece of tape along its length.*

Using a long piece of tape, create three or four fins about an inch long at one end of the rocket. Use scissors to trim them to a fun shape. The fins push the center of pressure back toward the base of the rocket. See "Rocket Stability" on page 7 to see why this is important.

Figure 1-28. *Form three or four fins from tape, evenly spaced around the base of the rocket.*

Figure 1-29. *You can trim the fins for a fancy shape.*

Use a small piece of clay to form a nose cone. It needs to block the end of the tube. A nose cone is usually used to reduce drag, which isn't terribly important at this point, but it will definitely add to the coolness factor of the rocket.

If you don't have any clay handy, you can fold the top of the rocket over and tape it shut.

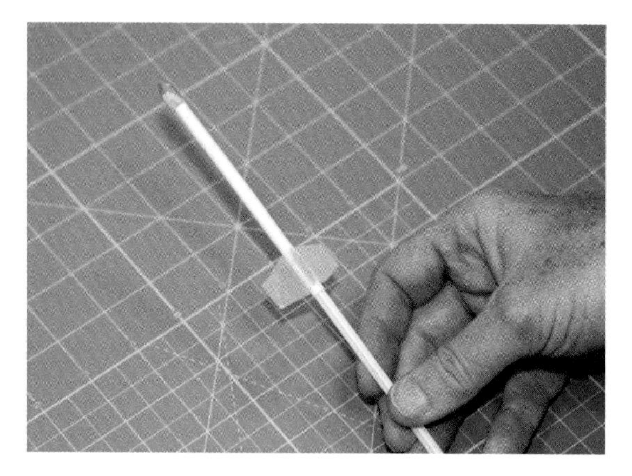

Figure 1-30. *The completed air rocket.*

It's time for the maiden voyage. Despite its simplicity, this air rocket will perform better than the match head or balloon rockets, easily flying 10 to 15 feet. Slip the rocket onto the straw, aim it a little less than 45 degrees from the horizontal for a really long flight, and give a sharp puff into the straw.

Figure 1-31. *A sharp puff into the straw will propel the air rocket 10–15 feet.*

Air rockets are a simple and inexpensive way to explore rocketry. We'll come back to them in Chapter 6. where we build much more powerful versions.

Model Rocketry Today

2

The Five Faces of Hobby Rockets

On October 4, 1957, the Soviet Union launched Sputnik, the world's first man-made satellite. It followed that with Vostok I, the first rocket-based human space flight, which put cosmonaut Yuri Gagarin into orbit on April 12, 1961. In the same time frame, the Unites States created NASA, the National Aeronautics and Space Administration, and announced the selection of America's first seven astronauts. Among the general public, interest in rocketry exploded. Unfortunately, it sometimes exploded literally, claiming the fingers, eyes, and occasionally lives of hapless would-be home scientists as they tried to build rocket motors in their garages and basements.

Fortunately, Orville Carlisle had invented a safe, reliable model rocket. He filed a patent application, which was granted in 1958. Just as concerned lawmakers were starting to clamp down on hobby rocketry, Harry Stine and Orville Carlisle founded Model Missiles, Inc., the first company to build and sell prepackaged model rockets and, more importantly, model rocket motors. While the business did not last, they were joined by Vernon Estes, who designed a machine to mass-produce rocket motors. This spawned Estes Industries, still the 800-pound gorilla in low-power model rocketry. Their prepackaged rocket motors and their paper, balsa, and plastic rockets earned a fabulous record for safety. This helped to reverse the legal trend against rockets, and allowed anyone with an interest to safely explore the hobby and the science behind building and flying model rockets.

Figure 2-1. *Three generations of rocketeers getting rockets ready for flight.*

This was pretty much the state of model rocketry until very recently. Sure, some companies came and went, and a few intrepid DIY types occasionally broke out of what has become known as low-power rocketry, but for the most part, hobby rocketry meant buying single-use black powder rockets and flying them from small launchers to a few hundred or a few thousand feet.

Like automobile designers, model rocket designers have always looked for higher performance. In the case of model rockets higher performance typically means larger payloads and higher altitudes.

One path was the use of solid rocket propellants that have more energy per volume and weight than black powder. A common chemistry used in professional rocketry, like the Space Shuttle's Solid Rocket Boosters, is ammonium perchlorate composite propellant (APCP). Motors using this propellant have total impulses approximately 2 to 3 times that of a black powder motor with the same fuel weight.

An example is the Aerotech F39. This motor has 2.5 times the total impulse (commonly referred to as total power) of an Estes D12 but still fits in the same size motor mount.

Figure 2-2. *The author prepares a J rocket for flight.*

Of course, in addition to a more energetic propellant, another way to get more performance is to use more of the propellant. For a time model rocketry was limited a bit by regulations that restricted the use of APCP motors, but that changed in 2009, when the National Association of Rocketry (NAR) and the Tripoli Rocketry Association (TRA) won a lawsuit that effectively opened the doors for the manufacture and distribution of higher-capacity APCP motors. While high-power rocketry had been around for a long time, this change allowed the easy, legal distribution of high-power motors.

Model rocketry had burst from its single-use, black powder beginnings. This paved the way for hobby rocketeers to safely build and fly rockets of almost any size. The extra power allowed rockets to scream to altitudes of a mile, then several miles, and even the edge of space. Hobby rockets, including rockets built by high-school classes, commonly break the 100,000-foot altitude barrier, sending back photos that show the darkness of space and the gentle curve of the Earth.

These ammonium perchlorate motors are not the small Estes rockets of a few years ago. Rockets that break the sound barrier or fly to the edge of space are not built from lightweight paper tubes with balsa fins. As their size, speed, and thrust climb, the construction of rockets changes to thick paper and plywood, then to fiberglass or carbon fiber, and eventually to metal. These are not the sorts of rockets you haul out to the local park for a casual Saturday morning launch. In the United States, most of these rockets require launch clearances from the Federal Aviation Administration (FAA), and for some, you need to file the plans for the individual rocket and flight profile before each flight.

Needless to say, these are also not the sorts of rockets you hand to a fifth grader, or allow a high-school student to launch unsupervised. There are even legal age limits on flying some of these rockets. We'll get into the details later, but for the most part, you must be 18 or older to fly rockets with F or larger motors without close supervision.

While all of this was going on, the DIY community began to develop two other kinds of rockets. The first is the water rocket.

Water rockets have been around for years; even back in the early days of space flight, stores carried small hand-pumped water rockets. But water rockets have changed dramatically in the ensuing years. Today's water rockets are usually built from soda bottles (and not always just one bottle!). Water rockets are pressurized from bicycle pumps, car tire pumps, compressors, or my own favorite, a scuba tank. They soar to hundreds of feet, often feature parachute recovery, and some even feature multiple stages! Unlike solid propellant rockets, water rockets pose no fire hazard, so they can be launched even when fire restrictions might leave other rockets on the ground.

How Safe Is Model Rocketry?

It's enough to give a parent pause. Your 11-year-old comes home from school with a permission form wanting to fly model rockets at school. Or a 14-year-old gets interested in a science class and wants to build and fly rockets. Is it safe?

In a word, yes.

Like anything, rockets can be misused and mishandled, and that can lead to accidents. The most common injury is a burn from doing things this book (and any other basic model rocket safety guide) will tell you *not* to do, like taking apart a motor or lighting it with a match. One study (*http://www.ncbi.nlm.nih.gov/pubmed/8040589*) found 18 burn injuries over an 18-year period.

Let's put one accident a year into perspective. Table 2-1 lists some other activities, with the number of deaths in the United States in a recent year.

Table 2-1. Deaths attributed to sporting activities

Activity	Deaths	Year
Skateboarding	42	2011
Bicycling	677	2011
Football	12	1990–2010 average
Water sports (drownings)	3,533	2005–2009 average

Compare that to an average of one burn per year requiring medical attention. Yes, model rocketry is safe—especially if you follow the rules.

Still, young children do not understand the potential consequences of some of their actions. That's why parents should supervise younger children. It's tempting to put an absolute age limit on supervision, but let's face it, I've met some 10-year-olds I trust with a model rocket, and some 30-year-olds I don't. Each parent should evaluate his or her own child. Here are some general thoughts, though.

First, model rocketry is suitable for any age, as long as a parent is directly involved in building and flying the rocket. I've had my own kids and grandkids at the launch site and pushing the launch button at the age of two.

A child should be 10 before building rockets without help, but still with supervision. This isn't so much a restriction based on rockets as it is based on the tools used for building rockets. Sharp tools are used to build rockets. The kids need to be old enough to handle them. Of course, if a parent does the cutting, younger kids can build rockets.

Children should be 12–14 before being allowed to build and fly rockets on their own.

What do the laws say? For most US states, there is no age limit on low-power rocketry. Here are a few exceptions:

- In California, you must be 14 to buy rocket motors up to D motors, and 18 for E or above.
- In New Jersey and Rhode Island, you must be 14 to buy rocket motors up to C motors, and 18 for D or above.

I hope there are a few folks from other countries reading this book. You should check your own local regulations. The only other country whose regulations I've seen is Canada, where you need to be 12 to buy rocket motors.

Air rockets also took off in popularity, powered by the same pressure systems but without the water. While a typical air rocket is a simple ballistic projectile, this is an area that is changing fast, as you'll see in Chapter 21 when we build a rocket plane that lifts off as an air rocket.

Rocketry is a big topic—too big to cover in a single book. This book covers aspects of rocketry open to people under the age of 18. There is one exception to this rule, in the last chapter. Mid- and high-power rocketry are more thoroughly covered in an upcoming book, *Make: High-Power Rockets*. Let's take a look at what all of this means by exploring the differences between low-power, mid-power, and high-power rocketry.

Solid Propellant Rocket Motors

You need to know a bit about rocket motors and how they are classified to really understand the differences between low-power rocketry, mid-power rocketry, and high-power rocketry, so let's take a look at solid propellant rocket motors.

Rocket motors are classified by their total power, measured in *newton-seconds*, abbreviated N-s. We'll talk more about what a newton-second really is in a moment, but for now it's enough to know that it is a measure of power, and that doubling the number of newton-seconds doubles the power of the rocket motor.

Is It a Rocket Engine or Rocket Motor?

If you want to start a passionate argument about something that really doesn't matter, go to a large rocket club meeting and ask whether the thing that propels a rocket is an engine or a motor, and why. There is a lot of bickering about which is correct.

There are two purely technical definitions, and they give different answers.

In one, a motor is something that takes fuel from an outside source and converts it to mechanical energy (electric motor), while an engine is something that has an integrated fuel system (car engine, rocket engine). Physics sites usually say an engine converts thermal energy to mechanical work (engine for a water pump—even if it's powered by an electric motor), while a motor converts any kind of energy to motion (rocket motor, electric motor); from this definition, all motors that use heat energy are engines, but not all engines are motors.

Estes Industries calls them rocket engines. Apogee Components, the Tripoli Rocketry Association, and the National Association of Rocketry (usually) call them rocket motors. Wikipedia decided it's a liquid rocket engine, and a solid rocket motor. Some references say either is fine. Some frustrated engineers are humorously calling them "whoosh generators" to avoid the debate.

I've tried to stick with "rocket motor" unless the specific part is almost always identified as an engine-something. So it's a rocket motor, but you hold it in place with an engine hook or engine block. I'm not really taking sides, here, just trying to be consistent!

While newton-seconds is the more precise way to give the power of a rocket motor, it's a lot more common to use a shorthand where a letter is assigned to a range of power. Table 2-2 shows the letter system used in model rocketry to designate motor power. An A motor has 2.5 N-s of total impulse. The total impulse doubles with each new letter, so a B motor is 5 N-s, a C motor is 10 N-s, and so forth. Eventually we run out of alphabet. A 2Z motor has twice the total impulse of a Z motor, and so on.

Table 2-2. Rocket motor classification

Class	Maximum total impulse	Example
1/4A	0.625	Low-power rocketry
1/2A	1.25	
A	2.5	
B	5	
C	10	
D	20	
E	40	
F	80	Mid-power rocketry
G	160	
H	320	High-power rocketry
I	640	
J	1280	
K	2560	
L	5120	
M	10240	
N	20480	
O	40960	
P	8.19E+04	Sidewinder air-to-air missile
Q	1.64E+05	
R	3.28E+05	
S	6.55E+05	
T	1.31E+06	
U	2.62E+06	Apollo launch escape rocket
V	5.24E+06	
W	1.05E+07	
X	2.10E+07	
Y	4.19E+07	
Z	8.39E+07	Mercury-Redstone
2Z	1.68E+08	
3Z	3.36E+08	Mercury-Atlas
4Z	6.71E+08	Gemini-Titan
5Z	1.34E+09	
6Z	2.68E+09	Space Shuttle
7Z	5.37E+09	
8Z	1.07E+10	Saturn V

Motors don't always have exactly the total impulse shown, of course. The table shows the upper limits for each motor class. If a motor has, say, 8 N-s of total impulse, it's still classed as a C motor. For the most part, motors up through E are manufactured right at the upper limit, so almost all C motors deliver 10 N-s of total impulse. Beyond E, though, the difference between successive categories is so large that it's common to find motors that are not right at the limit. If you hear a model rocketeer say she launched a rocket with a full J, she means the motor was right at the total impulse limit, delivering 1,280 N-s or very close to it.

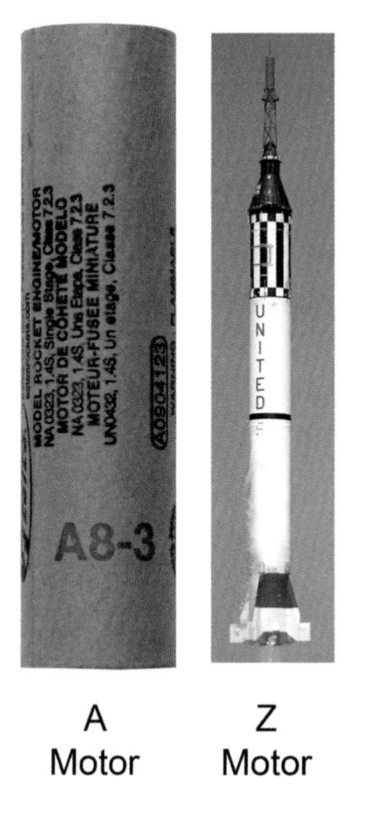

Figure 2-3. *Rocket motors from A (the Estes A8-3) to Z (the Mercury-Redstone that carried the first American astronauts into space). Not to scale.*

Doubling the power for each new letter gives a geometric progression. Take a close look at Table 2-2. The enormous Saturn V rocket that carried men to the Moon is an 8Z rocket, which puts it 28 steps beyond the E motors we'll use for some of the rockets in this book. Figure 2-4 puts this geometric progression in visual terms, showing the number of motors in the A–E classes that it takes to get equivalent power.

For example, it takes 16 A motors to deliver the same power as a single E motor, but it takes only 2 D motors to equal that same E motor.

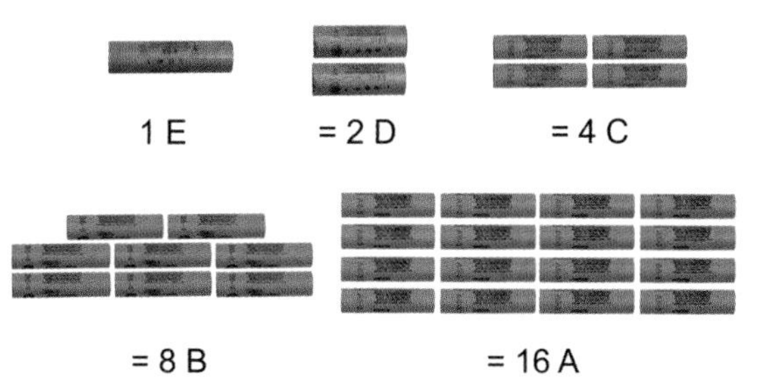

Figure 2-4. *Each letter represents twice the power of the previous letter, creating a geometric progression. Here it is in visual terms, from a single E motor to 16 A motors. Each pile of motors represents the same total impulse.*

So what does that mean compared to the mighty Saturn V? Well, the Saturn V was not a full 8Z rocket. It delivered roughly 8,733,800,000 N-s of total impulse. That's the equivalent of 3,493,520,000 A motors. Have there even been three and a half billion A motors produced? Stretched end to end, they would be 151,626 miles long—about 60% of the distance to the Moon. That's one way to get there, I suppose.

Some Common Motor Sizes

Figure 2-5 shows some common rocket motors. The ones on the left are designed for tiny rockets. We'll give them a try in Chapter 16. These motors come in 1/4A to A, maxing out at 2.5 N-s total impulse.

The next size of motor is far and away the most common size of solid propellant rocket motor. These are generally A, B, or C motors, although 1/2A and D are not unheard of. Most of the rockets in this book, including the introductory rocket in the next chapter, use this size motor.

The fat motor next in line is a D motor. This motor will easily lift larger rockets or heavier payloads.

Next in line is an E motor. It's pretty rare to *need* an E motor. They are frequently more of a *want*. You might really want to build an oversize rocket that gets a slow, impressive liftoff. You might want to send a sleek model soaring 4,000 feet in the air. This is a large, impressive motor for a model rocket.

The last motor is a reloadable motor. Unlike the black powder motors, this one uses ammonium perchlorate, just like the solid boosters on the Space Shuttle. The motor is cleaned and reloaded after each flight, ready to carry another rocket into the sky. We'll give reloadable motors a try in Chapter 23.

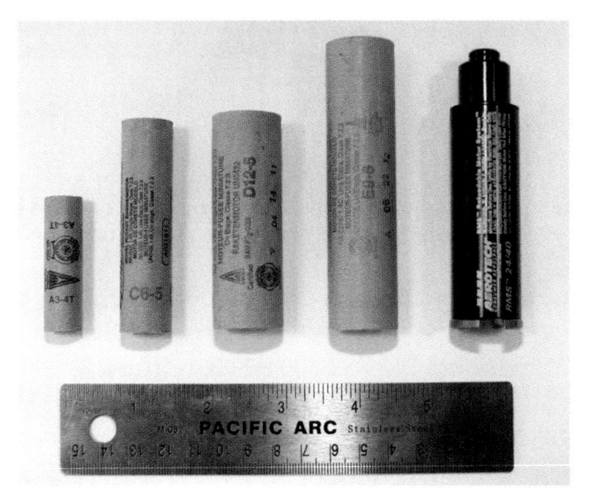

Figure 2-5. *From left to right: T-size motors come in 1/4A to A; standard motors usually come in A to C; the D motor; the E motor; and a reloadable motor that can take D, E, or F reloads.*

Reading Motor Labels

Motor labeling is pretty consistent, at least for single-use black powder motors like the ones we will use for most of this book. We'll take a look at labeling for reloadable motors in Chapter 23, when we put them to use.

Take a look at Figure 2-6. This is the B6-4, a pretty typical rocket motor.

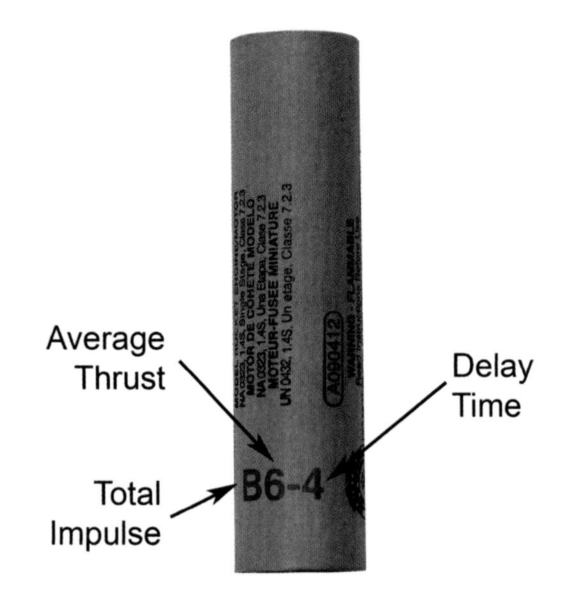

Figure 2-6. *Motor labels start with the total impulse letter, followed by the average thrust, and end with the delay time between burnout and the ejection charge.*

The B indicates the total impulse of the motor, which is larger than 2.5 N-s and no more than 5.0 N-s for a B motor. Like most motors below E, the B6-4 is designed to deliver the largest total impulse possible for this class of motor. But what is total impulse?

Thrust is how much force the motor develops—how hard it pushes on the rocket. While this could be measured in pounds of thrust, and sometimes is, it's more common to see units of *newtons*. One newton is about 0.225 pounds of force. That's an important number—it tells us how much thrust the rocket motor develops, and by extension, how heavy the rocket can be. But it's only part of the story. Obviously, if a rocket develops five newtons of thrust for one second, and another rocket develops five newtons of thrust for two seconds, then all other things being equal, the second rocket is going to fly higher and faster. And that's what *impulse* really is. Impulse is the average thrust multiplied by how long the thrust lasts. *Total impulse* is the impulse for the entire time the rocket motor fires. The first motor, delivering five newtons of thrust for one second, is a 5 N-s motor—a B-class motor. The second, developing five newtons of thrust for two seconds, is a 10 N-s motor—a C-class motor.

The number 6 in the B6-4 motor is the average thrust of the motor in newtons. If the motor jumped immediately to full power, stayed at exactly the same level, and dropped off to zero in a nice step, you could divide the total impulse by the average thrust to get the duration of the thrust for this motor, which would be 0.833 seconds. Plotting the thrust over time, we would get the plot shown in Figure 2-7.

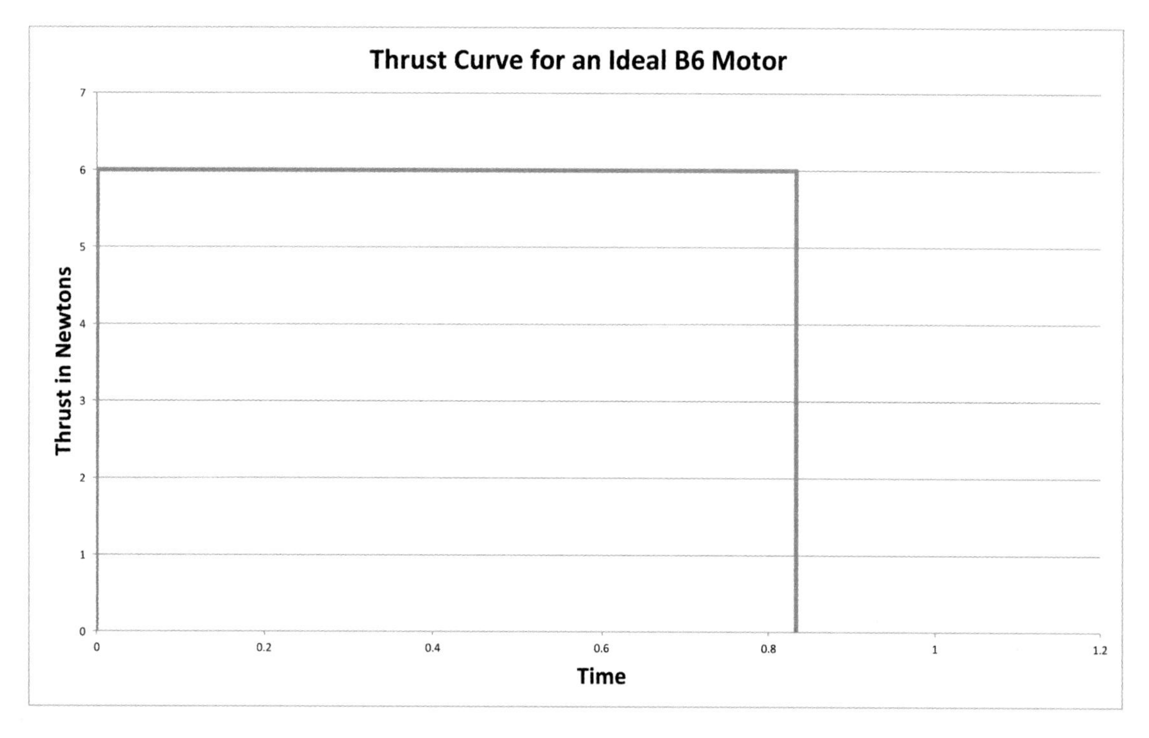

Figure 2-7. *If a motor jumped immediately to its average thrust and stayed there until burnout, the curve would look like this for a 5 N-s B6 motor.*

The plot for a real rocket motor rarely looks anything like this. Figure 2-8 shows the actual thrust curve for a B6-4 motor.

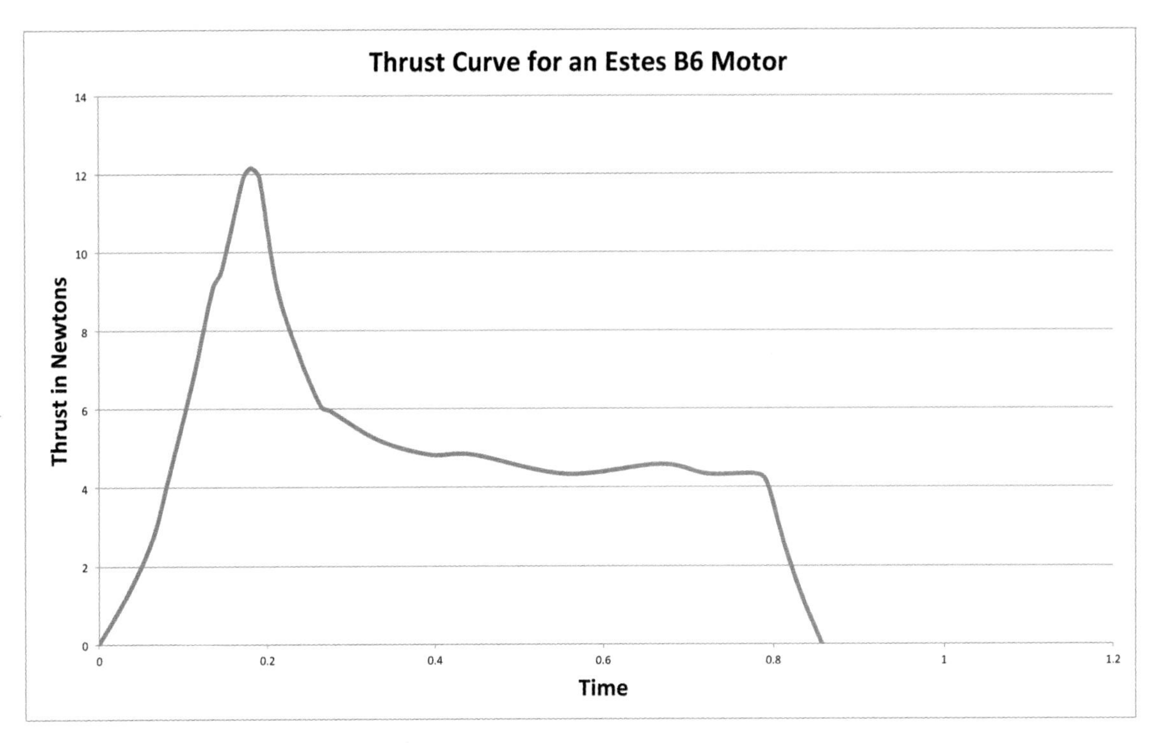

Figure 2-8. *Here is the thrust curve for an Estes B6 motor, created from actual test data. All Estes B6 motors have the same thrust curve; the difference is in how long the delay lasts and whether it has an ejection charge.*

Like many motors, it starts off with a spike in thrust that helps get the rocket off of the launch pad quickly. The motor then settles down to a fairly constant thrust until it burns out, when the thrust drops rapidly to zero. While it may not be as simple as the thrust curve from Figure 2-7, it's better for our rocket, since it delivers an extra punch at the start of flight to accelerate the rocket quickly to flight speed.

The last number in the B6-4, the 4, is the delay charge. This is the time in seconds the smoke charge burns. The smoke charge pours out smoke without producing any appreciable thrust, allowing the rocket to coast to apogee before the *ejection charge* fires to deploy the parachute.

Let's compare this to a similar motor, the B4-4. Like the B6-4, this motor has 5 N-s total impulse and a 4-second delay before the ejection charge fires. The average thrust is lower, though. A motor with perfectly constant thrust would fire for 1.25 seconds rather than 0.833 seconds. Figure 2-9 shows the actual thrust curves for both the B6-4 and the B4-4.

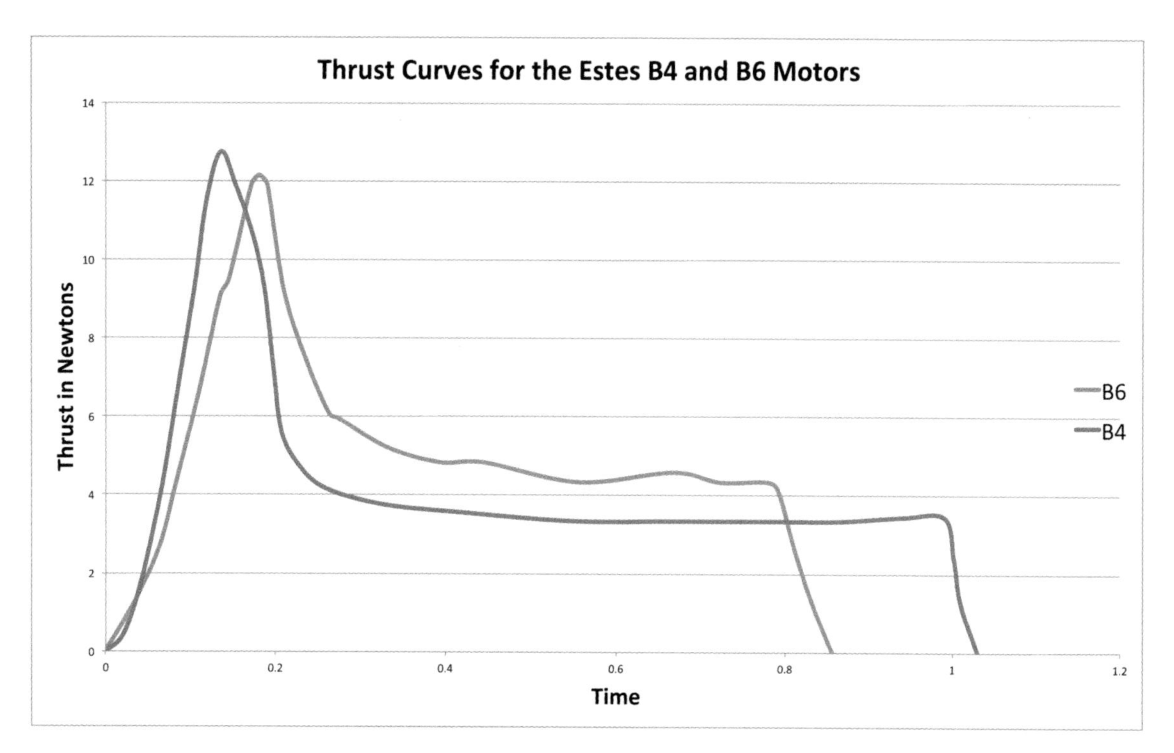

Figure 2-9. *Comparing the B4 and B6 motors, you see that the B6 has a higher average thrust, but burns out sooner than the B4.*

As you can see, the average thrust is lower for the B4-4, but it lasts longer. In general, you would select the B6-4 for a heavier rocket that needs the thrust to lift its weight, and the B4-4 for a lighter rocket.

There is one very special class of motor, those with 0 for the delay time. Figure 2-10 shows the B6-0.

Motors with a delay time of 0 have no delay smoke and no ejection charge. They are usually used as the booster motor on two-stage rockets. We'll see how they work and how they are used in Chapter 17, when we build a two-stage rocket.

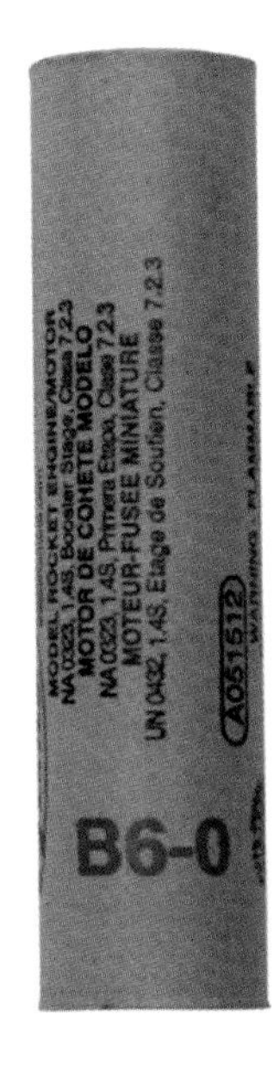

Figure 2-10. *The B6-0 does not have an ejection delay or an ejection charge. It is usually used as the first stage of a two-stage rocket.*

Storage and Disposal

Rocket motors are flammable, and need to be stored with that in mind. They are very unlikely to catch fire on their own, though. Black powder motors will burn hotly when ignited, but will not spontaneously combust. Still, they should not be stored in a garage, where large temperature swings might result in overheating or shorten the life of the propellant. Store the motors inside in a climate-controlled closet.

Don't stockpile motors. Several hundred motors stashed in your house could be a hazard to the building and anyone in it if a fire starts, including the firefighters who show up to help. There is also a legal limit—the National Fire Protection Association (NFPA) 1122, Code for Model Rocketry limits the total weight of model rocket motors stored in the living quarters of a residence to 25 pounds. NFPA codes are not legal codes in and of themselves, but most states use these codes as the basis for their laws. It's not completely clear if the 25-pound limit is the motor weight or the propellant weight, but even if it's the motor weight, that's 440 C6-5 motors—surely enough for anyone!

Two kinds of containers make sense for storing motors. The first is a metal container, such as a metal fishing tackle box. Metal containers help contain the flames should a motor catch fire. The second is an insulated container, such as a cooler. Insulation makes it less likely the motors will catch fire in the first place.

The best kind combines the characteristics of both. The one thing you do not want to use, though, is an airtight metal container, like some fireproof safes. True, they are designed to protect the contents from a fire, but an airtight container with rocket motors inside it—especially a strong one—is essentially a bomb. If, for whatever reason, the motors ignite while inside such a container, the pressure from the burning motors will build up until the container explodes, spewing pieces of the container and flaming gases. Burning motors would be bad. Exploding containers is worse.

There may come a time when you need to dispose of an unflown motor. My favorite way to destroy a motor is in a rocket, of course. Why not just fly it? Still, if this is not practical for some reason, soaking a black powder motor in water is a safe way to render it harmless. This chemically changes the black powder so it will not burn. After soaking it in water, dispose of it with your normal garbage.

Low-Power Model Rockets

Now that you have a basic understanding of rocket motors, we can take a look at the various kinds of rocketry, starting with low-power rockets.

Low-power model rockets are the typical rockets you see at most hobby stores. They are almost certainly what you see people flying at the local park or building and flying for grade-school science class. These rockets are usually made from lightweight paper tubes, balsa fins, and balsa or plastic nose cones. The paper tubes are thin—generally about the same thickness as the cardboard tube at the center of a roll of paper towels, but a bit stronger. The rocket motors typically use black powder for a propellant. They are usually single-use motors.

While this is called low-power model rocketry, don't assume these are underpowered toys. Low-power model rockets can soar well over 2,500 feet—about a half-mile. Extreme high-performance rockets can fly even higher. The current altitude record for an E rocket is 4,459 feet!

Figure 2-11. *Model rockets regularly take off from parks and small fields across the world.*

Model rockets are a great way for students and interested adults to learn about the science of rocketry. I have worked with numerous fifth-grade classes as they built and flew rockets. Science teachers in local third-grade classes also use rockets as a way to promote interest in science and engineering. They are not just for grade schools, though; model rockets are also used in many university science classes.

The reason rocketry has such a broad appeal is the vastly different ways you can approach the hobby. A third-grade class may simply build and fly the rockets, with little thought to how the rockets are really working. By middle school, students can start to practice their blossoming mathematics skills, using basic trigonometry to track the altitude of a rocket. With high school physics comes a basic understanding of the reason rockets fly as they do, along with the knowledge needed to write simple computer programs to predict the altitude of a rocket. Aerospace engineering in college introduces concepts like fluid flow and drag, as well as tools like wind tunnels. Model rockets are an ideal way to safely and inexpensively put these theoretical skills to use in real-world applications.

So what is a low-power model rocket? The distinction is a bit arbitrary these days. For this book, we'll draw the line at a model rocket with 40 newton-seconds or less of total impulse. This includes all rocket motors with a class of E or less. They usually weigh less than one pound, often just a few ounces.

So why draw the line here? There are two reasons. The first is that construction of rockets changes dramatically about the point where rockets start to use F motors. The rockets are bigger, heavier, and can potentially do more damage if they get out of control. The second is that sale of F motors is restricted to those 18 and older.

Mid-Power Rocketry

As the power of model rocket motors increases, the motors eventually get powerful enough to rip apart a rocket made with thin paper tubes and balsa wood fins. Most F and G rockets are made with thicker cardboard tubes that are physically closer to mailing tubes than paper towel rolls. The balsa wood fins are replaced by aircraft-grade plywood, almost always mounted using a double-wall construction technique for extra strength. These rockets are large enough that sales are restricted to people aged 18 or older, but they are not large enough that they require FAA clearance.

High-Power Rockets and Beyond

Rockets that weigh over 3.3 pounds (1.5 kilograms), have more than 4.4 ounces (125 grams) of propellant, use fast-burning propellant, or have heavier construction require FAA approval before flight. The propellant restrictions apply to some G and all H and above motors I'm aware of. This is the realm of high-power rocketry. Unlike low- and mid-power rockets, you can't just go down to the store or pop online and order H motors and beyond. You need to be certified through either NAR or Tripoli to fly high-power rockets.

High-power rocketry starts with the Level 1 certification. This is a relatively easy certification to obtain. Under supervision, you must build and successfully fly a rocket that uses an H or I rocket motor. The rocket will typically weigh a few pounds and reach an altitude of 2,000–3,000 feet, although there are lots of exceptions.

The Level 2 certification is a little more involved. It starts with a written exam that covers engineering, legal codes, and safety. After passing the exam, the candidate must build and successfully fly a rocket that uses a J, K, or L motor. These rockets usually weigh several pounds and can reach altitudes well over a mile high. They may still use heavy cardboard tubes, like Level 1 rockets, but most will sport fiberglass fins.

A complete engineering package is required for the Level 3 certification. This engineering package describes the design and construction of a rocket using an M, N, or O motor. It includes construction details and computer analysis predicting the flight characteristics of the rocket. As with the first two certifications, the applicant has to build the rocket and fly the rocket successfully. Level 3 rockets are generally either very large or fly very high. Lighter rockets will easily break the sound barrier—some can hit speeds of Mach 3. These are almost always built from fiberglass or carbon fiber. Larger rockets can weigh a hundred pounds or more, and might still be built from heavy cardboard tubes.

Rockets with P or larger motors are beyond the scope of even high-power rocketry. These are privately built research rockets (often called "amateur rocketry"), some of which can visit the lower reaches of space. Each flight requires the kind of engineering analysis used for a Level 3 certification, and the FAA must individually approve each flight.

Water Rockets

While solid propellant rockets were getting bigger and bigger, water rockets didn't really take off until about the 1970s, when soda manufacturers introduced the now ubiquitous two-liter plastic soda bottle. If you've ever had the misfortune of opening a soda bottle after it has been shaken, you know the pressurized contents can squirt out with some force. As we saw in the last chapter, that's the basic requirement for a rocket motor. Most of the soda's force is lost to the formation of foam, but imagine the thrust you could get from another source of pressure that's not hindered by the energy-absorbing foam…

You don't really have to imagine. Water rockets made from soda bottles are fun to build. Getting squirted with a little water on a hot summer day is also a treat.

Figure 2-12. *A typically wet liftoff of a water rocket.*

The current world record for altitude using pressurized air and water is a little over 2,000 feet. That's almost a half-mile, nearly out of sight for a rocket that size.

Chapter 11 covers construction of a launcher for water rockets, as well as construction of a rocket with parachute recovery.

Air Rockets

Air rockets take many forms, from stomp-rockets that use a foot-powered blast of air to rockets powered by bicycle pumps or compressors. Chapter 6 introduces the concepts with a launcher and some simple ballistic air rockets.

Figure 2-13. *An air rocket takes flight.*

Lest you think that's all there is to air rockets, take a look at Chapter 21, where we look at a rocket glider powered by the same air rocket launcher. Soaring a couple hundred feet into the air, this aircraft rises as a rocket, unfolds wings, and glides gracefully back to earth.

A Typical Model Rocket and Its Flight

The next chapter gives construction details for Juno, a very typical model rocket. Before we get into that, though, let's follow Juno through a flight to get an overview of how model rockets work. There are a lot of variations on Juno's classic design, of course, and we'll cover some of them in later chapters, but this rocket is a great place to learn the basics. This description also provides you with an overview that will help you understand how the pieces fit together over the next three chapters, as we build a rocket, build a launcher, and take to a field to fly the rocket.

As the launch approaches, we load an A8-3 motor into Juno and insert an igniter. The A8-3 is a smallish motor, but it's perfect for the maiden voyage of the rocket. After a shake-out flight, we'll load in a larger motor—perhaps a B6-4, or even a C6-5. Recovery wadding and a parachute are packed in, and the rocket slides down onto a launch rail. The launch lug—basically a short piece of reinforced drinking straw—holds Juno on the rail, while allowing the rocket to slide up and down easily. Figure 2-14 shows the major parts of the rocket, including the launch lug, recovery wadding, and parachute.

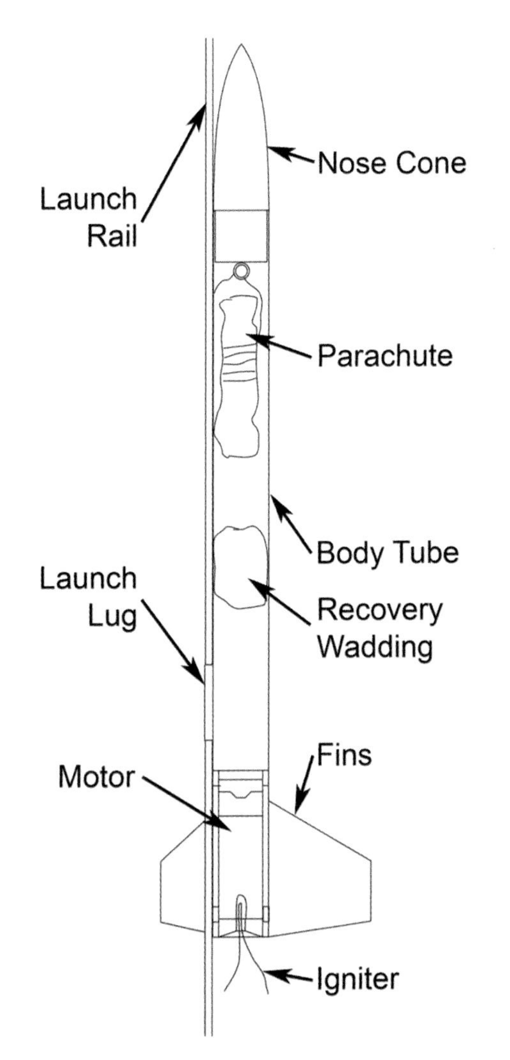

Figure 2-14. *The major parts of a model rocket.*

An electronic ignition system is connected to the igniter that is inserted into the prepackaged black powder rocket motor. Figure 2-15 shows the major parts of the motor. You will see these parts in action as we watch the rocket take flight.

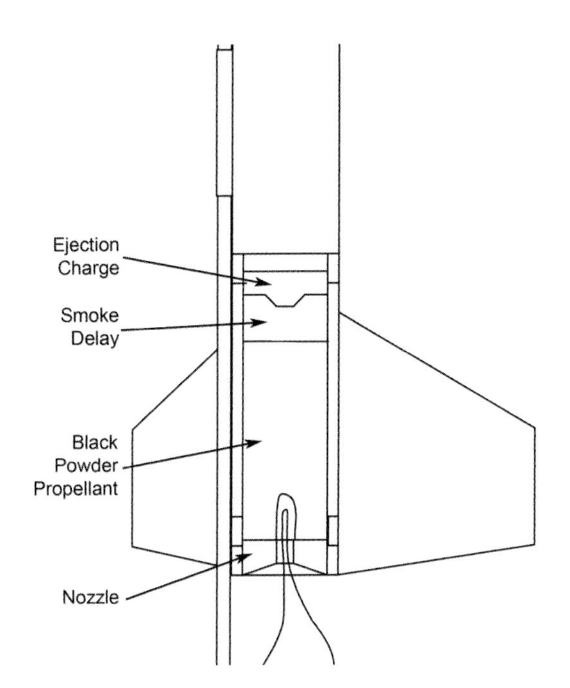

Figure 2-15. *The major parts of a typical rocket motor.*

With the igniter clips attached, the rocket is ready for flight.

Figure 2-16. *The rocket is ready to fly!*

After a short countdown, the firing button on our launcher is pressed and held. Twelve volts of electricity flow through the igniter wire, heating it to the point that the flammable pyrogen on the tip of the igniter combusts. That, in turn, starts the black powder burning in the rocket motor (Figure 2-17).

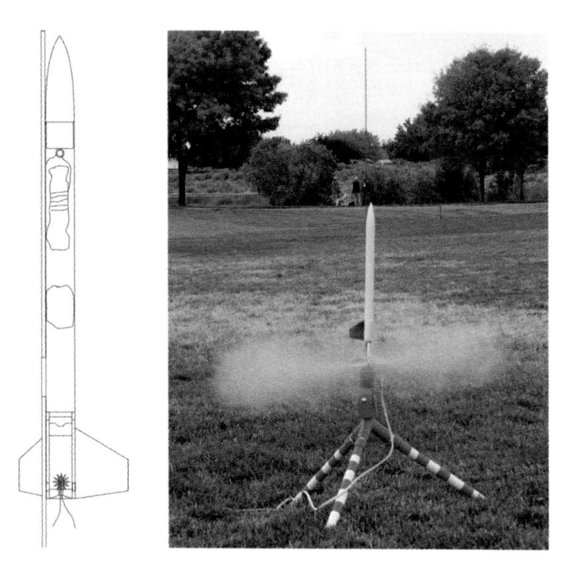

Figure 2-17. *The igniter lights the black powder propellant, and the rocket begins to lift off.*

When the black powder burns, it burns fast! Pressure builds up in the motor, forcing hot gases out of the rear of the rocket motor. The rocket starts to slide up the rail. It's not going fast enough yet for the fins to guide it, but the launch rail keeps the rocket on a straight course for the first three feet of flight. It only takes two-tenths of a second for the rocket to roar up the length of the launch rail! By then, it's traveling about 56 feet per second, fast enough for the fins to guide the rocket for the rest of the flight (Figure 2-18).

Three-quarters of a second into the flight, the black powder is exhausted. Juno is only about 70 feet in the air at this point, but it's reached its maximum speed, shooting upward at 160 feet per second—over one hundred miles per hour!

Figure 2-18. *As the motor fires, the igniter falls out. By the time the rocket reaches the end of the launch rail, it is traveling fast enough for the fins to provide guidance.*

The first portion of the flight is over. The black powder in the base of the motor has burned out, but it had just enough energy left to ignite a smoke charge at the top of the motor. This burns much more slowly than the black powder, and doesn't provide any significant thrust. It does create a smoke trail, though, making it easier to see the rocket as it coasts up to its maximum altitude, the *apogee* (Figure 2-19).

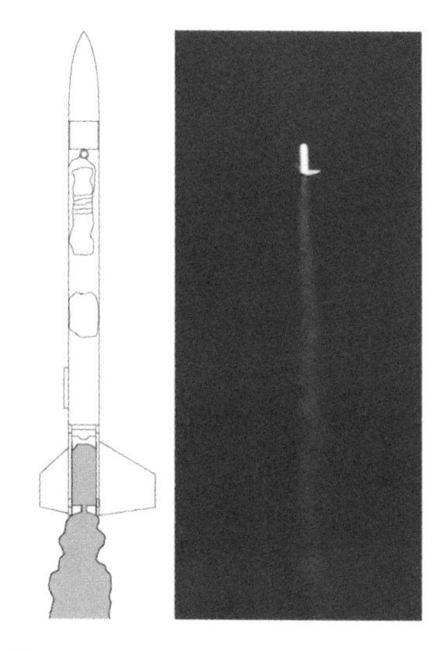

Figure 2-19. *The smoke delay starts to burn after the black powder fuel is exhausted. This gives the rocket time to coast to apogee, and makes it easier to see the flight.*

The smoke charge burns for three seconds—that's the 3 in the A8-3 motor designation. It's almost perfect timing. The rocket barely tips over after reaching an altitude of 350 feet, just as the smoke charge burns out.

The motor has one last section, the ejection charge, which fires as soon as the smoke delay stops. This is a small explosion, akin to a small firecracker, that blows the parachute out (Figure 2-20). The recovery wadding sits between the hot gases of the ejection charge and the delicate plastic parachute, protecting the chute from being scorched.

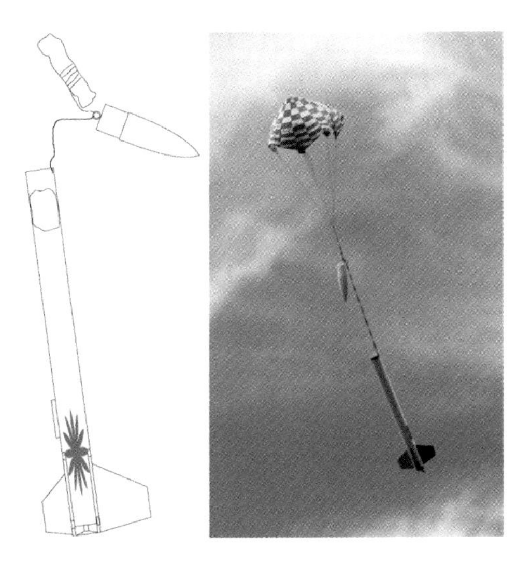

Figure 2-20. *The ejection charge fires when the smoke delay is complete. It pushes the parachute and recovery wadding out. The parachute deploys for a gentle return to earth.*

The parachute gives Juno a nice, slow descent. The flight so far has only lasted a little over four seconds, but it is a half a minute before the rocket finally returns to earth.

Doesn't that look like fun? Let's get started!

Let's Get Started!

The next three chapters show you how to build and fly a model rocket. You'll learn all sorts of sophisticated construction techniques, see how to build a great launcher, and learn how to run a safe flying field. The chapters are pretty detailed.

It may seem a bit daunting at first, but you don't really have to get that sophisticated that fast. There are a lot of great starter packages for model rocketry. Estes Industries (*http://www.estesrockets.com*) has a whole series of starter packages that retail for $30 to $50. They are called *launch sets*, and come with a launcher and one or two rockets that are really easy to put together. I've seen them on sale for as little as $20. You still need to add a rocket motor or two and some recovery wadding, but these launch sets let you build and fly a rocket the same day you get started. I'd recommend getting a package of the smallest rocket motors listed for the rocket in the launch set.

Quest Aerospace (*http://www.questaerospace.com*) has similar kits, called starter sets, that even come with a few motors and recovery wadding.

These starter sets are a great value, and the launchers they come with will work for most of the rockets in this book. They are a great place to start if you want to get flying quickly or are not sure if you are ready for the commitment you see in the next few chapters.

If you do decide to get going with a starter kit, be sure to skip ahead and read Chapter 5. That's the chapter that shows you how to select a launch site and prepare and fly a rocket. You can always go back to Chapters 3 and 4 once you get that first rocket in the air!

Juno: A Solid Propellant Rocket

3

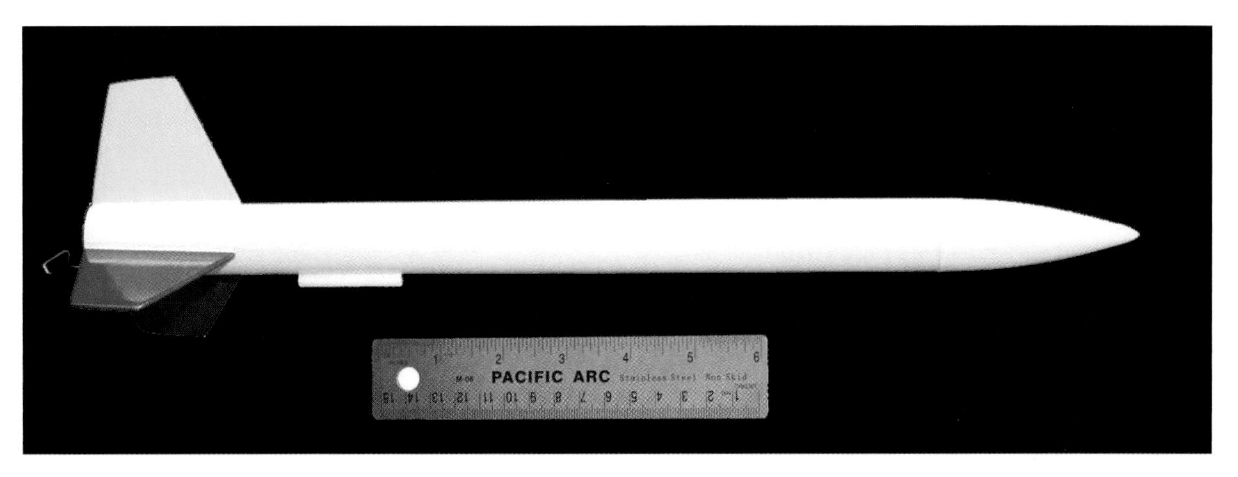

Figure 3-1. *Juno.*

The rocket we will build in this chapter is called Juno. See "What's in a Name?" on page 30 to learn a bit about where the name comes from. Juno is a workhorse, easy to build but very stable. It's also capable of reaching altitudes of 1,300 feet—over a quarter of a mile!

There are a lot of rocket motors out there. Table 3-1 shows the recommended motors for flying Juno, along with the projected altitudes. The altitude attained can vary a bit depending on the altitude you fly from, how you finish the rocket, how much glue you use, and even small variations in the motors themselves, but this should give you some idea of what to expect when you fly the rocket. If you are going out to buy parts now, buy a package of A8-3 motors. That's a great motor for the first flight.

Table 3-1. Recommended motors

Motor	Approximate altitude
1/2A6-2	60 ft
A8-3	200 ft
B4-4	520 ft
B6-4	550 ft
C6-7	1,280 ft

What's in a Name?

Names for rockets, both model rockets and big NASA rockets, range from serious to whimsical. The Apollo 11 lunar module that first landed on the Moon carried the classy moniker of Eagle, leading to those first words from the Moon: "Houston, Tranquility Base here. The Eagle has landed!" Apollo 10's lunar module had the fun name of Snoopy.

Like many of the rockets in this book, Juno is named after an asteroid. German astronomer Karl Harding discovered the asteroid in 1804. It's about 200 miles (320 kilometers) across at its widest point, but it isn't round. In another direction, it's only about 2/3 that size. Its orbit is fairly irregular, too. Juno's *perihelion* (closest point to the Sun) is about twice as far as Earth's orbit, while its *aphelion* (farthest distance from the Sun) is about 3.4 times Earth's orbit. It is probably the tenth-largest asteroid (estimates vary), accounting for roughly 1% of the total mass of the asteroid belt.

Like many of the bodies in the solar system, Juno's name comes from mythology. Juno was the highest-ranking Roman goddess, sometimes referred to as the Queen of Heaven. Her symbol is a star mounted on a scepter. All in all, a fine name for our first rocket!

Parts and Tools

Table 3-2 lists the parts you will need to build Juno. A list of additional tools and supplies is found in Table 3-3. Read through the construction description before you go shopping, though; there are lots of places where you can make substitutions, and some of the substitutions are a lot less expensive than the items shown in the tables.

Table 3-2. Parts list

Part	Description
BT-50 body tube, 12″	The main body tube for the rocket. This will be cut from a longer tube; they usually come in 15- or 18-inch lengths.
BT-20 body tube, 2 3/4″	The motor mount. This is precut in the Estes Designer's Special. It can easily be cut from a longer BT-20 body tube.
BT-50 size nose cone, 2 3/4″	Any BT-50 nose cone of balsa or plastic that is at least 2 1/2 inches long will do. Stick with balsa wood if you turn your own. Harder woods can turn a fun model rocket into a potentially dangerous projectile.
Screw eye (balsa nose cones only)	You will need a screw eye if the nose cone is balsa wood. This is screwed into the base of the nose cone to provide an attachment point for parachutes and shock cords. Plastic nose cones have an equivalent molded into the base of the nose cone.
3/32″ balsa fin stock	The balsa wood will be used for the fins. See "Fins" on page 34 if you intend to buy your own balsa at the local hobby store.
CR-20/50 centering ring (2)	Two spiral-wound centering rings are used to hold the motor tube inside the main body tube. There are lots of substitutes, from balsa wood to cardboard, but the commercial version is cheap and very easy to work with.
2 3/4″ engine hook	This is a small metal hook. They are too cheap to make it worthwhile manufacturing your own.
1/8″ launch lug	It's actually a little thicker, but the launch lug we will use is designed for a 1/8″ launch rail, so it's generally called a 1/8″ launch lug. Get the commercial version. You will need a 1 1/2"-long piece.
1/8″ shock cord, 18″	Made from rubber or elastic. This will come in the Estes Designer's Special, or you can buy some 1/8"-wide elastic from the fabric store. You will need a piece 18 inches long.
12″ plastic parachute	These come in kit form. If you're buying parts individually, though, skip this one. Buy some crochet string and some duck tape and get it in a colorful plastic sack from the store, instead.
Snap swivel (2, optional)	I like to attach parachutes and shock cords with a snap swivel, making it easier to move a parachute or payload bay from one rocket to another.

Table 3-3. Tools and supplies

Tool or supply	Description
Hobby knife	A hobby knife with several extra blades.
Cutting board	A cutting board or mat, approximately 18 x 24 inches.
Glue	Yellow wood glue and a small tube of plastic glue. The plastic glue is only needed if your plastic nose cone comes in two parts.
Sandpaper	Medium grit and extra fine grit sandpaper.
Sanding block	A small block of wood, about 0.75 x 1.5 x 2 inches, will do nicely.
Wood primer	Filling, sandable primer, or sanding sealer.
Paint	Enamel spray paint, acrylic paint, or model airplane dope. Practically anything will work as long as you don't mix paint kinds.
Paper	Used for fin guides, cutting guides, and shock cord mounts. Normal notebook paper or printer paper is fine.
Metal ruler	A metal-edge ruler is not essential, but is very handy. Get one about 15 inches long.

Tool or supply	Description
Pencil	Not a pen, but a pencil. You'll use it to mark body tubes.
Wax paper	Glue does not stick to wax paper. Use it to cover surfaces where glue might drip while parts dry.
Tape	Several kinds of tape are handy. At the very least, get painter's tape, used to mask areas while spray painting.

Where to Buy Parts

Most of the solid propellant rockets in this book can be built from the parts in the Estes Designer's Special, a collection of rocket components available by mail order from Amazon.com and many other retailers. You will need a few additional odd parts for some of the projects.

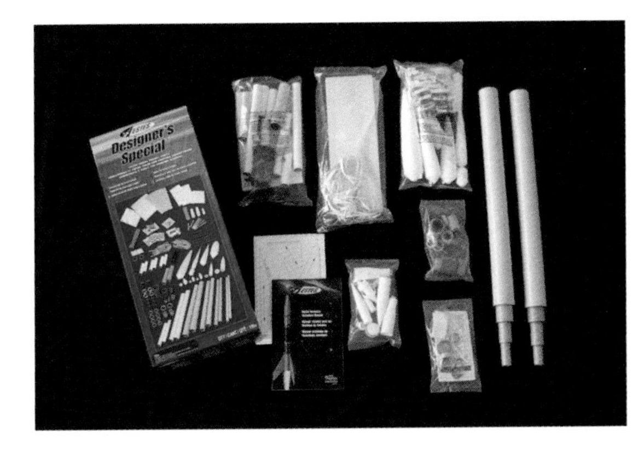

Figure 3-2. *The Estes Designer's Special*

All of the parts can also be purchased from mail-order houses and hobby stores. A quick Internet search will turn up a long list of hobby stores and model rocket companies with an online presence. Here are a few that I have dealt with over the years:

http://www.estesrockets.com
: The 800-pound gorilla of the low-power model rocket world, Estes Industries has been around since the dawn of model rocketry. It makes and sells the Estes Designer's Special, as well as various replacement parts, motors, and kits.

http://www.questaerospace.com
: One of the few competitors for Estes in the low-power model rocket arena, Quest has a nice online store featuring motors, parts, and kits.

http://www.amazon.com
: You can find the Estes Designer's Special, as well as motors and kits from many manufacturers, on Amazon.

Local hobby stores frequently carry model rocket kits and supplies. Most will have starter sets, kits, and motors. Some will have individual parts like body tubes and motors, too. There's nothing quite like the convenience of dropping into a local store and seeing all of the parts firsthand, or picking up an igniter or motor for a launch the same day. Try to support your local retailers so they are there to support you when you need it.

You will also find an extensive list of retailers, most with an online presence, in Appendix B.

What About Kits?

A quick online search or a trip to the hobby store will show a lot of rocket kits. Why not just buy a kit?

Why not, indeed. Go for it.

About half of the rockets I build are scratch built, and about half are built from kits—sometimes heavily modified, but still a kit. This book shows you how to build everything using nothing but raw parts, which is a lot of fun. However, there's a lot to be said for kits, chiefly that everything you need is included in one nice bundle.

There are equivalent kits available for most of the rockets in this book. Each chapter that shows a rocket design will also show one or two equivalent kits, if any are available.

Juno is a basic sport rocket, and there are a lot of equivalent kits, including the following:

Alpha
: From Estes industries, this was my first rocket many, many years ago. It's very similar to Juno. You can also buy the Alpha in bulk packs for schools or in starter sets with a launcher and a few motors.

Astra 1
: From Quest Aerospace, this rocket features through-the-wall fins, something generally not seen until rockets are quite a bit bigger. The Astra 1 is also available in bulk packs.

A Detailed Walk-Through of the Parts List

Since this is our first solid propellant rocket beyond the match head rocket, it's appropriate to take some time for a thorough look at the materials used to build model rockets.

Body tubes

Body tubes are the main structural element in the model rocket. They are used for the exterior body of most rockets, as well as for payload bays and engine mounts, and sometimes for decorative elements.

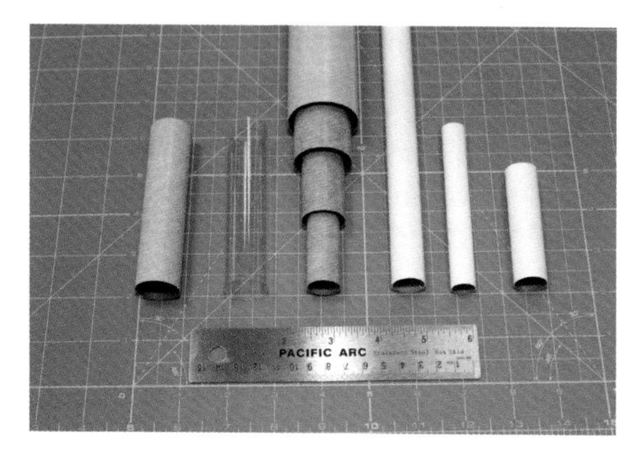

Figure 3-3. *Body tubes are generally made from spiral-wound paper. Seen here are the BT-5, BT-20, BT-50, BT-55, and BT-60. The clear payload tube is the same size as a BT-50.*

Body tubes are almost universally made from spiral-wound paper, usually brown but occasionally white or yellow. Some payload bay tubes are made from clear plastic, and a few rockets are built around other components entirely, but the paper body tube is the mainstay of model rocketry.

Body tubes come in a variety of diameters and lengths. The most common lengths are 18 inches and 15 inches. Tube diameters are most commonly labeled with BT- followed by a number. The BT- numbers originated as stock numbers for Estes Industries body tube sizes, but became a de facto standard. Table 3-4 lists the most common body tube sizes.

Table 3-4. Common body tube sizes

Size	Inside diameter (in)	Outside diameter (in)	Inside diameter (mm)	Outside diameter (mm)
BT-5	0.518	0.544	13.2	13.8
BT-20	0.710	0.736	18.0	18.7
BT-50	0.950	0.976	24.1	24.8
BT-55	1.283	1.325	32.6	33.7
BT-60	1.595	1.637	40.5	41.6
BT-70	2.180	2.217	55.4	56.3
BT-80	2.578	2.600	65.7	66.0

Metric or Imperial Units?

Table 3-4 shows the body tube diameters in both metric and imperial/US customary units. Throughout the book, you will see these systems of units used interchangeably. Why not just stick with one or the other?

This is a book about rocket science, so I'd prefer to stick with the metric system and ignore imperial/US customary units. There are two reasons for *not* doing that, though. The first is that a lot of people who read this book are still not familiar with the metric system, despite the fact that legally, the metric system has been "the preferred system of weights and measures for United States trade and commerce" since 1988. Quick: how big is an 18 mm tube? Oh, about 3/4 of an inch...I've used the metric system professionally for years, but I still visualize 3/4 of an inch better than 18 mm.

The other reason is that there is no standard within model rocketry. When you pick up a magazine or article about model rocketry, you'll see people switching back and forth between metric and imperial units with dizzying frequency, especially in high-power rocketry. You'll see people talking about their 3-foot-long 29 mm rocket (the body tube diameter) that reached 5,000 feet in altitude. The motor might have 1,000 newton-seconds of total impulse (a metric unit), with a maximum thrust of 250 pounds (an avoirdupois unit). Since you're going to see this constantly, you might as well get used to it here.

This book takes a pragmatic approach, embracing the schizophrenic nature of measurement in the United States by freely switching between metric and imperial. For the most part, I'll stick with the units I see used most often in hobby rocketry, occasionally giving both units when you might see both in articles and reference sheets. Your calculator no doubt enables conversions between the various units. It's how measurement is done in the real world.

The BT-20 body tube is exactly the right size to hold a typical A, B, or C model rocket motor. While the diameter varies a bit, most rocket motors are 0.7 inches in diameter and 2.75 inches long. Juno uses a short section of BT-20 for the motor mount.

Juno uses a BT-50 for the main body tube. This makes the rocket a little bigger than the motor mount, giving room for a motor retention clip that is internal to the rocket. The BT-50 is also the right size for motor mounts for D and E motors.

We will use several of the other body tube sizes in other rockets. The BT-5 is perfect for tiny motors and decorative trim, the BT-55 is just right for the main body tube on a D or E rocket, and the BT-60 is great for large, impressive-looking rockets.

Nose cones

Nose cones are usually made from plastic or balsa wood. They are never made from metal. I prefer working with balsa wood, which is easier to modify. Plastic nose cones are becoming more and more common, though, because they are cheap to make in large quantity.

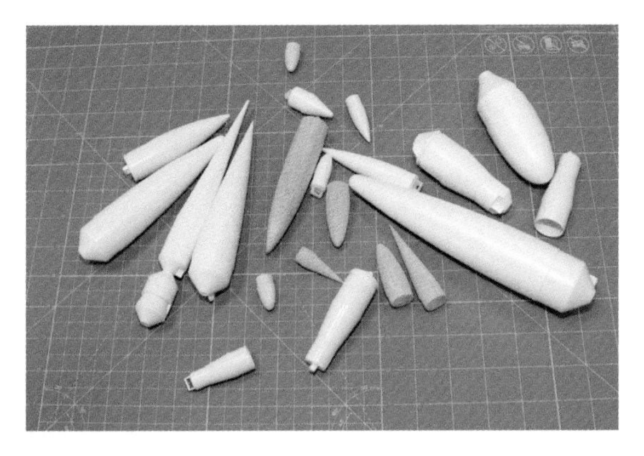

Figure 3-4. *Nose cones come in a variety of sizes, shapes, and materials. Most are plastic or balsa wood.*

Juno is designed for a plastic nose cone from the Estes Designer's Special. It's the one that fits the BT-50 body tube with a length of 2 3/4". That's the length of the part that is exposed when the nose cone is in the body tube.

Can You Safely Make Substitutions?

You might get a package of nose cones from another manufacturer, or Estes may change the Designer's Special. Is it safe to make changes?

Within certain limits, yes. We'll go over rocket design in Chapter 7, where you'll learn all of the things you need to know to evaluate a rocket design for stability. For now, these rules of thumb will keep your rockets stable if you vary from the design of the rockets in this book:

1. You can always use a longer, heavier nose cone. That increases the weight at the front of the rocket, bringing the center of gravity forward, making the rocket more stable.
2. You can always lengthen the body tube. That also moves the center of gravity forward.
3. You can add more fins or make them larger.
4. You can change the shape of the fins as long as you keep the surface area at least as big as that of the fins in the design. You can move them toward the base of a rocket, but not toward the nose cone. You can sweep the fins more, but only toward the base. All of these changes move the center of pressure toward the base of the rocket (or leave it unchanged), making the rocket more stable.

Screw eyes

You need some way to attach a parachute and shock cord to the nose cone of a rocket. A small screw eye is a perfect attachment point for a balsa nose cone. Plastic nose cones come with a premolded loop for this purpose, so you don't need a screw eye if you are using a plastic nose cone like the ones in the Estes Designer's Special.

Screw eyes are available at hardware stores. Look for small ones, about 1/2" to 3/4" long.

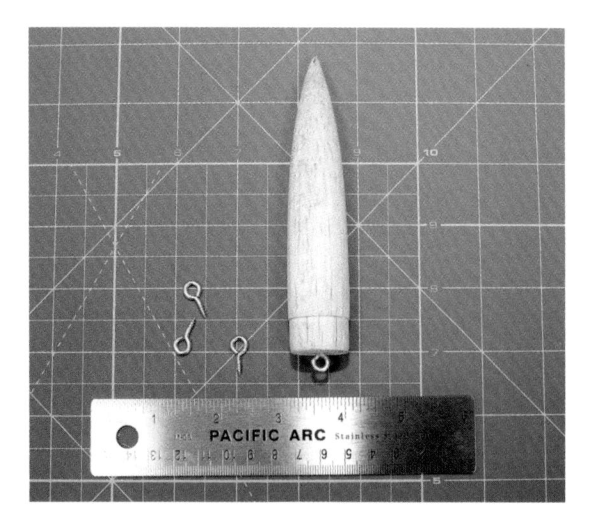

Figure 3-5. *Screw eyes are mounted in balsa wood nose cones to provide a place to attach shock cords and parachutes.*

Fins

Kits sometimes come with plastic fins, but almost all scratch-built rockets use balsa wood sheets for fins. Balsa wood is light, strong, and easy to work with, so it is likely to remain the material of choice.

Figure 3-6. *Fins are usually cut from sheets of balsa wood, although preformed plastic fins are fairly common.*

The fins for Juno are cut from 3/32"-thick balsa wood sheets. It's a great general-purpose thickness for small- to medium-size rockets flying on A, B, or C motors. Other common thicknesses are 1/16", which we will use for a high-performance rocket in Chapter 15; and 1/8", which we will use for larger rockets and rockets with D and E motors. Most sheets are three or four inches wide. Hobby stores generally carry sheets that are 36 inches long, but this varies a bit.

Like any natural material, balsa wood varies from one sheet to the next. For rocket fins, pick the densest, hardest balsa sheet in the bin. It is also important that the sheet not have twists or warps. Sight along the length of the sheet to see any bends and warps. A sharp, localized bend is not as bad as a long, gentle twist—you can always cut around a short imperfection.

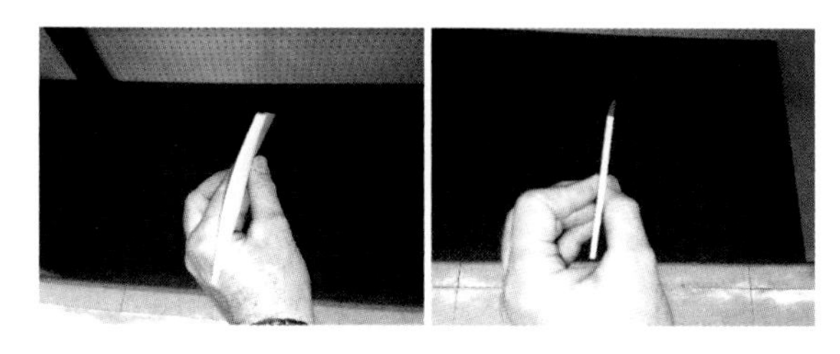

Figure 3-7. *Check balsa sheets for twists and bends by sighting along the length of the sheet. The sheet on the left is clearly not good fin material. The one on the right will work well.*

Centering rings

Centering rings are used to mount one body tube inside another. They are made from a variety of material, usually thick cardboard or spiral-wound paper. Juno is built with two wound paper rings—the green ones in the Estes Designer's Special.

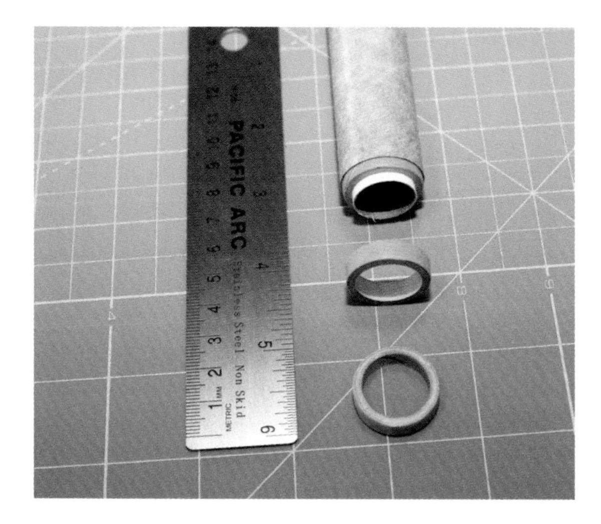

Figure 3-8. *Centering rings are made from cardboard or wound paper. You can see how the centering ring holds the BT-20 perfectly in the middle of the BT-50.*

Engine hook

There's a considerable amount of force driving the rocket motor forward as it launches, and another force that is trying to shove the motor out of the tube as the ejection charge fires. The rocket must be designed with some mechanism for holding the motor in place, preventing it from moving forward during the boost phase or backward when the ejection charge fires. It's nice if the motor retention system also makes it easy to install and remove the motor.

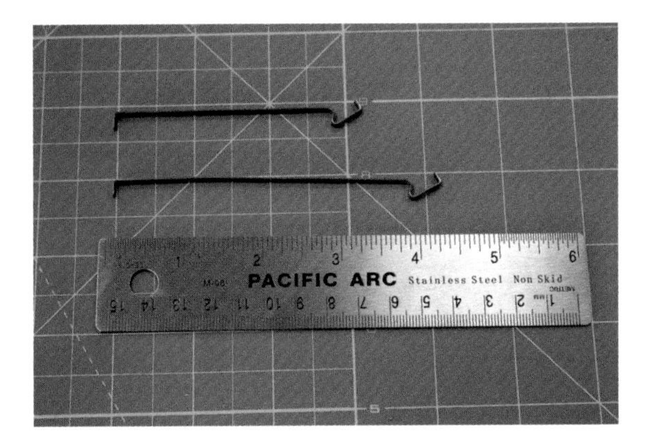

Figure 3-9. *The engine hook holds the rocket motor in place during the boost phase and when the ejection charge fires. Juno uses the shorter engine hook. The longer engine hook is for E-class motors.*

There are several ways to hold a rocket motor in the motor tube. By far the easiest and most reliable is the engine hook. That doesn't mean the engine hook is the only way to hold the motor in place, though. In later chapters, we'll see rockets with friction mounts, and even a couple of rockets that don't have a motor retention system at all—the motor is allowed to pop out of the rocket when the ejection charge fires.

For Juno, though, we'll stick with the engine hook. It's by far the best choice for this rocket. We'll revisit the choice when we build rockets where an alternative makes sense.

Launch lug

Launch lugs are thin tubes, usually made from paper reinforced with Mylar. There are two standard sizes. The smaller is about 5/32″ on the inside, designed to slide over a 1/8″ rod used as a launch rail on the rocket launcher. This is the size used on Juno. Rockets that weigh more than about 4 ounces use a slightly larger size (7/32″ diameter), which is designed for a 3/16″ launch rod. Occasionally you will find some other configuration, like the plastic rings with attached guides shown in Figure 3-10.

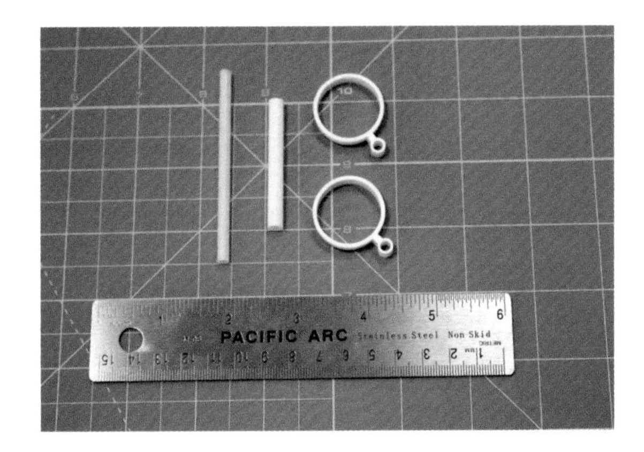

Figure 3-10. *The launch lug guides the rocket until it picks up speed.*

Shock cord

The shock cord is usually made from rubber or flat elastic. Rubber works great and is inexpensive, so that's what you usually find in kits and packages like the Estes Designer's Special. Flat elastic is much easier to find; you can buy a large quantity for a few dollars at the local fabric store. They both work well.

Shock cord is available in a lot of sizes. It's common to use 1/8"-wide shock cord for smaller rockets like Juno, switching to 1/4"-wide shock cord for larger rockets. The shock cord should be about one and a half times the length of the body tube of the rocket.

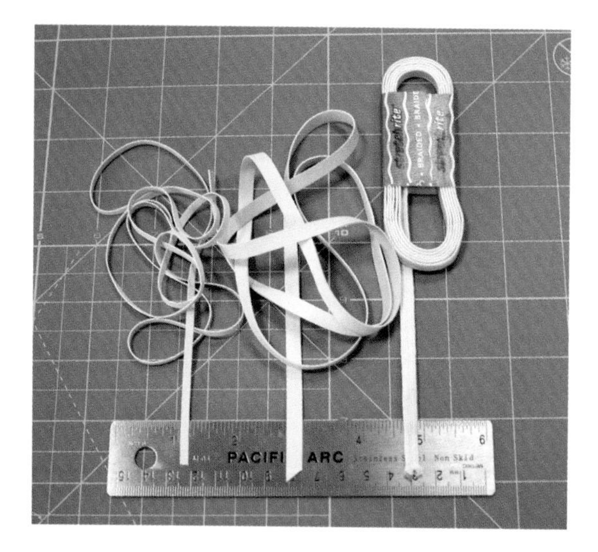

Figure 3-11. *The shock cord keeps the nose cone and body tube together after ejection.*

Parachute

Parachutes are usually made from thin plastic. Kits come with colorful printed designs, and I like them a lot. When they wear out, though, I rarely buy another one. Parachutes are very easy to make from grocery bags. An even better material is the thin metalized Mylar used for helium party balloons. Add some crochet string and a bit of tape, and you can create custom parachutes for practically nothing.

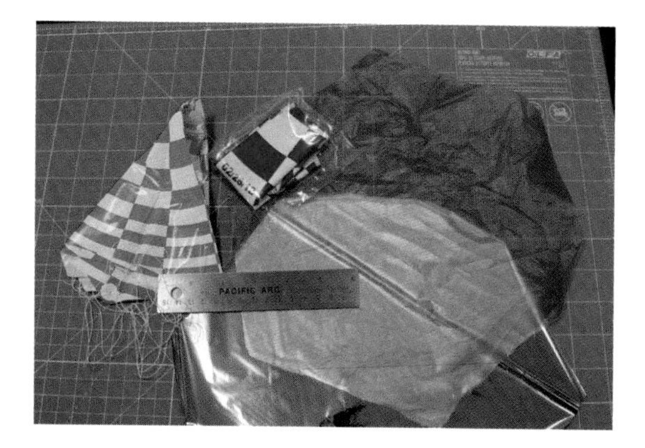

Figure 3-12. *Parachutes are usually made from thin plastic. Shown here are two preassembled parachutes from the Estes Designer's Special and two parachute canopies, one cut from a plastic bag and another cut from a Mylar balloon.*

Larger rockets that carry very heavy payloads, like egg lofters, use rip-stop nylon parachutes. We'll cover those later in the book.

Snap swivel

Snap swivels are optional, but I highly recommend them. Snap swivels make it easy to attach the parachute to the nose cone, and also make it easy to swap a parachute in the field if one gets torn.

Figure 3-13. *Snap swivels are a great way to attach parachutes to the nose cone.*

You can buy snap swivels from almost any place that sells fishing supplies. Look for snap swivels that are 1" to 1 1/2" long.

A Detailed Walk-Through of the Tools and Supplies

Let's take a look at some of the tools commonly used to build model rockets. You probably already have some of these items lying around the house from other hobbies and projects, and there are lots of places where alternatives will do. You can save some cash for rocket parts and motors by reading the descriptions for the tools carefully to see where you can make substitutions.

Hobby knife

You don't need fancy power tools to build model rockets. They are made from paper, balsa wood, and plastic: all materials that are easy to cut with a hobby knife. Hobby knives are available at almost any hardware store or hobby store. The set in Figure 3-14 is a pretty fancy one, but the only knife I used for Juno was the small one, and the only blade was the common #11 blade.

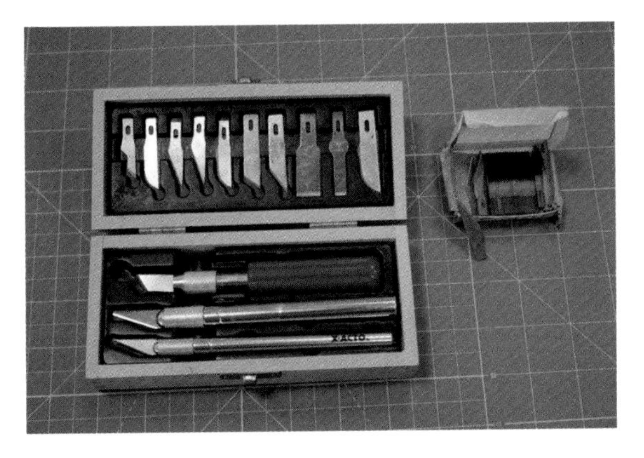

Figure 3-14. *A hobby knife is essential. A nice set is very handy.*

Get some extra blades, too. Knife blades dull quickly. I generally start a new rocket with a new blade.

Cutting boards

Cutting boards allow you to cut through wood or paper and into the surface below for a nice clean cut, all without dulling the tip of the knife. They come in a variety of sizes. I like the 18″ x 24″ size for most purposes.

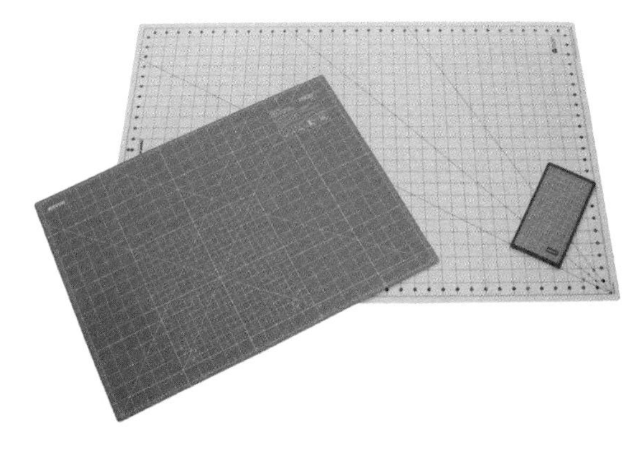

Figure 3-15. *Cutting boards protect surfaces.*

Cutting boards are available at fabric stores and craft stores. Check online for discount coupons for stores like Jo-Ann's craft store (*http://www.joann.com/*), which frequently has a 40% off coupon for one item.

Don't plan to share a cutting board with the fabric artists in the house, though. The model rocket cutting board will get a lot of abuse, collecting deep cuts, glue, and paint—all things that will make you very unpopular with fabric artists!

You can substitute an old piece of Masonite, a flattened cardboard box, or even an old piece of wood, but a cutting board is best. It is smooth, provides good support, and doesn't dull the knife blade as quickly as some of the alternatives.

Glue

Model rockets can be built almost exclusively with yellow wood glue. White glue will also work, but wood glue is quite a bit stronger—strong enough that it's the wood or paper that gives way on a properly glued fin, not the glue.

Figure 3-16. *Use yellow wood glue for almost all construction. You will need an occasional dab of plastic cement for fastening pieces of plastic nose cones.*

Model rocketeers are passionate about their choice of glue. Personally, I don't like a single glue. I use several glues and cements, depending on the material being glued. Still, if you have to pick one glue, use yellow wood glue.

Wood glue works great on porous surfaces, but not on plastics. Use plastic cement, superglue, epoxy, or household glue to glue plastic pieces together.

Cyanoacrylate (CA) glue, also called superglue, is a great choice in many situations. Some people use it as their primary glue for building rockets. There are several thicknesses of CA glue. Consider adding a small bottle of thin CA glue and another of thick CA glue, often called gap-filling CA glue, to your toolbox. Don't get a big bottle unless you plan to use it for a big project, though. The shelf life for CA glue drops once you open it.

Thin CA glue is great for strengthening parts. We'll use it later in the book to strengthen body tubes and thin balsa parts. Thick CA glue is best for gluing parts together, especially porous parts like paper and balsa wood.

Some of the components in later chapters will also use epoxy glue, but it's not needed for Juno. Epoxy is usually the strongest adhesive available, and unlike the other glues in this section, it will set quickly even if it's locked in areas that don't get good airflow.

Sandpaper

There is a wide variety of sandpaper. It's rated by *grit*, which indicates the size of the particles used in the sandpaper. The bigger the number, the smaller the particles and the smoother the sandpaper will be.

Figure 3-17. *Get at least one medium grit sandpaper and one fine grit sandpaper.*

Medium grit sandpaper is used to shape fins, remove burrs from plastic parts, and rough-sand balsa parts before finishing. A grit of 120 to 150 is about right.

The wood filler will be sanded between coats. A fine sandpaper is best for this job. I like 320 or 400 grit sandpaper for sanding the wood filler.

Sanding block

A lot of the surfaces on model rockets are large, flat areas that need to stay that way. Sanding with just finger pressure can easily etch small grooves into a flat fin. A sanding block holds the sandpaper flat, applying even pressure across the area being sanded.

Figure 3-18. *Use a sanding block to sand flat surfaces.*

Sanding blocks don't have to be fancy. I've used a piece of scrap wood that measures 3/4" x 1 1/2" x 2 1/4" for years. It's the perfect size for 1/16th of a sheet of sandpaper.

Wood primer

Balsa wood is very porous. Paint applied directly to the wood will be quite rough. There are a number of wood fillers available that prepare the wood for paint, creating a satin or glassy surface.

Figure 3-19. *Use some form of filler to prepare the wood for paint.*

While lots of products will do, my favorite when working with a brush is sanding sealer. It's getting a little hard to find, but is usually available at hobby stores that cater to model airplanes. There will be a section of paint called "Butyrate Dope" or "Hot Fuel Proof Dope," and somewhere in the rack will be a creamy, translucent jar labeled sanding sealer. If you can't find it locally, check online: you can order it directly from SIG Mfg. Co (*http://www.sigmfg.com*).

Dry-sandable filling primer works very well for spray-on applications. This is also a little hard to find. Be sure and get a filling primer, not just a standard primer. While wet-sandable primers will do, I prefer a dry-sandable primer like Rust-Oleum 2 in 1 Filler & Sandable Primer. I occasionally find this at Lowe's hardware store, but it's a specialty item and often out of stock. It's available from Amazon.com, too.

Some people skip this step. That works particularly well on rockets that have plastic fins. Sure, you can still see a little of the paper winding through the paint, but it doesn't have much effect on the way the rocket flies.

You can also use paint as a primer. It doesn't work quite as well, but you can paint a wooden part, let the paint dry, and sand the paint. Depending on the wood, you might have to repeat this process several times to get a smooth finish—or maybe you don't care if the finish is perfectly smooth. The rocket will fly either way.

Paint

Rockets should look cool. The only way to do that is with a good paint job. Almost any kind of paint that won't dissolve in water once it is dry will do.

Enamel spray paint works well, and it's what I use most of the time.

You might also want small (1/4 ounce) bottles of enamel paint in a few colors to paint details.

Figure 3-20. *Use almost any kind of non-water-based paint.*

You will need some paintbrushes. A nice wide brush, 1/4" to 3/16" wide, works really well. Brush cleaner is also a must. Paint thinner for the kind of paint you are using works well. Nail polish remover works fine for enamel paint.

Acrylic paint is my favorite if I'm painting the entire rocket with a brush. You can also use acrylic paint with an air brush. An air brush is a little more expensive to buy, and it will take a little time to get used to it, but it does save money compared to spray paint in the long run. You can also mix your own colors if you use acrylic paint and an air brush.

Model airplane paint, also known as butyrate dope, also works fairly well. It's a bit smelly, though, and needs to be used in a well-ventilated area. It also tends to shrink as it dries, which is good and bad. Shrinking on most surfaces smooths out small imperfections. Shrinking near the joint of a fin can cause the paint to pull away from the fillet.

Paper

Normal-weight writing paper is used for marking body tubes for cutting, for making some kinds of fin guides, and for shock cord mounts. Any scrap will do. Larger sheets are useful for making parachute patterns. You can paste several sheets of writing paper together, but a more practical idea is to use some of that leftover wrapping paper from the last holiday or birthday.

Pencil

There are several situations where we will mark guide lines on the rocket during construction. A normal pencil works best. Ballpoint pens tend to make indentations in the surface of balsa wood and body tubes, and often don't work on plastic nose cones. Felt tip markers work well, but tend to bleed through the paint.

Wax paper

Glue occasionally drips, but it will not stick to wax paper. Any time you are working over a surface you don't want to get glue on, especially while a part dries, cover it in wax paper. This is particularly helpful when gluing fins over a fin guide.

Metal-edge ruler

A metal-edge ruler is a great addition to your tool kit. It's really nice when you are cutting balsa wood fins or paper. A 15-inch ruler is a good length; they are available from office supply stores.

Tape

Several kinds of tape are useful. At a minimum, get a roll of painter's tape. It's available from the paint sections of hardware stores, and is usually blue or green. This stuff works a lot better than traditional masking tape for masking areas when you spray paint. Clear cellophane tape is also handy.

Building the Rocket

With parts and tools assembled, it's time to build a rocket!

Read all of the instructions before you start. I know. You hate directions. There is a reason, though. If you follow through step by step, you will build a perfectly good rocket, then discover that there are some cool options you could have used—but it's too late. Read through so you know what options you want before you start.

Cutting Body Tubes

Juno uses a 12-inch-long section of BT-50 body tube. Most body tubes are 18 inches long, while a few are 15 inches long. That means we'll need to cut the tube.

Measure 12 inches from the base of the tube and make a small guide mark with a pencil. Wrap a piece of paper around the body tube and match the edges. The paper forms a perfectly straight edge that is exactly perpendicular to the length of the body tube. Mark this line carefully with the pencil.

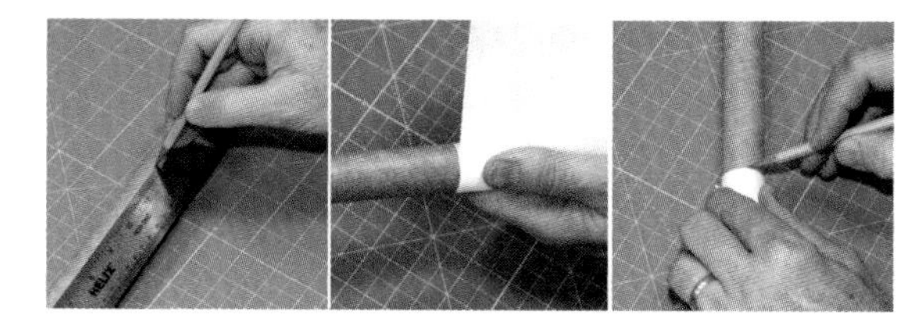

Figure 3-21. *Measure the location to cut, then use a piece of paper to extend the line around the body tube.*

Put something hard or thick into the tube, directly under the location to be cut. The long orange motor adapter from the Estes Designer's Special works well, as does a tube coupler, or a used D or E motor (not an unused one!). This gives a cleaner cut by supporting the tube as the hobby knife slices through.

Take your time, starting with a light cut that barely scores the body tube. Take extra care to create a nice, straight groove for the deeper cuts. I like to use the flat part of the blade for this cut—it makes it easier to get a straight cut.

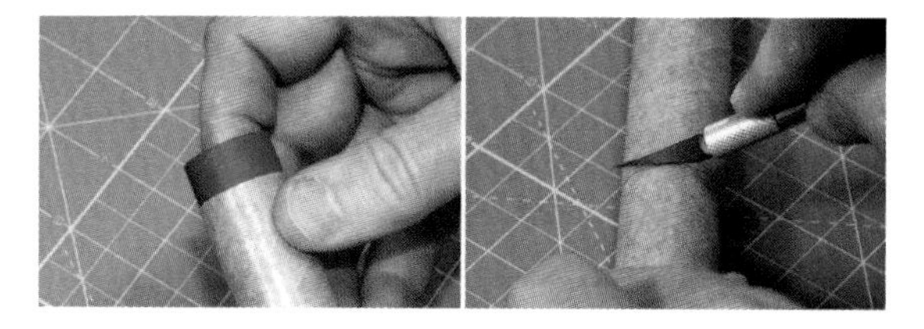

Figure 3-22. *Place a support under the cut location. Start with a shallow cut using the flat part of the knife blade.*

Switch to the point of the blade and increase the pressure after the guide cut is complete. Take your time. It should take three to five complete turns of the tube to make the cut. If you cut too deep, you could end up with a jagged edge.

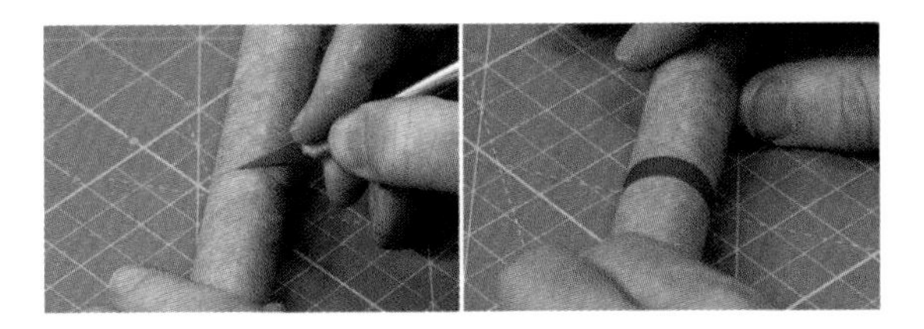

Figure 3-23. *Take your time, making the cut in three to five turns of the tube.*

If you do mess up, ending up with a body tube that is not perfectly flat or one that has some ragged edges, sand the body tube to flatten the end. Place a very fine grit sandpaper on a flat surface. Hold the tube perpendicular to the sandpaper and sand in a circular motion. You can check to make sure the end is still square by inserting a nose cone. Does it fit flush with the tube? If not, sand off the high spots.

The Estes Designer's Special comes with a precut BT-20 body tube for the motor mount. It's white, about 3/4" in diameter, and 2 3/4" long. If you are working from a longer piece of BT-20 body tube, repeat the above process to cut off a 2 3/4" piece.

Marking the Body Tube for Fins

The fins and launch lug must be straight for the rocket to slide off of the pad and fly straight. Marking the body tube is the first step in a perfect flight.

Figures 3-24 and 3-25 are fin marking guides. Use Figure 3-24 as a fin marking guide for Juno. Place the body tube on the guide, using the concentric circles to position the body tube in the exact center of the guide. Work around the rocket, making small marks at the base of the body tube to mark the locations of the three fins and launch lug.

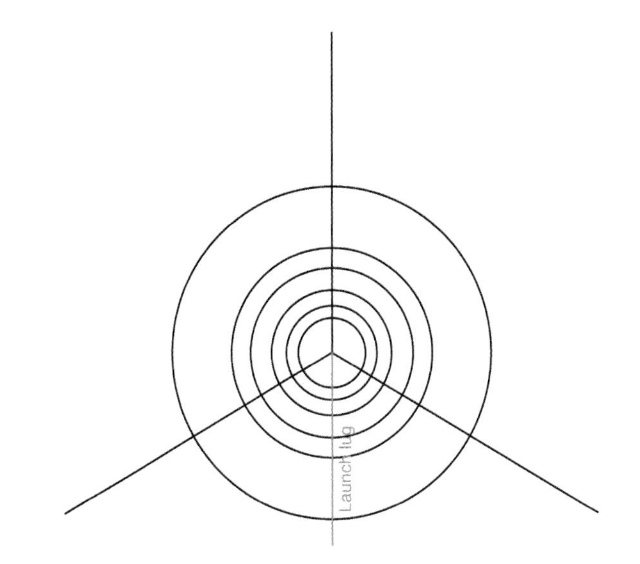

Figure 3-24. *Fin marking guide for three-finned rockets. Use this one for Juno.*

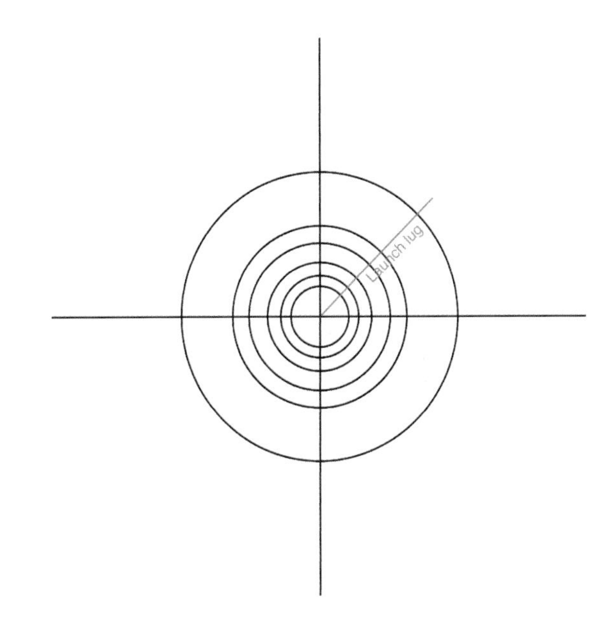

Figure 3-25. *Fin marking guide for four-finned rockets. Don't forget to mark the location for the launch lug.*

The next step is to extend the lines up the body tube. Find a convenient door jamb or the edge of a drawer, placing the body tube so it sits in the V shape formed by the wood. Rotate the tube until the guide mark is just next to the wood, then use your pencil to extend the guide lines about three inches for each fin and about six inches for the launch lug. Don't press hard-—

just hard enough to make a mark you can see clearly. Be careful not to mark the door jamb!

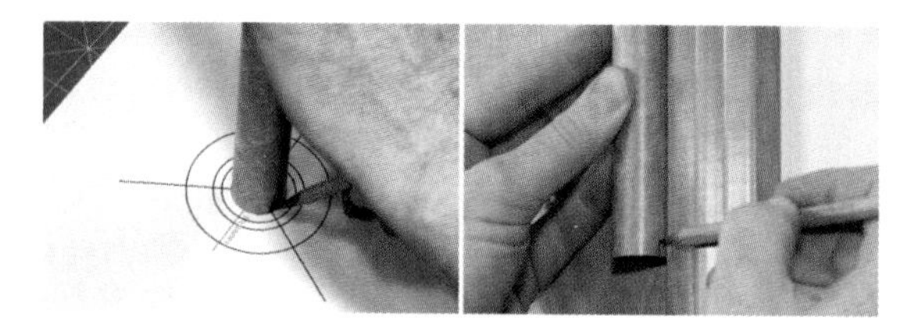

Figure 3-26. *Mark the position of the fins and launch lug using the fin guide, then extend the marks up the tube using a door jamb as a guide.*

Motor Mount Construction

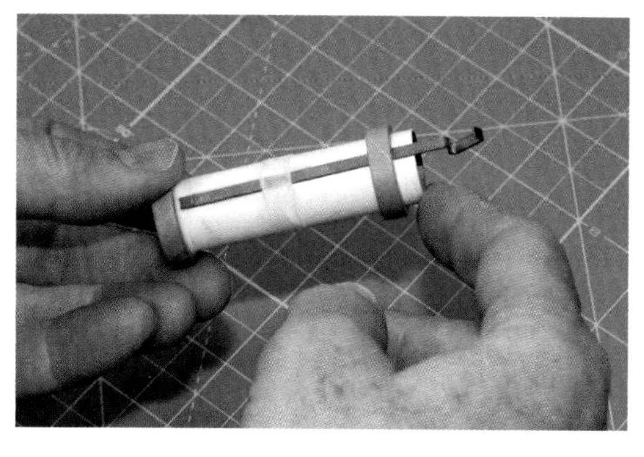

Figure 3-27. *The completed motor mount.*

The motor mount will slide into the body tube, holding the rocket motor in place during flight. It needs to be strong.

Start by measuring 1/4" from one end of the motor mount tube and making a mark 1/8" long, perpendicular to the length of the tube. Puncture the tube at this location with a hobby knife. Insert the flatter end of the engine hook into the slot.

The engine hook protrudes into the motor mount. It may not look like much, but this little tab of metal will easily hold a rocket motor in place during the thrust phase of flight.

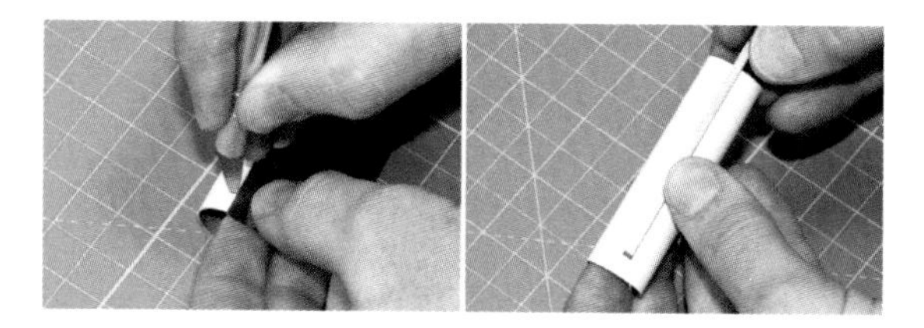

Figure 3-28. *Cut a notch in the motor mount and insert the engine hook.*

Cut a notch in one of the centering rings. The notch should be wide enough for the engine hook and about half of the depth of the centering ring.

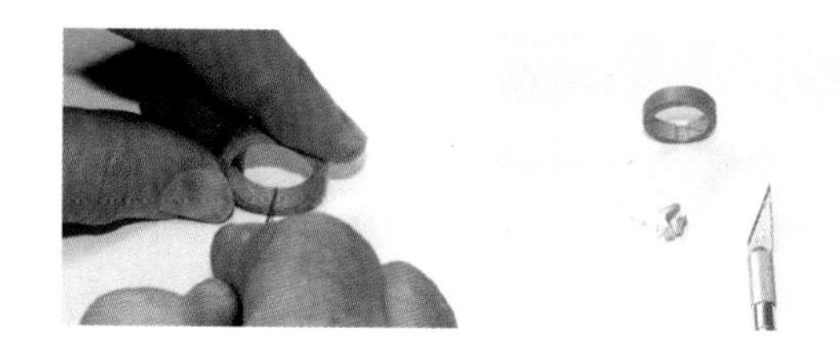

Figure 3-29. *Cut a notch in one of the centering rings.*

Test fit all of the pieces. Use your fingernail to score the edges of the centering rings if they don't slide easily over the motor tube. Use sandpaper if you must.

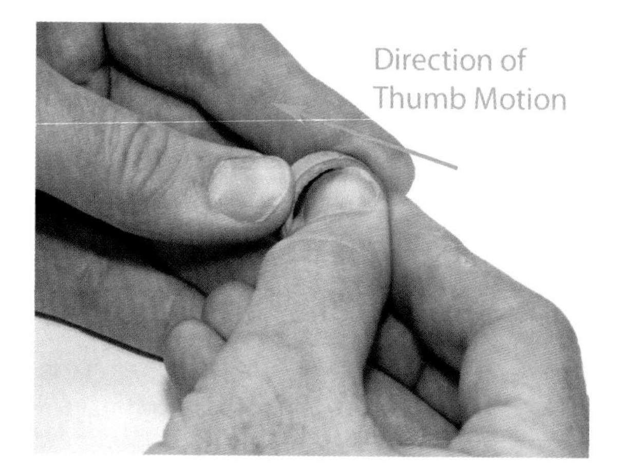

Figure 3-30. *Score the edge of the centering ring to make it easier to slide it onto the motor tube.*

Insert the engine hook. Place a bead of wood glue around the end of the motor mount, just above the slot for the engine hook, then slide the uncut centering ring onto the end of the motor mount. Wipe away excess glue with your finger, forming a nice fillet where the centering ring meets the motor mount tube.

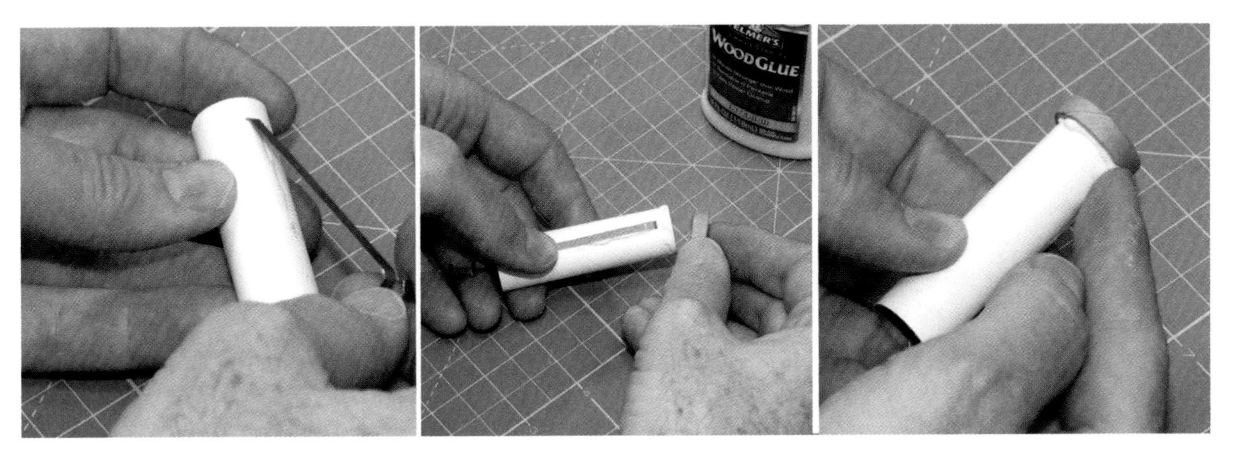

Figure 3-31. *Glue the uncut centering ring onto the end of the tube. Form a fillet using your finger.*

Fillets

You're going to see the word *fillet*, or sometimes *glue fillet*, a lot when building rockets. Filleting is a way to strengthen a joint. You form a fillet after the parts are already glued together by adding a bead of glue right at the joint where the two parts meet.

In most cases, especially when the part is on the outside of the rocket, you will want to smooth the bead of glue, since it will be too rough and nonaerodynamic right from the glue bottle. You have a great built-in fillet-smoothing tool handy at all times: just run your finger over the fillet, and then clean the excess glue off of your finger.

You might occasionally see bubbles in the glue. You can generally wipe these out with your finger. If there are any still there after the glue dries, either fill them in with a second coat of glue, or fill them in with primer when you are painting the rocket.

Wood Glue Messes

You're going to get wood glue on things—particularly your fingers—as you form fillets. That's really not a big deal. Wipe any extra glue from your fingers or other surfaces with a paper towel or other scrap of paper. Wash your hands in clear water to remove the glue, or use a damp paper towel to remove glue messes from the rocket or cutting board.

Wood glue cleans up well while wet, but is very tough once dry, so don't leave drips or smears for later!

The notch in the remaining centering ring gives a little more room for the engine hook to bend. Without some support, though, the engine hook will eventually work loose from the slot at the end of the motor tube. The Estes Designer's Special comes with a Mylar ring to support the engine hook.

Place glue under the engine hook, from the end that is inserted in the motor mount tube to about halfway down the tube. Glue the Mylar ring about halfway down the tube, then apply a liberal amount of glue over the engine hook between that ring and the forward centering ring. Don't apply glue below the ring.

Figure 3-32. *Secure the engine hook with glue and a band for stability.*

There are plenty of adequate substitutes for the Mylar ring, so don't bother ordering one if you are buying individual parts. A couple of turns of masking tape works well, as does a couple of turns of 1/2"-wide paper coated in wood glue.

Apply a bead of wood glue in a ring near the base of the motor tube. The glue should extend from 1/4" to 1/2" from the base of the motor tube. Slide the remaining centering ring onto the motor mount and form generous fillets on either side of the centering ring.

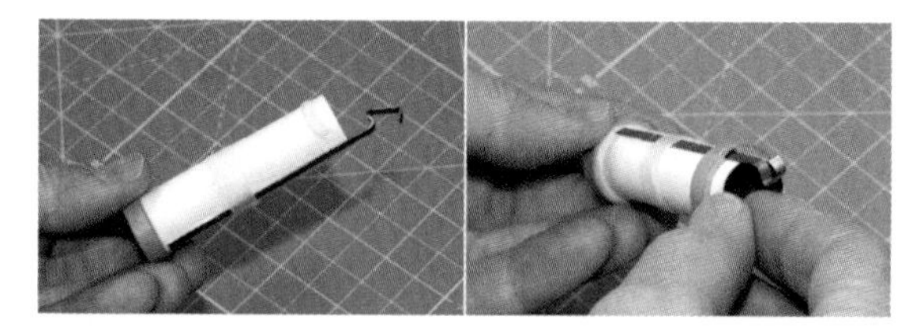

Figure 3-33. *Glue the remaining ring 1/4" from the end of the motor mount. Apply generous fillets of wood glue.*

Set the motor mount aside to dry. It's best to prop it up on its end so the glue doesn't drip off. Placing it on a small scrap of wax paper will make cleanup a lot easier if any glue does drip. The glue will set pretty quickly, but check it occasionally as the glue dries to make sure none has dripped. If any glue does drip, clean the drip with a damp paper towel or, if it is firm enough, with the hobby knife.

Cutting Fins

Take a look at a piece of balsa wood. You can clearly see the grain in the wood—long, parallel lines that were formed as the tree grew new rings. The wood is much stronger going across the grain than parallel to the grain (Figure 3-34).

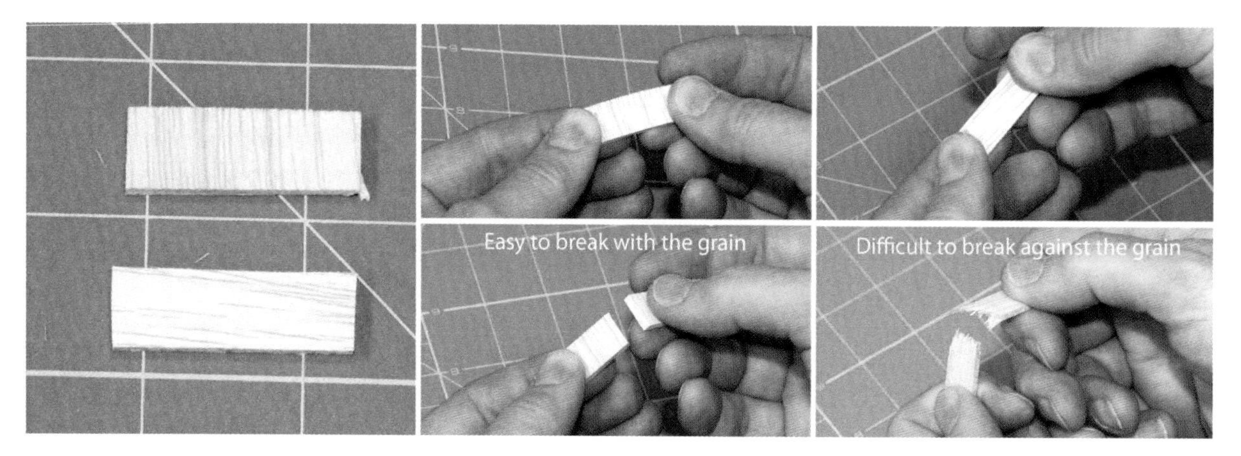

Figure 3-34. *Balsa wood is stronger perpendicular to the grain of the wood. It breaks easily and cleanly with the grain, but is harder to break against the grain.*

Cut two pieces of balsa from a 3/32"-thick plank, both about 1/2" wide and 1 1/2" long. Cut one with the grain parallel to the long axis, and one with the grain perpendicular to the long axis. Bend each piece until it breaks. It's not hard to break the piece with the grain running perpendicular to the direction of the bend, but the other piece is both more rigid and much harder to break.

If you think about it a bit, you'll understand why fins need to be cut so the grain of the wood extends out from the body tube, not running parallel to it. The general rule is to cut the fin so the grain runs parallel to the leading edge of the fin. Fins should be designed so there is never a line of grain that does not lead back to the body tube or a glue joint. If you need to cut a really fancy design, perhaps a concave curve, use multiple pieces of wood so the edges don't break off.

Take a look at the fins in Figure 3-35. *Root* indicates the part of the fin that is glued to the body tube, and in all cases, the nose cone is up. The lines with arrows indicate the direction of the wood's grain.

The fins for Juno are on the left. While the grain direction follows the general rule of cutting the fin so the grain is parallel to the leading edge of the fin, it would work just as well if the grain were perpendicular to the fin root. The middle fin shows a case where this is a bad idea, though. The bottom tip of the fin is very fragile, and is likely to break off even on a mild landing.

The rightmost fin presents an interesting case. There is no way to orient the grain of the wood so there is no fragile section. The normal rule of placing the grain parallel to the leading edge leaves a piece at the tip of the fin that is still very prone to breaking. The solution is to build the fin from two pieces of wood, as shown at the top. The second piece has the grain running vertically, where it is connected with glue to the rest of the fin. Wood glue is quite strong; if properly glued, the joint will be stronger than the surrounding balsa wood.

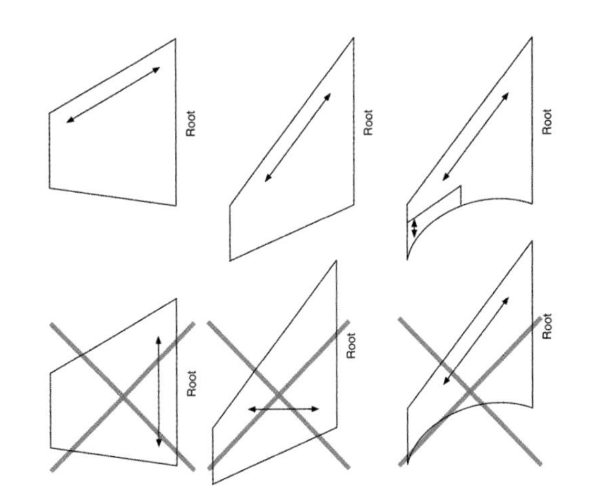

Figure 3-35. *Some dos and don'ts for fin grains.*

Figure 3-36 shows a fin-cutting guide for Juno. As in most fin guides, the direction of the grain is clearly marked. Juno fins fit together snugly on a sheet of balsa.

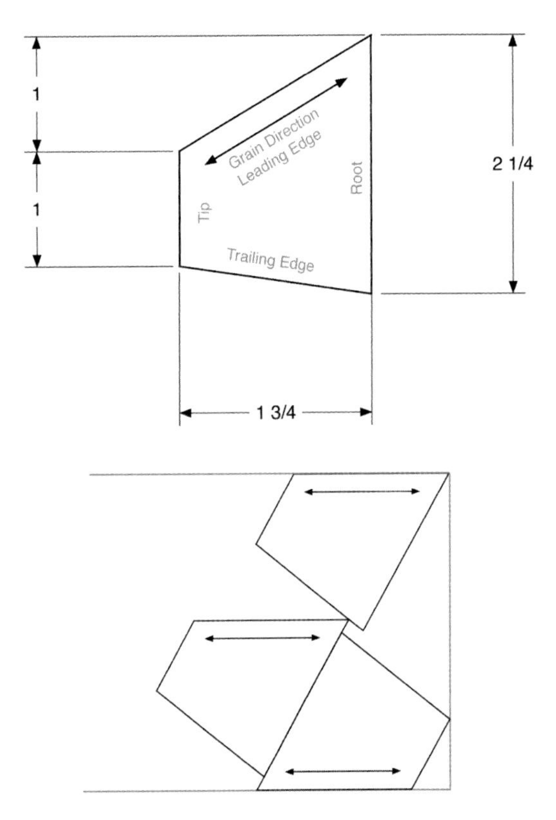

Figure 3-36. *A fin guide and one layout for efficiently cutting the fins from a plank of balsa.*

Trace the fin guide on a sheet of paper, cut it out, and use a pencil to draw the outlines of the fins on a 3/32"-thick plank of balsa wood. This is the middle-thickness plank from the Estes Designer's Special. Cut the fins using a hobby knife. Just like when cutting the body tube, start with a light guide cut, then make three to five additional cuts to cut all the way through the wood. While not essential, a metal ruler helps make nice straight cuts.

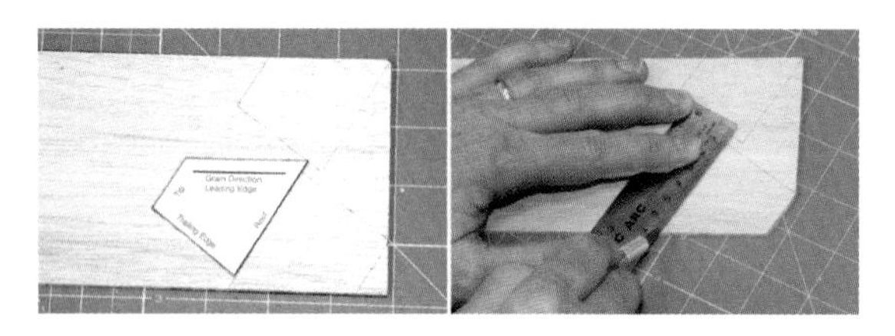

Figure 3-37. *Transfer the fin guide to the balsa wood, then cut out all three fins.*

Getting the Full-Size Patterns

This book is filled with rocket plans. Sometimes the pages of the book just aren't big enough for a full-size pattern. Even if they were, you may be reading an ebook version of the book, or may not want to cut the book up, or may not have access to a copier.

That's fine. You can find PDFs of all the patterns in this book at the author's website (*http://bit.ly/byteworks-make-rockets*). Print as many as you need.

The fins are not likely to be perfect, and that's OK. Press them together and sand them on a sheet of medium grit sandpaper, being careful to keep the stack of fins perpendicular to the sandpaper.

Sand until all fins are even.

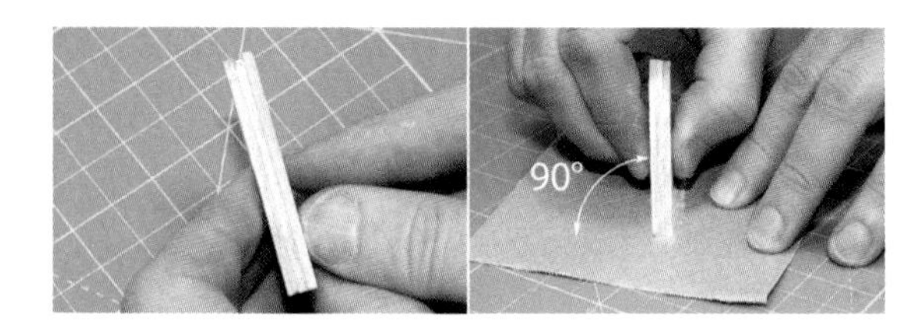

Figure 3-38. *Use sandpaper to even out the fins.*

There is nothing wrong with leaving the forward and trailing edges of the fin flat, but it will reduce the drag a lot if you round the edges. There are lots of alternatives to a simple semicircular rounding on the leading and trailing edges, but rounded fins are easy to create and work pretty well. Fin cross section is a topic we'll return to later.

Start by roughing out the rounded edge by sanding the fin at about a 45° angle from the sandpaper. Finish the shape by sanding by hand; your hand has some give and will help the sandpaper form a gentle curve. Be sure to round the leading and trailing edges and the fin tip, but do not round the fin root (the part that will be glued to the body tube). Figure 3-39 illustrates.

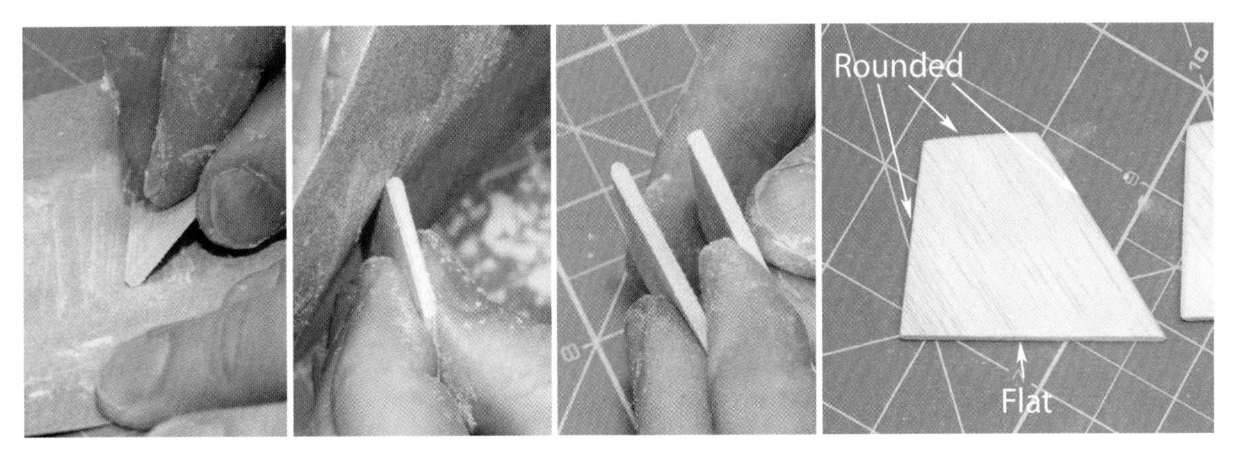

Figure 3-39. *Sand a rounded edge on the leading edge, trailing edge, and fin tip, but not the fin root.*

Now that the fins have their rough shape, it's time to smooth out the surfaces. Using a sanding block, as illustrated in Figure 3-40, sand the surfaces of each fin with the grain using medium grit sandpaper until the fin feels smooth.

Figure 3-40. *Use a sanding block to smooth the top and bottom surfaces of the fin.*

Close your eyes and feel the surface of the wood, then run your fingers along the rounded edges. Is everything smooth? If not, carefully touch up the roughness. Remember, it doesn't matter what the wood looks like—it's going to get painted soon. All that matters is what it feels like. Any roughness or irregularities will show up clearly once the surface is painted.

Gluing the Fins

Fins are glued to the body tube in two steps. Start by placing the cutting board on a flat surface that is low enough to allow you to sight down along the length of the tube when it is standing on end. Tape the fin guide under a piece of wax paper and center the body tube over the fin guide with the marked end down. Put a bead of glue onto the root of a fin, make contact with the base of the rocket, and press the fin onto the rocket, being careful to make sure the fin aligns exactly with the guide line (Figure 3-41).

Glue each fin in place, then sight along the length of the body tube from above to make sure each fin is perfectly straight and exactly follows the lines on the fin guide. Let the fins dry thoroughly, checking them occasionally to make sure they don't drift out of alignment.

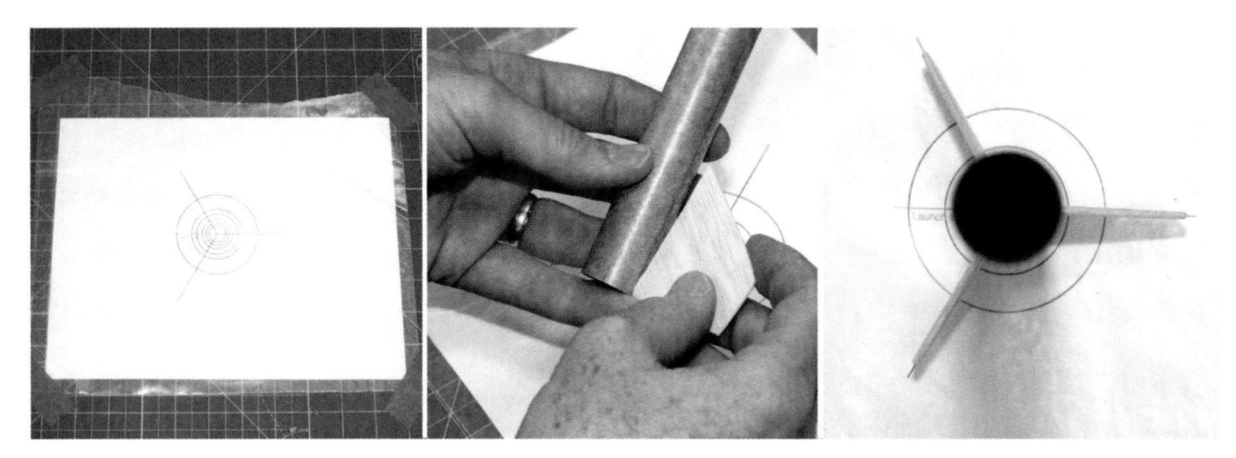

Figure 3-41. *Glue the fins over the fin guide to make sure they are aligned properly.*

Once the fins are completely dry, apply a fillet along each of the six edges where a fin meets the body tube. Hold and roll the rocket between your palms for a minute or two while the glue sets, then place it vertically on the fin guide to dry.

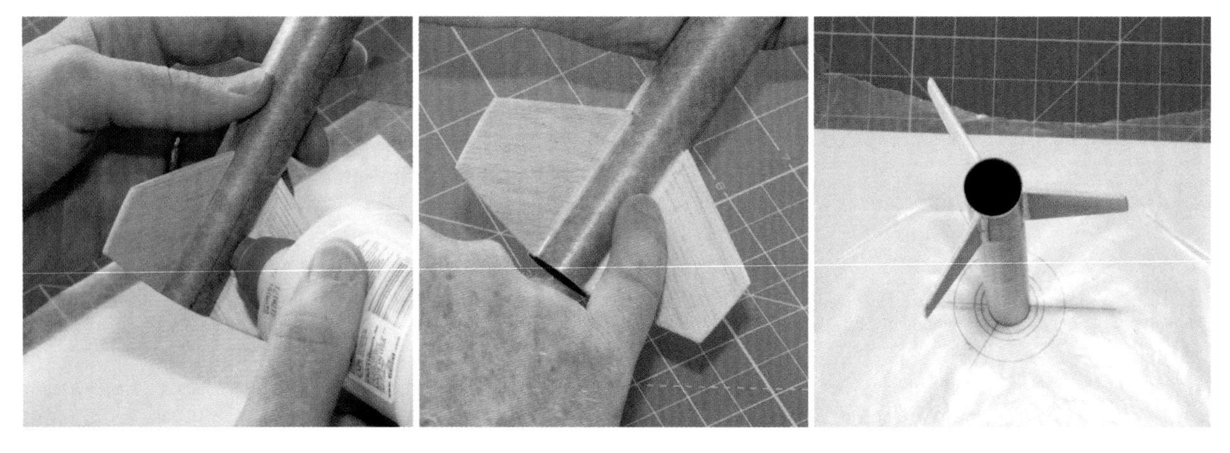

Figure 3-42. *Place fillets at all joints between the fin and body tube.*

Attaching the Launch Lug

Cut about a 1 1/2″ piece from a 1/8″ launch lug. They are generally three inches long, so you are cutting it in half. The exact length isn't really critical; anything from about an inch long to a full three inches will work, but cutting the launch lug in half gives you another launch lug for your next rocket.

Use a sharp hobby knife to saw through the launch lug. Sand the cut end on fine sandpaper if needed to remove any roughness.

Use a generous amount of glue to attach the launch lug three inches from the base of the rocket. Carefully sight along the length of the rocket to make sure the launch lug is perfectly aligned with the body of the rocket, as seen in Figure 3-43.

You can do all of this while the fin fillets are setting. Moving the rocket around as the fillets set helps prevent sags in the glue. Check the glue on the launch lug as the glue on the fin fillets sets, too, making sure there are no sags in the glue on the launch lug.

One final check is in order. As the glue starts to get tacky, but before it is completely set, carefully slide a launch rod through the launch lug to check the alignment. Straighten the launch lug if necessary, and then carefully remove the launch rod, making sure the launch lug does not twist as the launch rod is removed.

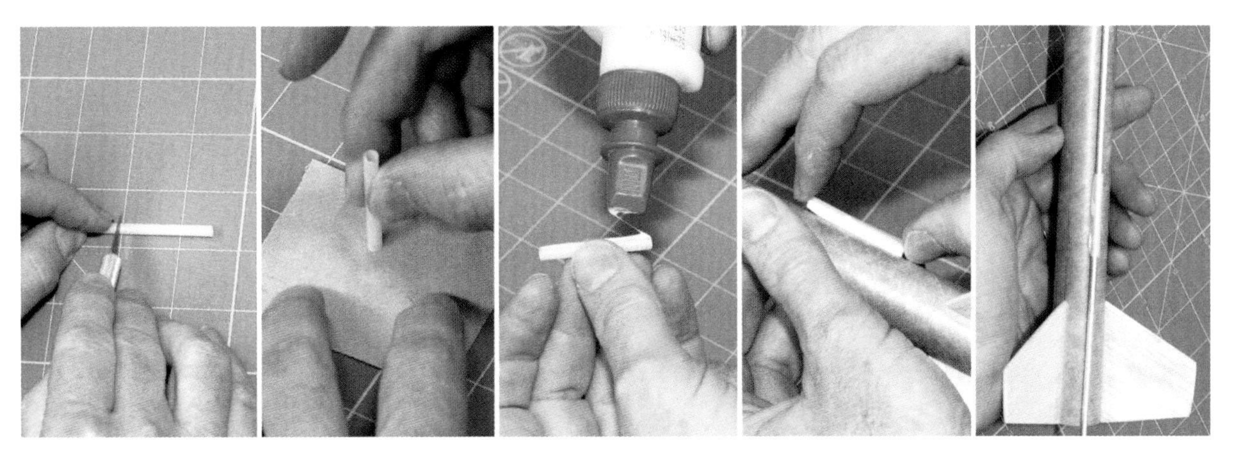

Figure 3-43. *Cut a 1 1/2" piece of launch lug and attach it 3" from the base of the rocket. Check the alignment carefully.*

Installing the Motor Mount

Test fit the motor mount in the rocket once the fins and motor mount are completely dry. Gently score the edge of the body tube or lightly sand the centering rings if the motor mount does not slip easily into the body tube.

The motor mount must be glued securely in the body tube. You will need to apply a generous amount of glue to the inside of the body tube, using your finger or a small stick, so that when you slide the motor mount into the tube, the glue gets smeared around underneath and in front of it.

Measure your finger against the motor mount, holding your thumb as a guide, so the tip of your finger almost reaches the end of the motor mount (Figure 3-44). This tells you how far into the body tube your finger needs to be when you apply the glue. Place a large dab of glue on the end of your finger and smooth this onto the inside of the body tube. Repeat as often as needed to get a thick ring of glue for the forward motor mount ring. Repeat this process with a thinner ring of glue about 1/4" into the body tube, and then slide the motor mount into the tube.

I like to align the engine hook with the launch lug; it makes it easier to connect the igniter wires to the rocket later when it is prepared for launch. In truth, though, it doesn't matter all that much.

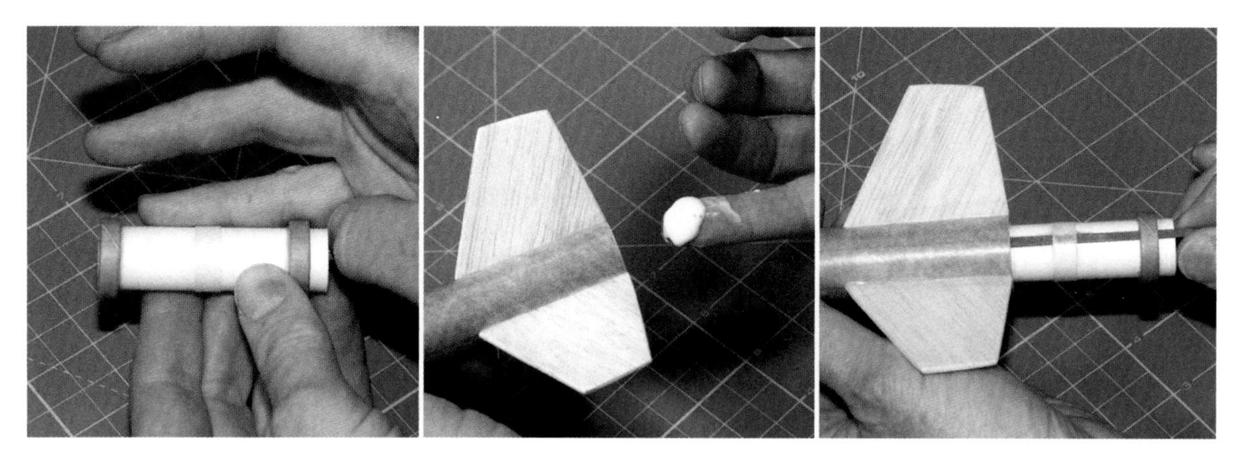

Figure 3-44. *Apply a generous amount of glue so the top centering ring will hit it when slid into the body tube. Apply a thinner ring of glue where the lower centering ring will sit.*

Slide the motor mount tube in until its end is even with the end of the body tube.

Hold the rocket upright while the glue sets. It will just take a couple of minutes.

Figure 3-45. *Slide the motor mount in until the end of the motor mount tube is flush with the end of the body tube, then stand the rocket up to dry.*

Once the glue is firmly set, turn the rocket over and dribble a glue fillet into the slot between the body tube and motor tube. The glue should cover the centering ring everywhere except where the engine hook pokes through; make sure there is no glue there, clearing away any drops that fall into this area. Stand the rocket so the fins are up while this fillet dries.

Shock Cord Installation

The rocket is popped apart with an explosive charge as it coasts to the top of its flight. This explosion forces the parachute out of the body tube and flings the nose cone away. The shock cord is an elastic cord connecting the nose cone, parachute, and body tube. It absorbs the shock of the explosion, keeping everything in one nice, neat package.

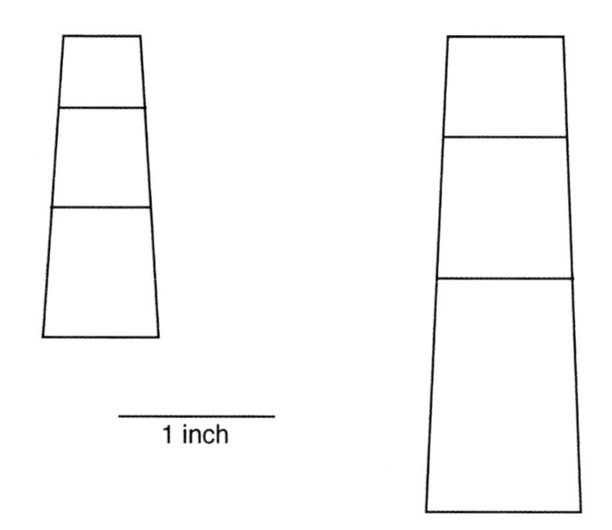

Figure 3-46. *Shock cord mount for 1/8" and 1/4" shock cords. Trace and cut from printer paper.*

The shock cord is fastened to the body tube with a tri-fold paper and glue mount. The Estes Designer's Special comes with some of these drawn on a sheet of stiff paper, but you can also cut one from normal writing paper.

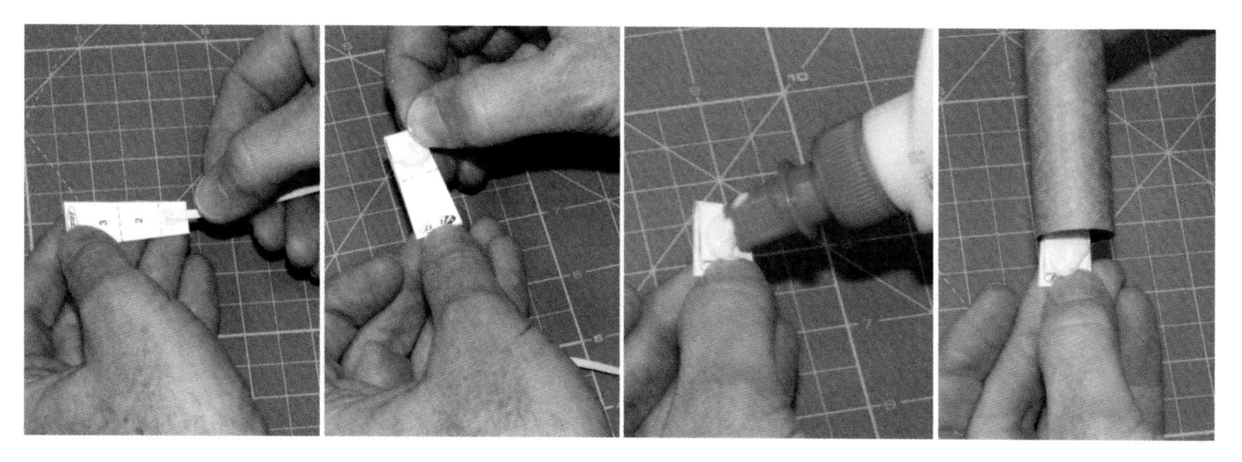

Figure 3-47. *Mounting the shock cord.*

Using either the Estes shock cord mount or one you cut yourself, apply wood glue to the smallest tab, lay the shock cord in the glue, and fold the paper over (Figure 3-47). Don't mash it down too much, or the glue will squirt out, leaving a dry gap.

Repeat this process with a second fold. Finally, apply glue to the inside edge of the paper and glue it inside the body tube, about one inch from the end of the tube. Make sure there is room for the nose cone to be inserted with at least a quarter-inch gap between the nose cone and the shock cord mount.

I generally have the end with the dangling shock cord pointed toward the nose of the rocket. Press down just enough for some of the glue to squirt out, and then smear this across the top of the paper to form a complete seal. Hold the mount in place until the glue begins to set.

Once the shock cord is dry, apply another coat of wood glue over the entire mount and let that dry, too.

Making a Parachute

The Estes Designer's Special comes with several prebuilt plastic parachutes. It's a nice little package, and you should use one of the 12-inch parachutes if you bought that package. Don't buy a parachute if you are using individual parts, though.

A bag from a retail store makes a great parachute, as does the metallic Mylar used for helium party balloons. Some of the plastic bags are pretty flimsy and made from plastic that sticks together. Use one of the slightly thicker ones that seems a little less like plastic food wrap.

You can make a nice parachute template using a protractor and a few sheets of printer paper taped together, or the back of a piece of wrapping paper. Use the protractor to draw six equally spaced lines radiating outward from a central point, as seen in Figure 3-49. These will be 60° apart.

Extend a nice, long line—about 8" long for a 12" parachute—from the center of the circle through each of the six points.

Measure 6 15/16" from the center of the circle along each line. Connect these lines around the outside edge of the parachute to form a hexagon. This is the outline for the finished parachute.

Parachute Math

Why 6 15/16"? And how long should this line be for other parachute sizes?

Finding the correct length is pretty simple with a little dab of trigonometry. Figure 3-48 shows a sketch of one of the six sections of the parachute. The line h is the one that is 6 15/16" long. We want a 12-inch parachute, so the line d is half that—-6 inches. The angle between the lines d and h is 30°, or half of the 60° you marked off to form each of the six sections.

From trigonometry, we know that the length of the line d (the adjacent line in trig-speak) divided by the length of the line h (the hypotenuse in trig-speak) is the cosine of the angle. You can get the value of the cosine of 30° from just about any calculator; it's about 0.866. All of this makes it easy to find the length of the line h, which is the one we marked to find the points that form the hexagon for the parachute:

$$\cos(30) = \frac{d}{h}$$

$$h = \frac{d}{\cos(30)}$$

$$h = \frac{6}{0.866} = 6.928$$

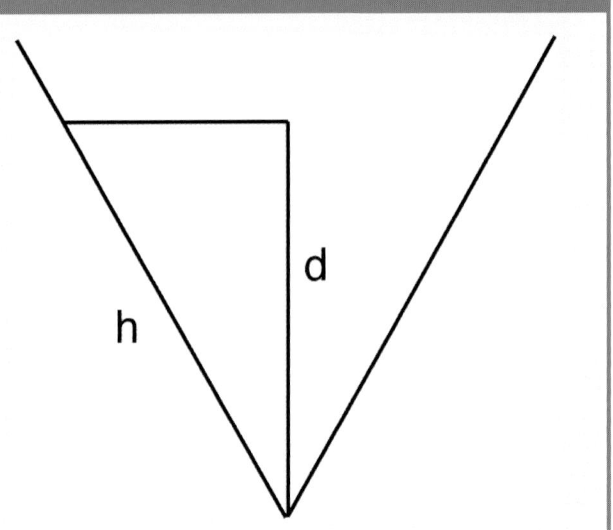

Figure 3-48. *Parachute trigonometry.*

You can use the same formula to find the length of the guide line for any size parachute.

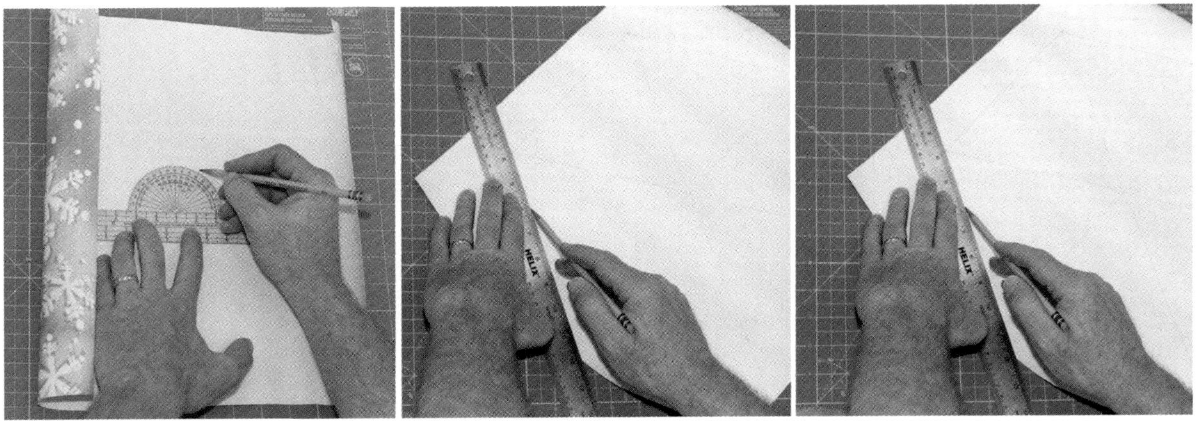

Figure 3-49. *Use a protractor and ruler to form a parachute template from an old piece of wrapping paper or several sheets of printer paper taped together.*

Use the template and a hobby knife to cut one or two layers of your chosen parachute material.

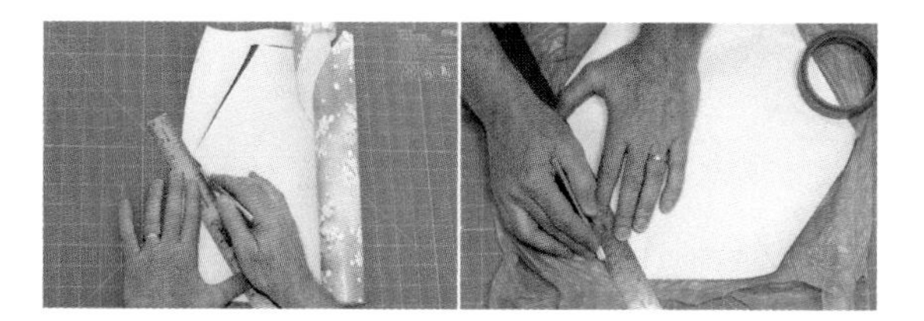

Figure 3-50. *Cutting the parachute.*

Cut three parachute shroud lines that are twice as long as the parachute's diameter—24", in this case—from crochet string. Any string that is thick enough not to get tangled in itself is fine; normal sewing thread is too thin.

Attach the shroud lines to the corners of the parachute with 1/2" square or round pieces of a sticky tape like duck tape. Form a loop under the tape to hold the string firmly in place as shown in Figure 3-51.

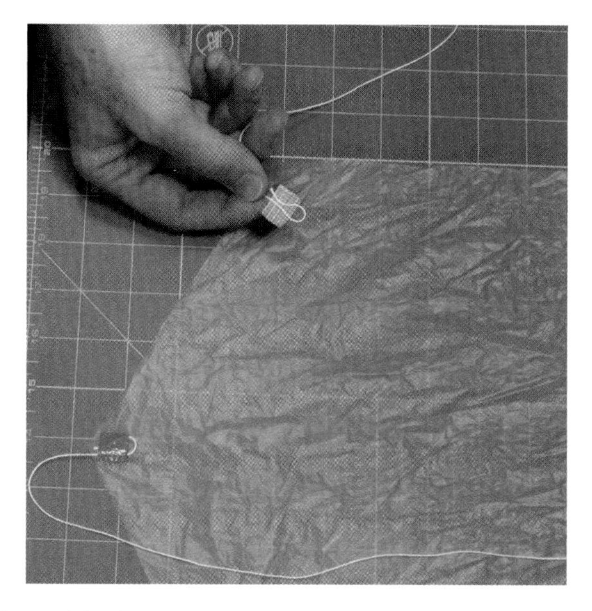

Figure 3-51. *Fasten the shroud lines to the parachute at the corners.*

One shroud line is attached across opposite corners of the parachute, while two others are attached to adjacent corners so they do not cross each other.

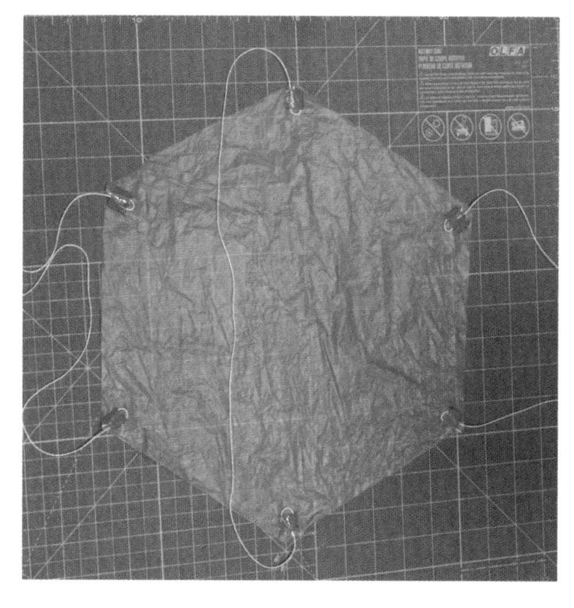

Figure 3-52. *One shroud line crossed from one corner to the opposite corner. The other two attach to the remaining adjacent corners.*

Apply baby powder to the finished parachute to keep the plastic and tape from sticking together when folded in the rocket (Figure 3-53). I like to store my parachutes open on a clothes hanger so they don't get permanent creases that may impede their snapping open."')))

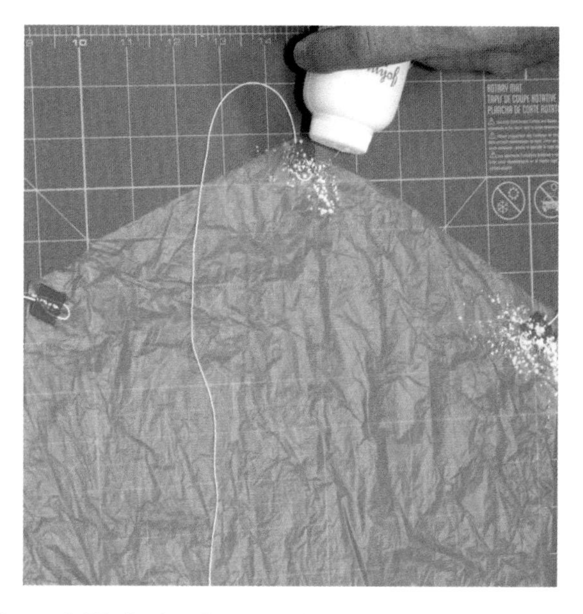

Figure 3-53. *Apply talcum powder to keep the top side of the tape from sticking when the parachute is folded.*

Attaching the Parachute and Shock Cord

You can attach the shock cord and parachute directly to the loop molded into the base of the nose cone, but I prefer attaching them using a snap swivel. The snap swivel attaches the parachute or shock cord to the nose cone. There are two main advantages to using a snap swivel.

First, rockets are made from standardized parts. It's pretty easy to swap a nose cone on Juno for a payload bay made from a BT-50 body tube. The shock cord has to be untied and retied if it is tied directly to the nose cone, but the change is easy if the shock cord is attached with a snap swivel. The same is true for parachutes. You might want to quickly change parachutes in the field, either due to damage or to change sizes as payload weight changes.

Rockets tend to twist a lot as they descend. A snap swivel can twist, too. The snap swivel generally won't keep up with the rapid twists of the rocket, but every little bit helps.

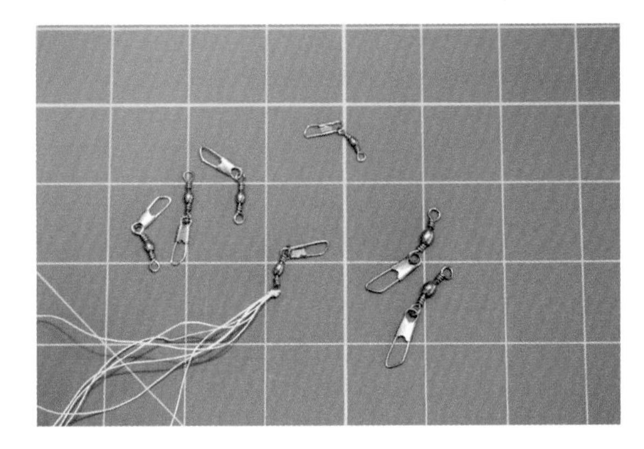

Figure 3-54. *Snap swivels make it easy to connect and disconnect parachutes and shock cords.*

The method for attaching a parachute is the same for both a nose cone and a snap swivel. Poke the shroud lines of the parachute through the opening, then loop the parachute around and pass it through the shroud line loop (Figure 3-55). Tighten the knot to secure the parachute.

Almost any knot will do for tying the shock cord to a snap swivel or nose cone, especially if you place a drop of glue in the middle of the knot. Don't cover the entire knot, though; just add a small dab where the end pokes through the knot. The glue is pretty rigid and makes a sharp edge, so covering the entire knot can lead to a broken shock cord.

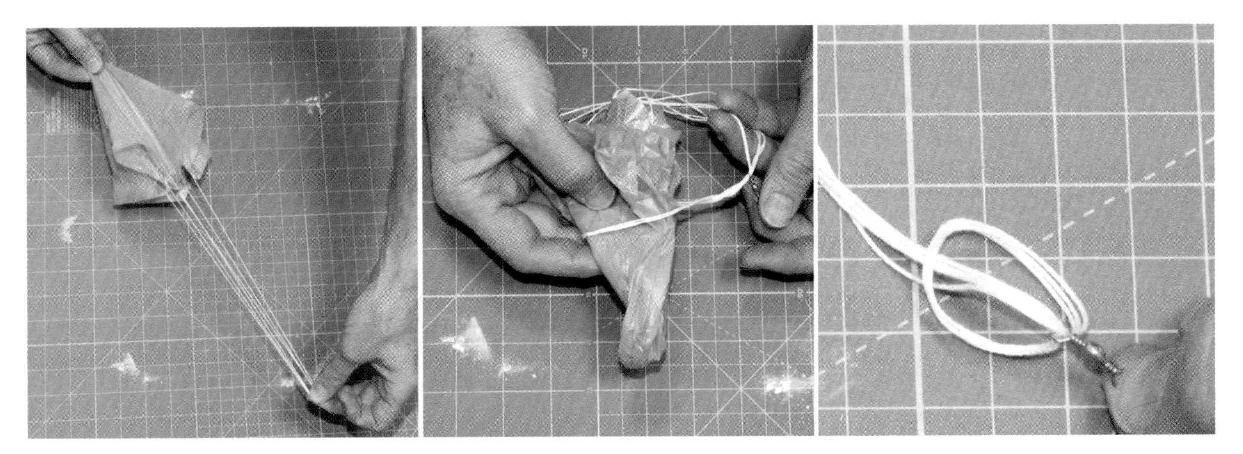

Figure 3-55. *Find the center of the shroud lines and loop the parachute through a snap swivel or the loop on a nose cone.*

The really classy way to attach a shock cord is to use a buntline hitch. Figure 3-56 shows what this looks like, along with some instructions on tying the knot from Wikipedia.

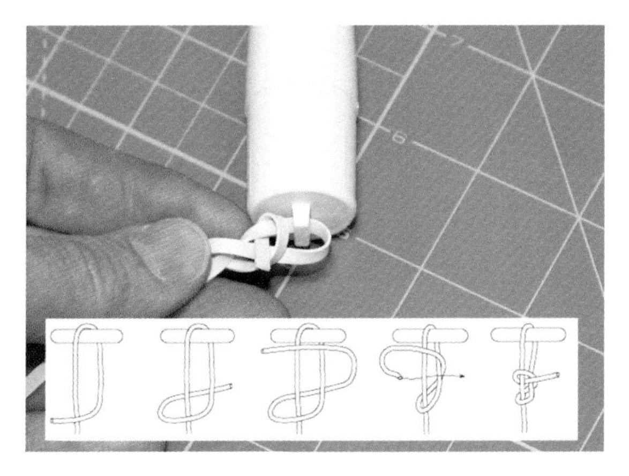

Figure 3-56. *The buntline hitch is a great way to tie the shock cord to a nose cone or snap swivel.*

Before You Paint

Your rocket is structurally complete. You can fly it if you like. If you go to a launch where there are lots of rockets, you will probably see several that are not painted. A poor finish does cause a bit more drag, but it's not that big a deal for a sport rocket. There is only one real problem with not painting the rocket...

It's ugly.

You've spent all of this time building a classy rocket. Take a little more time to make yourself proud!

There are two steps to take before applying the paint. Start with the nose cone. The one I picked out for Juno from the Estes Designer's Special gave me a pleasant surprise: the nose cone was molded vertically, so there is no *flash* running the length of the nose cone. There is another nose cone in the kit that does have flash. Take a look at the 6 1/2″ nose cone for the B-60; it's about 1 1/2″ in diameter, and there are two ridges running the length of the nose cone. The molding process causes this flash, and it's the norm, not the exception.

Look at Figure 3-57 carefully—the nose cone shown is a bit big for Juno, but it shows the flash well. You'll want to remove the flash using a hobby knife. You're not cutting the flash off; you are dragging the hobby knife backward to scrape it off. If the nose cone is still a bit rough after scraping, sand the entire nose cone with a very fine grit sandpaper. As with the fins, sand by feel, not sight. You will probably still be able to see the flash mark when you finish scraping and sanding, but if you can't feel it, you're done.

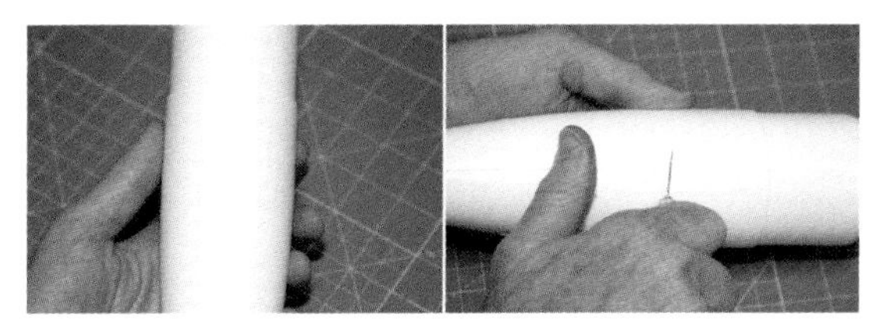

Figure 3-57. *Scrape any flash off of the nose cone, then sand with a very fine grit sandpaper if needed. You're done when it feels smooth.*

The balsa fins will have a very rough surface if they are painted right away. You can paint, sand, and repeat, but there are lots of great products for preparing the wood.

Sanding sealer is the primer of choice if you are using butyrate dope (model airplane paint). Butyrate dope tends to fill and shrink a bit on its own, and does a pretty good job of covering up small grooves left from the spiral winding process used to manufacture body tubes.

First, brush on a coat of sanding sealer. After it dries completely, sand the fins with fine grit sandpaper until they are smooth. You will probably need two or three coats of sanding sealer for a good finish.

Another option is a sandable filling primer, such as Rust-Oleum 2 in 1 Filler & Sandable Primer, which can be sprayed on. Unlike butyrate dope, spray enamel will show up every imperfection in the spiral winding of a body tube. A filler primer also works well on the body tube, filling in these gaps.

Shake the can well before you start to spray. There is a steel ball in the can that will rattle—if you don't hear it rattling, it's stuck in the paint. Keep shaking. Shake for at least another minute after the ball starts to rattle.

Spray the primer onto the wood and paper parts of the rocket, as seen in Figure 3-58. You can spray it on the nose cone, but the one used with Juno doesn't need it. After the primer dries, sand it until you see wood and paper showing through the primer. The intent is not to build up a thick layer of primer; it's to fill in the low-lying cracks and minor dings in the rocket. Repeat this process until the primer looks smooth when sprayed on, and give that final coat of primer a light sanding to smooth out any rough spots in the primer itself. At this point, you are looking at the finish, not feeling it, to see if it is smooth. Imperfections will show through dramatically.

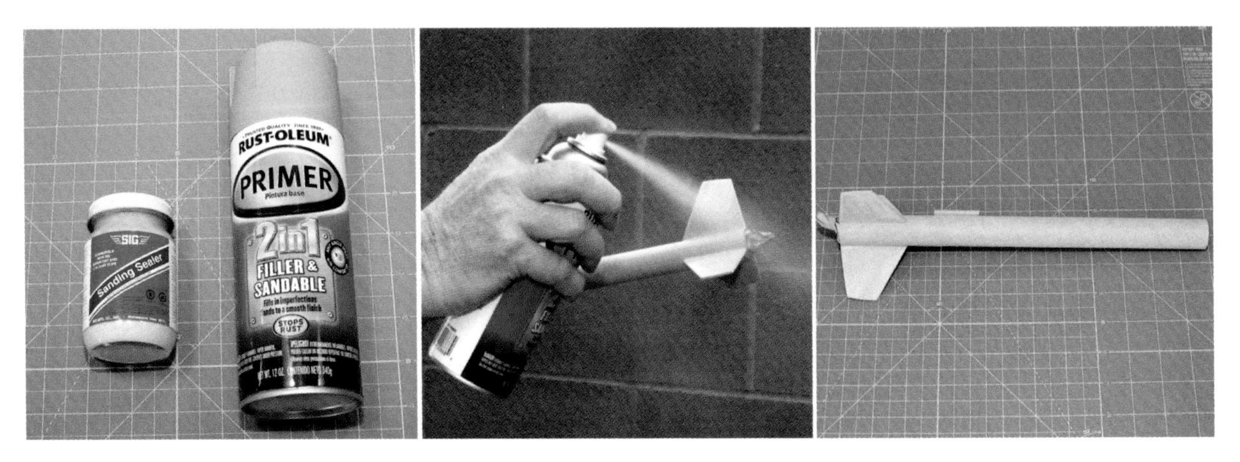

Figure 3-58. *Spray on primer, sand until wood and paper show through, and repeat until the last coat looks smooth.*

Painting the Rocket

Whether you spray paint or paint with a brush, start with the lightest color and progress to darker colors whenever possible. Dark colors cover light ones better than light colors cover dark ones. If you paint Juno as I did, you will start by spraying the body and nose cone white. Some paint will get on the fins. Just ignore that paint. It will get covered up later, and any attempt to tape the fins will just result in a visible ridge when the tape is removed and the new color applied over the line.

Shake the can for at least a minute after the ball starts to rattle, then apply a medium-thickness coat of paint. You will probably still see some gray from the primer in places. Resist the urge to take care of that. Let the paint set for 10 to 15 minutes, then apply another coat of paint. Repeat this process until the body tube and nose cone look perfect.

Tactics for Perfect Paint

When using spray paint, paint the hard-to-reach areas first. This includes the area around the launch lug and the edges of the fins. You will get coverage on a lot of the easy-to-paint areas in the process, and won't need to go back and add more paint to them. If you paint the easy-to-access areas first, you will inevitably repaint some of them when trying to paint the edges of the fins and other hard-to-reach places, increasing the chance of a drip or sag in the paint.

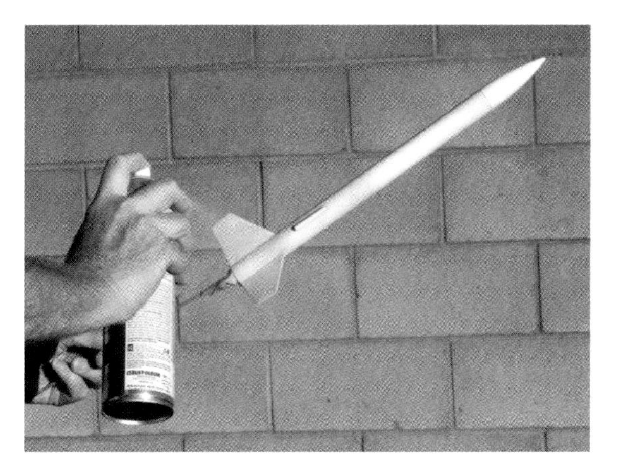

Figure 3-59. *Paint the nose cone and body tube white.*

Drips and runs in the paint are not just ugly; they also affect the aerodynamics of the rocket. If you get a drip or run, stop and let the paint dry completely. The next day, use the finest grit sandpaper you have to sand the paint until it feels smooth, and then start again with the painting process.

Let the first color dry for several hours, preferably overnight, then mask off the body tube using painter's tape and scrap paper, leaving two of the fins exposed. Press the tape at the edge down firmly with your fingernail.

When the rocket is taped, spray the two exposed fins with red paint. You might get complete coverage with a single coat of red paint, but don't push it. Two coats are much better than one coat with a drip.

Let the red paint dry thoroughly. You don't have to wait as long as before because you won't be applying tape directly to the paint. Make sure it is dry to the touch, though—test the paint on the taped surface so any fingerprints are on the tape, which you will remove, not the fins. Once the paint is dry, repeat the process to paint the final fin yellow.

You hold your breath, remove the tape, and...it's not perfect. What do you do?

One option is to wait until the paint dries thoroughly, sand it down, and try again. If the error is very small, though, another option is to touch it up with a brush. Some paint—particularly the expensive stuff sold for modeling—is available in both spray and bottle form, so you can buy the color you need in a small bottle. Even better (and probably cheaper) is to wrap a small bowl in plastic wrap and spray directly into the wrapped bowl. You'll collect a few drops of paint that can be applied with a brush for touch-ups.

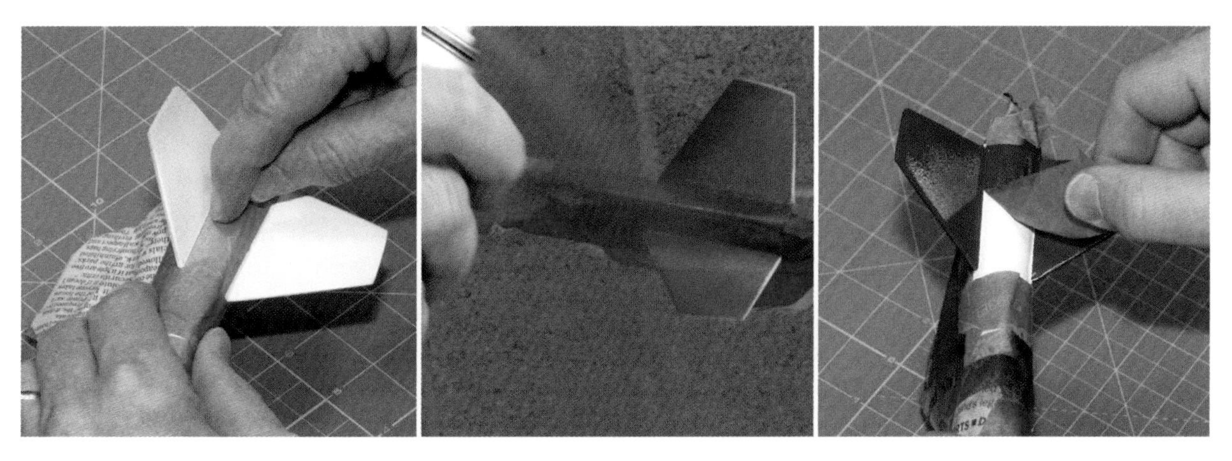

Figure 3-60. *Paint two fins red and one yellow.*

Figure 3-61. *Spray paint directly into a plastic wrapped bowl to get a few drops for touch-ups.*

Options

One of the really nice aspects of scratch building as opposed to kits is that you don't have to build exactly the same rocket I built. Here are a few suggestions. Read "Can You Safely Make Substitutions?" on page 33 if you would like to change Juno even more.

Alternate Fin Shapes

I like the fin shape on Juno a lot, but there are two other fins worth considering.

Swept fins look really cool. They are also good for stability, since they move the fin surface further to the rear of the rocket, pushing the center of pressure back, too. When you put standard fins on Juno, you need to use a stand to display the rocket, since the engine hook sticks out the back. Rockets with swept fins, however, can stand on their own.

For me, the problem with swept fins is that I live in a high mountain desert, which has a lot of hard surfaces and exposed rocks. Swept fins dent, chip, and break far more easily than the trapezoidal fins I selected.

Still, those swept fins look great, don't they? And you may live in a grassy area where they won't be subject to as much damage. If you'd like to try them, Figure 3-62 contains the pattern for swept fins.

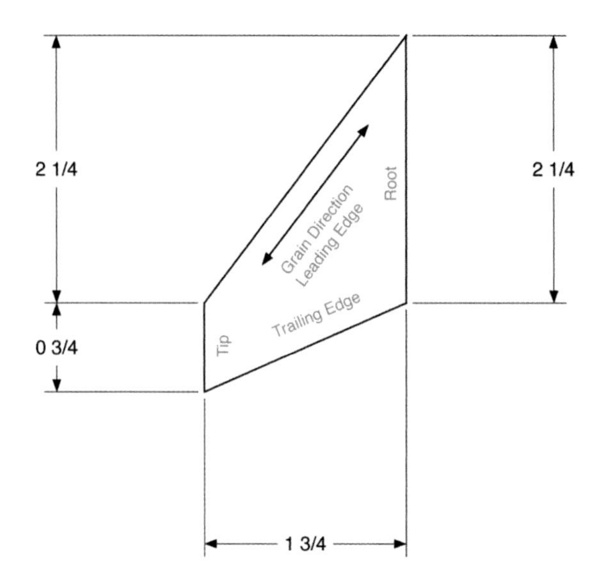

Figure 3-62. *Swept fin pattern for Juno.*

Another popular fin design is the oval. This is technically the best fin shape for subsonic flight, but imperfections in the fins and especially in the airfoil shape are likely to swamp any tiny advantage gained from the fin shape. Still, they look nice. The oval fins are a little harder to cut out and make, but give them a try if you are up for a challenge: Figure 3-63 contains the pattern.

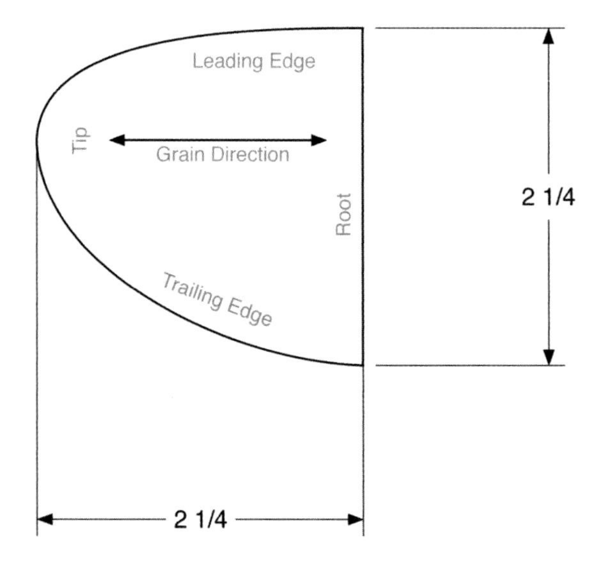

Figure 3-63. *Oval fin pattern for Juno.*

Juno uses three fins. Why not four? Well, mostly because four fins are not needed. High-performance rockets always use three fins, since a fourth fin produces more drag. Still, if you think four fins looks better, add another one. Your rocket won't fly quite as high, but it will be really tough to tell the difference, and it will be quite a bit more stable with all that extra fin area.

4 Launch Pads

Figure 4-1. *The mono and quad launchers.*

The launch pad is used to hold the rocket steady and keep it traveling straight for the first moments of flight, before it is traveling fast enough for the fins to work. It also needs to provide a reliable electrical ignition system for starting the rocket motor.

We'll see two launchers for solid propellant rockets in this chapter. The first is probably the one you will want to build. It's a single rocket launch pad and launch control system that satisfies the safety requirements for both NAR and Tripoli. We'll call this one the *mono launcher*, since it launches one rocket at a time. The second launcher is designed for schools or clubs, allowing up to four rockets to be set up at a time. We'll call this one the *quad launcher*. The rockets can be fired one after another, keeping the pace fast on launch day so you get the most from your classroom or club launch time. The quad launcher can also fly all four rockets at the same time. Known in rocketry circles as a "drag race," firing multiple rockets is a sight to behold!

What Makes a Good Launcher?

The primary design constraint on any rocket launcher is safety.

The launch pad itself must be sturdy. The rockets in this book will be up to one pound in weight (about half a kilogram). That's not a lot of weight, but the launcher does have to be strong enough to hold the rocket upright, even in a light wind.

The launch rod is used to guide the rocket as the flight starts. The launch rod needs to be 3 feet long. Launch rods for lighter rockets like Juno are made from stiff 1/8" diameter wire. Launch rods for heavier or more powerful rockets, like the egg lofters or D rockets we'll build later, are 3/16" in diameter, while launch rods for mid-range rockets are generally 1/4" in diameter and 4 feet long. The launcher should be able to hold all of these launch rod sizes. Both the mono and quad launchers use keyless chucks to securely hold all of these sizes and more.

While rockets are *never* flown in high wind, they are frequently flown in stiff breezes that can carry them quite a distance away from the launch site. While not essential, it's nice if a launcher allows the launch rod to tilt a bit from the vertical so the rocket can be pointed into the wind, flying upwind so the wind carries it back toward the launcher when the parachute deploys. The mono launcher has a tilt mechanism that allows the rocket to be securely positioned up to 30 degrees from the vertical—the maximum allowed tilt angle for model rockets.

This may surprise you, but in my opinion, the launch rod is the most dangerous thing at the launch site. Here's a 3-foot-long stiff metal wire sitting just below chest height, with preadolescent children frequently running excitedly about. A child tripping and catching a launch rod in the eye is about the scariest scenario I can imagine at the launch site, and I've seen a number of close encounters. There are three ways to minimize the risk from the launch rod. The first is to use a cap that sits on top of the launch rod until just before launch. That's a critical part of the launcher design, and an integral part of the mono launcher. A second method is to raise the launch rod above eye level. That's the technique used for the quad launcher. Finally, the launch site itself must be controlled. We'll deal with that issue in the next chapter.

What Is a Safe Launch Distance?

Most of the launch controllers you'll find at a hobby store have 17 feet of wire, not 30 feet. Is that acceptable? What is a safe distance?

Actually, it depends.

NAR says you should be 15 feet from a rocket that uses D motors or smaller, and 30 feet from rockets that use E motors or larger, or clusters of motors that have more power than a D motor-—for example, three C6-5 motors. The launch controllers with 17 feet of wire are designed for D motors or smaller and for flying under NAR rules.

Tripoli, on the other hand, requires a distance of 30 feet for any rocket launch, even a 1/2A motor.

I've used launchers with a 15-foot standoff for years, and never felt like there was a safety issue flying D or smaller motors. Still, that extra 15 feet of safety is never a bad thing. Some of the rockets later in the book also have clusters of motors or can handle E motors, so they will require a 30-foot distance.

Both the mono and quad launch controllers are designed with 35 feet of wire so you can stand at least 30 feet from the rocket. This lets you launch larger rockets and follow the Tripoli guidelines even for the smaller rockets. Alternate wiring for a 15-foot standoff will also be discussed. You can make the choice that's appropriate for the kind of flying you are planning. And one nice thing about both launchers is that you can quickly switch the launch controller from 17 feet to 35 feet of wire.

When the motor ignites, flame shoots downward. If not controlled, this poses both a fire hazard and a danger to the launcher itself. The launch pad should have a metal blast deflector plate to take the brunt of the initial rocket exhaust. We'll look at three alternatives that can be used with either the mono or the quad launcher.

The launch controller provides a strong electric current to the rocket igniter. This current heats up the igniter wire, which ignites a starter material that lights the motor. The launch controller must be capable of delivering 1 to 3 amps of current at 6 to 12 volts, depending on the igniter used. Ideally, it should be able to deliver that current to four igniters at a time to support clustering or drag races. It needs to do all of this while allowing the person launching the rocket to stand 30 feet from the rocket. Both launch controllers in this chapter will do the job.

I still remember Rob, my rocketry buddy from high school, proudly showing off his new home-brewed launch controller as we trundled off to an open field to launch some rockets. I noticed that it used a toggle switch, not the normal pushbutton switch I was used to, and that it had no safety key. Still, it was a really cool launch controller, getting power from a 12-volt car battery he lugged along.

After a few launches, Rob went out to set up another rocket. He slipped it down over the launch rod, squatted down, and carefully connected the igniter clips to the rocket. The rocket promptly took flight—Rob had forgotten to flip off the toggle switch!

Every launch control system should use a pushbutton launch switch for the launch button. It should spring back to an open position as soon as it is released. There must be an arming key that prevents the igniter from firing until the key is inserted. That key should be kept at the launch pad or in the launch control officer's pocket while rockets are set up. There must be an arming light that lights up when the rocket is ready to launch. The mono and quad launch controllers have all of these features.

By the way, Rob got quite a jolt of adrenaline, but was fine otherwise. The rocket did not scorch him. Model rockets are designed with safety in mind, so even a severe case of stupid caused him no real harm this time. In fact, perhaps it was a good thing—we all learned a valuable lesson about following safety guidelines that day.

The last requirement of a launch control system is the ability to deliver the required current to the igniter or igniters, and to not deliver too much current during the continuity test—that point when the arming key is inserted, and the arming light on the launcher lights up to indicate there is a complete circuit through the igniter, but before the launch button is pressed. For reasons we'll go into later, that means choosing the right wire and arming light. Most commercial launchers, such as the Estes launch controller, use 24-gauge wire and 6 volts with an incandescent arming light. That's fine for a single Estes igniter, but doesn't allow enough current for clusters of motors using Estes igniters or for lighting 12-volt igniters for reloadable motors. The Estes launch controller also dumps enough current through the igniter during a continuity check to fire a Quest Q2G2 igniter, so the rocket will fire as soon as the arming key is inserted! The lesson here is to only use a launch control system with igniters you *know* it is designed to handle. Both the mono and quad launchers are designed to handle from one to four igniters at a time, and will work properly with any of the igniters commonly sold for model rockets.

Varying the Designs

There are several alternatives listed for the launchers that can simplify construction or make it less expensive. The parts lists are for the most capable version of each launcher. Be sure to read through the directions before buying parts so you know exactly what options you want to use, and buy the appropriate parts.

There are also a number of ways to vary the design on your own. Be sure to read the theory sections first. There are some important details and a few simple calculations you need to perform to create a reliable launch controller.

Kits

One of the interesting things about launchers is that they seem to come in two varieties: starter kits and expensive.

Both Estes and Quest make nice portable, easy-to-assemble kits. Both are suitable for black powder rockets using D motors or smaller. Estes also makes a second version of its launch controller that has a longer launch wire, making it suitable for E rockets. Neither launcher will handle clustered rockets with Estes igniters, and neither can light igniters designed for composite motors. Even the Estes launch controller is not designed to work with some models of the Estes mid-power F and G motors.

At the same time, both of the launchers are inexpensive and easy to assemble and set up in the field.

I highly recommend either launcher if you are intent on getting into the field as fast as possible. I have one in my launch box, and carry it to almost every launch. Keep in mind that you will outgrow these launchers if you build all of the rockets in this book, though. You can stretch the time when you outgrow the launcher by getting the Estes E launch controller right away instead of the standard launch controller, but it still won't handle the composite motors near the end of the book.

The Estes launch controller cannot be used with the Quest Q2G2 igniter. Both are fine products, but they are not compatible. For reasons that will be explained in "Electrical characteristics of some common igniters" on page 80, when used with the Estes launch controller the Quest Q2G2 igniter will fire as soon as the arming key is inserted. This, as you can imagine, is undesirable.

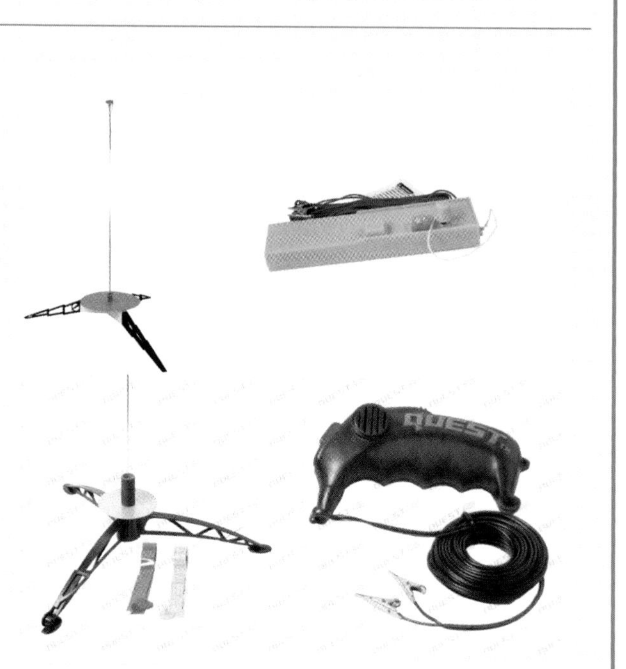

Figure 4-2. *The Estes launcher and launch controller (top) and the Quest launcher and launch controller (bottom).*

A Quick Lesson in Igniters

You need to know a few things about igniters to understand some of the design details for the launch controllers. You don't need to know much about them to fly rockets, though. This discussion gives you a lot of fun background, but it isn't really needed if you just want to build the launcher. Here's the simple version: buy either standard Estes igniters or Quest Q2G2 igniters for any single-use rocket motor. Estes sells two kinds of igniters. Don't buy the Estes Pro Series igniter. They are fine igniters that we'll use later, but for now you won't need them. We'll discuss these and other igniters when it is appropriate to use them. You can skip the rest of this discussion if you like.

Almost all single-use rocket motors use black powder for rocket fuel. Black powder is easy to ignite. The igniters are generally composed of thin nichrome wire coated at the end with pyrogen or some other material that will burn when the wire gets hot. The electric current from the launch controller heats the nichrome wire, which ignites the pyrogen. The burning pyrogen ignites the black powder from the rocket motor. These igniters work well with 6-volt ignition systems for a single igniter, and easily light black powder motors.

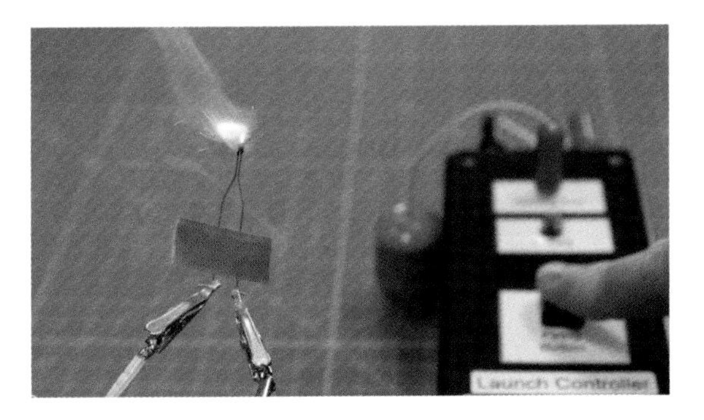

Figure 4-3. *Igniter firing. If you try this yourself, make sure you have a foot or so clear of flammable material. As you can see, the igniter does throw off some sparks, so wear gloves if you're holding it in your hand.*

The reloadable rocket motors we'll start using near the end of the book don't use black powder for a propellant. Just like the solid boosters on the Space Shuttle, reloadable rocket motors use aluminum perchlorate, sometimes mixed with other materials to vary the color of the flame or the amount of smoke. These are often referred to as composite propellants. These motors don't light as easily as black powder motors. They are usually ignited using 12-volt ignition systems and igniters with a bit more pyrogen, like the First Fire Jr. igniter or Estes Pro Series igniter.

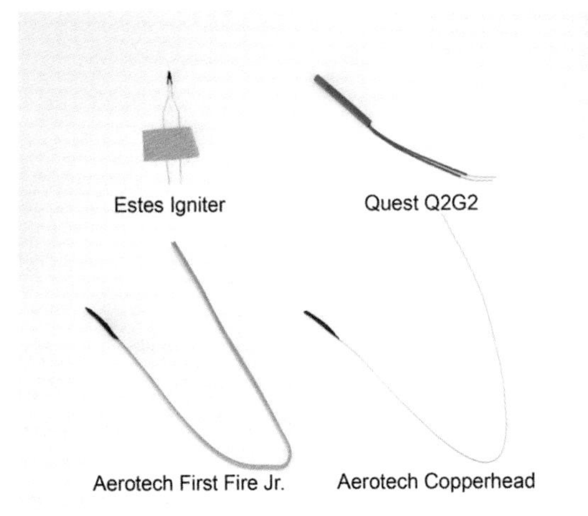

Figure 4-4. *Some common igniters. The Estes and Quest igniters are designed for black powder rockets, while the Aerotech igniters are used with composite motors.*

The trick with any of these igniters is to supply enough current at exactly the right time to heat the wire. Table 4-1 shows the ignition currents and voltages for the four most commonly used igniters for A through G motors.

Table 4-1. Igniter requirements

Igniter	Resistance	Recommended launcher voltage	Minimum current
Estes	0.67–0.73Ω	6 volts	1.0–1.2 amps
Quest Q2G2	0.48–0.56Ω	6 volts	0.55–0.6 amps
First Fire Jr.	0.53–0.67Ω	12 volts	1.75–1.8 amps
Copperhead	0.48–0.56Ω	12 volts	1.55–1.7 amps

The important point is the wide variation in the current required to light the igniter. The Quest Q2G2 only needs 0.55–0.6 amps at 6 volts, while the First Fire Jr. needs 1.75–1.8 amps at 12 volts. Generally, we want to supply about twice the required current to ensure fast, reliable ignition.

The wide range of current needed to fire an igniter is a real challenge for the launch controller designer. We'll see how the mono and quad launchers meet the challenge as the designs unfold.

The Mono Launcher

The mono launcher's base is made from 3/4" PVC pipe. The base unscrews for easy storage, and doubles as a launch rod storage tube. A wooden tilt mechanism allows the launch rod to be tilted so the rocket can be launched into the wind. A keyless chuck like the chucks found on electric drills firmly holds launch rods of any size.

Table 4-2 shows the parts list for the mono launcher, and Table 4-3 shows the tools you may need. Read the construction description before you go shopping, though; there are several options presented, and some of them change the parts or tools you need.

Figure 4-5. *The mono launcher with Juno loaded, getting ready for a launch.*

Figure 4-6. *The mono launcher packs and stores as a single long tube that holds the launch rods. It sets up quickly in the field with no tools.*

Table 4-2. Parts list

Part	Description
3/4″ schedule 40 PVC	This is used for the launch stand. You will need about 50″ total, cut into smaller pieces. Be sure to get schedule 40 or 80, not the thinner schedule 20.
90-degree side outlet elbow	This piece forms the top of the base.
PVC cap (2)	You will need two caps, one for the base of one leg and one for the launch safety cap. If you have a scrap of 1/2″ PVC, get one cap for 3/4″ PVC (the leg) and one for 1/2″ PVC (the safety cap). If you don't have any scrap 1/2″ PVC, just use 3/4″ PVC for the safety cap. It's not worth buying 10 feet of 1/2″ PVC just for the safety cap!
Adapter, male (3)	This adapter is threaded on one end to screw into a matching female adapter. The other end fits over the 3/4″ PVC pipe.
Adapter, female (5)	This adapter has female threads on one end. The other end slips over the PVC pipe. Three of these are used to form the cap of the launcher, while two more form the ends of two legs.
1x2 wood	This is used for the tilt mechanism. While it's called 1x2, the wood will actually be about 3/4″ x 1 3/4″. Get a good-quality piece of cabinetry wood, like poplar, not rough-cut white pine. You will need about 8″ total.
1/4-20 3″ carriage bolt	Carriage bolts are the ones with a small square of metal just under the head of the bolt. This will lock the bolt in place. The carriage bolt is used to tighten the tilt mechanism. Most 1/4"-diameter bolts have 20 threads per inch.
1/4-20 wing nut	Used to tighten the carriage bolt.
1/4″ washer	Used under the wing nut to protect the wood.
1/4-20 threaded rod	You will cut a 2″ piece from this rod.
1/4-20 nut and lock washer	Secures the tilt mechanism to the base.
1/2-20 3″ bolt	You will cut the threads from this bolt to form the mount for the keyless chuck. You need to end up with 1 1/2″ of thread. Buy the cheapest bolt you can find; the more expensive stainless steel bolts are very difficult to cut. One construction option avoids cutting the bolt.
1/2″ keyless drill chuck replacement	The keyless chuck is used to secure the launch rod. The launcher is designed for a chuck that screws onto a 1/2-20 threaded bolt. If the one you buy uses a different mount, you will need a different mounting bolt, too. Do an online search for "keyless drill chuck," particularly on Amazon.com. While specific models come and go, I've always found several under $10.
1/8″ music wire	This is the launch rod used for most of the rockets in this book. It should be 3 feet long.
3/16″ music wire	This is the launch rod used for the heavier and more powerful rockets near the end of the book. You don't have to get it right away.
Blast deflector	You have some options here. I like the Estes blast deflector plate, which has a nice central support to keep the deflector plate level. You can also use a piece of steel from the hardware store, as described in one construction option. You can even use a tin can—but not an aluminum can!

Table 4-3. Tools and materials

Tool	Description
PVC primer and cement	Purple primer and PVC cement are used to join the various pieces. They are frequently sold as a package, but may be purchased separately.
Primer	Used to treat the external surface of the wood before painting.
Furniture oil	Used to treat the internal wood surfaces so they are protected, but will not stick together. I used lemon oil.
Enamel paint	Spray paint works well.
Painter's tape	You will need painter's tape if you go for a fancy paint job like the one shown. It's also handy when painting the wood parts.
Drill	You will be drilling several holes in various parts, ranging from 3/32" to 1/2". A drill press is best, but a handheld drill will work.
Scroll saw	A scroll saw is used to cut the wood pieces to shape. One construction option avoids the use of the scroll saw.
Metal saw	You will be cutting a 1/2"-diameter bolt and a 1/4"-diameter threaded rod. There are lots of saws that can cut the metal; I used a cut-off saw. One construction option avoids cutting metal.
PVC saw	Almost any saw will cut PVC. I used the same cut-off saw that cut the metal. A hand saw works well. Specialized PVC cutters also work well.
Sander	You can sand the parts by hand, but a belt or disk sander is very nice, especially for shaping the tilt mechanism.
Epoxy glue	You're going to use about 2–3 cubic inches of epoxy resin to reinforce the head assembly. Get small bottles of glue from the hobby store; it will be much cheaper than buying the small tubes sold in grocery stores for quick repairs.
Rubbing alcohol	Use rubbing alcohol to clean up epoxy glue. It works great until the glue sets.
Small cup	A small disposable cup is used to mix the epoxy.
Stirring stick	A popsicle stick, plastic spoon, or plastic knife is needed to stir the epoxy.
Small bowl	A small bowl or large cup will be used to support and cool the head while the epoxy dries. If you're careful, it can even be reused.

Building the Base

The base is made from 3/4" PVC pipe and fittings. These are available from most hardware stores, although a few of the fittings may be hard to find locally. You can also find the fittings online at places like *http://www.lowes.com*.

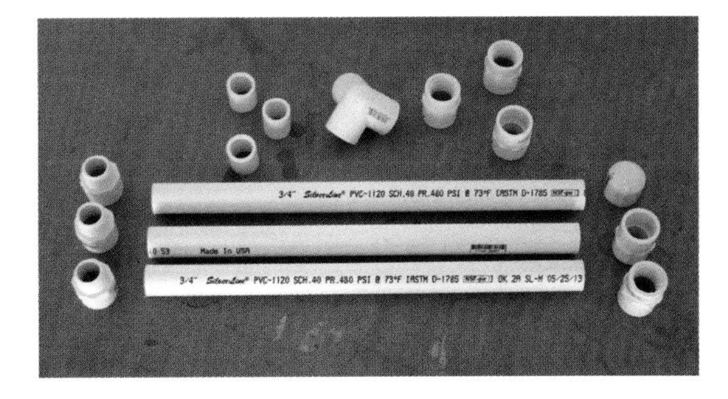

Figure 4-7. *PVC parts for the launcher base.*

The fittings include a 90-degree side outlet elbow, three male couplers, five female couplers, and an end cap. The legs and three shorter pieces to the left of the side outlet elbow are cut from PVC pipe.

Cut the three PVC legs from 3/4"-diameter schedule 40 PVC pipe. The legs are not all the same size. One leg will end with a PVC cap, while two will end with a coupler that allows the three legs to be screwed together for storage. The two shorter legs are 15" long, while the longer leg is 15 3/8" long.

You can use just about anything to cut PVC pipe. The hardware store will happily sell you a special pair of shears that cut PVC, and they work really well. You probably already have a pair if you live out West, in the land of sprinkler systems. Band saws work well, too. I used a cut-off saw that you will see later; it also cuts metal. Handheld saws also work well.

PVC is bonded with a three-step process. The first step is to clean the pipes. You can do this with soap and water or with specialty cleaners sold for the purpose. Take a look at the ingredients on specialty cleaners, though. For many products, including the Oatey brand shown here, the cleaner is just the primer without the purple dye. That purple dye is used so building inspectors can tell the pipes were primed. In other words, you can substitute the Oatey cleaner for the primer! In any case, make sure the PVC pipes are clean and dry before moving on to the second step.

The second step is to apply purple primer to both pieces where they will join. You are committed once the primer is applied, so test fit the parts before applying the primer, then test fit them again. The primer softens the PVC a bit. After the primer has soaked in for a moment, but before it dries, apply the cement to the inside of one piece, then insert the other and give it a 1/4-turn twist to smear the cement around. Hold the pipes in place for 30 seconds while the cement sets so the pipes don't slide around while everything is still loose. While the bond will set in a few seconds, it takes several hours to completely dry.

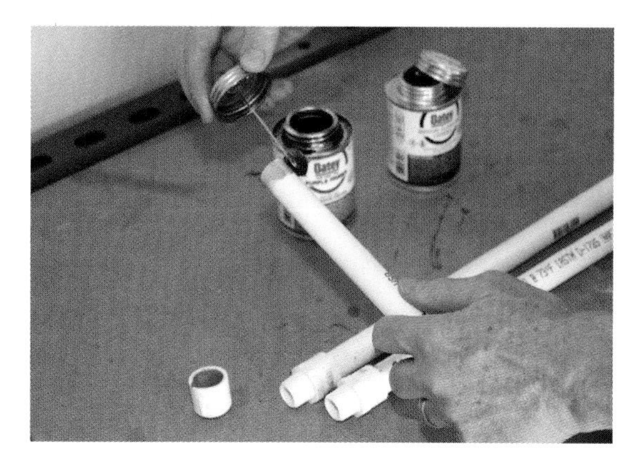

Figure 4-8. *After cleaning and a test fit, apply purple primer to both parts.*

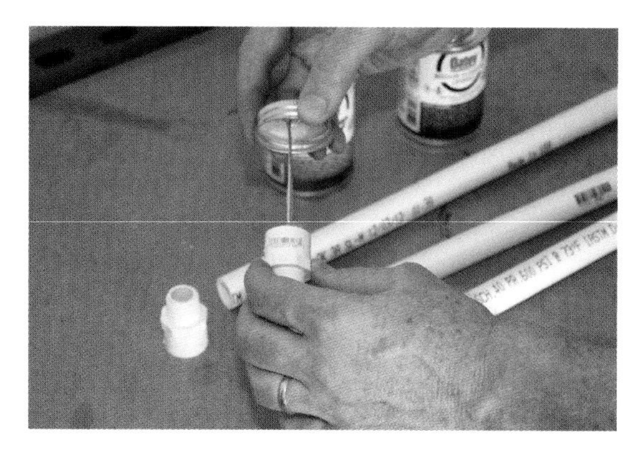

Figure 4-9. *Apply cement to the inside of one part.*

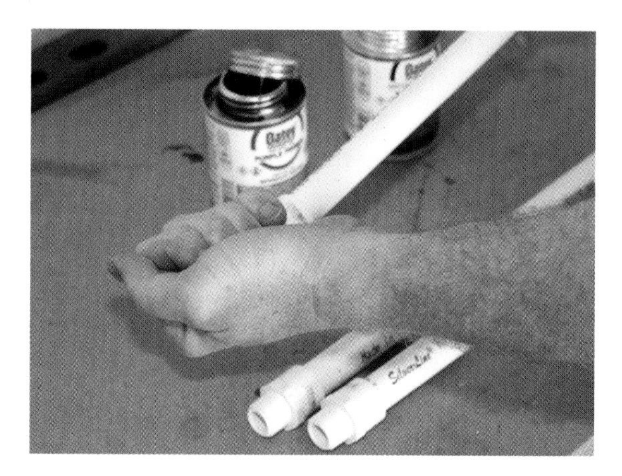

Figure 4-10. *Couple the parts, then do a 1/4-turn twist to smear the cement evenly.*

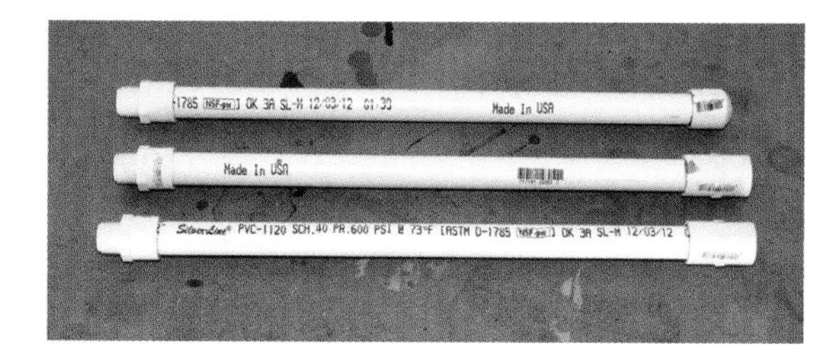

Figure 4-11. *The completed legs.*

The cap is made up of three female couplers, the 90-degree side angle elbow, and three 1"-long pieces of PVC pipe. One end of each coupler is designed to fit over the end of a pipe, and the three openings in the 90-degree side angle elbow are also designed to fit over a pipe. The three pieces of pipe are used to join the couplers to the side angle elbow.

As with the legs, apply purple primer to the pieces to be joined, then apply cement to the inside of the coupler or side angle elbow. Finally, insert the pipe and give a 1/4 turn to smear the cement.

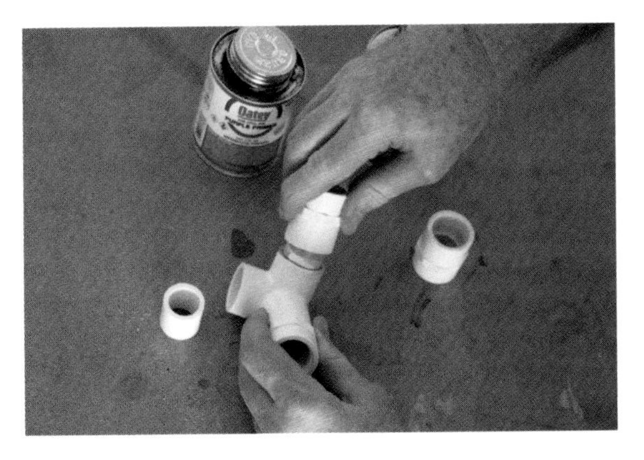

Figure 4-12. *Build the cap from the 90-degree side angle elbow, three female couplers, and three 1"-long pieces of pipe.*

Figure 4-13 shows the nearly completed head assembly for the launcher. Take note of the bolt sticking up through the base of the wooden tilt mechanism. This is cut from a piece of 1/4-20 threaded rod. It is mounted in the 90-degree side angle elbow. The PVC part won't stand up to long-term stress on this rod, so it needs to be reinforced.

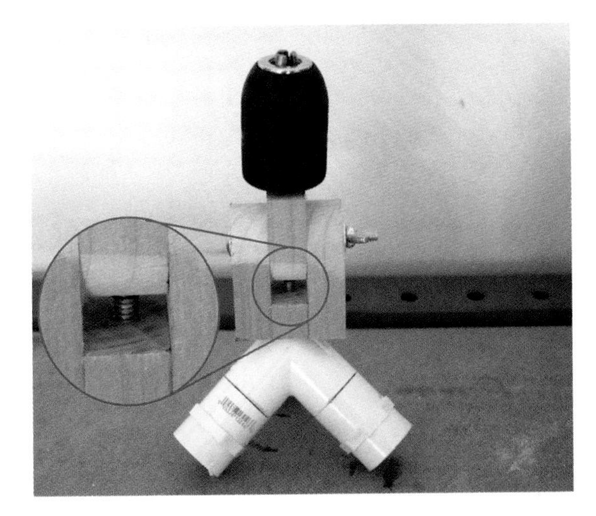

Figure 4-13. *The nearly completed head assembly.*

The solution is to fill the top of the head with epoxy resin, then drill a hole into the resin for the rod. The resin is quite strong, and will hold the rod securely.

Figure 4-14 shows the process. The bowl serves two purposes. First, it's an easy way to hold the head upside down while the epoxy is poured and hardens. The other is less obvious, but just as important. Epoxy can get quite hot as it cures. We usually don't notice the heat given off by epoxy resins because we're using small amounts for repairs, but large volumes of curing epoxy can get hot enough to melt plastic. The heat probably won't melt the PVC, but why take the chance? Add water to the bowl to help dissipate the heat. In my case, the water got pretty warm, too.

Mix enough epoxy to completely fill the side angle elbow, but don't pour in so much that it reaches the threads of the coupler. Use a paper funnel to keep the epoxy from covering the threads as the glue is poured in. The glue will set in a few minutes, but let the assembly dry overnight before proceeding.

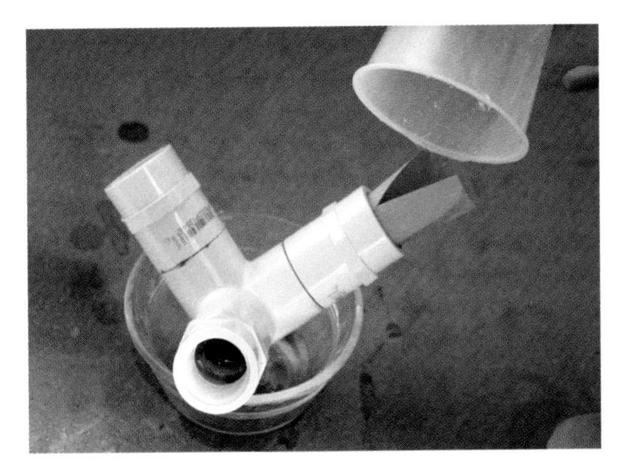

Figure 4-14. *Support the head assembly in a bowl and add enough epoxy to make a solid plug in the top part of the head. Add water to the bowl to keep the part cool as the glue dries.*

Once the epoxy is dry, drill a 1/4"-diameter hole 1" deep directly into the top of the head assembly. A drill press is by far the best choice for this step, since the hole needs to be perfectly vertical, but a hand drill will work if you are very careful. You also don't want to go too deep. Note the tape wrapped around the drill bit. This is a marker telling me how deep to drill the hole. Drill presses have calibration mechanisms, but this old-school method is easy and works with both a drill press and a hand drill.

Simplifying the Design

The mono launcher features a mechanism that allows you to tilt the launch rod and easily change between different sizes of launch rods. While these are great features, the tilt mechanism is the most involved part of the launcher to build, and the keyless chuck that supports different launch rod sizes is the most expensive part of the launcher. There is a much faster, less expensive alternative, although it means giving up the ability to tilt the rod or easily change launch rod sizes.

To make the simpler version, once the glue dries in the head assembly, simply flip it over and drill a 1/8"-diameter hole 1" into the top of the head. A drill press is really handy to get a straight hole, but careful use of a hand drill will work, too.

That's it—you're done! Insert the launch rod into the hole when you get to the launch site and add the blast deflector plate, and you are ready for launch. You can even get a little tilt by placing a stable rock or piece of scrap wood under one leg.

Figure 4-15. *Insert the launch rod directly into the hole in the launch head.*

What about multiple launch rod sizes? The easy solution is to build two head assemblies, one for 1/8" launch rods and one for 3/16" launch rods.

Figure 4-16. *Drill a 1" deep, 1/4"-diameter hole in the head assembly.*

Cutting the Metal Pieces

Two bolts need to be cut to size. The 1/2-20 bolt needs to be cut to 1 1/2" long, and the 1/4-20 rod needs to be cut to a length of 2". There are a lot of ways to cut the bolts. I used a cut-off saw, as seen in Figure 4-17. I have used cutting wheels on a Dremel tool for 1/4" threaded rods in the past, but it does take patience.

Whatever you use to cut the bolts, think ahead. They are going to get very hot. I used leather gloves to hold the bolt while it was cut, and had no problem.

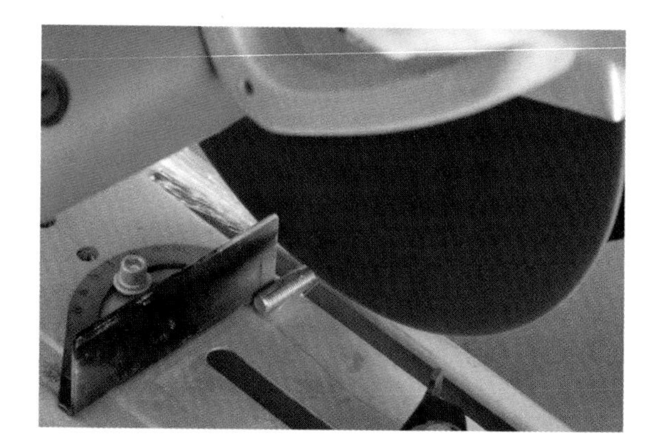

Figure 4-17. *Cutting bolts and threaded rod is easy with a cut-off saw. This type of saw also works well for PVC pipe.*

There may be some rough spots on the cut end that make it difficult to fit a nut over that end. Don't spend time grinding them off. One end of each piece will eventually be buried in glue. When the time comes, just make sure it is the irregular end that ends up in the glue.

Building the Tilt Mechanism

The tilt mechanism is constructed from 1x2 wood. While rough-cut pine is a popular choice for most shop projects, this is one project where a better-quality wood is a necessity. Most lumber yards and big-box hardware stores have a tiny section where they keep the good stuff. I like working with poplar for projects like this. It's soft enough to cut and shape easily, but it's hard enough to use for pieces that need to stand up to a little abuse now and then. You only need a piece about eight inches long, so spending a little more on good-quality wood shouldn't break the bank.

Like all wood, the 1x2 size refers to the size before the saw cuts the wood. The actual dimensions will be 1 1/2" by 3/4". The plans take that into account.

There are four pieces to cut. The bottom piece is 3/4" long, with a 1/4" hole drilled from top to bottom. This hole will fit over the 1/4" threaded rod to hold the assembly to the top of the PVC launcher head. Cut the piece using a table saw, band saw, or whatever other saw you have handy for making straight cuts.

Drill a 1/4" hole through the exact center of the wood.

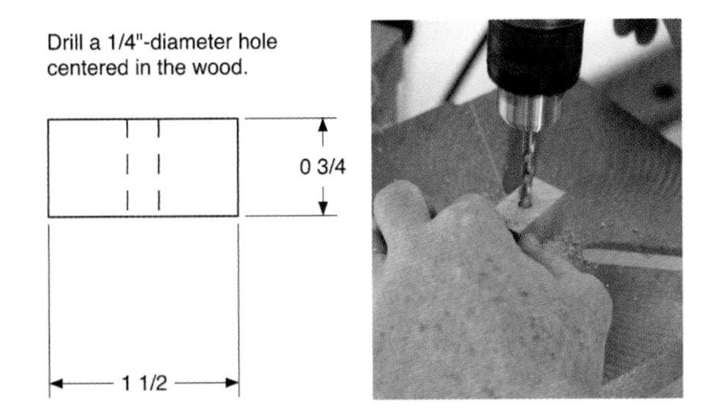

Figure 4-18. Cut the base from 1x2 stock, then drill a 1/4" hole through the center of the piece.

Cut two pieces for the sides of the tilt mechanism. The dimensions are shown in Figure 4-19. I think the easiest way to get a good result for the curves with hand tools is to cut just outside the desired curve with a jigsaw, then use a disk sander to form the final curve. Once both pieces are cut, drill a 1/4" hole at the center of the curved part.

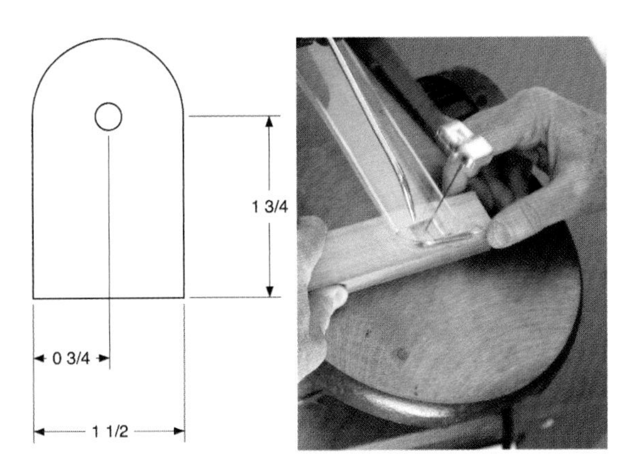

Figure 4-19. Cut two sides for the tilt mechanism, cutting slightly outside the desired shape.

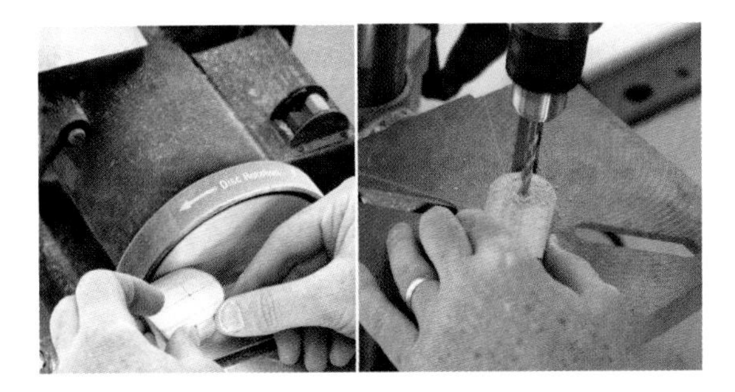

Figure 4-20. Sand the rounded edges of the sides to their final shape, then drill a 1/4" hole through the center of both parts.

The final piece is the top tilt mechanism. It has a 1/2" hole drilled partway through from the flat side. This is the hole that will eventually hold the 1/2-20 rod we cut from a bolt earlier. It's a lot easier to drill this hole before the rounded edge is cut into the bottom end of the piece.

Drill a 1/2"-diameter hole 3/4" deep, centered in the wood.

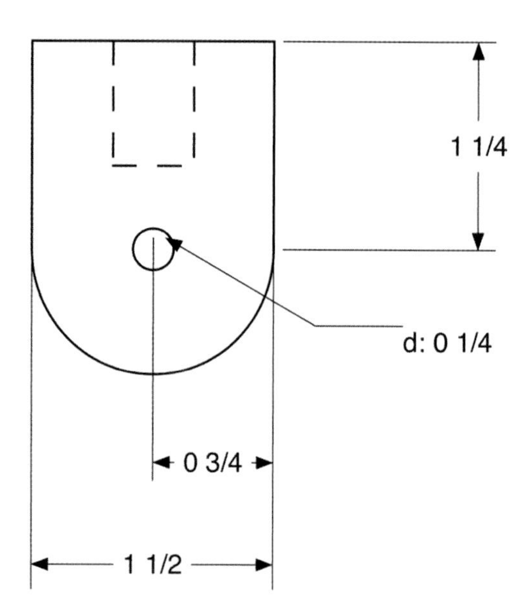

Figure 4-21. *Top piece of the tilt mechanism.*

The 1/2" hole doesn't go all the way through the piece, and another hole will come in from the side. Make sure the hole is the correct depth by marking the depth on the drill bit with a piece of tape. Drill down until the tape reaches the wood.

Figure 4-22. *Use a piece of tape to mark 3/4" of drill bit, then drill a 1/2" hole 3/4" deep in the piece.*

Once the hole is drilled, follow the same procedure used for the sides to form a rounded end for the piece, and then drill the final 1/4" hole through the center of the rounded portion.

It's time to glue the threaded rods into place. Mix up a small amount of epoxy glue and place some in the 1/2" hole in the top of the tilt assembly, and a bit more in the 1/4" hole in the PVC launcher head. You need enough glue to securely fasten the rods in place, but don't let any squirt out of the hole and fill the threads of the rods. If you have a problem, stop right away and clean the threads with rubbing alcohol.

The 1/4" rod should stick out of the PVC launcher head just far enough to go through the hole in the tilt assembly base and still leave room for a lock washer and nut to tighten the assembly to the base. The 1/2" rod should not stick out so far that the chuck cannot be screwed all the way down to make contact with the wooden piece that forms the top of the tilt mechanism.

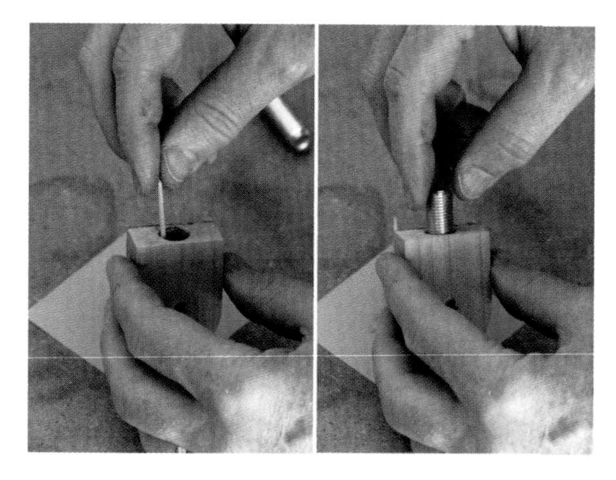

Figure 4-23. *Apply epoxy glue to the inside of the hole in the top part of the tilt mechanism, then insert the 1/2" threaded rod cut from a bolt earlier.*

Figure 4-24. *Repeat the process to glue the 1/4" rod into the PVC launch pad head.*

The bottom three pieces of the tilt mechanism are fastened together with 2" wood screws. If you've ever worked with small pieces of wood, you know the wood screws will split the wood unless there is a pilot hole. Pick a drill bit that is the same thickness as the screw would be if the threads were filed flat. Stack all three pieces in their final configuration, insert the carriage bolt through the two 1/4" holes in the side to make sure the alignment of the holes is correct, then drill two guide holes 2" deep. Finish up with a countersink bit to drill out the area for the top of the screws, then fasten the pieces together with the wood screws.

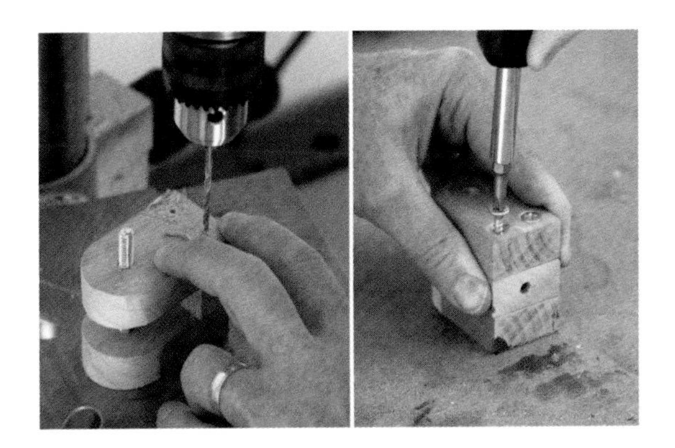

Figure 4-25. *Insert the carriage bolt as a guide, then drill two guide holes, followed by countersunk holes for the screw tops. Insert the screws to bind the three parts together.*

If the pieces are a little uneven—OK, *when* the pieces are a little uneven—finish by sanding the entire component on a disk sander or belt sander to get nice, flat surfaces.

Paint the outside of the pieces with a primer, then add color. I did not paint the surfaces that would need to rub against each other to tip the pad, though. I applied some lemon oil to the wood to protect it, but otherwise left the wood bare to avoid the risk of paint binding the pieces together.

Figure 4-26. *Paint the outside of the pieces, but leave the parts that will move against each other unpainted. Rub in furniture oil to protect the wood.*

Safety Cap

If you recall, I said that the launch rod is one of the most dangerous things at a launch site. The safety cap sits on top of the rod to protect anyone that falls on it.

Drill a 1/16" hole in the top of the remaining PVC end cap, then glue a 1 1/4" inch piece of PVC pipe in the cap to give the finished safety cap enough length to keep it on the launch rod. Set the piece aside; we'll finish it up when we build the launch controller.

Figure 4-27. *Drill a 1/16" hole in the remaining PVC end cap, then glue in a 1 1/4" piece of PVC pipe.*

Final Assembly

Plain PVC pipe looks pretty ugly. Spruce up the launcher a bit with some paint or plastic tape. I used enamel spray paint to cover the pipes with a nice, even white, then taped the pipes with painter's tape before applying the final orange color. The same orange appears on the wooden parts of the tilt mechanism.

Once the paint dries, fasten the tilt mechanism to the PVC launcher head by slipping the base over the 1/4" threaded rod. Drop on a lock washer, and then screw the nut in place to hold the base of the tilt mechanism on the launcher.

Figure 4-28. *Bolt the base of the swivel mechanism in place.*

Slip the top of the tilt mechanism between the sides. It should be a snug fit. Insert the carriage bolt, place a washer on the bolt, then tighten the wing nut. The wing nut will pull the square portion under the head of the carriage bolt into the wood. This prevents the bolt from twisting as the wing nut is loosened and tightened to adjust the launcher.

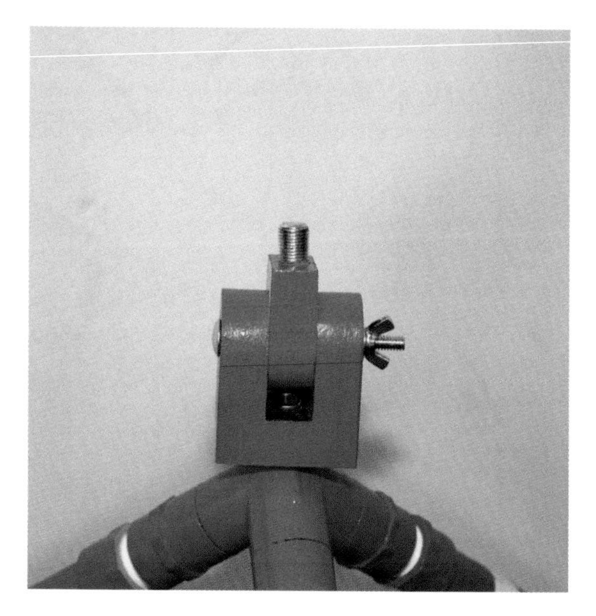

Figure 4-29. *Bolt the tilt mechanism in place.*

Screw the drill chuck onto the 1/2" rod, and the base of the launcher is complete! You can unscrew the legs and reattach them to form a long rod, making it easy to store the launcher in the corner of a closet. The launch rods store inside the PVC legs.

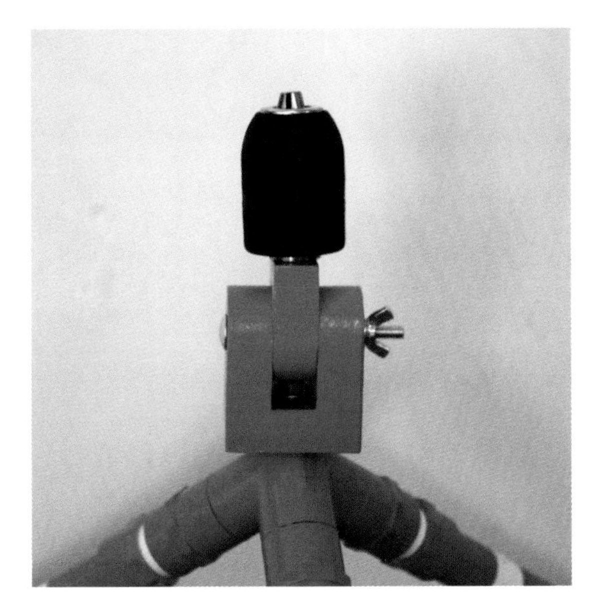

Figure 4-30. *Attach the chuck.*

Blast Deflector Plates

As the rocket lifts off, an intense flame will shoot downward. The flame is hot enough to melt plastic, burn through aluminum cans, and start fires if there is anything combustible in the way. The blast deflector plate is an essential part of the launcher. It takes the brunt of the flame, allowing the hot gas to dissipate and cool harmlessly before it reaches any delicate parts of the launcher or landscape.

There are a lot of great choices for a blast deflector plate. Figure 4-31 shows the three we'll look at in this section.

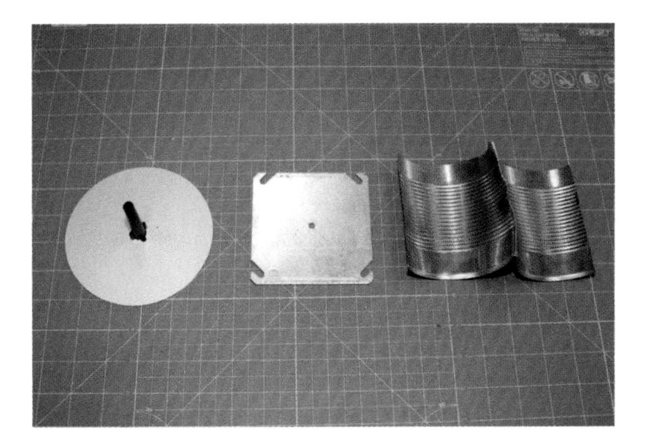

Figure 4-31. *Three of the many choices for blast deflector plates. From left to right: the Estes blast deflector, a blast deflector made from a metal plate, and a blast deflector made from a tin can.*

The Estes blast deflector plate

I admit I'm partial to the Estes blast deflector plate. I've scorched these plates with the exhaust from hundreds of rockets. They look nice, they're relatively inexpensive, and I really like the guide bar that slides over the launch rod to keep the deflector level. This is the one I've used to launch most of my E or smaller rockets, and it's the one pictured on the launchers in this chapter.

A steel plate from the hardware store

A good alternative to a commercial blast deflector is a steel plate from the local hardware store. The plate shown in Figure 4-31 is called a square box cover. It's a pretty thick steel plate; all you need to do to use it as a blast deflector is drill a hole in the center. I recommend a 13/64" hole. This is large enough so the plate can slide over the larger 3/16" rods used for D and E rockets and heavier rockets, but not so big to prevent using the plate with a standard 1/8" rod.

Be a little careful drilling the hole in the plate if you are not familiar with drilling metal. It's a different operation from drilling wood. If you go too fast, the drill bit can catch on the metal, spinning the plate. That can cause some painful scrapes and cuts. Take your time, drilling slowly through the metal with even pressure, and letting up on the pressure as the bit finally breaks through the plate. You don't see it in the pictures, but eye protection is a must with any power tool, and protective gloves are a must when drilling metal. Even better is to use a clamp and keep your hands completely away from the metal.

Figure 4-32. *You can use a metal plate for a blast deflector. Drill a 13/64" hole through the center of the plate.*

The cutouts at the corners may look odd, and might cause you to think this blast deflector is not cool enough for you to use. *Au contraire*. As programmers say, that's not a bug, that's a feature! Those notches are a great place to run launch wires. They grip the wires slightly, preventing them from slipping and pulling the igniter from the motor before it lights.

One minor problem with these plates is that the exhaust from the rocket can tilt them. The rocket is generally fast enough to get out of the way, and as you will see in the next chapter, it probably won't be close enough to the blast deflector plate to get hit anyway, but it's one of the reasons I like the Estes-style blast deflector.

These steel plates are the blast deflector of choice for the low-power pads at my local rocket club. While they also use Estes blast deflector plates, these are very cheap and last a bit longer than the Estes plates. After several hundred launches with powerful C14, D, and E motors, you can actually burn through a deflector plate. That happens more slowly with these, because they're made of thicker steel.

The tin can deflector

This classy little blast deflector is literally free. Snag the next tin can that's headed for the landfill, and start by cutting off the bottom. You can do that with a can opener on some cans, but metal shears or a Dremel tool with a cutting wheel will do the job quickly, too. After removing the bottom, cut the can from end to end.

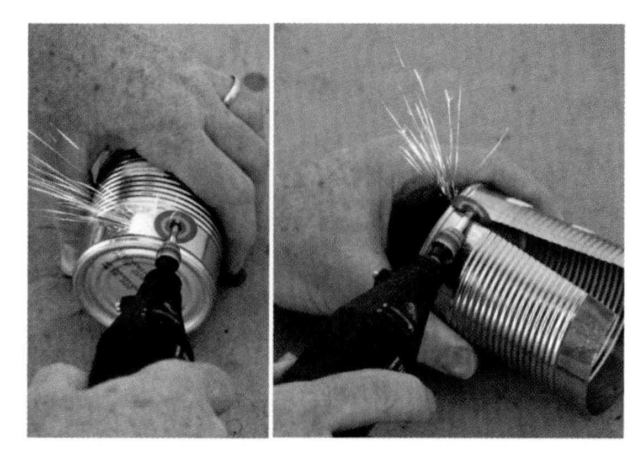

Figure 4-33. *Cut the bottom from a tin can, and then cut the can from top to bottom.*

Bend the can back on itself. A vise and rubber mallet are the best way to bend the can—my vise wasn't big enough, but the metal is pretty soft. You can impress your friends by bending it with your bare hands. Finish up the deflector by drilling a 13/64" hole through the bend.

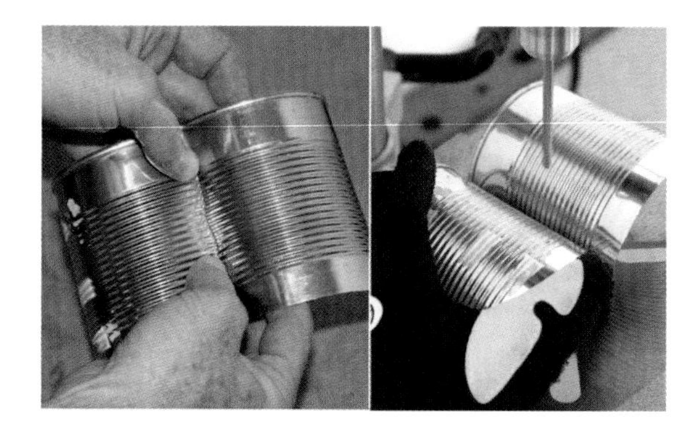

Figure 4-34. *Bend the can back on itself. Finish the blast deflector with a 13/64" hole through the center.*

These blast deflector plates won't last as long as the others, but then again, they can be replaced for free. They also have a cool curve to bend the flame.

Don't Use Soda Cans!

Don't use an aluminum soft drink can. The aluminum is too thin and has too low a melting point to serve as a blast deflector plate.

Launch Rods

Launch rods are readily available at hobby stores and hardware stores. Hobby stores call them piano wire, music wire, or hobby wire, while hardware stores have them in a section often labeled "metal by the piece." Either way, look for them in a group of vertical cardboard tubes.

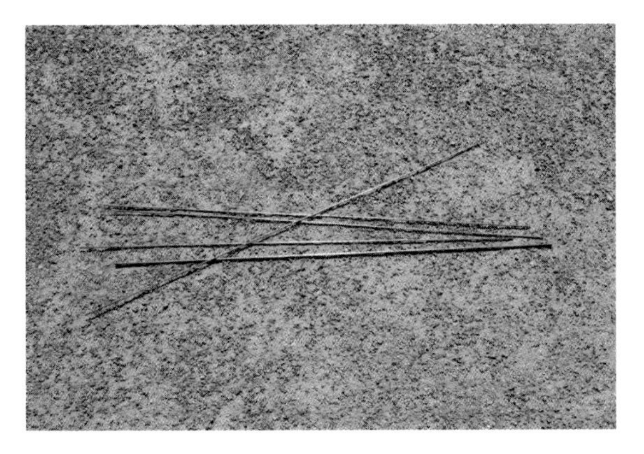

Figure 4-35. *Four 1/8" launch rods and one 3/16" launch rod. Each rod is 3 feet long.*

The rods are generally sold in 3-foot lengths, which is perfect. You might occasionally find them in 4-foot lengths, which is even better for thicker rods. I've also seen 6-foot rods, but that's too long—the rod is long enough that it will bend back and forth, causing rod whip. The rod can actually whip back and forth enough to damage the rocket.

While almost any metal rod will do, some are better than others. The music wire or piano wire rods are the best; they are quite rigid. Light aluminum rods and stainless steel rods are more flexible. They can be used, and I've seen them in commercial launchers, but the flexibility can be a problem. I recommend getting solid rods, not tubes. While it's possible to build a launch rod in sections using tubes, you have to be very vigilant about keeping the joint smooth, often sanding the joint to get it smooth enough for the launch lug to slide over the rod without catching. The tubes are not as strong as solid rods, either. The main reason for using shorter lengths is easy storage, but the mono launcher solves that problem—the legs themselves provide storage for the launch rods.

The launch rod may be a bit rough or even rusty when you buy it. That's not really a problem. Use fine sandpaper to sand down to smooth, bare metal, and then wipe the rod with some lightweight oil such as WD-40. You may need to repeat the process every few seasons, but a single launch rod will last long enough to pass on to your grandkids. That's something I can say with confidence, since my granddaughter is using a launch rod that predates my own children. If you want to hand the launch rod down to your great-grandkids, wipe it with oil after each launch day. The exhaust from rocket motors can cause the rod to rust faster. Wiping the rod with oil after a launch cleans any contamination from the rocket motor and protects the rod from normal corrosion.

The light rockets in the first part of the book all use a 1/8" diameter launch rod. The heavier rockets later in the book use a 3/16" launch rod. I'd recommend getting both right away, but there is no reason to buy a 1/4" launch rod just yet. There are months' worth of rockets to build and fly, and lots of crashes, repairs, experience, and fun to be had, before you build a rocket that needs a 1/4" launch rod.

The Mono Launch Controller

The mono launch controller described in this section is a 12-volt launch controller that allows you to stand 30 feet from the launcher. It has all of the standard safety features for a rocket launch controller, including an arming key, an arming light, and a pushbutton firing switch that returns to the open position as soon as it is released. Its 18-gauge wire carries enough current to the launcher to launch composite igniters like the First Fire Jr., Estes Pro Series, and Copperhead, yet the LED-and-resistor-based arming light cuts current enough to safely handle the sensitive Quest Q2G2 igniter. It can even handle cluster rockets. The Estes igniter needs a lot of current, but this launcher can deliver enough current to fire four Estes igniters if you use an external lead-acid battery. With the more sensitive Quest Q2G2 igniters, this launcher can fire a four-motor cluster rocket from the internal AA batteries, and a whopping 24 igniters using an external lead-acid battery (see Table 4-4).

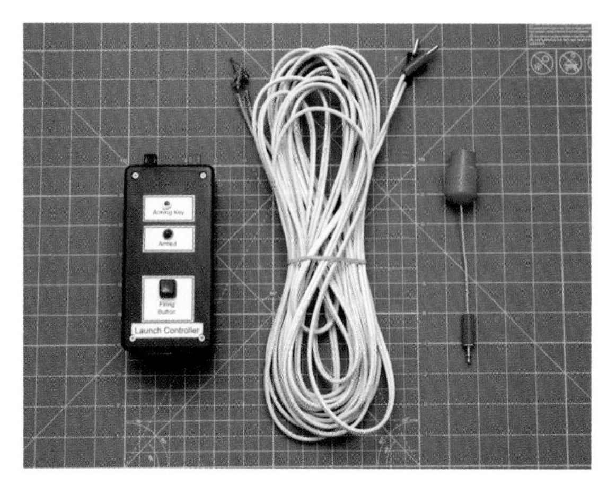

Figure 4-36. *The mono launch controller.*

Table 4-4. Maximum number of clustered motors with various igniters using the mono launch controller

Igniter	AA batteries	Car battery
Estes	1	6
Quest Q2G2	4	24
First Fire Jr.	1	4
Copperhead	1[a]	4

[a] The current supplied is slightly under the current recommended in the theory section, but testing shows the launch controller fires the igniter reliably.

How the Circuit Works

This theory section discusses exactly how the launch controller works. It's important to understand the workings of the launch controller system if you want to change the design of either launcher, but you can safely skip it if you build the launch controller exactly as described. The construction instructions, starting with "Building the Mono Launch Controller" on page 82, will include the most common alternatives for power and wiring, and explain which one you should choose based on the kind of flying you expect to do.

The basics of launch circuit design

Figure 4-37 shows a typical circuit for a launch controller. This one is the circuit for the mono launch controller. Let's walk through the circuit to see how it works.

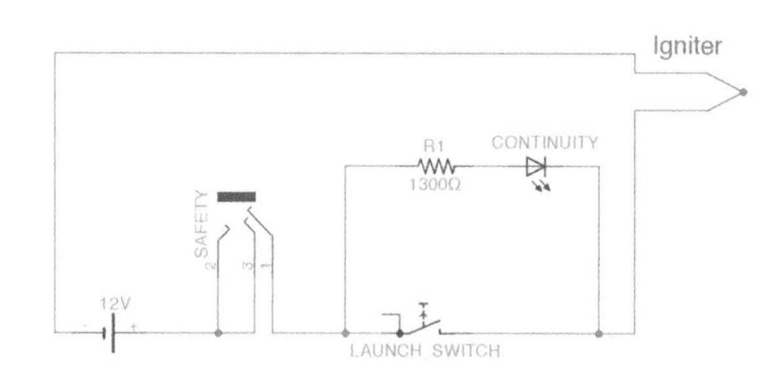

Figure 4-37. *A typical launch controller circuit.*

Power comes from a battery. This can be a 6-volt or 12-volt battery, depending on the number and kind of igniters. Following the positive flow of current as it exits the right side of the battery, we come first to a safety feature called the *arming key* or *launch key*. This is a switch—a complete break in the circuit—which is only closed when an external key is inserted (and turned, if it is an actual key switch). It's an important safety feature for the launch control system. Until this external arming key is inserted, no current can flow through the circuit, no matter how many times you press the launch button. As we'll discuss in the next chapter, this key must not be inserted until the launch area is clear and everyone is aware a launch is about to take place.

The current then splits. Following the upper branch first, we see a resistor and an LED. This is the arming light. Assuming the igniter is correctly connected and the arming key is inserted, current will flow through this resistor and LED, and the LED will shine, telling us the rocket is ready for launch. Other things can go wrong, as we'll see in the next chapter, but the glow of the arming light is our indication it is time for the launch.

Following the circuit around, we see the current flows through the igniter and back to the battery. That means there is current flowing through the igniter as soon as the arming key is inserted. Why doesn't the rocket launch right away? The simple answer is the resistor is cutting the amount of current flowing through the circuit, so there isn't enough current reaching the igniter to heat it significantly. We'll delve into a more complete answer in a moment, and see why the standard Estes launch control system will, in fact, fire a Quest Q2G2 igniter as soon as the arming key is inserted!

Going back to the branch in the circuit, the lower path goes to an open switch. This is the launch button. As soon as the launch button is pressed, the current bypasses the higher-resistance branch, dumping almost all of the current through the launch button and directly to the igniter. In fact, the arming light will go dark as soon as you press the launch button, and stay dark until the launch button is released. Assuming the igniter fires, it will probably stay dark; firing an igniter, especially in a rocket motor, usually burns through the igniter wire so it no longer completes the circuit.

Now let's go back and look at why the launch controller works. The launch controller is designed to dump as much of the battery's current through the igniter as possible to launch the rocket. It's the same idea as an electric heater or an old-fashioned incandescent light bulb: lots of current passing through a wire will heat the wire. Some kinds of wire, such as a light bulb filament, a heating element, or an igniter, heat up when enough current passes through the wire.

Let's state that again using the language of science and mathematics. We'll start with a basic equation from electronics to help us understand this circuit.

Rocket igniters need a specific amount of current to ignite. Ohm's law gives the current through a circuit:

$$I = \frac{V}{R}$$

where I is the current, V is the voltage, and R is the resistance. To get as large a current as possible, we want to have a high voltage and a low resistance.

Let's start by looking at the circuit when the launch button is pressed. This will tell us how much current is flowing through the igniter when we press the launch button.

As we'll see in a moment, the resistance in the mono launch controller when the launch button is pressed, not counting the igniter, is about 3.9 ohms, written as 3.9Ω. While igniters vary, a typical resistance for an igniter is 0.8Ω. The total resistance is 4.7Ω. This means the current flowing through the igniter is:

$$I = \frac{12.0}{4.7}$$

$$I = 2.6$$

So, the launcher is delivering 2.6 amps to a typical igniter. We'll look at igniters in more detail in the next section, but it turns out this is plenty.

Now let's analyze the circuit again, but this time before the launch button is pressed. That's important, because we want to make sure our circuit is not delivering enough power to launch the rocket as soon as the launch key is inserted.

The design for the mono launch controller calls for a resistor with a resistance of at least 1,200Ω, giving a total resistance for the launch controller of 1,203.9Ω. With a 0.8Ω igniter, the total resistance is 1,204.7Ω:

$$I = \frac{V}{R}$$

$$I = \frac{12}{1204.7}$$

$$I = 0.01$$

One one-hundredth of an amp is definitely not a lot of current. It won't heat the igniter much, which is why the rocket does not launch with the arming key inserted, even though current is flowing through the igniter.

Designing a launch controller is all about maximizing the current delivered to the igniter while the launch button is pressed, while at the same time minimizing the current flowing through the igniter when the launch button is not being pressed.

Electrical characteristics of some common igniters

We need to know two things about an igniter to design a launch controller. First, we need to know the resistance of the igniter so we can calculate the current through the igniter, both when the arming key is inserted and when the launch button is pressed. Second, we need to know how much current the igniter needs before it will fire, so we can make sure it won't fire when the arming key is inserted, but will fire quickly and reliably when the launch button is pressed. Table 4-5 shows the electrical characteristics of some common igniters from a series of excellent tests performed by J.R. Brohm in 2009. You can find the original article at *http://bit.ly/igniter-cont-tests*. The table shows the smallest and largest values that actually ignited a test igniter.((("Quest Q2G2 igniter"

Table 4-5. Electrical characteristics of common igniters

Igniter	Min volts	Max volts	Min amps	Max amps	Min resistance	Max resistance
Estes	0.70	0.80	1.00	1.20	0.58	0.80
Copperhead	0.75	1.20	1.55	1.70	0.44	0.77
First Fire Jr.	1.10	1.40	1.75	1.80	0.61	0.80
Quest Q2G2	0.60	1.60	0.22	0.32	1.88	7.27

Our launcher designs need to keep the current flowing through the igniter well below the minimum ignition current before the launch button is pressed, and deliver about twice the ignition current when the launch button is held down.

Resistance in the launch controller

The launch controller itself doesn't look like it would have much resistance once the launch button is pressed. After all, at that point it's just a battery and some wire. It turns out that's not quite the case, though. The resistance in the launch controller comes from two places: the wire and the batteries themselves.

Huh? The batteries? Yes, batteries have a resistance, called the *internal resistance*. Under the conditions of a typical launch controller, the internal resistance of a AA battery is about 0.4Ω, although that can vary with the temperature, the age of the battery, and the manufacturer. Of course, there is more than one battery in the launch controller, so for the four batteries used in a 6-volt launch controller, the internal resistance is 1.6Ω, and for the eight batteries used in the mono launch controller, the internal resistance is 3.2Ω.

The battery (or batteries) is the most significant source of resistance in the launch controller, so it makes sense to try to find ways to minimize that resistance. By far the easiest way is to change the kind of battery used. A car battery, for example, delivers 12 volts with an internal resistance of about 0.02Ω. The battery resistance to deliver 12 volts drops from 3.2Ω with AA alkaline batteries to 0.02Ω with a 12-volt lead-acid car battery.

The other source of resistance is the wire itself. Most electronics books and articles simply ignore the resistance of the wire. That's fine, since the resistances in the circuit are generally several hundred to tens of thousands of ohms, while the resistance in a short jumper wire on a breadboard is less than a thousandth of an ohm. Launch controllers are different, though. The resistance of the wire can't be ignored when we're using a few dozen feet of wire instead of a fraction of an inch.

The two most common wire sizes for launch controllers are 24 gauge and 18 gauge. We always use stranded copper wire. Aluminum wire is common, but it has almost twice the resistance of copper wire. There are several advantages to stranded wire, but the ones that apply to a launch controller are flexibility and durability. Stranded wire bends easier than a solid wire with the same resistance, so it's easier to deploy at the launch field and then wind up at the end of the day. It's also less likely to break after being bent, which will happen as the wire is wound and unwound.

The Estes Launch Controller

The Estes launch controller is probably the most commonly used launch controller in model rocketry. I've used one for years. It's rugged, simple, and very effective when used to launch black powder rockets using the Estes igniter. As long as you don't intend to launch anything larger than an E motor and are not planning to use clusters of motors, I highly recommend this launch controller.

I got quite a surprise the first time I connected one to a Quest Q2G2 igniter, though. Let's see why.

The Estes launch controller uses a #51 6-volt pilot light and four AA batteries, providing a 6-volt ignition system. The pilot light has a resistance of 30Ω, and it's the pilot light that cuts the current through the igniter when the arming key is inserted in the launch controller. The total resistance in the launch controller, not counting the igniter, is about 34Ω. The resistance in a Quest Q2G2 igniter is about 3Ω. Plugging these values back into our equations, we see that the current delivered to the igniter with the arming key inserted in an Estes launch controller is:

$$I = \frac{V}{R}$$

$$I = \frac{6.0}{37.0}$$

$$I = 0.16$$

That's not enough to hit the recommended current for firing this igniter, but it's close enough! Give the igniter a little time —a very little, from our perspective, just a fraction of a second —and it will ignite.

The Estes launch controller is a fine launch controller, but bear in mind that it is not suitable for all igniters.

Higher wire gauge numbers indicate thinner wire. Eighteen-gauge wire is thicker than 24-gauge wire, and also has a lower resistance. The resistance for 18-gauge wire is about 0.01Ω/foot, while the resistance for 24-gauge wire is about 0.04Ω/foot.

The mono launch controller uses 35 feet of 18-gauge wire, so the resistance is about 0.35Ω, right? Not necessarily. It does use 35 feet of wire, but it's two-conductor wire. The current goes out, then returns to the launch controller, for a total of 70 feet of travel, giving 0.7Ω of resistance.

And now you see where the resistance of the launch controller came from when we analyzed the launch controller circuit earlier. The resistance is 0.7Ω for the wire and 3.2Ω for the batteries, or 3.9Ω total.

Wouldn't it be nice to have the option of using a 12-volt car battery, though? That way the resistance drops to 0.72Ω, since the battery resistance drops from 3.2Ω to 0.02Ω. As you will see, the mono launch controller gives you that option. It has battery holders for eight AA batteries, but it also has a plug and cable so you can use power from an external lead-acid battery.

Clusters of rockets

Most rockets use a single motor, so they only need one igniter. Some use clusters of motors, though, and each motor needs its own igniter. When the current is split between multiple igniters, it divides evenly among the igniters. That means our launch controller needs to supply four times the current needed to fire a single igniter if we want to use it to launch four motors at once.

Summing it all up: Analysis of the mono launch controller

The mono launch controller is built with 35 feet of 18-gauge copper wire, for a total wire run of 70 feet. It can use either eight AA batteries or a single external 12-volt lead-acid battery. Table 4-6 shows the four igniters we've looked at, along with the various combinations of batteries required for either one or four igniters. The last column shows the recommended current for each igniter.

Table 4-6. Amps per igniter using the mono launch controller—the last column shows the recommended current to light the igniter

Igniter	AA batteries	AA batteries	Car battery	Car battery	Recommended current
	1 igniter	4 igniters	1 igniter	4 igniters	
Estes	2.55	0.72	7.89	3.08	2.00
Quest Q2G2	2.76	0.74	10.30	3.39	0.60
First Fire Jr.	2.59	0.73	8.22	3.13	3.00
Copperhead	2.76	0.74	10.30	3.39	3.00

Using AA batteries, the mono launch controller can launch a single motor using any of these igniters. The current delivered to the First Fire Jr. and the Copperhead is a bit lower than recommended, but field testing shows they fire just fine. In fact, the Estes launch controller only delivers 1.6 amps to an Estes igniter, which is also below the recommended 2 amps.

Clustering is a little different. The Quest Q2G2 igniter is definitely the right choice for clusters of black powder motors. You can use Estes igniters, too, but not with AA batteries; for clusters of Estes motors, switch to a 12-volt lead-acid battery, like a car battery.

Even a 12-volt lead-acid battery can barely handle clusters of Copperhead or First Fire Jr. igniters. You don't want to skimp on power with a clustered rocket—all of the igniters need to fire quickly, and at the same time. The mono launch controller will work with two or three motors using a lead-acid battery, but not with four.

Building the Mono Launch Controller

Most of the parts for the mono launch controller are available from Radio Shack, and most of the rest are easy to find at an office supply store. The parts list (Table 4-7) gives Radio Shack part numbers when appropriate. Of course, there are similar parts available from lots of other sources, so feel free to shop around.

The only parts that might be hard to find locally are the cigarette lighter socket and plug. These are used to connect an external 12-volt battery to the launch controller. The parts are readily available from many online retailers, such as Amazon.com. You can use some other form of connector, of course, but keep one important point in mind: as designed, the launch controller can use both the internal and external batteries. That means there is power on the external connector from the internal batteries at all times. Be sure the connector you use guards against accidental shorting. The cigarette lighter socket does this nicely.

Table 4-7. Parts list

Part	Description	Radio Shack part number
6x3x2 project enclosure	This is the black plastic enclosure that houses the launch box.	270-1805
Banana jack insulated binding post	These come in a package of four. We'll use one red one and one black one as a handy way to connect the wires that run from the launch box to the launcher.	274-0661
Banana plugs	The banana plugs give a quick way to attach the launch wire to the launch controller box. They come in a package of two; we'll use both.	274-0721
4 AA battery holder (2)	There are several styles of battery holders available. Because of the way we're going to use them, be sure to get two flat battery holders with an open top. If this specific part is not available, combine similar-style battery holders that hold two or three batteries.	270-0391
Ultra-high brightness 10 mm red LED	Any LED will work as an arming light, but this one is big, bright, and impressive!	276-0015
3-conductor stereo 1/8″ phono jack	This is the hole for the arming key.	274-0249
Solder-type mono 1/8″ phono plug	This is the arming key. It's also available in black, but who wants a black arming key?	274-0287
SPST soft touch momentary pushbutton switch	This is the firing button. It's a big, red switch that's easy to push.	275-1566
Micro 1 1/8″ smooth clips (2)	These are used to connect the launch wire to the igniters. They come in a package of 10.	270-0373
Battery clip	The battery clips connect the launcher to an external 12-volt battery. This part is perfect for connecting to small portable power supplies like the one shown in the text, but may be too small to clamp onto the terminal on a full-size car battery. Get a clip big enough for the expected use.	270-0344
1/4-watt resistor	The design isn't especially sensitive to the specific resistance used. Get the smallest resistor you can find that is larger than 1.2KΩ. Just make sure you get one that will handle at least 1/4 watt.	

Part	Description	Radio Shack part number
Marine-grade cigarette lighter socket 12 VDC	This socket connects the launch controller to an external 12-volt power supply. You can use several alternatives, but this is a nice, safe way to prevent the internal battery from shorting. It's available on Amazon, sold by Parts Express.	
8-32 3/4" flat-top machine screws, lock washers, nuts	You will need two 8-32 machine screws to fasten the cigarette lighter socket to the launch controller box. Get the kind with the flat top and conical head to match the opening in the socket (available at Ace Hardware or a similar store). Get matching lock washers and nuts.	
Marine-grade locking cigarette lighter plug 12 VDC	This plug is used in the cable that connects the external 12-volt battery to the launch controller. It's available on Amazon, sold by Parts Express.	
10-amp fuse	The cigarette lighter plug comes with a 5-amp fuse. Switch to a 10-amp fuse.	270-1015
Printable white label sheet (Avery 8165 or similar)	Use these to create labels for your box.	
42 feet of two-conductor 18-gauge wire	There are a lot of choices here, so shop around. Be careful to get copper wire. If one of the wires looks silvery, it's probably aluminum, and that won't do. I used lamp wire from the local hardware store.	

Figure 4-38. *Visual parts list for the launch controller box.*

The construction details discuss several alternatives to the base design. Be sure to read through the construction description to decide if you will use any of these modifications, since that could change the parts you buy. Additional tools and materials that you will need for this project are listed in Table 4-8.

Table 4-8. Tools and materials

Tool	Description
Soldering iron	Anything designed for electrical work will do, such as Radio Shack part 64-2808.
Solder	Solder designed for soldering electronics, such as Radio Shack part 64-009.
Wire strippers	Used for stripping wires, of course.
Dremel tool with cutting disk	You can get by with a hobby knife or even a file with some patience, but a Dremel tool will be easiest.
Hobby knife	Used to trim labels to decorate the launch controller.
Drill	A bench press is great, but a handheld drill will work fine. You will need the following drill bits: 1/8, 1/4, 5/16 and 3/8 inch. The sizes may change if you substitute any parts for the Radio Shack parts shown.
Glue	I recommend epoxy glue for this project, but thick CA glue or any other glue that will bond plastic will also work.

Building the launch box

Start by fitting the two battery holders in the bottom of the plastic project box—or at least, try to. They don't fit very well. There are four plastic nubs sticking up from the bottom, and they'll be in your way. Time for a Dremel tool.

Figure 4-39. *The launch box carries 12 V internally for high-voltage igniters, but also has external connections for lead-acid batteries for cluster launches.*

Cut or file the nubs until the bottom of the box is flat (Figure 4-40).

Figure 4-40. *Cut off the four plastic nubs at the bottom of the box.*

Use epoxy glue to glue the two battery holders to the bottom of the project box. The battery holders will still be at a slight angle due to the posts for the screws that hold the top of the box to the bottom, but they will fit flat on the bottom.

Make sure the wires on both battery holders are on the same side of the box, as in Figure 4-41.

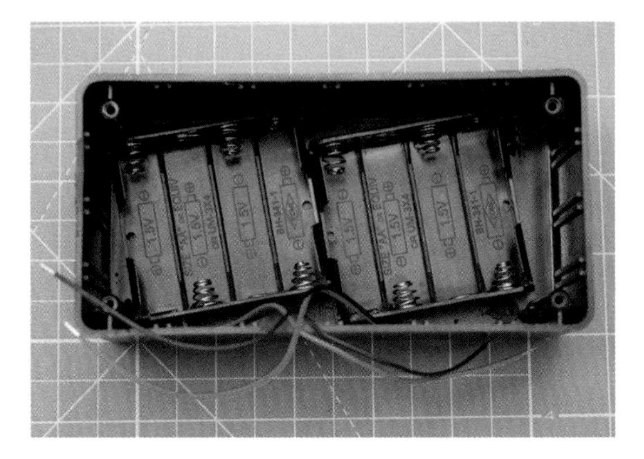

Figure 4-41. *Battery holders glued to the bottom of the project box. Twist them to make room where the cigarette lighter socket will protrude into the box.*

Drill holes in the locations shown in Figure 4-42, using the sizes shown in Figure 4-43. The size and location of the holes could change if you substitute parts for any of the Radio Shack parts shown in the parts list. See Table 4-9 for a list of the holes. If you do use different parts, mark any changes you need to make before you start drilling. Check the fit for each part as the holes are drilled.

Will Six Volts Work?

A lot of launch controllers use 6 volts instead of 12. That's half as many batteries! Why not use 6 volts in the mono launch controller?

Well, you can.

There are some limitations, though. A 6-volt launch controller will work fine with Estes igniters, and you can even launch cluster rockets if you switch to the Quest Q2G2 igniter. You won't be able to light igniters designed for composite motors; all of those require a 12-volt ignition system. That will limit you to E motors or smaller, but an E motor packs a lot of punch. Maybe you don't care about that limitation.

There's also an incompatibility between the 6 volts from the battery pack and the 12 volts from the external battery. The problem occurs when you connect an external 12-volt battery to the terminals that are also connected internally to 6-volt batteries. In effect, you're driving the batteries backward, which can damage the batteries, perhaps even causing them to leak. You could solve that by using a 6-volt external battery or by leaving off the connections for an external battery. You could also be careful to remove the AA batteries before using the launch controller with an external 12-volt battery, getting the best of both worlds.

If you're not sure, why not leave the decision for later? Build the launch controller as designed, but leave one of the battery holders empty. Connect a wire from the red wire to the black wire of the empty battery holder. If you decide to switch to 12 volts later, cut the wire and stuff in four more batteries. Again, though, be careful not to connect an external 12-volt battery when you have only four AA batteries installed internally.

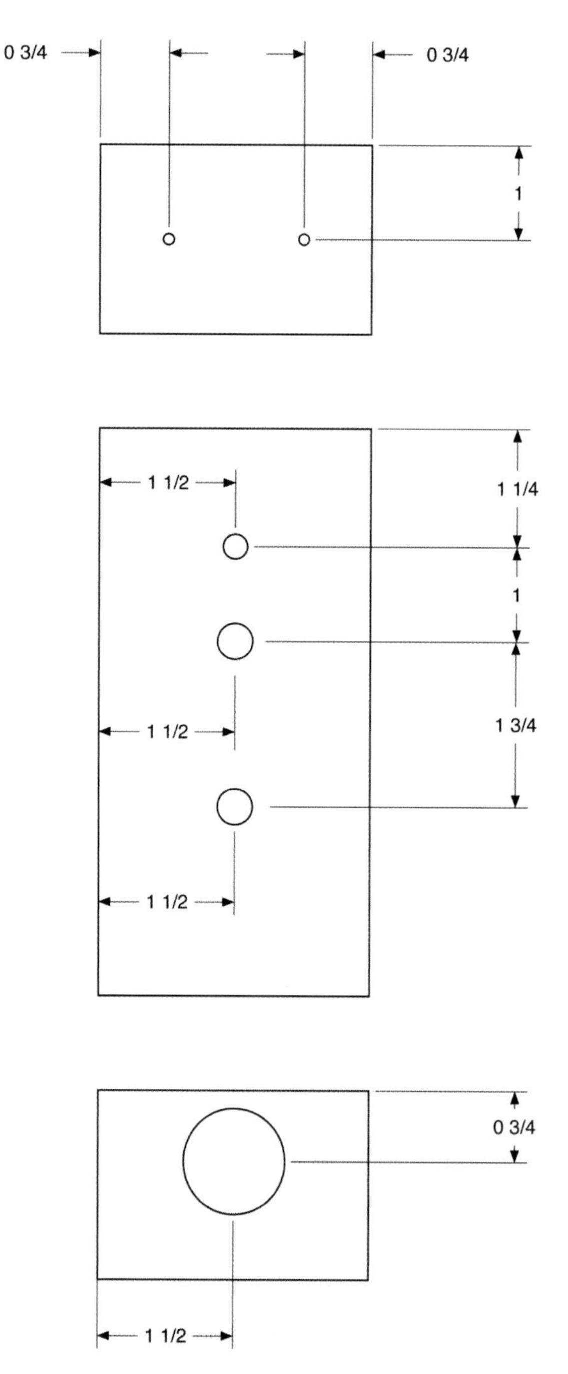

Figure 4-42. *Drill holes in the project box at these locations.*

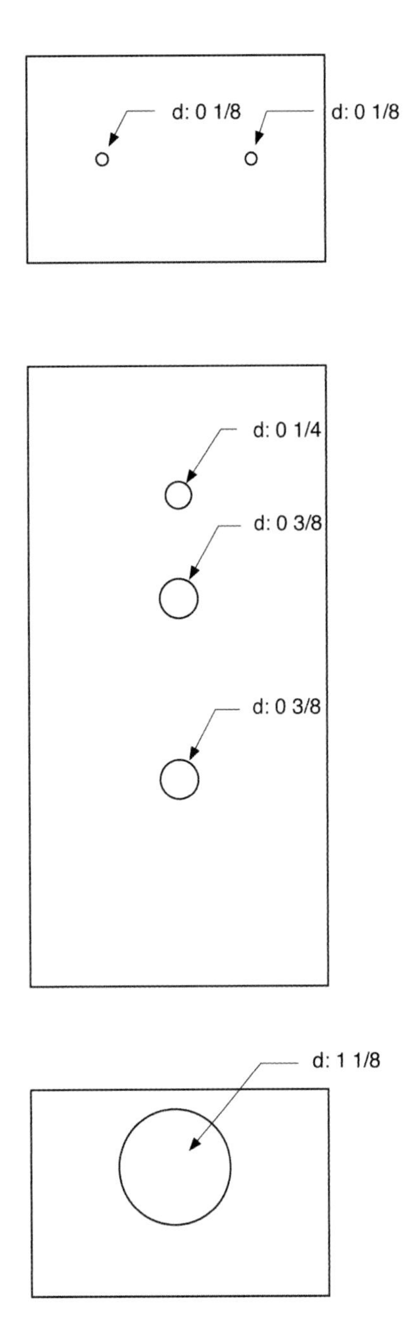

Figure 4-43. *Here are the sizes for each hole. The sizes may change if you substituted any parts for the ones shown in the parts list.*

Plastic Drilling Safety Tips

Standard bits will drill plastic, but you do need to be careful. Plastic is very soft, and tends to bind. This can grab the project box right out of your hand, spinning it around dangerously and damaging the box. And your hand.

Drill slowly to avoid these problems. Go extra slow just as the bit breaks through the plastic —this is when it is most likely to bind. Use gloves and clamp the box in place before drilling.

Table 4-9. Drill hole checklist

For part	Hole diameter	Location
Banana jack insulated binding post (launch wire connector)	1/8″	Two holes on one end of the box, both 1″ from the top of the box and 3/4″ from the side of the box.
Cigarette lighter socket	1 1/8″	One hole on the opposite end of the box from the banana jacks, centered horizontally and 3/4″ from the top edge of the bottom of the box.
LED (arming light)	1/4″	Centered in the top, 1 1/4″ from the end.
Phono jack (arming switch)	3/8″	Centered in the top, 2 1/4″ from the end.
Pushbutton switch (launch button)	3/8″	Centered in the top, 4″ from the end.

While the connections on the box are going to make perfect sense to you, they may not be obvious to others who may use your launch controller. A plain black launch controller is also pretty ugly.

Print some attractive stick-on labels for your launch controller using an inkjet printer, or, if you are the artistic type, draw some using colored markers. You can find templates for the labels shown in the photos at the author's website (*http://bit.ly/byteworks-make-rockets*). Place them on the box and trim the holes using a hobby knife.

Install the electrical components in the lid of the box. The LED needs to be glued into its hole. I applied glue from the bottom to keep the top looking neat. Check the LED before you glue it. The side next to the shorter lead wire is flattened slightly. The wiring will be a bit easier if this flattened side is oriented toward the pushbutton launch switch.

The phono jack and pushbutton switch each have a nut that can be removed. You then slip the device through the hole and tighten the nut on the other side to hold it in place. The pushbutton switch also has a small tab on the top. Use a hobby knife to carve a similar notch in the project box lid; this will keep the switch from twisting once it is installed.

Figure 4-44. Use stick-on labels to label and decorate the box, and add the components to the top of the box.

The cigarette lighter port that provides access to the external 12-volt power supply comes with four parts. The flexible rubber ring mounts under the surface plate using two 8-32 machine screws. You can get by with 1/2"-long screws, but I think it's easier to work with the longer 3/4" screws. Add two lock washers and nuts to hold the screws in place. You will need to trim away the posts at the edge of the project box to make room for the screws and nuts. Leaving the top 1/2" or so intact, use a Dremel tool or hobby knife to cut away enough of the posts for the nuts to fit. Once the surface plate is screwed in place, add the first two batteries to the battery holder. The socket will cover these once it is mounted, and will need to be pulled out a bit to change the batteries. With the first two batteries in place, insert the socket and use the plastic nut that comes with the socket to secure it in place.

Figure 4-45. Install the cigarette lighter socket. You will need to trim the corner posts to make room for the nuts. Install the first two batteries before inserting the socket; they can't be inserted with the socket in place.

Figure 4-46. Here is the front of the cigarette lighter socket with the plastic cover closed. Note that it sits about 1/16" above the lip of the box. The lid will add some height so the socket fits well, but we need all the room we can get inside the box!

Next, install the binding posts in the other side of the launch controller box. We'll make a launch cable in a moment that will plug into these posts. Figure 4-47 shows the bottom portion of the launch controller box at this stage, with all of the parts and two of the batteries installed.

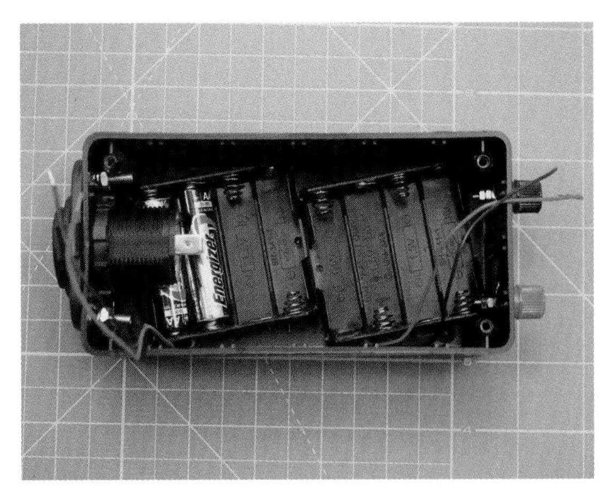

Figure 4-47. *Install the binding posts opposite the cigarette lighter socket.*

Figure 4-48 shows the circuit diagram for the launch controller. There is no printed circuit board, nor do we need one for the old-school point-to-point wiring we're about to do. While the circuit diagram is a nice blueprint to make sure you get all of the connections correct, follow the instructions carefully to learn some techniques that will make the wiring go smoothly.

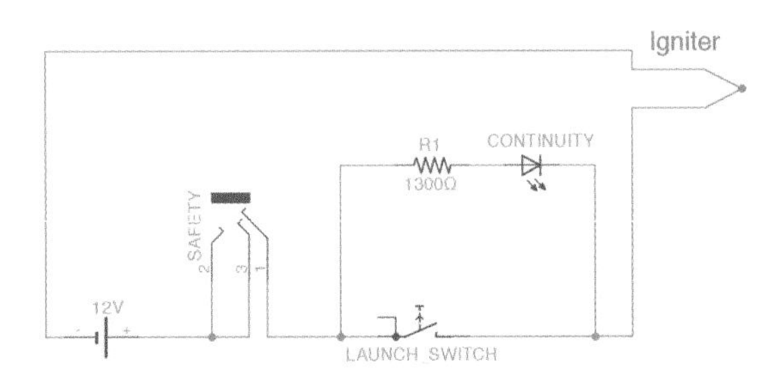

Figure 4-48. *Mono launch box wiring diagram.*

Start by soldering the red connector from one of the battery holders to the black connector on the other. This wires the two battery packs in series, delivering a total of 12 volts. Tuck the wires out of the way around the edge of the batteries. Covering the connection with a small piece of electrical tape doesn't hurt, but isn't really required.

Figure 4-49. *Solder the red wire from one battery holder to the black wire of the other.*

Cut two pieces of wire from the 18-gauge wire that will be used as the launch wire. One piece will connect the negative post for the external battery to the negative binding post used to connect the launch wire. This piece is 6 inches long. The other piece will connect the positive post for the external battery to the arming key. This piece should be at least 8 inches long.

The back of the cigarette lighter socket has two metal tabs sticking out. The one that is closest to the side of the socket is labeled with a small minus (–) sign molded into the plastic. Solder one end of the 6-inch wire to this terminal.

Twist the other end of the 6-inch wire together with the black lead from the battery pack. Solder both of these to the black terminal on the opposite side of the box.

Solder one end of the 8-inch-long wire to the positive terminal on the cigarette lighter socket. This is the terminal that is closest to the center of the socket, and is marked with a plus (+) sign molded in the plastic.

The phono jack we are using for an arming key is actually a stereo phono jack. The two connections that stick up both touch the side of the launch key. We'll wire those together, connecting them to the positive wire from the battery holders and the positive external battery terminal.

Splay the strands from the 8" 18-gauge wire and separate them into two groups, twisting these together to form two separate bundles of copper wire. Feed one through each of the top connectors on the phono jack, and then add the red (positive) wire from the battery pack to one bundle. Solder these in place.

Figure 4-50. *Connect the negative tab on the cigarette lighter socket and the black wire from the battery pack to the black terminal on the opposite side of the box from the cigarette lighter socket. Connect the positive terminal of the cigarette lighter socket and the positive lead from the battery pack to the two connectors sticking up from the phono jack on the lid of the launch controller.*

Cut a piece of wire 3 1/2″ long. If you have some handy, stranded 22-gauge connecting wire is easy to work with, and that's what is shown in Figure 4-51. You can also use another piece of 18-gauge stranded wire, though. Solder one end of this wire, along with the resistor, to the remaining connector on the phono jack. The value of the resistor isn't terribly important, but it must be at least 600Ω to protect the LED. I specified a 1.2KΩ resistor to give the LED plenty of protection from the 12-volt power supply, just in case you substituted a different LED. It should be at least a 1/4-watt resistor, though. A 1/8-watt resistor will technically do, but it would be cutting the tolerances really close, and there is no reason to push limits with the resistor.

I happened to have a 1.5KΩ resistor in my parts box, so I pressed it into service.

Unlike light bulbs, LEDs must be wired into the circuit in a specific direction. Solder the other end of the resistor to the positive side of the LED. There are two ways to tell the positive side from the negative side. On this particular LED, the easiest way is to look at the lengths of the wires coming from the bottom of the LED. The positive wire is the longer of the two wires. The other way to tell is to feel along the edge of the plastic base of the LED. There will be a flattened portion. This flat part is on the negative side of the LED.

Solder the wire to one of the connectors on the pushbutton launch switch, as seen in Figure 4-51. It doesn't matter which one you pick.

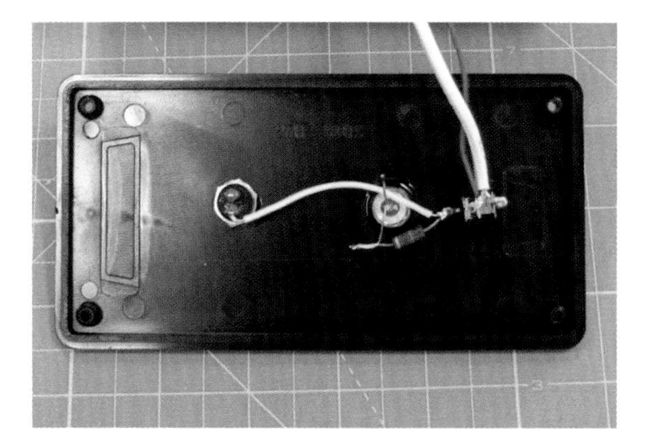

Figure 4-51. *Solder a resistor and a 2 1/2″ piece of wire to the remaining connector on the phono jack. Solder the other end of the resistor to the positive (long) wire of the LED, and the other end of the wire to the launch button.*

Cut two pieces of wire, one 1 3/4″ long, and the other 8″ long. Solder one end of the shorter wire to the remaining wire on the LED. Then solder both wires to the remaining connector on the pushbutton switch, as seen in Figure 4-52.

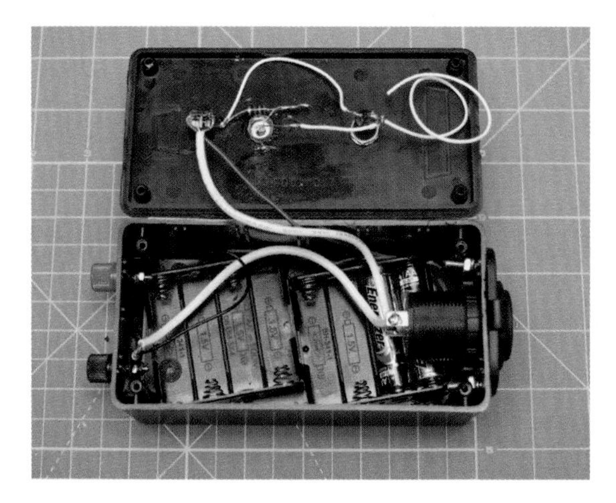

Figure 4-52. *Connect the remaining wire from the LED to the pushbutton switch, and connect an 8″ piece of wire to the same terminal on the pushbutton switch.*

Then solder the dangling end of the 8" wire to the positive (red) binding post, as shown in Figure 4-53.

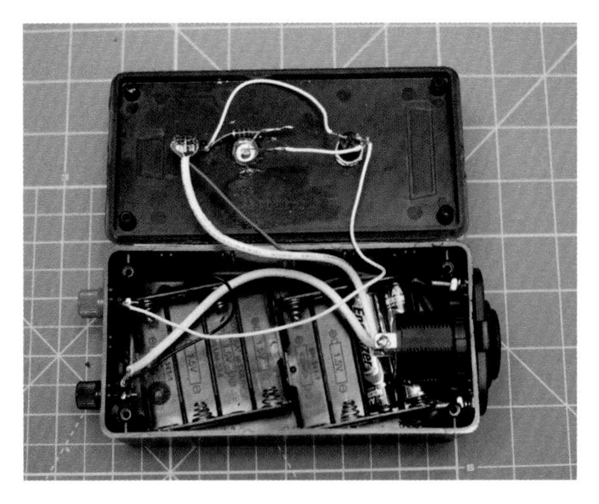

Figure 4-53. *Solder the 8" wire to the positive binding post.*

That's it—the launch controller box is done. Load it with eight AA batteries and screw on the lid with the screws that came with the project box.

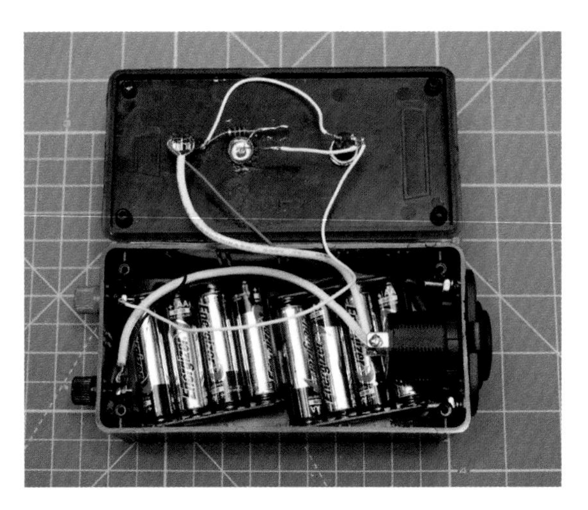

Figure 4-54. *Load the launch controller with eight AA batteries.*

Building the launch cable

The launch cable carries the current from the launch controller to the igniter, 30 feet away at the launch pad. While wire is generally not a significant consideration for power loss in an electrical circuit, this changes when we're building a launch controller. That's because there are really only three sources of resistance when the rocket is firing: the igniter itself, the batteries, and the long wire running from the launch controller to the launch pad. That's why the mono launch controller uses 18-gauge wire rather than the 24-gauge wire seen in starter launch controllers. For the same length of wire, the 18-gauge wire has about 1/4 the resistance of 24-gauge wire.

Alternative Launch Cables

Do we *really* need 35 feet of launch cable? Most people only have a 17-foot cable. And is 18-gauge wire really needed? Twenty-four gauge wire is easier to work with, and cheaper.

All of that is true. It really comes down to what sorts of rockets you intend to fly, and whether you intend to follow Tripoli or NAR launch rules. Our club, for example, flies under Tripoli rules, where the minimum launch distance is 30 feet for all launches. NAR allows a 15-foot distance for launches with a D motor or smaller.

Some of the rockets in this book will use clusters of motors, which are difficult to light with higher-resistance 24-gauge wire. Some of them also use more than 20 newton-seconds of total thrust, putting them out of the D class. Some also use reloadable composite motors, which require a 12-volt launch system that delivers a lot of current—again, 18-gauge wire does that better than 24-gauge wire. For all of these reasons, the mono launch controller is designed to use 18-gauge wire with a launch distance of 30 feet.

Still, most of the rockets in this book, and most of the rockets flown at launches I've attended, will work just fine with a 17-foot 24-gauge launch wire. The nice thing about the design of the mono launcher is that you can easily switch. Why not make both? The micro clips come in a package of 10, so you have plenty of extras. Buy an extra set of banana plugs and 17 feet of 24-gauge stranded wire and make a light-duty launch wire, too.

Figure 4-55. *The completed launch cable gives 30-foot standoff for large models and the current capacity needed for cluster launches.*

At first, it may seem a little difficult to find 18-gauge wire. Most electronics stores will have wire that is part copper and part aluminum, assuming they have 18-gauge wire at all. It's best to stay away from wire that uses aluminum, since aluminum has almost twice the resistance of copper. The surprise source is the hardware store. Most small electrical appliances use 18-gauge stranded copper wire for power cords. Hardware stores sell the stuff in spools of different sizes, often under the name "lamp cord." The wire is a pretty significant part of the cost of the launcher, so do a bit of shopping around before making a selection.

Assembly is pretty easy. Start by cutting a 35-foot length from the 18-gauge wire.

Trim off enough insulation to fill the tip of the banana plug with bare wire, plus about 1/8". Be sure to slip the plastic cover over the wire as shown in Figure 4-56. Take a close look at the metal part. It's actually two pieces. Unscrew the sleeve to reveal a small hole in the side of the main shaft of the banana plug. Push the wire in from the bottom of the banana plug and through this hole. Replace the sleeve to crimp the wire in place. Now solder the wire where it is inserted in the bottom of the banana plug. I know, the package says it's a solderless banana plug, but I also know I'm going to tug on mine a lot. I want that extra strength.

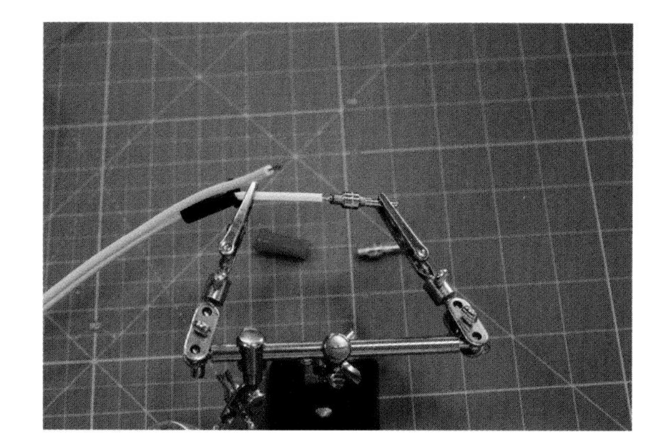

Figure 4-56. *Solder the banana plugs to one end of the wire.*

Screw the plastic sleeve onto the tip and repeat the process for the other banana plug.

Then solder two of the micro clips to the other end of the launch wire, as shown in Figure 4-57.

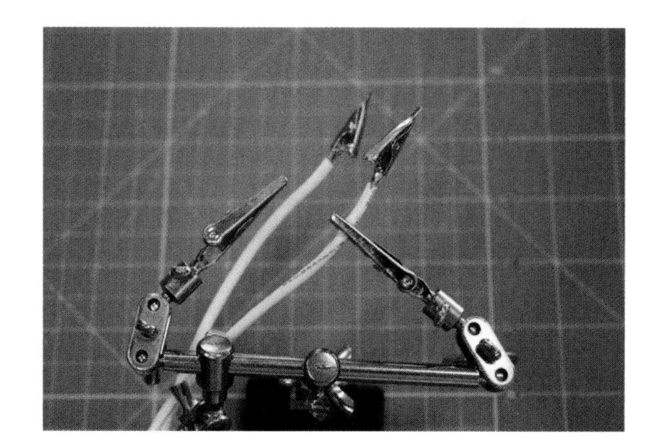

Figure 4-57. *Solder the launch clips in place.*

Building the arming key

You started the arming key in "Safety Cap" on page 73. That's when you built the safety cap that will sit on the launch rod until just before launch. Attaching the arming key to the safety cap encourages people to follow safe launch procedures. Leaving the safety cap/arming key on the launch rod between launches will help to prevent accidental launches. You must remove the safety cap from the launch rod *and* insert the launch key just before firing, so why not combine the two actions?

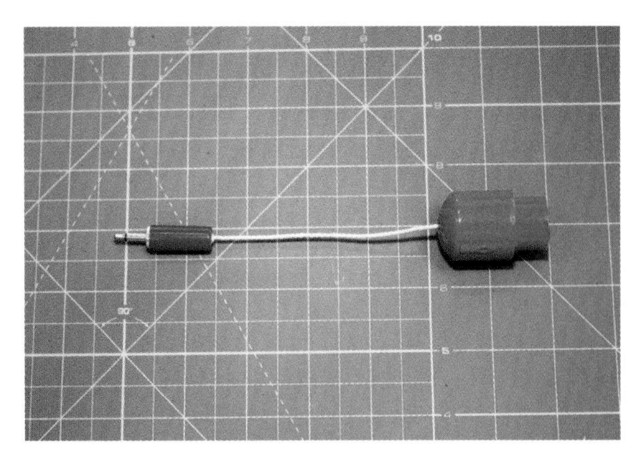

Figure 4-58. *The completed arming key with attached safety cap.*

The arming key is made from a phono plug. We want to complete the circuit in the launch control box by inserting the plug. To do that, the connectors in the plug need to be soldered together. Use a scrap piece of wire to connect the two connections on the plug. Solder the wire in place.

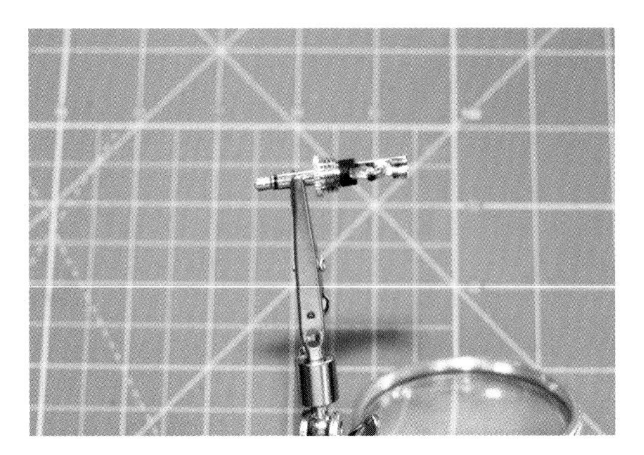

Figure 4-59. *Solder a piece of scrap wire to connect the two sides of the phono plug.*

Add a heavy string or some other cord to connect the pieces. Tie the string around the wire you just soldered in the phono plug, thread it through the plastic sleeve that covers the phono plug, then poke it through the hole you drilled in the safety cap when you built the launcher. Tie a knot in the string to keep it from pulling back through the hole in the safety cap.

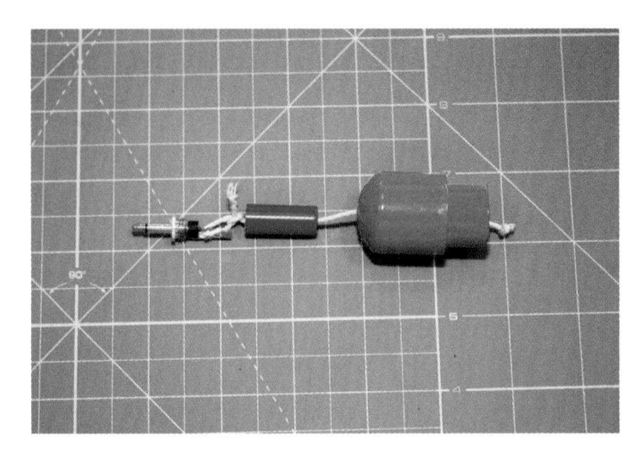

Figure 4-60. *Use a piece of string to connect the safety cap to the arming key.*

Screw the plastic sleeve onto the phono plug, pull the string until the knot rests against the bottom of the safety cap, and you have a completed launch rod safety cap and arming key.

Building the 12-volt wiring harness

Figure 4-61. *The completed 12-volt wiring harness.*

You may not need it right away, but eventually you might want to switch to an external 12-volt battery. A great way to connect the launcher to the external battery is with a short run of 18-gauge wire with hefty battery clips.

Caution: Lead-Acid Batteries Pack an Unexpected Wallop!

This section discusses connecting the launch controller to a 12-volt DC power source from a lead-acid battery such as a car battery. You might think all batteries are pretty harmless, but a 12-volt lead-acid battery can do some damage if it is mishandled.

As you saw in the theory section, a lead-acid battery has a very low internal resistance. This means it can dump a lot more current through a wire than eight AA batteries, which is why we want to use it. In fact, a 12-volt lead-acid battery can dump enough current through a wire to melt it in your hand. It can also melt jewelry, even melting a wedding ring right on your finger. People tend to pull away long before the metal melts, but severe burns have been reported.

This isn't meant to scare you away from a 12-volt power source. It's less dangerous than the 120 volts from an electrical outlet. It does, however, deserve more respect than you might ordinarily give small alkaline batteries, which generally do nothing more than overheat when shorted.

You might want to vary the design of this cable a bit. Mine is designed for use with the Black & Decker Portable Power Station. I'm a real fan of this device. It gives me a 12-volt DC power source for launching rockets in a nice, portable package. It has an air pump for water rockets and air rockets. It even has a 110-volt AC power converter for my telescope—but I digress.

Since I can put this power source right under a launch table or very close to where I stand, I decided to build my 12-volt wiring harness with a fairly short 6-foot run of 18-gauge wire. The battery clips from the parts list are perfect for connecting to the Black & Decker Portable Power Station, but may be a bit small for connecting to a standard car battery. If you intend to pop the hood up on your car and connect to the car battery for power—a great option if you can get your car close enough to the launch site—you may want to use a larger clip on one end. You might also want to use a bit more than 6 feet of wire. In short, think about how you will use this wiring harness before you build it.

Figure 4-62. *The Black & Decker Portable Power Station must have been designed for rocketry. It has 12-volt DC for the mono and quad launchers, a compressed air feed for water and air rockets, and 120-volt outlets for plugging in computers.*

That said, to build the harness shown, start by cutting a 6-foot length of the same two-conductor 18-gauge stranded copper wire used for the launch cable. Solder a set of battery clips to one end.

The other end uses a cigarette lighter plug to plug into either the mono launch controller described earlier, or the quad launch controller described later in the chapter. This connector isn't hard to install, but it's a bit different from most, so let's go through how to do it step by step.

We need to get at the insides of the connector. Three things hold it together: the metal end cap, which screws on; a small screw in the side of the case; and a rubber ring. Start by removing the rubber ring. Gently pry it off with a small screwdriver, as seen in Figure 4-63.

Figure 4-63. *Gently pry off the rubber ring, one of the three things that hold the connector together.*

Unscrew the metal end and remove the screw in the side of the plug. Disassemble the parts.

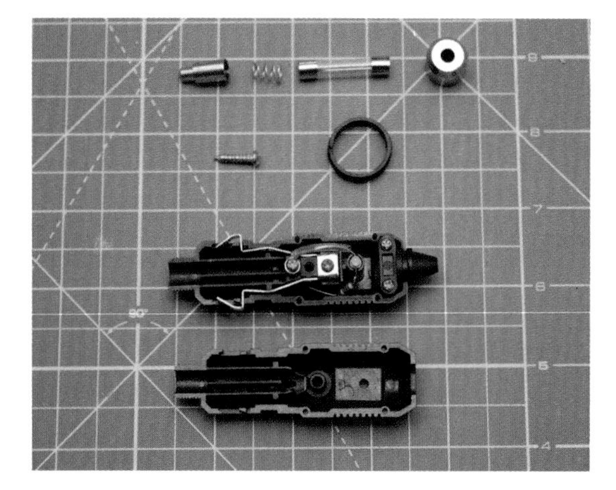

Figure 4-64. *Finish disassembling the plug.*

Loosen the two screws that hold the rubber sleeve in place and thread the free end of the 18-gauge wire through the sleeve. Cut a quarter inch from the wire connected to the black battery clip. This is the negative side of the wiring harness. Strip about 3/8″ of insulation from both wires and tin the tips of the wires to keep them from fraying, perhaps causing a short inside the connector. Screw them in place as shown in Figure 4-65.

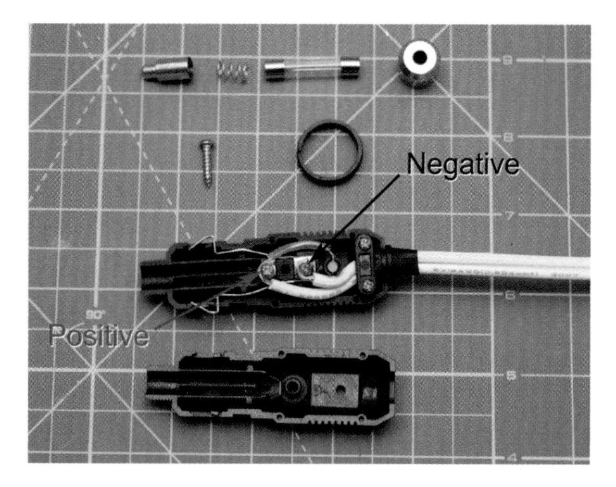

Figure 4-65. *Fasten the wire connected to the black battery clip to the screw closest to the entry point for the wire, and the wire connected to the red battery clip to the farther screw, then reassemble the plug.*

It is very important that the wire connected to the black battery clip is the short one, and that it's connected to the metal part that springs out on the side of the plug. The wire connected to the red battery clip connects to the small post that is eventually connected to the metal end cap, which is the positive connection on the plug.

Caution! Watch the Polarity!

Very Bad Things can happen with a 12-volt power supply if you reverse the polarity, connecting negative to positive and vice versa. If you remove the AA batteries from the mono launch controller and get the polarity backward, the LED arming light won't light, but the launch controller will perform acceptably otherwise. If you leave the batteries in, though, the reverse current could overpower the batteries, causing them to overheat. That could cause them to melt the plastic launch box or split, spilling caustic chemicals.

Once the wires are in place, reassemble the plug. This finishes your wiring harness.

Isolating the Power Sources

The text makes a big deal about how connecting the internal and external batteries backward can cause problems. Earlier, we also learned that connecting an external 12-volt battery to an internal 6-volt battery pack could cause similar problems. Why not do things differently?

Well, you can. I like this design because you have the option of using both the internal and external power supplies in unison, dumping even more current into the igniters. Still, if you prefer to avoid the possibility of making a mistake, there are three solutions.

The first was already mentioned: remove one or more of the AA batteries if you intend to use the launch controller with an external battery. It only takes a moment.

The second way is to build the launch controller either for internal batteries or external batteries, but not both.

The third is to isolate the two batteries so the external battery can't overpower the internal battery when they are connected backward. The traditional way to do this in a circuit like the one in the mono launch controller is to use a diode. This isn't a very good idea in a launch controller, though, because the diode will add 3–4 ohms of resistance to the circuit. That's enough to prevent the external battery from being useful for cluster launches. A better solution is to use a single-pole, double-throw switch to isolate the two batteries. Flip the switch one way and the internal batteries supply the power; flip it the other way and the external battery is in use.

I prefer using both batteries at once, though, so I'm willing to exercise a bit of care to get the benefit of their combined current.

The Quad Launch Pad

The quad launch pad is designed for clubs and schools. It handles up to four rockets at a time. That allows four club members or students to prepare their rockets at the same time, cutting preparation time so more rockets can fly in a limited time span.

It's also a lot of fun for an event called a "drag race," where multiple rockets are launched at once. "The Quad Launch Controller" on page 98 shows how to build a launch controller that can launch one, two, three, or four rockets at the same time.

Table 4-10 shows the parts list for the launcher shown in Figure 4-66. There are several places where you can make substitutions, so read through the construction description before buying parts. Table 4-11 shows some tools and supplies you can use. Again, lots of alternatives are possible.

Figure 4-66. *The quad launch pad.*

Table 4-10. Parts list

Part	Description
2x4, 56 inches long	The main body of the launcher is made from construction-grade 2x4 lumber. It's cheap, readily available, and works just fine. The lumber yard will probably even cut the wood for you. Just be careful to pick a nice, straight piece—construction-grade lumber is not the same as cabinetry wood!
2x4, 36 inches long (4)	These are the legs. I'm tall, and built this launcher for adults. The tips of the launch rods are six feet above the ground. You may want to adjust the leg length, but keep the tips of the launch rods at or above eye level. Shorter people can reach up to slide the rocket over the top of the rail, and everyone will appreciate not having to squat to set up a rocket.
Workforce sawhorse brackets	These are the metal parts that hold the legs together. This particular brand is available at Lowe's hardware stores, including the online store at *http://www.lowes.com*. There are other brands that will do the same job.
#10 1 1/2" round-headed wood screws (20)	These are used to fasten the legs into the sawhorse brackets. The number needed may vary with other brands of bracket. I only used 16, opting not to screw in the crossbar so the pad was more portable.

Part	Description
1/2″ keyless drill chuck replacement (4)	The keyless chuck is used to secure the launch rod. The launcher is designed for a chuck that screws onto a 1/2-20 threaded bolt. If the one you buy uses a different mount, you will need a different mounting bolt, too.
1/2-20 4″ bolts (4)	These hold the drill chucks to the launcher. If you get a different brand of drill chuck, make sure the bolts match. These may be hard to find locally, but I found some at Ace Hardware. They are also available online at *http://www.boltdepot.com*, where they're listed as Hex bolts, Zinc plated grade 8 steel, 1/2"-20, 4″ bolts. Two- to three-inch-long bolts will also work, but require a bit more drilling.
#3 screw eyes (4)	These are used to run the launch wire on the back side of the pad. A #3 screw eye is about an inch in diameter. The size isn't that critical, as long as the micro clips for the launch controller will fit through the hole.
1/8″ music wire (4)	This is the launch rod used for most of the rockets in this book. It should be 3 feet long.
3/16″ music wire	This is the launch rod used for the heavier and more powerful rockets near the end of the book. Our club sets up these pads using three 1/8″ rods and one 3/16″ rod, but we have plenty of alternate rods, including some 1/4″ rods. Schools may not need a heavier rod at all. Consider how you will use the rods and buy what you need.
Blast deflector (4)	You have some options here. I like the Estes blast deflector plate, which has a nice central support to keep the deflector plate level. You can also use a piece of steel from the hardware store, as described in "Blast Deflector Plates" on page 75. You can even use a tin can—but not an aluminum can!
White shipping labels (Avery 8165 or similar)	Use these to create the pad numbers.

Table 4-11. Tools and materials

Tool	Description
White primer	The wood is very porous, so it needs a primer before applying paint. I used white wood primer, which doubled as the base color.
Latex paint	Use the same semi-gloss paint you would use for a kitchen or bathroom. It's readily available and works great. Many hardware stores now have small sampler-size containers of paint for about $3. Go crazy. I obviously did. You might pick school colors, or you might stick with the color-coded scheme seen in the photos.
Painter's tape	You will need painter's tape if you go for a fancy paint job like the one shown.
Drill	You will be drilling four 1/2″ holes through the 2x4. A drill press is best, but a handheld drill will work if you are very careful. You will also be drilling several 3/32″ pilot holes for the wood screws and widening the holes in the sawhorse brackets with a 3/16″ bit. A handheld drill works best for these holes, but a drill press can be used, too.
Inkjet printer, computer	Used to create the pad numbers.
Paintbrush	For the paint, of course.
Clear paint	Either a clear varnish or clear spray paint to protect the labels on the launcher.

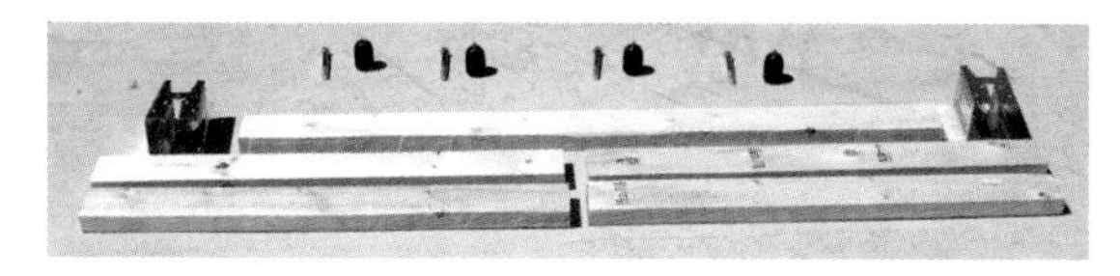

Figure 4-67. *Visual parts list for the launcher. The paint, labels, screw eyes, and small wood screws are not shown in this photo.*

If the lumberyard did not cut the wood to length, start by cutting the 2x4 boards into four legs and one crossbar. The crossbar should be 56 inches long. You could do the "stretch limo" version to handle more rockets; the design can be extended easily. If you do, add 12 inches for each additional rocket. Realize, though, that the launch controller cannot handle an unlimited number of simultaneous launches, nor do NAR and Tripoli safety guidelines allow them without additional clearances. You can set up a launcher to launch 10 or so rockets one after another with simple extensions of the design, though—that and a really long 2x4.

The legs shown are 36 inches long. I stand 6′ 4″, so perhaps I went a bit overboard, but the idea is to get the top of the launch rod above eye level. As long as the shorter people who will use the launcher can reach 6 feet in the air to slide the rocket over the tip of the launch rod, this length is fine. Still, I'd recommend shortening the legs to 30 inches if the launcher will be used at an elementary school. Whatever length you choose, cut four legs.

A Lower-Cost Alternative Design

This launcher uses drill chucks to hold the launch rods, and that works really well. The drill chucks are the most expensive and difficult part to find, though. An alternative is to drill a 1/8" hole 2" deep into the top of the wood. Stick the launch rod directly into this hole. Add a 3/16" hole about 1" away from the 1/8" hole for the thicker launch rods. If you decide you want to switch to drill chucks later so you can handle any size rod, just drill through the original holes as described in the instructions and install the drill chucks.

Mark the top of the crossbeam 10 inches from one end on top of the 2x4. This is where you will drill the first hole for the 1/2-20 bolts used to secure the drill chucks. Mark the locations for the other three launch rods at 12-inch intervals. This should leave you with 10 inches between the last launch rod and the other end of the crossbeam. You will notice in Figure 4-68 that I didn't trust the 2x4 to be cut square, and you shouldn't, either. Use whatever is handy that you know forms a 90-degree angle. Prop it up against the side of the 2x4 as you drill the 1/2" holes all the way through the wood at the locations marked for the launch rods.

Finally, drill a 1/8" hole near the end on the top of the beam. This is used as a storage location for the launch key while the pads are being set up.

Figure 4-68. *Drill four 1/2" holes all the way through the wood at the locations of each launch rod.*

Getting By with Shorter Bolts

Four-inch-long 1/2-20 bolts are a bit hard to come by. You can get by with shorter bolts. Make sure they are at least 2" long for good mechanical stability, though.

Start by drilling four 1/2" holes, just like the instructions show. Measure the length of the bolts and the size of the bolt head, and drill a larger hole partway through that is wide enough for the bolt head and just deep enough that the bolt, when dropped in, sticks out of the other end of the wood by 1/2".

For example, let's say you are using a 2"-long bolt with a 1" hex head. The 2x4 isn't actually 2x4 at all—that's the distance between the saw blades, not the size of the wood. It will generally be about 3 1/2" tall. In this case, you would drill a 1" hole two inches into the wood in exactly the same place as the 1/2" hole that goes all the way through. When the bolt is dropped in, it will stick out 1/2".

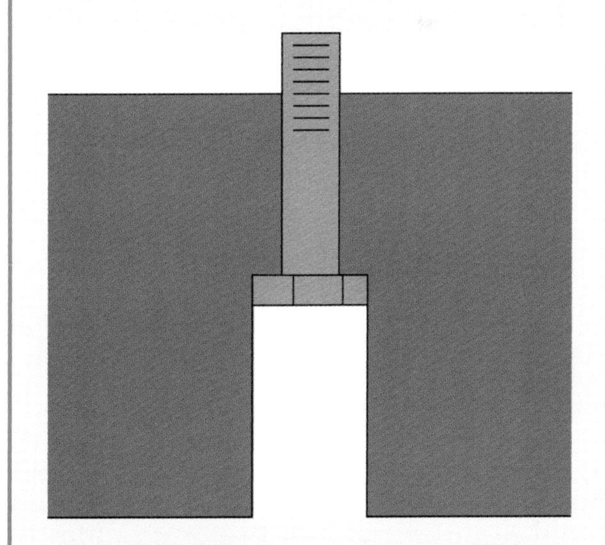

Figure 4-69. *Cutaway drawing showing how to use shorter bolts to mount the drill chuck.*

Paint all of the wood parts with primer. It took two coats before I was satisfied. Add any detail paint you like. The launcher in the picture is color-coded, with 11-inch-wide colored sections that go completely around the crossbar. Add printed labels if you like. You can get the pattern for the labels shown at the author's website (*http://bit.ly/byteworks-make-rockets*). I added a coat of clear spray paint over the labels to protect them from wear and tear.

Next we will attach the sawhorse brackets to the legs. I selected nice, heavy wood screws, and they are a bit too thick to fit through the holes in the sawhorse brackets. Drill the holes out with a 3/16" bit so the wood screws will fit through. Slide the legs into the sawhorse brackets (actually, pound them in; it's a tight fit), then install the wood screws to hold the legs in place.

The lumber will split if you drive the screws in with no preparation. Drill 3/32" holes into the wood to match the location of the holes in the sawhorse brackets. Add the wood screws to keep the legs in place.

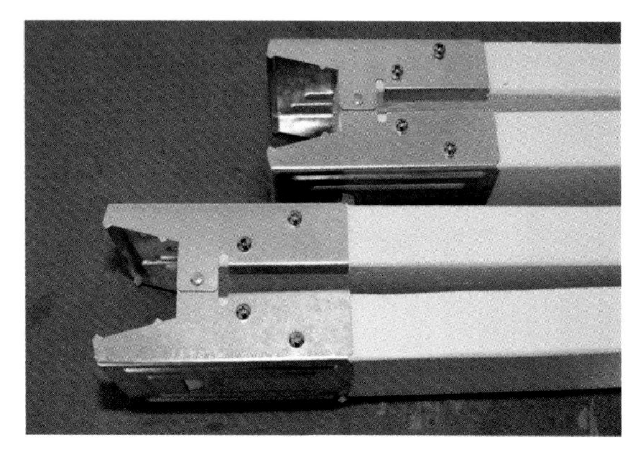

Figure 4-70. *Install the legs in the sawhorse brackets with the wood screws.*

You can add four more screws to fasten the legs to the crossbeam, but I didn't. The legs are pretty secure just from being pinched by the sawhorse brackets, and I expect to move the launcher around a lot. If you have the luxury of storing the launcher near where it will be used, and have the storage space, you might want to add those four additional screws. The launcher will probably last longer that way.

The last step is to shove the bolts through the holes and thread the drill chucks onto the bolts. Add the screw eyes to the back of the launcher; these are used to hold the launch wire so the weight of the wire doesn't pull the igniter out of the rocket motor.

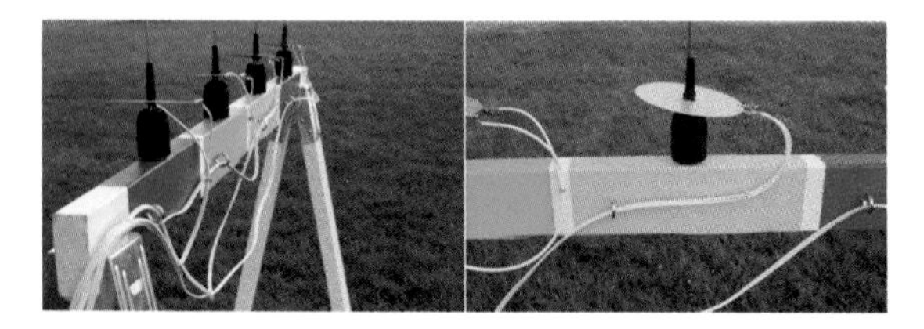

Figure 4-71. *Install screw eyes on the back side of the launcher to hold the cables near the launch pads.*

Launch Rods and Blast Deflectors

The launch rods and blast deflectors for the quad launch pad are the same as for the mono launch pad. See "Launch Rods" on page 77 for a discussion of launch rods, and "Blast Deflector Plates" on page 75 for three alternatives for blast deflector plates.

The Quad Launch Controller

The quad launch controller is a 12-volt launch controller designed for use with the quad launch pad. It allows you to stand 30 feet from the launcher as you launch one, two, three, or four rockets in succession or simultaneously. This launch controller can handle any commonly used igniter, from the sensitive Quest Q2G2 to the 12-volt igniters used in composite motors. All igniters can be used in cluster rockets using an external 12-volt lead-acid battery like a car battery, and clusters can be launched using the Quest Q2G2 igniter with AA batteries.

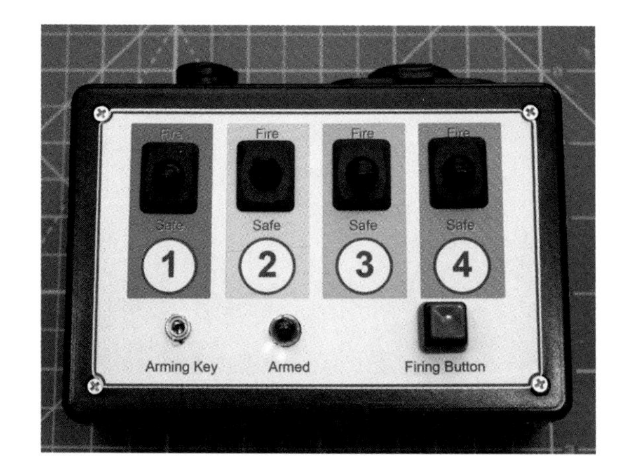

Figure 4-72. *The quad launch controller.*

Table 4-12 shows the maximum number of simultaneous ignitions with various igniters using the quad launch controller. These can take the form of simultaneous launches of rockets with a single motor or launches of clustered rockets with multiple motors.

Table 4-12. Maximum number of simultaneous ignitions using the quad launch controller

Igniter	AA batteries	Car battery
Estes	1	6
Quest Q2G2	4	24
First Fire Jr.	1	4
Copperhead	1[a]	4

[a] The current supplied is slightly under the current recommended in "Electrical characteristics of some common igniters" on page 80, but testing shows the launch controller fires the igniters reliably.

The parts you'll need to build the quad launch controller are listed in Table 4-13. Additional tools and materials you may need for this project are listed in Table 4-14. Read the construction details before buying parts. There are several options you can consider, and you will want to make those decisions before purchasing.

Table 4-13. Parts list

Part	Description	Radio Shack part number
6x4x2 project enclosure	This is the black plastic enclosure that houses the launch box.	270-1806
4 AA battery holder (2)	There are several styles of battery holders available. Because of the way we're going to use them, be sure to get to flat battery holders with an open top. If this specific part is not available, combine similar-style battery holders that hold two or three batteries.	270-0391
Ultra-high brightness 10 mm red LED	Any LED will work as an arming light, but this one is big, bright, and impressive!	276-0015
3-conductor stereo 1/8" phono jack	This is the hole for the arming key.	274-0249
Solder-type mono 1/8" phono plug	This is the arming key. It's also available in black, but who wants a black arming key?	274-0287
SPST soft touch momentary pushbutton switch	This is the firing button. It's a big, red switch that's easy to push.	275-1566
SPDT panel mount paddle switch (4)	These switches are used to select the launch positions, allowing you to fire any combination of rockets from the pad.	275-0648
Micro 1 1/8" smooth clips	These are used to connect the launch wire to the igniters. You need 8, but the package comes with 10, so just get one package.	270-0373
4-40 1/2"-long machine screws and nuts (4)	These are used to mount the 9-pin connector. Lock washers are nice, but they will have to come from the hardware store.	64-3011 and 64-3018 or similar
Battery clip	The battery clips connect the launcher to an external 12-volt battery. This part is perfect for connecting to small portable power supplies like the one shown in the text, but may be too small to clamp onto the terminal on a full-size car battery. Get a clip big enough for the expected use.	270-0344
9-pin connector	Any connector designed for 8 or more pins will do as long as it can handle 10 amps. These are hard to find, so either visit an electronics supply house or plan on ordering from the Internet. You can find the connectors shown at *http://www.newark.com*, part numbers 44F8388, 44F8389, 73K4032, 52K3325, and 52K3323. Some of these parts are the male and female pins. Order a few extra in case you mess up the wiring and need to redo a few connections; the pins sometimes get mangled when making a change.	
1/4-watt resistor	The design isn't especially sensitive to the specific resistance used. Get the smallest resistor you can find that is larger than 1.2KΩ. Just make sure you get one that will handle at least 1/4 watt.	
Marine-grade cigarette lighter socket 12 VDC	This socket connects the launch controller to an external 12-volt power supply. You can use several alternatives, but this is a nice, safe way to prevent the internal battery from shorting. It's available on Amazon, sold by Parts Express.	

Part	Description	Radio Shack part number
8-32 3/4" flat-top machine screws, lock washers, nuts	You will need two 8-32 machine screws to fasten the cigarette lighter socket to the launch controller box. Get the kind with the flat top and conical head to match the opening in the socket (available at Ace Hardware or a similar store). Get matching lock washers and nuts.	
Marine-grade locking cigarette lighter plug 12 VDC	This plug is used in the cable that connects the external 12-volt battery to the launch controller. It's available on Amazon, sold by Parts Express.	
10-amp fuse	The cigarette lighter plug comes with a 5-amp fuse. Switch to a 10-amp fuse.	270-1015
147 feet of two-conductor 18-gauge wire	There are a lot of choices here, so shop around. Be careful to get copper wire. If one of the wires looks silvery, it's probably aluminum, and that won't do. I used lamp wire from the local hardware store.	
22-gauge hookup wire	You will need a few inches of insulated hookup wire to wire the box. The instructions show black and red wire, but the color isn't important, since all of the wire will eventually be hidden inside the box.	
Printable white label sheet (Avery 8165 or similar)	Use these to create labels for your box.	

Table 4-14. Tools and materials

Tool	Description
Soldering iron	Anything designed for electrical work will do, such as Radio Shack part 64-2808.
Solder	Solder designed for soldering electronics, such as Radio Shack part 64-009.
Needle-nose pliers	Used to crimp the pins and manipulate small wires.
Wire strippers	Used for stripping wires, of course.
Hobby knife	Used to trim labels to decorate the launch controller.
Drill	A bench press is great, but a handheld drill will work fine. You will need the following drill bits: 1/8, 1/4, 5/16, 3/8, 9/16, and 7/8". The sizes may change if you substitute any parts. The larger bits will probably be spade bits; see the construction description for information on these bits.
Glue	I recommend epoxy glue for this project, but thick CA glue or any other glue that will bond plastic will also work.

Building the Launch Box

Start by drilling the holes in the launch box. The easiest way to do this is to print the decorative labels you can find at the author's website (*http://bit.ly/byteworks-make-rockets*), attach them to the box, and drill right through the marked spot on each label. Sure, you might mess up one of the labels, but after the holes are drilled, you can just peel it off and replace it with a new one.

Print the labels on white label paper, using a standard inkjet printer. Office supply stores have lots of labels. You're looking for the full-sheet white labels. You can trim the labels with scissors or a hobby knife; I like using a hobby knife myself.

Figures 4-73 through 4-76 show the locations and sizes for each of the holes. Remember, if you substituted parts, the sizes of the holes may change too. The holes for the 9-pin connector and the paddle switches are large, and are not drilled in one step.

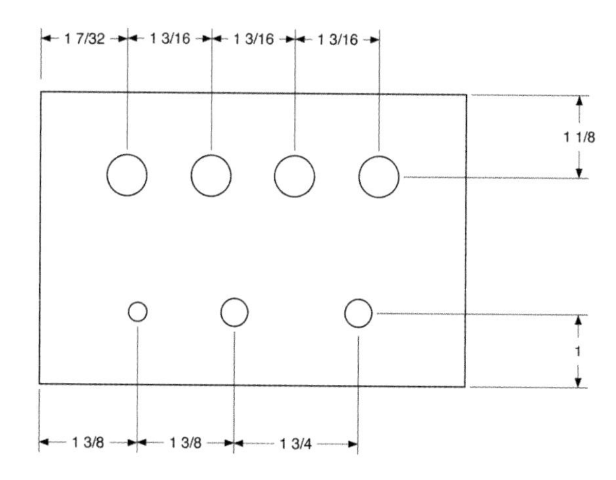

Figure 4-73. *Positions of the holes in the top.*

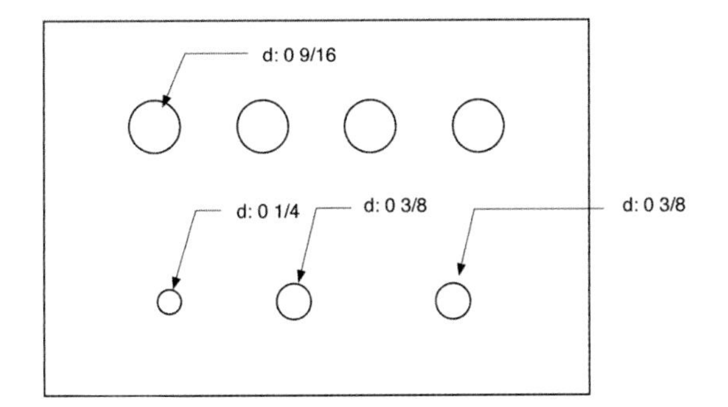

Figure 4-74. *Sizes of the holes in the top. The holes in the top row are the same size.*

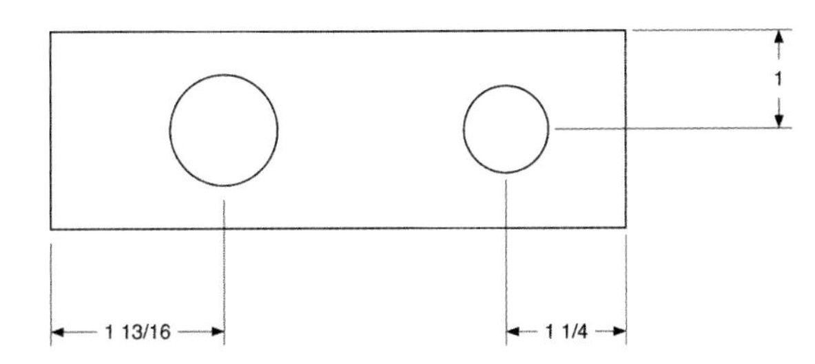

Figure 4-75. *Positions of the holes in the side.*

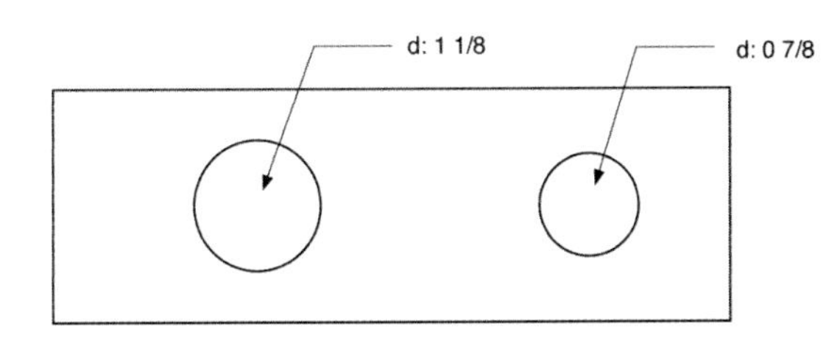

Figure 4-76. *Sizes of the holes in the side.*

Start by drilling the holes for the arming key, arming light, and firing button. While standard bits work well, keep in mind that plastic is soft and tends to bind as you drill. Work very slowly, especially as the drill bit breaks through the plastic. This will reduce the chance that the bit will bind in the plastic, grabbing the project away from you and spinning it around willy-nilly.

Large holes like the 9/16″ holes for the paddle switches, the 7/8″ hole for the 9-pin connector, and the 1 1/8″ hole for the cigarette lighter socket are drilled with paddle bits (also known as spade bits). I like Irwin Speedbor Blue-Groove spade bits for drilling plastic. They have a funny burr on the outside end of the cutting edge. This burr cuts a nice, even circle in the plastic as the bit nears the project box. The bit will punch out a plastic disk without you ever needing to drill through all of the plastic.

Switch to a 1/8″ bit and drill a pilot hole for the four paddle switches, the 9-pin connector, and the cigarette lighter socket. This gives a precise guide hole for the paddle bits as they start through the plastic. Switch to the paddle bits after drilling the pilot holes. Again, drill very slowly.

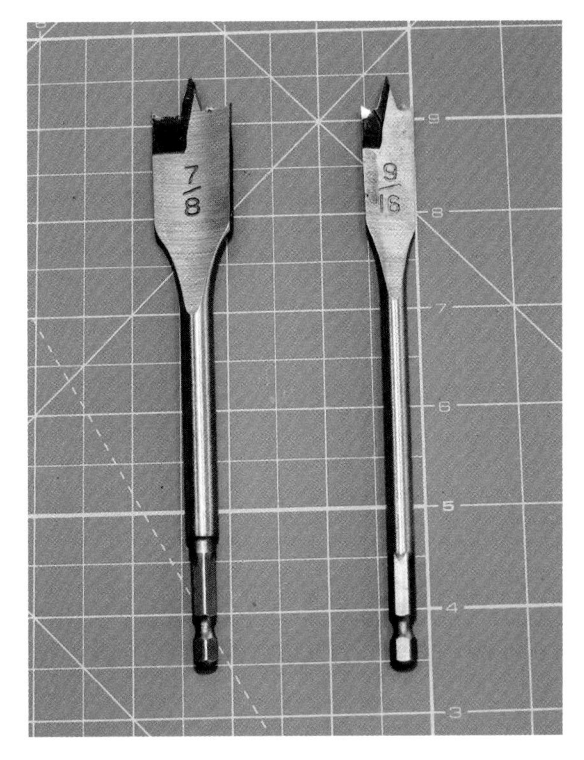

Figure 4-77. *Use paddle bits to drill the larger holes. These are two of the three Irwin Speedbor Blue-Groove bits mentioned in the text.*

Using the 9-pin connector as a guide, drill four 1/8″ holes for the machine screws that will hold the connector in place. Do the same for the cigarette lighter socket, drilling two 5/32″ holes to accommodate the 8-32 machine screws that will hold it in place.

The paddle switches are not actually round, and they won't fit through those 9/16″ holes you just drilled. The switches have a small bulge in each side to keep them from rotating once they are mounted. Use a grinding wheel on a Dremel tool, a file, or a piece of sandpaper to elongate the holes until the paddle switches just fit through. You want to end up with an oblong hole, just like the shape of the switch, so it won't rotate.

Figure 4-78 shows the complete wiring diagram for the quad launch controller. There is no printed circuit board. Instead, the wires are connected directly from part to part. Use the circuit diagram as a blueprint as you wire the launch controller.

The Six-Volt Option

You could build the launcher with a single battery holder and just use four batteries, but that will limit the launcher. See "Will Six Volts Work?" on page 85 for a description of this option.

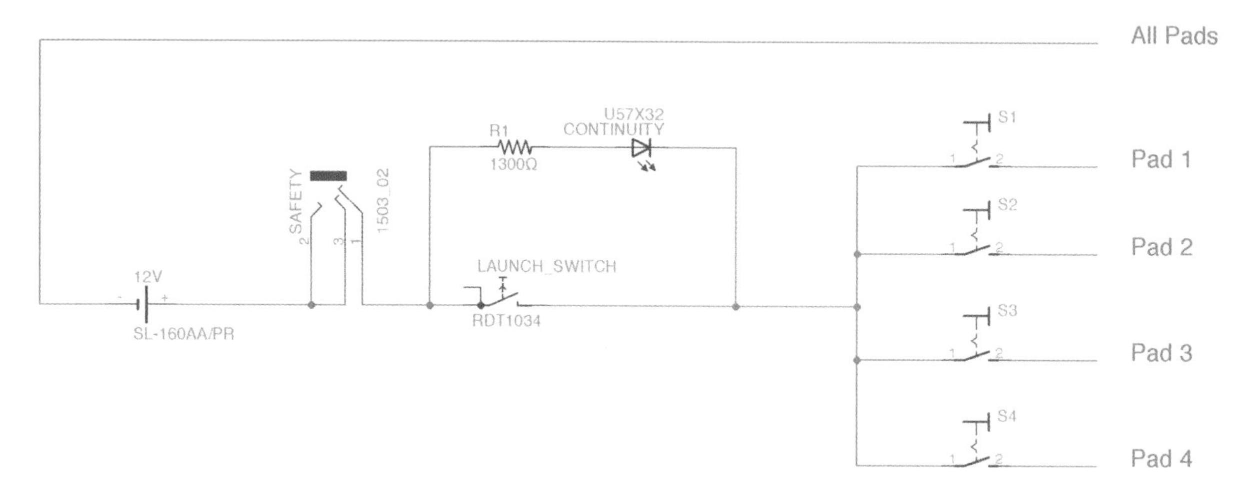

Figure 4-78. *Wiring diagram for the quad launch controller. The connections on the right continue to the launch pad, where they connect to the igniters.*

Use epoxy glue to fasten the two battery holders to the bottom of the project box as shown in Figure 4-79. The battery holders are positioned to give plenty of room for the cigarette lighter socket once it is inserted into the 1 1/8" hole. Test fit the socket to make sure the battery holders are far enough out of its way.

Solder the red wire from one battery holder to the black wire from the other. Use electrical tape to fasten the wire out of the way on the bottom of the box.

Figure 4-79. *Glue the battery holders in the box.*

The quad launch controller is built with a circular 9-pin connector. While I like this connector, there are alternatives that work, too. One example is the 9-pin D-Sub connector from Radio Shack, part numbers 276-1537 and 276-1538. Really, any connector with at least 8 pins that can handle 10 amps will do, although I recommend it be a connector that can't be plugged in upside down.

Take your time assembling the 9-pin connector. There are 16 connections to make and 18 pins to install, and a bit of artful carving of insulation is required to create the complete connector. We'll start with the easy half, which is the connector for the launch box; we'll leave the external connector for the next section, when the launch wire is assembled.

Four of the connectors will be bound together and connected to the negative side of the power supply. Cut four 2" lengths of 22-gauge hookup wire. Strip one end and fasten four of the male pins for the 9-pin connector to the wires. A little patience and a pair of needle-nose pliers works just as well as a crimping tool. While I like the crimp-style pins, I don't count on the crimps to hold the wires. Solder the connections, too.

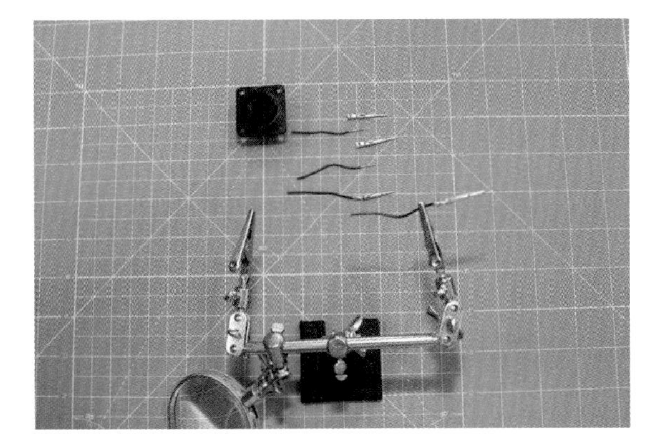

Figure 4-80. *Solder 2" connector wire to four of the male pins.*

Insert the pins into the 9-pin connector in pin positions 2, 4, 6, and 8. These are the top, bottom, left, and right holes. Push the pins in until you hear or feel them click into place.

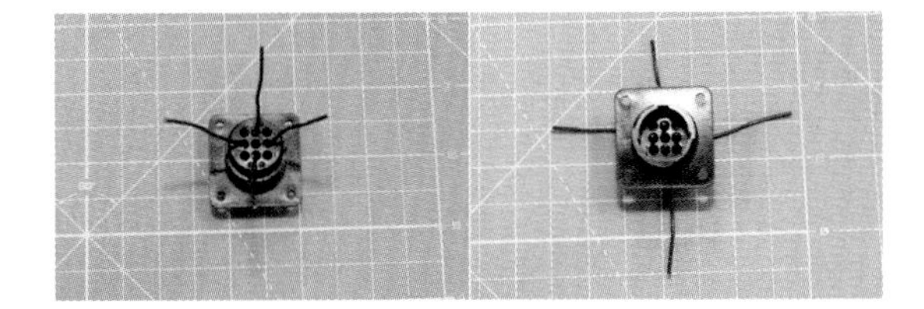

Figure 4-81. *Insert the four pins into the even-numbered pin holes.*

Repeat this process with four 8"-long wires. Install these wires in the corner holes, which are holes 1, 3, 7, and 9.

Push an empty pin into the middle position, which is pin 5.

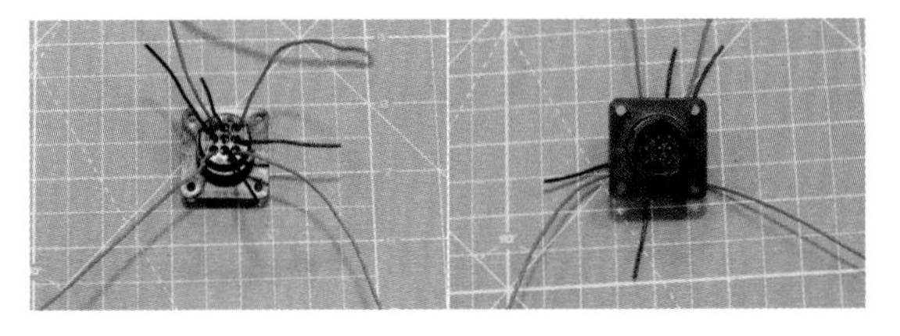

Figure 4-82. *Install four 8" wires in the corner pins, then push an empty pin into the center hole.*

Strip about 1/4" of insulation from all four black wires from the 9-hole connector. Cut a 4" piece of 18-gauge wire and strip about 1/4" of insulation from one end. Pull the unattached black wire from the battery holder through the hole for the 9-pin connector. Bundle the black wire from the battery holder, the 4" piece of 18-gauge wire, and all four of the black wires from the 9-pin connector, twisting them together. Solder this collection of wire, then protect it with some electrical tape.

Figure 4-83. *Form a bundle of the black wire from the battery holder, the four black wires from the 9-pin connector, and a 4" piece of 18-gauge wire. Solder these together.*

Pull all of the wires through the hole for the 9-pin connector and mount it to the project box using four 1/2"-long 4-40 machine screws and matching nuts. I had some lock washers handy, too, but they are not essential.

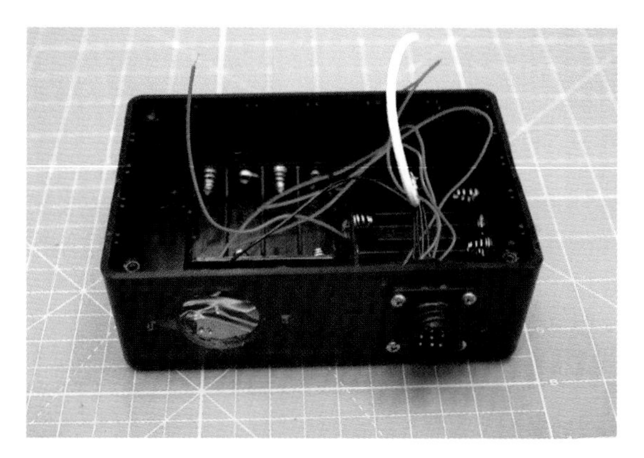

Figure 4-84. *Mount the 9-pin connector with four 4-40 machine screws.*

The cigarette lighter socket that provides access to the external 12-volt lead-acid battery has four parts, not counting the mounting screws. Mount the cover plate —the part with a central hole and two screw holes— with two 3/4" 8-32 flat-top machine screws with matching nuts and washers. The flexible rubber part goes under the cover plate, with the tab sticking up through the matching slot in the back of the cover plate. Shove the socket itself in place and secure it with the plastic nut.

Figure 4-85. *Secure the cigarette lighter socket with two 8-32 machine screws.*

Glue the LED into its hole in the lid. It's easy to bump the LED once the launcher is assembled, so use a good glue and be generous with it. Glue on the bottom of the lid won't show up once the launch controller is assembled, so it doesn't hurt to use a little extra. I used epoxy glue, but any good-grade glue that will bind plastic should work.

Mount the phono jack, the pushbutton switch, and the four paddle switches in the lid using the provided nuts.

Figure 4-86. *Glue the LED in place, then mount the remaining components with the nuts that come with the parts.*

Strip about 1/2" of wire from the loose end of the 18-gauge wire attached to the wiring bundle on the 9-pin connector. Solder it to the negative connector on the cigarette lighter socket. The negative connector is the one closest to the outside of the socket. It's also marked with a small minus (–) sign molded into the plastic.

Cut an 8" piece of 18-gauge stranded wire. Strip 1/2" from one end and solder it, along with the red wire from the battery pack, to the positive terminal on the cigarette lighter socket. Like the negative terminal, the positive terminal is marked with a character molded into the plastic; this time it is a plus (+) sign.

The phono jack that serves as the arming key has three connections. The two that stick straight up are connected internally to the barrel of the phono plug when the plug is inserted in the jack. That's used to support stereo sound, but of course we really don't care. We're going to connect both of these connections to the positive side of the power supply. Strip about 1/2" of insulation from the unattached end of the 8" piece of 18-gauge wire. Separate the strands of wire into two bundles and twist them together. As shown in Figure 4-87, feed these strands through the two connections on the phono jack that stick up in the same direction.

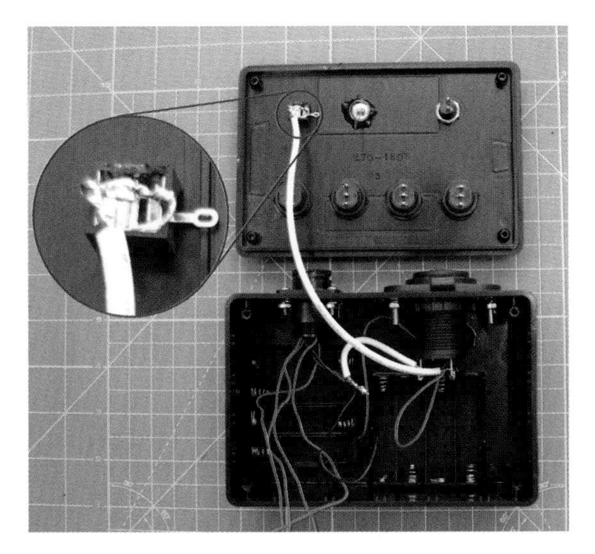

Figure 4-87. *Solder the 18-gauge wire from the 9-pin connector to the negative terminal of the cigarette lighter socket. Use an 8" piece of 18-gauge wire to connect the top two terminals of the phono jack to the positive terminal of the cigarette lighter socket. Solder the red lead from the battery pack to the positive terminal of the cigarette lighter socket, too.*

Cut a 3" piece of 22-gauge connection wire and solder it and the resistor to the remaining terminal of the phono jack. The resistor should be at least a 1/4-watt resistor. The specific resistance is not terribly critical as long as a few ground rules are followed. It must be at least 600Ω for the LED specified in the parts list; I recommend at least 1.2KΩ, mostly to guard against substitutions with a different, more sensitive LED. Pick the lowest-rated resistor you have that is at least 1.2KΩ. I had a 1.5KΩ resistor in my parts box, so that's what you see here.

LEDs have a polarity, which means it matters which end is connected to the positive voltage and which end is connected to the negative voltage. There are two ways to tell the positive side of the LED from the negative side. On this particular LED, the easiest way is to look at the lengths of the wires coming from the bottom of the LED. The positive wire is the longer of the two wires. The other way to tell is to feel along the edge of the plastic base of the LED. There will be a flattened portion. This is on the negative side of the LED. Solder the other end of the resistor to the positive side of the LED.

Solder the other end of the connecting wire to the firing button.

Figure 4-88. *Connect the remaining terminal of the arming switch to the positive side of the LED (the long wire) using a 1.2KΩ or larger 1/4-watt resistor, and to the firing switch using 22-gauge connecting wire.*

Use short runs of connecting wire to daisy-chain the remaining lead on the LED, the other terminal of the firing switch, and the center terminals on each of the four paddle switches.

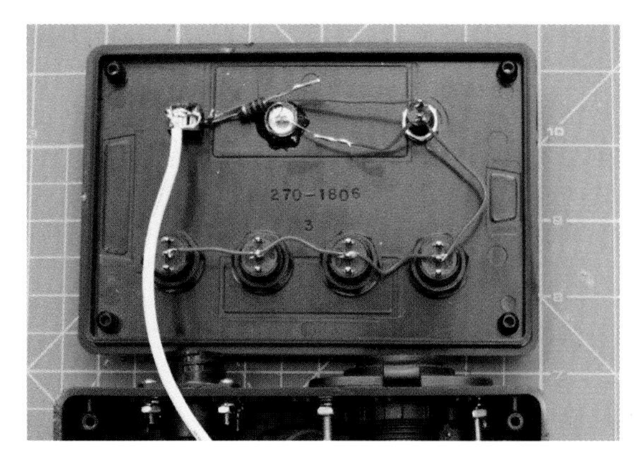

Figure 4-89. *Connect the remaining lead on the LED, the remaining terminal on the firing switch, and the center terminals on the four paddle switches.*

Solder the four wires from the 9-pin connector to the terminals on the paddle switches that are closest to the LED, pushbutton switch, and phono jack. Which wire goes to which switch is important. Figure 4-90 shows the pin numbers, looking at the connector from inside the box.

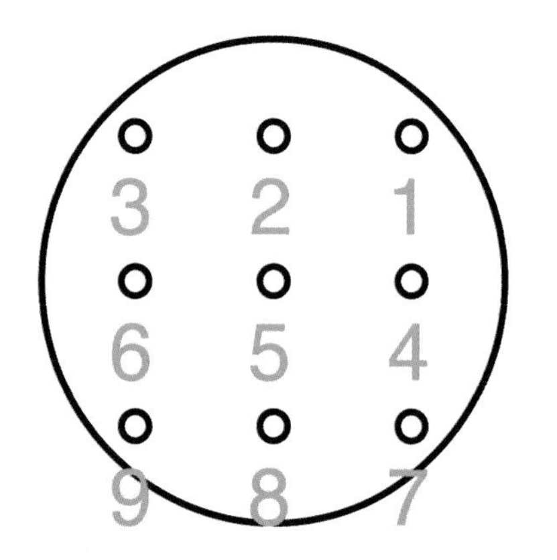

Figure 4-90. *Pin outs as seen from inside the launch controller box.*

You can flip the top of the box over to see which paddle switch corresponds to a particular launch pad. Connect the wires from the 9-pin connector to the paddle switches as outlined in Table 4-15.

Install eight AA batteries and screw the lid on. The launch box is complete.

Table 4-15. Wiring legend for connecting the 9-pin connector to the paddle switches

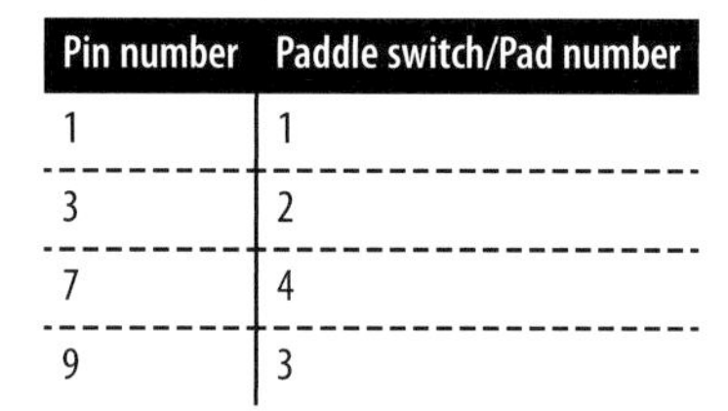

Pin number	Paddle switch/Pad number
1	1
3	2
7	4
9	3

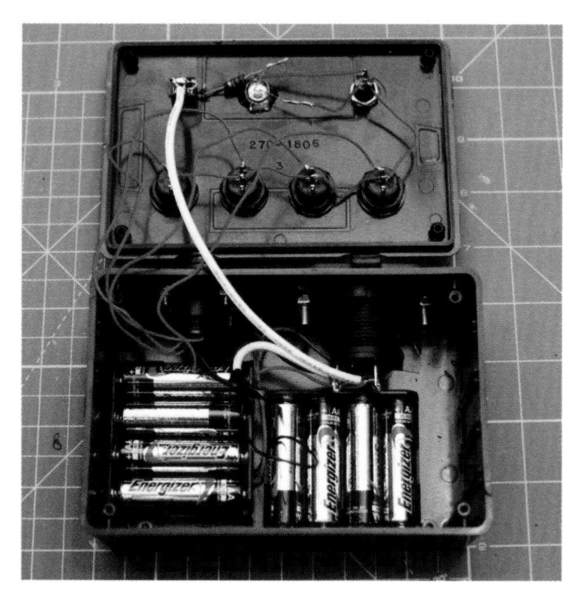

Figure 4-91. *Connect the 9-pin connector to the paddle switches following Table 4-15, then install the batteries.*

Building the Launch Wire Harness

The wiring harness connects the launch controller to the igniters for the various rockets on the launch pad. It is designed to give the proper current to each igniter while still allowing you to stand 30 feet away from the launch pad when launching a rocket.

There are alternatives. You can create a 17-foot harness rather than a 35-foot harness, and you can also use cheaper and more flexible 24-gauge wire rather than the 18-gauge wire specified in the parts list. I don't recommend the thinner, shorter wire for this launcher, though—it sort of misses the point. The technical reasons for these choices were discussed back in "How the Circuit Works" on page 78. It's also very nice to have a 30-foot clearance for a launcher that will allow four rockets to fly at once. Come on, you know you want to try it! When you do, you'll find it's easier to keep track of all those rockets if you have a bit more distance between you and the pad. It also gives you more time to duck.

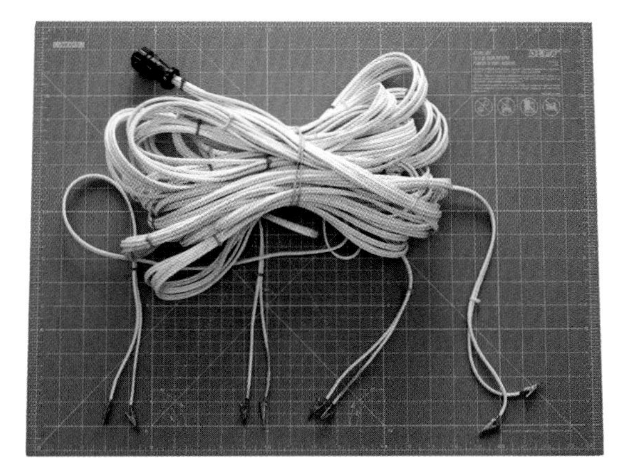

Figure 4-92. *The completed launch harness for the quad launcher.*

If you can find an eight-conductor wire, so much the better, but I didn't find any prebuilt cables with that many conductors. Oddly, seven-conductor wire is fairly common. You might be tempted to use that, or even five-conductor wire, and wire the negative side of the battery to all four pads. Keep in mind, though, that this will dump the current from all four pads into a single wire. It will work, but it does cut the current delivered to each igniter. You'll need to do the calculations from "How the Circuit Works" on page 78 to decide if that's really what you want to do. These instructions assume you went with 18-gauge lamp wire, commonly available from hardware stores.

The wire runs are not all the same length. The wiring harness will come up along the launch pad leg next to pad 4, then run down the back side of the launcher. The wire for pad 1 is 3 feet longer than the wire for pad 4. Cut the wires to lengths of 34 feet, 35 feet, 36 feet, and 37 feet. Put a piece of tape at the end of each wire and label them for pads 4, 3, 2, and 1, respectively. Strip about a quarter inch of insulation from each end of each wire and twist the strands together. Solder the female connectors from the 9-pin connector to the eight conductors at one end of the wires.

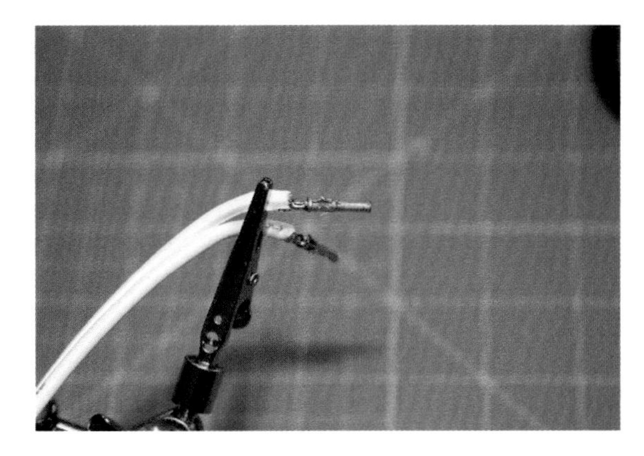

Figure 4-93. *Cut wires to 34, 35, 36, and 37-foot lengths. Solder one of the pins from the 9-pin connector to one end of each of the cables.*

Assemble the 9-pin connector by pushing the pins into the outside holes on the connector. The pin numbers on the connector are molded into the plastic on both sides of the connector. Two wires from each pad get wired into the connector; it doesn't matter which of the two wires goes to which pin, as long as the pins for each pad are placed in the two holes assigned to that pad. Table 4-16 shows which pins are used by which pad. The numbers may seem odd, but if you look at the pin layout on the connector, you will see that the pin outs are designed to keep the wires from each pad next to each other so there is a minimum of tangling as the wires are inserted into the connector.

Table 4-16. Wiring legend for connecting the 9-pin connector to the paddle switches

Pin numbers	Pad number
1, 2	1
3, 6	2
4, 7	4
8, 9	3

The insulation on the wire may be too thick to allow all of the wires to fit into the connector. In this case, use a hobby knife to carefully thin the insulation right next to the connector. You don't want to bare any wires and cause a short, but you do want the pins to fit.

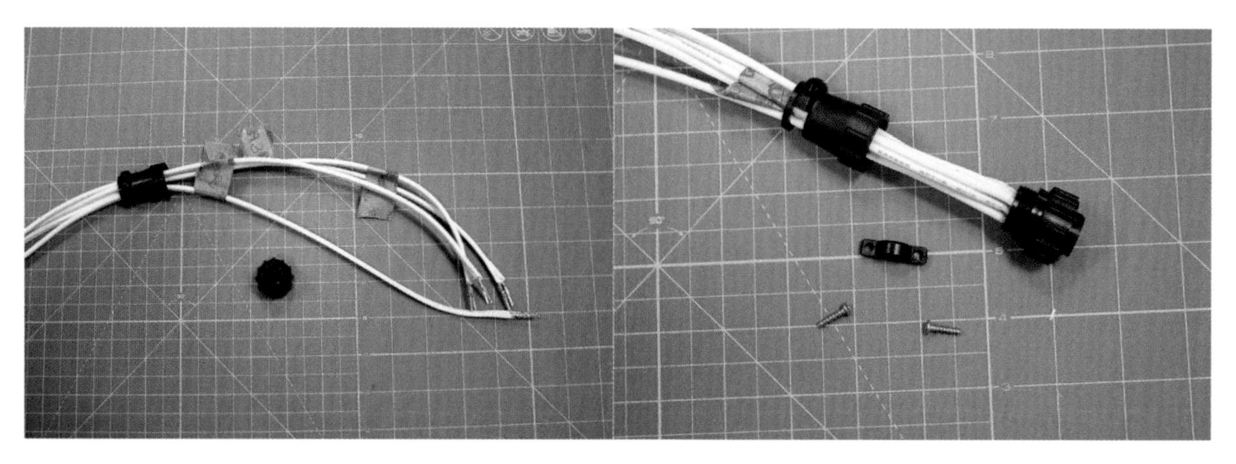

Figure 4-94. *Insert the pins into the connector as shown in Table 4-16.*

Slide the plastic sleeve for the connector over the four cables, then insert the pins into the connector. Insert a pin into pin 5, too, even though there is no wire connected. Once all of the pins snap into place, screw the plastic cover onto the connector and attach the fitting to secure the wires.

This is a lot of cable, so I used cable ties about every foot to combine the individual wires into a complete mounting harness. By a fortuitous happenstance, my box of cable ties had bundles in red, green, blue, and yellow. Taking a look at the launcher, it was obvious that I had to alternate between all of them!

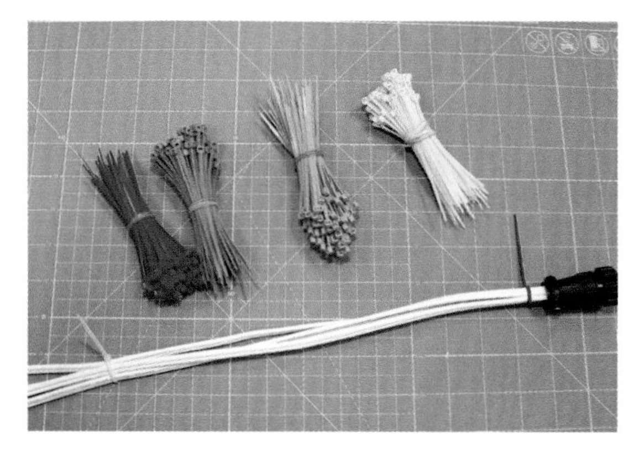

Figure 4-95. *Use cable ties to form a cable from the individual wires.*

Solder eight micro clips to the ends of the wires that are closest to the launcher. Separate the wires for a length of about six inches so the micro clips can easily be fastened to the separate ends of an igniter. I used wire ties to keep the wires from splitting too far back, and color-coded the wire ties to match the launch pad colors. Be sure to use a label of some kind to indicate which wire goes to which pad if you are using chromatically challenged wire ties.

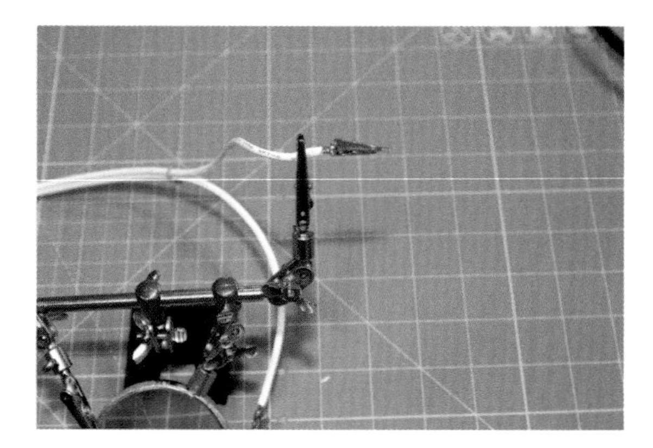

Figure 4-96. *Solder micro clips to the ends of the cable that will connect to the launcher.*

The Arming Key

The arming key for the quad launch controller is the same as the arming key for the mono launch controller, except that a colored streamer of ribbon is used instead of a launch rod cap. See "Building the arming key" on page 91 for detailed assembly instructions.

Testing the Launch Controller

It's time to see if you've got all of the wires connected to the right places. Connect the launch cable to the launch box and insert the arming pin. Flip the paddle switch for pad 1 to the Fire position. Connect the two micro clips for pad 1 together, and make sure none of the other micro clips are touching another clip. The Armed light should be lit, indicating there is a complete connection through pad 1.

Now flip the paddle switch for pad 1 to the Safe position and flip the switches for pads 2 through 4 to the Fire position. The arming light should be off, indicating there is no complete circuit to any pad.

Repeat this process for the other three pads. Check your wiring carefully if they don't behave as expected. It's possible a pin was inserted into the wrong hole in a connector, or there may be a short in one of the wires. Changing the pin positions is not easy, but it can be done. There is a special tool available from electronic supply houses to release the pins, or you can release them using a long, stiff needle.

Once you've finished the bench testing, it's time for the real thing. It's off to the flying field to launch a few rockets!

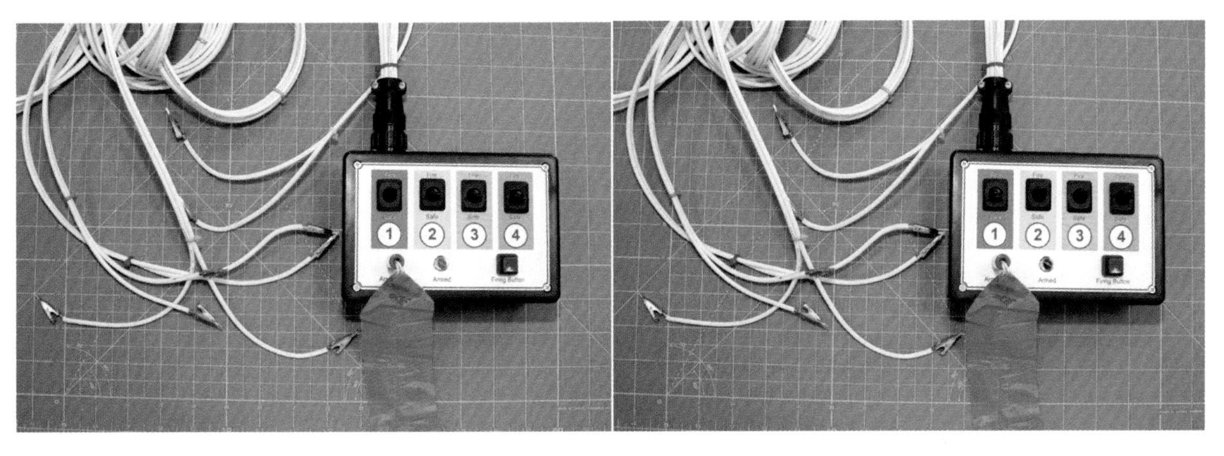

Figure 4-97. *Testing the completed launch system.*

Flight Operations

5

The day has finally arrived when you're ready for a launch. You've built Juno and the mono launcher, or kits that are similar to them, and you're ready for a flight. Let's take a look at where you should fly, how to get ready for the flight, and what to expect when flying and recovering your rocket.

Picking a Launch Site

The first step is to pick a place to fly your rocket. Local clubs are a great source of information about locations to fly. In most cities and states in the US, you can fly just about any of the rockets in this book in a public field or any field where the owner gives permission.

Getting permission to use a field may be pretty simple or very hard, depending on where you live. Most schools are happy to let you fly rockets on unused athletic fields. Rockets are very educational, after all. If you're under 18, it might help to have an adult ask for you, and to promise to have an adult present when flying. You might also know someone who runs a local farm or owns some unused land. Some parks may work well, too, although a lot of them have way too many trees to be useful for flying rockets. Check with the local authorities or other model aviators to make sure it's OK to use parks.

Local rocket clubs are a great source of information on local laws and attitudes about rockets. They may already have permission to use some great flying fields. We'll talk more about clubs in Chapter 22, but if you are having trouble finding a field, you might want to skip ahead to that chapter to find out how to link up with a rocket club.

Finally, look around for other model aviation enthusiasts, especially model airplane pilots. I live a few miles from Maloof Air Park, a local area set aside for model aviation. I've used the paved helicopter pads as a launch site on many occasions.

The ideal launch site will be a large, flat, grassy area. How large depends on the size of the motors you will be using. Table 5-1 is from the National Association of Rocketry (NAR) Model Rocket Safety Code (*http://www.nar.org/MARmrsc.html*). It shows the minimum allowed field sizes for various motor sizes. Remember, these are minimum sizes; a bigger field is always better. The site should not be near an airport or a major road. While a stray rocket is very unlikely to damage a car, I've seen a report of a water rocket that startled a driver, causing an accident. The accident resulted in regulations that hampered rocketry activities in the city where the accident occurred.

Table 5-1. Launch site dimensions

Installed total impulse (N-sec)	Equivalent motor type	Minimum site dimensions (ft)
0.00–1.25	1/4A, 1/2A	50
1.26–2.50	A	100
2.51–5.00	B	200
5.01–10.00	C	400
10.01–20.00	D	500
20.01–40.00	E	1,000
40.01–80.00	F	1,000
80.01–160.00	G	1,000
160.01–320.00	Two Gs	1,500

The first flight of any rocket should be on a low-power motor. This lets you check out the rocket and its construction before applying too much power. The first flight for Juno should be on an A8-3 motor, so we need a field 100 feet across. An unused football or baseball field is absolutely perfect. B motors are fine, too, but a single football field or baseball field is a bit too small for a C motor. A football field is about 200 feet wide when you include the team and coaching areas at the side, and it is 300 feet from one goal post to the other. You'll need a field 400 feet wide in all directions for a C motor. Keep your eyes open: an athletic field field that is technically too small might have enough surrounding open space to make it just right.

Other areas that make great flying fields are large vacant lots, pastures that are not currently occupied by animals, and parks.

You also need to keep an eye on the weather. You want as little wind as possible. Flying rockets is absolutely not allowed if the wind speed is over 20 miles per hour, even if it is gusting to those speeds. Sophisticated wind meters are great, but you can simply look around and get a good idea of how fast the wind is blowing. Table 5-2 is the Beaufort wind scale (land conditions): it was invented in 1805 by Francis Beaufort, an officer in the Royal Navy, to regularize weather reports at sea when no wind speed instruments were available. It uses empirical observations—the movement of waves, trees, and smoke—to estimate wind speed. The scale runs from absolute calm to hurricane speeds, but clearly you won't be flying rockets if you see the wind breaking twigs off the trees!

Table 5-2. Beaufort wind force scale

Description	Wind speed (mph)	Indicators
Calm	Less than 1	Calm. Smoke rises vertically.
Light air	1–3	Smoke drifts, but leaves and wind vanes are stationary.
Light breeze	4–7	You can feel the wind on your skin. Leaves rustle and wind vanes move.
Gentle breeze	8–12	Leaves and small twigs move. Light flags are extended.
Moderate breeze	13–17	Dust and small pieces of paper are blown around. Small branches move.
Fresh breeze	18–24	Moderate-size branches move. Small leafy trees begin to sway.
Strong breeze	25–30	Large branches move. Power lines begin to whistle in the wind. Empty plastic trash bins tip over.
High wind	31–38	Whole trees are in motion. It is difficult to walk against the wind.
Gale	39–46	Some twigs break from trees. Cars veer on road.

Finding a day when the wind is calm enough for flying rockets is a real problem in some seasons and locations. One thing to try is getting up very early in the morning. The wind is almost always calmer just after sunrise than later in the afternoon.

Packing for the Launch

Once you've found a field and picked a day with low wind, you need to haul your rocket stuff to the field. Most of us are not lucky enough to live somewhere where we can step out of the back door and fly a rocket, so we need to make sure we take everything we need to the launch site.

The key to making sure you arrive at the launch site with everything you need is a checklist. I've heard every excuse there is for not using a checklist, from "I'm too smart to need one," to "I can always go back for something I forgot," to "Only newbies need checklists." Right. The last excuse is particularly silly. As Atul Gawande points out in *The Checklist Manifesto: How to Get Things Right* (Picador, 2011), the more professional an activity becomes, the more likely you are to see a checklist. I carry a checklist with me when I teach scuba diving. When I was flying, a book of checklists for various situations was strapped to my left leg. Airline pilots have checklists for flying their planes. NASA uses checklists for preparing and flying its rockets. And yes, I use a checklist to make sure I'm taking everything I need to the launch. It's not a sign of weakness or forgetfulness to use a checklist; it's a sign of professionalism in action, making sure you get all of the details right. We'll develop three checklists now that you can adapt to your own needs.

Setting up a good range box is the first step in getting ready for launch day. The range box should contain everything you need to get a rocket ready for flight or make quick adjustments or repairs. We'll walk through what you might want to pack in your own range box, but keep in mind that you may want to modify this list, depending on local conditions and what kinds of rockets you fly.

I use a plastic fishing tackle box for my range box. It's only the second one I've owned. I used a metal launch box until I left for college. I didn't go to a college where I could take a range box with me, so after college I bought a new one. Figure 5-1 shows my range box.

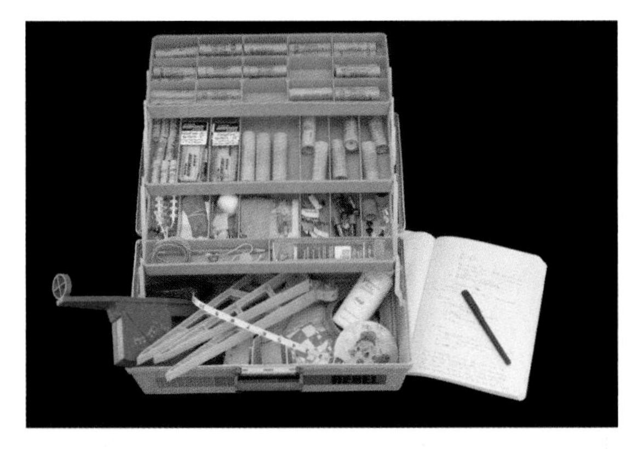

Figure 5-1. *A typical range box. A range box contains all the items needed to prepare a rocket for flight and make minor repairs and adjustments.*

Your range box should contain motors, igniters, and launch plugs. Even if you get your rocket ready before leaving for the field, you may want to fly it again. Having extra motors makes that easy. Extra igniters are important, since you occasionally get a misfire and need to install a fresh igniter.

Your range box should also have recovery wadding, spare parachutes or streamers, and talcum powder for packing parachutes. I like to use a small can of baby powder; it works great and comes in a convenient size.

Be sure to include sandpaper and painter's tape or masking tape. You may need to adjust the fit of a nose cone or payload by sanding it if it is too tight, or adding tape if it is too loose. Tape is also useful for holding in igniters, securing payloads, and tightening the fit on friction-fit motors. (Juno doesn't have a friction-fit motor, but some of the rockets later in the book do.) Masking tape sticks a bit better, but painter's tape works well, too.

My range box also has spare batteries, a 150-foot tape measure for measuring out tracking baselines, and an altitude tracker. This is a cheap, effective way to find the approximate altitude of a model rocket, but not a necessity in your own range box. The other thing people commonly measure is the flight time, especially for gliders. You probably have a stopwatch built into your phone, though, so you may not need a separate one in the launch box.

I've been using the mono launcher since I started flying cluster rockets and E motors, but I also carry an Estes launcher as a spare. It's a great launcher for rockets that have a single motor, size D or smaller, and it is very portable.

You might want to add a small flashlight for peering into body tubes to check the condition of shock cord mounts and so forth. Spare launch lugs and some thick CA glue are nice for fast field repairs. A small pair of pliers is handy for removing stubborn motors after a flight. Larger range boxes may include a multimeter for tracking down problems with the launcher or a small tool set.

Range Box Content

- Rocket motors
- Igniters
- Motor plugs
- Recovery wadding
- Parachutes
- Streamers
- Sandpaper
- Painter's tape
- Pencil
- Spare launch key
- Spare batteries
- Talcum powder
- Altitude tracker
- Tape measure
- Miscellaneous hardware
- CA glue (superglue)
- Spare launch lugs
- Flashlight
- Small tool set or pliers

You'll also want to take a few personal items. These can vary depending on how long you will be at the launch site and how remote the launch site is. Our local high-power launch site is pretty remote, so we pack just about anything we might need. Look over my list and see what makes sense for your launch site.

Personal Items to Take to the Launch

- Journal for taking notes
- Sunglasses
- Sunscreen
- Hat
- Sweatshirt
- Camp chair
- Shade structure
- Snacks or lunch
- Water
- Trash can
- Binoculars
- First aid kit
- Camera

Finally, don't forget your rockets and launcher!

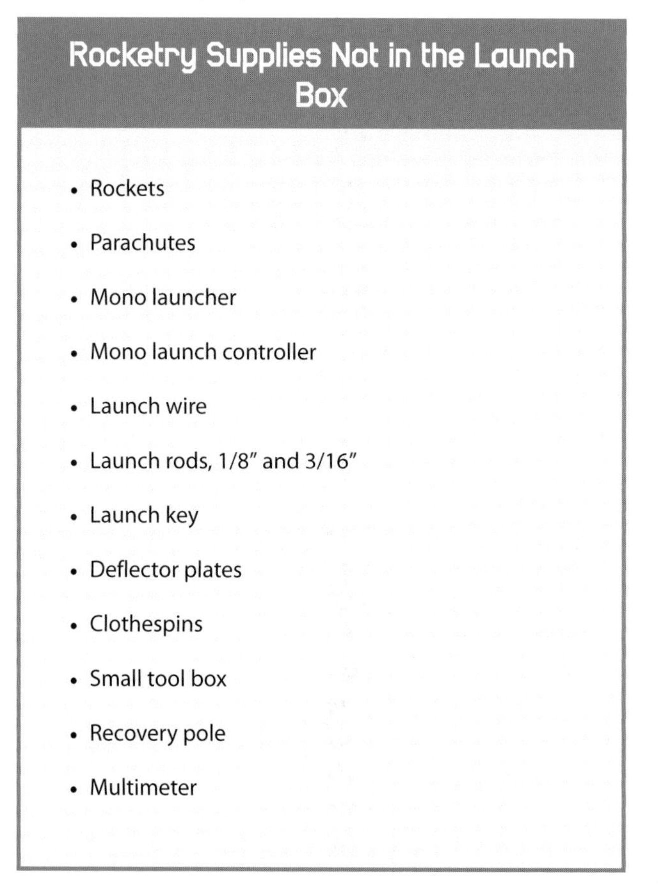

Rocketry Supplies Not in the Launch Box

- Rockets
- Parachutes
- Mono launcher
- Mono launch controller
- Launch wire
- Launch rods, 1/8" and 3/16"
- Launch key
- Deflector plates
- Clothespins
- Small tool box
- Recovery pole
- Multimeter

Most of the items are pretty obvious. It might seem like overkill to list the various parts of the launcher, but before I started using a checklist, I left every one of those items behind at one time or another. Use the checklist, and make sure you have each of those items!

It may not be obvious why you need clothespins. They are used to hold the rocket away from the blast deflector plate. There are a lot of things you can substitute. We'll cover this in more detail when we set up the launcher later in the chapter.

The recovery pole is used to get rockets out of trees. I use a long extendable aluminum pole designed for washing windows. I've saved more than one rocket from a tree using my recovery pole! As I'll mention later, though, you should never try to get a rocket down from a power line, particularly not with an aluminum pole.

Range Operations

The rules and procedures for setting up and running a model rocket range were originally developed and passed around by Harry Stine and other rocket professionals from White Sands Missile Range and various NASA operations. They were designed with safety in mind, and they have worked remarkably well over the years. Learning and applying them will make your rocket range just like those used by NASA and other professional rocket scientists, just as our rockets are smaller versions of the rockets they fly.

NAR Safety Code

Let's start with the National Association of Rocketry (NAR) safety code. It's the bible for safe operation of low-power solid propellant rockets. There are two variations on this safety code, one for high-power rockets and one for radio control gliders. All three safety codes can be found in the NAR website's safety section (*http://www.nar.org/safety.html*), and the Model Rocket Safety Code is reprinted in its entirety in Appendix D.

The safety code itself is shown in bold. In some cases, the reason for the rule may not be obvious to a beginner, but there is always a reason. I've added some commentary after some of the rules in a normal typeface. This commentary is not part of the safety code; it's just there to help you understand the reasoning behind the rules.

1. **Materials. I will use only lightweight, non-metal parts for the nose, body, and fins of my rocket.**

   Model rockets are designed to be lightweight and to crumple if they hit something at high speed. We'll explore the physics of impacts in Chapter 10, but for now, keep in mind that one of the reasons model rockets are so safe is that their light weight minimizes the energy of impact, and their construction spreads that energy over a longer period of time by crumpling if they hit something hard. That's the same technique used in modern cars to make them safer in a collision. That's why we don't use extremely strong body tubes or sharp rigid points on the nose cones of the rockets.

2. **Motors. I will use only certified, commercially-made model rocket motors, and will not tamper with these motors or use them for any purposes except those recommended by the manufacturer.**

   Back before Estes and other companies started making safe, reliable rocket motors, many garage scientists were making their own. I don't have an accurate count of the number of eyes, fingers, and lives that were lost in the process, but there were quite a few. Making rocket motors is something that kills even professionals. It's not something you should try yourself without extensive training and access to proper facilities. A book or Internet article is not proper training, and your garage is not a proper facility.

   On average, about once a year someone makes a trip to the emergency room after being burned by a model rocket. Invariably, the burn was caused by breaking this rule. I once heard of a candidate for a Darwin Award (*http://www.darwinawards.com*) who dumped the contents of a rocket motor into a paper cup and lit them with a match. A rocket motor is safe if used properly. If you are incapable of using one properly, please find another hobby—we don't want you messing ours up by providing fodder for a well-intentioned legislator who wants to add restrictions.

3. **Ignition System. I will launch my rockets with an electrical launch system and electrical motor igniters. My launch system will have a safety interlock in series with the launch switch, and will use a launch switch that returns to the "off" position when released.**

   Remember my friend Rob from Chapter 2? He broke this rule and was lucky not to end up in the emergency room himself.

   Why don't we use pyrotechnic fuses, the kind you light with a match? We use electrical ignition systems because fuses are not particularly reliable, and because it is extremely hazardous (to the point of foolhardiness) to try to cancel a countdown once a fuse is lit. It could spell disaster if a low-flying airplane came into view or a child ran into the launch area once a fuse was lit.

   The safety interlock is used to make sure the rocket doesn't go off accidentally. The safety interlock key, also called the launch key or arming key, should be on or near the launch pad—never in the launch controller—when anyone is working on the rocket. That's why the safety interlock key is connected to the launch rod cap in the mono launcher, and why a hole is provided on the quad launcher for storing the safety interlock key. The launch button should be a momentary pushbutton switch—at the very least, it *must* return to the off position when released, as an extra guard against an accidental launch. If you don't think all of these precautions are reasonable, talk to Rob.

4. **Misfires. If my rocket does not launch when I press the button of my electrical launch system, I will remove the launcher's safety interlock or disconnect its battery, and will wait 60 seconds after the last launch attempt before allowing anyone to approach the rocket.**

   We do get misfires on occasion. These happen when an igniter lights, but the rocket fuel is not ignited—at least, not right away. Just like a smoldering campfire, it's possible for a heat source to still be active after the launch button is released. Giving the rocket time before approaching it is just a precaution to make sure any smoldering heat sources die out, to avoid accidents.

5. **Launch Safety. I will use a countdown before launch, and will ensure that everyone is paying attention and is a safe distance of at least 15 feet away when I launch rockets with D motors or smaller, and 30 feet when I launch larger rockets. If I am uncertain about the safety or stability of an untested rocket, I will check the stability before flight and will fly it only after warning spectators and clearing them away to a safe distance. When conducting a simultaneous launch of more than ten rockets I will observe a safe distance of 1.5 times the maximum expected altitude of any launched rocket.**

   This is the rule that drove the design of the launchers in the previous chapter. It keeps you and everyone else a safe distance from the rocket while it is being launched.

I would encourage you to make sure everyone is not only paying attention, but standing up and facing the rocket for a launch. If something goes wrong, they can then move quickly out of the way.

For untested rockets, our club announces "Heads up!" before the launch. This is a warning to spectators to pay special attention in case something goes wrong.

6. **Launcher. I will launch my rocket from a launch rod, tower, or rail that is pointed to within 30 degrees of the vertical to ensure that the rocket flies nearly straight up, and I will use a blast deflector to prevent the motor's exhaust from hitting the ground. To prevent accidental eye injury, I will place launchers so that the end of the launch rod is above eye level or will cap the end of the rod when it is not in use.**

   The launch angle is common sense. Rockets are not projectiles to be fired at something. They also don't fly well in a horizontal direction.

   The blast deflector plate is designed to prevent fires and damage to the launcher. Take a look at a used blast deflector plate. I've seen them with scorch marks that go through the metal plate. I've even seen holes burned in really old ones that have been used for hundreds of launches. A blast deflector plate is a necessity, not a cute design addition.

   You'll recall that the mono launcher has a launch key that is also a cap for the launch rod. We'll cover procedures in a moment, but the combination makes it easier to follow this rule.

7. **Size. My model rocket will not weigh more than 1,500 grams (53 ounces) at liftoff and will not contain more than 125 grams (4.4 ounces) of propellant or 320 N-sec (71.9 pound-seconds) of total impulse.**

   This rule pretty much defines "model rocket." Yes, there are rockets larger than this. They are made from different materials, and need different rules to fly safely. They require FAA approval before they can be launched. While there is nothing wrong with flying high-power rockets, they are the topic of another book planned for release in 2015, *Make: High-Power Rockets*.

   This rule also covers a class of rocket that fits between low-power rockets and high-power rockets. I called these "mid-power rockets" in Chapter 2, and that's a pretty common term. Mid-power rockets are powered by a motor or motors with more than 40 N-s of impulse, which puts them in the F or G category. The rockets generally have stiffer body tubes, harder nose cones, and plywood or fiberglass fins to withstand the power of these larger motors. You also have to be over 18 to buy the motors. Mid-power rockets have a lot more in common with high-power rockets than low-power rockets, and are covered in *Make: High-Power Rockets*. This book and the techniques it describes assumes your rocket is made from thin, collapsible cardboard tubes and balsa fins, weighs one pound or less, and uses motors with 40 N-s or less total impulse.

8. **Flight Safety. I will not launch my rocket at targets, into clouds, or near airplanes, and will not put any flammable or explosive payload in my rocket.**

   Duh. But it has to be said.

   The rule about flying a rocket into a cloud is to prevent your rocket from hitting low-flying airplanes that are in or over the cloud.

9. **Launch Site. I will launch my rocket outdoors, in an open area at least as large as shown in the accompanying table, and in safe weather conditions with wind speeds no greater than 20 miles per hour. I will ensure that there is no dry grass close to the launch pad, and that the launch site does not present risk of grass fires.**

   High winds can cause a rocket to *weathercock*, tipping it into the wind. In severe cases, the rocket can fly almost level to the ground, or even impact with the motor still burning. We'll look at this more thoroughly in Chapter 7.

   A model rocket has been known to start at least one grass fire. The area around the launcher should be clear of anything that will burn easily. We'll discuss techniques for fire prevention in a moment.

   The accompanying table is Table 5-1.

10. **Recovery System. I will use a recovery system such as a streamer or parachute in my rocket so that it returns safely and undamaged and can be flown again, and I will use only flame-resistant or fireproof recovery system wadding in my rocket.**

    There are lots of ways to recover a model rocket. If you're working through the book, building the rockets without reading the science sections, stick with the recovery methods listed for each rocket. See Chapter 10 if you want to experiment.

11. **Recovery Safety. I will not attempt to recover my rocket from power lines, tall trees, or other dangerous places.**

    To the best of my knowledge, there have been exactly four deaths associated with model rocketry as described in this book over the years. Every one of them was caused by someone trying to retrieve a model rocket from a power line. If a rocket lands across a power line, it is lost. Period. Only the power company can change that.

    All other deaths associated with nonprofessional rocketry that I am aware of were caused by people making their own motors. That's not model rocketry.

The Roles of the RSO, LCO, Flier, and Spectator

There are four roles at any model rocket launch. The obvious one is the flier; that's you. Less obvious but still important is the role of spectator, which is a more active role than you might think. We'll get to that in a moment. There are two other roles you need to understand, though. If you're headed out on your own or with a couple of friends to launch a rocket or two, you might be the one filling the other two roles, too. If you're a new member of a club, headed off to fly rockets for the first time at the club, you'll definitely not be filling either role.

The most important role at any launch is the *Range Safety Officer* (RSO), sometimes called the *Launch Safety Officer* (LSO). The RSO is the person ultimately responsible for the safety of everyone at a launch. It is the responsibility of the RSO to reduce risks to people and property from model rockets from the time the site opens until the last person leaves.

In a club, the RSO will be a senior member who has seen it all—or, if he's really good, perhaps has prevented a few things from happening so he *hasn't* seen it all! The RSO is the ultimate authority for safety at a launch. He may decide that a rocket looks unstable, and should not be flown. He might decide that the weather is unacceptable for the launch conditions and close the launch site, temporarily or for the day. Whatever decision the RSO makes is final. No one can override the RSO on a safety decision. Once the RSO makes a call, you have two choices: follow it or leave.

Some ranges in the United States follow NAR launch rules, and some follow Tripoli launch rules. In other countries, ranges may operate using rules appropriate to their local situation, laws, or culture. Whatever set of rules you are using at a launch, though, the RSO is responsible for making sure they are followed. The Model Rocketry Safety Code serves as a checklist for the RSO when setting up, running, and tearing down a launch site.

The second role is the *Launch Control Officer* (LCO), sometimes called the *Range Control Officer* (RCO). With one flier or a small launch with a single rocket launcher and a few fliers, the RSO and LCO are frequently the same person. With two or three fliers who know each other well, the roles might even pass from one person to the next, with the individual flier assuming the roles of RSO and LCO. With two or more pads, though, someone needs to be responsible for safety, and that's an important enough role that a second person takes charge of running the range. That's the LCO.

The LCO is the most visible role at the launch site. The LCO directs where the pads will be set up, where spectators will stand, and where the rockets will be prepared for flight. It's the LCO who will tell people when they can approach the launch pads and which pads are clear to fly. The way my local club runs its high-power launchers, the LCO is even the person who does the countdown and presses the launch button, although the LCO may allow the flier to perform those two steps if she likes.

The RSO can step in at any time to handle a situation that is getting out of hand—that's his job—but if things are running smoothly, the LCO may do her job well enough that you have to ask to figure out who the RSO is.

You might think a *spectator* doesn't really have anything to do. You would be wrong.

Spectators at a launch are responsible for following the safety rules for the launch. The RSO or LCO can, and will, stop a launch if spectators are behaving in an unsafe manner. I've watched an RSO stop the launch countdown until spectators stood up and faced a launch so they would know, and be able to react, if something went wrong.

Spectators have another important role. It might be a spectator that sees an unsafe condition before the RSO or LCO is aware of it. Maybe a small child breaks loose from the crowd and heads for the launcher, or perhaps a model glider from the nearby airplane flying field wanders overhead. Anyone, including a spectator, can yell "Hold!" at absolutely any time to halt a countdown. The RSO will take over from there, figuring out why the hold was called and resolving the situation or shutting down operations if the situation cannot be resolved.

Fire Prevention

Model rockets are built to be safe. The blast deflector plate catches or deflects virtually all of the blast from a model rocket until it is high enough in the air that there is no fire danger. Still, things happen. I remember one cone-shaped rocket I built in high school from a complementary kit. It was a really cool but odd-looking rocket. I'd made it a bit too quickly and glued the launch lug on slightly crooked. I did the countdown and pressed the firing button. The rocket moved up a couple of inches, and then lodged on the launch rail! It stayed there as the motor went through the thrust, coast, and ejection phases. The rocket didn't budge. Then it turned a little brown around the top and burst into flame! Rob and Steve, my two high-school rocket buddies, stood there with me as we watched it burn to a crisp, staying well away until the fire was completely out.

This could have been a big problem. It wasn't because we were following rule 9 of the NAR safety code. As was our habit, we were launching from the center of an unused gravel road, well away from anything flammable.

Damp grass, gravel, dirt, or small paved areas are all perfect places to set up a launch pad. An area covered with dried grass or weeds is definitely not a good place to set up a launch pad—at least, not without a little site preparation. You need to make sure there is nothing flammable for a few feet in any direction from the center of the launch pad. If there is, you have three choices.

The first, and possibly the easiest, is to move the launch pad. There may be a better spot just a few feet away.

The second possibility is to clear the area of dried grass and weeds. That's what our club does. We have access to an area of high-desert prairie that will someday be covered by houses. For now, though, it's covered in sagebrush, low cactus, and grass. About once a year the area is cleared with a backhoe, scraping off all of the weeds to leave bare dirt. This is cleared of any new growth through the year. After some record recent rains, for example, the club president sent out an email asking everyone to bring a shovel to the next launch to clear out new growth.

Figure 5-2. *A view from behind the LCO table toward the 100- and 200-foot high-power launch pads at the Albuquerque Rocket Society launch site. The ground has been scraped to reduce fire danger, and there are fire extinguishers by every pad and at the LCO table. There is also a fire beater out of the view.*

The last possibility, again a good one, is to invest in a cheap tarp. You don't need a large one—five feet on a side will do—but the tarp can be placed over the grass to tamp it down and prevent igniter wires, flame from the rocket exhaust, or really poorly constructed rockets that burn on the pad from igniting any dry grass.

Figure 5-3. *Use a tarp at least five feet square under a launcher in a field of dried grass.*

Here in the West we sometimes have fire restrictions. Don't fly solid propellant rockets if there are fire restrictions in place. Check with your local rocket club, though. Even in severe fire restrictions, our club has had an area where we could fly. This was only approved, however, because of the extraordinary measures the group takes to prevent fires, and even then only after a review by the local fire department.

Site Setup

In this section I'm going to assume that you are setting up a site for yourself or a couple of people to use the mono launcher, or for a small group to use a single quad launcher. If you are trying to set up and run a launch site for a larger group of people or with multiple pads, refer to Chapter 22. Or, better yet, join a rocket club and see how it operates its range.

Mono Launcher

It's launch day. The weather is fine, you've found a suitable field and have permission to use it, and your rockets are ready to fly. Perhaps you have a couple of friends who will be flying with you. It's time to set up the site.

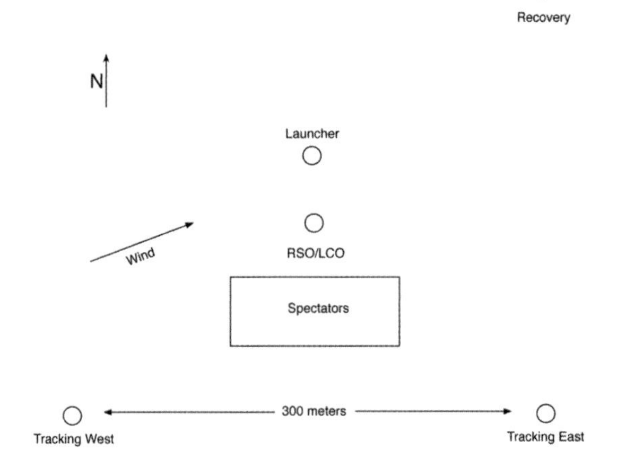

Figure 5-4. *Typical site layout for a single-launcher rocket range.*

Pick a spot for the launcher that is at the center of the site, or upwind from the center of the site. It should not be too close to the edge of the site, though. The rocket should land in the designated area. In general, the launcher should be at least half of the designated distance shown in Table 5-1 in each direction.

In the Northern Hemisphere, the best place for the LCO and spectators is south of the launch pad. That puts the sun at everyone's back, making it easy to see the rocket, which makes for a safe and enjoyable launch. If something goes wrong with the stability of the rocket, it is more likely to turn into the wind than any other direction, so if the wind is coming from the south, shift everyone around a bit in the direction of the sun—toward the east in the morning, or west in the afternoon.

If you're launching in a large field, or if the wind is a little high, you might preposition a recovery crew to chase down a rocket. If you don't have a designated recovery crew, though, be sure and tell everyone present that only the flier should approach the rocket unless the flier specifically says it's OK to fetch the rocket. If a recovery crew is used, only the flier or the recovery crew should fetch the rocket.

You might want to track the altitude of your rocket with one or two trackers. Like spectators and the LCO, the trackers should have their backs to the sun. Generally a site is set up with Tracking West and Tracking East positions, about 300 meters apart and 150 meters south of the launch pad. See Chapter 8 for more on setting up and running tracking stations.

Quad Launcher Operations

The quad launcher is ideal for small groups of fliers, small clubs, or schools. The site is set up much like the mono launcher site. In general, though, there will probably be more people around, so the range needs to be more organized. If you're setting up the quad launcher for yourself or for a couple of fliers, follow the site setup for the mono launcher.

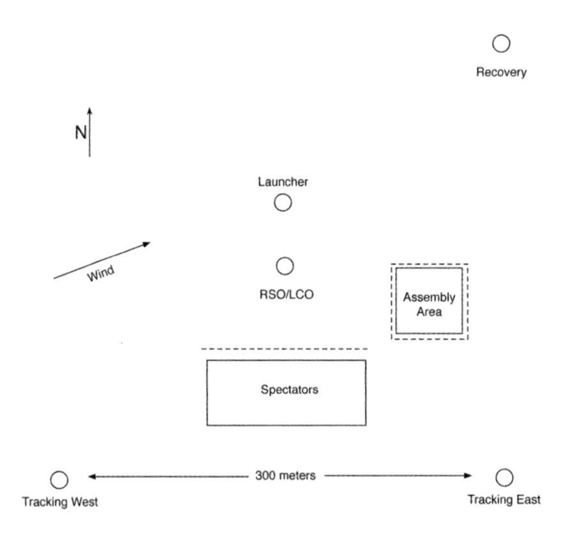

Figure 5-5. *Site setup with a larger number of fliers and spectators, but still one launcher. Dashed lines are great places for roped off areas to keep people in place.*

Once again, set up the site so the sun is behind the spectators and the LCO. Try to set up a table for the LCO and a separate table for preparing rockets, especially if the group will have young children around. I've had an excited Cub Scout step right on top of a rocket I had laid out on the ground for launch preparation; using a table makes damage a lot less likely.

If the group will be large, consider setting up flags or tape to rope off the launch area, LCO table, and rocket preparation area. Make sure spectators stay out of those areas. You can also set up a spectator line, and ask them to stay behind it. If you're at a school athletic field, ask spectators to stay in the stands, just as they would for a game.

If the number of people who will be around is large enough or likely to be loud enough that someone speaking in a loud voice might not be heard, make sure you have a PA system or megaphone handy. Before each launch, ask spectators to stop what they are doing, stand up, and face the rocket. The RSO should hold the launch until all spectators comply. The LCO should announce the rocket, which pad it will fly from, and the approximate altitude ("Juno will fly from pad 3 on a C motor. It will go really high, folks!" is fine—the spectators may not know what 2,000 feet really looks like for a rocket flight), then give the countdown.

If the number of fliers is large, ask each flier to fill out a flight card like the one in Figure 5-6. When a flier's rocket is ready, the flier brings the rocket and flight card to the LCO table. The LCO will assign a pad to each of up to four fliers, make sure the launch key is removed from the pad, then announce for each, "The pads are safe. Fliers may install their rockets." Keep the flight cards at the LCO table until after the flight so the LCO can announce the name of the rocket, name of the flier, motor, and expected altitude.

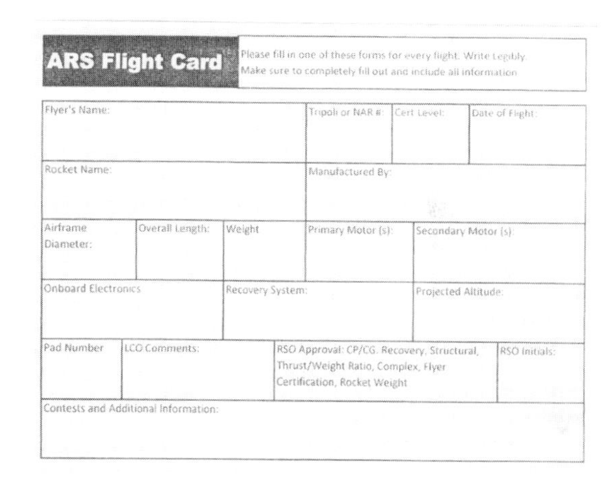

ARS Flight Card

Please fill in one of these forms for every flight. Write Legibly.
Make sure to completely fill out and include all information

Flyer's Name:			Tripoli or NAR #:	Cert Level:	Date of Flight:
Rocket Name:			Manufactured By:		
Airframe Diameter:	Overall Length:	Weight	Primary Motor (s):	Secondary Motor (s):	
Onboard Electronics		Recovery System:		Projected Altitude:	
Pad Number	LCO Comments:		RSO Approval: CP/CG, Recovery, Structural, Thrust/Weight Ratio, Complex, Flyer Certification, Rocket Weight		RSO Initials:
Contests and Additional Information:					

Figure 5-6. *This is the flight card used at Albuquerque Rocket Society launches.*

Fly all of the rockets and remove the launch key before announcing once again that the pads are safe and allowing the fliers to go recover their rockets. No one should be allowed to head downrange until all of the rockets have launched—anyone heading downrange after their own launch is not paying attention to the next launch.

The LCO and RSO should not be involved in flying rockets of their own when managing a large group. If the LCO and RSO have rockets of their own, the group should occasionally switch out the roles, and have other people take over the positions.

Repeat the process until all of the rockets have flown.

Preparing the Rocket

The range is set up. It's finally time to prepare your rocket for launch. Let's walk through the steps for preparing Juno for launch. Other rockets in the book are similar, but there are minor differences. That's why there is a separate section on launch preparation for each rocket, along with separate checklists for preparing and flying the rocket. All of them follow this basic outline, though.

Before you start, make sure you are working in an area where interruptions are unlikely. If someone stops to chat, politely ask the person to wait a moment. I've messed up rockets before because some well-intentioned or curious person talked with me during setup. If you are interrupted, after the interruption, restart at the top of the checklist and follow it down, making sure each step has been completed, until you get to the point where you left off. Don't just jump back in the middle to the step you think you were on! That's how mistakes are made.

Installing the Motor

I like to start by installing the motor. The recovery wadding and parachute will slide into the body tube from the nose cone end of the rocket. If the motor is in place, they can't slide too far.

Select the motor for the flight. For the first flight it should be an A8-3 motor, one of the lowest-power full-sized motors available. When you look at the motor, you'll see that one end has a clay nozzle, and the other has a clay or paper cap. The nozzle will always be right at the end of the rocket motor, but the cap may be recessed on A and B motors.

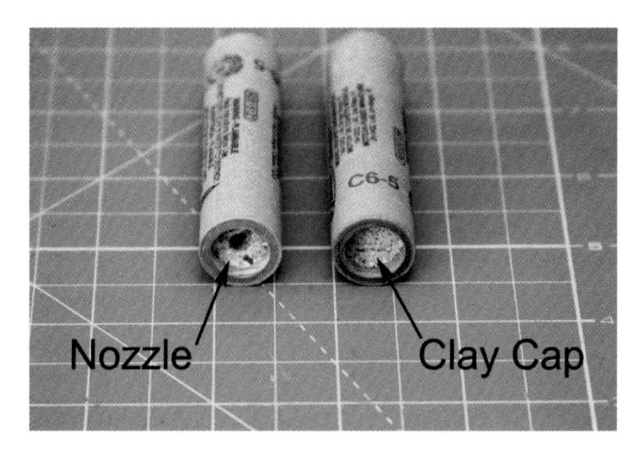

Figure 5-7. *Install all motors so the nozzle is toward the bottom of the rocket and the clay or paper cap is toward the parachute.*

Move the engine hook to the side and slide the motor in so the nozzle is at the bottom of the rocket. Push the motor in until the clip snaps in place, which should prevent the motor from sliding to the rear.

Figure 5-8. *Push the engine clip aside and slide the motor in until the clip snaps in place to prevent the motor from sliding to the rear.*

> **Other Motor Mount Systems**
>
> While Juno uses a motor hook, some rockets do not. If you bought a kit that uses friction fit motors, skip ahead to Chapter 15 for information on properly fitting a motor using friction fit. A few rockets eject the motor and descend without it. See Chapter 16 for a discussion of featureweight recovery, one recovery method that generally ejects the motor.

Recovery Wadding

Recovery wadding is used to protect the parachute from the hot gases in the ejection charge. It sits between the motor and the parachute and shock cord.

There are two main kinds of recovery wadding. Estes Industries, Quest, and others sell packets of recovery wadding that looks like light brown or blue toilet paper. Don't worry: the light brown color is a fire retardant, hidden a bit by blue dye in the Quest version. The recovery wadding will char a bit when it's hit by the ejection charge, but it won't burn. Wad up three or four sheets of recovery wadding and insert it into the launch tube.

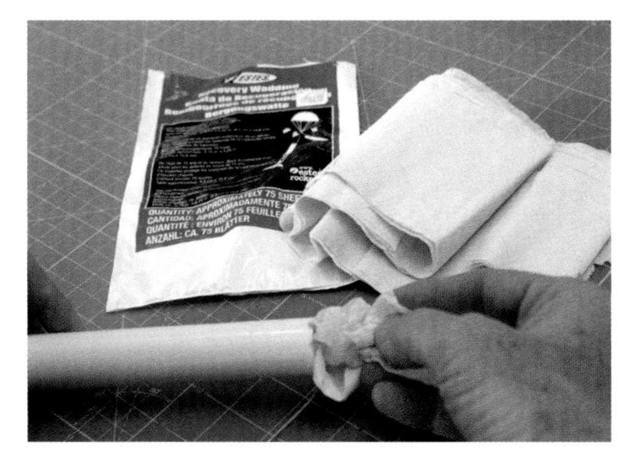

Figure 5-9. *Recovery wadding is essentially flameproof toilet paper, great for rockets and Mexican food. Wad up three or four sheets and stuff it deep into the body tube. Use more for larger tubes—there is a table on the package.*

A cheaper alternative for clubs and for very active fliers is popularly known as dog barf. It's fiber insulation available from the hardware store made from recycled newspapers, usually used to insulate homes. It comes in a huge bag and, like recovery wadding, has been treated with a fire-resistant chemical. You'll see fliers with plastic bags full of the stuff at many launches. Stuff enough dog barf in the body tube to fill it to about twice the diameter of the tube. If your tube is one inch in diameter, pack about two inches of wadding into the tube (you'll use two inches for Juno). While you don't want to pack it too tightly, you don't want to dump it in too loosely, either. Pack it in lightly so compressing the dog barf does not absorb the energy of the ejection charge.

Figure 5-10. *An alternative to recovery wadding, dog barf is the popular name for flame-resistant house insulation. Loosely pack the tube about two tube diameters deep.*

Don't Use Cotton Balls!

Never use cotton balls or other flammable material for recovery wadding. Flammable recovery wadding is a serious fire hazard.

Packing the Parachute

I've seen more than one rocket, including a couple of my own, fall from the sky without a parachute attached. Sometimes this happens because a worn shock cord breaks. Sometimes it happens because the knot securing the shock cord or parachute is not tight, and slips. Sometimes it happens because the snap swivel was not fastened, or wasn't attached at all. A NAR study (*http://www.nar.org/pdf/launchsafe.pdf*) reports that 72% of all flight failures are recovery system failures. This is where you need to slow down and pay attention!

Before packing the parachute, stretch out the shock cord and parachute and give both a visual inspection for wear. Make sure the knots are secure. If you are using snap swivels, attach both of them now and make sure the clips are completely closed.

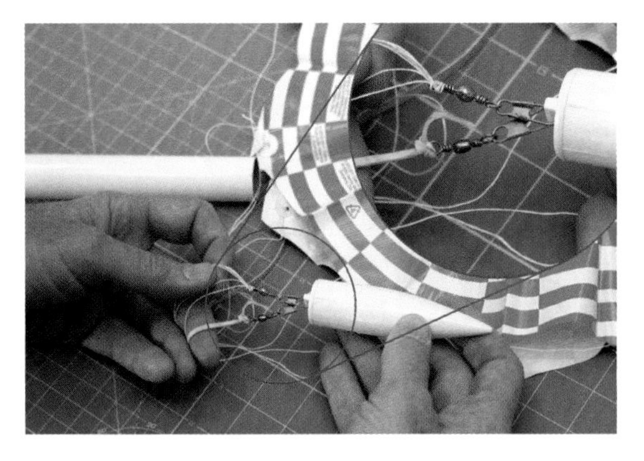

Figure 5-11. *Attach the parachute and shock cord. Make sure knots are secure and snap swivels are fully closed.*

Tug on the shock cord to make sure it is securely fastened. The shock cord mount may be too far down in the body tube to see clearly, so tugging is the only way to make sure it has not weakened from use. If you are using rubber or elastic shock cords, take a look at the stretched shock cord. Is it showing signs of cracking? If so, replace it before flight.

A lot of parachute plastic will stick to itself. Parachutes, especially plastic ones, also tend to get stiff in cold weather. You want your parachute to snap open quickly once it is ejected from the rocket. A generous application of talcum powder helps a lot. The talcum powder also makes it easier to see the rocket when the parachute opens, adding to the puff of smoke from the ejection charge itself. Another thing that helps parachutes open quickly is to store them on a clothes hanger so they "remember" their open shape, not their creased and folded shape.

Figure 5-12. *The best way to store plastic parachutes is hanging open. Use tabs formed from tape to keep them separate while hanging.*

Start by laying the parachute out with the top side down and sprinkling plenty of talcum powder onto the underside of the parachute. Spread it around a bit.

Figure 5-13. *Begin by covering the underside of the parachute with talcum powder.*

Fold the parachute in half so there is a shroud line at the outside edge on each side of the parachute. Add more talc to about 2/3 of the exposed part of the parachute.

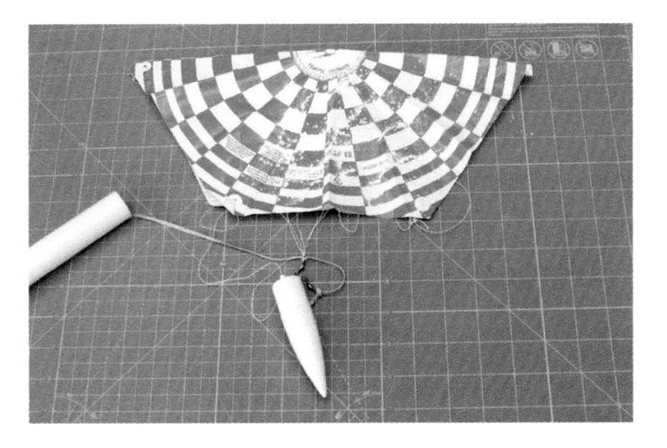

Figure 5-14. *The right two panels of the parachute are covered with talcum powder. Fold the left third over first, then fold the right third over.*

Fold the unpowdered third over, then the remaining third, so the parachute shroud lines match up at the opposite corners. Add powder to half of the exposed surface.

Figure 5-15. *Powder the left side of the panel, then fold to bring all of the shroud lines together.*

Continue folding the parachute in half and adding powder until it forms a long pencil-like shape about half of the width of the body tube.

Figure 5-16. *Continue folding until the parachute is about half the width of the body tube.*

Fold the pencil-like shape in half, then loosely wrap the shroud lines around the parachute to keep it from expanding too early.

Insert as much of the shock cord as you can, then slide the parachute and any remaining shock cord into the body tube. Slide the nose cone into position.

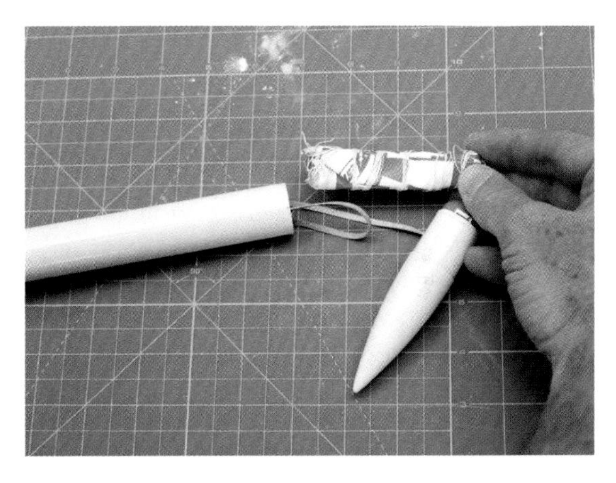

Figure 5-17. *Fold the parachute in half, then wrap the shroud lines loosely around the parachute. Insert as much of the shock cord as possible, then follow with the parachute.*

The parachute must not be tight! Remember, the nozzle end of the rocket motor has a hole in it, and the pressurized gas from the recovery charge can squeeze out there, too. I've seen rockets plummet to earth with the parachute still in the body tube, only partially ejected, because the parachute was packed too tightly.

The nose cone itself should slide in and out easily, but should not be loose. Turn the rocket over. If the nose cone falls out with a gentle shake, it's too loose. Add some tape around the shoulder of the rocket to tighten it a bit. If it's too tight, sand the shoulder a bit.

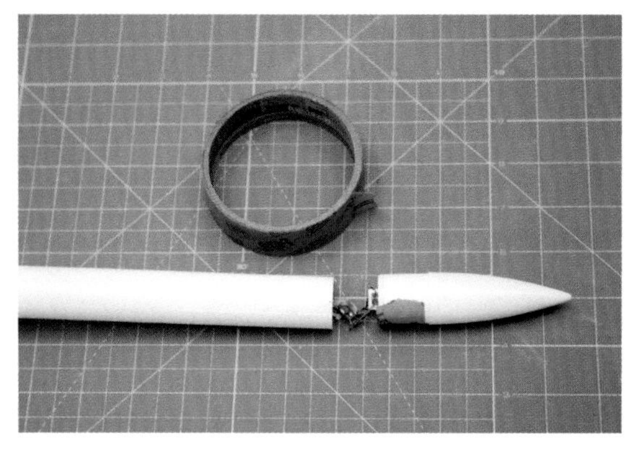

Figure 5-18. *If the nose cone is loose enough that it falls out when the rocket is turned upside down and shaken gently, add tape to the shoulder to make it tighter. Sand the shoulder if the nose cone is too tight to slip in and out of the body tube easily.*

Installing the Igniter

While there are certainly some variations, there are two main types of igniters used in black powder motors in model rocketry today. The two igniters are similar, though there are slight differences in the way they are installed. Both work extremely well.

The Estes igniters are preformed, with nichrome wire *legs* held apart by a strip of paper. This helps prevent shorts. Always do a visual check of the igniter, though, to make sure the leads do not touch, especially right near the tip of the igniter. If they do, spread the wire gently. A short is one of the two common causes of igniter failure—everything looks right, and you even get a good continuity light, but the current goes through the short instead of through the tip of the igniter, and it doesn't light.

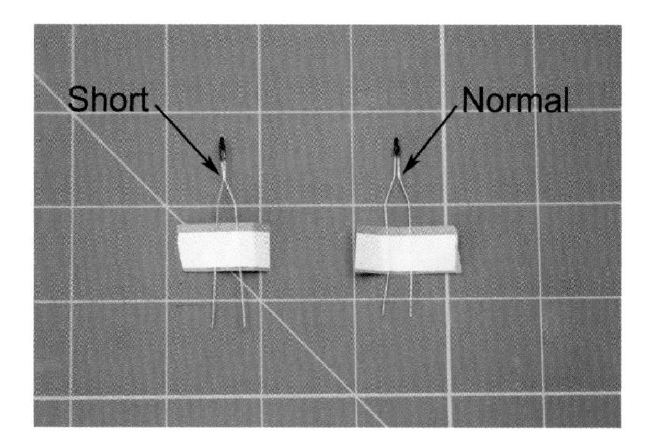

Figure 5-19. *Make sure the igniter wires do not touch, forming a short. If they do, spread the wires slightly.*

Insert the igniter into the motor's nozzle as far as it will go. The tip of the igniter should touch the black powder just inside the nozzle. Bend the igniter away from the launch lug and insert the small plastic plug that comes with the motor as far as it will go into the nozzle. The plastic plug will hold the igniter in place and allow the motor to build up a little pressure before it starts to move the rocket.

Figure 5-20. *Making sure you don't short the wires, insert the plastic igniter cap to keep the igniter in place.*

Quest igniters use insulated wire to prevent shorts. The igniter tip is protected in a red plastic sleeve. Remove the red plastic sleeve before using the igniter.

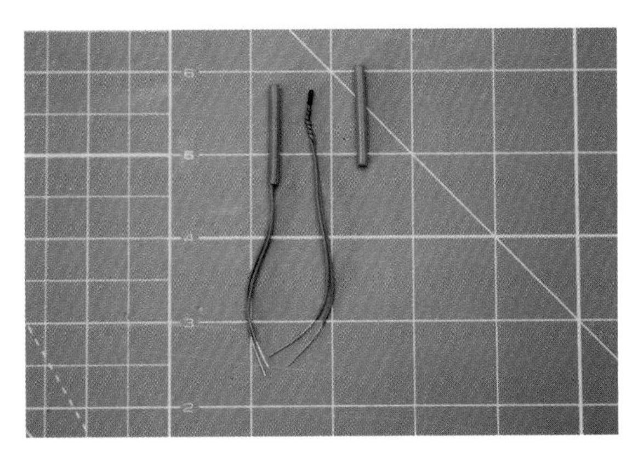

Figure 5-21. *Quest igniters look different, but do the same job. Remove and discard the red protection sleeve just before installing the igniter.*

Insert the igniter all the way into the motor so the tip touches the black powder. Add a piece of tape across the motor to hold the igniter in place. (You can use tape with an Estes igniter too, if you lose the plastic plug.) An alternative is to stuff a small piece of recovery wadding into the nozzle to hold the igniter in place.

Figure 5-22. *Use tape to secure the igniter in place.*

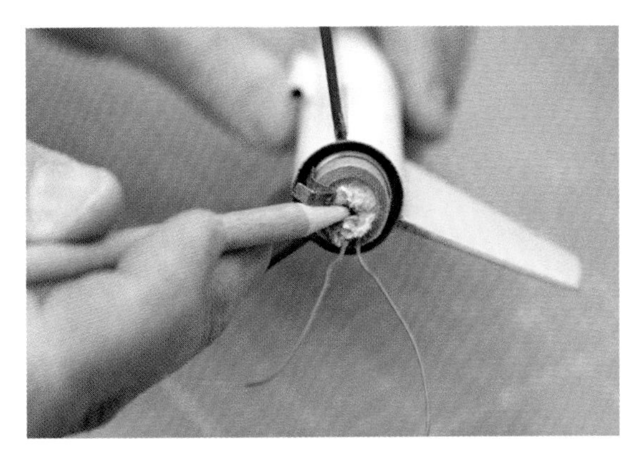

Figure 5-23. *You can also stuff recovery wadding or dog barf into the nozzle to secure the igniter.*

Flight Preparation Checklist

You're not going to reread this entire preparation description each time you get a rocket ready for flight, but you can and should run through a checklist as you work. Here's a flight preparation checklist for Juno.

Flight Preparation Checklist

1. Install the motor.
2. Insert recovery wadding.
3. Check the shock cord and parachute for wear or loose knots.
4. Attach the parachute and shock cord.
5. Insert the parachute and shock cord; make sure they are not too tight.
6. Insert the igniter.

The Launch

There is a formal check-in procedure at most large launches and many club launches. The RSO or a designated official does a quick safety check of the rocket. The RSO Safety Checklist shows what the RSO will be looking for. There may be local variations based on club experience or local flying conditions.

The RSO Safety Checklist

1. Are the fins securely attached, with no obvious cracks, especially at the fin root?
2. Is the launch lug securely fastened and parallel to the launch direction?
3. Push on the rocket motor to make sure it will not "fly through" the model.
4. Pull on the rocket motor to make sure it is restrained in the model.
5. Check nose cone fit. It should not fall out under its own weight or be too tight.
6. Is the recovery wadding there? Is it flameproof?
7. Does the rocket look stable?
8. Is the motor powerful enough for the rocket?

It's time to load the rocket onto the launcher for the launch. I will assume you are using the mono launcher and flying with a small group of friends. There are some obvious changes in procedure if you are flying in a club setting—and if the differences are not obvious, the LCO or a club member can help.

The launcher is set up, the launch cord extends back to the LCO area, and your rocket is ready. Start by clearing all nonessential people from the launch area. No one should be within 15 feet of the launch rail unless they are helping you or supervising.

Approach the launcher and remove the safety cap from the launch rod. Slide your rocket onto the pad, then immediately replace the safety cap. From this point forward, absolutely nothing except the hand of the person who removes the safety cap before launch should be over the top of the launch rod.

Launch Rod Safety

The safety cap can turn a serious injury into a minor mishap. You don't want to slip or stumble and impale yourself on the launch rod. Just imagine, for one moment, tripping and hitting the launch rod with your eye. Seriously, take a moment. Imagine it.

Now you'll never forget to put the safety cap in place!

Is there any wind? If it's a steady wind, you might want to angle the launch rod into the wind by 10 or 15 degrees. The rocket will fly high and into the direction of the wind, then get carried back toward the launcher after the parachute ejects. It's both fun and practical to see how close to the launch pad you can get the rocket to land. Our club even has an unofficial "closest to the pad" contest at each launch.

The igniter wires should not touch the blast deflector plate. If they do, raise the rocket slightly and use a piece of tape wrapped around the launch rod or a clothespin to hold the rocket off of the blast deflector plate. Sliding an old motor case with the nozzle removed over the launch rod before the rocket is placed on the pad works well, too.

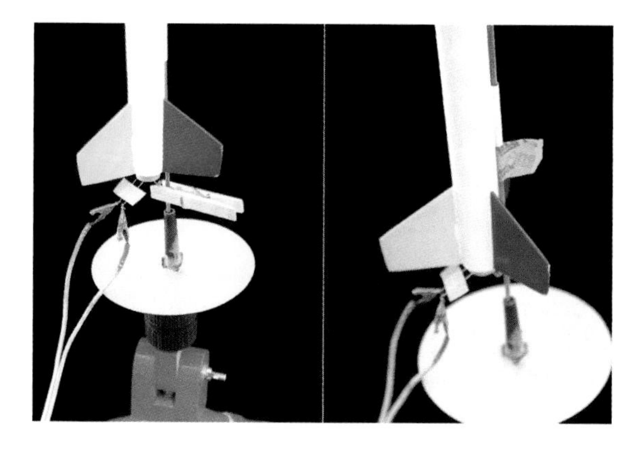

Figure 5-24. *If needed, use a clothespin or tape to support the rocket so the launch clips do not touch the blast deflector plate.*

Check the igniter clips. They tend to get dirty after a few launches. Clean them off with a piece of sandpaper if they look blackened.

Clip one igniter clip to each end of the exposed igniter wire. Wrap excess wire around the clip.

If anyone came into the launch area with you, clear them out of the launch area. Remove the launch key and return to the launch pad.

Insert the launch key. The continuity light should come on. If not, remove the launch key, replace it on the launch rod, and check for problems. Did an igniter clip come loose? Are the launch clips too dirty? Are the batteries in the launcher dead?

Check the skies for low-flying aircraft or clouds low enough that the rocket might go into the clouds.

Alert the spectators. Check to make sure everyone is standing up and facing the rocket. Yes, you can tell Mom to stop talking to her friend and pay attention, or leave the launch area. If you're the RSO, it's your duty.

In a voice loud enough for everyone to hear, count down from five. When you get to zero, press and hold the launch button. The rocket may take a second or two to launch. Don't let up on the launch button for three or four seconds, or until you hear the rocket lift off.

Immediately remove the launch key from the launch controller. I know, you're watching the rocket flight. That's good! You can remove the launch key by feel. Watch the rocket until it lands, then go to the launcher and place the safety cap back on the launch rod.

Now you can go recover the rocket.

Recovery

Hopefully your rocket recovery is nice and simple, and the rocket came down close to the pad. Don't try to catch it! Juno isn't likely to hurt you when it lands, but you could easily stumble or grab too hard and hurt the rocket. Let it land, and then pick it up.

Unless you are on a recovery team or the flier of a rocket says it's OK, you should not recover another flier's rocket. There's a lot to learn from a recovery. Moving the rocket is like contaminating a crime scene. You want to leave everything as is so the flier can analyze what went right and what went wrong with the flight.

What if your rocket lands in a tree, on top of a building, or on a power line? It may be safe to recover the rocket from a tree or building, but don't put yourself at risk. A rocket on a power line, though, is lost. Only the power company should recover a rocket from a power line.

Be sure to notify the power company if your rocket lands on a power line. They might get it back for you. They might also remove the rocket to keep it from attracting curious people who might be tempted to try to get the rocket down themselves. If nothing else, they may want to visually check the rocket to make sure the weight of the rocket will not cause structural problems with the power line, or to make sure the rocket is not in danger of causing a short.

Flight Checklist

1. Clear spectators from the launch zone.
2. Remove the launch key from the launch rail, slide the rocket onto the launch rail, and replace the launch key on the launch rail.
3. Attach the igniter clips.
4. Get the launch key from the pad.
5. Insert the launch key. Make sure the continuity light is on.
6. Check for clear skies—no aircraft or low clouds.
7. Make sure spectators are alerted. Count down from five and launch.
8. Remove the launch key and place it on the launch rod.
9. Recover the rocket.

Figure 5-25. *The author recovers a rocket from a rocket-eating tree using a long extendable window-washing pole. Of course, you should not use a pole to recover a rocket from a power line. Hook the parachute with the pole and lift the rocket off.*

Power Lines Are Dangerous!

There have been four deaths associated with model rocketry. In each case, the death was caused by someone trying to recover a rocket from a power line. Rockets on a power line are lost. Don't try to get them down.

Once the rocket is recovered, return to the launch area. You may want to repack it for another flight, or you may have more rockets to fly.

Be sure to pick up any trash before you leave the site. You may not be able to find the recovery wadding, but it will disintegrate in the first rain. The plastic plug that holds the igniter in place is often hard to find, too, and may no longer be the same shape it started in! Make an extra effort to find the igniter and put it in the trash. That's especially important on an athletic field or in a park, where someone might step on the igniter wire or fall on it, injuring themselves on the wire. The igniter is usually pretty easy to find, since it's attached to the igniter clips. It's usually in two pieces after the launch, though.

Whenever possible, don't leave anything behind at the launch site, especially igniter wires. Not only is it polite, but it makes it more likely you will be invited back to use the same field again.

Compressed Air Rockets

6

Figure 6-1. *Compressed air rockets are fast and easy to build, and appropriate for all ages.*

Compressed air rockets use a puff of pressurized air to blast a rocket from a tube. As I mentioned in Chapter 1, some purists may feel that compressed air rockets are not really rockets at all, but these newcomers to the model rocketry scene are really gaining in popularity. Why?

First, compressed air rockets are very fast and easy to build, and can be built by anyone old enough to handle tape and scissors. A hot glue gun is the tool of choice for gluing the parts together, but you can even use tape instead of glue to make the rockets easy to build with younger kids.

Compressed air rockets are fun. Setup time for a launch is less than a minute if you have a source of compressed air, like the emergency tire pump you will see in the launch section. The short time from construction to launch is great for anyone, but especially for kids who might not yet have the attention span for solid propellant rockets.

All of the science of aerodynamics, stability, and tracking applies just as much to air rockets as to solid propellant rockets. The principles of center of pressure and gravity still apply. You can carry payloads, launch gliders, or track the altitude of air rockets using the same techniques used for solid propellant rockets.

It's also easier to get ignorant administrators to approve compressed air rockets. Let's face it; schools and clubs sometimes have trouble getting parents and administrators on board with solid propellant rockets, despite their fantastic safety record. While they deserve the same level of safety precautions as solid propellant rockets, air rockets just don't sound as scary as solid propellant rockets to someone asked to approve an activity.

Air rockets are getting a lot better, too, which is making them more popular with traditional model rocket fans. I've seen air rocket launchers and rockets show up at our club launches, taking flight right alongside massive J- and K-powered rockets.

My personal experience with air rockets is pretty similar to that of a lot of people who started with solid propellant rockets. I really didn't think that much of them at first, but air rockets have become my preferred choice for parties and any other large group activities where people come to a launch with no preparation to build and fly a rocket. And after sending air rockets to 244 feet in the air (roughly the same height reached by a solid propellant rocket on an A motor) for essentially no cost, I've become a fan of air rockets for any activity where I need to launch a rocket many times. Finally, an air rocket launcher is a perfect way to launch a glider, as you will see in Chapter 21.

Being Safe Under Pressure

The air rockets from this chapter and the water rockets in Chapter 11 both use air pressure rather than burning chemicals for propulsion. This might cause you to think you don't need safety precautions with air rockets. That's simply not true. Like solid propellant rockets, air rockets are safe if built and used properly. They do not pose a fire hazard. You are, however, working with high-pressure systems, and that poses new safety concerns.

There are two safety issues with air rockets, and both are addressed by the design of the launcher and the launch procedures described in this chapter.

The first is impact. The air rockets in this chapter are light and made from soft materials, but they still leave the launch pad with considerable speed. You should not point an air rocket at anything, nor should you be near the air rocket launcher when it is under pressure.

The second is the launcher or rocket bursting under pressure. The launcher has been overdesigned to prevent ruptures, and I have never heard of a report of a launcher like this one bursting. Still, we always stay at least 15 feet form the launcher if it is under pressure. I have seen rockets burst under pressure. The most common failure is a nose cone that is blown off. The second most common failure is a rupture in the body tube. Neither is dangerous if you are more than a few feet from the launcher, since the rockets are made from lightweight, soft materials like paper, foam, and tape. There really isn't much danger to your skin, even if you are very close to the rocket. Even a small piece of paper could hurt your eyes, though, so again, we stay away from the launcher when it is under pressure. Even if the rocket bursts, the fragments are too large and light to fly more than a few feet from the launcher—and you are likely to blink when you hear the pop, anyway. As long as you observe the 15-foot minimum launch distance, you should be completely safe from anything that might go wrong on the launch pad.

Building an Air Rocket Launcher

There are a lot of air rocket launchers out there. The essential feature in all of them is a way to deliver a large volume of air in a short period of time. That usually means that the launcher has a pressure chamber to hold air and some way to pump air into that pressure chamber. The pressure chamber must be able to hold the air under reasonably high pressure, usually 50 to 75 pounds per square inch (psi). When it's time to launch the rocket, the launcher needs some way to deliver the air from the pressure chamber in a short period of time, dumping it into the rocket to fling it from the launcher. The air is typically sent through a tube that also serves as a launch rod. The air rocket slides over the tube, and is propelled a few dozen to a few hundred feet into the air when launched.

You see all of these features in Figure 6-2. The big, mostly yellow tube is the pressure chamber. Made from PVC pipe and wrapped in duck tape for extra strength, the pressure chamber is filled from the clear plastic hose to the left using any air source that can pump up a tire.

The launch controller is an electrically operated system. When the red button atop the controller is pressed, the green sprinkler valve opens, delivering all of the air from the pressure chamber to the rocket through the white PVC pipe. That's where the rocket sits—but not for long! With a soft pop, the rocket will fly high into the air. The air rockets are soft enough and light enough that we can allow them to return to earth ballistically, although you'll see an air rocket glider in Chapter 21.

This classic launcher was developed by Rick Schertle and first seen in *MAKE* magazine, volume 15. It is the nucleus for a small but growing community of air rocket enthusiasts. You can join that community at *http://airrocketworks.com*.

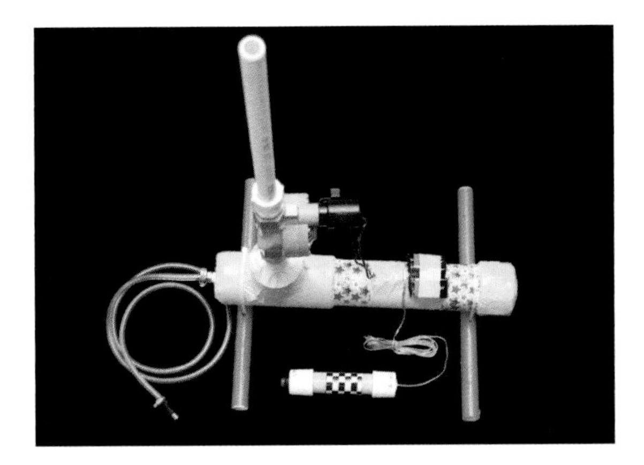

Figure 6-2. *The air rocket launcher uses any air source that can pump up a tire.*

Our launcher is built from common PVC and sprinkler parts available at local hardware stores (see Table 6-1). You can also find the parts online if your local hardware store isn't particularly well stocked—Table 6-1 lists some mail-order locations, and you can find details about these outlets in Appendix B.

Shop around a bit. Prices vary quite a lot, and availability can change from time to time, too. A list of additional supplies you'll need to build your launcher is included in Table 6-2.

Table 6-1. Parts list

Part	Description	Part number
2″ x 1/2″ PVC bushing	The smooth outer shoulder glues into the 2″ PVC pressure chamber. The threaded 1/2″ hole holds the brass parts that connect to the hose.	Lowe's 51013
1/2″ x 1/4″ brass bushing	Joins the hose barb to the 2″ x 1/2″ PVC bushing.	Lowe's 127954
1/4″ brass hose barb	Fastens the pressure hose to the 1/2″ x 1/4″ brass bushing.	Lowe's 74486
2″ 90° PVC tee	This is the part of the pressure chamber that joins the main pressure chamber, pressure cap, and launch assembly.	Lowe's 23908
2″ schedule 40 PVC	This forms the main pressure chamber. PVC pipe comes in several thicknesses; be sure to get schedule 40 pipe.	Lowe's 23832
2″ schedule 40 PVC cap	Forms the end of the pressure chamber.	Lowe's 23900
2″ x 3/4″ schedule 40 PVC slip bushing	Attaches the pressure chamber to the launch assembly.	Lasco 437248
Orbit sprinkler valve	Other sprinkler valves will work, but check the fittings. This is the model shown.	Lowe's 50161
3/4″ PVC adapter (2)	This adapter slip-fits over a pipe on one end and threads into the sprinkler valve on the other. Get two.	Lowe's 23856
3/4″ x 1/2″ PVC slip adapter	This adapter fits over a 1/2″ PVC pipe on one end, and a 3/4″ PVC pipe on the other end.	Lasco 437101
3/4″ schedule 40 PVC pipe	You will need two pieces, one 3″ long and one 4″ long.	Lowe's 23971
1/2″ schedule 40 PVC pipe	You will need three 13″ pieces and one or more 10″ pieces. The 10″ pieces are used as a tool when building the rockets. You really only need one, but it's nice to have several if a group of people will build the rockets.	Lowe's 23830
3/4″ schedule 40 end cap (2)	Get two 3/4″ PVC end caps.	Lowe's 23896
Pushbutton switch	This is the launch button.	Radio Shack 275-646
Wire	You will need 17 feet of two-conductor 24-gauge stranded copper wire.	Radio Shack 278-1301
9-volt battery holders (2)	Clips to the top of a 9-volt battery.	Radio Shack 270-324
9-volt batteries (2)	These are standard 9-volt batteries.	
Schrader valve	This is a standard tire valve, available from many auto supply stores.	Amazon Slime 2080-A Rubber Tire Valve
Hose clamp (2)	You will need two worm gear hose clamps, one for each end of the clear plastic fill tube.	Amazon Precision Brand M4P Micro Seal, Miniature Partial Stainless Worm Gear Hose Clamp, 7/32"–5/8″
Hose	You will need 15 feet of 3/8″ clear vinyl hose.	Lowe's 443064
Zip ties	You will need two 12″ zip ties to attach the supports to the pressure chamber.	Amazon 100 Pack 12″ Clear Nylon Cable Ties

Table 6-2. Additional materials and supplies

Part	Description	Part number
Teflon tape	Available from hardware stores, Teflon tape is a thin, white tape with no sticky side. It's used on threaded high-pressure joints.	Lowe's 25010
PVC cement and primer	PVC is joined using this two-step primer and cement.	Lowe's 150887
Duck tape	The entire system will be wrapped in duck tape. Get some fun colors!	Lowe's 503443

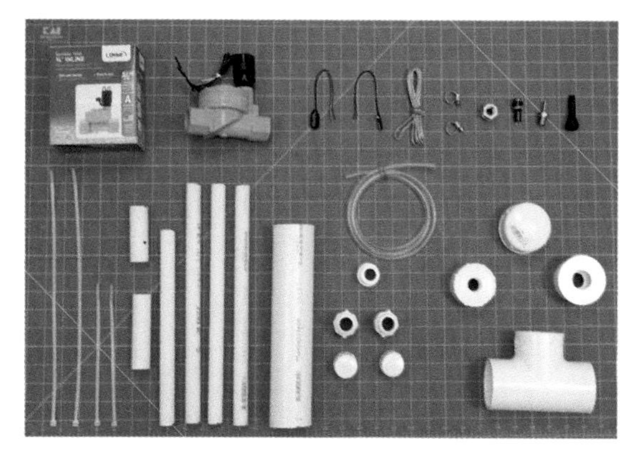

Figure 6-3. *Visual parts list for the launcher.*

Figure 6-4. *The older Maker Shed kit.*

Equivalent Kit

The design shown in this chapter is a slight modification of a kit that used to be available through Maker Shed. This kit is currently being revamped to make it simpler to build and to make the pressure components from metal parts rather than PVC. While the design shown in this chapter is a great one, you should check out the new kit once it is available. I've used prototypes of the new launcher, and it will be much easier to build and maintain. Getting the kit also means you don't have to visit multiple stores, possibly paying multiple mail order shipping charges, to get all of the right parts. The new kit will be announced on *http://airrocketworks.com*, among other places.

Figure 6-5. *Here is a prototype of the new air rocket launcher.*

Estes Industries has a line of air rockets that use a flexible bulb to deliver air to launch a rocket. You stomp on the bulb to deliver a blast of air to launch the rocket. These are popularly known as *stomp rockets*, although that particular name is a registered trademark of another company that has a nice line of air rockets, called Stomp Rocket®.

Toy stores frequently carry stomp rockets and hand pump rockets from a variety of manufacturers.

The bane of any high-pressure system is a leak. Leaks make it difficult or impossible to pressurize the system, and quickly rob a pressurized system of the power we need to launch our rockets. Preventing leaks starts with the first parts we assemble on the launcher. Do it right, and the launcher will go together with no leaks. Rush ahead, or try to make do without the proper glue or tape, and the system will leak.

All threaded parts, whether they are metal, PVC, or a combination, must be sealed with Teflon tape. This is frequently called plumber's tape, since plumbers need to keep threaded joints from leaking under pressure, too. Teflon tape is a thin, white, stretchy tape with no sticky side. You can find it at just about any hardware store.

Wrap a little more than one turn of the tape around the brass bushing that screws into the 2" x 1/2" threaded PVC bushing—just enough to overlap the starting point. Wrap the tape so screwing the part in tightens the tape rather than unwinding it. Cut the Teflon tape smoothly with scissors or a hobby knife. It's easy enough to just pull the tape apart, but that will leave wrinkles that can cause leaks. Take the time to do a neat job.

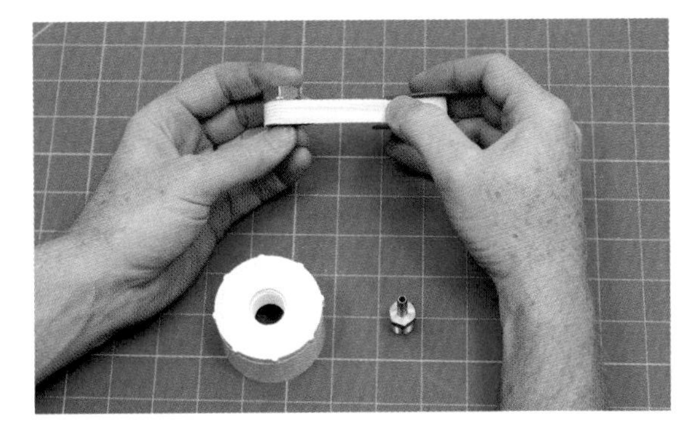

Figure 6-6. *Wrap Teflon tape around all threaded joints before assembly.*

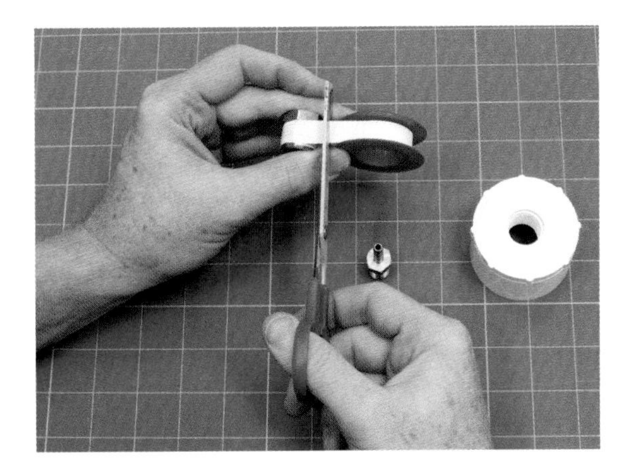

Figure 6-7. *Always cut Teflon tape; never tear it. Cutting leaves a nice, smooth end that seals well. Tearing leaves thick spots that might leak.*

Screw the brass bushing into the PVC bushing. Use a wrench to tighten the parts. If you have to disassemble the parts for any reason, remove the old Teflon tape and apply a new layer.

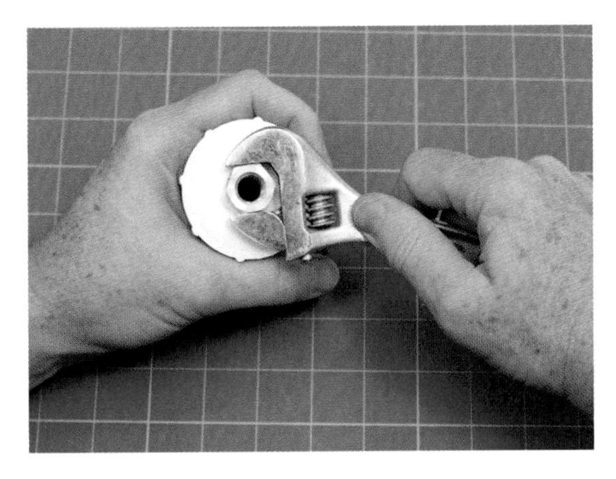

Figure 6-8. *Threaded parts need to be tight. Use a wrench to tighten them.*

Finish the pressure cap by installing the brass hose barb. As with the bushings, use Teflon tape and tighten the parts with a wrench.

Figure 6-9. *The completed pressure cap.*

PVC is a marvelously cheap, adaptable construction material. You first saw it in Chapter 4 when we built the mono launcher. If you are cutting your own PVC parts, refer to "Building the Base" on page 67 for some tips.

Begin by cutting the PVC parts. The cut pieces are shown in Figure 6-3. You will need three sizes of PVC for this project. In all cases, you should buy schedule 40 PVC. Cut three 13"-long pieces and one 10"-long piece from 1/2"-diameter PVC. Cut a 3"-long piece and a 4"-long piece from 3/4" PVC. Finally, cut a 10"-long piece from 2"-diameter PVC.

The 1/2"-diameter pipe and 3/4"-diameter pipe are pretty common sizes that you can use in lots of projects. The 3/4"-diameter PVC was also used in the mono launcher, for example. The 2"-diameter PVC is a bit unusual, though. PVC is not expensive, so you can buy one of the 10-foot long sections sold at most hardware stores. Ask at the store first, though—many hardware stores have a stack of scraps, and you might be able to find a piece that's pretty close to the right length. If nothing else, getting a shorter piece will save you from storing 9 feet of unused PVC pipe for several years!

As you'll recall, joining PVC parts is a three-step process. The very first step is always to dry fit the parts, since you won't get a second chance once the primer and cement are applied. Once you are satisfied with the fit, apply purple primer to both parts. After a few seconds, apply the cement, and then join the parts quickly, twisting to spread the cement evenly.

The purple primer stains virtually everything it touches. Wear gloves unless you want that cool plumber's look of purple fingers. Don't worry, though; it wears off in a day or two. The cement smells bad, and can't do you any good if you breathe heavy concentrations of it. Make sure you set up your work area in a well-ventilated place, like a garage with the garage door open.

Glue the pressure cap assembly into one end of the 2" PVC tee as shown in Figures 6-10 and 6-11. Be sure to glue it into one of the two ends, and not into the top.

Figure 6-10. *Apply purple primer to one end of the PVC tee and the outside of the pressure cap bushing.*

Figure 6-11. *Apply PVC cement over the primer in the PVC tee, then twist the pressure cap into place.*

The bulk of the pressure chamber is built from a 10″ length of 2″ schedule 40 PVC pipe, capped on the far end. This gives a nice, large volume of air for a fast launch of our rockets. Glue the pipe into the opposite end of the 2″ PVC tee, then glue a 2″ PVC cap to the other end of the pressure chamber.

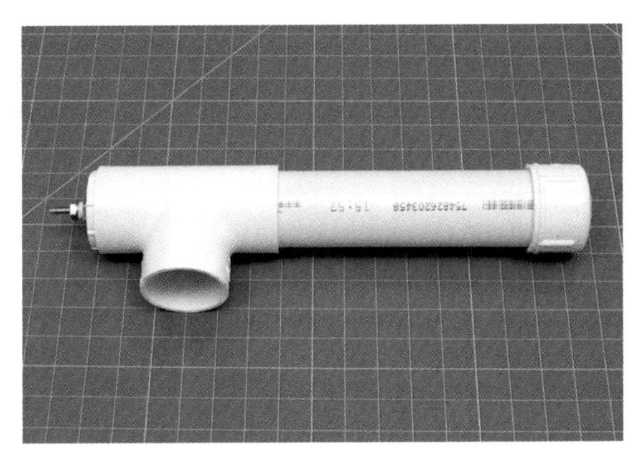

Figure 6-12. *Glue the 10″-long 2″ PVC pipe and cap in place.*

Glue a 2″ x 3/4″ PVC slip bushing into the top opening of the tee, completing the pressure chamber. The slip bushing has a hole in the middle for a 3/4″ PVC pipe and slips into the end of the 2″ PVC tee. Be careful if you are buying your own parts. Most bushings are threaded in the middle; you want one that is smooth. The smooth connection is called a *slip connection*.

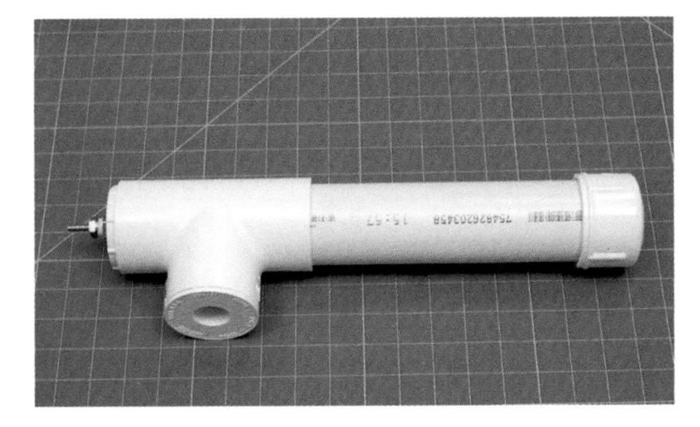

Figure 6-13. *With the slip bushing in place, the pressure chamber is complete.*

Sprinkler systems use electric valves designed to work with pressures up to 75 psi. They snap open when you apply a voltage. The Orbit valve shown here is just one of the many that will work. It is designed for 24-volt sprinkler systems, but testing shows we can use it with two 9-volt batteries, too.

Using Teflon tape and pliers, attach the two 3/4″ PVC threaded adapters to the ends of the sprinkler valve.

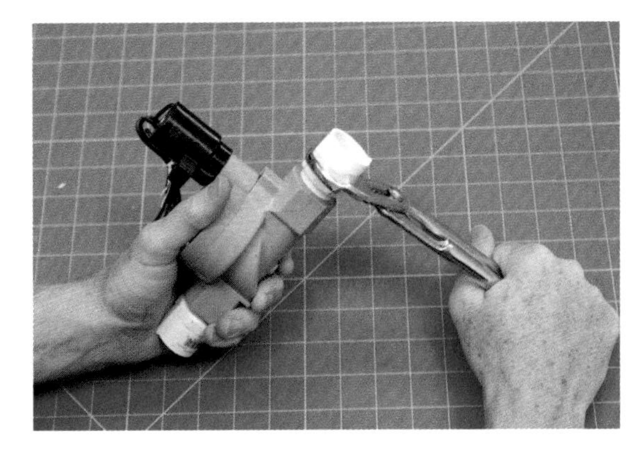

Figure 6-14. *Attach the two PVC adapters to the sprinkler valve.*

The valve has an "in" side and an "out" side. The direction of flow is marked with an arrow. The "in" side gets attached to the pressure chamber, and the "out" side is attached to the rocket launch tube. Glue a 3″ piece of 3/4″ PVC pipe to the adapter screwed into the "in" side of the valve.

Figure 6-15. *Glue a 3″-long piece of 3/4″ PVC pipe to the "in" side of the valve.*

The opposite side of the launch assembly attaches to a 13″ length of 1/2″ PVC pipe. This is the pipe the rocket will slide over. Use the 3/4″ to 1/2″ PVC adapter to fit the launch tube into the larger 3/4″ PVC fitting. Glue the launch tube and adapter into place on the "out" side of the valve.

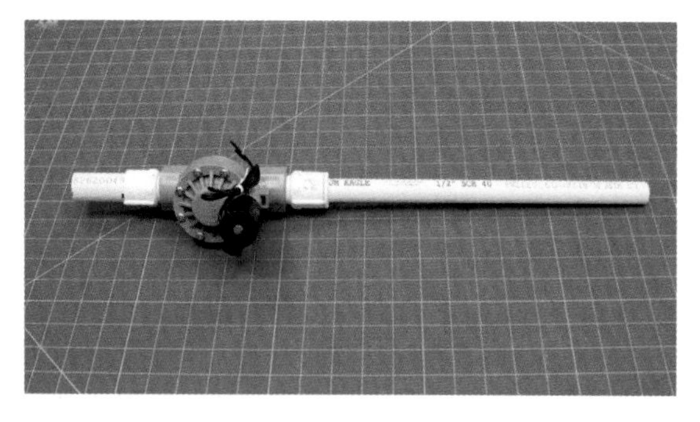

Figure 6-16. *Glue the launch tube into the 1/2" x 3/4" adapter, then glue the adapter into the 3/4" fitting on the "out" side of the valve.*

Make sure the valve is turned in the direction shown in Figure 6-17, and glue the 3/4" pipe attached to the "in" side of the launch assembly into the top of the pressure chamber assembly.

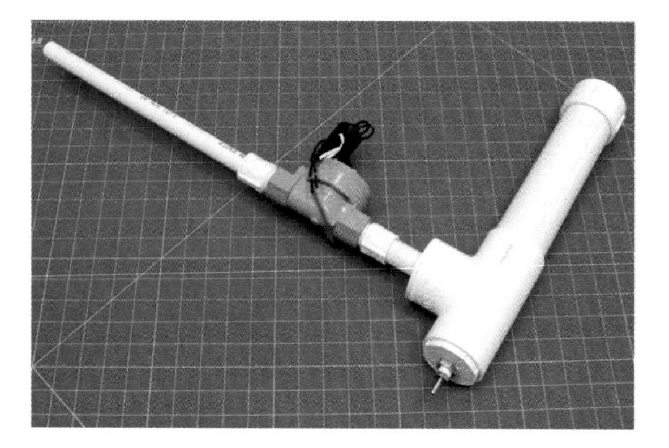

Figure 6-17. *Glue the launch assembly to the pressure chamber assembly.*

Build the stand for the launcher from two of the 13"-long 1/2"-diameter PVC pipes. Drill two 1/4"-diameter holes 2 3/4" apart, centered in the pipes. That makes each hole 5 1/8" from the end of the pipe.

These holes will eventually be threaded with the long zip ties. Don't fasten them yet, though.

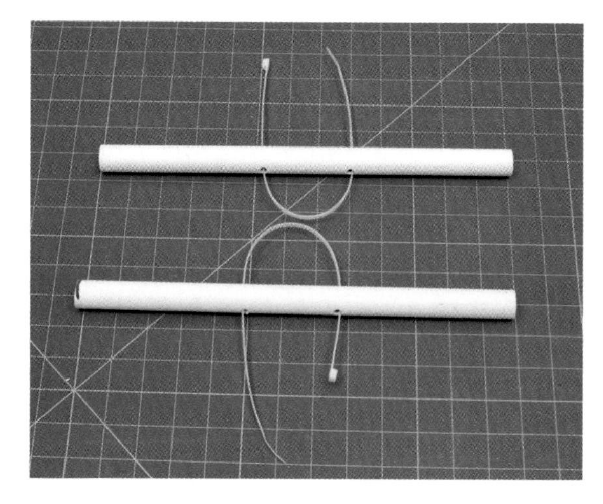

Figure 6-18. *Drill two 1/4" holes in the remaining 13" PVC pipes. The holes are centered along the pipe, 2 3/4" apart.*

Two 3/4" end caps will form the ends of the launch button assembly. Drill a 1/8" hole in one of the caps. This should be large enough for the launch wire to fit through the hole. Drill a 1/2" hole in the other cap. This should fit the pushbutton launch switch.

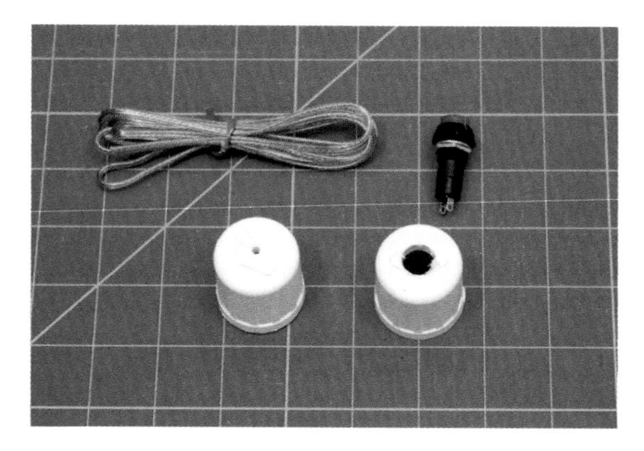

Figure 6-19. *Drill a 1/8" hole in one end cap and a 1/2" hole in the other.*

The rocket is fired using an electrical ignition system that allows the launch controller to stand a bit away from the launcher as the rocket lifts off. The launch controller is made from a short length of 3/4" PVC pipe capped at each end. One end has a launch button, and the launch wire runs out the other end, connecting in series with two 9-volt batteries and the sprinkler valve. When the launch button is pressed, the valve opens. If the pressure chamber has been filled and there is a rocket on the launch tube, the rocket will be popped into the air.

Solder the launch wire to the launch button. Use electrical tape or heat shrink tubing to insulate the connections.

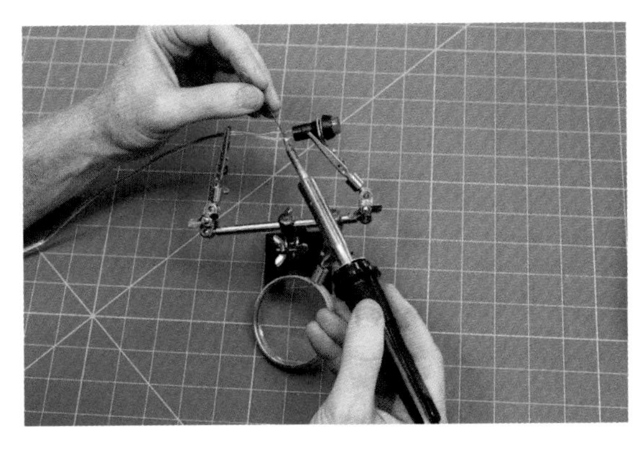

Figure 6-20. *Solder the launch wire to the launch switch.*

The launch button has a nut on the bottom. Remove that nut, slip the wire and launch button through the 1/2" hole in one of the end caps, and refasten the nut. It's a tight fit—I found that a pencil eraser was an effective way to attach the nut.

Tie a knot in the wire so that if anyone tugs on the wire, the knot will be pulled against the hole in the bottom end cap, and they will not pull the soldered connections off the launch button. Tying the knot about 5" from the soldered connection works well.

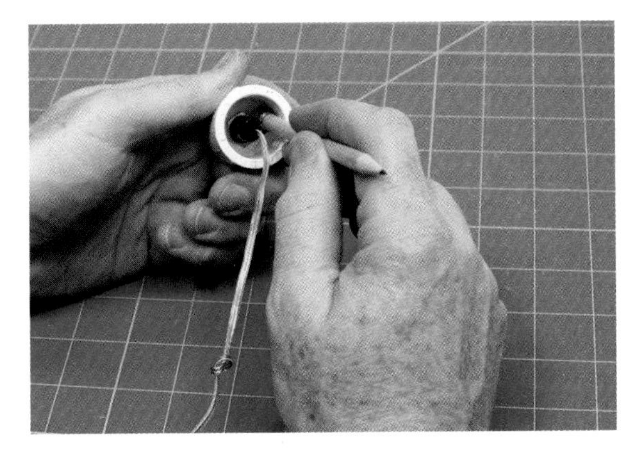

Figure 6-21. *Install the launch button in one end cap. The knot in the wire will pull up against the hole in the bottom end cap, so tugging the wire doesn't put strain on the soldered connections.*

Thread the wire through the 4" length of 3/4" PVC pipe and the other end cap, then press the parts together. Don't glue end caps to the pipe. If something breaks, you'll be able to pull the parts apart to fix them.

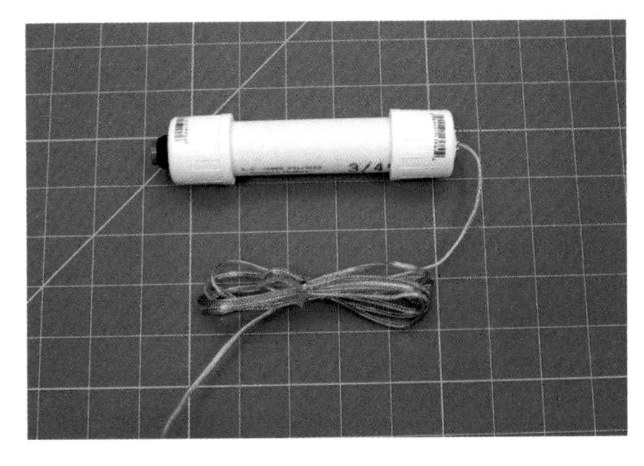

Figure 6-22. *Assemble the PVC pipe and the other end cap to complete the launch controller. Do not glue the end caps.*

What? No Safety Key?

The air rocket launcher doesn't have a safety key. Why?

Air rockets are a bit different from electrically ignited solid propellant rockets. The heavy work of pumping up the rocket by hand, or the noise from a compressor, is perfectly adequate to let everyone know the rocket is being prepared for flight. That's one purpose of the safety key in a solid propellant launcher—it's easy to forget you left it armed.

The other purpose of the safety key in a solid propellant launcher is to give you some way of disarming the rocket before approaching the launcher, and being sure there is no chance of accidentally or mistakenly pushing the launch button. We do that with an air rocket by bleeding off the air pressure before approaching the rocket.

Solder the wires for the battery holders, valve, and launch controller. Solder one red wire and one black wire from the battery holders together. This connects the batteries in series, so they deliver a total of 18 volts to the valve when the launch button is pressed. Solder one of the remaining battery wires—it doesn't matter which one—to one of the leads from the sprinkler valve. Cut a 17-foot length of two-conductor 18-gauge wire to use as a launch wire. Solder the other sprinkler valve wire to one of the leads from the launch wire, and the other lead from the launch wire to the remaining battery wire.

Use heat shrink tubing or electrical tape to protect the exposed wires.

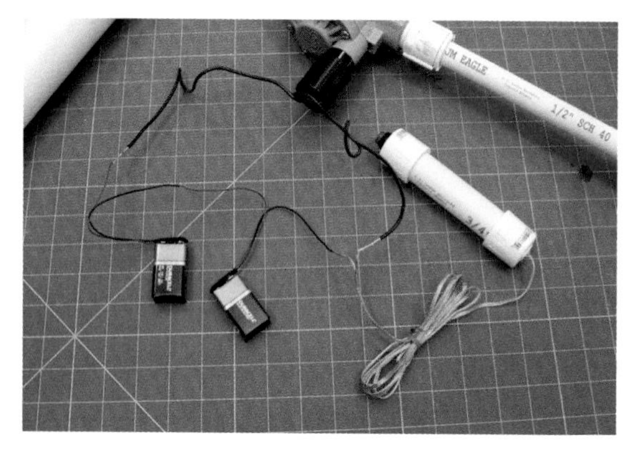

Figure 6-23. *Solder the wires for the valve, batteries, and launch wire as shown.*

At this point, pressing the launch button should operate the valve. You will hear a soft click in the valve when the launch button is pressed. If not, stop and double-check all of the wiring, and make sure the batteries are fresh.

Wrap the pressure chamber and fittings up to the valve in duck tape. This forms a protective layer that will tend to hold pieces together rather than letting them fly apart if any part of the system ruptures under pressure. The components are designed for about three times the pressure we will be using, so a rupture is extremely unlikely, but never pass up a chance for an extra layer of safety. Besides, duck tape is available in all sorts of fun colors, and using colored tape is simpler than painting the launcher!

You can wrap the two 13" launcher stands and the launch controller in duck tape for decoration, too.

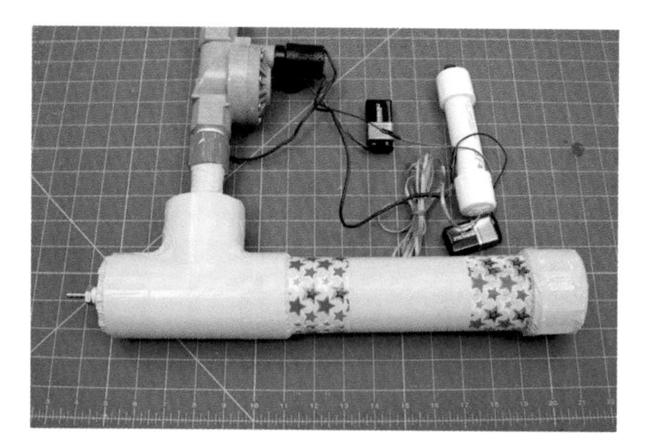

Figure 6-24. *Wrap all of the high-pressure parts in duck tape for added safety.*

Fasten the batteries and extra wire to the top of the pressure chamber with a little extra duck tape. Put some of the launch wire under the duck tape, too, to protect the soldered connections from occasional tugs.

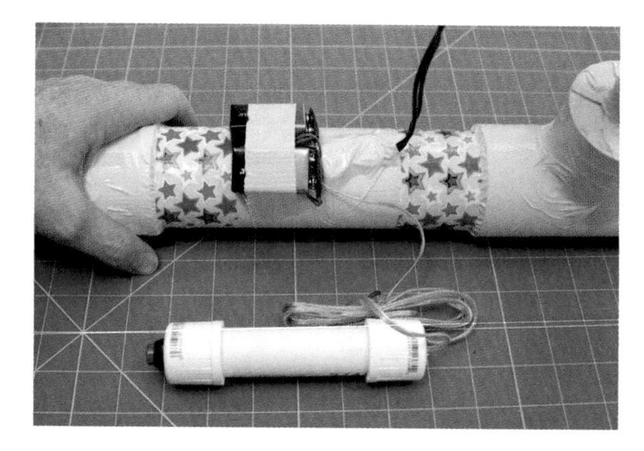

Figure 6-25. *Tape the battery in place. Tape extra wire to the top of the pressure chamber.*

The valve used on most tires is called a Schrader valve. The Schrader valve from the parts list is covered in a thick rubber coating. We don't want the coating. Use a hobby knife to cut it off. Take your time and be precise near the end. You need to cut the rubber until the underlying brass part is exposed. Trim all ridges until the fitting is smooth so the valve does not leak when inserted in the plastic hose.

Figure 6-26. *Remove the thick rubber coating from the tire valve.*

The valve is a one-way valve designed to allow you to pump air into a tire but not let the air back out. Our use is a little different. We need to be able to dump air from the system if there is a problem. The easiest way to do that is to convert the tire valve from a one-way valve to a connector that allows air to flow both ways. The valve portion is screwed into the housing, and there are special tools for unscrewing it. If you don't want to buy a special tool, and don't have anything else handy that can reach deep enough into the valve to get a good grip for unscrewing it, try pulling it out. Grab the pin in the center of the valve with a pair of needle-nose pliers and pull. You will have to pull very hard! Another alternative is to drill the pin out from the back with a 3/32″ drill bit.

Now the system will stay pressurized while the pump is attached, but you can easily depressurize the system by detaching the pump. An added advantage is that it is now impossible to store the launcher accidentally with pressure in the system. The pressure chamber will expel any extra air as soon as the pump is disconnected.

It's time for final assembly. Cut 15 feet of 3/8″ clear vinyl hose. Use the hose clamps to attach the plastic hose to the pressure chamber and tire valve. Use long zip ties to attach the support legs to either end of the pressure chamber. Your launcher is ready to go. It's time to build some rockets!

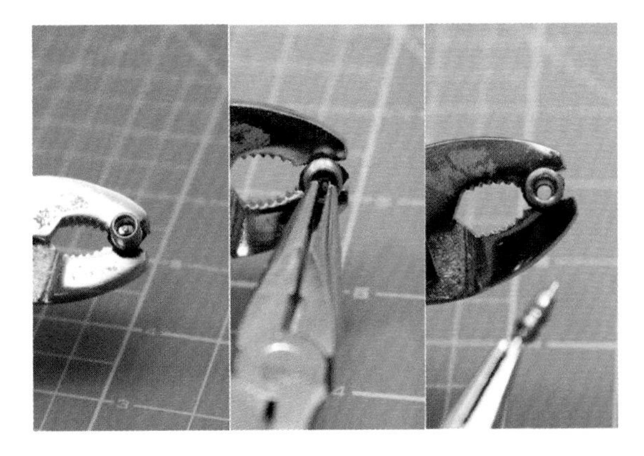

Figure 6-27. *Remove the pin from a Schrader valve by unscrewing it, pulling it out with a pair of needle-nose pliers, or drilling it out from the back with a 3/32″ drill bit.*

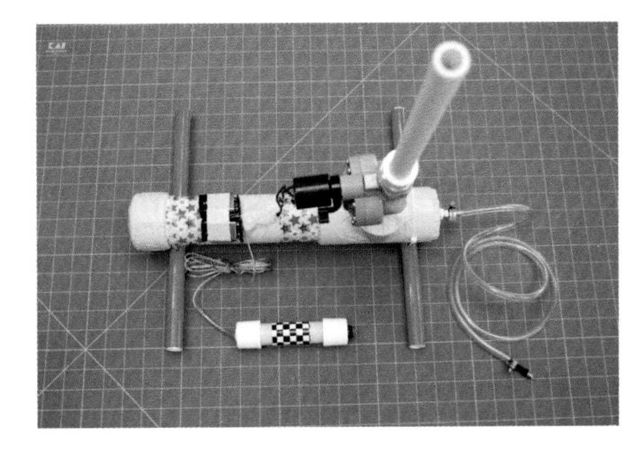

Figure 6-28. *The completed air rocket launcher.*

Building Ballistic Air Rockets

Let's learn to build some simple ballistic air rockets. Before we do, though, I want to point out something that may not be obvious at first glance. While a lot of this book may seem to deal with just solid propellant rockets, there are a lot of sections that have just as much to do with air rockets as any other kind of rocket. Stability, aerodynamics, and tracking are the same for any kind of model rocket. There is also a more sophisticated air rocket later in the book—check out Chapter 21, where you will find a rocket glider that is launched from the air rocket launcher.

There are a lot of ways to build air rockets, as an Internet search will show. They all work great. The air rockets we will create here are fast, cheap, and easy to build. They are also very colorful. These rockets are great at parties or other gatherings, because everyone there can quickly build and fly their own rocket.

Figure 6-29. *Air rockets are fast and easy to make—and very colorful!*

Don't get too hung up on the parts list (Table 6-3), since there are a lot of different but equally good ways to build the rocket. Read through the construction description to get some ideas for alternatives.

Table 6-3. Parts list

Part	Description
Card stock	The body tube of the rocket is made from 50 lb to 65 lb paper (card stock). It can be made from regular paper, too, but only if it is reinforced with duck tape. Get a package with various colors from the scrapbook section of your local craft store.
Duck tape or packing tape	You can get by without strong tape if you use card stock, but even with card stock, you can build a stronger rocket with duck tape.
Foam sheets, cardboard, or heavy pasteboard	You'll need something stiff for fins. Foam also works well for nose cones.
Cellophane tape	Great for initial fastening of body tubes or for sealing card-stock body tubes.
Hot glue gun	For attaching the fins. You can also do this with tape, if you don't have a hot glue gun.

Figure 6-30. *Basic construction materials are available from hobby and craft stores. Here you see foam sheets on the left, card stock on the right, cellophane tape, and several rolls of colorful duck tape. There are lots of possible substitutes; read through the construction details for some ideas.*

Air rockets consist of three basic parts, and one of those is more or less optional.

The body tube of the air rocket provides the basic structure that everything else is attached to. It's also the source of the thrust, and must stand up to 75 pounds per square inch of pressure from the air rocket launcher. That's enough pressure to turn a poorly constructed paper rocket into confetti, something I've done once or twice. It's also enough power to blow the nose off of a rocket if it is not attached well.

The fins provide stability. You need three or more fins for stable flight. They should be about two square inches each for a typical 9"-long body tube, and should be at or very near the bottom of the rocket.

Most air rockets also have a nose cone, but you can make a very adequate air rocket by just folding the end of the body tube over and taping it. That's something you will do anyway. It's pretty difficult to make a nose cone strong enough, and to attach it well enough, to withstand 75 psi. Instead, we'll build the rocket so the nose cone doesn't need to withstand the pressure. The nose cone is there for decoration and better aerodynamic performance.

Your basic tool for building an air rocket is a piece of PVC pipe left over from building the launcher. This is why you cut a 10"-long piece of 1/2" PCV pipe when you built the launcher. It wasn't used in the launcher, but you will use it now to make body tubes. For a rocket party, you will want to have about one of these forms for every 3–4 people. Use a piece of sandpaper or a grinder to round the ends slightly so it is easy to slide the paper on and off the tube.

You will use this guide both for making body tubes and as a fin guide. Use the fin guides in Figures 3-24 and 3-25 to mark the PVC tubes as if you were going to glue either three or four fins to the PVC. Permanent markers work well for this, and colored permanent markers are even better. You can mark the three-fin guides in one color and the four-fin guides in another.

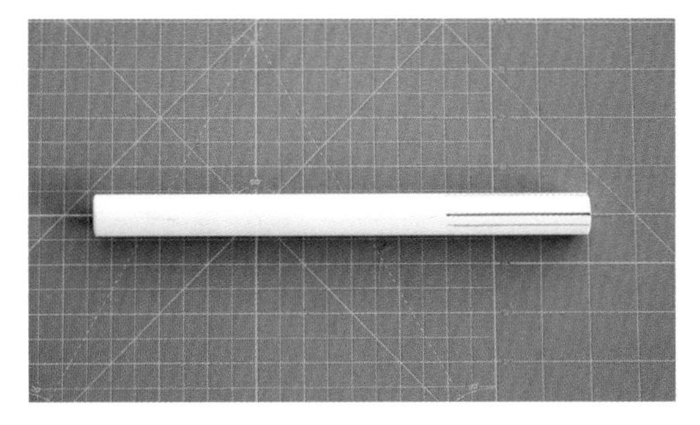

Figure 6-31. *The body tube tool is made from a 10" piece of 1/2" PVC pipe. Mark it for both three and four fins using different colors of permanent marker.*

With your body tube tool in hand, start constructing your rocket by cutting a piece of paper or card stock that is 4" x 9". If you use paper, you will want to cover it with duck tape or shipping tape, so the color won't matter. If you are using card stock, get some of the colorful heavy paper from the scrapbooking section of your local craft store. At alternative is to use 50 lb to 65 lb white paper from the office supply store and color it, print on it, or wrap it in duck tape.

Once the paper for the body tube is cut, put two small pieces of cellophane tape on one of the long ends, then turn it over so the tape is sticky side up. Begin rolling the body tube around the body tube tool, starting with the untaped side. Continue to roll until the paper hits the tape, which will stick to the paper to form the tube.

The ideal body tube will be loose enough to slip on and off the PVC tool easily, but will still be a fairly close fit. We want the rocket to get off of the pad easily, but we don't want to let too much air escape along the side.

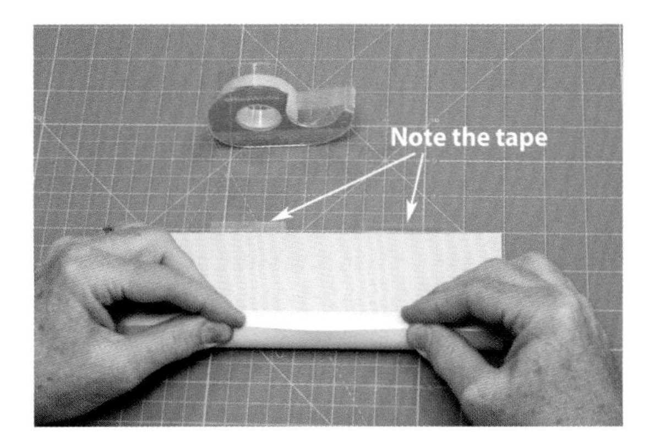

Figure 6-32. *Place two pieces of tape at one edge of the paper, and then turn it over so the tape is sticky side up. Roll the paper around the PVC tool to form the body tube.*

Leave the paper tube on the PVC body tube tool while you finish applying tape and crimp the end to form the nose bulkhead. For a card-stock rocket that will not be wrapped in duck tape, add additional short pieces of tape to the sides of the tube to close it completely. If you plan to use duck tape, now is the time. Duck tape is almost a necessity for thin paper rockets, but it's definitely optional for card-stock rockets. Using duck tape on a card-stock rocket will make the rocket stronger, though. It will stand up to a lot more flights, and won't be as prone to damage from damp grass.

One strategy for a party is to premake the body tubes. You will still need some PVC tools to attach the fins, but that won't take as much time, so you won't need as many tools.

Figure 6-33. *Body tubes are easy to make and can be very colorful!*

Closing the end of the body tube is probably the most critical step in the construction of the air rocket. This is what seals the end of the rocket so the 75 psi of pressure from the launcher powers the rocket into the air, rather than blowing the nose cone off of the top of the rocket!

Slip the body tube up so it sticks about 3/4″ off of the end of the body tube tool, then fold one end over with your thumb. Repeat the process from the other side, and then from the left and right, so you fold the edge over the end four times. This forms a nice, solid bulkhead at the end of the body tube.

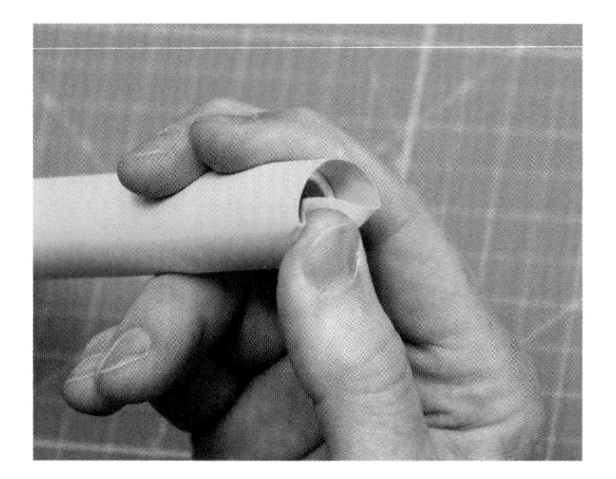

Figure 6-34. *Fold the body tube over the top of the body tube tool to form a solid cap at the end of the tube.*

Wrap a piece of cellophane tape about 5″ long across the top of the tube. Repeat this two or three more times from slightly different angles to form a nice, tight seal.

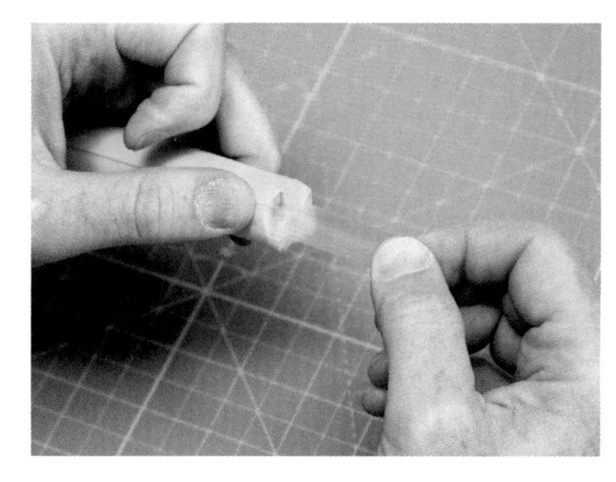

Figure 6-35. *Wrap several pieces of tape across the top of the rocket to seal the body tube shut.*

Finally, add one layer of tape all the way around the top of the tube. This strengthens the tape strips running across the top of the tube.

Figure 6-36. *Wrap one layer of tape over the ends of the tape that form the seal on the top of the rocket.*

My tool of choice for attaching fins and nose cones is a hot glue gun. It works well, because the glue sticks to duck tape, cellophane tape, paper, cardboard, and foam. You can tape the fins on, though. Use a strong tape, like duck tape or packing tape, and run a strip along both sides of the fin where it joins the body tube.

I like to use colored sheets of foam for fins. You can find these in craft and hobby stores. The sheets are easy to cut with scissors or a hobby knife. They are also flexible enough to fold over if the rocket happens to hit something. Balsa wood fins don't do a lot of damage if they hit something they are not supposed to hit. Foam fins do even less—and tend to survive the impact themselves, whereas balsa wood fins frequently break.

You can use just about any stiff material for fins, though, including heavy pasteboard, corrugated cardboard, or foam core.

Cut three or four fins, each about 2" x 2". Figure 6-37 shows a few possible fin patterns, but there is nothing magical about them. The rocket will fly well as long as the side of the fin that attaches to the body tube is flat, the fin is about two square inches in size, and all of the fins are the same shape. Trace three or four copies of one of the fins on a foam sheet using a pencil, then cut out the fins with a hobby knife or scissors—or just cut a couple of 2" squares from foam core and cut across the diagonal to form two triangular fins.

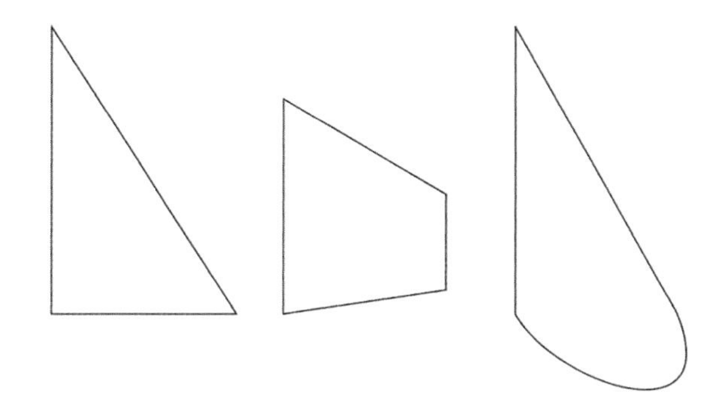

Figure 6-37. *Some suggested fin patterns.*

Once the fins are cut, slide the body tube up on the body tube tool to expose several inches of the fin-marking guide. Apply a bead of glue from the hot glue gun along the edge of the fin that will attach to the body tube. Sight along the body tube to make sure you are gluing the fin on straight, and press the fin in place. Hold it in place until the glue sets, which won't take long. Repeat the process for the other fins.

Figure 6-38. *Glue the fins in place using the fin markings on the body tube tool as a guide.*

I've seen the fins rip right off a rocket from the acceleration generated by the compressed air launcher. Add an extra fillet of glue along each joint where the fin meets the body tube. That's a total of six fillets for a three-finned rocket. If you are using duck tape instead of hot glue, this is the location where the tape is applied—again, a total of six pieces for a three-finned rocket.

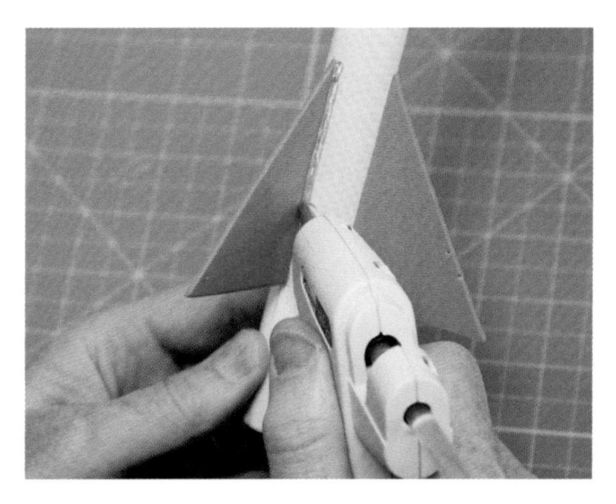

Figure 6-39. *Add a fillet of hot glue where the fin joins the body tube for extra strength.*

At this point, your rocket is ready to fly. I like to add a nose cone, though. The rocket will look more like a rocket, and if the nose cone is well made, the rocket will fly quite a bit higher.

Nose cones are made from a circle with a pie wedge cut out. The bigger the wedge that is cut out, the flatter the nose cone will be. The template you will see in the construction details that follow is the leftmost template from Figure 6-40.

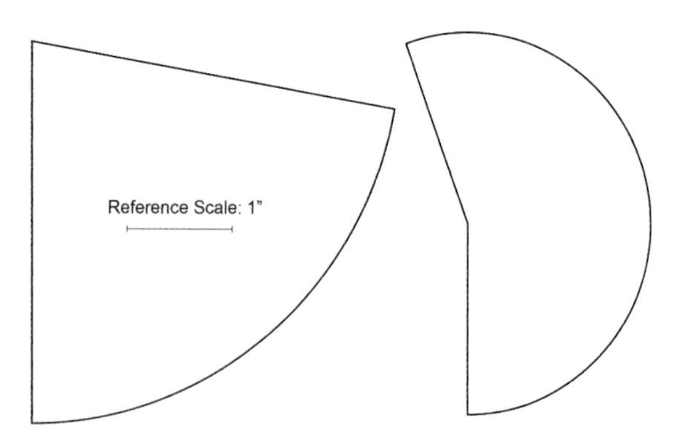

Figure 6-40. *Two nose cone templates. The one on the left makes a long, thin nose cone, while the one on the right makes a short, wide one that hangs over the edge of the body tube.*

I really like foam sheets for nose cones. These air rockets will return to earth the same way they leave it, screaming in nose first. The nose will take a beating. While you can make it from paper, expect a paper nose cone to get bent on each flight. A foam nose cone will fold over to absorb the impact of landing, which will make it far less likely to do damage to itself or anything it hits.

Mark the nose cone pattern on a foam sheet with a pencil, then cut out the nose shape with a hobby knife or scissors.

Roll the foam over on itself in the shape of a very tight nose cone. This convinces the foam it wants to roll up instead of lie flat, and will help when you start to glue the nose cone together. Test fit the nose cone to make sure it is the right size. The most aerodynamic size will just match the size of the body tube, but the easiest size to work with overlaps the body tube slightly.

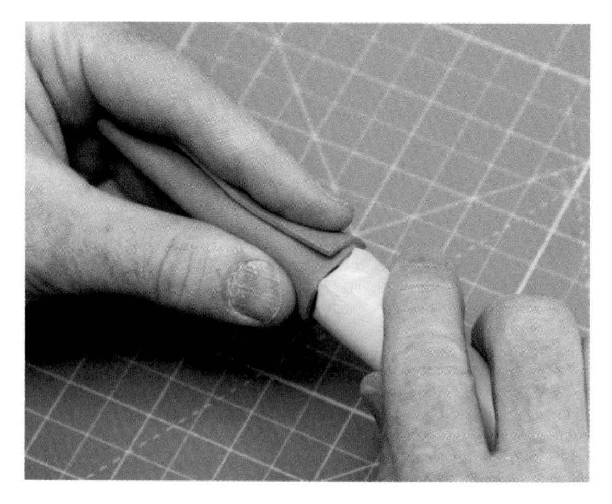

Figure 6-41. *Roll the nose cone until it starts to hold the shape of a cone, then test fit to make sure it is the correct size.*

Tack the base of the nose cone in place with a small spot of glue from a hot glue gun, as shown in Figure 6-42, or with a strong tape like duck tape. Hold the nose cone firmly closed until the hot glue dries, then repeat the process near the tip of the nose cone. Once that joint dries, the nose cone will retain its shape while you run a bead of glue down the entire joint. Once again, you can use duck tape or some other strong tape instead of the hot glue gun.

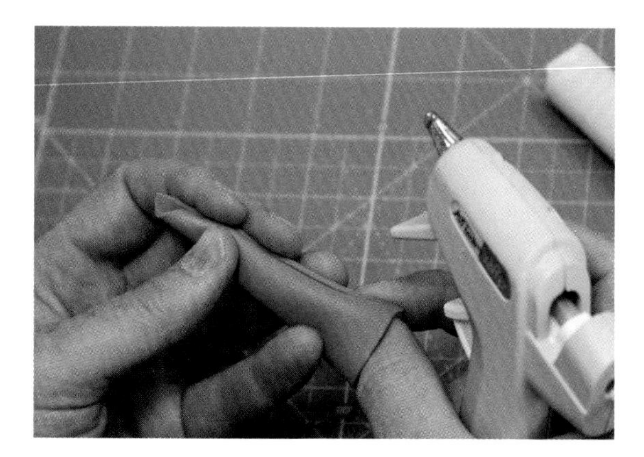

Figure 6-42. *Place a small dab of hot glue near the base of the nose cone and hold it in place while the glue sets. Add a second dab near the nose, then run a bead along the entire joint.*

Run a bead of glue just inside the base of the nose cone, then hold it in place at the top of the body tube while the glue sets. Check the alignment of the nose cone carefully! It's easy to glue it on slightly tilted. You want the point of the nose to be centered on the rocket. Alternately, tape the nose cone in place with some colorful duck tape or plastic tape.

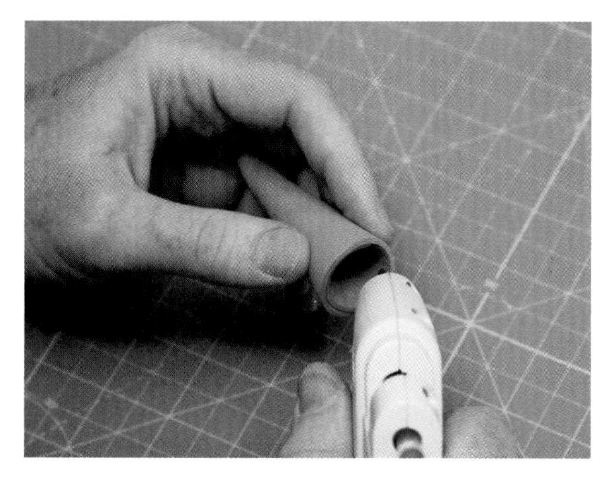

Figure 6-43. *Add a bead of glue at the base of the nose, and then press it into place on the body tube.*

Your air rocket is ready to fly! Figure 6-44 shows the rocket we just built in the middle. The top rocket doesn't have a nose cone; the body tube is simply folded over and taped. The lower rocket is one of my favorites. It's wrapped in colorful duck tape and sports far more interesting fins and nose cone.

These rockets will fly 50 to 150 feet in the air. You will learn more about aerodynamics in Chapter 13, but these rockets offer a nice preview. The top two rockets both fly to about 150 feet. The bottom one only gets to about 50 feet. The difference is the nose cone. All that extra area on the yellow nose cone adds significantly to the drag—the force of the air pushing on the rocket to slow it down. The difference in altitude is staggering.

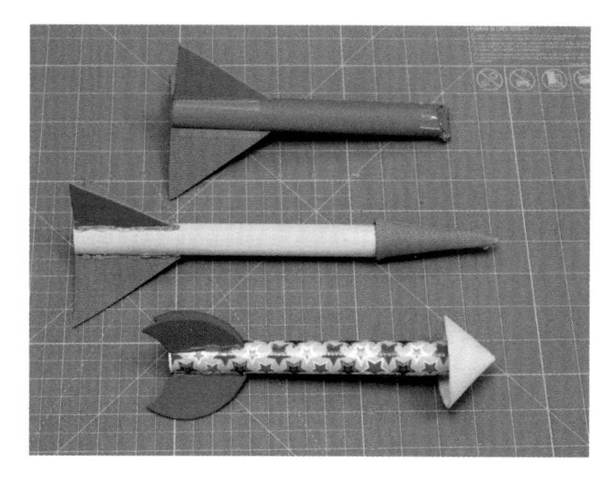

Figure 6-44. *Our completed air rocket in the middle, with two others alongside.*

Flying Air Rockets

The first step in launching any rocket is to decide on the safety rules and other requirements for the launch. You saw that in Chapter 5, when we went through launch practices for solid propellant rockets. Air rockets are simpler to set up and operate, and they don't go quite as high as most solid propellant rockets, so the procedures are a lot simpler. Still, these rockets take off at a high speed, so it's important to treat them with some respect.

There is no generally accepted air rocket safety code like the ones for low-power solid propellant rockets, high-power rockets, or rocket gliders. Here is one based on the low-power rocket safety code, adapted for our use.

Air Rocket Safety Code

1. Build rockets from lightweight materials such as paper, tape, foam, and balsa wood.
2. Do not use hard plastic nose cones unless the rocket is equipped for parachute recovery.
3. Never stand or lean over the rocket or launch tube when the launcher is pressurized.
4. Never launch a rocket at people or animals.
5. Launch air rockets outdoors in a clear area with at least 100 feet on a side.
6. Do not pressurize the system above the rating for the launcher. (The launcher in this chapter is rated at 75 psi.)

As launch day approaches, the first step is to pick an appropriate launch area. It doesn't need to be as large as a launch area for solid propellant rockets. Pick a site that's about half as far across as the maximum expected altitude for the rockets. A 100-foot site is perfect, although something a bit smaller will do. To put that in familiar terms, that's 1/3 of the length of a football field. Try to avoid trees, rooftops, or other rocket-eating obstructions.

Place the launcher either in the middle of the field or toward the direction of any wind, so the wind will carry the rocket into the middle of your launch area. Connect whatever you will use to pump up the launcher. A bicycle pump will work fine, but it will be a fair amount of work. I like to use the Black & Decker Portable Power Station. It has a 12-volt lead-acid battery for launching solid propellant rockets and a great portable compressor for launching air rockets and water rockets. That's the big orange device you see beside the launch pad in Figure 6-45.

Move all spectators back at least 15 feet from the launcher. They will be able to see more from a slight distance—that rocket takes off fast! They will also have more time to react if the rocket shreds, creating a burst of confetti.

Figure 6-45. *The launch pad is set to go with this portable compressor.*

Working Well with Pressure

The launch system is very reliable, and should fire and release all pressure reliably, too. If it does not release all of the pressure, be sure to depressurize the system before handling or moving it.

To manually depressurize the system, remove the pump from the valve. Assuming you followed the construction details for the launcher, this will immediately depressurize the system. If you did not remove the pin from the middle of the Schrader valve, though, the system will stay pressurized. In that case, push the pin in the middle of the valve and hold it down while the air rushes out.

Slide the rocket over the launch tube. Pressurize the system to between 40 psi and 75 psi, then turn off the compressor. Make sure all spectators are at least 15 feet away and are aware a launch is about to take place. Check for low-flying airplanes. My favorite air rocket launch site is frequently populated by low-flying model airplanes; be sure to check for them, too. Count down from five so the spectators are aware of the impending launch, then press the launch button just as you hit zero.

As you know from Chapter 5, checklists are a big part of any complicated discipline—even one where people are highly skilled at the task they are doing. Here is a flight checklist for air rockets that summarizes the steps we've just taken.

Flight Checklist

1. Clear spectators from the launch zone.
2. Slide the rocket onto the launch tube.
3. Pressurize the launcher to 40–75 psi.
4. Check for clear skies—no aircraft.
5. Count down from five and launch.

7 Rocket Stability

Remember the first balloon flight from Chapter 1? The balloon flipped crazily in the air, using most of its energy to twirl itself around. Imagine what that would be like with a model rocket. As soon as it left the launch rod, an unstable rocket would start to fly like the balloon, careening around in random directions, perhaps running into a spectator or flopping onto the ground with the motor still burning. It's a good thing we build models from lightweight, collapsible materials and test them for stability before flying, because an unstable rocket can be pretty scary. Fortunately, rocket instability is easy to avoid once you understand what makes a rocket stable.

We're going to look at several ways to see if a rocket will be stable before flight. Most of these work even before you start buying parts. They use a technique called *static stability* for the analysis. Basically, we find the center of pressure and center of gravity for the rocket. We looked at this idea briefly back in Chapter 1: that the center of gravity is the balance point for the rocket, and the center of pressure is the place where the force from the wind is balanced. I'll expand on these ideas here to show exactly how these are calculated. You'll also learn where the center of pressure needs to be in relation to the center of gravity to build a stable rocket.

It might surprise you, but a rocket really can be too stable. An overstable rocket can weathercock, or turn into the wind. In many ways, this is scarier than an unstable rocket; an unstable rocket flips around randomly, but the overstable rocket that weathercocks badly can end up flying almost level with the ground. Model rocketeers call this a *land shark*, as if it is hunting for prey. High wind and slow rockets make the problem a lot worse. All of this has to do with the way the rocket behaves when it is moving in the air, which is also moving. This kind of stability is called *dynamic stability*. We will look closely at dynamic stability later in the chapter, too.

How Much Can You Skip?

This is the first chapter that is very heavy on theory. Do you really need all of this information? Well, maybe.

You can safely skip this entire chapter if you are building the designs from the book with little or no modification and following all of the recommendations regarding flying the various rockets.

If you want to design some of your own rockets, but don't want to get too deep into the science and math behind the process, skim the chapter for key ideas and spend some time learning to use one of the simulators. The simulators are very, very good at analyzing static stability, and they can even help a bit with dynamic stability.

Read this chapter carefully if you really want to understand the math and science behind stability. This will help you understand what the simulators are doing, their strengths and limitations, and how to coax the utmost from the rockets you build. This is what I hope you will do. This is a book about rocket science, after all!

What Is Stability?

Imagine for a moment a marble in a bowl, as seen in Figure 7-1. If the marble is placed in the center of the bowl, with nothing pushing it around, it will just sit there. If you push the marble to one side and then let it go, it will roll toward the center of the bowl. The marble's momentum will probably keep it rolling past the center and up the other side of the bowl. It will then roll back to the center again, and probably past it again. Eventually, friction between the marble and the bowl will slow the marble to a stop, and it will end up right back in the middle of the bowl. We say the marble is *stable* —or, if we want to impress people with our technical vocabulary, we might say the marble is *positively stable*. Once it stops, it is in *equilibrium*, in this case *stable equilibrium*.

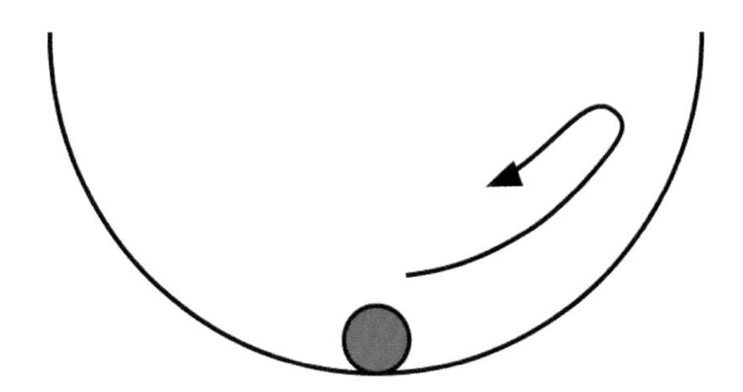

Figure 7-1. *A marble in a bowl is positively stable. If disturbed, it will eventually roll back to its original position.*

Now imagine the bowl turned upside down. With a lot of care, we can balance the marble on top of the bowl. The marble is still in equilibrium, but this time it's in *unstable equilibrium*. The slightest disturbance will move it away from the center, just a bit. The marble will then roll quickly off of the bowl, never to return. The fact that the marble keeps on going, without returning to its original position, means the marble is *unstable* or *negatively stable*.

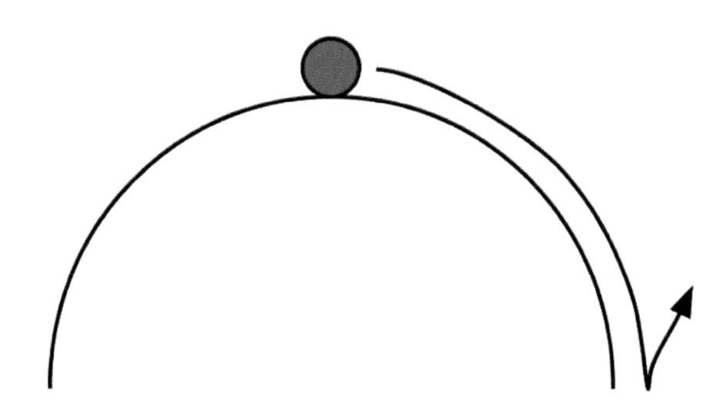

Figure 7-2. *A marble balanced on top of an upturned bowl is negatively stable. The slightest disturbance will send it careening off the edge.*

Lay the marble on a large, flat table, and the marble is *neutrally stable*. Give the marble a push, and it just keeps going, rolling smoothly until friction gradually slows it to a stop.

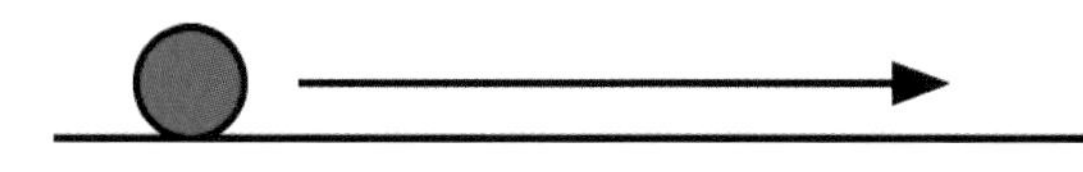

Figure 7-3. *A marble on a flat surface is neutrally stable. If disturbed, it will roll until friction or some other force stops it.*

The key to stability is building a system that, when disturbed, returns to its original state. It may oscillate a bit in the process, like the marble rolling back and forth in the bowl, but a stable system is designed to return to the state it was in before something bumped it out of that state.

The same kinds of stability can occur in a model rocket. Anything tumbling in air or a vacuum will rotate around its center of gravity, which is the place where the object balances. Toss a pencil into the air, and it will rotate pretty much around the center (although the weight of the eraser might throw things off a bit). Put a tiny dab of clay on the sharp end of the pencil to counterbalance the eraser, and the pencil will rotate about a point closer to the center. Either way, a pencil is neutrally stable.

If the object is moving through the air, or air is moving past the object, the air pressure pushes around the object's center of pressure. For the pencil, that's right at the middle. If you add feathers or fins on one end, as with an arrow or a rocket, the wind pushes more on the end with the feathers or fins, moving them away from the flow of the air. The fins don't weigh much, so the center of gravity for the pencil is still more or less in the middle, but the wind pushes more on the fins, pushing them away from the flow of the air. The pencil is now a stable system that wants to orient itself so the fins point away from the direction of air flow.

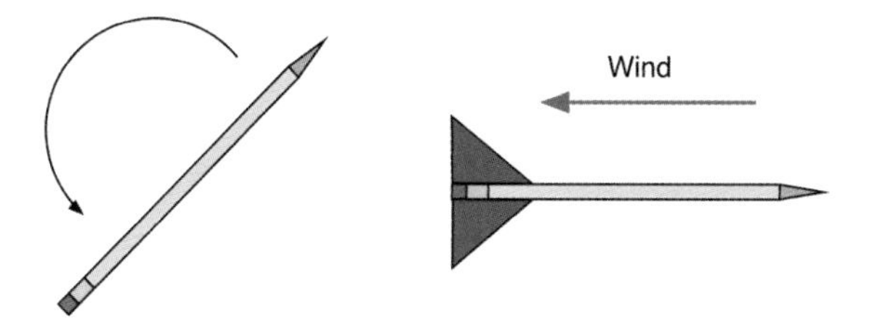

Figure 7-4. *A pencil tossed into the air will flip around the center of gravity, roughly the center of the pencil. Add fins and it orients itself to the direction of air flow.*

This is a very important point, so let's step back and look at it in an entirely different way with a little thought experiment. Let's say you tape a long ribbon to a baseball, and throw it. What will be up front, the heavy baseball or the light, flapping ribbon? The baseball, of course. No matter how hard you try, the ribbon will not lead the way. Once again, the point is that the heavier part of the projectile ends up in front, while the part with the highest drag ends up in back. To make a rocket more stable, we add weight to the nose or area to the fins, making sure the fins are behind the center of gravity.

In a stable rocket, the center of gravity needs to be in front of the center of pressure by at least the diameter of the body tube. With a BT-50 body tube, like the one we used for Juno, the center of gravity needs to be about one inch in front of the center of pressure. We can move the center of gravity forward by reducing the weight at the base of the rocket, or adding weight to the nose of the rocket. We can move the center of pressure toward the rear of the rocket by making the fins larger, moving them toward the rear of the rocket, or adding more of them—as long as the fins we add are behind the center of gravity!

This distance between the center of pressure and center of gravity, measured in body tube widths, is called the *caliber* of the rocket. Going back to the example of Juno, when the center of pressure is one body tube width behind the center of gravity, the caliber of the rocket is 1. If the center of pressure is 1.5 body tube widths behind the center of gravity, the caliber is 1.5. For most rockets, the ideal caliber is between 1 and 2. More is not necessarily bad, especially in a rocket that comes off the pad fast, but it's something to be aware of. We'll see why when we look at dynamic stability, later in the chapter.

Incidentally, the actual caliber for Juno varies with the motor selected, since the different motor weights move the center of gravity a bit. For an A8-3, Juno's caliber is about 1.75. A heavier C6-7 motor moves the center of gravity toward the aft end of the rocket, changing the caliber to 1.3. Both values are right in the range of 1 to 2 calibers that we want for a model rocket.

Static stability is all about making sure the center of pressure is at least one body tube width behind the center of gravity. Dynamic stability looks at the way the rocket actually moves when it is in the air. A side wind as the rocket comes off the launch rod or a gust of wind as the wind direction changes at altitude can rotate the rocket in the air. When the caliber of the rocket is in the 1 to 2 range, it will act a lot like the marble inside the bowl. The rocket will continue in more or less the same direction, but the fins will push the rocket back away from the flow of the air. The angular momentum of the rocket will rotate it past the ideal point, though. Once it swings far enough, the wind will reverse the direction. Eventually, friction causes the rocket to settle down so it is pointed in the direction of flight again.

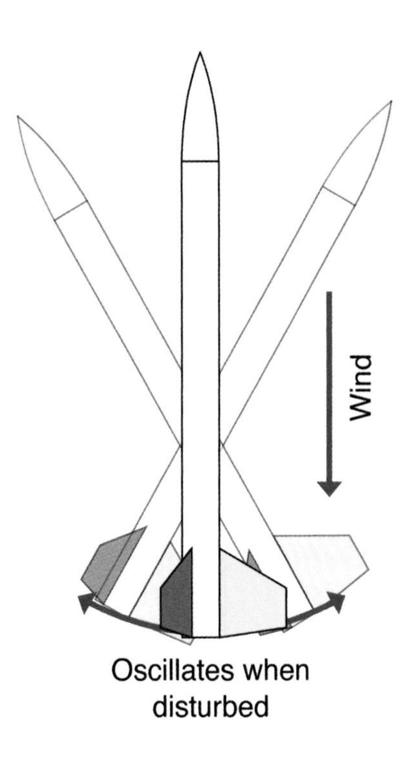

Figure 7-5. *A stable rocket will oscillate from side to side when disturbed, settling down as friction dampens the oscillation. This is greatly exaggerated for clarity. Typical oscillations are a degree or two.*

In an overstable rocket, the center of pressure is well behind the center of gravity—the rocket has a higher caliber. An overstable rocket doesn't oscillate, at least not as much as one that has a caliber in the 1 to 2 range. A very stable rocket that is traveling slowly compared to the wind speed will pop around and face the new wind direction if the wind direction changes a bit.

Let's think about what happens to an overstable rocket coming off the launch pad. Let's assume the rocket is traveling at a slow 20 mph, below the recommended minimum launch speed of 30 mph. Let's also assume the wind is traveling at about 20 mph, the highest allowed for model rocket flights. As it leaves the launch rod, the rocket will rapidly rotate to about 45° from the vertical, since the air flow it is experiencing has 20 mph pointed down due to its speed coming off the launch rod and 20 mph sideways from the speed of the wind. The motor is now pushing at a 45° angle, not vertically. It's still experiencing the sideways wind, though. As the speed increases, the wind matters less and less, but the rocket will turn to the side more than just 45° in those first few seconds of flight. As the rocket slows, gravity will pull it back to earth along an arc—maybe causing the rocket to hit before the recovery charge fires! That's one way to get a classic land shark.

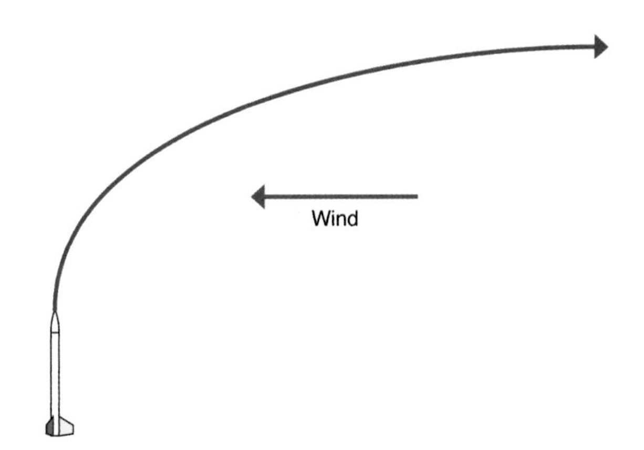

Figure 7-6. *An overstable rocket that has a slow liftoff speed compared to the wind speed can weathercock, turning dramatically into the wind. In severe cases, the rocket can end up going sideways in powered flight.*

This is why we generally don't want a rocket that is overstable. That doesn't mean a rocket with a caliber of 3, 4, or even higher is necessarily going to cause problems. If it is a fast rocket or the launch rod is long, it might come off of the pad doing 100 mph. A little 20 mph side wind won't cause much of a problem. On a very calm day, even a slow rocket won't have an issue. What this means is we have to be very aware of the caliber and launch speed of our rockets, and not fly overstable rockets—especially those that lift off slowly—on windy days. You may see warnings on some rocket kits not to fly them in high winds, or to only fly them on calm days. Now you understand why.

Dynamic stability also has a big impact when we're trying to fly high or fast. Think about it for a moment. If you want to get the maximum altitude, you don't want a rocket to weathercock at all. That wastes energy moving the rocket sideways when you want it to go up. If the rocket is pushed away from vertical, you want it to oscillate back and forth, but to keep going straight. You want a small caliber.

If you want high speed, though, all that oscillating is wasting energy. If you're trying to get speed, and you don't care if the rocket happens to fly 20° from the vertical, increase the caliber as much as possible. The rocket may bend a bit into the wind, but it won't waste a lot of energy by oscillating back and forth. Oscillations will increase the drag, slowing the rocket down.

Experimenting with Stability

It's too dangerous to build and fly unstable rockets just so we can learn some lessons about stability. However, we can examine unstable rockets using a very simple wind tunnel. Figure 7-7 shows our experimental setup. It's made from a household fan, a straw, some tape, and a long needle. My needle is pushed into a cork, but that's not strictly necessary.

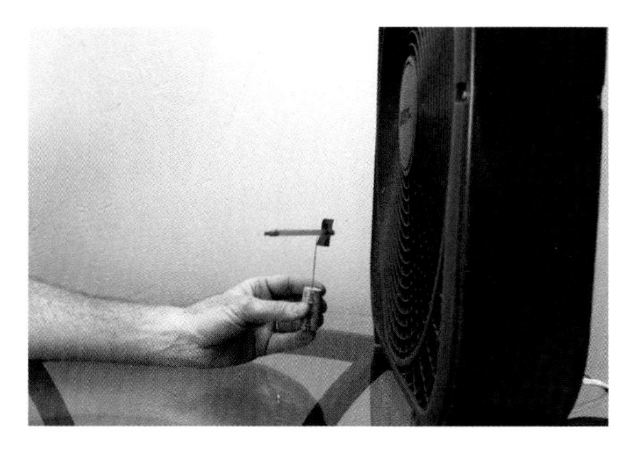

Figure 7-7. *You can learn a lot about stability with a fan, a straw, some tape, and a needle.*

Build the rocket from a soda straw, adding a flap of tape near the rear. Use another piece of tape to form a crude nose cone. That's important, since we don't want wind whistling down the center of the body tube, which would cause the tube itself to act a bit like a fin.

Find the center of gravity by balancing the rocket on your fingertip, and then push the needle through the bottom of the body tube roughly at the center of gravity. Make the hole fairly loose so the rocket can twist easily. Push the needle to make an indentation into the top of the body tube, just enough so the rocket spins fairly freely on the top of the needle.

Hold the rocket so it is sideways to the direction of flow from the fan and turn the fan on. The rocket will quickly rotate so the nose is pointed into the wind. That's a stable rocket!

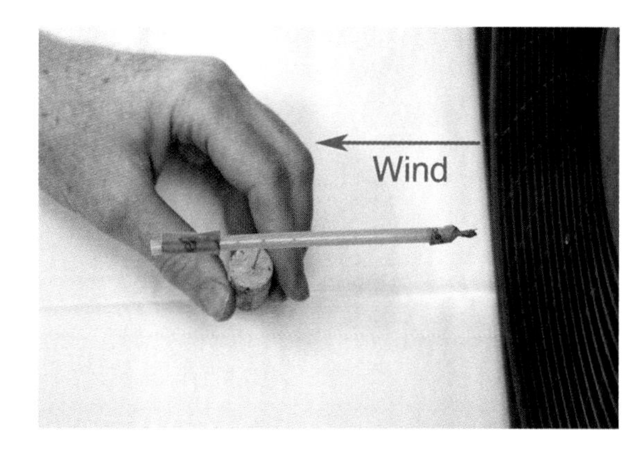

Figure 7-8. *A stable rocket will spin around to face the wind.*

Make another hole fairly near the fins. Hold the rocket sideways to the air flow and turn the fan on. If you find just the right point, the rocket will stay sideways in the wind. You can see in Figure 7-9 that mine actually stabilized with the fins a little closer to the fan than the nose cone. That would sort of work if there was no motor, but the motor would be pushing the rocket in another direction. With the motor pushing one way and the flow of air trying to stabilize the rocket in another direction, the rocket will flip about randomly. It's unstable.

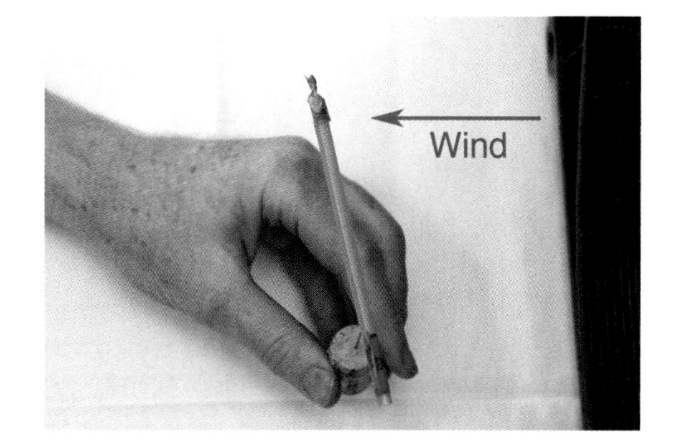

Figure 7-9. *An unstable rocket will spin or turn sideways to the wind.*

Even more odd is a semistable rocket. You may have to put several holes in the straw to find the right position, but there is a point where you can make the rocket stay in either position. If it starts off pointed into the wind, it will stay that way, but if you push it to the side a bit, the rocket either will flip around until it is sideways on to the wind or, if you get everything just right, it will start to spin. If it the rocket stabilizes, it is semistable; if it spins, it is a classic unstable rocket. If you were to launch a real rocket with the same characteristics, the rocket would tumble in the air, even after the motor stops pushing.

Finally, make a hole near the nose of the rocket to make it overstable. That should be a good thing, right? When you turn the fan on, the rocket certainly pops right around to face the wind. Watch the tail, though. It will quiver, vibrating back and forth due to slight changes in the direction of the wind from the fan. That's one characteristic of an overstable rocket. If such a rocket comes off the pad slowly, and in a high wind, it will tend to rotate into the wind, perhaps becoming a land shark.

Checking Stability with the String Method

We're going to cover several ways to check the stability of a rocket. Most methods depend on finding the center of pressure and the center of gravity to make sure the caliber of the rocket is at least 1. There are two methods that don't require finding the center of pressure, though. Too many people try the first method at one point or another, although it really should never happen: fly the rocket and see if it works. Yes, it's the ultimate way to check, but it's also dangerous to the rocket and perhaps the things (and people) around it.

There is an experimental way to check stability without actually flying the rocket. It's called the string test.

Start by finding the center of gravity for the rocket by balancing the rocket on your finger. You need to do this with the motor and parachute installed.

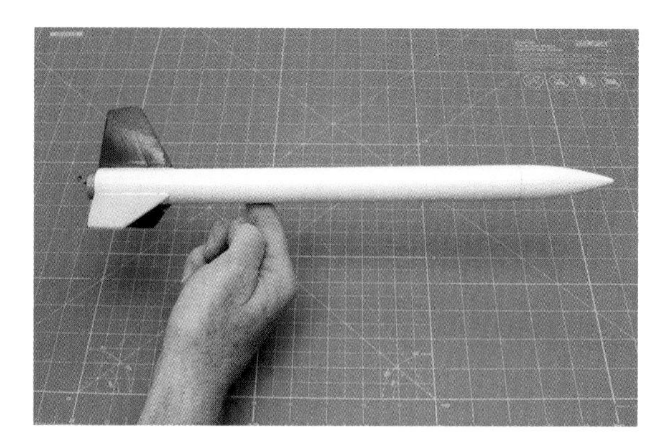

Figure 7-10. *Balance a rocket on your finger to find the actual center of gravity.*

Tape a long piece of string to the rocket right at the center of gravity. The string should be as long as your height plus your arm length. Hold the end of the string above your head; the rocket should be near the ground. If you found the balance point, it should also stay fairly level.

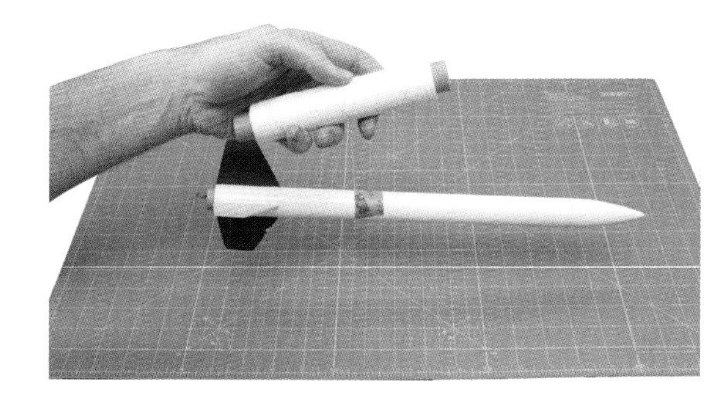

Figure 7-11. *Juno balancing on a string, getting ready for the string test.*

Find a place with very little wind. A grassy area on a very calm day is great. The center of a gymnasium is even better. A living room with fragile lamps sitting around is emphatically not the best choice.

Now spin around like a discus thrower (or a ballerina, if you like), increasing your rotational speed until the rocket is actually flying. Look carefully: is the rocket flying straight?

If the rocket straightens out and points "into the wind" (i.e,. if it stabilizes in the direction you're rotating in), it's a stable rocket. If it flips about, don't despair. Stop turning, straighten the rocket out, and try again. A stable rocket will often flip as if it is unstable if it does not start off going straight. The rocket is also going pretty slowly. With a 5-foot-long string, spinning once every two seconds, the rocket is only going about 10 mph. That is just barely enough speed for the fins to begin to work. If the rocket fails the string test, it may still be stable. Check it using one of the methods in the rest of the chapter.

The string test is a very conservative test. Many perfectly stable rockets will fail the string test, but if a rocket passes this test, it is stable. Juno actually fails the string test. As we'll see from the more exact methods, though, Juno is quite stable.

Checking Stability Experimentally

Almost all methods for checking stability depend on finding the center of gravity and center of pressure and checking to make sure that the center of gravity is at least one body tube width in front of the center of pressure. Let's look at some ways to find the center of pressure and center of gravity.

Finding the Center of Gravity Experimentally

You already found the center of gravity experimentally when you did the string test; just balance the rocket on your finger. The place where it balances is the center of gravity.

Finding the Center of Pressure Experimentally

You can also use the balance method to find the center of pressure. This time, though, you want a two-dimensional cardboard cutout of the rocket. The cutout uses weight to represent the cross section of the rocket that is seen by the wind.

Take the image of Juno from Figure 3-1 and trace it onto pasteboard, cardboard, or some other material that is stiff enough not to bend under its own weight. Don't take everyday paper and try to strengthen it with tape or anything like that; uneven weight distribution from overlapping tape will throw off the experiment. Balance the cutout just as you balanced the full-size rocket. Take a look at the result in Figure 7-12. The cross section is, roughly at least, what the wind blows on. Each square inch of paper represents an area that the air will push against. Areas that are farther away from the center of pressure exert more torque, just as they do when finding the center of gravity. The balance point is the approximate center of pressure.

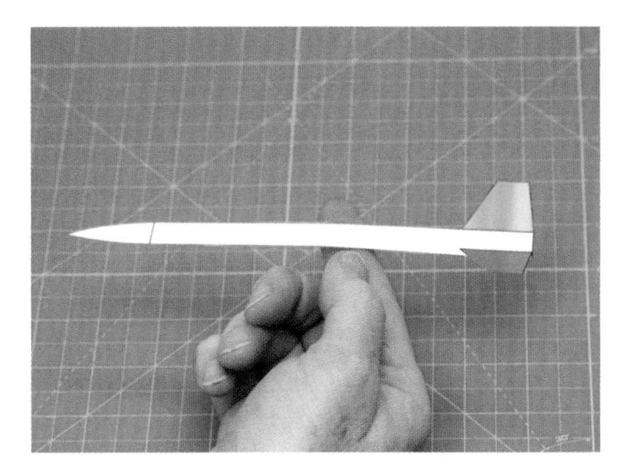

Figure 7-12. *A cutout of a model rocket shows the approximate center of pressure. It's a bit conservative, but if this test says the rocket is stable, it will be stable.*

Well, it's one center of pressure, anyway. As it happens, the center of pressure varies depending on which way the rocket is pointing. When the rocket is headed directly into the wind and gets nudged to the side just a bit, the fins actually work a lot better than when the rocket is sideways on to the wind. A nearly head-on rocket will have a center of pressure that is farther back than a rocket heading into the wind sideways. That means the balance test is a little overly conservative, which is not a bad thing.

Based on the cutout test, the center of pressure for Juno is 9 1/4" from the nose. Based on the balance test, the center of gravity is 10" from the nose. Since the body tube is about 1" in diameter, the caliber is 3/4. For a stable rocket we want a caliber of 1 or more. Once again, this simple test indicates Juno is unstable, as it will for most rockets with a caliber of less than 2. Like the string test, this method is overly conservative. If the rocket passes this test, it is stable, but if it fails, you need to move to a more accurate test. Fortunately, we're about to do just that.

Checking Stability with a Rocket Simulator

These days, the most common way to find the center of pressure and center of gravity for a rocket is with a rocket simulator. Rocket simulators are specialized computer programs that combine a computer aided design (CAD) program tweaked specifically for drawing model rockets with a physics simulation package that can calculate the altitude the rocket will achieve with various motors, payload weights, and weather conditions. We'll return to the topic of simulating the rocket flight in Chapter 14. For now, we're going to use a rocket simulator to find the center of gravity, center of pressure, and caliber of a rocket.

There are several rocket simulators available. They all have individual advantages and disadvantages. The most talked about rocket simulator is called RockSim. Sold by Apogee Components, it's a really nice simulator that I have used a lot. You can find it and download a free trial at the Apogee website (*http://bit.ly/apogee-rockSim-v9*). While RockSim is a fine simulator, it also costs over $100, so we're going to use another rocket simulator here. It doesn't have all of the features of RockSim, but it has all of the features we need for now, and the price is right. OpenRocket is an open source rocket simulator you can download for free from the OpenRocket repository on SourceForge (*http://openrocket.sourceforge.net/download.html*). The installation instructions vary a bit depending on whether you are using a Linux, Macintosh, or Windows computer. Go to the site, download the program, and install it for your particular machine. OpenDoc is written in Java, so be sure you install Java for your platform if you have not already done so.

Most of the rockets in this book have simulation files available at the author's website (*http://bit.ly/byteworks-make-rockets*). These rocket simulators don't work especially well with gliders, odd rockets like the Nicomachus helicopter rocket we'll build in Chapter 19, water rockets, or air rockets, so you won't find simulation files for those rockets at the site. You will find Juno, though. Go ahead and download the Juno simulation file, called *Juno.rkt*. This is actually a RockSim file, but it will load in either RockSim or OpenRocket.

Once you load the *Juno.rkt* file, you'll see a screen like the one in Figure 7-13. Take a look at the area that's circled in the screenshot, near the top-right corner of the drawing. There's the center of gravity and the center of pressure, calculated for us!

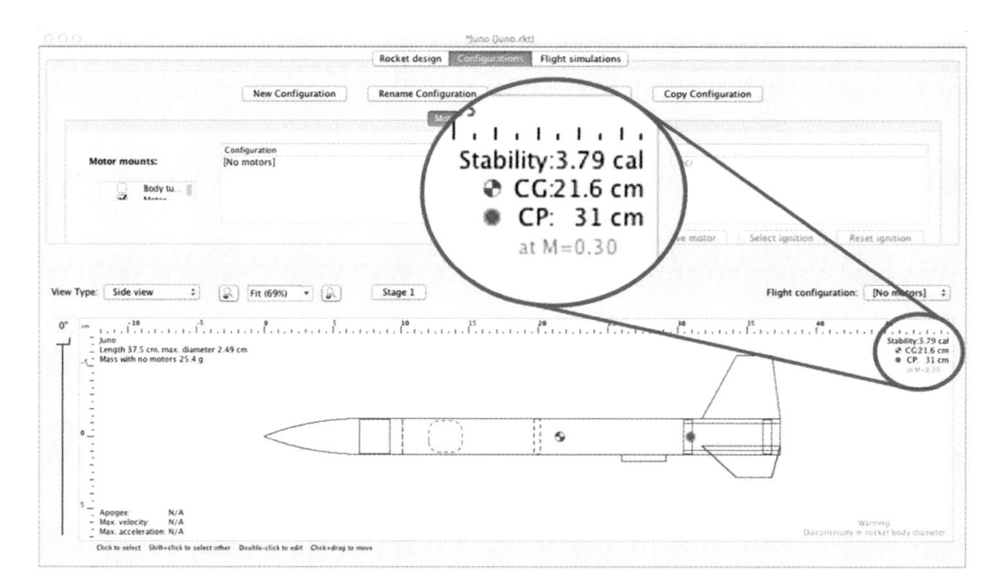

Figure 7-13. *As soon as you load a file in a rocket simulator, it will show you the center of gravity, center of pressure, and caliber. OpenRocket calls the caliber "Stability."*

Unfortunately, this calculation does not take a motor into account. Let's load one. Click on Configurations, then on the New Configuration button to create a new rocket configuration. This will add a line with the ominous configuration of "[No motors]."

Click on the None under Motor Mount. Once you do, the "Select motor" button is enabled. Click that button and you will see a dialog like the one in Figure 7-14.

Select the motor you want to load—I loaded an Estes A8 motor—then set the ejection delay charge. This one is set to 3 seconds. Click the OK button.

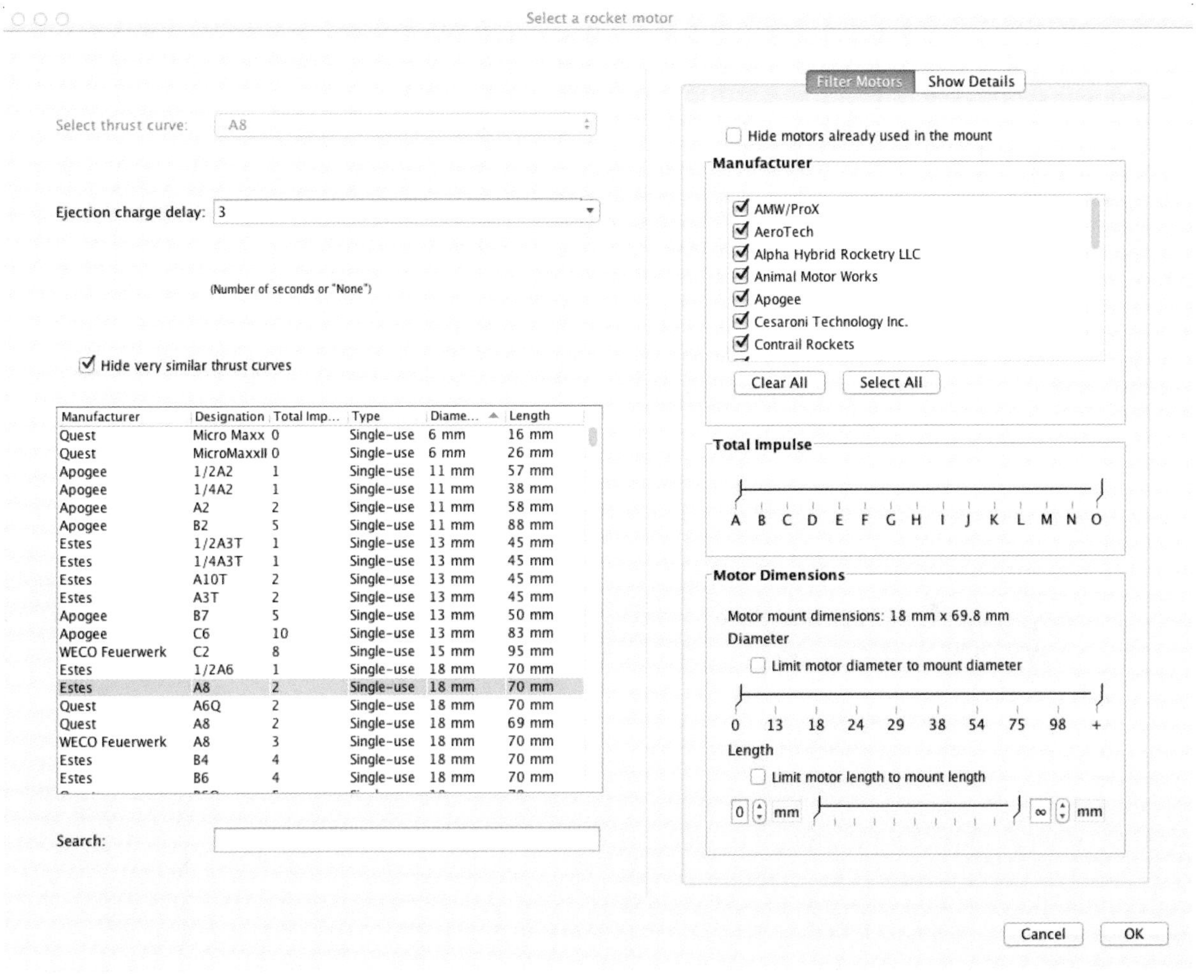

Select a rocket motor

Select thrust curve: A8

Ejection charge delay: 3

(Number of seconds or "None")

Hide very similar thrust curves

Manufacturer	Designation	Total Imp...	Type	Diame...	Length
Quest	Micro Maxx	0	Single-use	6 mm	16 mm
Quest	MicroMaxxII	0	Single-use	6 mm	26 mm
Apogee	1/2A2	1	Single-use	11 mm	57 mm
Apogee	1/4A2	1	Single-use	11 mm	38 mm
Apogee	A2	2	Single-use	11 mm	58 mm
Apogee	B2	5	Single-use	11 mm	88 mm
Estes	1/2A3T	1	Single-use	13 mm	45 mm
Estes	1/4A3T	1	Single-use	13 mm	45 mm
Estes	A10T	2	Single-use	13 mm	45 mm
Estes	A3T	2	Single-use	13 mm	45 mm
Apogee	B7	5	Single-use	13 mm	50 mm
Apogee	C6	10	Single-use	13 mm	83 mm
WECO Feuerwerk	C2	8	Single-use	15 mm	95 mm
Estes	1/2A6	1	Single-use	18 mm	70 mm
Estes	A8	2	Single-use	18 mm	70 mm
Quest	A6Q	2	Single-use	18 mm	70 mm
Quest	A8	2	Single-use	18 mm	69 mm
WECO Feuerwerk	A8	3	Single-use	18 mm	70 mm
Estes	B4	4	Single-use	18 mm	70 mm
Estes	B6	4	Single-use	18 mm	70 mm

Search:

Filter Motors | Show Details

Hide motors already used in the mount

Manufacturer

AMW/ProX
AeroTech
Alpha Hybrid Rocketry LLC
Animal Motor Works
Apogee
Cesaroni Technology Inc.
Contrail Rockets

Clear All | Select All

Total Impulse

A B C D E F G H I J K L M N O

Motor Dimensions

Motor mount dimensions: 18 mm x 69.8 mm

Diameter

Limit motor diameter to mount diameter

0 13 18 24 29 38 54 75 98 +

Length

Limit motor length to mount length

0 mm ∞ mm

Cancel | OK

Figure 7-14. *Select a motor from the motor selection dialog. Don't forget to select the proper ejection charge delay.*

With the motor loaded, the center of gravity shifts rearward. You can read off the center of gravity, center of pressure, and stability from the panel that shows the rocket's design. Stability is the rocket caliber, which you recall should ideally be between 1 and 2. It's sitting at 1.74, a very nice value. If you repeat these steps with a few other motors, you'll see the center of gravity shift around a bit, which will also change the caliber. It will stay between 1 and 2 for any motor that will fit in this rocket, though.

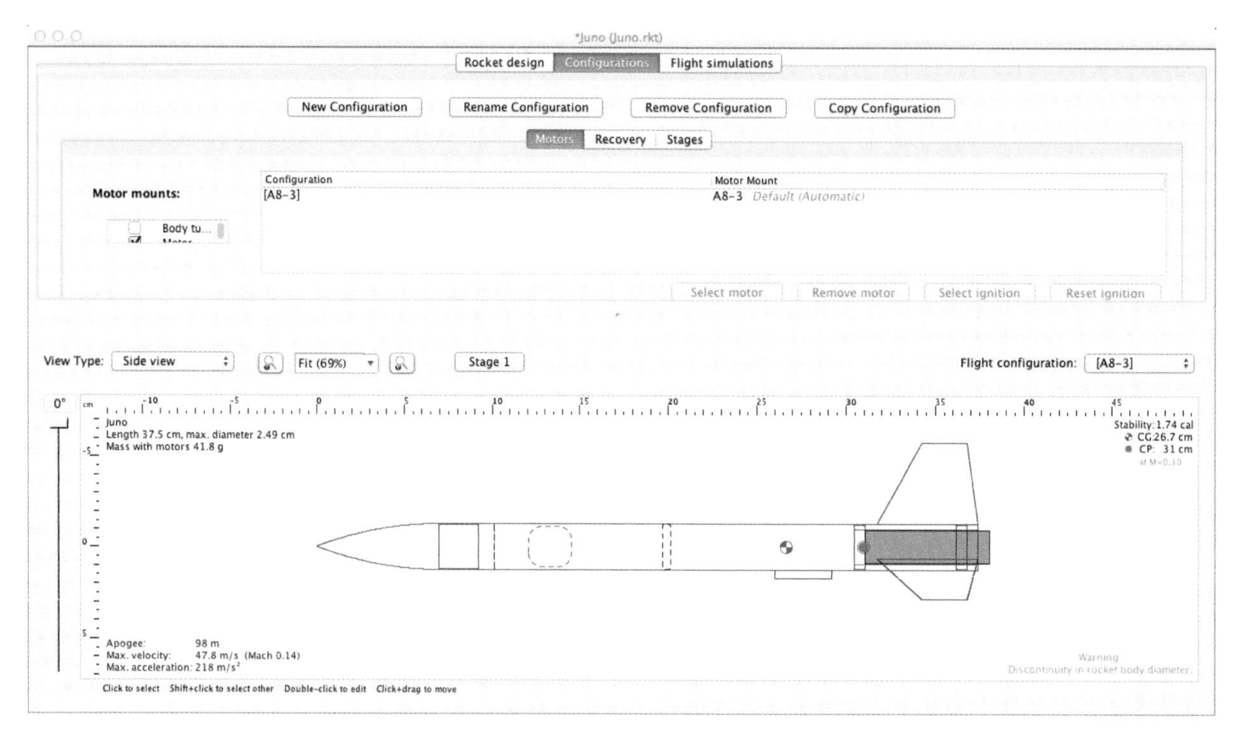

Figure 7-15. *The center of gravity shifts for each motor, so check the stability of a design with all of the motors you plan to use.*

Of course, the whole purpose of the rocket simulator is to simulate a rocket flight. We won't cover this function in detail until Chapter 14, but who can resist a sneak peek? Click the "Flight simulations" tab, and then select the simulation you just created. Click the "Run simulation" button. After a very short delay, you'll see all of the important flight parameters filled in for you. There are several options you can explore. You can also change the units from the Preferences if you prefer imperial units to metric units.

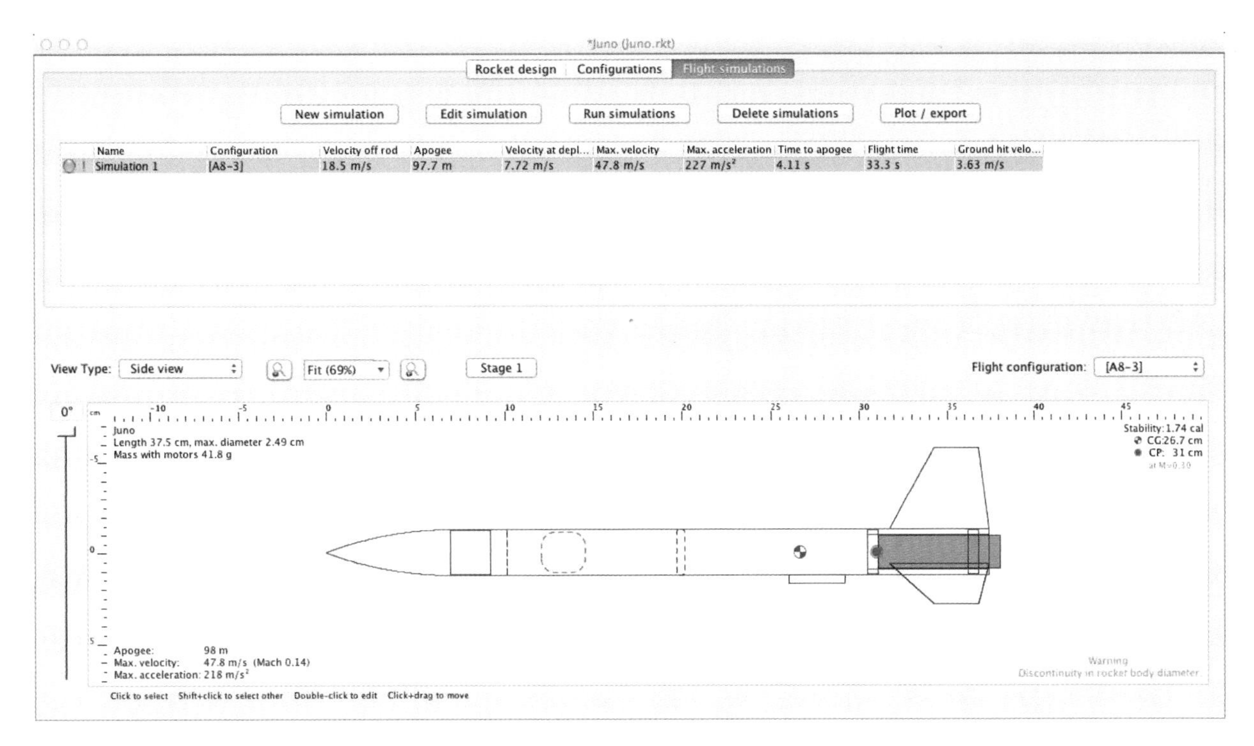

Figure 7-16. *Run the simulation to see the projected altitude, speed, and other parameters.*

Checking Stability with Math

The rest of the chapter deals with how to find the center of gravity and center of pressure using math. It's the way the rocket simulators do it, so working through the math will give you a much better feel for how the simulators work and what their limitations are. It also gives you a way to calculate the center of pressure and center of gravity if the rocket simulator isn't working for your particular design. Still, you can skip all of this if it doesn't interest you.

Finding the Center of Gravity with Math

Still here? Good! Let's see how to calculate the center of gravity by hand. The procedure is fairly easy, but it can be a bit tedious.

Let's start off with a very, very simple "rocket" and find the center of gravity. You can even follow along by building the rocket and testing to make sure the center of gravity matches the theory.

Our test rocket will be a soda straw with a ball of clay at each end. The soda straw is very light compared to the clay, so we can ignore its weight. Start by cutting a soda straw so it is 12 cm long. The length isn't that critical, but 12 cm will make some of the math we're going to do a lot easier. If you prefer to work in inches, make it 12 half-inches long. This will be a bit longer than the example shown, but working in half-inches will make the math easier. Roll some clay out into a long strip and cut it in thirds. Roll one of the pieces into a ball and stick it on one end of the straw, then roll the other two pieces into a second ball and put it on the other end of the straw. Find the center of gravity by balancing the straw on your finger, then measure its location. It should be 1/3 of the length of the straw, approximately 4 cm, from the larger ball of clay.

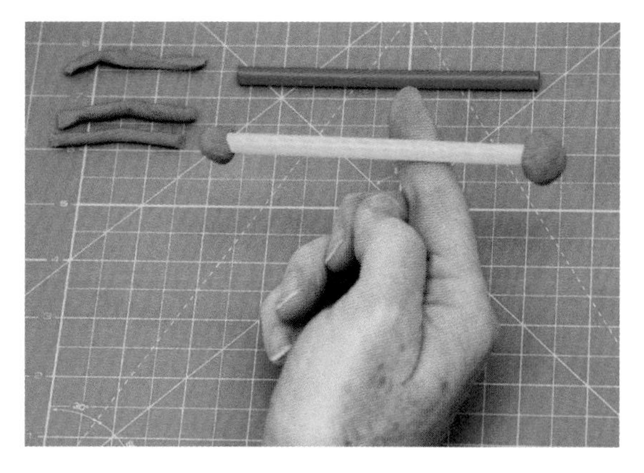

Figure 7-17. *The "rocket" with one clay ball twice the weight of the other balances 1/3 of the way from the heavier weight.*

To do the same thing with math, we choose a point called the *datum*. The datum is an arbitrary starting point. For a rocket, we typically use the tip of the nose cone for the datum. For our clay and straw rocket, we'll use the middle of the larger lump of clay.

The center of gravity is the sum of the weight of each object on the rocket, multiplied by its distance from the datum, divided by the total weight of the rocket:

$$CG = \frac{\sum_{i=1}^{n} w_n d_n}{\sum_{i=1}^{n} w_n}$$

It turns out that we don't need to know the precise weights of the balls of clay to do this calculation, only their relative weights. To make this easier to follow, though, let's assume the smaller ball of clay is about a half-inch in diameter, which would be about two grams, and the larger one is twice that, or four grams. We chose the larger ball of clay for our datum, so it is 0 cm from the datum, while the smaller ball of clay is 12 cm from the datum. Our equation therefore becomes:

$$CG = \frac{4 \times 0 + 2 \times 12}{4 + 2} = \frac{24}{6} = 4\,\text{cm}$$

This tells us the center of gravity is 4 cm from the large clay ball, which is 1/3 of the way along the length of the 12 cm straw, just like the balance test told us to expect.

Doing the calculation for a real rocket is a bit more complex. What we really need to do with any shape is sum up the mass and distance from the datum for each individual source of weight. Carried to an extreme, that means we need to do the calculation for every molecule in the object. We clearly need a shortcut! For pieces made up of one material, like the balls of clay, we can find the center of gravity for each piece, then treat the piece as if all its weight is at the center of gravity for that piece. That's what we did by measuring from the center of the ball of clay. Just looking at the ball of clay, you can probably convince yourself that its center of gravity is at the center of the sphere. It's not so easy to see where the center of gravity is for a nose cone, though. This is a situation where we turn to an advanced mathematical tool called calculus. It gives us a way to actually sum up the contributions from each infinitesimally small part of an object.

As you might imagine, other rocketeers have already done the calculation for common rocket parts. I'll show you how it's done for a sphere, but you can find the center of gravity for some common nose cone shapes in Figure 7-18 so you don't have to do the calculus each time you want it for one of these shapes. The centers of gravity shown there are for solid shapes, but the results for hollow plastic nose cones are very close. Of course, once you know the technique, you can figure out the center of gravity for any odd shape that comes along.

Figure 7-18 shows the top of each nose cone. As you know, nose cones have a portion below this part that sticks out, called the *shoulder*; it slides into the body tube. When calculating the center of gravity, you treat that as a separate part. It's a cylinder, with a center of gravity halfway between the bottom and top.

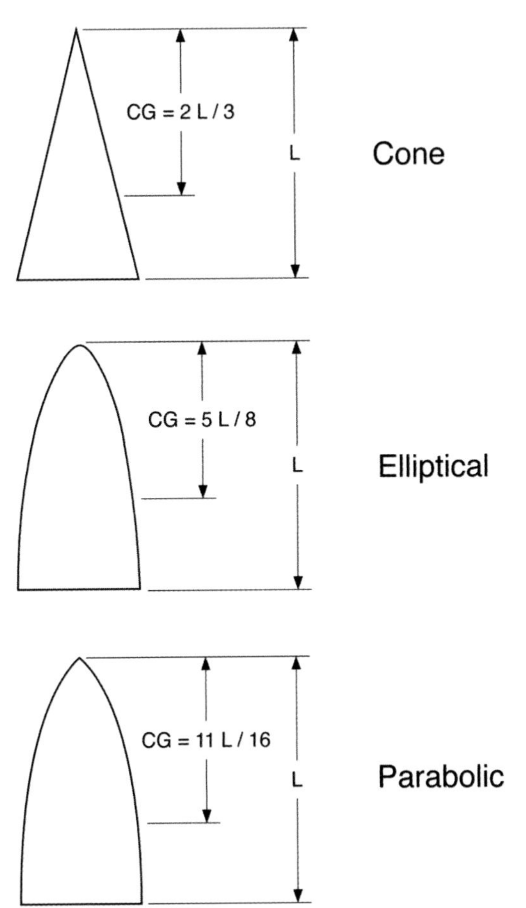

Figure 7-18. *The center of gravity for a few common nose cone shapes.*

To find the center of gravity for the sphere, let's start by defining a coordinate system. Our sphere will have a radius of 1 and will be centered at the origin, as shown in Figure 7-19. We'll slice that sphere into very thin sections, parallel to the y-axis. Of course, the sphere is three-dimensional, so each slice is a disk (think of the way a hard-boiled egg slicer makes disks of egg).

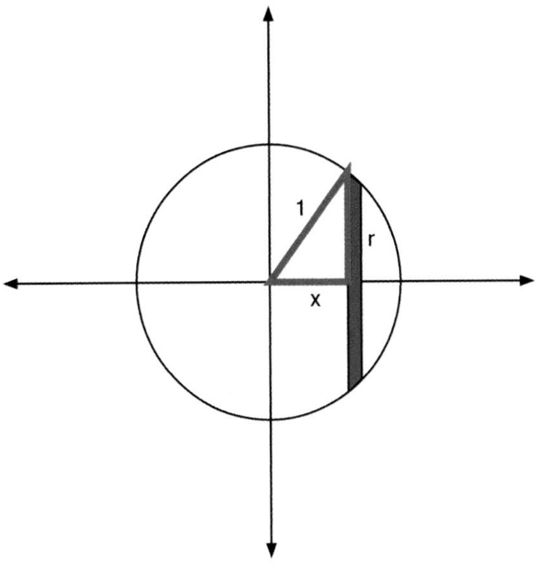

Figure 7-19. *One slice from a sphere, showing r, x, and the hypotenuse, which is 1.*

At each point along the x-axis, the radius of a disk from the sphere is found using the Pythagorean theorem:

$$r^2 + x^2 = 1$$

$$r = \sqrt{1 - x^2}$$

The volume of the sphere, which is proportional to the weight we need in order to find the center of gravity, is approximately the sum of the areas of the disks multiplied by the thickness of each disk. The volume is approximate because the edges of the disk are formed from the side of the sphere, so they are curved. The thinner the disks, the better the approximation is. Here's the volume of one disk, where the thickness is a small value, dx:

$$V = \pi r^2 dx$$

$$V = \pi (1 - x^2) dx$$

Calculus gives us a way to sum infinitesimally thin slices of the sphere, so the result is exact instead of an approximation. Multiplying by the distance from the center of the sphere and summing all of the slices from –1 to 1, then dividing by the total weight, we get:

$$CG = \frac{\int_{-1}^{1} \pi\,(1 - x^2)\,x\,dx}{\int_{-1}^{1} \pi\,(1 - x^2)\,dx}$$

Evaluating the integral in the numerator gives 0, indicating the center of gravity is at the origin, just as we expected. We don't even need to evaluate the integral in the denominator to find the total weight.

The center of gravity for fins can be found using a similar method. Let's start with a simple triangular fin like the one shown in Figure 7-20.

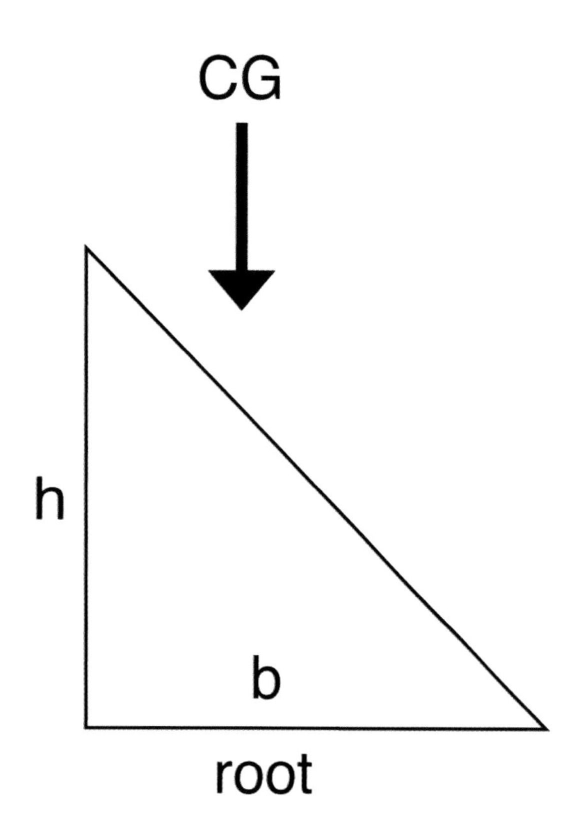

Figure 7-20. *This triangular fin has the base of the triangle along the x-axis and the height along the y-axis.*

The equation for the leading edge of a fin with the base *b* along the body tube and x-axis and the height along the y-axis is:

$$y = h - \frac{h\,x}{b}$$

This time we slice thin sections from a plane, so the center of gravity (CG) is:

$$CG = \frac{\int_{0}^{b} \left(h - \frac{h\,x}{b}\right) x\,dx}{\int_{0}^{b} \left(h - \frac{h\,x}{b}\right) dx}$$

After a bit of calculus, we find the center of gravity for a triangular-shaped fin is:

$$CG = \frac{1}{3}\,b$$

or 1/3 of the way from the base of the fin.

A lot of fins, like the ones on Juno, look pretty complicated. You can divide them into smaller chunks, though, and treat each individual piece as either a triangle or a rectangle. We just found out that the center of gravity for a triangle is 1/3 of the way from the base. The center of gravity for a rectangle is its center, 1/2 of the way from the base. Figure 7-21 shows how to break up Juno's fin. We can use these facts to find the overall center of gravity.

At this point you can see how you would find the center of gravity for a rocket like Juno. You know how to find the center of gravity for each individual part. You can find the weight of each part by checking catalogs, online references, or even weighing the parts on a kitchen scale. With the weight and center of gravity of each part, you can find the center of gravity for the entire rocket using the same technique used for the two clay balls. There are more parts, so there is more arithmetic involved, but the idea is exactly the same. Fortunately, it's rare to have to do all of these calculations by hand, because we have rocket simulators.

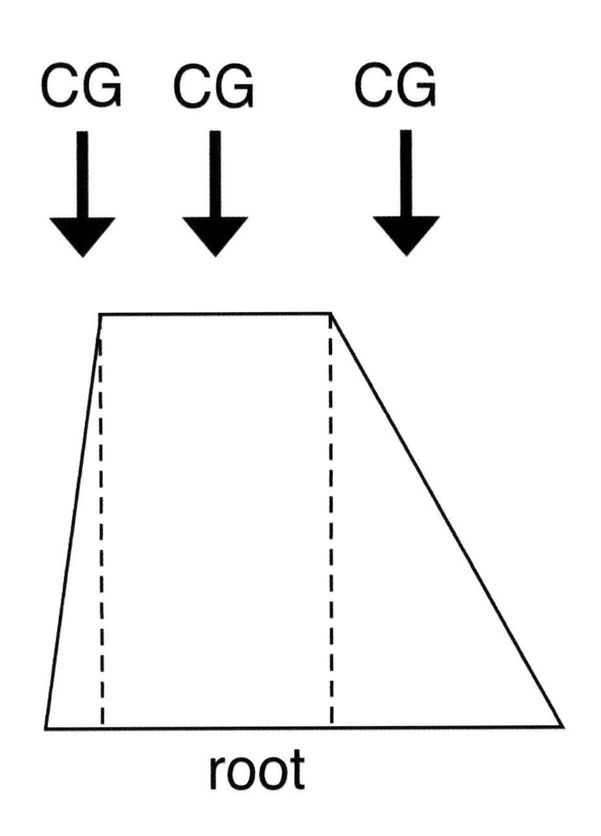

Figure 7-21. *Complex fins can be split into simpler shapes to find the center of gravity. Here, Juno's fin is split into two triangles and a rectangle, all of which are easy to deal with. The three separate centers of gravity for the parts are shown.*

Finding the Center of Pressure with Math

At the 1966 NARAM, a national gathering of rocketeers sponsored by the National Association of Rocketry, James and Judith Barrowman presented a paper as an R&D project. Titled "The Theoretical Prediction of the Center of Pressure," this is perhaps the most famous paper on rocketry produced by a model rocketeer. It shows how to calculate the center of pressure for a rocket using techniques that are similar to the method we just used to calculate the center of gravity. As with the center of gravity, we break the rocket up into component parts and find the center of pressure for each part, then reassemble the parts mathematically to get the overall center of pressure. Let's see how this works. Before we do that, though, it's important to know that this method, which is also used by both the RockSim and OpenRocket simulators, has some limitations.

Limitations of the Barrowman Method for CP (Center of Pressure) Calculation

1. The angle of attack—the angle between the axis of the rocket and the direction of flight—must be close to zero. Angles below 10° generally are fine.
2. The speed of the rocket must be less than the speed of sound. Aerodynamics change in the transonic region, and stay different once you get past the transonic region. This change starts at about Mach 0.8, which is about 880 ft/s, or 600 mph. Fortunately, model rockets rarely get that fast.
3. The air flow must be smooth, and can't change quickly.
4. The rocket is thin compared to its length.
5. The rocket is symmetrical about its center axis.
6. The rocket is rigid. All rockets bend a little, but it can't bend much.
7. The fins are thin plates.

Figure 7-22 shows the rocket with the component parts we'll use. The parts are the nose cone, body tube, conical shoulders, boat tails, and fins.

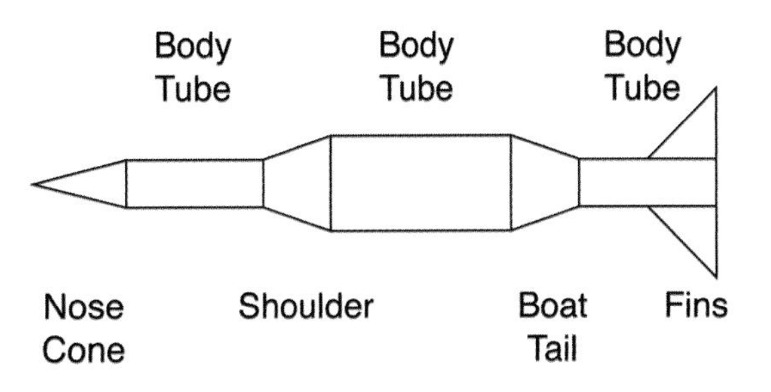

Figure 7-22. *The major components of a rocket for the purpose of calculating the center of pressure are shown here.*

When a stable rocket is tilted relative to the flow of air, the pressure from the air tries to straighten it out. The part we care about is called the *normal force*, which pushes sideways on the rocket. Force toward the rear of the rocket is drag, and doesn't matter in terms of straightening out the rocket.

The trick, then, is to find the size and location of the normal force for each component of the rocket. Once we know the magnitude of the normal force and where it is located on the part, we can use the same method we used for finding the center of gravity to find the overall center of pressure.

The normal force is:

$$N = \frac{1}{2} C_{N\alpha} \rho v^2 \alpha A_r$$

where:

- N is the normal force acting on the rocket.
- α is the *angle of attack* of the rocket, which is the angle between the rocket's direction and the direction of the wind. This is 0° if the rocket is flying perfectly straight.
- $C_{N\alpha}$ is a dimensionless coefficient that takes the shape of the component into account. It does depend on the angle of attack, which we assume is small enough that it will be constant.
- ρ is the density of air.
- v is the speed of the rocket.
- A_r is a reference area. We will use the area at the base of the nose cone.

Does the equation make sense? Let's work through the terms and see.

The equation is telling us that the normal force varies with the air pressure—thicker air gives a larger normal force. That makes sense; more air will give more force, and fins serve no purpose in a vacuum, where the air density is zero.

The normal force goes up with the square of the velocity. If the rocket isn't moving, the fins don't do any work. That also makes sense, and tells us why we need a launch rod. The fins don't work until the rocket picks up some speed.

If the angle of attack is zero, there is no normal force. As the angle goes up, the normal force gets larger. That makes sense, too. You might wonder if it really goes up linearly, though. Will 90° give twice the normal force of 45°? Well, no—but for small angles, the equation works. That's why the method is restricted to small angles.

Finally, the normal force goes up if the rocket is bigger. That makes sense, too. If there is more rocket, the force from the air will be larger.

It is possible to find the normal force, coefficient force, and locations for various rocket parts using some of the same general kinds of techniques used to find the center of gravity, coupled with numerical analysis. But we won't do that here. There are some references at the end of the chapter that will point you to more detailed papers if you are interested. For now, we'll just take a look at the results.

For all nose cones that come to a point and vary smoothly from one end to the other, the normal coefficient is:

$$C_{N\alpha} = 2$$

While the size of the force is the same for all nose cones, the location is not. The locations of the center of pressure for a few common nose cones are shown in Figure 7-23.

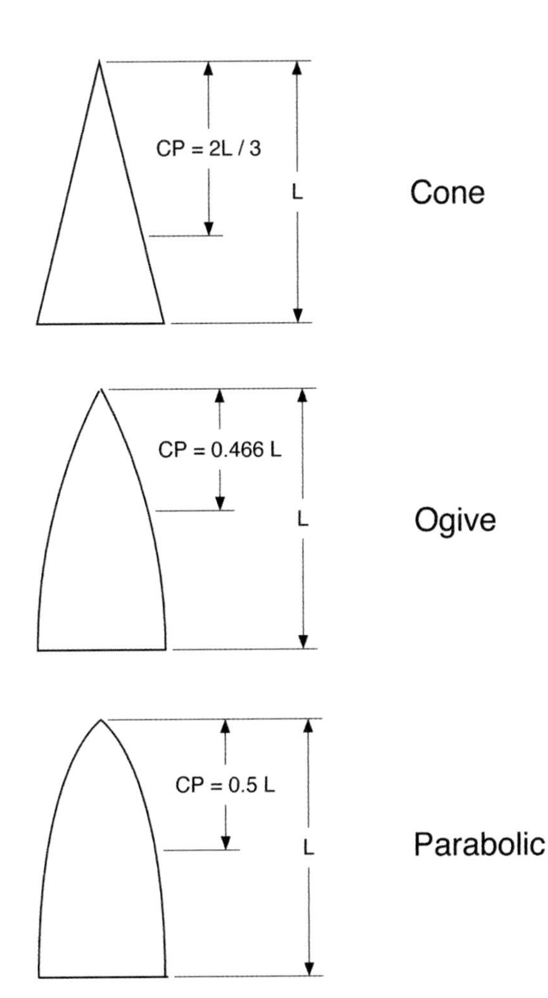

Figure 7-23. *Locations of the center of pressure for a few common nose cone shapes.*

The body tube is simple. It turns out that the body tube doesn't contribute to the normal force, at least at low angles of attack.

Figure 7-24 shows the location of the normal force for a conical shoulder.

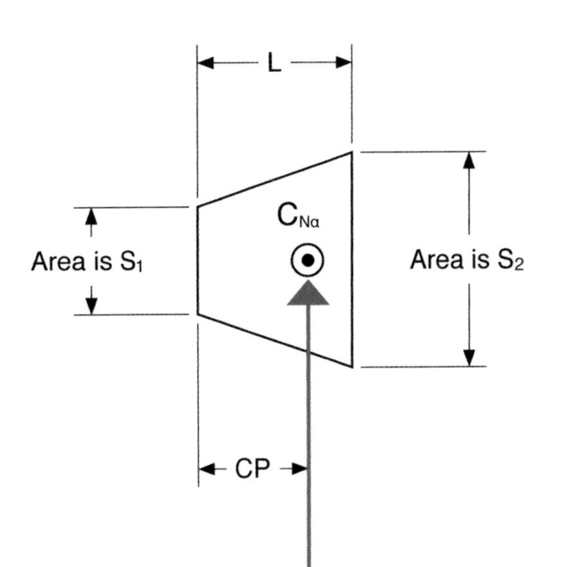

Figure 7-24. *Location of the center of pressure on a conical shoulder, along with a definition of the variables used to compute it.*

Conical shoulders have a normal force coefficient of:

$$C_{N\alpha} = \frac{8}{\pi d^2}(S_2 - S_1)$$

where:

- d is the diameter at the reference area, A_r. By convention, this is the diameter at the base of the nose cone.
- S_1 is the cross-section area at the top of the shoulder.
- S_2 is the cross-section area at the bottom of the shoulder.

Of course, if you know the diameter of the body tube that attaches to the top and bottom of the shoulder, it is easy enough to find the area using:

$$S = \frac{\pi d^2}{4}$$

In this case, the diameter is the diameter of the body tube.

The location of the center of pressure for the shoulder is:

$$CP = \frac{L}{3}\left(1 + \frac{1 - \frac{d_1}{d_2}}{1 - \left(\frac{d_1}{d_2}\right)^2}\right)$$

where:

- d_1 is the diameter at the front of the shoulder.
- d_2 is the diameter at the base of the shoulder.
- L is the length of the shoulder.

Figure 7-25 shows the location of the normal force for a boat tail.

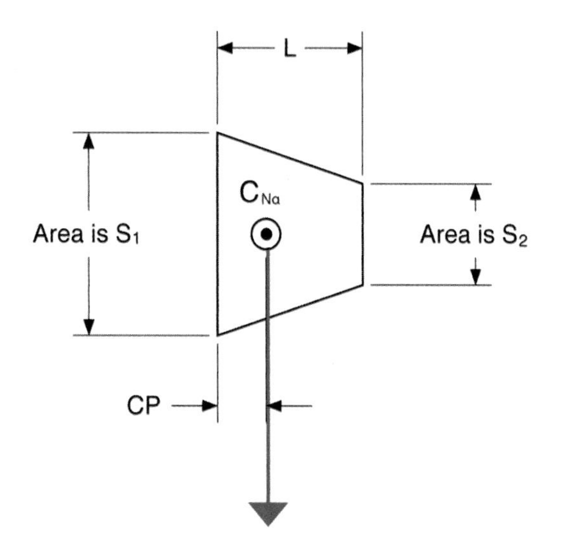

Figure 7-25. *Location of the center of pressure on a boat tail, along with a definition of the variables used to compute it. The direction of the normal force is reversed compared to other components, meaning the force pulls, rather than pushes, when the rocket is not flying straight.*

A boat tail is essentially a shoulder that is turned around, so it will come as no surprise that the equation for the normal force is just the equation for the shoulder with the areas reversed:

$$C_{N\alpha} = \frac{8}{\pi d^2}(S_1 - S_2)$$

The surprise comes when you realize that the value will be negative! That means the normal force is backward, pulling on the rocket instead of pushing. A boat tail ahead of the center of gravity actually acts like a negative fin, pulling the nose of the rocket back in line while the fins push the rear toward the center line.

The location of the center of pressure is the same as for the shoulder.

Fins come in a lot of shapes. There are two ways to deal with this. One way is to start with the equation for the fin and do the calculations necessary to find the actual center of pressure and normal force. The other is to estimate a bit. Estimation actually works pretty well in this situation. Figure 7-26 shows the idealized fin shape for doing center of pressure calculations, and Figure 7-27 shows a few examples of converting a fin to the idealized shape. The idea is to create a fin that has the same area as the original, but is formed using a simplified shape with a leading edge, a trailing edge, and a fin tip that is parallel to the body tube. Set b to 0 for a triangular fin.

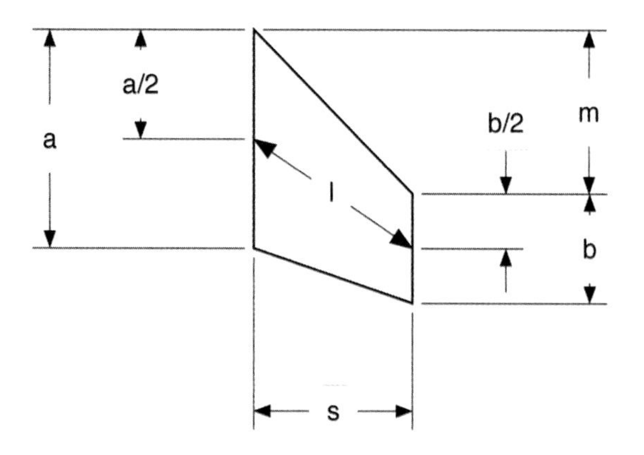

Figure 7-26. *The idealized fin for center of pressure calculations, along with the dimensions used in the equations.*

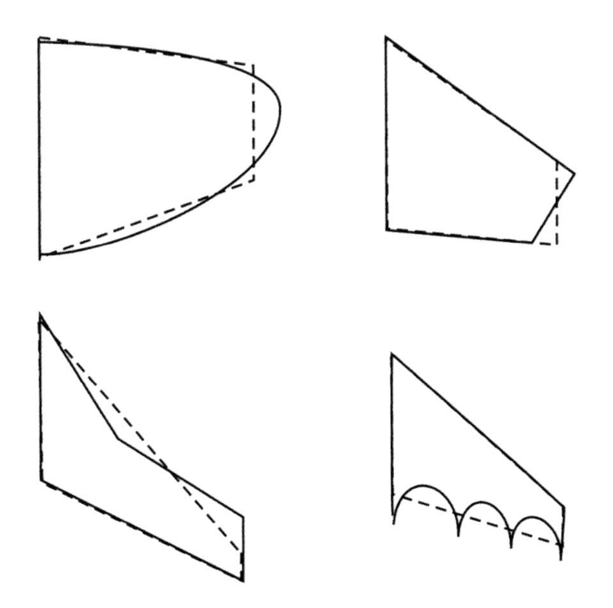

Figure 7-27. *Fins that don't fit the ideal pattern can be changed so they do. Keep the overall area the same, but match the original shape as well as you can with the idealized pattern. Here are a few fins with the idealized equivalents shown with dashed lines.*

For the idealized fin, the normal force coefficient is:

$$C_{N\alpha} = K_f \frac{4n\left(\frac{s}{d}\right)^2}{1 + \sqrt{1 + \left(\frac{2l}{a+b}\right)^2}}$$

where the lengths a, b, and s are defined in Figure 7-26, and n is the number of fins. This equation only works for rockets with three, four, or six fins. If the rocket has more than six fins, the fins start to mask each other—the wind flow past one fin causes the air to behave differently when it hits the fin behind the first one.

The length l is shown in Figure 7-26, but it's not one we usually measure. The length l is called the semichord. It's the length from the center of the root to the center of the fin tip. With a little geometry, we see that it is:

$$l = \sqrt{s^2 + \left(m + \frac{1}{2}(b - a)\right)^2}$$

K_f is an additional coefficient that accounts for the interference between the body tube and the fins. For three or four fins, K_f is:

$$K_f = 1 + \frac{r}{r+s}$$

For six fins, K_f is:

$$K_f = 1 + \frac{1}{2}\left(\frac{r}{r+s}\right)$$

In this case, r is the radius of the body tube and s is the semispan, as shown in Figure 7-28.

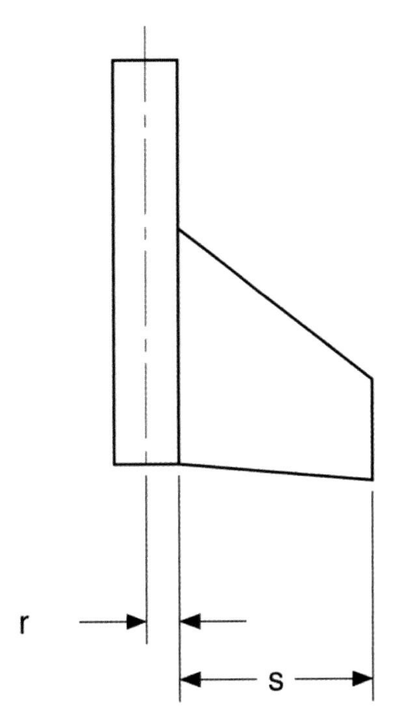

Figure 7-28. *The radius r and semispan s for calculating the fin interference coefficient.*

The location of the center of pressure is given by:

$$CP = \frac{m(a + 2b)}{3(a+b)} + \frac{1}{6}\left(a + b - \frac{ab}{a+b}\right)$$

Once all of the individual components are known, the overall center of pressure is given by:

$$CP = \frac{\sum_{i=1}^{n} CP_i C_{N\alpha i}}{\sum_{i=1}^{n} C_{N\alpha i}}$$

Let's see how this works for Juno. It's a pretty simple calculation, since only the nose cone and fins contribute to the center of pressure. As we saw earlier, the body tube is ignored at low angles of attack.

The nose cone is an ogive that is 2.75" long. $C_{N\alpha}$ is 2, while the location for the center of pressure is:

$$CP = 0.466\,L = 0.466 \times 2.75 = 1.282\text{ in}$$

The fins are already in the idealized shape. Figure 7-29 shows the fin with the dimensions for the various lengths on the ideal fin from Figure 7-26.

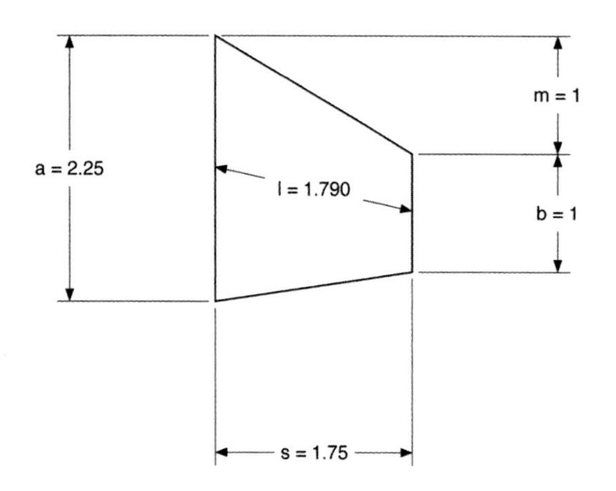

Figure 7-29. *Juno's fin with the dimensions needed to calculate the center of pressure labeled.*

The interference coefficient is:

$$K_f = 1 + \frac{r}{r+s} = 1 + \frac{0.488}{0.488 + 1.75} = 1.219$$

We will also need the semichord length:

$$l = \sqrt{s^2 + \left(m + \frac{1}{2}(b - a)\right)^2}$$

$$l = \sqrt{1.75^2 + \left(1 + \frac{1}{2}(1 - 2.25)\right)^2} = 1.790\text{ in}$$

The normal force for the set of fins is:

$$C_{N\alpha} = K_f \frac{4n\left(\frac{s}{d}\right)^2}{1 + \sqrt{1 + \left(\frac{2l}{a+b}\right)^2}}$$

$$C_{N\alpha} = 1.219 \frac{4 \times 3\left(\frac{1.75}{0.976}\right)^2}{1 + \sqrt{1 + \left(\frac{2 \times 1.790}{2.25 + 1}\right)^2}}$$

$$C_{N\alpha} = 18.904\text{ N}$$

The location of the center of pressure, measured from the front of the fins, is:

$$CP = \frac{m(a + 2b)}{3(a+b)} + \frac{1}{6}\left(a + b - \frac{ab}{a+b}\right)$$

$$CP = \frac{1(2.25 + 2)}{3(2.25 + 1)} + \frac{1}{6}\left(2.25 + 1 - \frac{2.25 \times 1}{2.25 + 1}\right)$$

$$CP = 0.863\text{ in}$$

When we add the individual contributions, we need to remember that the center of pressure for the fins is relative to the front of the fins. The fins themselves are 12.5" from the front of the rocket, so we need to add that value to the CP (center of pressure) for the fins. Summing the individual contributions for the center of pressure, we get:

$$CP = \frac{\sum_{i=1}^{n} CP_i\, C_{N\alpha i}}{\sum_{i=1}^{n} C_{N\alpha i}}$$

$$CP = \frac{1.282 \times 2 + (12.5 + 0.863) \times 18.904}{2 + 18.904}$$

$$CP = 12.207\text{ in}$$

The location of the center of pressure is in inches, as were all of our units. Converting to centimeters, it's 31.006 cm, which is a good agreement with the rocket simulator's calculation of 31 cm.

Further Reading

This chapter showed you all of the methods commonly used in model rocketry to determine if a rocket is stable. It also showed you some of the strengths and weaknesses of each method. But we could only scratch the surface of this topic. Here's a list of references that will help you to learn more about rocket stability. All are available for free online, and many point you to still more detailed information:

- "The Theoretical Prediction of the Center of Pressure" (*http://bit.ly/barrowman_report*), James and Judith Barrowman's original 1966 report on calculating the center of pressure
- "Stability of a Model Rocket in Flight" (*http://bit.ly/rocketStabilityFlight*), an article by Jim Barrowman on rocket stability
- "Calculating the Center of Pressure of a Model Rocket" (*http://bit.ly/calcCenterPressure*), a reworking of Jim Barrowman's original article as a Centuri (now Estes) technical report
- The "Rocket Stability" section of the Apogee Components website (*http://bit.ly/apogee-stability*), with links to over a dozen different articles on rocket stability from Apogee's fine series of newsletters

How High Did It Go? 8

After watching a rocket roar into the sky, seeing the smoke trail, and watching the rocket descend gracefully under a parachute, the first thing most people want to know about the flight is how high the rocket went. The second thing they want to know is how fast it went. We'll investigate several methods to find out how high the rocket went in this chapter, and one method to find out how fast it went.

There are basically two practical ways to track the altitude of a model rocket. The first is to track the rocket visually, measuring the angle of the rocket from the launcher and using trigonometry to find the altitude. This is the method NAR uses most often for official tracking for contests and records for low-power rockets. The second method is to put an altimeter in the rocket. This finds the rocket's altitude using a tiny barometer and accelerometer. Using an altimeter is the most accurate way to find the altitude of the rocket, and it's also the favored way to track the altitude for high-powered rockets, which often go out of sight.

The Single-Axis Tracker

Let's start with a single-axis altitude tracker—essentially, and sometimes literally, a protractor to measure an angle. This is a very simplified form of the kind of tracker used by NAR for contests. Keep in mind that while this tracker is great for giving a reasonable idea of the altitude of a rocket flight, it is not adequate for NAR-sanctioned contests. We'll show a more sophisticated tracker, one that satisfies all of the NAR requirements, later in the chapter. Still, this first tracker is cheap, easy to build and operate, and reasonably accurate.

The idea is very simple, and very familiar to anyone who has taken a trigonometry class. If you know the distance from the launch pad to the tracker, and you know the angle between the ground and the rocket's highest point, you can find the altitude of the rocket. We do make the simplifying assumption that the rocket went straight up. The altitude A is:

$$A = d \tan \theta$$

where d is the distance between the tracker and the launcher, and θ is the angle between the ground and the rocket.

The equation is pretty easy to use even if you don't know much about trigonometry. The tangent function, abbreviated as *tan*, is available on most calculators. Table 8-1, later in the chapter, also shows the altitudes for various angles.

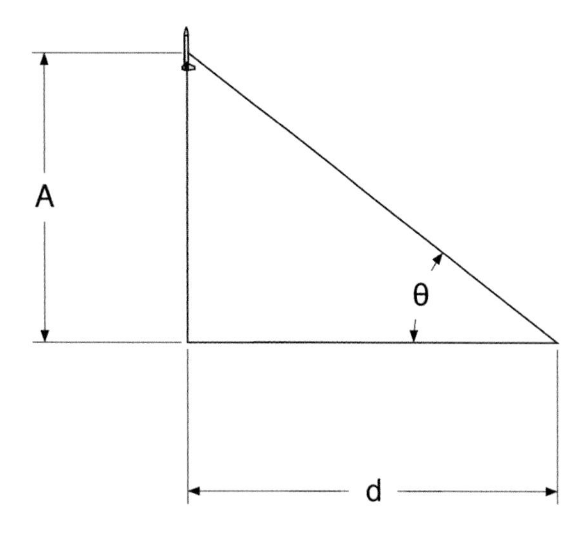

Figure 8-1. *Finding the altitude when you know the angle θ and baseline distance d is a simple calculation on a calculator.*

Building a Single-Axis Tracker

It is extremely easy to build a single-axis tracker. All you need is a protractor, a string, and a weight. The protractor should be as large as possible. Figure 8-2 shows one you can copy and paste onto foam core, cardboard, tin, or plywood. You can get a full-size template for the protractor and tracker at the author's website (*http://bit.ly/byteworks-make-rockets*).

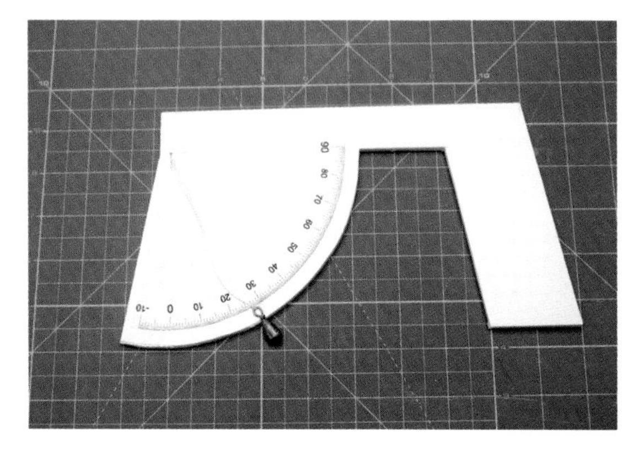

Figure 8-2. *You can make a single-axis tracker from a free template and a few cents' worth of parts.*

Thread the string through the marked hole at the center of the protractor and tie the weight to the other end. The weight can be a fishing weight like the one shown, a ball of clay, or any other convenient dense material.

Make sure the string is long enough to extend past the edge of the protractor. Tie a knot or use tape or glue to fasten one end of the string at the center of the semicircle, and fasten the other end securely to the weight.

The tracker is used to measure the angle between the launcher and the apogee of the rocket. Sighting along the top of the tracker to the launcher should give an angle of 0°. If not, the tracking position is uphill or downhill from the launcher. The best thing to do in that case is move. If moving isn't practical, and the difference is small, adjust the final angle by subtracting the angle to the launcher from the angle to the rocket's apogee.

When tracking the rocket, sight along the top of the protractor and follow the rocket to its highest altitude. Once there, clamp the string in place with your thumb so you can turn the tracker to read the angle without changing the angle.

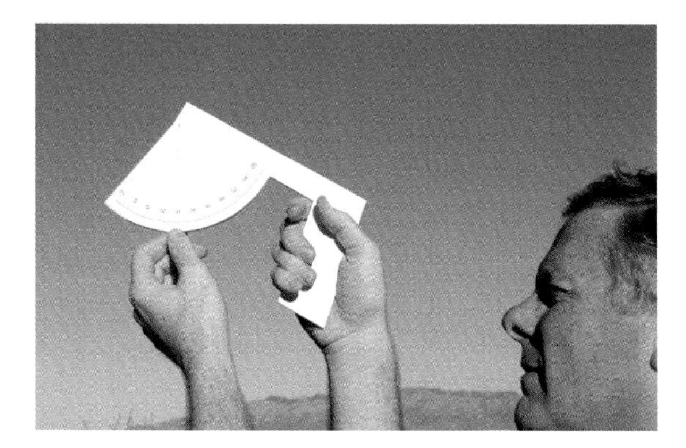

Figure 8-3. *Track the rocket to apogee, then clamp the string in place so you can turn the tracker to read the angle.*

A nice alternative to the homemade tracker is the tracker from Estes Industries shown in Figure 8-4. It uses a clever trigger mechanism to lock the measurement in place, and has great sights for accurate tracking.

Figure 8-4. *This tracker from Estes is rugged and sports a trigger-based locking mechanism.*

Math Behind Single-Axis Trackers

The math in the first part of this section is appropriate for anyone who has had a trigonometry course, and can be followed by anyone who is adventurous and has had algebra. There is an error analysis section later that uses calculus. You can skip all of this and just use Table 8-1 to look up the altitude, but even if you don't dig into the math, there are some interesting facts about where tracking error comes from that make it worth skimming this section.

The math used to find the altitude of the rocket relies on the definition of the tangent of an angle. The definition states that the tangent of an angle is the length of the opposite side of the triangle over the length of the adjacent side, provided the angle between the two sides is 90°.

From there, it's a simple rearrangement to get the formula for the altitude of the rocket:

$$A = B \tan \theta$$

where A is the altitude, B is the length of the *baseline*, which is the distance from the launch pad to the tracker, and θ is the angle measured by the tracker.

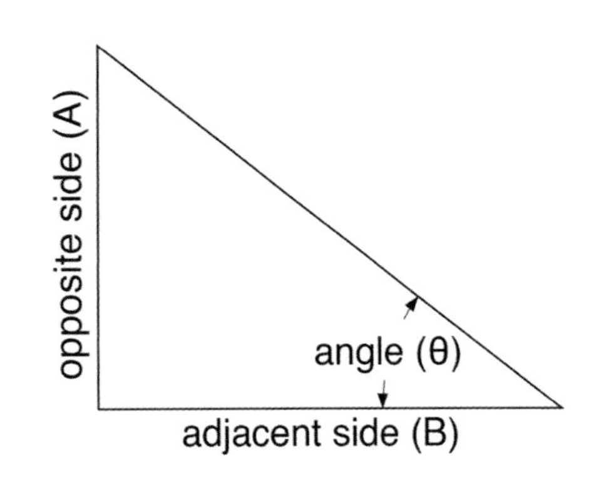

Figure 8-5. *The tangent of an angle is defined as the length of the opposite side divided by the length of the adjacent side.*

Plugging in the data and solving the equation will give you a number, but how accurate is that number as a representation of the rocket's altitude? Error analysis is both a well-known area of science and engineering, and a frequently ignored one. Look at it this way: if I tell you my rocket flew to 500 feet on an A motor, you might doubt my calculations. After all, the predicted altitude for Juno on an A motor is about 350 feet. Now, if I tell you I tracked my Juno to 500 feet plus or minus 100 feet, you'll have more confidence that I didn't mess up, simply because you understand I'm telling you the error was large. You might reasonably ask why I don't work to get a lower error. That's where understanding the math behind error analysis helps. It will show you, very clearly, which part of the measurement has the biggest effect on the error in the altitude for the rocket.

Let's start by looking at what we mean when we give an error on a number. When I say the altitude is 500 feet plus or minus 100 feet, that's usually written as 500 ± 100 feet. The 100 is one *standard deviation*. A standard deviation describes the way many random samples fall about the middle of the data. In this case, it's telling us that our attempts to measure something are likely to average to the correct value, and we're more likely to be close to the correct value than really far away from it. Saying the standard deviation is 100 feet says there is a 68.2% chance the actual altitude was in the range of 400 feet to 600 feet, and a 95.4% chance it was in the range of 300 feet to 700 feet.

Table 8-1, later in the chapter, lists the error for various altitudes. Maybe that's good enough for you. If so, feel free to skip the rest of this section. If you want to dig into the math, though, get ready for a little calculus.

The first thing we need to do is look at the equation and decide where error can creep in. It's pretty clear we could be off a bit measuring the length of the baseline. Of course, we may not measure the angle accurately, either. These are measurement errors. One way to figure out the error is to make the same measurement multiple times and calculate the standard deviation. Another is to estimate the error. That's what is done later for Table 8-1. We assume our measurement for the baseline is within 2%, and the angle is ±2°.

Can you do better than that? Maybe. Measure the angle to the top of a fixed object, like a flagpole, 10 or 20 times and find the standard deviation to see. Be sure to do it quickly, since a rocket will not give you time for a careful measurement. One way to do that would be to start off pointed away from the flagpole, then turn and take the measurement right away. Do the same for measuring a 500-foot baseline from the flagpole. Make a chalk mark each time to measure the distance and see how close they are. Yes, it's painstaking. Welcome to engineering. If you want to get a better answer, it often takes more work.

There are two other sources of error that aren't as obvious. They both have to do with the assumption that the angle between the baseline and the rocket at apogee is 90°. There are two reasons that might be off. The first is that the tracker might be uphill or downhill from the launch pad. Try measuring the angle to the launch pad itself. It should be 0°. If not, you're uphill or downhill. For our calculations, we'll assume you are approximately level with the launcher, and use 2° for this error.

The other source of error shows the real problem with a single-axis tracker. The rocket might arc toward the tracker, making the measured altitude much higher than the actual altitude, or it might arc away from the tracker, making the altitude appear too low. The cause of the rocket's arc is usually the wind, so we can reduce this error by placing the tracker 90° from the direction of the wind. However, even if we do this, the error never goes away completely. Table 8-1 assumes the rocket is within 10° of the vertical.

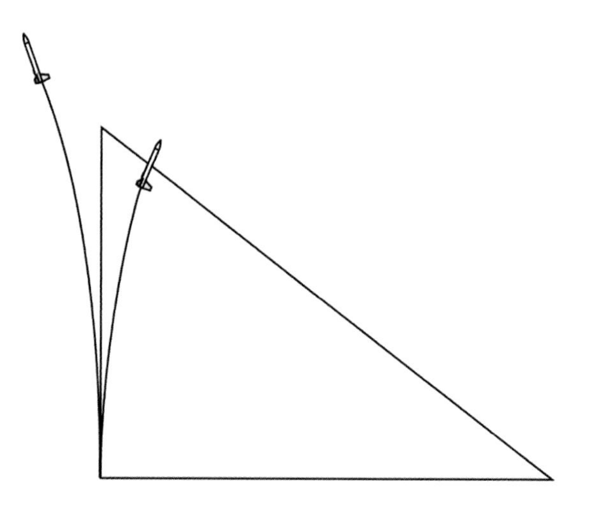

Figure 8-6. *The actual altitude can be off quite a bit if the rocket doesn't really go straight up. The drawing shows the error from just a 10° difference from a vertical angle. The tracker will report both altitudes as being at the tip of the triangle.*

Now we can go back to our original equation and figure out how much these errors contribute to the overall error in altitude. First we have to make a change in the equation, though. Since the angle of the rocket might not be exactly 90° from the baseline, we need a new equation that includes the second angle.

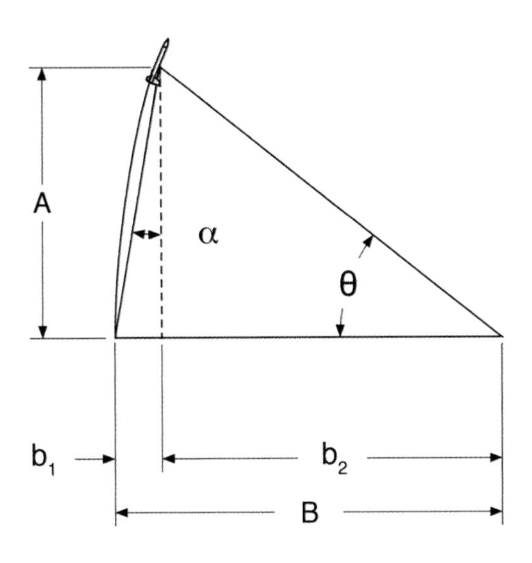

Figure 8-7. *The new tracking equation will use an angle* α*. Dropping a line from apogee splits the baseline into two components,* b_1 *and* b_2*, that add to B.*

We need to develop an altitude equation that assumes α is not zero. A little algebra will do the trick. Start with the fact that the distance from the point under the rocket to the launcher and to the tracker is equal to the measured baseline:

$$B = b_1 + b_2$$

We can use the definition of the tangent to find the relationship between the short baselines, the angles, and the altitude:

$$\tan\alpha = \frac{b_1}{A}$$

$$\tan\theta = \frac{A}{b_2}$$

Solving for A gives us a relation between the angles and small baselines:

$$A = \frac{b_1}{\tan\alpha} = b_2 \tan\theta$$

We need the value for b_2 so we can find the actual altitude, so we need to solve this equation for b_2:

$$b_2 = \frac{b_1}{\tan\alpha\ \tan\theta}$$

Using the fact that

$$b_1 = B - b_2$$

and doing a bit of rearranging, we end up with:

$$b_2 = \frac{b_1}{\tan\alpha\ \tan\theta}$$

$$b_2 = \frac{B - b_2}{\tan\alpha\ \tan\theta}$$

$$b_2 \tan\alpha\ \tan\theta = B - b_2$$

$$B = b_2(1 + \tan\alpha\ \tan\theta)$$

$$b_2 = \frac{B}{1 + \tan\alpha\ \tan\theta}$$

$$\frac{A}{\tan\theta} = \frac{B}{1 + \tan\alpha\ \tan\theta}$$

$$A = \frac{\tan\theta}{1 + \tan\alpha\ \tan\theta} B$$

As a quick sanity check, if the rocket really does fly straight up, α is 0, so tan α is 0, and the altitude equation reduces to the original one we had before we started the error analysis.

With an altitude equation that has all of the sources of error included, it's time to figure out the relative contributions of the errors. This involves a bit of calculus, since we will need to take some derivatives. If calculus is not something you've had a chance to master yet, skim over this section and check out the results.

Assuming the errors are relatively small compared to the measured values, we can get the square of the overall error by adding the square of the contribution of each individual error, multiplied by the square of the derivative of the equation with respect to the variable whose error we're looking at.

Huh?

Let's try it with math. It's so much clearer that way. The error in the altitude, indicated as σ_A, is:

$$\sigma_A^2 = \sigma_\theta^2 \left(\frac{\partial A}{\partial \theta}\right)^2 + \sigma_\alpha^2 \left(\frac{\partial A}{\partial \alpha}\right)^2 + \sigma_B^2 \left(\frac{\partial A}{\partial B}\right)^2$$

The fractions are partial derivatives. They indicate we should take the derivative. The partial derivatives are:

$$\frac{\partial A}{\partial \theta} = \left(\frac{\sec^2 \theta}{1 + \tan \alpha \tan \theta} - \frac{\tan \alpha \tan \theta \, s\sec^2 \theta}{(1 + \tan \alpha \tan \theta)^2}\right) B$$

$$\frac{\partial A}{\partial \alpha} = - \frac{\tan^2 \theta \sec^2 \alpha}{(1 + \tan \alpha \tan \theta)^2} B$$

$$\frac{\partial A}{\partial B} = \frac{\tan \theta}{1 + \tan \alpha \tan \theta}$$

This simplifies a lot because α is supposed to be 0°. Making that substitution, the partial derivatives become:

$$\frac{\partial A}{\partial \theta} = B \sec^2 \theta$$

$$\frac{\partial A}{\partial \alpha} = - B \tan^2 \theta$$

$$\frac{\partial A}{\partial B} = \tan \theta$$

This gives a final equation for the error in altitude of:

$$\sigma_A = \sqrt{\sigma_\theta^2 (B \sec^2 \theta)^2 + \sigma_\alpha^2 (B \tan^2 \theta)^2 + \sigma_B^2 (\tan \theta)^2}$$

Let's try an example to see how this would work. For our example, we will assume we are tracking a rocket we expect to reach an altitude of about 350 feet, so we'll use a 300-foot baseline. That's the value for B. Assuming a 2% measurement error, that gives B as 300±6 feet.

Let's say the tracker returned an angle of 50°. This means θ is 50±2°.

The angle between the baseline and the line from the launcher to the rocket should be 0°, since we hope the rocket goes straight up. We're allowing for a 10° error, to account for the arcing path of the rocket and the possibility that the tracker might be a bit uphill or downhill from the launcher. The means α is 0±10°.

The calculated altitude in feet is:

$$A = B \tan \theta$$

$$A = 300 \tan 50$$

$$A = 358 \text{ ft}$$

While most calculators are quite happy to use degrees or radians, the error in an angle needs to be in radians for the math to work. That means the error in each angle must be multiplied by π/180. The error is:

$$\sigma_A = \sqrt{\sigma_\theta^2 (B \sec^2 \theta)^2 + \sigma_\alpha^2 (B \tan^2 \theta)^2 + \sigma_B^2 (\tan \theta)^2}$$

$$\sigma_A = \sqrt{\frac{2\pi^{2}}{180} (300 \sec^2 50)^2 + \frac{10\pi^{2}}{180} (300 \tan^2 50)^2 + 6^2 (\tan 50)^2}$$

$$\sigma_A = \sqrt{642.37 + 5530.23 + 51.13}$$

$$\sigma_A = 78.9 \text{ ft}$$

So the altitude is 358±79 feet. The error tells us we have an approximation of the altitude, but it's only good to about 20% or so. The error analysis tells us more, though. It tells us that we need to work on making sure the rocket is vertical to get a good altitude measurement with a single-axis tracker, since the overwhelming majority of the error comes from the error in the angle between the baseline and the line from the rocket to the launcher. Knowing the error is useful. Knowing where the error comes from justifies the work of going through the math, because it tells us how to reduce the error.

Using a Single-Axis Tracker

Tracking the altitude of a rocket with a single-axis tracker is pretty easy. The first step is to lay out the baseline. This is usually done using a tape measure. The location for the tracker should be perpendicular to the wind direction to minimize the error in the launch angle. The sun should be behind the tracker so it doesn't get in the eyes of the person doing the tracking. The length of the baseline should be roughly the same as the expected altitude of the rocket. The tracker should also be at the same elevation as the launcher—it should not be uphill or downhill from the launcher.

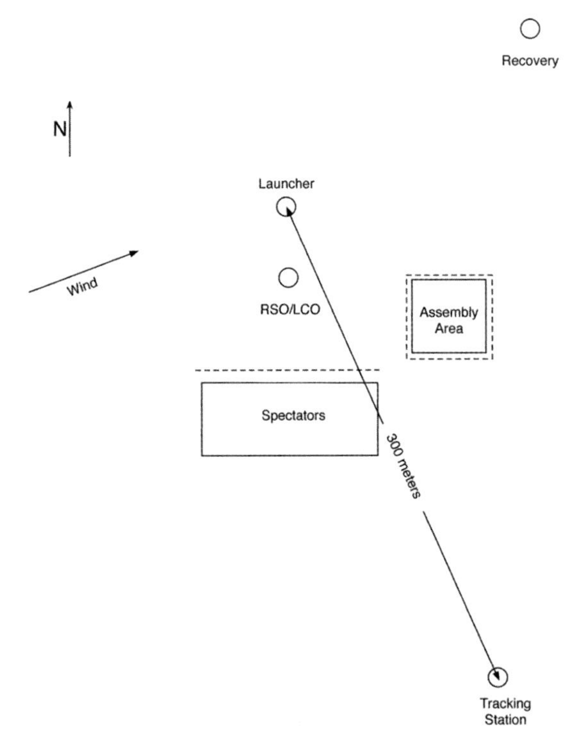

Figure 8-8. The tracking station should be perpendicular to the wind, generally to the south so the sun is behind the tracker, and about as far away from the launcher as the expected altitude of the rocket.

Official altitude measurements at NAR-sanctioned contests are done in meters, so the baseline is laid out in meters, too. Those of us in the US usually have better access to tape measures that measure feet, though, so it's common to see measurements in feet for less official results. Either will work just fine.

You might not have a tape measure long enough to measure the baseline. If not, measure the distance from the launcher to the end of the tape measure, then measure the distance from that point to the end of the tape measure, and so on until you get the desired total distance. Be sure to measure in a straight line by sighting from your endpoint to the person standing at the last measured location and on to the launcher.

Figure 8-9. The baseline is often longer than your tape measure. Here, the author and his lovely assistant are setting up a 300-foot baseline with a 100-foot tape measure. The end of the first 100 feet is located, and then the tape measure is held there and extended another 100 feet. The tape measure is in a line with the launcher, seen behind the post in the background.

With the baseline measured, track the rocket to either its highest altitude or the ejection charge. It is usually easier to see the ejection charge than the rocket, particularly at higher altitudes, so the ejection charge is usually the better choice. Let the string and weight hang free as you track the rocket. Once you see the ejection charge, or when the rocket reaches apogee, clamp your thumb onto the string. Keeping your thumb on the string, bring the tracker down and read the angle. From there, you can either find the altitude using trigonometry, or look it up in the table in the next section.

Altitude Tables for Single-Axis Trackers

Table 8-1 shows the altitude of a rocket for various angles and baselines. Let's see how it is used.

Find the baseline length along the top row of Table 8-1. Scan down to find the row showing the angle you measured with the altimeter. The table shows the altitude and standard deviation in the error for the altitude. The error calculation assumes the following measurement errors:

- Angle between the apogee and true vertical from the launcher: ±10°
- Baseline length: ±2% of the measured length
- Tracker angle: ±2°

The table works for either meters or feet. If the baseline is measured in feet, the altitude is in feet; if the baseline is measured in meters, so is the altitude.

Table 8-1. Altitude table for a single-axis altitude tracker

	100	200	250	300	400	500	1000
1	2±3	3±7	4±9	5±10	7±14	9±17	17±35
2	3±3	7±7	9±9	10±10	14±14	17±17	35±35
3	5±4	10±7	13±9	16±11	21±14	26±18	52±35
4	7±4	14±7	17±9	21±11	28±14	35±18	70±35
5	9±4	17±7	22±9	26±11	35±14	44±18	87±35
6	11±4	21±7	26±9	32±11	42±14	53±18	105±35
7	12±4	25±7	31±9	37±11	49±14	61±18	123±36
8	14±4	28±7	35±9	42±11	56±14	70±18	141±36
9	16±4	32±7	40±9	48±11	63±14	79±18	158±36
10	18±4	35±7	44±9	53±11	71±15	88±18	176±37
11	19±4	39±7	49±9	58±11	78±15	97±19	194±37
12	21±4	43±8	53±9	64±11	85±15	106±19	213±38
13	23±4	46±8	58±10	69±11	92±15	115±19	231±38
14	25±4	50±8	62±10	75±12	100±16	125±19	249±39
15	27±4	54±8	67±10	80±12	107±16	134±20	268±40
16	29±4	57±8	72±10	86±12	115±16	143±20	287±41
17	31±4	61±8	76±10	92±13	122±17	153±21	306±42
18	32±4	65±9	81±11	97±13	130±17	162±22	325±43
19	34±4	69±9	86±11	103±13	138±18	172±22	344±45
20	36±5	73±9	91±12	109±14	146±19	182±23	364±46
21	38±5	77±10	96±12	115±14	154±19	192±24	384±48
22	40±5	81±10	101±13	121±15	162±20	202±25	404±50
23	42±5	85±11	106±13	127±16	170±21	212±26	424±53
24	45±6	89±11	111±14	134±17	178±22	223±28	445±55
25	47±6	93±12	117±14	140±17	187±23	233±29	466±58
26	49±6	98±12	122±15	146±18	195±24	244±30	488±61
27	51±6	102±13	127±16	153±19	204±26	255±32	510±64
28	53±7	106±13	133±17	160±20	213±27	266±34	532±67
29	55±7	111±14	139±18	166±21	222±29	277±36	554±71

	100	200	250	300	400	500	1000
30	58±8	115±15	144±19	173±23	231±30	289±38	577±75
31	60±8	120±16	150±20	180±24	240±32	300±40	601±80
32	62±8	125±17	156±21	187±25	250±34	312±42	625±85
33	65±9	130±18	162±22	195±27	260±36	325±45	649±90
34	67±10	135±19	169±24	202±29	270±38	337±48	675±95
35	70±10	140±20	175±25	210±30	280±40	350±51	700±101
36	73±11	145±21	182±27	218±32	291±43	363±54	727±107
37	75±11	151±23	188±29	226±34	301±46	377±57	754±114
38	78±12	156±24	195±30	234±36	313±49	391±61	781±121
39	81±13	162±26	202±32	243±39	324±52	405±65	810±129
40	84±14	168±28	210±34	252±41	336±55	420±69	839±138
41	87±15	174±29	217±37	261±44	348±59	435±73	869±146
42	90±16	180±31	225±39	270±47	360±62	450±78	900±156
43	93±17	187±33	233±42	280±50	373±67	466±83	933±166
44	97±18	193±35	241±44	290±53	386±71	483±89	966±177
45	100±19	200±38	250±47	300±57	400±76	500±95	1000±189
46	104±20	207±40	259±50	311±61	414±81	518±101	1036±202
47	107±22	214±43	268±54	322±65	429±86	536±108	1072±215
48	111±23	222±46	278±58	333±69	444±92	555±115	1111±230
49	115±25	230±49	288±61	345±74	460±98	575±123	1150±246
50	119±26	238±53	298±66	358±79	477±105	596±131	1192±263
51	123±28	247±56	309±70	370±84	494±113	617±141	1235±281
52	128±30	256±60	320±75	384±90	512±121	640±151	1280±301
53	133±32	265±65	332±81	398±97	531±129	664±162	1327±323
54	138±35	275±69	344±87	413±104	551±139	688±173	1376±347
55	143±37	286±75	357±93	428±112	571±149	714±186	1428±373
56	148±40	297±80	371±100	445±120	593±160	741±200	1483±401
57	154±43	308±86	385±108	462±129	616±173	770±216	1540±431
58	160±47	320±93	400±116	480±140	640±186	800±233	1600±465
59	166±50	333±100	416±126	499±151	666±201	832±251	1664±502
60	173±54	346±109	433±136	520±163	693±217	866±272	1732±543
61	180±59	361±118	451±147	541±176	722±235	902±294	1804±588
62	188±64	376±128	470±160	564±192	752±255	940±319	1881±638
63	196±69	393±139	491±174	589±208	785±278	981±347	1963±694
64	205±76	410±151	513±189	615±227	820±303	1025±378	2050±757
65	214±83	429±165	536±207	643±248	858±331	1072±414	2145±827
66	225±91	449±181	562±227	674±272	898±363	1123±453	2246±907

	100	200	250	300	400	500	1000
67	236±100	471±199	589±249	707±299	942±399	1178±498	2356±996
68	248±110	495±220	619±275	743±330	990±440	1238±549	2475±1099
69	261±122	521±243	651±304	782±365	1042±487	1303±608	2605±1216
70	275±135	549±270	687±338	824±406	1099±541	1374±676	2747±1352
71	290±151	581±302	726±377	871±453	1162±604	1452±755	2904±1510
72	308±169	616±339	769±424	923±508	1231±678	1539±847	3078±1694
73	327±191	654±382	818±478	981±574	1308±765	1635±956	3271±1912
74	349±217	697±435	872±543	1046±652	1395±869	1744±1086	3487±2173
75	373±249	746±497	933±622	1120±746	1493±995	1866±1244	3732±2487
76	401±287	802±574	1003±718	1203±861	1604±1149	2005±1436	4011±2871
77	433±335	866±670	1083±837	1299±1004	1733±1339	2166±1674	4331±3348
78	470±395	941±790	1176±987	1411±1184	1882±1579	2352±1974	4705±3948
79	514±472	1029±944	1286±1180	1543±1416	2058±1888	2572±2359	5145±4719
80	567±573	1134±1147	1418±1433	1701±1720	2269±2293	2836±2866	5671±5733
81	631±710	1263±1421	1578±1776	1894±2131	2526±2841	3157±3552	6314±7103
82	712±902	1423±1804	1779±2255	2135±2706	2846±3608	3558±4510	7115±9019
83	814±1181	1629±2363	2036±2954	2443±3544	3258±4726	4072±5907	8144±11814
84	951±1612	1903±3224	2379±4030	2854±4836	3806±6448	4757±8060	9514±16120
85	1143±2326	2286±4652	2858±5815	3429±6978	4572±9305	5715±11631	11430±23262
86	1430±3641	2860±7282	3575±9102	4290±10923	5720±14563	7150±18204	14301±36408
87	1908±6481	3816±12962	4770±16203	5724±19444	7632±25925	9541±32406	19081±64812
88	2864±14597	5727±29193	7159±36491	8591±43790	11455±58386	14318±72983	28636±145966
89	5729±58420	11458±116839	14323±146049	17187±175259	22916±233678	28645±292098	57290±584195

Dual-Axis Tracking

If you worked through the error analysis section for the single-axis tracker earlier in the chapter, you saw that the biggest problem with getting an accurate reading from a single-axis tracker is that the rocket might not go straight up. We can eliminate that source of error using a *theodolite*. A theodolite measures two angles, one from the horizon to the apogee of the rocket, as before, and a second from another theodolite to a point under the rocket at apogee. The angle from the horizon to the rocket is called the *elevation angle*, while the angle from the rocket to the other theodolite is the *azimuth angle*. Instead of locating the rocket somewhere along a line and assuming the rocket is right above the launcher, the dual-axis tracking system figures out the position of the rocket in three dimensions. If you wanted to, you could even find how far the rocket is from the launcher when it hits apogee.

The biggest problem with dual-axis trackers is that they are hard to come by. Dual-axis tracking also involves more people. Doing it well requires at least six people, and doing it at all takes at least three. The calculations are more difficult, too, so you can't just plug an angle and distance into a simple formula, or look up the answer in a table. While we'll walk through the math, in practical use you'll need to use a short computer program to find the altitude of your rocket. Of course, I'll give you that program.

Building a Theodolite

The theodolite we will build is a simple yet fairly accurate way to track the altitude of a rocket. It meets all of the requirements from the NAR Pink Book for use in contests. It is designed to mount on a standard camera tripod. Camera tripods can be very cheap or very expensive, but even the cheapest camera tripod is likely to be more stable and far more portable than a tripod you build from scratch.

Figure 8-10. *A pair of theodolites can locate a rocket in three dimensions, getting rid of the biggest source of error from a single-axis tracker.*

Building a theodolite is no harder than building one of the more sophisticated model rockets. It takes a few materials and parts you may not have on hand, though. Look through the parts list (Table 8-2) and tools list (Table 8-3) carefully.

Don't be afraid to make substitutions, especially for the level, the indicator wire, and the brass tubes used as joints. You will need to adjust the plans a bit if you make substitutions, but there is nothing magic about the particular parts shown here. If you have something in your parts bin that will work, use it.

The parts list and directions that follow show how to build one theodolite. If you need two (you do actually need two theodolites to track a rocket), be sure to double the quantities of all parts.

Table 8-2. Parts list

Part	Description
1/4″ plywood	This is used for most of the structural parts. Visit your local hardware or woodworking store; most will have sheets of plywood about 2 ft x 3 ft that have a nice smooth surface on both sides. This is enough to make two trackers. Get cabinetry-grade hardwood, not the rough grade used for construction.
3/4"-thick wood	You will need two 1″ squares. One piece will need to be sanded down to a thickness of about 0.6″.
1/16″ piano wire	This is used for indicators. A little thicker or thinner won't hurt, so don't run out and buy more if you already have something that is close. This is sometimes called music wire or simply steel wire. You can find it at most hardware stores and hobby stores.
Photo paper, 2 sheets	The indicator dials will be printed from an inkjet printer. My tests showed photo paper works a lot better than card stock or plain paper, but it's not essential.
1/4-20 nut	Most photo tripods thread into a nut that is 1/4″ in diameter and has 20 threads per inch. This tracker is designed to sit on top of a photo tripod, and this nut is what will hold it.
Bull's-eye level	Also called a surface level, this type of level is usually circular with a bubble that is in the center of the level when the surface is level. The plans show Ace Hardware part 24539. You can substitute, but a few dimensions may change if you do, so be sure you have the level before you start cutting.
1/8″ plywood	This is used for a small ring to mount the level. You may need a different thickness if you use a different level. While I used plywood, you could also use balsa wood. There won't be much stress on the part.
Red seed bead	This is the top of the rear sight. I used a small red seed bead from an art kit. You can substitute anything that is about 1/32″ to 1/16″ in diameter and easy to see against black paint and blue or cloudy skies.
1/32″ plywood	This is used for the cross hairs on the front sight. If you make a substitution, pick something thin and strong.
BT-60 body tube coupler	This forms the outside of the front sight. A little larger or smaller won't hurt.
Precision metals 5/16″ brass tube	See the next item.

Part	Description
Precision metals 9/32" brass tube	A well-stocked hobby store will have a section with a dizzying variety of brass, copper, or aluminum tubes. Pick two sizes that are around 1/4" to 1/2" in diameter, where one tube slides smoothly inside the other. These will be used to create smoothly turning parts for the tracker. If you don't have access to a well-stocked hobby store, check *http://www.onlinemetals.com* or other online stores. If you are ordering online, allow about 1/64" difference between the outside diameter of the smaller tube and the inside diameter of the larger tube.
2"-long 3/16-20 bolt (4)	These will hold the two plates together. The nuts and washers that follow should fit this bolt.
3/16-20 nut (4)	
3/16-20 wing nut (4)	
3/16" lock washer (4)	
3/16" flat washer (4)	
Spring (4)	You are looking for four small springs, about 1 1/2" long, that will fit over the 3/16-20 bolt. The ones shown are Prime-Line 0.047" x 11/32" x 1-1/2" compression springs from Ace Hardware.
Small wood screw (3)	These are used to mount the level. I used 1/4" #2 pan head sheet metal screws from Ace Hardware. Anything that will hold the level in place will do, even glue.
Felt	A 4" square of felt to help form a smooth mechanism.

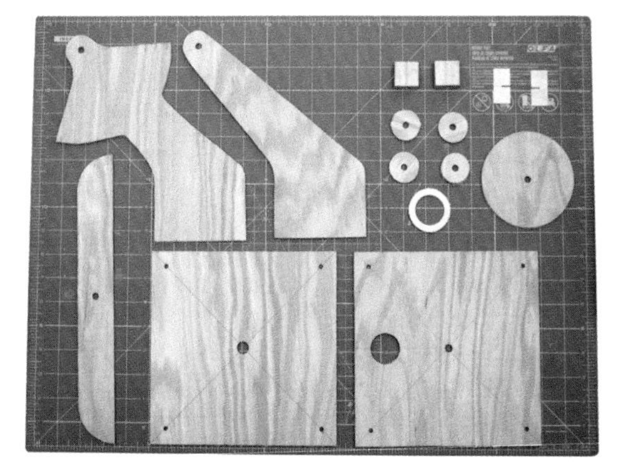

Figure 8-11. *Wooden parts are cut from 1/4" plywood, with a few small, specialized parts cut from 3/4" hardwood and 1/8" plywood.*

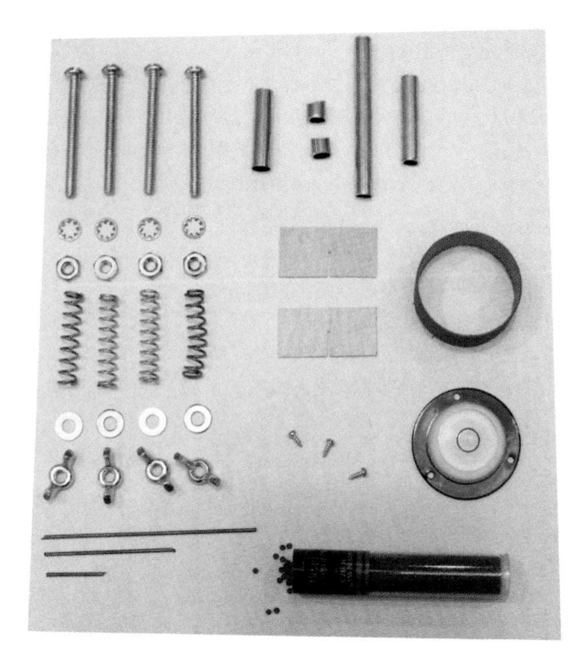

Figure 8-12. *Various additional hardware is used. The two small wooden parts are cut from 1/32" plywood, but they can be cut from other thin, strong materials.*

Table 8-3. Tools and materials

Tool	Description
Epoxy glue	You can get by with a slow-drying epoxy (say, 30-minute epoxy). It's nice to have both a slow- and a fast-drying epoxy, though. I used 5-minute and 30-minute epoxy for this project.
Sealer	This will be used to prepare the wood. You can use the same sealer you used for rocket fins, but it might be cheaper to use a wood sealer from the hardware store.
Spray glue	Used to attach the printer indicators to the wood. You can get by with wood glue, but spray glue makes life a lot easier.
Semi-gloss polyurethane spray	This is used to seal and protect the indicator dials.
Plastic wrap	We'll use a bit of plastic wrap to protect the level while adding wood putty.
Wood putty	Any wood putty or plastic filler will do. It won't show on the completed project, so it doesn't have to match the wood.
Drill and bits	You will need drill bits to fit the bolts, the level, and both sizes of brass rods.
Metal cutting tool	A metal cutting wheel on a Dremel tool, a hacksaw, or a cut-off saw will work.
Scroll saw	You will need to cut the plywood into various shapes. A scroll saw backed up by a sander works well.

Tool	Description
4" and 1 1/2" hole cutters	You can get by with a scroll saw, but these specialized drill bits make cutting out the various disks a breeze.
Table saw	Once again, you can get by with the scroll saw, but a table saw works well for making the large, straight cuts for the square pieces and the sight arm.
Level	This is used to make sure various parts are aligned.

Cutting the parts

Most of the parts are cut from 1/4" plywood. Many of the parts are circular. You have a choice about these: you could cut the parts using a scroll saw, but an easier way is to use the specialized hole saws used to cut large holes in wood, like the one shown in Figure 8-13. This is not the cheapest way to do it, but if you have the tools available, this is a great time to use them. These hole saws also have a nice side benefit: they drill a 1/4" guide hole in the exact center of the disk with no mess or measuring.

The large circular base should be about 4" in diameter. A 4 1/4" hole saw will leave a center piece that is just about a perfect 4" circle. A little smaller or larger is OK, but you will need to adjust the size of the upright supports if the circle size is not 4".

Hole saws cut all the way through the wood. Unfortunately, that isn't really good for the surface underneath the wood! To save the surface, put a piece of scrap wood under the piece you are cutting, as seen in Figure 8-13.

Figure 8-13. *You can cut the circular pieces with a scroll saw, but it's a lot easier with a hole saw like this one, seen cutting the 4" center disk.*

Cut four additional 1 1/4" disks with a 1 1/2" hole saw or scroll saw. The size is not absolutely critical, but again, the side supports will need to be adjusted if you use a different size. The rounded edge at the top of the side pieces should be the same diameter as the disks you cut.

Once all five disks are cut, drill a 9/32" hole in the exact center of each disk. These holes should accommodate the smaller of the brass tubes, so change the hole size if you used a different size than shown in the parts list. The fit will be very tight, which is a good thing, but don't be afraid to ream the hole out if it seems too tight. You can ream the hole by pushing the drill bit in and out several times.

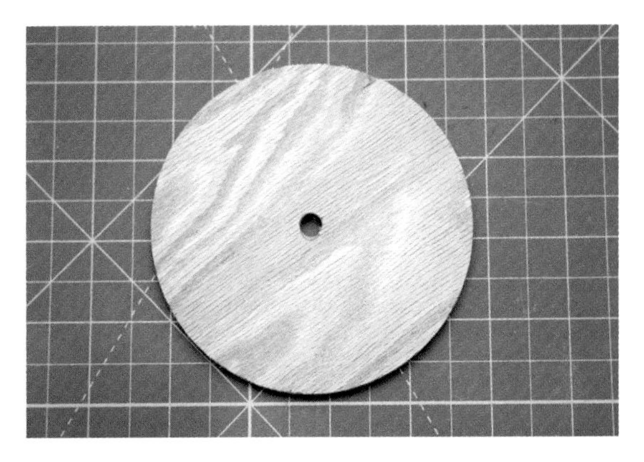

Figure 8-14. *Cut a 4" disk from 1/4" plywood. Drill a hole 9/32" in diameter to accommodate the smaller brass tube.*

Figure 8-15. *Cut four 1 1/4" disks from 1/4" plywood. Drill a 9/32" hole in the center of each disk.*

Cut two side supports from 1/4" plywood (a scroll saw works well here). Figures 8-17 and 8-18 show the dimensions for the side supports, which you can adjust if the size of other parts changes. However, if you make adjustments, be sure the supports will hold the sight bar far enough from the center of the theodolite so you can measure angles up to 90°. That's why the supports have the odd jog to the side.

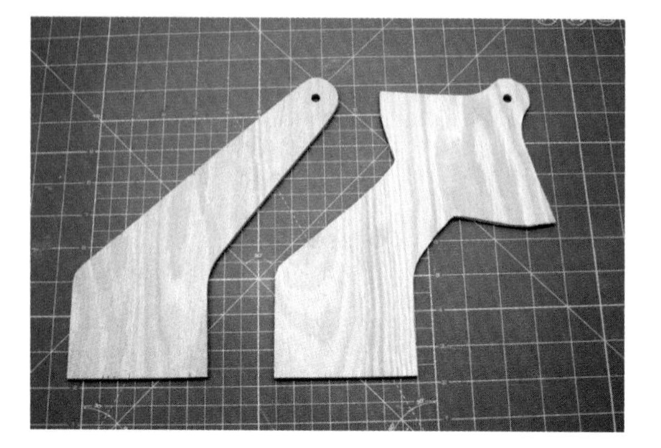

Figure 8-16. *Cut two side supports from 1/4" plywood.*

The support with the wing-like structures is designed for the elevation angle protractor.

You can find full-size plans for the side supports and all other parts online at the author's website (*http://bit.ly/byteworks-make-rockets*).

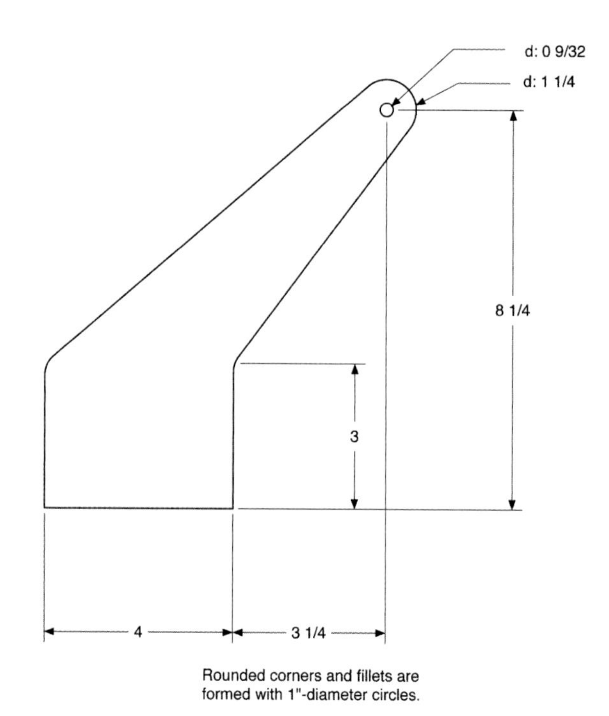

Figure 8-17. *Cut one of this style of side support from 1/4" plywood. Full-size plans are available at the author's website (http://bit.ly/byteworks-make-rockets).*

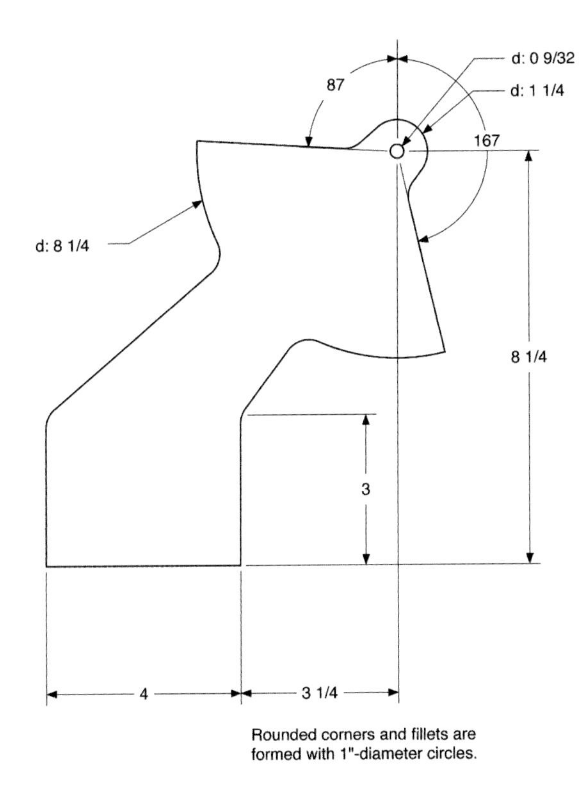

Figure 8-18. *Cut one of this style of side support from 1/4" plywood. Full-size plans are available at the author's website (http://bit.ly/byteworks-make-rockets).*

Cut two 8" square plates from 1/4" plywood. Carefully locate the center of each plate by drawing a line from one corner across to the other in both directions. Place an additional mark 7/8" from each corner. These are the locations where the holes for the connecting bolts will eventually be drilled.

Figure 8-19. *Draw lines from corner to corner to locate the center. Add cross marks 7/8" from each corner to mark the position of holes for bolts.*

You will need a surface level to make sure the tracker is perfectly level before each use. The tracker has an indicator that will sweep around the top plate, and it reaches almost to the edge of the plate. The level needs to be mounted flush with the surface of the top plate, unless it is small enough to fit in a corner between the indicator and the connecting bolts. That means you'll be drilling a hole through the plate to accommodate the level, and mounting it from the bottom. This way the top of the level will be flush with the top of the plate.

There are a lot of different kinds and sizes of surface levels. You don't have to use the same bull's-eye level from the parts list. However, if you make a substitution, you will have to make changes in the next few steps to accommodate the level you pick.

The hole for the level will be right below the zero point for the indicator dial. Find the location using a 45° angle to draw the vertical line. Measure 2 3/4" from the center and make a cross mark. This is the location of the center of the hole for the level. You can also use a protractor or even a folded sheet of paper if you don't have a 45° angle handy.

Figure 8-20. *Use a 45° angle to draw a line from the center of the top piece to the edge. Mark a location 2 3/4" from the center along this line for drilling the hole for the level.*

Drill 3/16"-diameter holes at the marks 7/8" from the edge of each corner. Drill these holes in both the top and bottom plates. Drill an additional 1 1/4" hole for the level.

Drill a 5/16" hole in the exact center of the top plate. This hole needs to be the same size as the larger of the brass tubes, so if you used different-sized tubes, adjust the hole accordingly. While it's always nice to be precise, precision really matters with this hole. Get it as close to the center as possible. Use a drill press for this hole if you possibly can. Try to get the hole as vertical as possible if you are using a hand drill instead of a drill press.

Measure the outside diameter of the 1/4-20 nut that will mount the tracker to the camera tripod. The one I used fit into a 1/2" hole. Drill a hole in the center of the lower plate to accommodate the nut.

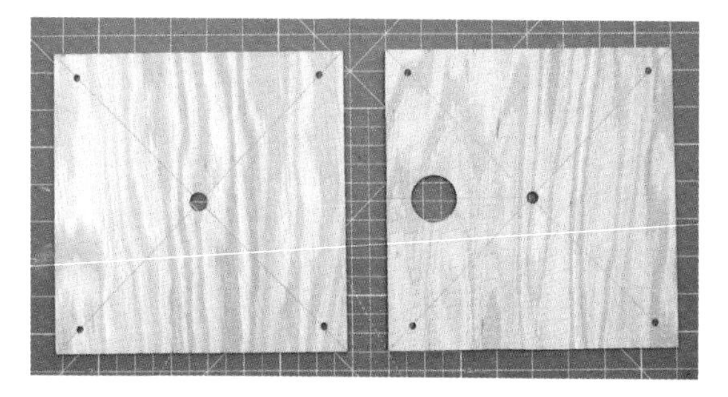

Figure 8-21. *Drill the holes in the bottom (left) and top (right) plates for the bolts, tripod nut, central brass tube, and level.*

Cut the sight bar from 1/4" plywood. The sight bar is 12" long and 1 1/2" tall. I rounded the corners using 3"-diameter circles because I thought it looked better, but that's not essential. Drill a 9/32" hole in the center to accommodate the smaller of the brass tubes.

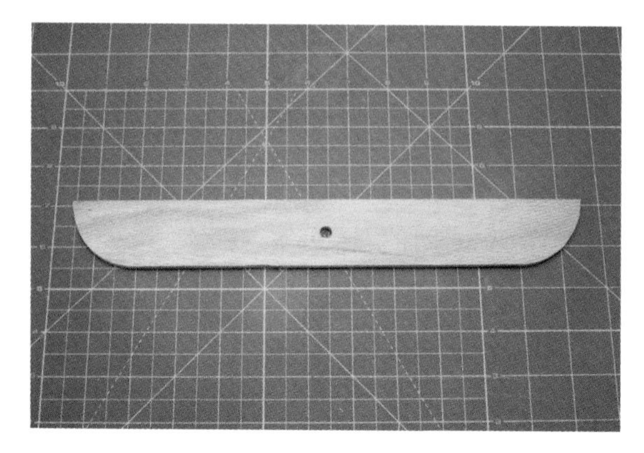

Figure 8-22. *Cut the sight bar from 1/4" plywood.*

The level will poke through the top of the plywood. Test fit the level to double-check the height with the specific level and plywood you are using. In my case, the level protruded from the top by about 1/8"—easily enough to interfere with the smooth movement of the theodolite, since it caught the needle as the device turned. To solve this problem, cut a shim from an appropriate thickness of wood. I used a 2" hole saw to cut a disk from 1/8" plywood, then used a 1 1/4" drill bit to drill out the hole for the level.

Drilling a hole in the center of a circular disk can be tricky. There will be a 1/4" starter hole left by the scroll saw, so finding the center is easy enough. The disk will tend to turn easily while drilling out the center hole, though. I suggest you hold the disk with a large pair of pliers or a clamp while you're drilling the center hole.

Of course, the shape of the outside of this disk is not critical. You could also cut a 2"-square piece, or cut the disk with a scroll saw. The only really critical aspect of this part is that the hole in the middle be the same size as the level.

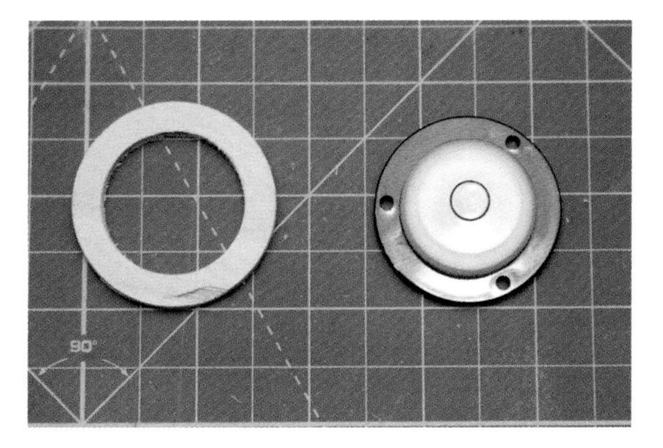

Figure 8-23. *Cut a spacer from 1/8" plywood. This will lower the level enough that it will not interfere with the theodolite's horizontal needle.*

Cut two 1"-square pieces from 3/4" thick clear wood (no knots or other major imperfections). One of these pieces will fit between the two sides to support them. We'll call this the top block; the other piece is the bottom block that fits below the top plate of the theodolite. The top block needs to be sanded to match the thickness of three sheets of plywood, since there are three pieces of plywood—one sight arm and two disks—at the top of the theodolite. If your plywood is really 1/4" thick, you won't need to do any sanding, since the three 1/4"-thick pieces of plywood will be the same thickness as the piece of wood. If the plywood is more like 1/5" thick, like mine was, you will need to sand the wood quite a lot to get it thin enough.

Drill a 9/16" hole through the center of the top block. This hole will fit the thinner of the brass tubes. Drill a 5/16" hole in the bottom block to fit the larger brass tube.

The angle of the holes through these wooden blocks is critical, so use a drill press if at all possible. If you don't have a drill press, make several wooden blocks and pick the one with the most precisely vertical hole. It's worth taking the time to get this step right, since the smooth movement of the theodolite depends on a perfectly aligned hole.

Figure 8-24. *Cut two 1"-square pieces of 3/4" wood. Sand one to the thickness of three sheets of plywood and drill a perfectly vertical 9/32" hole through the center. This is called the top block. Drill a 5/16" hole through the other. It is called the bottom block.*

Cut a 3/4" piece from a BT-60 body tube coupler. This forms the outside of the forward sight. Mark the tube coupler as if you planned to install four fins, but put the marks on the inside of the tube. These marks show where the cross hairs will be glued.

It's not absolutely critical that you use a BT-60 body tube coupler. A BT-55 coupler will work fine, as will a BT-70 coupler. You might even find some other stiff tube about the right diameter that will work well as a forward sight. I'd stay away from body tubes unless you plan to reinforce them, though. The forward sight will get bumped from time to time, and a body tube won't stand up to as much abuse as a tube coupler.

Cut two pieces of very thin plywood to 3/4" high, with the length matching the inside diameter of the tube coupler you selected. I used 1/32" aircraft plywood from a hobby store. Use a hobby knife to cut slots in the center of each piece so they can be assembled in a cross. This will form the cross hairs for the forward sight.

Finally, cut the brass tube to length using either a Dremel tool with a cutting blade, a chop saw, or a hacksaw. You will need three pieces from the thicker tube. Two should be the same thickness as the plywood; these will be used in the upright supports as strong, smooth bushings. Cut another piece that will fit through one of the 1"-square pieces of wood and the upper plate of the theodolite. For 1/4" plywood, it will be 1 1/4" long. Use sandpaper or a grinding stone on a Dremel tool to smooth off the ends so the smaller brass tube slides into and out of these pieces easily.

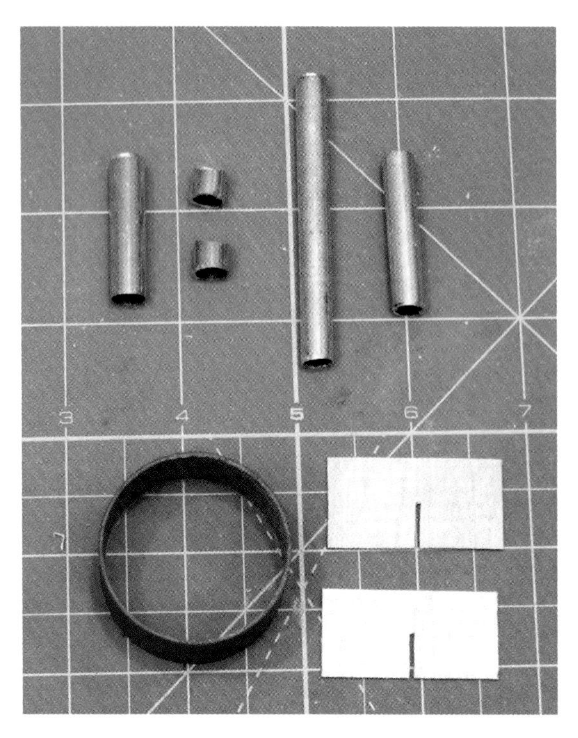

Figure 8-25. *Cut the pieces for the forward sight and the brass tubing as described in the text.*

Cut one piece of the thinner tube to match the thickness of seven layers of plywood (that is, the thickness you get by compressing the wood slightly). Small differences in the wood thickness will add up when you layer seven pieces of wood together. The best way to get a tube the right length is to assemble the top portion of the theodolite, push the tube through until it is flush with one end, and then mark the tube where it sticks out. Compress the wood enough that the sight arm can be moved easily, but will stay in place when you release it. Now cut the tube, sand the ends, and check to make sure it slides into the two small bushings easily and rotates smoothly.

Figure 8-26. *Cut the smaller brass tube so it goes through all of the top pieces of the theodolite.*

The last piece of thin tube needs to be long enough to go through both 1″ blocks and two pieces of plywood. Once again, assembling the parts and marking the tube is the most accurate way to cut the tube to the right length. Unlike the top piece, though, this piece can be a little too long or short without any real effect on the operation or appearance of the finished theodolite. Any excess tube will just stick out between the two plates. Sand the ends, and make sure the part slides into the longer thick brass tube and turns freely.

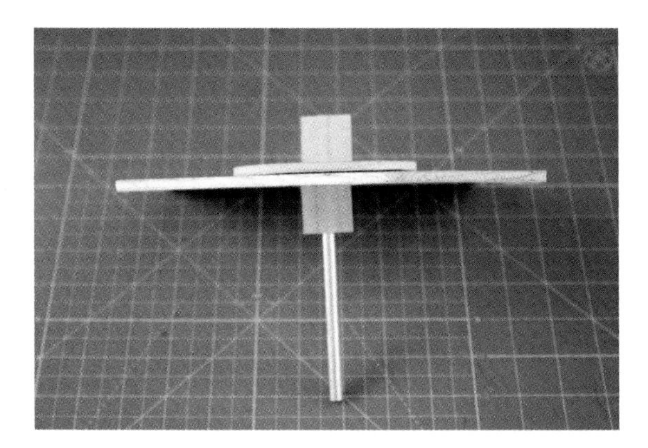

Figure 8-27. *Cut another length from the thinner brass tube that fits through the top block, the bottom block, the 4″ disk, and the upper 8″-square plate of the theodolite.*

First steps in assembly

There are some odd nooks and crannies in the theodolite that make it difficult to prime and sand once all of the pieces are fastened together. The solution is to assemble some of the small parts, then prime and paint the wood. Once the paint dries, the pieces can be glued together and a little touch-up paint used to cover up any small areas of bare wood.

Start by installing the 1/4-20 nut in the lower plate. This is the nut that serves as an attachment point with the tripod. You will want this nut to be flush with the bottom of the plate so the tripod bolt will screw down against the nut. If the nut is recessed, the tripod bolt will pull on it as you tighten it into place. Since a threaded bolt can exert a lot of pressure, this would either splinter the wood or break the glue joint if the nut itself is not taking the pressure.

Place a small piece of wax paper or plastic wrap on your work surface. Push the nut into place and apply a generous amount of fast-curing epoxy around the sides of the nut. Be sure you don't get any inside the threads! Reinforce the wood by spreading glue in a thick layer about 1/2″ over the wood that surrounds the nut on both the top and bottom sides of the plate.

Place the assembly over the wax paper to dry. Make sure the nut is pushed all the way down so it is in contact with the wax paper. Double-check to make sure there is no epoxy in the threads. If there is, you can clean it out with a Q-Tip and rubbing alcohol, or pull out the nut and substitute another. A flashlight might help when checking the threads for glue.

Figure 8-28. *Use epoxy glue to fasten the 1/4-20 nut in place. This is the nut the tripod will screw into.*

Print the two protractors from the online files at the author's website (*http://bit.ly/byteworks-make-rockets*). The quality of the paper you use really matters on these, especially if you paint the theodolite black as in our example. After trying normal inkjet printer paper, card stock, and photo paper, I opted for photo paper. It's thick enough that the black paint does not show through. You can get by with card stock, especially if the paint color is light, but we're going to glue and treat the paper in such a way that normal inkjet paper just won't do.

Place the 4″ central disk in the center of the 360° protractor. Cut a 2"-long piece of 1/16″ piano wire to use as a horizontal angle needle. Drill a 1/2″ hole into the edge of the 4″ circular disk, and then use a ruler to align the needle against the protractor. The needle should point straight out from the center point of the disk, just barely touching the base of the 1/2-degree marks on the protractor. Glue the needle in place with a small shim under the tip so it won't rest directly on the protractor as the disk rotates. I used epoxy because that's what I was using on almost everything else, but wood glue will work, too. There isn't a lot of strain on this needle.

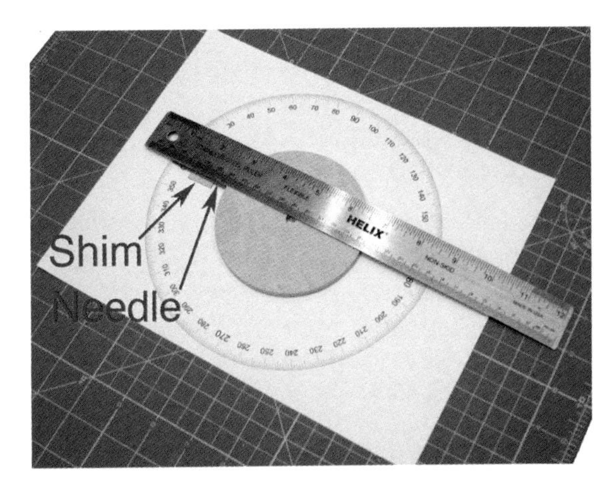

Figure 8-29. *Glue an indicator needle in place on the 4″ disk.*

While the glue dries, repeat the process for one of the four small disks. The needle will need to be 3 3/8″ long this time, also made from 1/16″ piano wire. Set both of the disks with their needles aside to dry.

Figure 8-30. *Glue a 3 3/8" needle made from 1/16" music wire into one of the four small disks, using the 100° protractor as a guide.*

The level has a cute beveled edge around the top. That will be a weak spot under the protractor, so we need to fill it in with wood putty.

Begin by gluing the 1/8"-thick plywood ring into place so the center of the ring matches up exactly with the center of the hole in the top plate. The ring gets glued to the bottom of the plate, although the plate is reversible up to this point. Let the glue dry enough so you can work with the piece. While wood glue will work just fine, if you use 5-minute epoxy, you can proceed in about 10 or 15 minutes.

Cover the level in plastic wrap. This prevents the wood putty from adhering to the level. Push the level into place, making sure it is flush against the wood ring, then screw it into place with three #2 1/4"-long wood screws. At least, that's what I used. Any screw that will fit is fine, of course.

Flip the board over and press wood putty into the area around the top edge of the level, filling in the area left by the bevel.

Allow the wood putty to dry overnight. Remove the level and use a sanding block to sand the wood putty smooth.

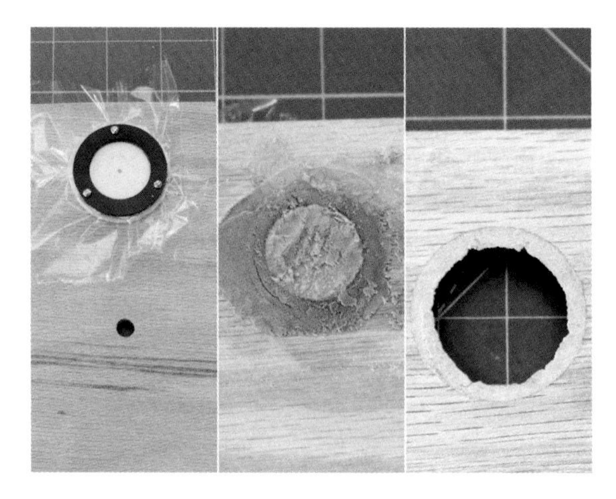

Figure 8-31. *Glue the spacer disk to the bottom of the top plate, then fasten the level in place with a protective layer of plastic wrap. Use putty to fill in the gap on the top side. After it dries, remove the level and sand the putty flush with the wood.*

Once the needle on the central 4" disk is dry, flip the disk over. Cut a felt disk slightly smaller than the wooden disk and glue it in place with a thin layer of wood glue. This gives a smooth surface that will allow the theodolite to rotate easily, but stop right away when it is in position.

Once the glue dries, trim the excess felt away from the central hole.

Figure 8-32. *Glue felt to the bottom of the central disk.*

Use small amounts of wood glue to fasten the cross hairs to form the forward sight. Glue the wood to the paper, but don't bother gluing the wood where the two pieces cross. Glue isn't needed in the middle, and a stray glob of glue will be very annoying when using the finished theodolite.

Figure 8-33. *Glue the cross hairs in place.*

Finishing

Dry fit all of the parts so you understand how they fit together to form the final theodolite. The theodolite will be held together by glue. Mark the areas that will eventually be glued and tape them off. There is a 1"-square area at the base of each of the vertical supports where the top block will be glued. (One of them is on the back side of the support in Figure 8-34.) The top block gets painted on the top and on two of the faces adjacent to the top, but the bottom and sides are taped. The bottom block is taped on top, but painted on the other five sides. The edges of the four small disks need to be painted, but only the two outer disks need to be painted to the center, and then only on one side.

The most complicated taping job is the top of the 4″ disk. Tape off the area where the top 1″ block and the two vertical supports will be glued. These should align with the indicator needle. Absolute precision is not critical; it won't affect the accuracy of the instrument. Still, it needs to look straight.

Once all of the areas that will be glued are taped, apply a coat of sealer to the wood. I used wood sealer, but the same sanding filler you use on rocket fins will work, too. After sanding—and perhaps replacing some of the tape that gets roughed up—paint the wood.

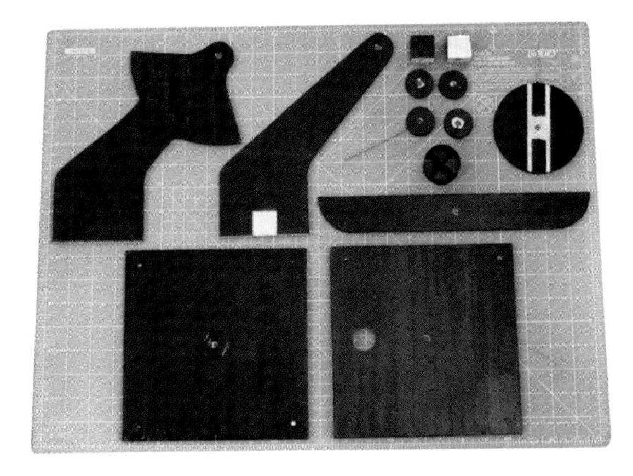

Figure 8-34. *Finish the wood pieces, being careful to leave places that will be glued as bare wood.*

Installing the protractors

Glue the protractors in place. It is very important that the center of each protractor is directly over the hole in the wooden piece. With that exception, it just needs to look straight. It won't affect the accuracy of the instrument if the protractors are rotated by a degree or so.

Spray glue works really well for attaching the protractors, but a thin layer of wood glue will work, too.

Once the glue sets, spray the protractors and the wooden surface on the same side as the protractors with a light coat of clear varnish. This needs to be a *light* coat, because a thick coat might cause the ink from the inkjet printer to bleed. Once the first layer of varnish sets, add a nice, thick top coat.

Let the paint and varnish dry very thoroughly before proceeding. The paint could bind if the disks are clamped in place before the paint is absolutely dry. Let it dry at least overnight.

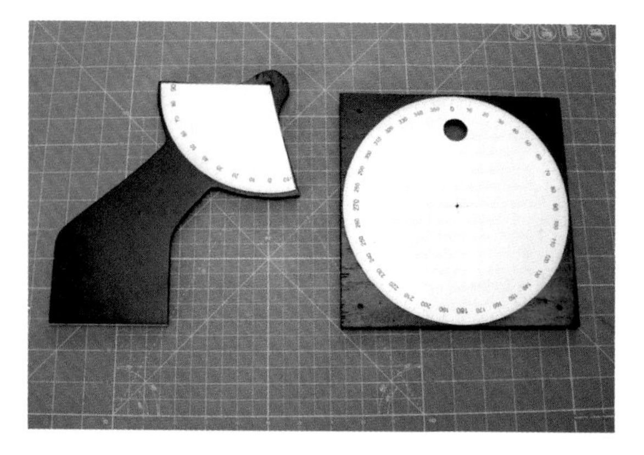

Figure 8-35. *Glue the protractors in place, and then cover with a layer of varnish.*

Final assembly

The thick brass tube will be glued into the lower wood block, so rough up the entire outside of the tube with sandpaper. Also rough up about 1″ of the long, thin tube. Part of it will be glued into the top block, but the lower part needs to stay smooth so it turns easily when inserted in the lower block.

Dry fit the parts. The thick brass tube should be flush with the top surface of the plate. It's OK if it sticks out of the bottom of the bottom block a bit, or doesn't quite reach the end. The thinner brass tube should be flush with the top of the top block on the 4″ disk, extending out of the bottom of the 4″ disk. Make sure they are cut and drilled straight enough that the 4″ disk turns smoothly in a 360° circle. It's not too late to sand or trim the parts, or even replace one completely, but it will be after the next step.

Use fast-drying epoxy glue to fasten the wooden blocks to the bottom of the top plate and to the top of the 4″ disk and to glue the brass tubes in place. Double-check the fit of the parts before the glue sets. The 4″ disk must rotate freely and without wobbling. If the disk binds or wobbles, adjust the pieces to get rid of the problem. Remove the pieces and start over if you have to, but don't let the glue set until the pieces are perfectly aligned.

Figure 8-36. *Glue the 1″ wood pieces in place.*

Allow the glue on the 4″ disk to set thoroughly. Dry fit the side supports, making sure the holes in the top portion align. Use fast-drying epoxy to glue the side supports in place. A clamp is very handy at this point, but patience and a steady hand will work, too.

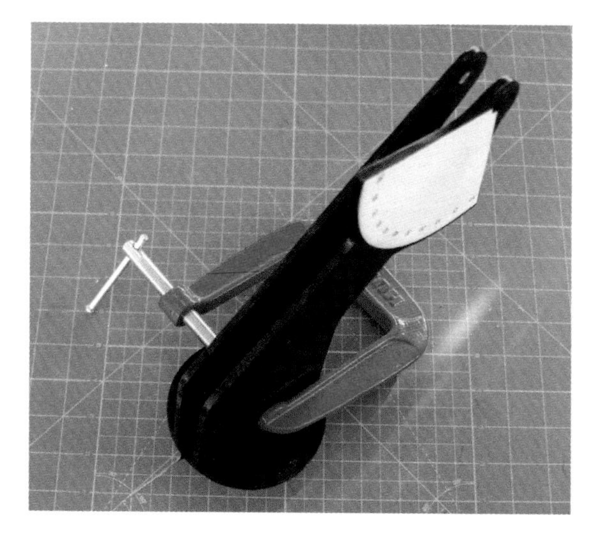

Figure 8-37. *Glue the side supports in place.*

Use a level to make sure the bubble level is perfectly aligned with the top surface of the plate, and then screw the level in place from the bottom. Be sure to check the alignment both horizontally and vertically.

Figure 8-38. *Install the level, checking its alignment with another level.*

Fasten the top and bottom plates together with 2"-long 3/16" bolts. Use a lock washer on each bolt, and then slide it through the top plate. Add a nut and tighten it. This keeps the bolt in place as the wing nut is used to tighten and loosen the bolt for fine-tuning the plates when leveling them. Repeat this process on all four corners.

Add a spring and washer to each bolt, and then slip the bottom plate into place. Add four wing nuts, tightening them just enough to hold the bottom plate in place.

Figure 8-39. *Place a lock washer on the bolt, insert it into the top plate from the top, and then add and tighten a nut. Add the spring, place the bottom plate on the bolts, and then add a wing nut.*

Cut a piece of 1/16″ music wire to use as the rear sight. It should be about 1/2″ longer than half the diameter of the forward sight. Glue a red bead or some other small, colorful object to the end of the wire and let it set.

Drill a small hole about 1/2″ deep in the top of the sight bar. This hole should be about 3/8″ from the end. Use a toothpick to shove some fast setting epoxy glue to the hole, and then sink the rear sight into the hole. Before the glue sets, put the forward sight right next to the rear sight. Make sure the red bead aligns exactly with the center of the cross hairs. Once the glue sets enough to handle the piece, glue the forward sight in place at the opposite end of the sight bar. Prop the piece up while it dries.

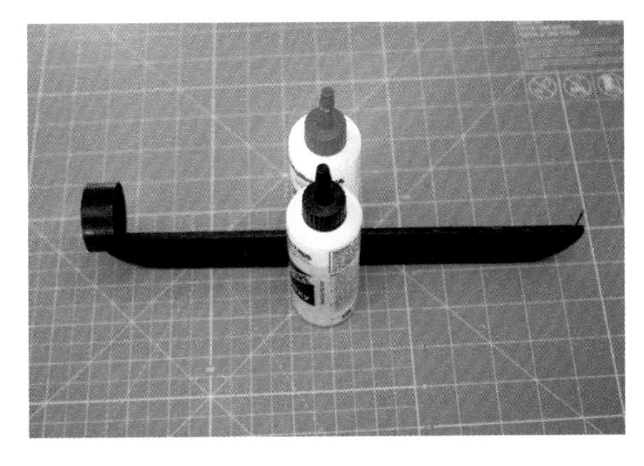

Figure 8-40. *Glue the rear and forward sights in place on the sight bar. Make sure the bead on the rear sight is exactly the same height as the center of the cross hairs.*

Make sure all of the paint and glue has dried completely before moving on to the last step in assembly. Tacky paint will dry solid and weld the top pieces together. It's also good to have a little time when assembling the top portion of the theodolite, so use slow-cure epoxy if you can.

Begin by dry fitting all of the parts to make sure the fit is good. The two short brass bushings cut from the larger-diameter brass tube go in the holes in the side supports. The remaining brass tube gets shoved through one of the circular disks, a side support, another circular disk, the sight bar, a third circular disk, the other side support, and the final circular disk, as shown in the final assembly in Figure 8-41.

Figure 8-41. *The disk with the indicator needle is on the left, followed by the winged side support, another disk, the sight bar, another disk, a side support, and one final disk.*

Make sure all parts fit and rotate smoothly. It's not too late to replace a bad piece or sand a piece that almost fits. Read through the next steps of the assembly before you do anything, to make sure you understand them completely. Have everything you need laid out where you can reach it easily, because once you start, you will need to finish before the glue sets.

Scuff the end 1/8" and middle 1/4" of the long brass tube with sandpaper so the glue has something to cling to. Apply a small amount of epoxy to the inside of the disk with the indicator, and insert the long brass tube. Make sure there is no glue at all on the inside surface of the disk or brass tube. Slide the assembly into place, and slide on the next wooden disk. Apply a small amount of epoxy to the brass tube and slide on the sight bar. It's OK if the sight bar gets glued to the wooden disk.

Slide on the third wooden disk and shove the tube through the bushing on the final side support. You will be sliding the components back and forth to get them through the various holes. Be sure no epoxy has worked its way onto the part of the brass tube that fits through the bushing. If it has, wipe it away with rubbing alcohol.

Put a small amount of epoxy inside the final wooden disk and slide it into place. Putting the epoxy inside the disk and sliding the disk over the brass tube should guarantee there is no epoxy on the inside surface of the disk or on the part of the brass tube that goes through the bushing. Keep a close watch for leaks.

Check the movement of the mechanism. The disks should be tight enough that the sight bar will stop when released, but they should be loose enough that the sight bar can be moved up and down easily.

Working quickly but very accurately before the glue sets, place a bubble level on top of the sight bar and make sure the sight bar is perfectly level. While it is level, rotate the sight bar relative to the disk until the needle points exactly to zero degrees. It is critical that this be done accurately. This is how the theodolite reports its vertical angle; you don't want it to report anything but 0° if the sight bar is level.

Remove the level and let the glue dry.

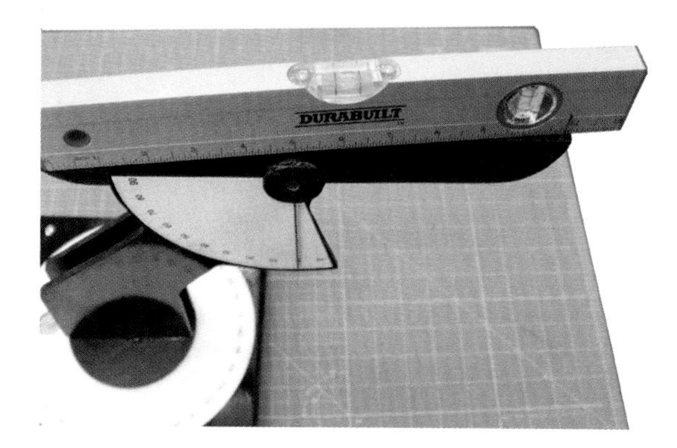

Figure 8-42. *Use slow-cure epoxy and the remaining brass pieces to fasten the sight bar in place. The indicator needle should point to zero when the sight bar is perfectly level.*

Check the movement of the mechanism when the glue is just about at its cure time. For example, if you are using 30-minute epoxy, check the movement 25 minutes after you mixed the glue. Does the sight bar move smoothly? Does moving the sight bar move the indicator needle? Are they moving together with no slipping? If there is a problem, quickly disassemble the components and clean them with rubbing alcohol. Figure out why the problem occurred and, after correcting the problem, repeat the assembly process.

Math Behind Dual-Axis Tracking

If you recall, the biggest source of error with a single-axis tracker occurs because the rocket doesn't necessarily go straight up, so the angle between the tracker, launcher, and rocket is not exactly 90°. Another way of looking at this is that we don't precisely know the length of the baseline to the point exactly under the rocket. With two dual-axis trackers, we no longer assume the rocket's apogee is exactly over the launcher. Instead, we measure the position using the two horizontal angles. Knowing the distance between the trackers and the angles between them and the rocket at apogee, we can find the actual baseline length. Then we can calculate the altitude from either tracker. In practice, we calculate the altitude from both trackers and take the average, since there will probably be some error that causes the two trackers to report slightly different altitudes.

There are several ways to do the math to find the altitude from these angles. Two are approved by NAR for use in contests. Let's take a look at one of them in detail, and the other briefly, to figure out what dual-axis tracking is all about.

The vertical midpoint method

The vertical midpoint method is the most obvious way to find the altitude using two theodolites. Figure 8-43 shows the two trackers, T_1 and T_2.

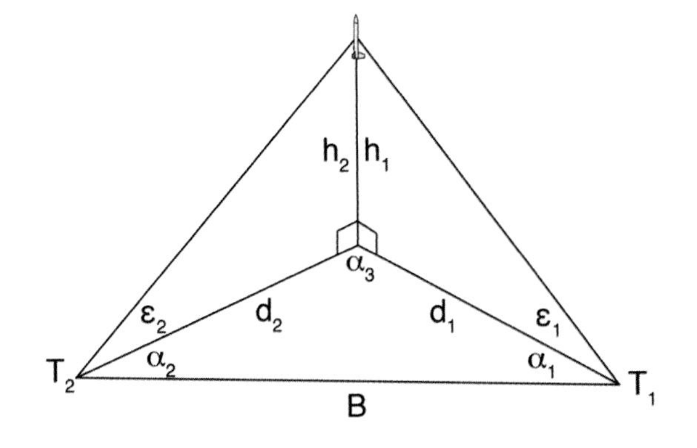

Figure 8-43. *Definition of variables for the vertical midpoint method. In an ideal world h_1 (the altitude calculated for the first tracker, T_1) is the same as h_2 (the altitude calculated by the second tracker), but in practice these are rarely the same.*

When the rocket hits apogee, both trackers report two angles. One angle is the horizontal angle between the tracker and the rocket, labeled α_1 for the first tracker and α_2 for the second tracker. Each tracker also reports the elevation angle of the rocket at its peak. This is labeled ε_1 for the first tracker and ε_2 for the second tracker. The distance between the trackers is B.

We can use the law of sines to find the distance from each tracker to the point just below the rocket. The law of sines relates the sine of an angle and the length of the side of a triangle opposite the angle to the other angles and sides of the triangle. Looking at the triangle on the ground, this gives:

$$\frac{d_1}{\sin \alpha_2} = \frac{d_2}{\sin \alpha_1} = \frac{B}{\sin \alpha_3}$$

All of those values are known except α_3, the angle right under the rocket. We can use the fact that the sum of the angles in a triangle is always 180°, and a few trig identities, to get that term into a more manageable form:

$$\alpha_1 + \alpha_2 + \alpha_3 = 180$$

$$\sin \alpha_3 = \sin (180 - \alpha_1 - \alpha_2) = \sin (\alpha_1 + \alpha_2)$$

The baseline for the first tracker, then, is:

$$d_1 = \frac{B \sin \alpha_2}{\sin (\alpha_1 + \alpha_2)}$$

Now that we have an accurate baseline for the elevation angle, we can solve for the altitude as if we were using a single-axis tracker:

$$A_1 = \frac{B \sin \alpha_2}{\sin (\alpha_1 + \alpha_2)} \tan \epsilon_1$$

The altitude measured by the second tracker is calculated the same way, giving:

$$A_2 = \frac{B \sin \alpha_1}{\sin (\alpha_1 + \alpha_2)} \tan \epsilon_2$$

These are usually different, since the angles are never measured with perfect accuracy. Since we don't know which one is correct, the only fair thing to do is report the altitude as the average of the two:

$$A = \frac{A_1 + A_2}{2}$$

We used error analysis to find the expected error for the single-axis tracker. NAR uses a slightly different way to get a handle on the error for altitude contests and records. If the two altitudes differ by more than 10%, the result is discarded, and the rocket must be flown again. This is called an *open track*. If the difference is 10% or less, it's called a *closed track*, and the result is the average of the two altitudes rounded to the closest meter. If the result is exact to a half meter—say, 112.500 meters—the result is rounded up (here, to 113 meters).

The equation used to calculate the error is:

$$C = \left| \frac{A_1 - A_2}{2A} \right|$$

If C is less than or equal to 0.1, the track is closed; otherwise, the track is open and the rocket must be reflown.

The geodesic method

This method finds the altitude using a slightly different procedure. It's more complicated to understand, but it gives a closed track more often than the simpler vertical midpoint method.

The geodesic method starts by assuming the two theodolites will not get exactly the same result. That means the line from one tracker to where it thinks the rocket is located won't actually touch the line from the other tracker to the position of the rocket. It finds the closest distance between the two lines, and reports the altitude as the midpoint of that line.

We won't go through all of the math here. Check online or in a textbook on analytic geometry for ways to find the distance between two lines if you would like to work out the derivation of the equations.

Here's the math used by NAR for finding the altitude using the geodesic method:

$$f = \sin \epsilon_1 \sin \epsilon_2 - \cos \epsilon_1 \cos \epsilon_2 (\cos \alpha_1 \cos \alpha_2 - \sin \alpha_1 \sin \alpha_2)$$

$$d_1 = B \frac{\cos \epsilon_1 \cos \alpha_1 + f \cos \epsilon_2 \cos \alpha_2}{1 - f^2}$$

$$d_2 = B \frac{\cos \epsilon_2 \cos \alpha_2 + f \cos \epsilon_1 \cos \alpha_1}{1 - f^2}$$

$$A = \frac{d_1 d_2}{d_1 + d_2} (\sin \epsilon_1 + \sin \epsilon_2)$$

$$C = B \left| \frac{\cos \epsilon_2 \sin \epsilon_1 \sin \alpha_2 - \cos \epsilon_1 \sin \epsilon_2 \sin \alpha_1}{A\sqrt{1 - f^2}} \right|$$

Once again, if C is less than or equal to 0.1, the track is closed.

The vertical midpoint method and the geodesic method might give slightly different results. How is that handled in a real contest? If one result is closed and the other is open, the closed result is used and the other discarded. When both methods give closed results, the general practice seems to be to use the geodesic method for all results. This is a detail left to the contest director, though. That person might pick the method with the lowest value for C, or the higher of the two closed altitudes. Based on the current NAR Pink Book, as long as the directors are consistent, they can use whatever method they choose. Of course, that should be decided before the competition, not after!

Using Dual-Axis Trackers

Theodolites are usually set up to the south of the launch pads so the sun won't get in the eyes of the trackers. They should be at a distance of 50% to 200% of the expected altitude for the rockets they are supposed to track. This way, the theodolites won't have to measure overly small or overly large angles, which can make it harder to get an accurate altitude measurement. It's common to use a baseline of 300 meters, which allows fairly accurate tracking of rockets that will hit altitudes of 150 meters to 600 meters—about 500 feet to 2,000 feet. That's a pretty good range of altitudes.

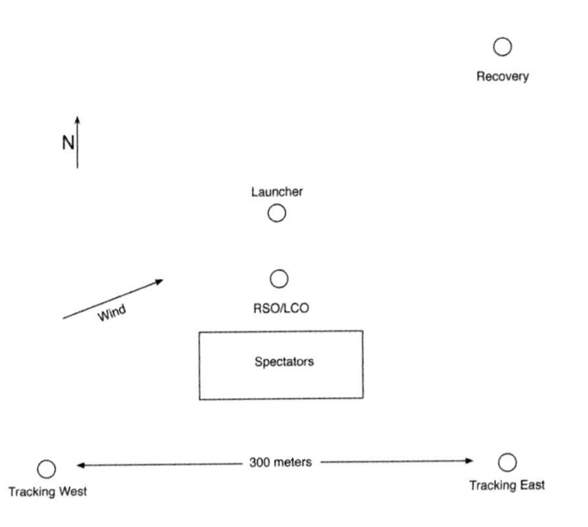

Figure 8-44. *The trackers should be set up so the sun is behind them. The baseline should be 50% to 200% of the expected altitude of the rockets.*

The trackers need to communicate with each other and with the LCO so they can agree on when the rocket reaches apogee. Unless they are as loud as Mom used to be when she shouted from the back door for me to get my tail back home, they will need walkie-talkies, or some similar method for communicating.

It takes at least three people to track a rocket with a pair of theodolites; six people are better. With just three people, the trackers should agree beforehand that they will track the rocket until the ejection charge fires. The two trackers then report their angles back to the LCO, who calculates the final altitude. You could even get away with having the trackers write down the angles and report them later. The problem with this method is that the rocket may arc over at apogee and start back down before deploying the parachute. It would be nice if the actual apogee could be recorded. Unfortunately, the rocket might arc in such a way that the two trackers disagree on when it reaches the highest point.

You can avoid this problem with six people. One person is the LCO, who announces the launch. Two trackers operate the theodolites. Three additional people operate the walkie-talkies and watch the rocket. As soon as one of the trackers thinks the rocket has hit apogee, she calls out to stop the track. The trackers then lock in the angle on the theodolite, and the people working with each tracker report the results back to the LCO table for calculation. In a well-oiled group, the result is available very quickly, and the LCO can announce the altitude to the spectators.

Using more than two theodolites

It's pretty tough on a competitor to have a perfect flight, only to find that one of the trackers got the sun in his eye or had some other issue that caused an open track, forcing the competitor to fly again. Larger contests will try to use more than two theodolites to track the rocket. If one messes up, the others can still get a result.

One popular method shows how this is done. Four theodolites (call them A, B, C, and D) can be positioned in a huge square around the launch area. Before the launch, each pair of theodolites reports the angles to each of the other theodolites. During the flight, they each report the angle to the rocket's apogee. You end up with a total of six different pairs of theodolites (A-B, B-C, C-D, D-A, A-C, B-D) providing data! If they all succeed, that gives six different closed tracks and, in most cases, six slightly different altitudes. All of the closed tracks are then averaged for the final contest altitude.

Program for Dual-Axis Trackers

There are lots of programs around for reducing the angles from the trackers and reporting the altitude. There is one in Appendix A, written in a fairly generic modern BASIC, that works well. This one is also built into techBASIC. You can find a similar program in the NAR Pink Book. A quick online search will turn up several more.

Run the program from Appendix A. It will ask for the baseline distance and the angle from the tracker to the rocket. It then asks for the azimuth and elevation angles —the horizontal angle between the tracker and the rocket and the angle between the horizon and the rocket—for both trackers. The program then reports the altitude using both tracking methods.

Here's some sample output from the tracking program in Appendix A:

```
Baseline length: 300
Tracker 1 azimuth: 90
Tracker 1 elevation: 45
Tracker 2 azimuth: 50
Tracker 2 elevation: 40
```

```
Vertical Midpoint Method:
  Altitude: 374.574097
  Closed track; C = 0.045513

Geodesic Method:
  Altitude: 380.048248
  Closed track; C = 0.063234
```

Altimeters

Visually tracking a rocket from the ground with a protractor or theodolite is cheap and fairly accurate. There is a way to track the flight electronically, though. Modern altimeters use a barometer to sense the changes in air pressure as the rocket ascends and descends. They are often supplemented by an accelerometer that senses the motion of the rocket. I've even flown small sensor packages that could sense the magnetic field and rotation of the rocket.

The advantage of altimeters is that their accuracy does not depend on multiple people doing a perfect job. They work even if you lose sight of the rocket, and they give a lot more information. Since the altitude is updated constantly, you can also figure out how fast the rocket went and what forces the payload had to endure.

Altimeters do have disadvantages, though. While some are remarkably small and light, they do add weight to the rocket. Altimeters also require a vent hole leading outside the rocket so the barometer works. Finally, while they are not terribly expensive, there is always the chance they will be lost. To my knowledge, no one has ever lost a theodolite during a model rocket launch!

Most model rocket altimeters are designed for high-power rocketry. Many high-powered rockets use small pyrotechnic charges to deploy the parachute rather than an ejection charge built into the motor. The altimeters for high-power rockets are generally powered by a fairly heavy external battery and have electronics to fire at least the pyrotechnic charges.

The tiny MicroPeak altimeter is one of my favorite altimeters for smaller rockets. It can't fire pyrotechnic charges like the high-power rocket altimeters, but it's a lot smaller and uses a tiny coin cell battery. The whole altimeter, with the battery, fits into a BT-20 nose cone, and only weighs 1.9 grams. Yes, that includes the battery! They cost about $50, and are available from several of the online rocket stores listed in Appendix B, such as Apogee Components (*http://www.apogeerockets.com*).

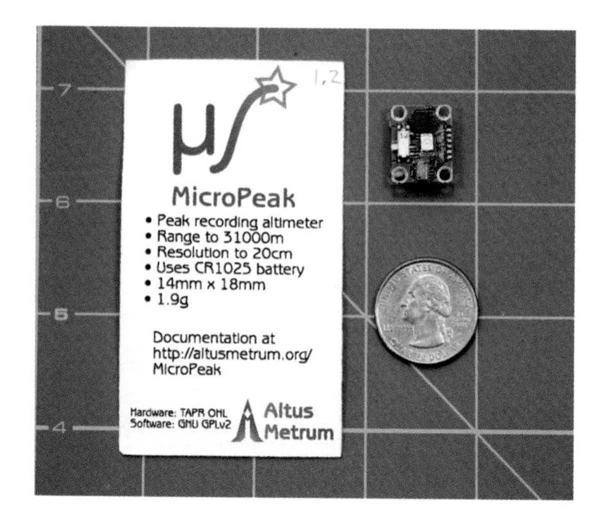

Figure 8-45. *The tiny MicroPeak altimeter weights just 1.9 grams and fits into nose cones for BT-20 body tubes.*

We're going to build a high-performance rocket named Hebe (pronounced HEE-bee) in Chapter 15. Let's take a look at one flight of Hebe with a C6-7 motor to get an idea of what this altimeter can do.

The MicroPeak fits into BT-20-size plastic nose cones. Stuff a little cotton in the top for padding, then tape the bottom securely to the nose cone. Don't let the pressure from the ejection charge get to the altimeter. One way to prevent this is to seal the opening in the base of the nose cone with a strip of paper covered in epoxy.

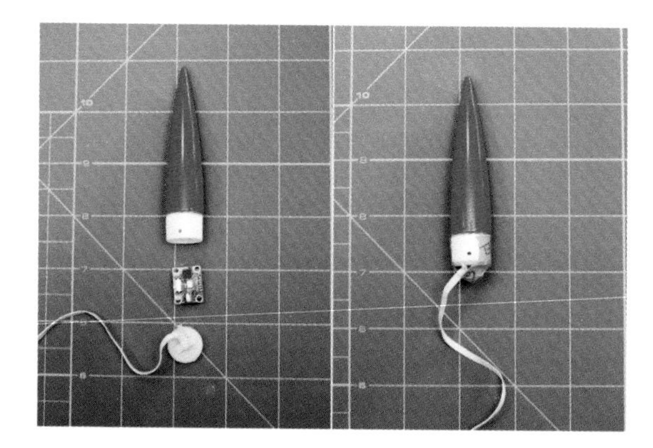

Figure 8-46. *Note the hole in the side to vent the nose cone so the barometer gets accurate pressure readings. This must be lined up with a similar hole in the body tube.*

After the flight, the altimeter flashes an LED with the altitude in decimeters. It flashed 6-10-7-5 for this flight. Ten flashes indicate a zero, so the altitude was 607.5 meters, or 1,993 feet. I had that answer as soon as I popped the altimeter out of the nose cone and counted the flashes.

Back at my computer, I hooked the altimeter up to a small device that downloads the flight data to a free program. The main display is a configurable plot, showing the acceleration in blue, the speed in green, and the altitude in red. You can clearly see the motor thrust by looking at the blue line in Figure 8-47. You can even see the rocket bouncing around as the streamer flapped a bit. It's also worth taking a close look at the acceleration right after motor burnout. Why does it go negative by so much? That may be an overreaction of the sensor to the switch from high forward thrust to a slight negative thrust.

The altitude also shows an interesting—and for this accelerometer, unusual—anomaly. The flight ends at an altitude of about 230 meters. The rocket absolutely did not land on a hill that high above the launch point!

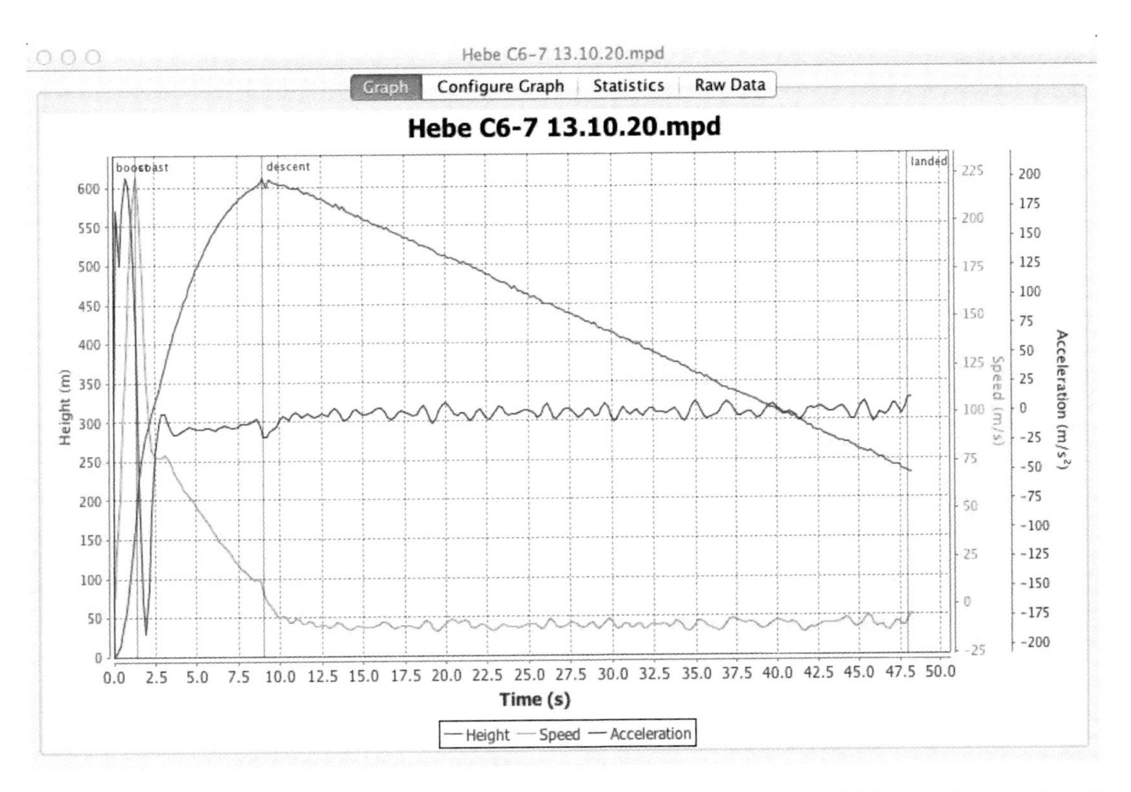

Figure 8-47. *This plot shows the acceleration, speed, and position of Hebe under power from a C6-7 motor. Model rocket altimeters almost always have support software that will show the flight like this.*

The Statistics tab gives a lot of great information about the flight. For example, it's pretty cool to see that the maximum speed was just over 500 mph, almost 2/3 the speed of sound! You can also see the descent rate, which is a bit fast on the streamer. That was a conscious choice for this flight. It's a small, light rocket, so coming down a little fast wasn't going to hurt anything. It was also going really high. There is a good chance the rocket would have been lost if it had been on a parachute.

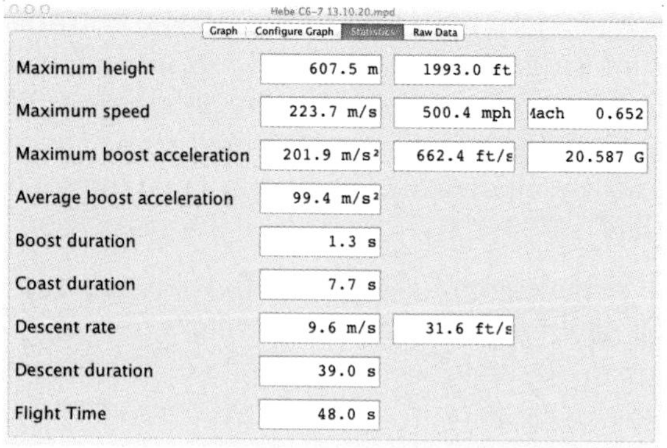

Hebe C6-7 13.10.20.mpd

Graph | Configure Graph | Statistics | Raw Data

Maximum height	607.5 m	1993.0 ft	
Maximum speed	223.7 m/s	500.4 mph	1ach 0.652
Maximum boost acceleration	201.9 m/s²	662.4 ft/s	20.587 G
Average boost acceleration	99.4 m/s²		
Boost duration	1.3 s		
Coast duration	7.7 s		
Descent rate	9.6 m/s	31.6 ft/s	
Descent duration	39.0 s		
Flight Time	48.0 s		

Figure 8-48. *Here are the vital statistics from flying Hebe on a C6-7 motor.*

There is a configuration tab that lets you turn various plotting features on and off or switch to imperial units. There is also a raw data tab that shows the actual values recorded by the accelerometer and barometer. You can offload the data with an Export command to run your own analysis.

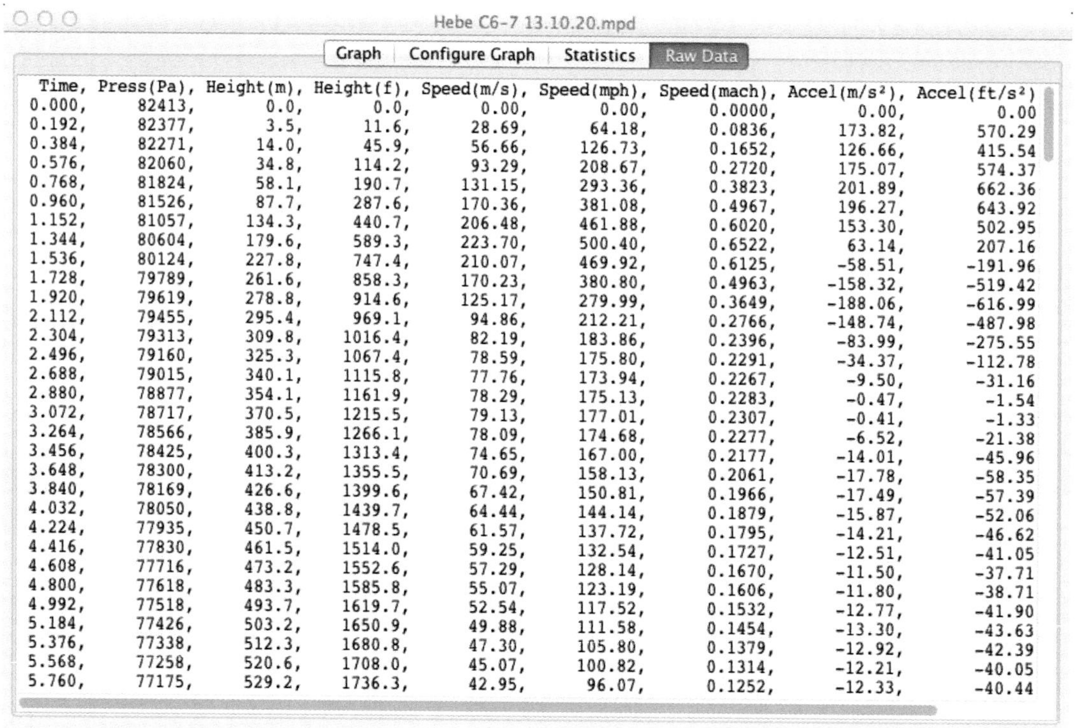

Hebe C6-7 13.10.20.mpd

Graph | Configure Graph | Statistics | Raw Data

Time,	Press(Pa),	Height(m),	Height(f),	Speed(m/s),	Speed(mph),	Speed(mach),	Accel(m/s²),	Accel(ft/s²)
0.000,	82413,	0.0,	0.0,	0.00,	0.00,	0.0000,	0.00,	0.00
0.192,	82377,	3.5,	11.6,	28.69,	64.18,	0.0836,	173.82,	570.29
0.384,	82271,	14.0,	45.9,	56.66,	126.73,	0.1652,	126.66,	415.54
0.576,	82060,	34.8,	114.2,	93.29,	208.67,	0.2720,	175.07,	574.37
0.768,	81824,	58.1,	190.7,	131.15,	293.36,	0.3823,	201.89,	662.36
0.960,	81526,	87.7,	287.6,	170.36,	381.08,	0.4967,	196.27,	643.92
1.152,	81057,	134.3,	440.7,	206.48,	461.88,	0.6020,	153.30,	502.95
1.344,	80604,	179.6,	589.3,	223.70,	500.40,	0.6522,	63.14,	207.16
1.536,	80124,	227.8,	747.4,	210.07,	469.92,	0.6125,	-58.51,	-191.96
1.728,	79789,	261.6,	858.3,	170.23,	380.80,	0.4963,	-158.32,	-519.42
1.920,	79619,	278.8,	914.6,	125.17,	279.99,	0.3649,	-188.06,	-616.99
2.112,	79455,	295.4,	969.1,	94.86,	212.21,	0.2766,	-148.74,	-487.98
2.304,	79313,	309.8,	1016.4,	82.19,	183.86,	0.2396,	-83.99,	-275.55
2.496,	79160,	325.3,	1067.4,	78.59,	175.80,	0.2291,	-34.37,	-112.78
2.688,	79015,	340.1,	1115.8,	77.76,	173.94,	0.2267,	-9.50,	-31.16
2.880,	78877,	354.1,	1161.9,	78.29,	175.13,	0.2283,	-0.47,	-1.54
3.072,	78717,	370.5,	1215.5,	79.13,	177.01,	0.2307,	-0.41,	-1.33
3.264,	78566,	385.9,	1266.1,	78.09,	174.68,	0.2277,	-6.52,	-21.38
3.456,	78425,	400.3,	1313.4,	74.65,	167.00,	0.2177,	-14.01,	-45.96
3.648,	78300,	413.2,	1355.5,	70.69,	158.13,	0.2061,	-17.78,	-58.35
3.840,	78169,	426.6,	1399.6,	67.42,	150.81,	0.1966,	-17.49,	-57.39
4.032,	78050,	438.8,	1439.7,	64.44,	144.14,	0.1879,	-15.87,	-52.06
4.224,	77935,	450.7,	1478.5,	61.57,	137.72,	0.1795,	-14.21,	-46.62
4.416,	77830,	461.5,	1514.0,	59.25,	132.54,	0.1727,	-12.51,	-41.05
4.608,	77716,	473.2,	1552.6,	57.29,	128.14,	0.1670,	-11.50,	-37.71
4.800,	77618,	483.3,	1585.8,	55.07,	123.19,	0.1606,	-11.80,	-38.71
4.992,	77518,	493.7,	1619.7,	52.54,	117.52,	0.1532,	-12.77,	-41.90
5.184,	77426,	503.2,	1650.9,	49.88,	111.58,	0.1454,	-13.30,	-43.63
5.376,	77338,	512.3,	1680.8,	47.30,	105.80,	0.1379,	-12.92,	-42.39
5.568,	77258,	520.6,	1708.0,	45.07,	100.82,	0.1314,	-12.21,	-40.05
5.760,	77175,	529.2,	1736.3,	42.95,	96.07,	0.1252,	-12.33,	-40.44

Figure 8-49. *This is the raw data captured by the MicroPeak accelerometer during Hebe's flight.*

The MicroPeak isn't the only small altimeter out there, and more are being added all the time. Shop around on the Internet a bit if you decide to buy an altimeter. Check Appendix B for a list of online stores.

Comparing the Methods

This chapter shows three different ways to find the altitude of a rocket. Chapter 14 will show three different ways to predict the altitude before a flight. How do they stack up? Let's take a look at the data from some actual rocket flights (Table 8-4) to find out. These flights are from the payload conversion of Juno flying with a MicroPeak altimeter. On this particular day, we didn't have theodolites set up because it takes too many people, but of course they would have been more accurate than the single-axis tracker. The flights were also simulated using OpenRocket, RockSim, and the 1D simulation from Chapter 14. The weather conditions were taken from a nearby airport (not too near!) so the simulators had a chance to accurately predict the altitudes. All altitudes are in meters. Error estimates for the single-axis tracker come from Table 8-1. Errors from the mean are the standard deviation from all of the observed and predicted altitudes.

The point here isn't to determine which device was better. All altitude predictions and measurements have error, and we really have no way of knowing the *exact* altitude. While I would certainly rank the single-axis tracker last for accuracy, the result we got from it is not far off from the other methods. The simulations tend to disagree, too, but again, not by all that much.

Table 8-4. Juno's altitude

Method	A8-3	B6-4	C6-5
1D simulation	57	156	315
RockSim	59	173	395
OpenRocket	57	150	326
Altimeter	64	169	392
Single-axis tracker	59	146	419
Mean	59±3	159±12	369±46

The point is that we have three different ways to measure altitude, each with its own advantages, disadvantages, and typical error. It's an engineering decision to pick the best method for measuring altitude for a given flight, based on budget, the number of people available, and the accuracy needed. As the rocket scientist, you get to make that choice.

Payloads

9

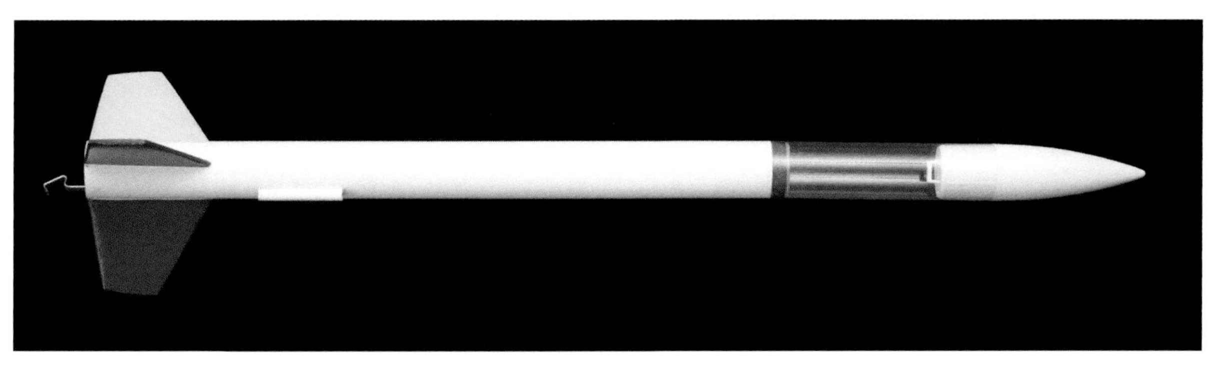

Figure 9-1. *Payload conversion of Juno. Most model rockets can be quickly converted to carry a payload.*

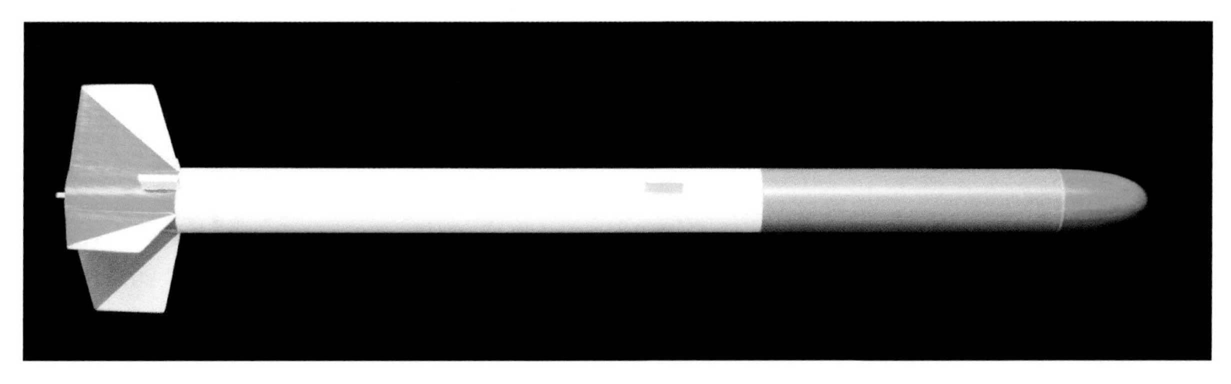

Figure 9-2. *The basic Ceres booster and payload. We'll look at two variations of the booster and five payload bays in this chapter, and another booster in Chapter 18.*

Flying the rockets we've seen so far is fun, but at some point you will want to step it up a notch and build rockets that carry something. It might be something whimsical, like some candy to pass out as "space candy" to the kids at a launch; it might be for a contest like the popular egg-lofting competitions; or it might even be an electronics payload to measure the altitude and acceleration of the rocket or the temperature and humidity in the atmosphere. Whatever it is you want to lift, you need a payload rocket.

Before we get too far into the planning stages, though, there is one rule about model rocket payloads that you need to be aware of. *Model rockets are not allowed to carry live vertebrates as a payload.* There is no scientific reason to pack a mouse into a model rocket. An egg is just as challenging technically, and the failures are entertaining instead of tragic. If it has a spine, it's a spectator, not a payload. Insects are allowed; I can personally say that grasshoppers, crickets, and bees all fly quite well in model rockets. Better than eggs, in fact. I've never lost an insect, but I have scrambled a few eggs.

This chapter covers two payload rockets, giving you a good fleet of payload vehicles. One of the payload rockets is a Tinkertoy rocket that has lots of variations. We'll look at two different boosters in this chapter, and another later, in Chapter 18. The chapter also shows how to build five different payload capsules for different purposes, and some obvious alterations that can be made on them for still other uses. You will find out everything you need to know to fly an electronic altimeter to find out how high a rocket really went; an egg for the challenge of lifting and recovering a delicate, heavy payload; and a camera to take high-definition video of a rocket flight.

Some of the rockets later in the book also have payloads, but they still depend on information from this chapter. Chapter 15 introduces high-performance rockets, and uses the same altimeter you will see in this chapter to figure out how high the rocket really goes. Chapter 18 shows two cluster rockets, both of which can carry payloads.

Liftoff Weight

Most of the rockets in the book list specific motors that will work in them. That works fine, because the weight and drag of the rockets were already known. Sure, there will be variations depending on how much effort you put into finishing your rocket, what kind of paint you use, and so forth, but those are not big enough differences to affect which motors will work.

Double the weight of the rocket by putting an egg in the payload bay, though, and things change. Your motor may not be powerful enough to lift the rocket. Worse, it may be powerful enough to lift the rocket, but not quite powerful enough to get it up to speed before it gets to the end of the launch rail. The result is called a *land shark*, where the rocket flops horizontal and powers off in a dangerous direction.

Or, perhaps the motor has plenty of power. The rocket lifts off, soars into the sky, and the delay charge begins to burn. The rocket reaches apogee, arcs gently, and turns over. The delay charge is still burning. The rocket screams to earth, impacting the ground. The ejection charge fires, deploying the parachute across the grass. That's called a *lawn dart*, since the rocket can easily stick in soft ground like the old lawn dart toys.

Or maybe you get lucky and the parachute deploys a few dozen feet off of the ground. Just in time, right? Well, maybe. The rocket may be moving fast enough to shred the parachute or break or rip loose from the shock cord. Even if the parachute and shock cord survive, the shock cord can rip through the body tube of the rocket. That's called a *zipper*.

I've seen all of these things happen. When you start adding payloads, you also have to start selecting the motors specifically for each flight.

There are three things to consider when picking a motor for a rocket:

1. Is the motor powerful enough to lift the rocket?
2. Will the rocket be traveling fast enough to be stable when it reaches the end of the launch rod?
3. Will the delay charge deploy the recovery system near apogee?

Using Charts to Find the Proper Motor

The first two questions are really a mater of the thrust available for the rocket. If the motor has enough thrust, it will lift the rocket, and lift it quickly enough to reach an adequate speed for the fins to work before the rocket leaves the launch rail. It turns out that the thrust needed when using a 3-foot launch rail is 11 times the weight of the rocket. We'll see why it is 11 times the weight of the rocket in "Using Math to Cheat Mother Nature" on page 209.

This means we need to know the thrust of the rocket motor. We should be able to get that from the motor's label, right? A B6-4 motor delivers 6 newtons of thrust, after all. Well, yes, that's the *average*, but most motors actually deliver quite a bit more thrust initially, during that first 0.13 seconds the rocket is on the launch rail. Take a look at the thrust curve for the Estes B6-4, shown in Figure 9-3.

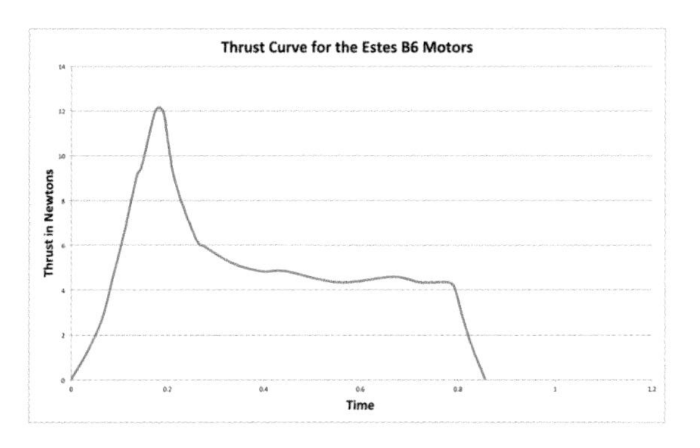

Figure 9-3. *The thrust curve for the Estes B6 motor shows that the maximum thrust is 12 newtons, about double the average 6 newtons from the motor designation.*

You can see that the maximum thrust is about twice the average thrust. That lets the B6-4 motor lift a rocket that is twice as heavy as you would expect based on the motor designation alone. That's also why you need to look at the thrust curve for the motor, not just the motor designation, when picking a motor for a payload rocket.

Thrust curves are available from several sources. Manufacturers have thrust curves on their websites, and sometimes even include them with the documentation that comes with a motor. You can also find a great summary of model rocket thrust curves at ThrustCurve.org. These files are generally designed for simulators, but you can see the thrust curves, too. Start with the Browser button on the main page, and navigate to the motor you would like to examine. There are a number of ways to find a particular motor, so I won't describe a specific one here. Once you've selected a motor, you will see a page like Figure 9-4.

thrustcurve.org
MOTOR PERFORMANCE DATA ONLINE
Search
Guide
My Stuff
Outbox

Estes B6

Manufacturer:	Estes Industries
Entered:	May 25, 2006
Last Updated:	Sep 12, 2012
Mfr. Designation:	B6
Brand Name:	B6
Common Name:	B6
Motor Type:	SU
Delays:	0,2,4,6
Diameter:	18.0mm
Length:	7.0cm
Total Weight:	15g
Prop. Weight:	6g
Cert. Org.:	National Association of Rocketry
Cert. Date:	Mar 25, 1995
Average Thrust:	5.0N
Maximum Thrust:	12.1N
Total impulse:	4.3Ns
Burn Time:	0.9s
Case Info:	SU 18x70
Data Sheet:	link

Click here

Don't like the units? Login to set your preferences.

Simulator Files

Format	Source	Updated	Contributor	Options
RockSim	user	Jun 25, 2008	John Coker	
RASP	cert	Mar 31, 2008	Mark Koelsch	

Have data or other info on this motor? Please login to contribute it.

thrustcurve.org MOTOR PERFORMANCE DATA ONLINE
Browser • Glossary • Manufacturers • Contribute
Search • Statistics • Simulators • Updates

Figure 9-4. *General data for a motor from http://www.thrustcurve.org. Click the View Data button to see a plot of the thrust curve, or examine this page to find the maximum thrust.*

This page lists the maximum thrust, but you need to make sure that the maximum happens soon enough—or, at least, that the thrust reaches 11 times the weight of the rocket you want to lift soon enough. So what is "soon enough"? As a general rule, you want the maximum thrust in the first 0.2 seconds after ignition. That's based on a 3-foot-long launch rod and an acceleration large enough to hit 30 mph by the time the rocket gets to the end of the rod. The time will vary with other launch rod lengths.

The only way to see if the maximum thrust happens soon enough is to look at the data. Click on the View Data icon on the right under Options. This particular motor has files for two simulators. It doesn't matter which one you pick. Figure 9-5 shows the data for RockSim. It's pretty similar to the curve shown in Figure 9-3.

Table 9-1 shows the maximum thrust for some common motors, including the ones used in this book. The delays are not shown, since the thrust does not change with the delay charge; a C6-0 and a C6-7 have the same thrust curve.

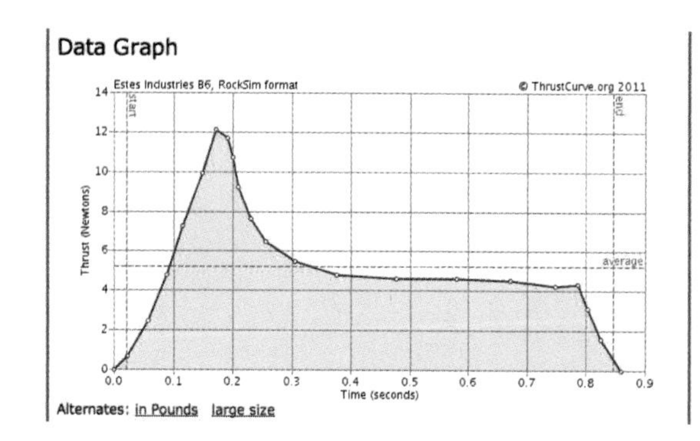

Figure 9-5. *A thrust curve from ThrustCurve.org.*

Table 9-1. Maximum thrust for some common motors

Motor	Maximum thrust in newtons	Maximum liftoff weight (ounces)
Estes 1/4A3 T	5.0	1.6
Estes 1/2A3 T	7.6	2.5
Estes A10 T	12.6	4.1
Estes A8	9.7	3.2
Estes B4	12.8	4.2
Estes B6	12.1	4.0
Estes C6	14.1	4.6
Estes D12	33.3	10.9
Estes E9	19.5	6.4
Quest A8	11.1	3.6
Quest B4	14.4	4.7
Quest C6	15.5	5.1
Quest D8	28.9	9.5

Thrust curves for model rocket motors are usually shown in newtons, but occasionally in pounds of force.

Model rocket weights are usually given in ounces, grams, or kilograms. Figure 9-6 shows a chart that will let you quickly pick a motor based on the maximum thrust. Find the weight of the rocket along the horizontal axis, and then go straight up or down to the line. Follow the line to the left or right to read off the minimum required thrust in newtons or pounds that you need to safely launch the rocket.

For example, let's say we're getting an egg lofter ready to launch. Our handy kitchen scale tells us the rocket, fully loaded with a parachute, motor, and egg, weighs 9 ounces. Figure 9-6 has 8 and 10 ounces along the x-axis. Picking a point about halfway between them and following it up to the line, then left to the y-axis, we see that we need about 27 or 28 newtons of thrust to lift our rocket. That rules out many of the smaller motors, but the D12, with 33.3 newtons of maximum thrust, will work just fine.

What if you don't know the exact weight of your rocket? A good estimate will work, too. This chapter shows the finished weight of each of the boosters and payload bays. Add in the weight of the motor and payload, and you have a pretty good estimate of the launch weight. You can get the weight of the motor from the manufacturer's website. You can come up with a pretty good estimate for the weight of most payloads with some thought or a few Internet searches. For example, it's not hard to use the Internet to find the average weight of a Grade A egg. True, recovery wadding and the parachute weigh something, too, but they don't contribute much to the total weight of a rocket carrying a heavy payload.

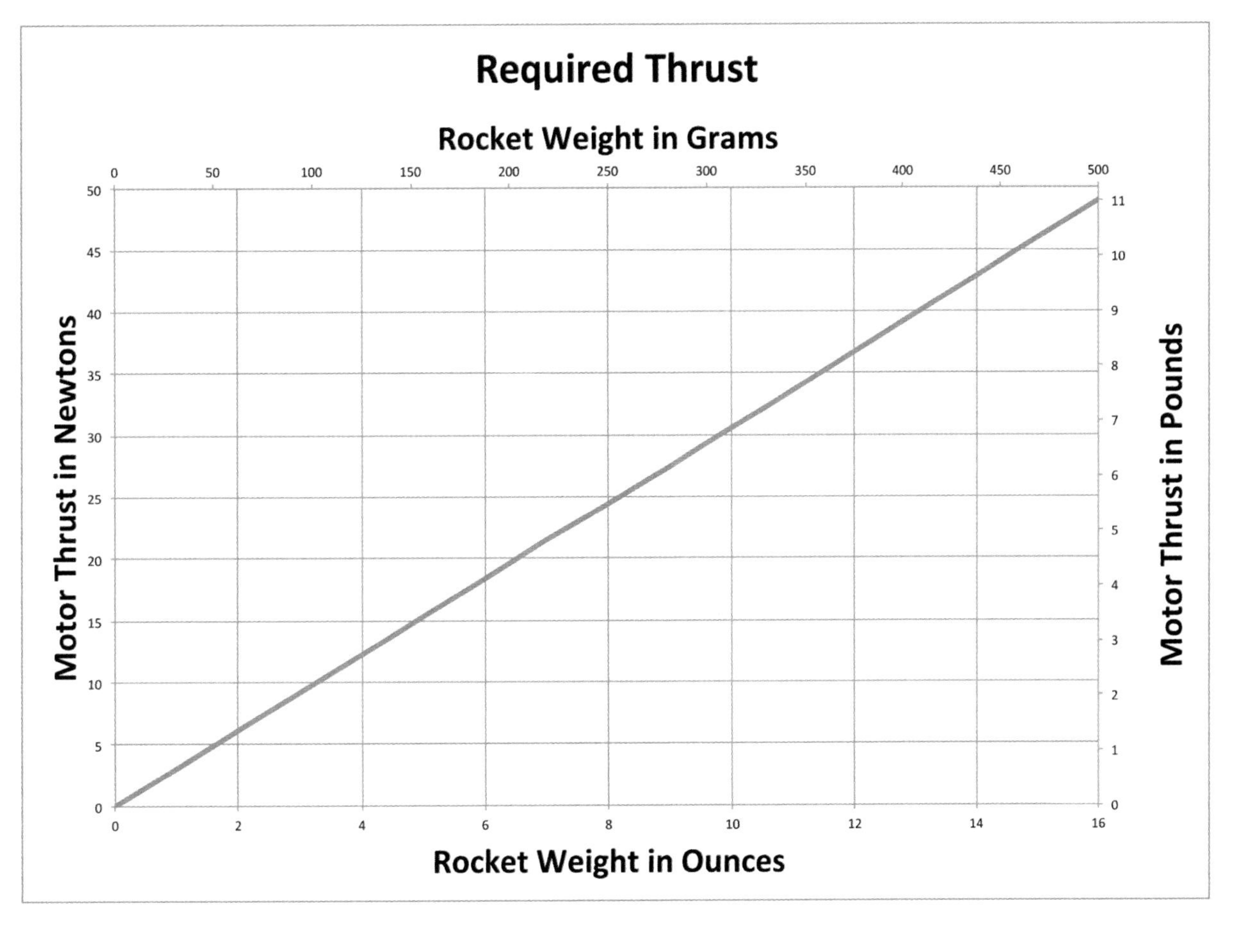

Figure 9-6. *Use this chart to find the safe thrust for a rocket of a given weight. Use the maximum thrust for the motor from the thrust curve for the first 0.2 seconds of thrust.*

Picking a Delay Time

Picking a delay time is a little trickier. We can ignore drag for the first fraction of a second of flight. It's still there, but the average speed of the rocket on the launch rail is only about 15 miles per hour—not enough for drag to be a huge impact on the dynamics of the rocket. That changes as the rocket powers up to speed, though. Hebe, a high-performance rocket you will see in Chapter 15, can hit 500 miles per hour. That's 33 times the average speed on the launch rail, and since the force of drag is proportional to the square of the velocity, that means Hebe is experiencing over 1,100 times more drag force at peak speed than the average aerodynamic drag force it experiences on the launch rail. A streamlined high-performance rocket like Hebe will not have the problems with drag that an irregular rocket like Nicomachus from Chapter 19 does. Hebe therefore needs a longer delay time than a rocket with more drag, even if you add enough weight so they weigh the same.

The most accurate way to pick a delay time is to fly the rocket and see what works. That's a little hard on rockets and spectators, though, so we generally use simulations to pick the delay time. Simulations are not quite as accurate as the real thing, but they are close enough to pick reasonable ejection delay times. Simulations are covered in more detail in Chapter 14. For this chapter, each rocket will have a table showing what motor to use for various payload weights.

Using Math to Cheat Mother Nature

In "Using Charts to Find the Proper Motor" on page 206, we used a thrust-to-weight ratio of 11 to determine the safe launch speed for a rocket. Why that number?

The answer comes from some basic physics. Understanding the physics will also help you cheat the system a bit, flying rockets that are not "safe" under standard launch conditions.

This section is a bit long to be a sidebar, but it does go into some mathematical details you can safely skip if you don't find them interesting. Stick with Figure 9-6 and you will be fine.

Still here? OK, let's look at the physics. After all, physics is phun!

Starting from rest, the distance a body travels under constant acceleration in a given amount of time is:

$$d = \frac{1}{2} a t^2$$

where *d* is the distance traveled, *a* is the acceleration, and *t* is the time. Traveling in a straight line, the speed after that amount of time is:

$$v = a t$$

With a little bit of algebra, we can solve for the acceleration needed to reach a specific speed in a certain distance. Basically, we solve both equations for *t*, then set them equal to each other, and then solve for *a*:

$$t = \sqrt{\frac{2d}{a}} \qquad t = \frac{v}{a}$$

$$\sqrt{\frac{2d}{a}} = \frac{v}{a}$$

$$\frac{2d}{a} = \frac{v^2}{a^2}$$

$$a = \frac{v^2}{2d}$$

This gives the acceleration needed to reach a specific velocity in a given distance. Both simulations and experience from hundreds of thousands of model rocket launches teach us that a rocket needs to be traveling at about 30 miles per hour, or 13.4 m/s, for the fins to keep it stable in flight. A 36-inch launch rod is about 0.91 meters long. Substituting 13.4 m/s for the velocity and 0.91 meters for the length, we find the acceleration needed to launch a model rocket, which is 98.66 m/s^2.

The thrust of a rocket motor is usually measured in newtons, which is a measure of force. Returning to basic physics, we have:

$$F = m a$$

where *F* is the force, *m* is the mass of the object, and *a* is the acceleration. The acceleration from gravity is 9.8 m/s^2, so the downward force on the rocket from gravity is 9.8×*m* newtons. The force needed to get enough acceleration to reach an appropriate launch speed by the end of the launch rail is 98.66×*m* newtons. The rocket motor must overcome both the force from gravity and the inertia from the mass of the rocket, so it needs 108.46×*m* newtons for a clean launch.

Another force comes from the weight of the rocket. While we sometimes use "weight" and "mass" interchangeably in everyday conversation, they are not really the same thing. The *mass* is the resistance of the object to movement, while the *weight* is the force exerted by gravity. The rocket weighs 9.8×*m* newtons, so the thrust-to-weight ratio is:

$$\frac{thrust}{weight} = \frac{108.46\, m}{9.8\, m} = 11.07$$

That's where the thrust-to-weight ratio of 11 comes from.

That's the physics, but we measure thrust in newtons and the weight of the rocket in grams or ounces. Of course, a gram is actually a unit of mass, so it does have to be converted into a force, but we're stuck with grams as the "weight" on a scale because that's the way scales are calibrated in the metric world. You convert newtons of force to grams of mass by dividing by the acceleration of gravity and then converting kilograms to grams, so one newton is the force exerted by 102 grams at the surface of the earth. Dividing by 11, a rocket with 1 newton maximum thrust can safely launch a rocket whose mass is 9.28 grams. The conversion for ounces is similar; a rocket motor delivering 1 newton of thrust can lift a rocket that weighs 0.327 ounces. You get the actual safe launch weight by multiplying 9.28 grams or 0.327 ounces by the actual maximum thrust of the rocket motor.

You might object that this is an approximation. There is friction from the launch rail and the air that we've ignored, and the motor doesn't deliver its maximum thrust the entire time the rocket is on the launch rail. You could also argue that some rockets are more stable than others, and may not need as much speed, or may need more.

All those objections are valid. But both rocket simulations using computers and practical experience from hundreds of thousands of rocket flights show that this approximation is good enough. As long as you are within about 10% of the listed weight, your rocket should fly fine.

So why is this science important? After all, we know the answer: multiply the maximum thrust by 9.28 grams or 0.327 ounces. We could also stick with using Figure 9-6. Who cares about the details? But stop and think for a moment. What if the thrust-to-weight ratio is only 8, rather than 11? Does that kill the rocket design? Not necessarily; you may just need a longer launch rod! And, with the science to back you up, you know how to decide just how long that launch rod needs to be.

You may see people successfully launching rockets that seem to defy these calculations. The Ceres A booster, which we'll meet later in this chapter, is one example. These numbers are based on a speed that works for almost any model rocket, even when wind speeds approach 20 miles per hour. That's a moderate breeze, enough to move small branches, cause flags to flap, and raise small whitecaps on water. You probably won't want to fly rockets in winds that high, since the chance of losing one goes up with the wind speed, but the rocket should still be designed to be stable at those speeds. After all, a gust of wind may hit just as you press the launch button, and you don't want your rocket to become a land shark. Still, with calm winds and large fins, it's possible to launch with a lower thrust-to-weight ratio. That's where simulators and test flights become important. In general, don't launch rockets unless they have a thrust-to-weight ratio of at least 11. For rockets with a lower thrust-to-weight ratio, or even rockets right at a thrust to weight ratio of 11, only fly in calm conditions.

What happens if the thrust-to-weight ratio is too low? If it is very low—below 1—the rocket won't move at all. Beyond that, it all depends. If the thrust-to-weight ratio is so low that the rocket is not traveling fast enough for the fins to work, the rocket will be unstable. If the rocket is going fast enough for the fins to work, but is still going very slowly compared to the current wind speed, the rocket will weathercock, turning sharply into the wind. Exactly when these things happen depends on the actual thrust-to-weight ratio, the design of the rocket, and the wind speed. If you keep the liftoff speed to 30 mph or higher, launch in wind speeds of 20 mph or lower, and make sure the rocket is stable, the rocket should fly fine.

Juno Payload Conversion

Back in Chapter 3, we built our first solid propellant rocket: a great sport flier called Juno. You might also remember the basic stability rules from that chapter. Adding weight at the nose of a rocket makes it more stable. That means we can do a payload conversion on just about any model rocket. Let's see how to do that with Juno.

Figure 9-1 showed Juno with the payload bay installed. Table 9-2 shows the motors you can use for the Juno payload conversion. Note that the motor selection depends on the payload weight. Heavier payloads require a shorter ejection delay.

Table 9-2. Recommended motors

Motors	Approximate altitude	Payload weight
A8-3	120–280 ft	0–1 oz
B6-4	390–660 ft	0–1 oz
C6-7	1,100–1,400 ft	0–1 oz
C6-5	690 ft	2 oz

Table 9-3 lists the parts you'll need for the Juno payload conversion.

Table 9-3. Parts list

Part	Description
BT-50 clear payload tube	This is the clear plastic tube from the Estes Designer's Special. Clear plastic payload tubes are also available separately from Estes and several other manufacturers.
BT-50 tube coupler	Used to form the shoulder of the payload bay. If you are ordering parts, you can substitute a BT-50 balsa nose block for the tube coupler and balsa fin stock.
1/8″ balsa fin stock	This will be used to form a bulkhead at the base of the payload bay.
Screw eye	You will need a small screw eye for the base of the nose section. These are available at hardware stores.

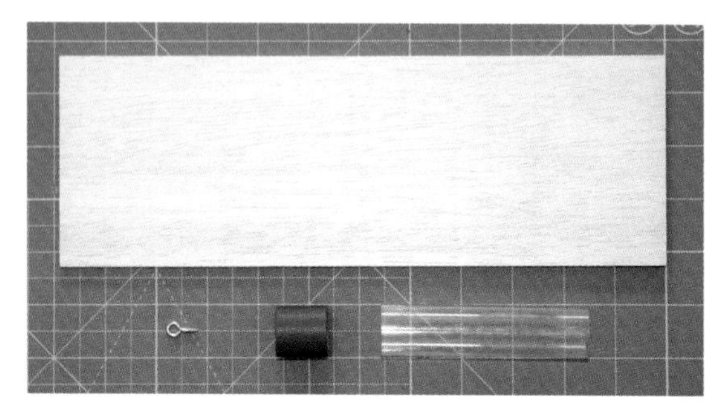

Figure 9-7. *Visual parts list for the Juno payload conversion.*

Equivalent Kit

The Estes Payloader II™ is a nice equivalent to the Juno payload conversion. The Skybird from Custom Rockets is similar, but with a nice twist: its payload bay is a clear red plastic.

Construction

Refer back to Chapter 3 for basic construction techniques, tools, and materials. That's when you built Juno. If you built it without the snap swivel, start by adding one to the shock cord and parachute. That will let you quickly switch between the payload bay and the original Juno design.

Cut two disks from the 1/8″ balsa stock. Each disk should just barely fit inside the tube coupler. Glue the disks together with the grain from one disk perpendicular to the grain from the other; this increases the strength of the disks. Glue the completed assembly into one end of the tube coupler.

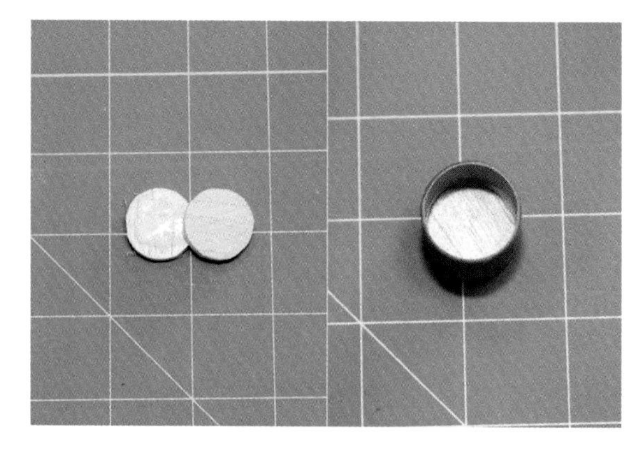

Figure 9-8. *Glue two 1/8″ balsa disks together, then glue the assembly into the bottom of the tube coupler.*

Glue the tube coupler into the base of the clear payload tube. The tube coupler is 1″ long; glue about 1/3 of it in the tube. Wood glue does not work especially well on the clear plastic payload tube. If you have some handy, use a small dab of epoxy instead. If you don't have epoxy handy, rough up the portion of the inside of the clear plastic tube that will be glued to the tube coupler with sandpaper. That will give the wood glue something to embed itself in, so it won't break free from the slippery plastic as easily.

Screw the screw eye into the bulkhead formed by the balsa disks, remove it, squirt some glue into the hole, and reinsert the screw eye.

That's it for construction. Detach the nose cone from the rocket and slide it into the top of the payload tube, then attach the shock cord to the screw eye of this payload bay. You can use the same technique to convert just about any model rocket to carry a payload.

Figure 9-9. *Glue the tube coupler into the payload tube and the screw eye into the bulkhead to complete the payload bay.*

Flying the Juno Payload Conversion with an Altimeter

There really is no difference between flying Juno and flying the Juno payload conversion, other than any preparation you might need to do to get the payload ready. Here's one example that might be both fun and interesting.

The MicroPeak altimeter from Altus Metrum is a tiny little thing, but it has capabilities that will surprise you. That's the altimeter lying on the card to the left of the Juno payload bay in Figure 9-10. The battery is on the back side of the altimeter. It's a barometric altimeter that finds the altitude of your rocket based on pressure changes, the same way many altimeters in airplanes work. That means you need to let the pressure in the payload bay change as the rocket flies. You can do that by drilling a 1/16" hole in the payload bay. In general, the hole should be at least one body tube diameter in length from the base of the nose cone, but the shoulder of the nose cone is long enough that it pretty much has to be anyway. I put the hole near the bottom of the payload bay, well away from the turbulence from the transition around the base of the nose cone.

Figure 9-10. *The MicroPeak altimeter and USB adapter.*

The altimeter itself is very easy to use. Just before the flight, flip the tiny switch on the altimeter. It will flash the altitude from the previous flight in decimeters using an LED. For example, if the pattern is 1 flash, 4 flashes, 2 flashes, and 1 flash, the maximum altitude was 142.1 meters. The device then flashes a series of short pulses, followed by a long pulse, and then settles down to an occasional flash to let you know it is on.

Drop it into the payload bay, stuff some cotton in to keep it from rattling around, and fly the rocket just like normal.

After the flight, turn the altimeter off and back on. Count the flashes. Right away, you know how high the rocket went.

Back at your desk, download the free software from *http://www.altusmetrum.org/MicroPeak/* and install it on your computer. It will show you how to connect to the MicroPeak USB adapter and move the data from the flight computer to your desktop computer. Figures 9-11 and 9-12 show the flight data from the Juno payload conversion, flying on an A8-3 motor.

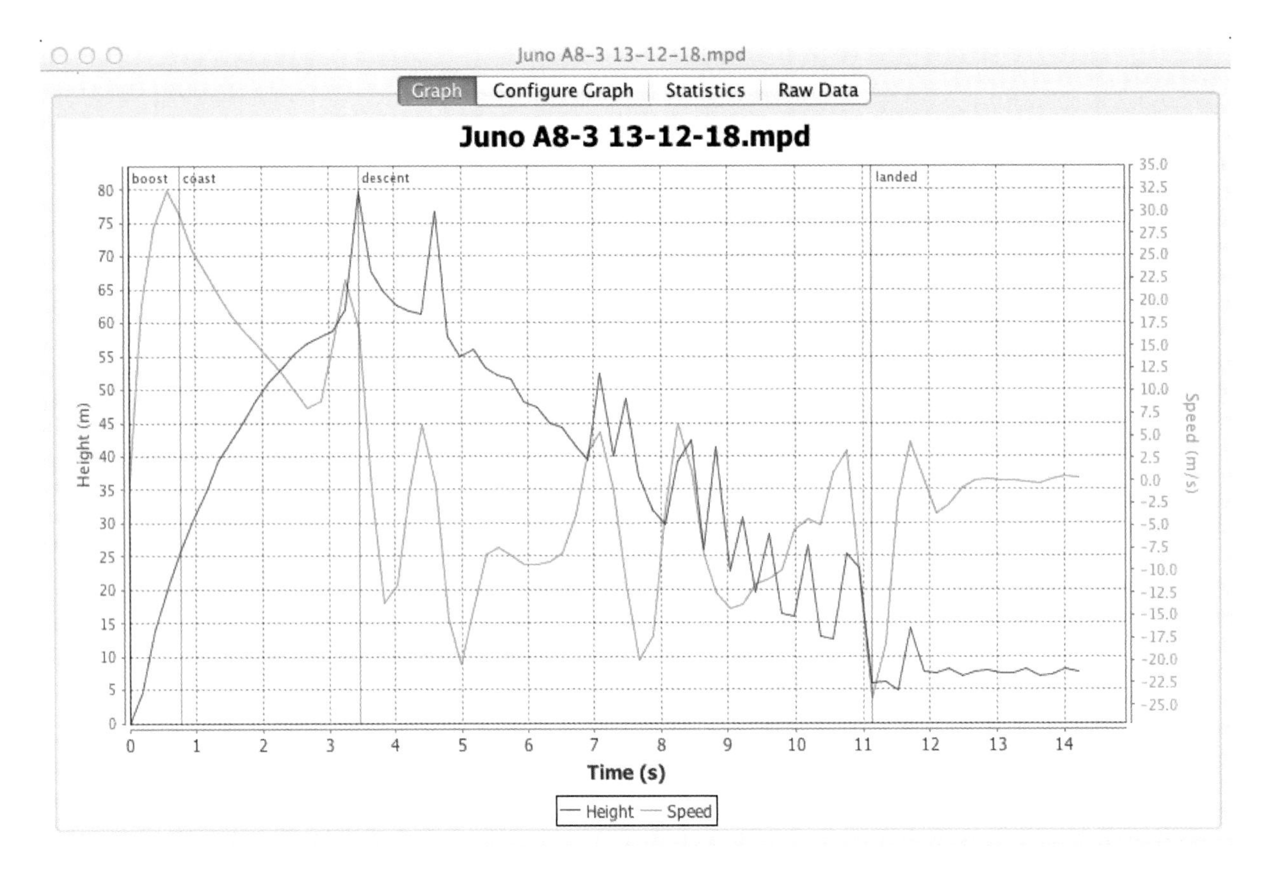

Figure 9-11. *Data from a MicroPeak altimeter flown on an A8-3 motor.*

You can see some places where the altitude and speed seem to jump in an unreasonable way. This is normal with real sensor data. Most of the cause is the rocket itself bouncing around on the shock cord.

Maximum height	63.8 m	209.4 ft	
Maximum speed	32.2 m/s	72.1 mph	Mach 0.094
Maximum boost acceleration	93.1 m/s²	305.4 ft/s²	9.494 G
Average boost acceleration	43.5 m/s²		
Boost duration	0.8 s		
Coast duration	2.7 s		
Descent rate	7.3 m/s	24.1 ft/s	
Descent duration	7.7 s		
Flight Time	11.1 s		

Figure 9-12. *The summary page from the data shows basic flight statistics.*

Before we move on to our next rocket, here's a complete launch preparation checklist for the Juno payload conversion with a MicroPeak altimeter, and a launch checklist to use before each launch.

Launch Preparation Checklist

1. Insert recovery wadding.
2. Attach the parachute and shock cord.
3. Insert the parachute and shock cord; make sure they are not too tight.
4. Install the motor.
5. Insert the igniter.

Flight Checklist

1. Clear spectators from the launch zone.
2. Remove the launch key from the launch rail, slide the rocket onto the launch rail, and replace the launch key on the launch rail.
3. Turn on the altimeter and place it in the payload bay.
4. Attach the igniter clips.
5. Get the launch key from the pad.
6. Insert the launch key. Make sure the continuity light is on.
7. Check for clear skies—no aircraft or low clouds.
8. Count down from five and launch.
9. Remove the launch key and place it on the launch rod.
10. Recover the rocket.
11. Check the altitude.

The Ceres Payload Rocket

Unlike the rockets built so far in this book, the Ceres payload rocket isn't just one monolithic design. NASA rockets like the Saturn IB carried many different payloads, including the Earth-orbiting Apollo flights and ferrying astronauts to Skylab. Some of those same payloads were also carried on the Saturn V. The Apollo command service module and the lunar module flew on both the Saturn IB and Saturn V. We will follow suit here, with three boosters and five payload bays.

The boosters are the Ceres A, Ceres B, and Ceres C. They look pretty much the same. The external design is identical, with four large fins for stability and an 18"-long BT-60 body tube. The difference is the propulsion systems. The Ceres A flies on the same A, B, and C motors you have used on other rockets in this book, meaning that you don't have to add to your existing stock of motors. The Ceres B uses D or E motors. That gives you a lot of power for heavy payloads or high flights. The Ceres C is a cluster rocket that uses three A, B, or C motors. It is covered in Chapter 18, where you will see why the cluster rocket with three C6 motors can lift even heavier payloads than an E motor, which has 33% more total impulse.

There are five payloads for these rockets. Any payload can be flown with any rocket. That's a total of 15 possible combinations! Three payloads are general-purpose payload bays like the one on the Juno payload conversion. Two others are highly specialized payloads. One carries a small video camera, while the other carries an egg.

Ceres

Ceres is the largest of the asteroids in the asteroid belt, making up a full one-third of the total mass of the asteroids. It's large enough that gravity has pulled it into a spherical shape, unlike most asteroids, which are irregular lumps. At nearly 600 miles in diameter, it's about one-quarter of the size of our moon. It's the only asteroid large enough to also be classed as a dwarf planet. It seems appropriate that the largest rocket in this book be named for the largest asteroid.

Ceres is probably made up of a rocky core with an icy outer layer. It may even have a very thin atmosphere. While it seems unlikely, the presence of ice on Ceres has led to some speculation that there may be primitive life there.

Back in 1766, Johann Titius, drawing on earlier work by Johannes Kepler, noticed that there was a regular pattern to the distance between planets, except for something missing between Mars and Jupiter. Johann Bode got involved, and they came up with what became known as the Titius–Bode law predicting planet positions. The discovery of Uranus in 1781, right in the orbit where they predicted the next planet should be, convinced a lot of people there should be something between Mars and Jupiter. A search was organized in 1800. Ceres was discovered on January 1, 1801 by the Italian astronomer Giuseppe Piazzi, exactly where the "missing planet" should be. While the Titius–Bode law itself was later discredited, it did lead to an active and successful search for asteroids.

Ceres is named for the Roman goddess of agriculture.

Unlike the rockets in the book, the payloads have descriptive or whimsical names. The texting generation will have no trouble seeing why the clear payload is called ICU or why the camera is called ICU2. For the non-texting generation... well, think about it a moment.

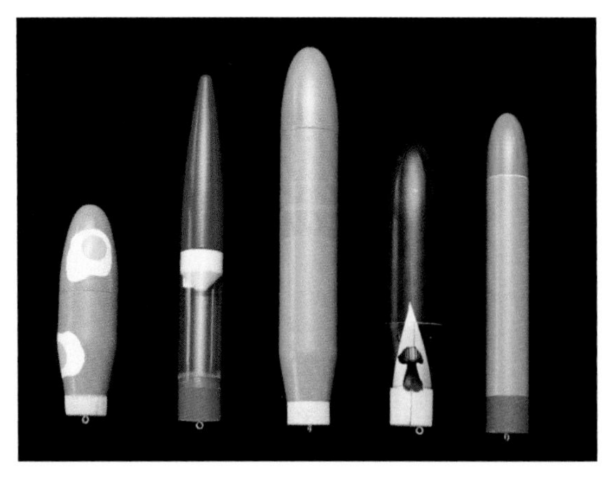

Figure 9-13. *From left to right: Over Easy (egg lofter), ICU, Fat Man, ICU2 (camera), and Thin Man.*

You can build any one of the Ceres boosters and the Thin Man payload bay with the parts in the Estes Designer's Special. Table 9-4 lists the additional parts you will need to build all of the boosters and payload bays. All of the parts except the camera are usually available from *http://jonrocket.com*, although they are occasionally out of some of the balsa parts. You can also get the balsa parts from *http://balsamachining.com*, and pick up the other parts from *http://estesrockets.com* or other online stores. Complete parts lists for each booster are also given at the start of each construction section.

Table 9-4. Parts list

Part	Description
BT-60 body tube (2)	You only have one left in the Estes Designer's Special. You will need one more for each booster you build.
BT-70 body tube	One section will be used for the egg lofter, while a longer piece will be used for the Fat Man payload bay.
BT-80 body tube	A small section will be used for the ICU2 payload.
1/8" balsa fin stock	The supply in the Estes Designer's Special will run short. Pick up another sheet from your local hobby store, or you can mail order a sheet when buying the other parts.
BT-60 clear payload tube	This is used for the ICU payload bay. This is a larger version of the payload for the Juno payload conversion. You can find these individually, or get a package of several clear payload bays from *http://estesrockets.com*.
BT-70 nose cone (2)	Get two of these. One is used for Over Easy and one for the Fat Man. The nose cone for Over Easy must be balsa; a plastic one will not work.
BT60–BT70 transition (2)	This is the piece you see at the bottom of the larger payload bays that transitions from a BT-60 to a larger body tube. Get two, one for Over Easy and one for the Fat Man. The transition for Over Easy must be a balsa transition.
BT-80 nose cone	For the ICU2 payload. You can get a second if you like, and make an even bigger version of the Fat Man payload. The nose cone for ICU2 must be balsa.
BT60–BT80 transition	For the ICU2 payload. This must be a balsa transition.
Y2000 mini camcorder	This is the camera for the ICU2 payload. Actually, any small camera will work. Do some shopping on the Internet and see if there is a newer one with features you like better. You will also need a micro SDHC memory card. You may already have one for something else. If you buy one, get a small one. A rocket movie doesn't take a lot of space, and there is a risk of losing the camera, so you don't want to spend a lot of money on a big memory card.
Snap swivels	It's possible to tie and untie the parachutes and shock cords as you switch payload bays, but it's a lot easier to use snap swivels. You will need one snap swivel for the shock cord on each booster, plus one snap swivel for any parachutes you plan to use.
Screw eye (5)	You will need a small screw eye for the base of the nose section for each payload you build.

The Ceres A Booster

The Ceres A booster can lift any of the payloads, but it's a fairly underpowered rocket for lifting payloads of any weight. Its main advantage is that the rocket can be built and flown with a single standard motor. With thrust-to-weight ratios running as low as 7.2, you should use a longer launch rod when available, and launch only on very calm days. That said, I have test flown the Ceres A with most of the payloads, including the Over Easy payload. All were flown with 3-foot launch rods. The Over Easy payload with the Ceres A booster has the lowest thrust-to-weight ratio in the book. Tables 9-5 through 9-9 list the recommended motors to use with the Ceres A for each payload.

Table 9-5. Ceres A/Thin Man recommended motors

Motors	Payload weight	Liftoff weight	Approximate altitude
C6-5	0 oz	4.7 oz	410 ft
C6-3	1 oz	5.7 oz	270 ft

Table 9-6. Ceres A/Fat Man recommended motors

Motors	Payload weight	Liftoff weight	Approximate altitude
C6-3	0 oz	5.6 oz	280 ft
C6-3	1 oz	6.6 oz	220 ft

Table 9-7. Ceres A/ICU recommended motors

Motors	Payload weight	Liftoff weight	Approximate altitude
C6-3	0 oz	5.0 oz	390 ft
C6-3	1 oz	6.0 oz	270 ft

Table 9-8. Ceres A/ICU2 recommended motors

Motors	Payload weight	Liftoff weight	Approximate altitude
C6-3	0.6 oz	5.8 oz	290 ft

Table 9-9. Ceres A/Over Easy recommended motors

Motors	Payload weight	Liftoff weight	Approximate altitude
C6-3	2 oz	7 oz	180 ft

Table 9-10 lists the parts you will need to build the Ceres A booster.

Table 9-10. Parts list

Part	Description
BT-60 body tube	Use the full 18"-long body tube. Some companies sell 34"-long body tubes. Cutting one in half and using 17″ works fine, too.
BT-20 motor mount	You can use a precut 2 3/4″ motor mount tube, or cut your own from a longer piece of BT-20 body tube.
3/32″ balsa fin stock	The balsa wood will be used for the fins and for launch lug standoffs.
1/8″ launch lug	You will use two pieces, each 1″ long.
Engine hook	Use the shorter (2 3/4″) engine hook.
Engine sleeve	You can use tape if none are available.
1/8″ shock cord	You will need about 30″ of shock cord.
Snap swivel	You will need one snap swivel for the shock cord.

Construction

Refer back to Chapter 3 for basic construction techniques.

There are no body tubes to cut if you are starting with a precut 2 3/4″ motor tube. If you are starting with a BT-20 body tube, cut a 2 3/4″ piece from the tube for the motor tube.

The main body tube is an uncut 18″ BT-60 body tube.

You will need four fins cut from 3/32″ balsa and rounded on all edges except the root edge. Figure 9-14 shows the full-size fin pattern.

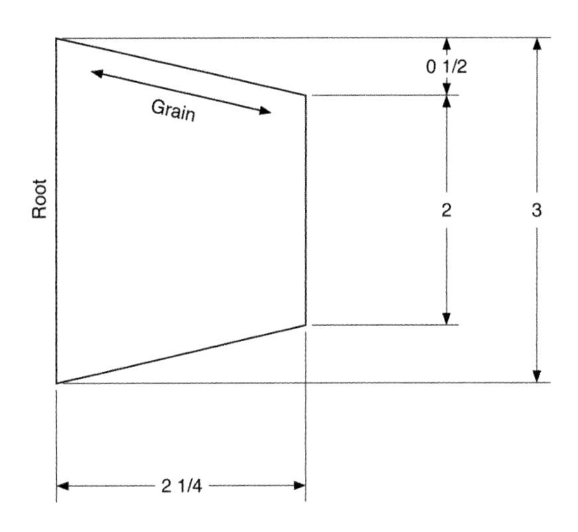

Figure 9-14. *Fin pattern for the Ceres booster. Cut four fins from 3/32″ balsa for the Ceres A, and four fins from 1/8″ balsa for the Ceres B and Ceres C.*

You will need two launch lug standoffs. Looking at some of the payloads, you can see that the body tube for the payload bay is larger than the body tube for the booster. The launch lug needs to be mounted 1/2″ away from the body tube to allow space for the payloads. Cut two 1″ x 1/2″ standoffs from 3/32″ balsa. Make sure the fin grain is parallel to the short side of the standoff so it runs from the body tube to the launch lug, as shown in Figure 9-15.

If you look closely at the recommended motor charts, you'll notice that the Ceres A is a bit underpowered for some of the payloads. You will want to use a longer launch rod if one is available. Long 1/8″ launch rods tend to be a little flexible, and can create *rod whip*, where the launch rod flaps back and forth violently as the rocket travels rapidly up the rod. If you have a long launch rod, it may be a 3/16"-diameter rod. Be sure and swap out the smaller launch lugs for 3/16" launch lugs if you plan to use a longer, 3/16" launch rod.

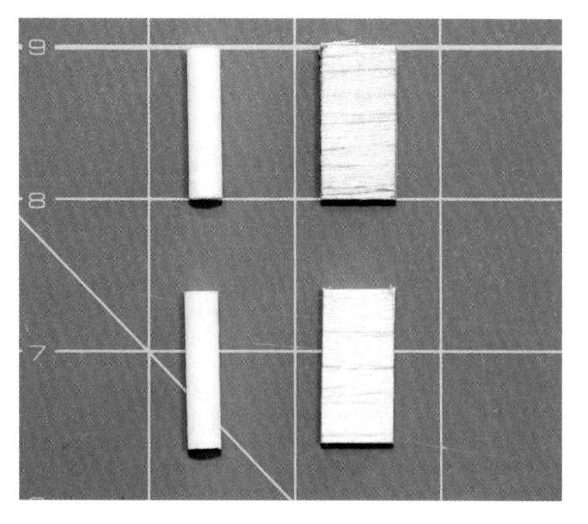

Figure 9-15. *Cut two 1" x 1/2" launch lug standoffs from 3/32" balsa. Cut two 1"-long launch lugs for a 1/8" launch rod.*

The motor mount is cut from one of the thick paper motor mount sheets in the Estes Designer's Special. If you are buying individual parts, you can either buy BT-60 to BT-20 centering rings or cut your own from 1/8" balsa.

With the parts cut, it's time to assemble the rocket. Start with the motor mount. It is assembled just like the motor mount for Juno, so we won't go over it in detail. You may want to save the plastic engine sleeves for the Ceres C booster. Tape works well, too; just coat it with a layer of glue to add strength. That's what you see in the assembled motor mount in Figure 9-16.

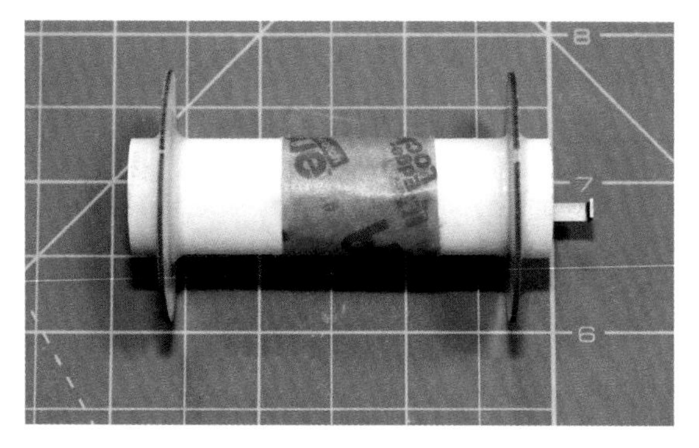

Figure 9-16. *The assembled Ceres A motor mount.*

While the motor mount dries, use the fin guide in Figure 3-25 to mark the body tube for four fins and a launch lug. Attach the four fins flush with the bottom of the body tube. Attach a launch lug on a standoff two inches from each end of the tube. Set the assembly aside to dry.

Once the parts are dry, mount the motor mount in the base of the body tube, and attach the shock cord near the upper end of the body tube using the standard paper mount. Since you will be switching between a variety of payloads, use a snap swivel at the end of the shock cord.

That completes the construction of the booster. It is shown in Figure 9-17 with the completed Thin Man payload.

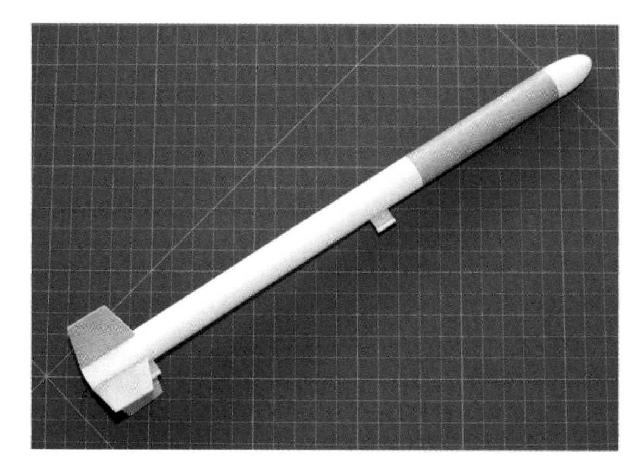

Figure 9-17. *The Ceres A booster with the Thin Man payload.*

Flying the Ceres A

The Ceres A can be flown with any of the payloads described later in this chapter. The payload may require some changes in the launch checklist, but the booster itself is very standard. Here are the launch preparation and flight checklists for the Ceres A.

Launch Preparation Checklist

1. Insert recovery wadding.
2. Attach the parachute and shock cord.
3. Insert the parachute and shock cord; make sure they are not too tight.
4. Insert the motor.
5. Install the igniter.
6. Install the payload.

Flight Checklist

1. Clear spectators from the launch zone.
2. Remove the launch key from the launch rail, slide the rocket onto the launch rail, and replace the launch key on the launch rail.
3. Attach the igniter clips.
4. If needed, start the payload (for cameras or other electronics).
5. Get the launch key from the pad.
6. Insert the launch key. Make sure the continuity light is on.
7. Check for clear skies—no aircraft or low clouds.
8. Count down from five and launch.
9. Remove the launch key and place it on the launch rod.
10. Recover the rocket.

The Ceres B Booster

If you are only going to build one of the three boosters, build this one. The powerful D and E motors will lift all of the payloads well. The only downside is that D and E motors have age restrictions in some states, and you must use a launcher with a 30-foot launch wire if you're igniting E motors. If you are under 18, you may need to get an adult guardian to buy the motors. Discuss the rocket and your plans with that adult, and plan some extra time to allow her to make the purchase if she agrees.

Table 9-11 through Table 9-15 list the recommended motors to use with the Ceres B for each payload.

Table 9-11. Ceres B/Thin Man recommended motors

Motors	Payload weight	Liftoff weight	Approximate altitude
D12-7	0-1 oz	5.1-6.1 oz	940-1,200 ft
D12-5	2-5 oz	7.1-10.1 oz	400-740 ft
D12-3	6 oz	11.1 oz	340 ft
E9-6	0-1 oz	5.9-6.9 oz	1,500-1,800 ft

Table 9-12. Ceres B/Fat Man recommended motors

Motors	Payload weight	Liftoff weight	Approximate altitude
D12-7	0 oz	6.3 oz	860 ft
D12-5	1-2 oz	7.3-8.3 oz	460-700 ft
D12-3	3-4 oz	10.3 oz	140-270 ft
E9-8	0 oz	7.1 oz	1,500 ft

Table 9-13. Ceres B/ICU recommended motors

Motors	Payload weight	Liftoff weight	Approximate altitude
D12-7	0-1 oz	5.6-6.6 oz	840-1,000 ft
D12-5	2-4 oz	7.6-9.6 oz	440-660 ft
D12-3	5-6 oz	10.6-11.6 oz	120-360 ft
E9-8	0 oz	6.5 oz	1,700 ft

Table 9-14. Ceres B/ICU2 recommended motors

Motors	Payload weight	Liftoff weight	Approximate altitude
D12-7	0.6 oz	6.6 oz	840 ft
E9-6	0.6 oz	7.3 oz	1400 ft

Table 9-15. Ceres B/Over Easy recommended motors

Motors	Payload weight	Liftoff weight	Approximate altitude
D12-5	2 oz	7.8 oz	640 ft
E9-6	2 oz	8.5 oz	1200 ft

Table 9-16 lists the parts you will need to build the Ceres B booster.

Table 9-16. Parts list

Part	Description
BT-60 body tube	Use the full 18"-long body tube. Some companies sell 34"-long body tubes. Cutting one in half and using 17″ works fine, too.
BT-50 motor mount	You can use a precut 3 1/4″ motor mount tube or cut your own from a longer piece of BT-50 body tube. Note that these are the fatter motor mount tubes that are about 1″ in diameter, not the smaller, roughly 3/4″ motor mount tubes used in most other rockets in this book.
1/8″ balsa fin stock	The balsa wood will be used for the fins and for launch lug standoffs.
3/16″ launch lug	You will use two pieces, each 1″ long.
Engine hook	Use the longer, 3 1/4″ long engine hook, not the shorter 2 3/4″ engine hooks used elsewhere.
Engine sleeve	You can use tape if none are available.
1/4″ shock cord	You will need about 30″ of shock cord.
Snap swivel	You will need one snap swivel for the shock cord.

Construction

Other than using larger parts in a few places, the construction of the Ceres B is identical to the construction of the Ceres A. Refer back to "Construction" on page 217 for step-by-step instructions. I'll note the differences here.

The Ceres B uses 1/8″ balsa for all fins and launch lug standoffs. In general, use 1/8″ fin stock for D motors or above, or for cluster rockets that have a total impulse of 20 newton-seconds or more.

D and E rockets are also generally launched from heavier 3/16″ launch rods, so you need to use the fatter launch lugs designed for 3/16″ rails.

The motor mount is built just like the one for the smaller A to C motors, but it's made from a slightly longer BT-50 body tube. The motor tube is 3 1/4″ long. The engine hook is also longer (3 1/4″, as opposed to the 2 3/4"-long hook used in the Ceres A).

Finally, we use a thicker shock cord, moving from the 1/8″ shock cord appropriate for smaller motors to a 1/4″ shock cord.

This basically means the only thing that didn't change is the main body tube. All of the other parts got a little thicker, a little longer, or a little fatter.

Figure 9-18 shows the motor mounts for the Ceres A, Ceres B, and Ceres C lined up for comparison.

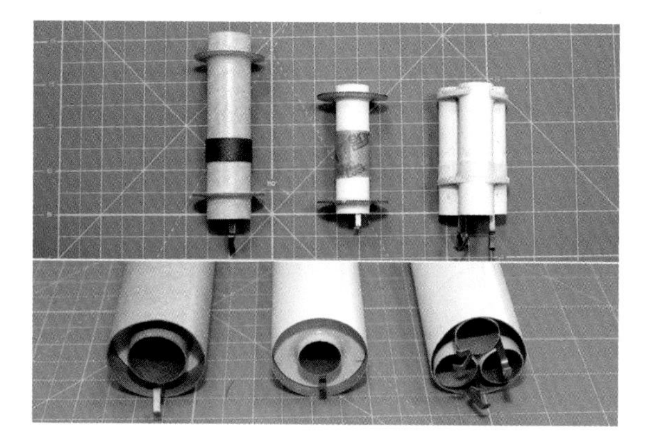

Figure 9-18. *The motor mounts for the Ceres rockets, both outside the body tube and being test fit into the body tubes. From left to right, the motor mounts are for the Ceres B (D and E motors), the Ceres A (C motors), and the Ceres C from Chapter 18.*

Flying the Ceres B

The Ceres B can be flown with either D or E motors. The motor mount is designed for an E motor, which is a half-inch longer than a D motor. So how do you use a D motor?

There is a bright orange paper tube in the Estes Designer's Special, as well as in many D–E motor mount kits. This is used as an adapter for D motors. Insert the adapter first, then the D motor, then fly like you normally would.

You should always save a few spent motors for various uses, such as supports when cutting body tubes. Another use is as an adapter—1/2″ cut from an expended D or E motor case makes a fine motor adapter.

Figure 9-19. *The E and D motors. The D motor is shown with the orange motor adapter.*

There is one other special consideration when flying the Ceres B booster.

E motors are large enough that everyone must be at least 30 feet from the rocket when it is launched. That includes the launch control officer, so you need a launcher with at least a 30-foot launch cord.

With that exception, flying the Ceres B is the same as flying the Ceres A. Launch preparation and flight checklists follow.

Launch Preparation Checklist

1. Insert recovery wadding.
2. Attach the parachute and shock cord.
3. Insert the parachute and shock cord; make sure they are not too tight.
4. Insert the motor.
5. Install the igniter.
6. If you are using an E motor, make sure the launch cord is at least 30 feet long.
7. Install the payload.

Flight Checklist

1. Clear spectators from the launch zone.
2. Remove the launch key from the launch rail, slide the rocket onto the launch rail, and replace the launch key on the launch rail.
3. Attach the igniter clips.
4. If needed, start the payload (for cameras or other electronics).
5. Get the launch key from the pad.
6. Insert the launch key. Make sure the continuity light is on.
7. Check for clear skies—no aircraft or low clouds.
8. Count down from five and launch.
9. Remove the launch key and place it on the launch rod.
10. Recover the rocket.

The Thin Man and ICU Payload Bays

These two payload bays are built identically. The only differences are the body tubes and nose cones used, and you can actually use either nose cone on either payload bay. These are also very similar to the payload conversion for Juno that we looked at earlier in this chapter.

Tables 9-17 and 9-18 list the parts required to build the Thin Man and ICU payload bays.

Table 9-17. Parts list for the Thin Man payload bay

Part	Description
BT-60 body tube	Cut a piece from a larger tube. You have enough left in the Estes Designer's Special for an 8" payload bay, which is what is shown.
2 1/2" BT-60 nose cone	Use a lightweight nose cone for flights with the Ceres A. You can use a longer one with the Ceres B.
1/8" balsa	Used to form a bulkhead at the base of the payload.
BT-60 tube coupler	Used to form the shoulder of the payload bay.
Screw eye	Used to attach the shock cord and parachute.

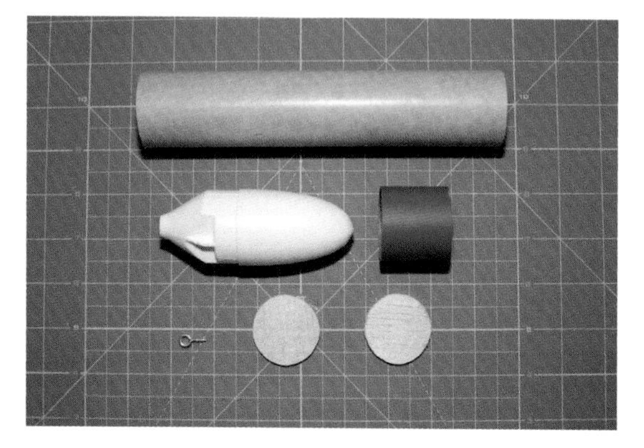

Figure 9-20. *Visual parts list for the Thin Man payload bay.*

Table 9-18. Parts list for the ICU payload bay

Part	Description
Clear BT-60 payload tube, 4" long	Clear payload tubes are available from various retailers as the "Clear Payload Section Assortment," Estes part 003171.
6 1/2" BT-60 nose cone	Any BT-60 nose cone will do; the one shown is the longer one in the Estes Designer's Special.
1/8" balsa	Used to form a bulkhead at the base of the payload.
BT-60 tube coupler	Used to form the shoulder of the payload bay.
Screw eye	Used to attach the shock cord and parachute.

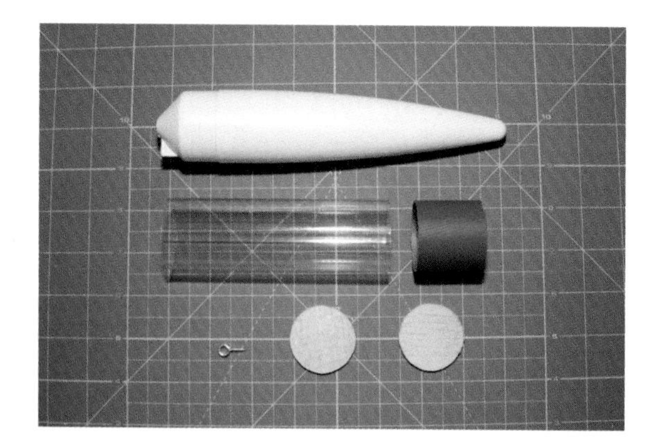

Figure 9-21. *Visual parts list for the ICU payload bay.*

Construction of the Thin Man and ICU Payload Bays

Begin by cutting the BT-60 body tube or clear payload tube to the desired length. There is nothing magic about the 8" length shown for the Thin Man; if you need a longer one for a specific payload, feel free to extend it. If you do extend the tube, though, stick with the Ceres B or Ceres C boosters—the payload is already a little heavy for the Ceres A.

The tube coupler in the Estes Designer's Special is 3" long. If that's the one you are using, cut it in half. You can use one piece for each payload bay.

Using 1/8" balsa, cut two disks that fit snugly inside the tube coupler. These will form a bulkhead at the lower end of the tube coupler, where we will eventually install a screw eye to attach the shock cord and parachute. Glue the two pieces together so the grain is perpendicular. Glue the finished assembly into one end of the tube coupler.

Glue about 1/2" of the tube coupler into the payload bay body tube. Wood glue works fine for the paper tube, but I recommend epoxy for the plastic tube. If you don't use epoxy glue for the clear tube, be sure to rough up the inside with sandpaper so the glue will stick.

Once the bulkhead is dry, screw the screw eye into the center of the bulkhead, remove it, squirt some wood glue into the hole, and screw the screw eye back into place.

The nose cone does not get glued in. You will remove it to insert payloads into the payload bay. Do you recall how we added a payload bay to Juno, using the original nose cone as the nose cone for the payload bay? You can also do a reverse payload conversion with either of these payload bays, flying one of the Ceres boosters with just the nose cone.

The plans show the short, squat nose cone on the Thin Man and the long, sleek nose cone on the ICU. You can certainly swap them, or use either one if you only build one of these payloads. That long nose cone really does look nicer, doesn't it? It just looks more rocket-like. However, the short, round nose cone is actually the better one technically: it's lighter, which is really important if you are using the Ceres A booster.

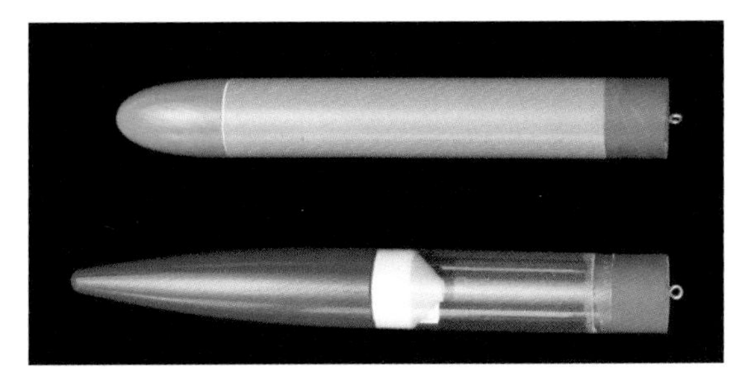

Figure 9-22. *The Thin Man (top) and ICU (bottom) payload bays*

Flight Preparation

You will need to modify your flight preparation checklist to accommodate your payload. How you change the checklist depends on the particular payload you use, though.

The only special consideration for the payload bay is making sure the payload does not pop out at apogee. For light payloads, just make sure the nose cone is snug, adding tape around the shoulder of the nose cone if needed. For heavier payloads, wrap a strip of tape around the joint where the nose cone attaches to the body tube.

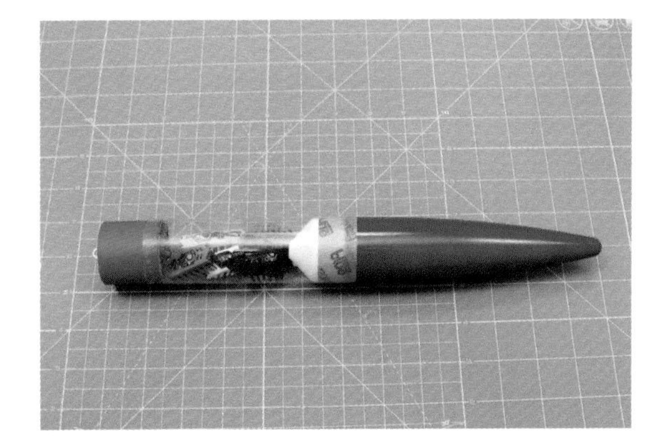

Figure 9-23. *Taping the payload bay shut could prevent a payload from breaking out of the capsule when the ejection charge fires.*

The Fat Man Payload Bay

The Fat Man payload bay is a little different from the ones we've built so far. This is the first payload bay that uses a tube size that differs from the rocket itself. It's also why we added launch lug extenders to the rocket. The parts you'll need to build this payload are listed in Table 9-19.

Table 9-19. Parts list for the Fat Man payload bay

Part	Description
BT-70 body tube	BT-70 body tubes come in various lengths, depending on the source. The design shows 8" cut from a longer tube.
BT-60 to BT-70 transition	This is a balsa piece that has a shoulder on each side, one for the BT-60 used as the main body tube of the booster, and one for the BT-70 payload tube.
BT-70 nose cone	There are several BT-70 nose cones available from various manufacturers. This one is fairly short and light, but any will do if you use the Ceres B or Ceres C boosters.
Screw eye	For attaching the parachute and shock cord.

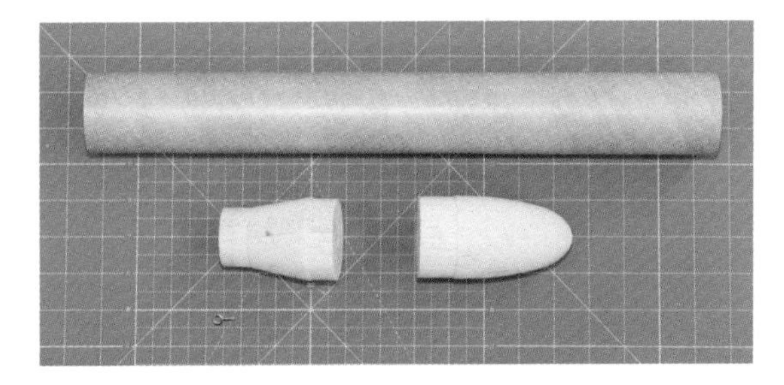

Figure 9-24. *Visual parts list for the Fat Man payload bay.*

Construction

Cut the BT-70 body tube to the desired length. The design shown uses an 8" section of tube, but you can make it longer or shorter. Stick with the Ceres B or Ceres C boosters if you use a longer body tube, though. Like the Thin Man payload, this payload bay is already pretty heavy for the Ceres A.

Use wood glue to glue one end of the transition into the payload tube.

Screw the screw eye into the base of the transition—the part that will slide into the BT-60 body tube—then remove it, squirt glue into the hole, and screw it back in.

That's it. The payload bay is ready for finishing.

While this version of the payload bay uses a BT-70 body tube, you can build a similar version using a BT-80 body tube. I've used BT-80 body tubes with rockets that carried iPhones to record data; see Chapter 7 of *Building iPhone and iPad Electronic Projects* (O'Reilly) for details.

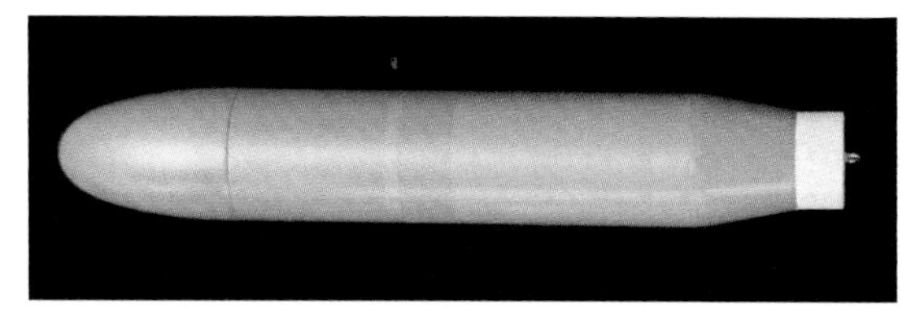

Figure 9-25. *The Fat Man payload bay.*

Flight Preparation

As with the Thin Man and ICU payloads, modify your flight checklist as appropriate for the payload you are carrying. Secure the nose cone to the payload tube with tape for heavier payloads.

The Over Easy Payload Bay

The Over Easy payload bay is my personal favorite. It's fun and different to build, and egg lofting is tons of fun at the launch site. Everyone wants to see what happened to the egg!

The parts list for the Over Easy (Table 9-20) is almost the same as that for the Fat Man, but construction is very, very different. The only additional part is some padding for the egg. The construction details discuss some alternatives.

Table 9-20. Parts list for the Over Easy payload bay

Part	Description
BT-70 body tube	BT-70 body tubes come in various lengths, depending on the source.
BT-60 to BT-70 transition	This is a balsa piece that has a shoulder on each side: one for the BT-60 used as the main body tube of the booster, and one for the BT-70 payload tube.
BT-70 nose cone	Like the transition piece, the nose cone must be balsa.
Screw eye	Used to attach the parachute.
Some sort of padding	See the construction description for ideas.

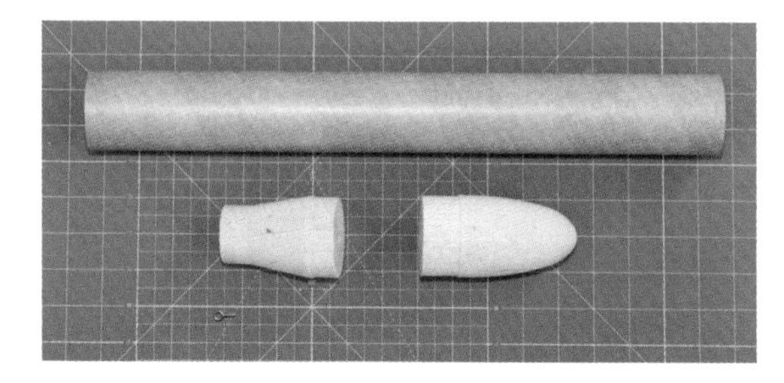

Figure 9-26. *Visual parts list for the Over Easy payload bay.*

Over Easy Construction

The egg will be cradled in a cavity we will hollow out of the balsa nose cone and transition. The two pieces will actually butt up against each other; the payload bay tube is just used to hold them together.

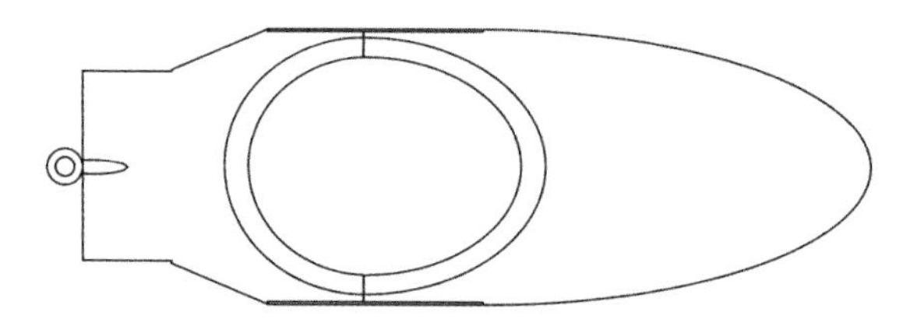

Figure 9-27. *This cutaway drawing shows how the egg fits inside a capsule carved from the balsa parts. The space between the egg and balsa will be filled with padding.*

Start by measuring the shoulder length of the two balsa parts. Mine had a total length of 1 13/16". Yours may vary; take the time to measure the particular pieces you get. Cut a length of BT-70 body tube to match this length.

The next step works well with a thin kitchen knife like a paring knife. You are going to slice the balsa parts through the middle repeatedly until both the nose cone and the transition piece are cut into eight roughly equal parts. Figure 9-28 shows the first cut through the nose cone.

Warning: Watch Your Fingers!

About halfway through the cut, the wood will suddenly split. That's fine as far as the cut goes. To avoid getting blood all over the balsa parts, think about how you are holding the knife. No fingers should be under the cutting edge of the blade. That way, when the part gives way and splits, you won't suddenly force the knife into the cutting pad on top of your fingers.

Figure 9-28. *The first cut through the nose cone. Watch your fingers!*

Continue cutting each piece in half until there are eight total wedge-shaped pieces. Repeat the process with the transition piece. Figure 9-29 shows the final eight pieces of the transition piece.

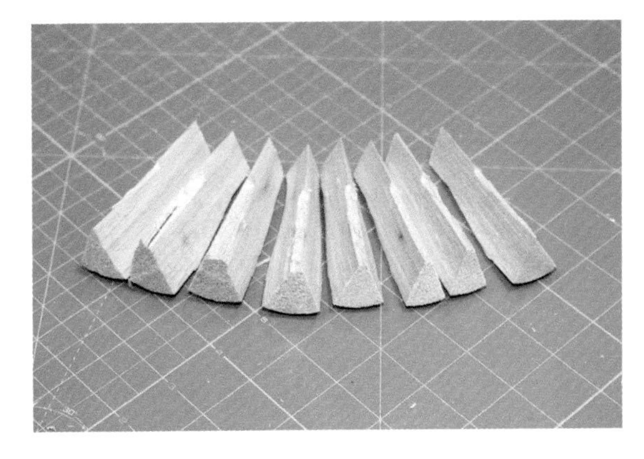

Figure 9-29. *Here are the eight finished pieces of the transition piece.*

You will be pulling these pieces apart and fitting them back together many times as you carve the egg cavity. Do yourself a favor and piece them all together now, then number the pieces on both the top and bottom of each end. You will carve the numbers off of the nose section; replace them with numbers inside the cavity as you carve. A rubber band is handy for holding the pieces together.

Figure 9-30. *Number the ends to make it easier to fit the parts together during carving.*

The size of the cavity depends a little on the thickness of the padding you select. I like to use the 1/8"-thick foam padding that frequently comes as packing material when I mail order parts. Thin bubble wrap works well, too. You can even use several layers of napkin or paper towel, but crumple the paper first so it doesn't lie too flat. Whatever you intend to use, get it now and wrap the egg the way you intend to wrap it during flight.

Wrap the Egg

Rotten eggs smell so bad, they are a cliché for a bad smell. You don't want that smell permeating your rocket after a mishap. Wrap the egg in plastic food wrap or seal it in a plastic bag before packing it. That way, if the egg cracks, the remains might stay safely separated from your rocket.

With the egg safely entombed, start carving the balsa parts to form a cavity for the egg. A hobby knife works well. You can get specialized blades for a hobby knife with the cutting edge on the side; that works even better. Test fit the parts frequently as you near completion to make sure you are carving off enough wood—but just enough! Figure 9-31 shows the parts once they are complete. You can, and should, assemble the entire capsule, holding it together with tape or rubber bands, before proceeding to the next step—gluing the parts back together. Make sure the egg fits perfectly before you glue anything.

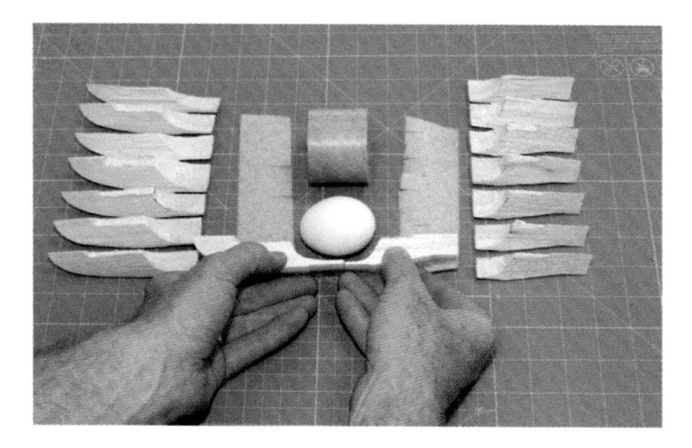

Figure 9-31. *Carve the pieces so the egg, with padding, will fit in the cavity.*

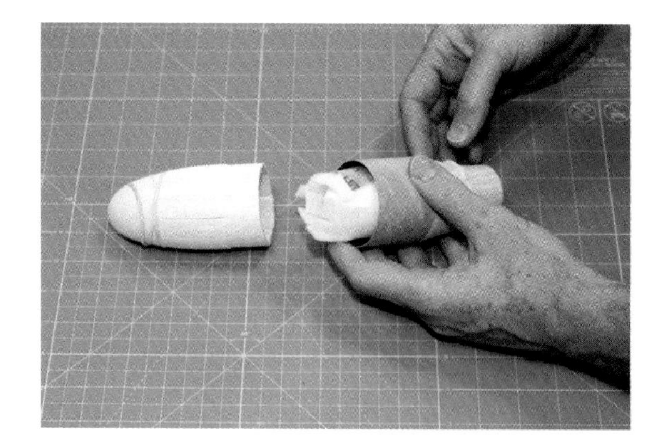

Figure 9-32. *Test fit the parts frequently, especially as carving is nearly done. You want a good fit for the egg that doesn't squeeze it or let it rattle around. Use tape and rubber bands to hold things together during test fits.*

Once the carving is complete and you have checked the fit, use wood glue to glue the parts of the nose cone and transition piece back together, then glue the transition piece into the bottom of the payload tube.

The balsa pieces I got from JonRocket.com also came with some 1/2″ wood dowels. These can be used to strengthen the wood where the screw eye is inserted. I've build and flown a lot of egg capsules like this one, and have never used a hardwood dowel or had a screw eye pull loose, but it's a good precaution. If you decide to use the dowel, drill a 1/2″ hole about 1/2″ into the base of the transition piece, then glue the dowel into place. Saw off any excess. Once the glue dries, install the screw eye just like you would in plain balsa.

Which brings us to the last step. Screw in the screw eye, remove it, squirt some glue in the hole, and then reinsert the screw eye.

All that's left is finishing and painting to create a really nice egg lofting capsule.

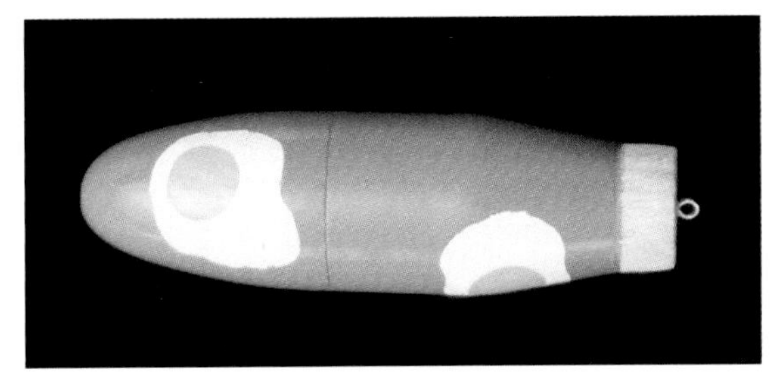

Figure 9-33. *The Over Easy payload bay.*

Flight Preparation

There are some special steps for flying the Over Easy payload.

As launch day approaches, make sure you have a fresh egg that can be sacrificed for a test of your engineering prowess. I like to prepare the egg before leaving for the launch site, but you may have to prepare it in the field for contests. Make sure you have plastic wrap, padding, and tape for the packing job.

Wrap the egg in plastic wrap, then in the padding material you have selected. Insert the egg into the capsule, and place the nose cone in place.

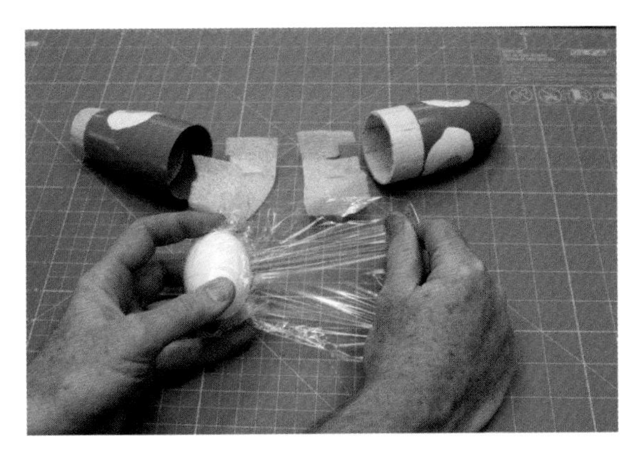

Figure 9-34. *Wrapping the egg in plastic wrap can save a capsule from an untimely, stinky retirement from service.*

The egg is a heavy payload. I've seen the shock of the shock cord pulling back on the capsule pop the nose cone off and eject the egg from the capsule, launching it in a ballistic return to earth. Splat! It's the real-life version of Angry Birds. To avoid this mishap, be sure to fasten the nose cone securely to the capsule. Even though it's a bit ugly, and spoils the effect of a nice paint job, I use a ring of tape around the nose cone joint.

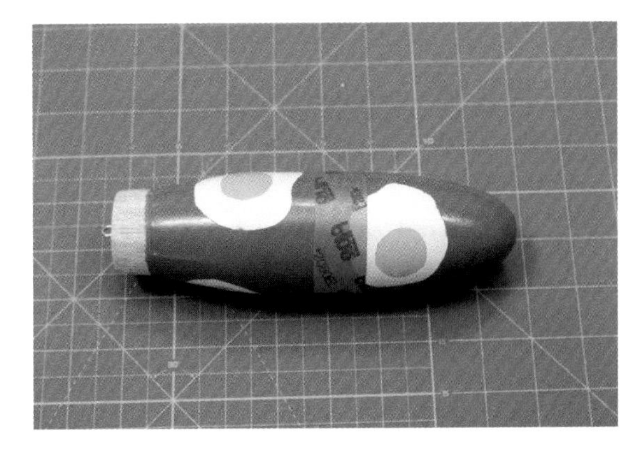

Figure 9-35. *Eggs should only be airborne inside rockets or birds. Use a little tape around the joint in the payload bay to prevent the egg from taking flight on its own.*

After the launch, remove the egg and unwrap it. Are there any cracks? Congratulations! You're ready for breakfast!

Rocket Eggs

Keep your eggs cool in a cooler until just before launch, then return them to the cooler right after recovery. After five successful flights, combine:

- 5 eggs
- 5 tablespoons of semi-skimmed milk
- A pinch of salt

Stir the ingredients with a fork. You can substitute whole milk or cream, but add something; it adds texture to the finished egg. Melt a pat of butter in a pan. As soon as the butter starts to bubble, add the eggs. Once the eggs start to solidify, increase the heat to high and begin to shake the pan a bit, folding the eggs over. Don't stir them; fold gently. You're trying to trap moisture in the egg, not beat it to death. Serve as soon as there is no liquid running around in the pan. They may not look done, but the eggs continue to cook after you remove them from the heat, and will be perfect by the time you eat them.

Add pepper or other garnishes at the table.

The ICU2 Payload Bay

This fun payload bay launches your very own spy camera into the sky. You can use it to create entertaining movies of your exploits, to do inexpensive aerial photography, or just for the sheer challenge of getting movies from a rocket.

There are a lot a ways to build this payload bay, so read through the description and think about your goals. The design shown is set up to create a movie of the rocket flight, pointing down along the tube of the rocket as it launches into the air. It points the camera down from the edge of the transition piece, so the transition piece needs to be big to give you room to mount the camera. The result is a dramatic movie of the rocket flight.

Other alternatives don't need such a fat payload bay, and are much better at aerial photography. You can shoot sideways through the tube, then rig the parachute so the camera points down during the long, slow descent back to earth; or you could point the camera straight up. Either way you do it, use two parachutes: one for the camera pod and another for the booster. The result is a pretty boring liftoff and flight, but lots of frames on the descent. The movie isn't that great—there is enough jerking around that you could make your audience sick. Or maybe that's your goal. If it's not, plan to grab still frames from the movie. This is a great way to get aerial photos.

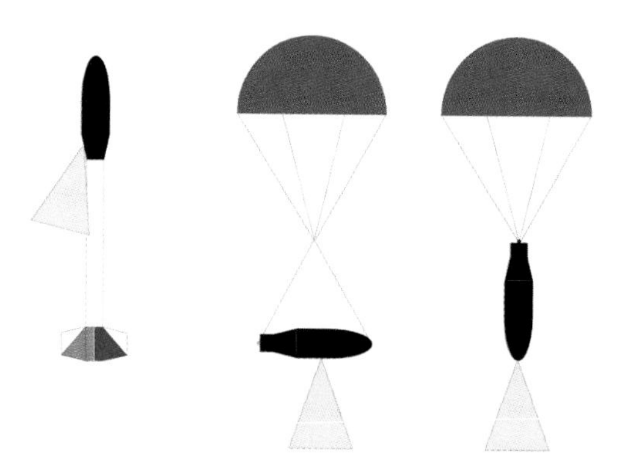

Figure 9-36. *Three of the many ways to orient the camera. The first orientation is what this payload will deliver. The side-view and top-view cameras are great for aerial photography, but not as good for dramatic launch movies.*

Table 9-21 lists the parts you'll need to build this payload. In addition to the standard tools and materials, a Dremel tool or other carving device will come in handy.

Table 9-21. Parts List for the ICU2 payload bay

Part	Description
BT-80 body tube	You will need a piece about 3" long, but measure the specific nose cone and transition piece you buy to find the exact length.
BT-60 to BT-80 transition	This is a balsa piece that has a shoulder on each side: one for the BT-60 used as the main body tube of the booster, and one for the BT-80 payload tube.
BT-80 nose cone	The nose cone can be balsa or plastic.
Screw eye	Used to attach the shock cord and parachute.
Camera	The Y3000 is shown, with its cousin the Y2000 in the visual parts list, but any small video camera will do. They are changing constantly, so shop online before selecting a camera.
Memory card	Most small cameras don't come with a memory card. Buy a small one—you don't need a large one for a single movie, and the larger ones are pretty expensive. There is a chance the camera and card will be lost, so keep your investment small.

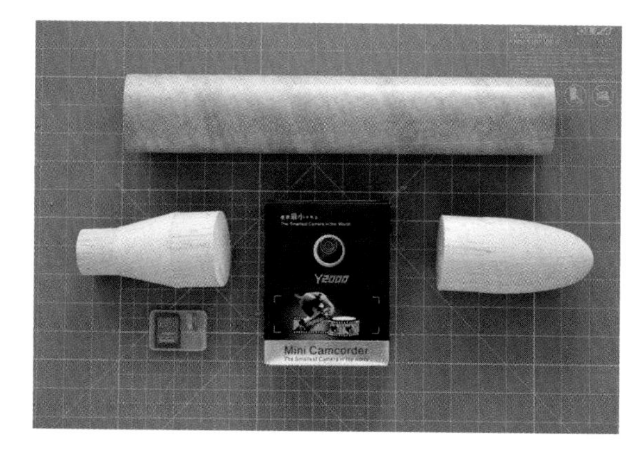

Figure 9-37. *Visual parts list for the ICU2 payload bay.*

Construction

The camera will be housed in a hollow cavity in the transition piece. Figure 9-38 shows the finished carving.

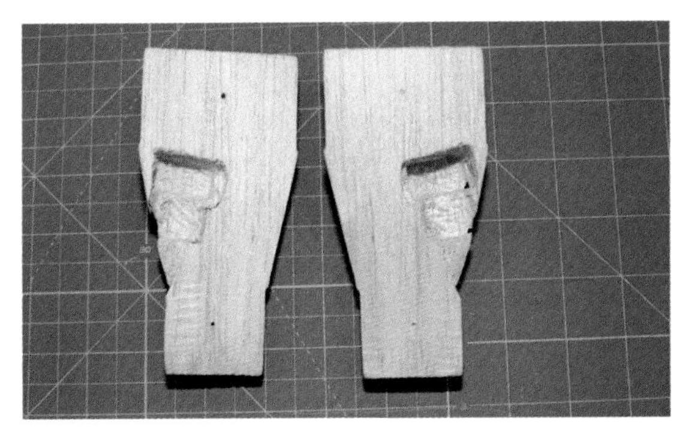

Figure 9-38. *The completed, carved-out cavity that will hold the camera.*

Split the transition piece in half. Use the technique described in "Over Easy Construction" on page 224, being careful to hold the knife so you cut the wood, not your fingers.

Spend some time with the camera before you start to carve out the cavity. All cameras have different fields of view. You need to check the field of view on your specific camera so you can plan ahead and get exactly what you want in the frame. Personally, I like to see some of the rocket in a lift-off movie, so I plan the field of view so a bit of the fins and body tube are visible. That also lets you catch some of the smoke and flame from the rocket motor. Others don't like to see the rocket at all. It's your camera, and your choice.

Begin by taking a movie and raising the camera to two or three measured distances above something with a known dimension. Engineering paper, as seen in Figure 9-39, is a perfect choice.

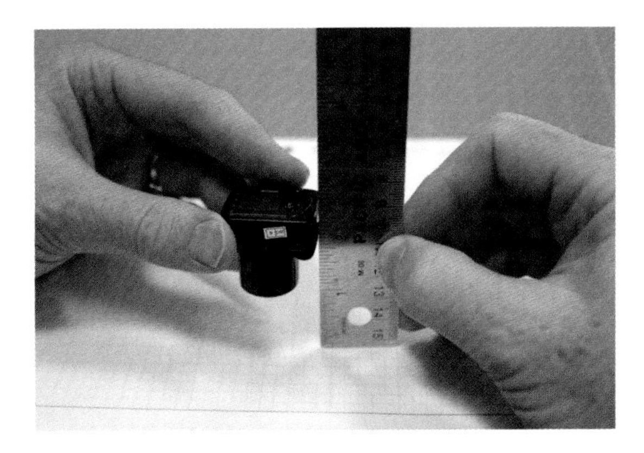

Figure 9-39. *Do a test movie at several distances over a piece of graph paper.*

Use the test movie to see what the field of view is at various distances. Plot the results on a piece of graph paper.

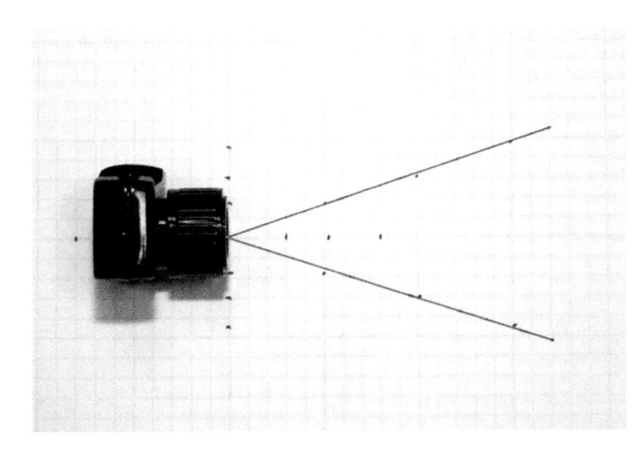

Figure 9-40. *Plot the field of view to create a physical representation.*

Cut out the field of view. Use it and the camera to decide on the specific location and orientation of the camera. Don't let the camera's normal orientation get in the way of a creative fit. In Figure 9-41, the top of the camera is on the lower side.

Figure 9-41. *Use the field of view to decide on the position and direction of the camera.*

Once you decide on the proper position for the camera, trace the outline of the camera and the field of view on the transition piece. There are a number of ways to carve the cavity. A Dremel tool with a spherical cutting head works really well, but you can do the job with patience and a hobby knife. Work slowly, checking the fit as you go, until the camera sinks into the wood so that the lens is centered on the split in the wood. Then clean out the field of view for the camera.

Figure 9-42. *After outlining the camera, slowly carve out a cavity until the camera lens sits centered on the split in the wood.*

The next step is to carve the cavity on the other side of the transition piece. You want them to match perfectly, so you need some way to form an image of the cavity you already have on the other piece of wood. A simple, effective way to do this is to put some colored powder in the existing cavity, place the two pieces together, turn the assembly over, shake gently, and remove the piece with the cavity. The colored powder will form the outline of the camera cavity on the other side. Trace it with a pencil and start carving, doing frequent test fits to make sure the pieces line up perfectly.

So where do you get colored powder? The spice cabinet works well. I used cinnamon, as seen in Figure 9-43.

Figure 9-43. *Use colored powder to form the outline of the cavity on the uncarved side of the transition.*

Continue carving until the pieces fit together, cradling but not squeezing the camera. Depending on the orientation you pick, the corners of the camera may poke through, as they do in Figure 9-44. That's not really a problem. The payload bay will work fine as long as there is enough wood left to hold the camera in place.

Figure 9-44. *The camera, nestled securely in the payload bay. The camera isn't quite symmetrical, so neither are the locations of the corners that poke out.*

Don't glue the pieces back together. The easiest way to hold the pieces together during flight is to put a ring of tape around both the upper and lower shoulders of the transition piece. A slightly higher-tech way of doing it is to countersink two screws, one in each shoulder.

Mount the screw eye in the normal way. Insert it, then remove it. Add some glue in the hole and reinsert the screw eye.

This is a large, heavy piece of wood, so it's not a bad idea to use a hardwood dowel to reinforce the screw eye hole, as discussed in "Over Easy Construction" on page 224. The only peculiarity is that you can't center the screw eye, since the crack in the wood runs right through the spot it would normally occupy. Instead, mount it in the center of one of the halves of the transition piece.

Measure the lengths of the shoulders on the transition piece and the nose cone. Cut a length of body tube just long enough to cover the shoulders. The nose cone and transition piece should just touch when inserted into the body tube. My body tube was 3" long, but measure before you cut in case there is some variation in the nose cone and transition piece you buy.

Glue the nose cone into the body tube, then finish as desired.

Figure 9-45. *The ICU2 payload bay.*

Flight Preparation for the ICU2

You will want to start the camera just before flight and turn it off immediately after recovery. Here's the flight checklist for launching the ICU2 payload with the Ceres A or Ceres B booster.

Proceed with the launch in the normal way, then recover the rocket and turn off the camera as quickly as you can.

Flight Checklist

1. Clear spectators from the launch zone.
2. Remove the launch key from the launch rail, slide the rocket onto the launch rail, and replace the launch key on the launch rail.
3. Attach the igniter clips.
4. Start the camera.
5. Secure the transition piece halves with tape (or screws, if that's what you used).
6. Insert the camera pod into the rocket.
7. Make a funny face looking up at the camera.
8. Get the launch key from the pad.
9. Insert the launch key. Make sure the continuity light is on.
10. Check for clear skies—no aircraft or low clouds.
11. Count down from five and launch.
12. Remove the launch key and place it on the launch rod.
13. Recover the rocket.

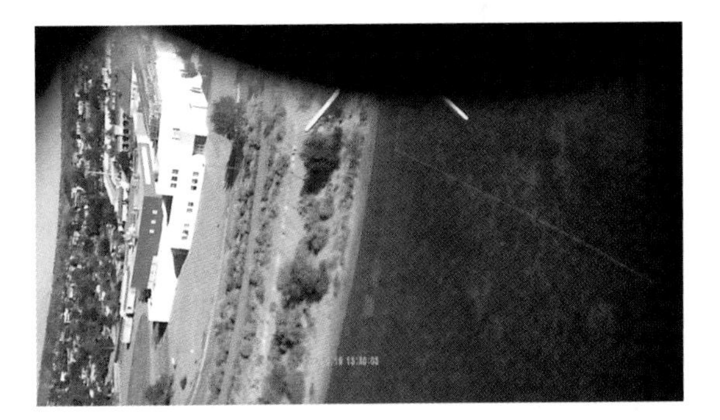

Figure 9-46. *This frame from a flight movie shows a nearby middle school. The rocket is just tipping over. In a few more frames, the parachute pops out.*

Selecting Parachutes and Streamers

10

As you know by now, parachutes are the most common way to recover a solid propellant model rocket. Chapter 12 will cover ways to use parachutes with water rockets. Streamers are often used with lighter rockets. But how can you tell when to use a streamer and when to use a parachute? When you use a parachute, how big should it be? That's what this chapter is all about.

There are many ways to answer this question, and we'll look at two of them. The first is to use tables or graphs to pick from common parachute sizes based on the weight and expected altitude of your rocket. That's fast, easy, and works perfectly well for the vast majority of cases where you just want to pick a reasonable parachute for a sport launch. The other way to pick a parachute—or even to design one for a special purpose—is to understand the math and physics behind parachute selection. We'll take a look at that a little later in the chapter.

How Fast Should a Rocket Fall?

There is a basic question that has to be answered before we can begin to pick an appropriate recovery mechanism: how fast should a rocket fall? It's not a question with an obvious answer, nor is there one correct answer.

The most common recovery speed cited for low-power model rockets is around 3.5–4.5 meters per second, which is about 11–15 feet per second, or 8–10 miles per hour. That's good enough for most purposes, but we'll return to that question in the physics section.

That said, there are many reasons to return a rocket to earth faster or slower than this general rule would allow. You would want a slower speed for a parachute duration event—that's a fun contest where you see who can keep a rocket in the air the longest. You might pick a slightly higher speed when flying from a small field or when the wind picks up so you have a better chance of recovering the rocket. We'll leave the issue of how fast is too fast for "Parachute Physics" on page 238, because you need to do a little math to really answer that question. If you intend to work strictly from the tables, without fully understanding the science, don't go up or down more than one parachute size from the recommended size.

Streamers Versus Parachutes

A *streamer* is a long ribbon of material that flaps in the wind to slow the rocket as it falls from altitude. Streamers are typically made from plastic strips, crêpe paper, or strips of Mylar. Streamers are generally used for light rockets that fly really high, or are being flown from a small field.

Why Speed Matters

Why do we care so much about the descent speed? Because in parachute selection, size matters, but speed matters even more. The damage done by a rocket impact depends on many things, like the shape of the rocket (and the object the rocket hits) and how rigid both objects are. By far the biggest factor, though, and the one we have the most control over when selecting a parachute, is the energy of the impact. Energy is generally measured in joules (J), and for a moving object, the energy is:

$$E = \frac{1}{2} m v^2$$

where m is the mass and v is the speed (or velocity). If you double the speed of descent, the energy quadruples—and roughly speaking, so does the amount of damage. Controlling the speed of a descending rocket is crucial.

As an extreme example, let's take a look at the Ceres B carrying a camera, returning from a flight to 1,400 feet without a parachute. It's a full ballistic return, perhaps because we mistakenly used an E9-0 first-stage motor instead of the appropriate E9-8 motor. The rocket is traveling 209 ft/s when it hits the ground, and weighs about 6 oz. After converting some units and calculating the energy, we find that the impact energy is about 345 J. We'll return to how to calculate this in the physics section, and define a joule then; for now we're just interested in the result so we can compare this to some familiar impacts. For example, the energy from being hit by a 90 mph baseball is about 117 J. How about the energy from a 9 mm pistol? It's about 578 J just as the bullet leaves the barrel of the gun. The impact energy of dropping an 8 lb bowling ball on your foot from a distance of one meter, a bit over three feet, is 35 J.

If we use a parachute that returns the rocket to the ground at 15 feet per second, we cut that impact energy from 345 J to 1.7 J. We've converted a dangerous projectile, with more than half the impact energy of a bullet, into a safe, soft model rocket.

As an aside, I'd much rather be hit by the Ceres B with an energy of 345 joules than a 90-mile-per-hour fastball. Why? Because the Ceres B is designed to crumple on impact, so the energy is absorbed over a much longer period of time. That's the same technique used in modern cars to make them safer in crashes.

You've already used parachute recovery, but we have not covered streamers. Streamers will be used later in two of our rockets, though. The high-performance Hebe in Chapter 15 uses a streamer because it can reach very high altitudes of over 2,000 feet, and it's nice to get it back after the flight. The pop pod on the Icarus glider in Chapter 20 will also use a streamer, this time because we want the pop pod out of the sky as quickly as possible so we can enjoy the glider flight.

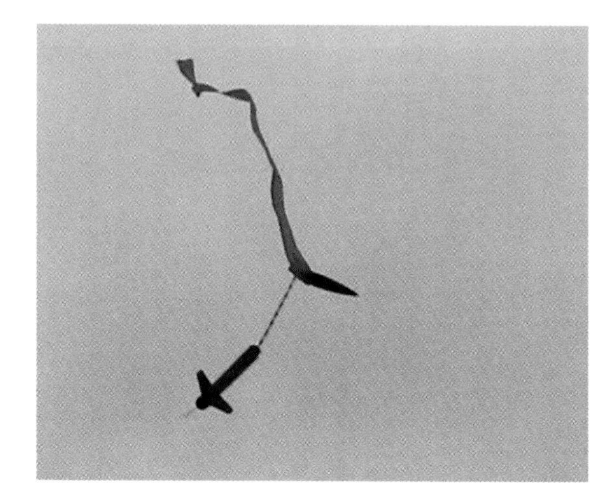

Figure 10-1. *Hebe returning on a streamer. Streamers are great for light rockets, especially high fliers or on small fields.*

So when is it OK to use a streamer? In general, the rocket should weigh less than 30 grams, which is about one ounce. The streamer won't be able to slow the rocket down enough if it is any heavier. Again, though, there are exceptions. See "Parachute Physics" on page 238 for some thoughts on using streamers.

Streamer Design

Streamers come in all sizes. The streamers we use in this book are made from the 32"-long plastic strip included with the Estes Designer's Special. It's 1" wide, so it is 32 times as long as it is wide. Is that the best shape?

Well, no. In general, streamers should be about 10 times longer than they are wide, but if you're restricted to a specific width, the longer the streamer is the better it will perform. You might be tempted to fold the streamer in half so it essentially forms two streamers. This actually seems to perform worse than a single streamer that is longer, probably because the air flow over one half of the streamer interferes with the air flow over the other. This is an area ripe for further testing, but unless you're really trying to do the tests, it's probably best to stick with a single long streamer rather than doubling it over.

As for material, stiffer streamers seem to work better, so metal-coated Mylar is very popular in competitions. One good source of thin Mylar sheets for parachutes and streamers is used helium party balloons. (Helium easily seeps between the molecules of a latex balloon; it doesn't seep as easily through metal-coated Mylar, so those balloons last longer.)

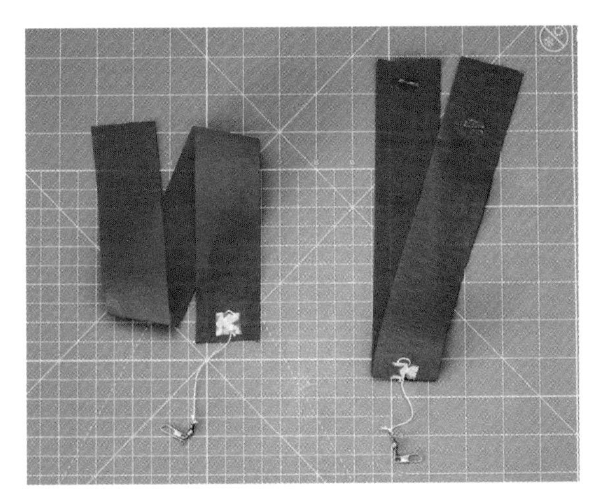

Figure 10-2. *A long streamer performs better than the same length of material doubled over.*

A crinkled material has more drag than a smooth one, so fold the Mylar streamer many times to crease it. It's tedious, but it will help increase the drag and slow the descent of your model.

Crêpe paper, in the form of party streamers, is another great streamer material. One of the nice things about crêpe paper is that it is pre-crinkled, which adds to the drag of the streamer. A paper streamer won't last as long as a Mylar or plastic streamer, but they are pretty cheap. A roll of 200 feet of crêpe paper costs a couple of dollars. At that price, who cares if you have to replace the streamer every few flights?

Keep your eye out for odd materials, too. I once found a package of metallic Mylar ribbon that was crinkled like crêpe paper. It has a very high drag, it is brightly colored, and it flashes in the sunlight. It's my favorite streamer material by far, and I found it in a shop's Christmas decoration section several years ago.

Thoughts on Parachute Material

Thin plastic and Mylar are the mainstays for low-power rocketry parachute material. You saw how to make your own parachutes from Mylar or just about any thin plastic in "Making a Parachute" on page 51. The Estes Designer's Special comes with a collection of various-sized premade parachutes, though, so you may not have bothered to make your own.

I like plastic parachutes, and I use them a lot on light rockets. Plastic is not a particularly strong material, though, and it tends to wear out quickly and tear easily under a heavy load. For rockets that weigh 4 ounces or more, I prefer something a bit stronger. When model rocketeers want a strong parachute, they almost always choose ripstop nylon.

Ripstop nylon parachutes are amazingly strong—strong enough to be used for parachutes that carry people. They are also heavier and stiffer than thin plastic parachutes, so they don't work well for small rockets. But for larger rockets that carry a payload, like the Ceres, they are a very good alternative to plastic parachutes.

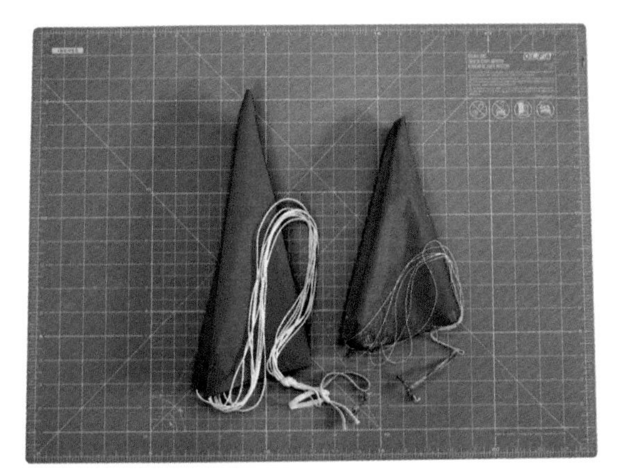

Figure 10-3. *A homemade 18" ripstop nylon parachute and a commercial 24" parachute from Estes. Both are equipped with a short loop of shock cord material to help absorb the impact of opening.*

You can make your own ripstop nylon parachutes from material purchased at a fabric store. Use strong nylon line for the shroud lines. A great choice for 18″ to 36″ parachutes is braided mason line, available from most hardware stores. Another option is to buy a premade parachute.

Parachute Selection Tables

Selecting a parachute can be pretty simple. Weigh your rocket to determine the recovery weight, then check Table 10-1 or Table 10-2 (depending on whether you used ounces or grams). The tables show the preferred size of parachute for different ranges of weight.

Recovery Weight of a Rocket

The recovery weight of the rocket isn't the same as its weight before the flight. The rocket will weigh less—sometimes substantially less—after the propellant in the motor burns away. You can weigh the rocket using a spent motor casing and the parachute you plan to use. Still another way to get the weight of your rocket is to weigh the rocket, parachute, and motor before launch, then subtract the propellant weight listed in the manufacturer's motor specification information. We'll work through an example of this later in the chapter.

The absolute best way to find the recovery weight of the rocket is to fly it and then weigh the rocket. Yes, that does seem a little facetious, but it is not. I've done exactly that on many occasions to get the best weight for doing physics calculations.

What if you don't have a scale? Throughout this book, I'll give you the weights for all of the rockets we build. These are finished weights, including glue and paint. While your own rocket will weigh a bit more or less because of differences in finishing, the weights from the book will definitely be close enough for parachute selection. Kits generally list the weight of the finished rocket, too.

For your own designs, you might get the weight from OpenRocket or some other simulator, or you might just compare the rocket to the ones in this book and then pick one that is closest. None of these methods are exact, but they will all get you close enough to pick an appropriate parachute.

Table 10-1. Minimum and maximum recommended weights in ounces for various parachute sizes

Parachute size (diameter)	Minimum weight	Maximum weight
6.25″	0.4 oz	0.7 oz
9″	0.9 oz	1.5 oz
12.5″	1.7 oz	2.9 oz
15″	2.5 oz	4.1 oz
18.75″	3.9 oz	6.4 oz
24″	6.4 oz	10.6 oz

Table 10-2. Minimum and maximum recommended weights in grams for various parachute sizes

Parachute size (diameter)	Minimum weight	Maximum weight
6.25″	12 g	20 g
9″	25 g	42 g
12.5″	49 g	81 g
15″	71 g	117 g
18.75″	110 g	183 g
24″	181 g	299 g

You can see that there are a few places where the weights overlap. For example, a rocket that weighs 2.8 ounces will return safely under either a 12.5″ parachute or a 15″ parachute. If you want the rocket to come down a bit quicker because you're in a small field, or so it will be less vulnerable to wind or nearby rocket-eating trees, pick the 12.5″ parachute. If you are in a big, open field on a calm day and want to watch the rocket return to earth a bit more slowly, pick the 15″ parachute.

There are also a few situations where it will be hard to get an ideal parachute. A rocket weighing 1.6 ounces, for instance, falls outside the maximum weight for a 9″ parachute, yet it is below the minimum weight for a 12.5″ parachute. Or the table might tell you to use a parachute of a specific size, but you don't have that parachute on hand. (The Estes Designer's Special comes with parachutes that are 6.25″, 12.5″, 18.75″, and 24″ across.)

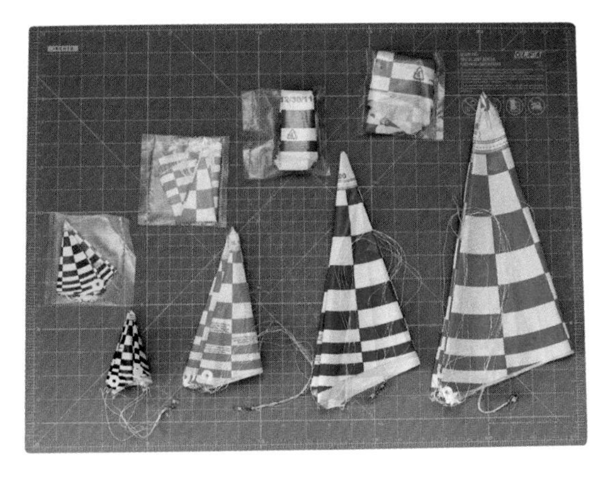

Figure 10-4. *The four parachute sizes in the Estes Designer's Special are typical for model rockets. From left to right you see the 6.25", 12.5", 18.75", and 24" parachutes.*

If you don't have the parachute size listed in the table, you can make one that fits your needs, or go either up to the next-larger size so the rocket comes down a bit slower, or down to the next-smaller size so the rocket comes down a bit faster. In general, it's safe to go up or down one parachute size to speed or slow the descent. Be cautious when going down to a smaller parachute, though. It's extremely unlikely that your rocket will damage anything on the ground, since the rockets are built from lightweight materials for safety. It's very possible, though, that the rocket itself will sustain damage, such as a broken fin or a slightly bent body tube. Look around at the likely landing spots. Will your rocket land in soft grass or on hard pavement or rocks? Does it have delicate swept fins, or is it a scale model with fragile decorations? If the landing area is hard or the rocket is delicate, going down a parachute size increases the risk of damage to the rocket. Of course, damage can occur with the correct parachute size, too; it's just less likely with a slower descent.

It's not completely out of the question to go up or down two parachute sizes, but there are issues with a parachute that is way too big or way too small. A rocket with a parachute that is too large might not have enough weight to pull the chute open properly, especially in a crosswind. While a large, unopened parachute might work as a streamer, it won't do what you intended, so there is no point in using it. A parachute that is too small will probably not slow the rocket down enough; almost certainly, a chute that is too small will result in damage. That can ruin a lot of work, and it will cost you money to replace the rocket—all of which could have been avoided with the right parachute. If the chute is extremely undersized, the rocket might even damage whatever it lands on. In NASA-speak, that is not an option. (See the physics section for more on this issue.)

In short, you should not go up or down more than one parachute size without some careful thought about the reason for the change in size, and without making sure there is no danger to whatever the rocket might land on. Breaking your rocket is sad, but to put it bluntly, it's your problem. Damaging something on the ground is not an option.

If you are starting to get the idea that selecting the correct parachute is a bit of an art, you are right. The tables will get you started, but they don't provide guarantees. You also need to weigh the possibility of damaging your rocket or losing it in the wind.

Figure 10-5 gives the same information in a more informative, but slightly more complicated way. It shows how fast a rocket will descend for a given weight in ounces and a given parachute size. For example, it shows that a 16-inch parachute carrying a 3-ounce rocket will return to earth at about 12 feet (3.66 meters) per second. The shaded area shows the optimal recovery speeds of 3.5 to 4.5 meters per second.

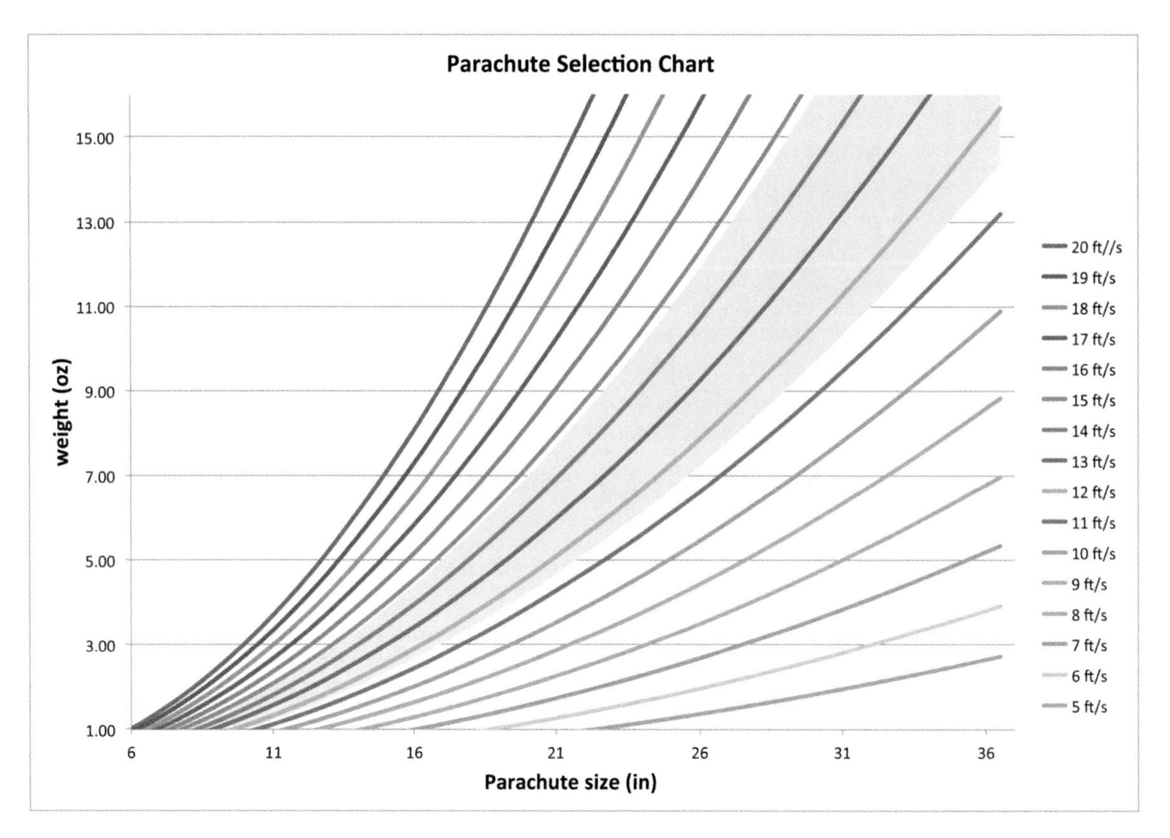

Figure 10-5. *Select the size of the parachute on the x-axis and the weight of the rocket on the y-axis. The line at the intersection shows the approximate descent speed.*

Parachute Physics

This section on parachute physics looks into two topics. The first is the energy of a falling body. It's important to understand the concept of the energy of the falling rocket if you want to step outside the narrow confines of using a table to pick a parachute. You can't make a reasonable choice about the acceptable descent rate of a rocket without understanding this topic.

The second topic is the physics and math behind the parachute tables. You might learn this just for fun, but knowing how it all works also gives you the tools to understand how to select the best parachute or streamer for a particular purpose. Be sure to read this section carefully if you want an edge designing a rocket for that next parachute duration contest!

If you're content to use the tables for selecting a parachute, though, you can safely skip the rest of the chapter.

The Energy of a Falling Body

I'm going to make a clean break with the information you find on parachute selection in most books by stating that the recovery speed is not that important. The important factor is the recovery *energy*. What's the difference? Speed is a measure of how fast an object is moving. Energy is a function of speed plus the mass of the moving object.

Right now, air molecules traveling about 505 meters per second—approximately the cruising speed of an F-15 supersonic jet fighter—are constantly bombarding you. Yet you're not hurt. This is because for all their speed, air molecules weigh next to nothing: about 4.65 $\times 10^{-23}$ grams each. Their total energy, even at such a high speed, is not enough to push you around or hurt you—at most, you feel fast-moving molecules as warm air, slower-moving molecules as cold air.

Parachute Spill Holes

Take a close look at some of the round parachutes the next time you are flying rockets with a large group of people. The chances are pretty good that the larger nylon parachutes have a hole in the middle. You might even see holes cut in some of the small plastic parachutes used on model rockets the size of the ones in this book. Why would anyone cut a hole in a parachute?

The answer, amazingly enough, is that it makes the parachute more stable.

The round canopy of a parachute traps some air underneath as the rocket descends. The slightest imbalance causes air to spill out of the side of the parachute that is highest up. That causes the parachute to move to the side, raising that side even more. The parachute continues to move to the side until it swings so far that the rocket swings it back like a pendulum, and the parachute swings back the other way. The parachute continues to rock back and forth, or perhaps to twist if it isn't perfectly round. Cutting a hole in the middle of the parachute lets some of the excess air escape from the center of the parachute, reducing the rocking motion.

You might think the hole would make the parachute less efficient. It does, but not nearly as much as you probably expect. Cutting a spill hole in the middle of a round parachute with a diameter 20% that of the whole parachute only reduces the total area of the parachute by 4%. That would be a 2.4″ hole in the middle of a 12″ parachute. The descent rate goes up even less. According to a RockSim calculation, putting a 2.4″ hole in the 12″ parachute used with Juno only increases the descent rate by 2.5%—a small enough difference that you can ignore it for the purpose of picking a parachute.

Another odd fact is that the spill hole can actually *reduce* the descent rate! That's because all of that oscillation spills a lot of air. If the oscillation without a spill hole is big enough to reduce the efficiency of the parachute by a lot more than 2.5%, adding a spill hole with a diameter 20% that of the parachute will bring the rocket down more gently.

Why not give it a try? Do some test flights with parachutes with and without a spill hole (20% of the diameter of the parachute is a good place to start) and see what you think.

On the other hand, while a bowling ball dropped from a height of one meter is traveling over a hundred times more slowly than a typical air molecule, I guarantee that if it lands on your foot, you'll feel something much more than a warm breeze. That's because the mass of a bowling ball compared to that of an air molecule is enormous: an increase of 2×10^{26}. The energy of impact of a bowling ball is correspondingly increased.

As we look at energy in this section, keep in mind that the whole point of using a rocket recovery system is to avoid damage to people, property, or the rocket when the rocket lands. That damage is caused by energy, not speed.

Energy comes in many forms. The kind of energy associated with motion is called *kinetic energy*. The equation for kinetic energy is pretty simple. It is:

$$E = \frac{1}{2} m v^2$$

where m is the mass of the object and v is its velocity. For our purposes, velocity is the same as speed.

The reason this equation is important is that stopping a moving body requires energy, too, and the energy it takes to stop something is exactly the kinetic energy of the moving body. For a rocket falling from the sky, the energy to stop the rocket might be supplied by grass bending out of the way, a bump on a fin that bends it momentarily, or any other sort of movement. In the end, most of this energy ends up as minuscule amounts of heat, which almost immediately dissipates away.

The problem, though, is that the energy required to stop the rocket can also do damage to the rocket. It might chip some paint, break a fin, or crumple the body tube. The energy needed to stop a rocket could also damage things on the ground. This might amount to nothing more than poking a hole in some dirt, but a rocket coming down with a great deal of energy could, in theory, do quite a bit of damage if it hit something more delicate or important than dirt.

Let's take a look at the equation for kinetic energy with this in mind. We can see that the amount of energy goes up as the mass of the rocket increases. If one rocket weighs twice as much as another, and they are going the same speed, the rocket that weighs twice as much will have twice the kinetic energy. This means we will have to engineer a slower descent rate for heavier rockets.

The other factor in the equation is the velocity. The kinetic energy of a moving body is proportional to the *square* of the velocity. If two rockets weigh the same, but one is moving twice as fast as the other, the faster rocket has four times the energy. If it is moving three times as fast, it has nine times the energy. This has impacts—sometimes literally—far outside rocketry. Think about automobile speed limits for a moment. Fifty-five mph may seem insufferably slow on the highway, but let's see what happens if we increase the speed to 78 mph. The kinetic energy equation tells us that for the car traveling at 55 mph, the kinetic energy is 3,025 multiplied by the mass of the car. At 78 mph, the energy is 6,084 multiplied by the mass of the car. The energy has doubled, even though the speed only went up two-fifths, from 55 to 78. That means the damage from a crash, both to the car and to the people in it, will be twice as bad at 78 miles per hour as it would be at 55.

Exactly the same thing happens with our rockets. Rockets are a lot smaller than cars, but they can go a lot faster. As the speed increases, the energy of the collision between the rocket and what it hits goes up even faster. Hebe, a high-speed rocket described in Chapter 15, can travel at speeds up to 500 mph!

There are different units for kinetic energy, just like there are different units for speed and mass. While we can work in feet per second and ounces, most of the time people switch to the metric system when dealing with energy. Energy is measured in *joules* in the metric system. In terms of kinetic energy, 1 joule is the energy of 1 kg (about 2.2 lb) moving at 1 m/s (about 2.2 mph). Table 10-3 shows the energy of some of the rockets from this book alongside some other moving objects.

Table 10-3. Kinetic energy of a few rockets and miscellaneous moving objects (speed is in meters per second, mass is in kilograms, and energy is in joules).

Object	Speed	Mass	Kinetic energy
Nitrogen molecule in air	505	4.6×10^{-26}	6×10^{-19}
Hebe landing	4.5	0.0262	0.3
Ceres B with ICU2 payload landing on 24″ 'chute	3.9	0.1715	1.3
Ceres B with ICU2 payload landing on 18.75″ 'chute	7.8	0.1715	5.2
Bowling ball dropped from 1 meter	4.4	3.6	35
90 mph baseball	40.2	0.145	117
9 mm bullet	380	0.008	578
Ceres B in flight with ICU2 payload	96	0.2076	957
Hebe in flight	224	0.0387	971
100 lb boy riding a bike at 15 mph	6.7	45.4	1,019
Pro football defensive back	8.0	90.3	2,890

Some of what you see in Table 10-3 might seem a little scary, but keep in mind that two different objects with the same kinetic energy can do a very different amount of damage when they hit something. One reason is the shape of the object itself. A hard, pointed object like a dart is more likely to puncture something it hits than, say, a round object like a baseball. A rigid object like a baseball is more likely to do damage than an object that crumples or gives, like a rubber ball. That's why we make rockets out of materials like paper tubes and balsa nose cones, not aluminum tubes and steel nose cones. The paper tubes will crumple, spreading the impact over more time, just like the rubber ball bending as it hits spreads the impact over more time so it does less damage than the baseball.

A few brave souls have actually tested flying a model rocket directly into a window. The rocket crumpled, destroying the rocket itself, but the window was unharmed. While I have not repeated that particular experiment myself, and would not suggest that you do it either, I have certainly propelled a baseball through a window. I know from experience that it will go through quite nicely!

The lesson we can take away from this is that energy is important, but so is proper construction. The baseball has about one-eighth the kinetic energy of Hebe in flight, yet because of its density and rigidity, it can do more damage. We need to build our rockets from appropriate materials so the rockets absorb the energy of any unintended impacts, not the objects they hit. Trust me, the rocket costs less to replace than a window.

So what energy is safe? Assuming you are following all of the normal rules for building low-power rockets, like using thin paper tubes, balsa fins, and balsa or plastic nose cones, we can use the maximum values for speed and mass as an upper limit. Low-power rockets generally weigh one pound or less. (You are allowed to build them heavier—but when you do, you are generally building mid-power rockets that use F or G motors, and heavier construction material. That's not what we're discussing at the moment.) The maximum recommended recovery speed is 4.5 meters per second. Converting 1 pound to 0.45 kilograms, we find that the maximum recovery energy should be:

$$E = \frac{1}{2} m v^2$$

$$E = \frac{1}{2} 0.45 \times 4.5^2$$

$$E = 4.6 \text{ joules}$$

Four and six-tenths joules will give the rocket a bit of a thump, but won't break anything on a well-made rocket unless you are unlucky enough to hit a fragile part on a rock or sidewalk. It also won't do much damage to anything the rocket lands on, but it might sting a bit—which is one reason we never try to catch a rocket. The other reason is that a careless grab can also damage the rocket.

Getting back to the topic of recovery speed verses recovery energy, there is still one reason why the actual speed is important. As we build bigger rockets, we tend to make them from stronger materials, precisely because they need to be stronger to withstand the buffeting they get during flight. Extremely thin fins can actually break in flight because of the forces exerted on them by air turbulence. A large rocket like the Ceres is built from thicker balsa than a smaller rocket because it has more mass, exerting more force on the fins as they guide the rocket through the air. Whether we're making the fins thicker because of careful engineering, because everyone else seems to, or simply because our rockets broke when we used thinner materials, we have to build rockets to withstand a certain flight speed, and that means making heavier, faster rockets stronger.

This is why we can't make a decision about how to recover small rockets by just looking at the rocket's kinetic energy; we need to factor in the fin thickness as well. We need to keep the recovery speed down to a reasonable level to protect the rocket, and anything it might hit.

Parachute Aerodynamics

Parachutes obey the same laws of aerodynamics as rockets, which we will cover in Chapter 13, but we use them in a dramatically different way. In general, we think of aerodynamics as a tool to let us decrease drag, allowing our rockets or airplanes to slip more easily through the air. With a parachute or streamer, we're using aerodynamics in exactly the opposite way. We want as much drag as we can get so we can slow the descent of our rocket.

The same fundamental equation governs drag, whether it is for parachutes or rockets. The force from drag is:

$$F_d = \frac{1}{2} \rho v^2 A C_d$$

where:

F_d
: The force due to drag.

ρ
: The density of air.

v
: The velocity of the rocket.

A
: The area of the parachute.

C_d
: The coefficient of drag.

We're trying to find the descent speed for a rocket, so we actually want to find the velocity of the rocket, not the force of drag. Let's rearrange the equation as:

$$v = \sqrt{\frac{2 F_d}{\rho A C_d}}$$

Now let's take a closer look at the terms in this equation, using Juno as an example.

Once the parachute opens, the rocket will quickly reach terminal velocity—the point when the force of drag pushing up on the rocket exactly balances the force of gravity pulling down on the rocket. That means we know the force of drag; it's the same as the mass of the rocket in kilograms times the acceleration of gravity. It's Newton's Second Law of Motion:

$$F = m a$$

You might know the mass of the rocket in grams or the weight in ounces rather than the mass in kilograms, but it's easy enough to convert. Divide by 1,000 to convert from grams to kilograms. Multiply by 0.02835 to convert from ounces to kilograms. The acceleration of gravity is 9.8 m/s^2.

Once all of the fuel is expended, Juno coming down from a flight on a C6-7 motor weighs about 1.3 oz, so it has a mass of 0.0373 kg. The force on the rocket, then, is 0.366 N, and at terminal velocity that's also the force of drag.

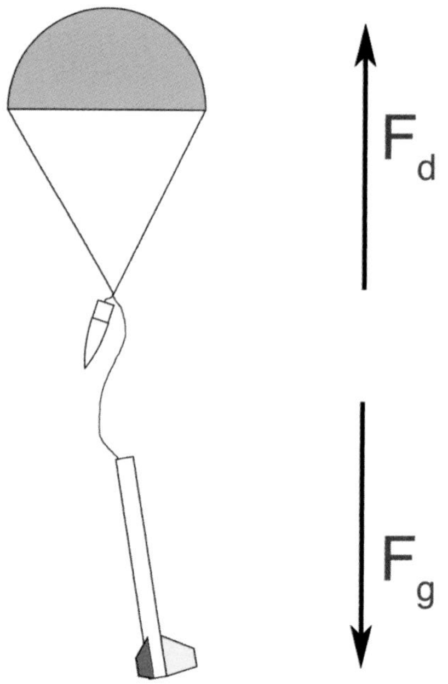

Figure 10-6. *At terminal velocity, the force from gravity pulling the rocket down is exactly balanced by the force of drag.*

The density of air is about 1.22 kg/m^3. This is at sea level, though, and altitude does play an important part in reducing the density of the air. Here in Albuquerque, where the altitude is about 5,000 feet above sea level, the density of air is about 1.03 kg/m^3. You can get as precise as you like with this term, adding in the effects of altitude, temperature, and humidity if you like. The error from using a reasonable approximation is pretty small compared to other errors, though, so it's not worth the effort to get more precise unless you put the same effort into precision everywhere else in your calculations. All of the tables from the first part of the chapter were calculated with an air density of 1.2 kg/m^3. That value is good enough for anything but an advanced simulator, or someone getting ready for a rocket competition. In those cases, you will want all the precision you can get.

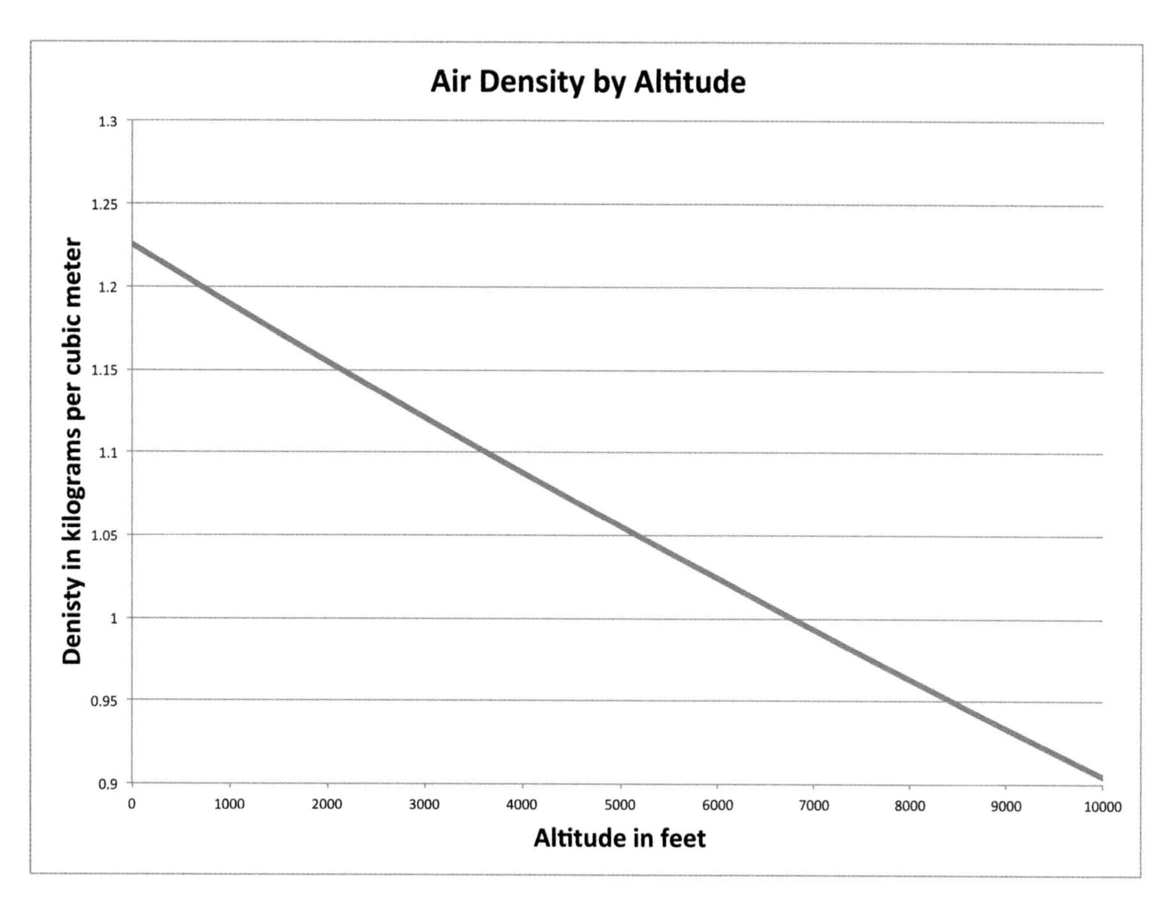

Figure 10-7. *Air density varies with altitude, temperature, and humidity. This plot shows the effect of altitude.*

The area is the area of the parachute. Parachutes come in a variety of shapes. For a standard hexagonal six-string parachute like we usually build for model rockets, the area of the parachute is:

$$A = \frac{\sqrt{3}}{2} d^2$$

where *d* is the distance across the parachute, as shown in Figure 10-8.

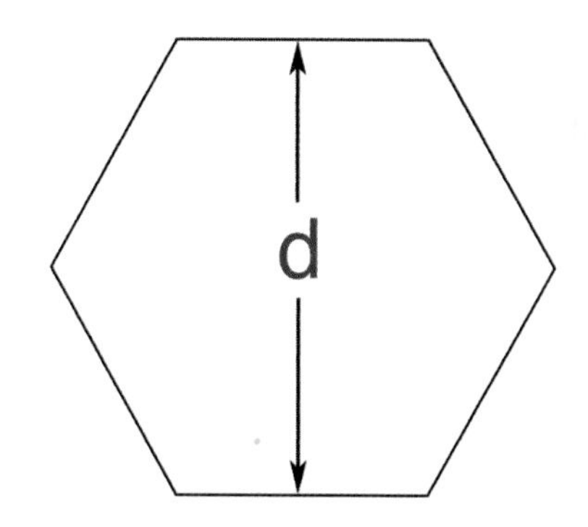

Figure 10-8. *The distance across the parachute is measured from one side to the opposite side, not from one corner to the opposite corner.*

Remember to convert the size of the parachute from inches to meters, since all of the other values used in this section are in the metric system. Table 10-4 shows common parachute sizes in inches and meters, and the area in square meters.

Table 10-4. Common hexagonal parachute sizes

Inches	Meters	Area
6.25	0.159	0.022
9	0.229	0.045
12.5	0.318	0.087
15	0.381	0.126
18.75	0.476	0.196
24	0.610	0.322

The last term we have to be concerned with is the coefficient of drag, C_d. The generally accepted value for the coefficient of drag for a model rocket parachute is 0.75. That's without the rocket, which does add a bit more drag, but it's a small amount compared to the parachute drag, so it's usually ignored.

Let's see how we can put all of this information to use.

Example: Find the speed of descent for a rocket

Let's say we want to know how fast the Ceres B will come down with an egg payload if it uses an 18.75″ parachute. We'll start with the weight of the rocket itself. From Table 9-1, we know the liftoff weight with a D12-5 motor is 7.8 oz. The propellant will be burned by the time the rocket is coming down, though, so we need to subtract the propellant weight from this liftoff weight.

Taking a look at the Estes data sheet for the motor, which you can find on the Estes website (*http://www2.estesrockets.com/pdf/Estes_Engine_Chart.pdf*), we see the propellant mass for the D12-5 motor is 24.93 g. Converting the liftoff weight and propellant mass to kilograms, we find the recovery mass:

$$m = 7.8\,\text{oz}\frac{0.02835\,\text{kg}}{\text{oz}} - 24.93\,\text{g}\frac{\text{kg}}{1000\,\text{g}}$$

$$m = 0.22113\,\text{kg} - 0.02493\,\text{kg}$$

$$m = 0.1962\,\text{kg}$$

We are considering the Estes 18.75″ parachute that came with the Estes Designer's Special. From Table 10-4, we know that the area of the parachute is 0.196 m^2. Using 1.22 for the density of air, 0.75 for the coefficient of drag, and 9.8 for the acceleration of gravity, we can now find the descent rate:

$$v = \sqrt{\frac{2\,F_d}{\rho\,a\,C_d}}$$

$$v = \sqrt{\frac{2\,m\,g}{\rho\,a\,C_d}}$$

$$v = \sqrt{\frac{2 \times 0.1962 \times 9.8}{1.22 \times 0.196 \times 0.75}}$$

$$v = 4.6\frac{\text{m}}{\text{s}}$$

That's a bit over the top end of the desired descent rate of 3.5 to 4.5 m/s, but easily close enough for our purposes. It also agrees nicely with Table 10-1 and Figure 10-5.

Example: Find the proper parachute for a duration event

We're getting Juno ready for a local parachute duration event, where the goal is to keep the rocket in the air as long as possible. The rocket must be flown on B motors, and we'd like to know what parachute to use. On the one hand, we want to keep the rocket up as long as possible, but thermals—rising air due to uneven air temperatures—are common at this time of year, so we don't want to make it so light the rocket is lost. After some thought, we've decided on a fairly slow descent rate of 3 meters per second—a little below the normal descent rates of 3.5 to 4.5 meters per second, but not too slow.

We start by weighing Juno on a kitchen scale with a B4-4 motor and a 12″ parachute installed. We find the liftoff weight of 1.7 oz. Just as in the previous example, we subtract the propellant weight—this time, 8.33 g—and convert to kilograms, giving a recovery mass of 0.0399 kg.

We're looking to calculate the area of the parachute, so we'll have to rearrange the drag equation a bit. Solving for A, we get:

$$A = \frac{2\,F_d}{v^2 \rho\, C_d}$$

Plugging in the various values, we find the desired area to be:

$$A = \frac{2 \times 0.0399 \times 9.8}{3^2 \times 1.22 \times 0.75}$$

$$A = 0.095\ \text{m}^2$$

Checking back to Table 10-4, we see this is between a 12.5" and a 15" parachute. But remember, we're preparing for a competition, which is worth a little extra effort. So, we'll build a parachute that's exactly the right size. Using the formula for the area of a hexagonal parachute, we know the area is:

$$A = \frac{\sqrt{3}}{2} d^2$$

Solving for d, we have:

$$d = \sqrt{\frac{2\,A}{\sqrt{3}}}$$

$$d = \sqrt{\frac{2 \times 0.95}{\sqrt{3}}}$$

$$d = 0.33\ \text{m}$$

$$d = 13\ \text{in}$$

Example: Find the coefficient of drag for a streamer rocket

Streamers are a bit trickier to plan for than parachutes. That's because there is no generally accepted value for the coefficient of drag of a streamer. We can use the drag formula and a test flight to find the value for a particular rocket and streamer combination, though. The data for this example is from a real test flight of Hebe from Chapter 15, flown on an A8-3 motor and using the 1" x 32" plastic streamer from the Estes Designer's Special.

In this case, we know, or will measure, everything but the coefficient of drag, so we need to rearrange the drag equation like this:

$$C_d = \frac{2\,F_d}{\rho\, v^2 A}$$

We can get the force of drag by remembering that the rocket descends at terminal velocity, so the force of drag matches the force from gravity. Weighing Hebe after the flight gave the descent mass as 30 g. The force of drag, then, is:

$$F_d = m\,a = 0.030 \times 9.8 = 0.294\ \text{N}$$

This particular flight was in Albuquerque, New Mexico, where the ground is about 5,000 ft above sea level. For that reason, we'll use an altitude-adjusted air density of 1.03 kg/m^3 rather than the sea-level value of 1.22.

One easy way to get the terminal velocity of the rocket is to measure the distance of the descent and divide it by the time of the descent. The distance of the descent is the altitude where the streamer deployed, which you can find using the techniques in Chapter 8, and time of the descent can be found with a stopwatch. In this particular experiment, though, a MicroPeak altimeter was installed in Hebe's hollow plastic nose, as described in Chapter 15. The descent speed reported by the altimeter was 8.8 m/s. Yes, that's a bit faster than the recommended parachute recovery speed—but remember, it's the energy, not the speed, that is important. Hebe doesn't use swept fins, so the fins are less likely to break on landing. Most of the impact will be absorbed by the motor and engine block, which were designed to withstand some force.

Figuring out the area needed for the drag equation is a bit tricky. We want to be able to scale our results, but that's tricky, too. The drag from a streamer depends on the aspect ratio (length divided by width), the stiffness of the material, and the roughness of the material. With so many variables, we really won't be able to use these results for any other streamer except the same kind of plastic streamer used on this flight. The streamer is one inch wide; we'll estimate that it flaps back and forth a bit more than this and call the area two square inches. That's 0.00129 m^2.

Crunching the numbers, we find:

$$C_d = \frac{2 F_d}{\rho v^2 A}$$

$$C_d = \frac{2 \times 0.294}{1.03 \times 8.8^2 \times 0.00129}$$

$$C_d = 5.7$$

How can we use this value? One way is to prepare the rocket for flight at sea level, perhaps to get ready for a rocket competition at another location. It's also a good way to estimate the recovery speed for a lighter or heavier rocket. Unfortunately, there really isn't any way to scale the results to other streamers. Unlike parachutes, which get their drag from a more or less circular canopy, streamers get their drag from flapping in the breeze. Each streamer size and material will need a separate measurement. Perhaps that is why streamer recovery of model rockets is such a popular science fair topic!

Water Rockets 11

Figure 11-1. *Water rockets are great summer fun.*

Water rockets have been around a long time. Small hand-pumped water rockets were a staple when I was a kid. Larger water rockets started to blossom in the 1970s, when soda manufacturers introduced the two-liter plastic soda bottle. Little did they know that they were packaging soda in a model rocket body!

The basic idea is pretty simple. Place a small amount of water in the soda bottle. Turn the bottle upside down, and pump in pressurized air to about 60–75 psi. When the rocket is released, the pressurized air at the top of the bottle pushes the water out quickly, providing thrust. Once all of the water is out, the pressurized gas itself pushes out, adding still more to the thrust, just like a balloon flying around as it releases air pressure. It's the same principle as with any rocket—material ejected through a nozzle moves the rocket in the opposite direction. Unlike with solid propellant rockets, though, there is no burning rocket fuel.

This chapter shows how to build a launcher and water rocket based on a two-liter soda bottle. The nozzle is the neck of the bottle, normally covered by the cap. "Equivalent Kit" on page 250 mentions a kit that uses a different nozzle.

The Launcher

Oddly, the first step in building a water rocket launcher is to find the right soda bottle. That's because the diameter of the neck of the bottle is critical to the design of the launcher. While most soda bottle necks are about the same size, some are a bit too wide or too narrow for our purposes. The main tube of the launcher needs to slide easily into the soda bottle, but it can't be loose.

There is also a wide flange running around the outside of the bottle's neck. The exact location of that flange changes according to which company made the bottle, and its location will also change the design of the launcher slightly. We'll be using the heads of zip ties to hold the rocket in place until it is time to launch. These must be placed in exactly the right spot to hook over the flange on the bottle's neck.

Water Rocket Safety

I hope you're supplementing this book by poking around on the Internet. If you are, though, you are quickly going to find that people are pretty cavalier about safety when building and flying water rockets. That's a big mistake.

There are two very real safety concerns when flying water rockets, and both are addressed by the design of this launcher.

The first is impact. Even the simplest water rocket weighs in at about a pound when it is full of water, often a little over. Water rockets quickly hit speeds of about 100 mph. They can do a lot of damage if they hit you (or anything) during the powered phase of flight. There have been at least two cases of people having their hands broken by water rockets. Please don't be the first person to break something more serious. Never approach the launcher when it is under pressure. Always use a launcher like the one in this chapter that lets you depressurize the system if something goes wrong.

The second hazard is from the PVC itself. If you add too much pressure, it will rupture, just like any other material would. The odd thing about polyvinyl chloride (PVC) is that it ruptures in two very different ways. If the PVC pipe is filled with a liquid, like water, it will swell slightly and spring a leak. The excess water will shoot out of the hole. If it is filled with a gas, though, it does something very different when it bursts. Instead of springing a small leak, the pipe will fragment, spraying pieces in all directions.

This launcher is designed so the PVC is the least likely part to rupture. The soda bottle and the air line leading to the launcher will both split open at considerably lower pressures than the PVC. Even if the PVC does burst, the launcher and launch operations are designed so you are far enough away to minimize any risk.

Still another factor is the number of threads on the neck of the bottle. Some bottles have two threads, while some have three. This also changes the positioning of the bottle on the launcher.

Figure 11-2. *Soda bottles have a wide, flat flange. We will hook the heads of some zip ties over this flange to hold the rocket in place until launch time.*

It's tempting to just tell you to buy brand X, which has the right diameter neck and the right flange location. Unfortunately, it's not that easy. Bottle specifications can vary slightly even among the same brand of soda, and they might not always be uniform around the world (or even in different regions of the United States). It's just something you will have to discover by trial and error. In my area, I've found that Pepsi, Orange Crush, 7 Up, and the off-brand grocery-store colas work well, while one-liter bottles and Coca-Cola products have necks that are too small.

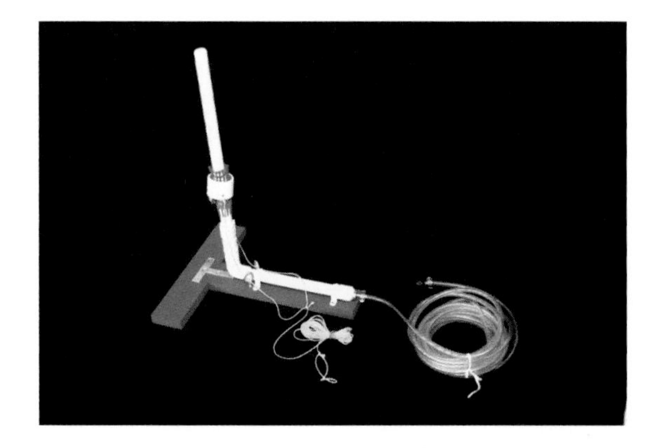

Figure 11-3. *The water rocket launcher uses any air source that can pump up a tire.*

Tables 11-1 and 11-2 list the parts and additional supplies you'll need for this project. All of the parts for the launcher should be available from a well-stocked hardware store, and I've listed some online sources as well. Shop around a bit. Prices vary quite a lot, and availability can change from time to time, too. Also, don't be afraid to make substitutions. While the choice of the PVC pipe, soda bottle, and O-ring are all critical, you can easily substitute most of the rest of the items. If you find something that looks like it will work, it probably will.

Table 11-1. Parts list

Part	Description	Part number
Clear pine or hardwood, 1x3	You will need two pieces of 12"-long wood to serve as the launcher base. The size can vary a bit. The actual dimensions of 1x3 lumber will be 3/4" x 2 1/2".	Lowe's 83707
T-plate	Joins the two pieces of wood securely into a T shape.	Lowe's 315655
3/4" tube strap (2)	Clamps the PVC to the base.	Home Depot C4724
1/4" brass hose barb	Fastens the pressure hose to the 1/2" x 1/4" brass bushing.	Lowe's 74486
1/2" schedule 40 PVC (at least 2 feet)	Get a little extra, say 3 feet.	Lowe's 23966
1 1/2" PVC	This piece is used for the release mechanism. The neck of the soda bottle you use can vary enough that 1 1/2" PVC won't fit. See the text for details. You only need a piece about 1 1/2" long. The hardware store may have some scraps or smaller pieces; be sure to ask, unless you really need the other 9 feet of PVC!	Lowe's 23830
1/2" schedule 40 PVC cap	Forms the end of the pressure chamber.	Lowe's 23937
1/2" schedule 40 PVC slip elbow	Forms the 90° angle.	Lowe's 23867
1/2" schedule 40 PVC adapter (slip/female)	This adapter has female threads on one end and a slip connection on the other.	Lowe's 23861
1/2" schedule 40 PVC adapter (slip/male)	This adapter screws into the one above, and has a slip joint on the other side for inserting a PVC pipe.	Lowe's 23855
1/2" schedule 40 PVC coupling	This looks like the previous two adapters, but both sides have slip joints.	Lowe's 23849
19/32" x 0.014" brass tube	Any stiff, waterproof tube that fits snugly inside the PVC pipe will work. Take a piece of PVC pipe with you to a good hobby store and try the various brass and aluminum tubes to find one that fits.	

Part	Description	Part number
9/16" ID, 13/16" OD O-ring	The size of this O-ring is pretty critical, but don't trust the online descriptions. A lot of folks seem to be math challenged. Lowe's lists this as a 7/8" O-ring, for example, but if you look at the detailed specifications and convert to sixteenths of an inch, you see that their part is really 13/16".	Lowe's 174788
Screw eyes (2)	The size isn't all that critical, as long as the launch string will fit through easily.	Lowe's 605303
#18 mason line	This is a thick, sturdy string. It's available in lots of fun colors.	Home Depot 65405
5/16" clear vinyl tubing	Get 20 feet for a 60 PSI launcher, or 40 feet for a 75 PSI launcher. See the instructions for a discussion.	Home Depot 714422
1 1/2" hose clamp	Used to bind the zip ties to the PVC coupler	Home Depot 6720595
Tire valve	Search the Internet or local auto supply stores.	Amazon Slime 2080-A Rubber Tire Valve
Hose clamp (2)	Two worm gear hose clamps, one for each end of the clear plastic fill tube.	Amazon Precision Brand M4P Micro Seal, Miniature Partial Stainless Worm Gear Hose Clamp, 7/32"–5/8"
Zip ties	You will need 15–20 zip ties, depending on the size. Get some with fairly large heads. They should be about 4" long, although the length is not critical.	
Tent stakes	Use two tent stakes or other supports to hold the launcher in place when you pull the launch string.	

Table 11-2. Additional supplies

Part	Description	Part number
Teflon tape	Available from hardware stores, Teflon tape is a thin, white tape with no sticky side. It's used on threaded high-pressure joints.	Lowe's 25010
PVC cement and primer	PVC is joined using this two-step primer and cement.	Lowe's 150887
Epoxy glue	Quick-setting epoxy is best, but any epoxy works.	

Equivalent Kit

Several manufacturers have water rocket kits. Quest has one I like a lot. Many of these kits, including the one from Quest, feature manufactured nozzles that reduce the size of the opening. Give them a try!

Other Launchers

The launcher in this chapter is based on a design from the US Water Rockets team. There are several other good launcher designs. Some are just alternatives, while others are specialized launchers for more advanced water rockets. Be sure to visit the US Water Rockets website (*http://bit.ly/uswr-cable-tie-launcher*) to get ideas and to see just how far —and how high!—you can go with water rockets.

Building the Launch Tube

Bottle sizes vary, so start by establishing the length of the piece that fits up into the bottle. This piece has a number of functions. It's the launch rod, guiding the water rocket for the first few inches of flight until it picks up enough speed for the fins to function. It's also a snorkel that prevents the water from draining down into the launcher while still providing a way to pump air into the bottle.

Shove a 1/2" schedule 40 PVC pipe into the bottle until it is a couple of inches from the end. Mark the pipe about 1/2" from the opening of the bottle and cut the pipe.

Working with PVC

We've covered working with PVC in previous chapters. See "Building the Base" on page 67 for tips on cutting and gluing PVC pipe.

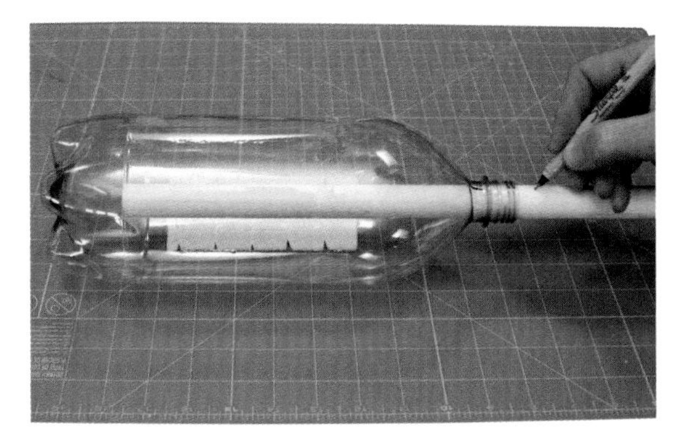

Figure 11-4. *Cut a piece of PVC pipe so it is long enough to reach about 2" from the end of the bottle, and sticks out 1/2" from the neck.*

While you are at it, cut the remaining 1/2" PVC parts. You will need three more pieces cut to 1", 4", and 9" long.

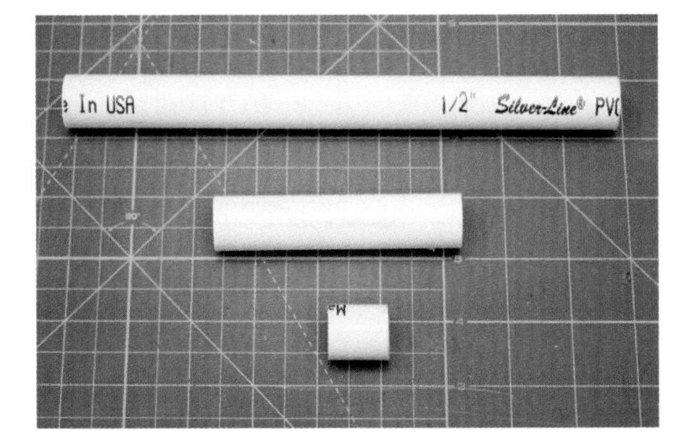

Figure 11-5. *Cut three additional lengths of PVC, 1", 4", and 9" long.*

One end of the PVC pipe will slide in and out of the bottle. While not essential, it's nice to sand this end so it is rounded a bit. This makes it easier to slide the bottle onto the launcher.

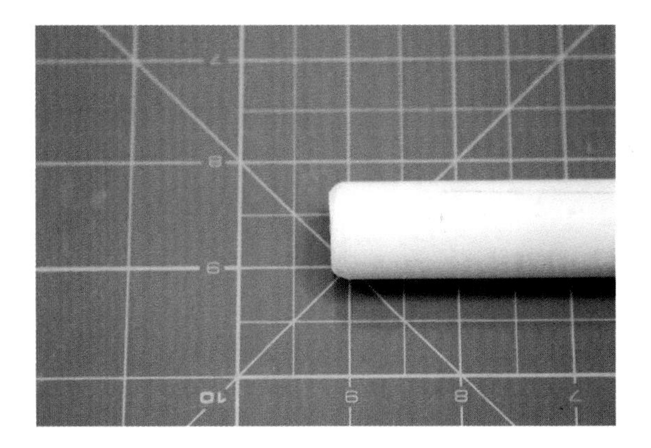

Figure 11-6. *Using sandpaper, round off one end of the PVC pipe that slides into the bottle.*

Dry fit a 1/2" PVC adapter. This piece slips over the non-rounded end of the PVC pipe and has male threads on the other end. Find a spot on the bottle's neck, about halfway down, where it is not changing diameter. The neck of the bottle shown here is about an inch long, so I picked a spot 1/2" from the end of the adapter. Cut the pipe at this spot.

This cut will serve as the seat for an O-ring, and that O-ring is the critical component in the entire launcher for holding pressure until the launch. Make sure this cut is absolutely straight. Sand the surfaces flat if necessary.

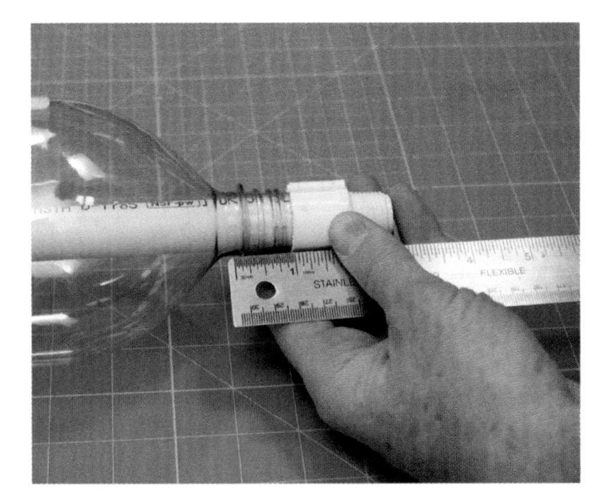

Figure 11-7. *Find a spot about halfway along the neck of the bottle and cut the pipe at that location. Make sure the ends are perfectly flat.*

The O-ring needs to sit against a tube that fits just inside the PVC pipe. Any stiff, waterproof material will work.

I've heard rumors there are some PVC parts that will work, although I've never found one. Some people make their own tube from heavy paper laminated with epoxy. An empty lip-balm tube works well, too. Personally, I went shopping at the local hobby store and found a 19/32" brass tube that was a perfect fit. Whatever you use, cut about a 1" length of the material to serve as the O-ring seat.

The size of the O-ring itself is critical. You will want an O-ring with an outside diameter of 13/16", an inside diameter of 9/16", and a thickness of 1/8". I found these in the plumbing section of my local Lowe's outlet. They are also available online, but read the descriptions carefully. The dimensions of the part are mislabeled on at least two websites, including the Lowe's website. The decimal dimensions in the specifications section are correct, though.

The O-ring will eventually wear out, allowing the water to leak. The same thing happens with the washers in a hose or faucet. Keep an extra O-ring or two with your flight kit so you can replace the O-ring if it leaks on launch day.

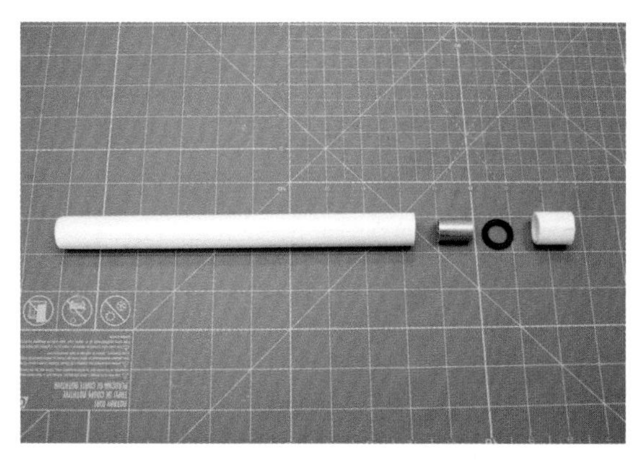

Figure 11-8. *The water needs to be held inside the bottle, as the pressure will be pumped up to as much as 75 psi. We'll form an O-ring seat between the two pieces of pipe using a stiff tube that fits inside the PCV pipe.*

We'll use epoxy glue to fasten these parts, since we need a glue that will bond well to both PVC and brass, and will be waterproof and dry completely in an area that doesn't have good air circulation. Rough up the ends of the brass tube so the glue will bond, but don't rough up the middle, where the O-ring will sit. Glue the pieces together so the gap between the PVC parts is exactly the width of the O-ring. It won't hurt anything if a little epoxy oozes into the gap; just wipe it out with a cotton swab soaked in rubbing alcohol.

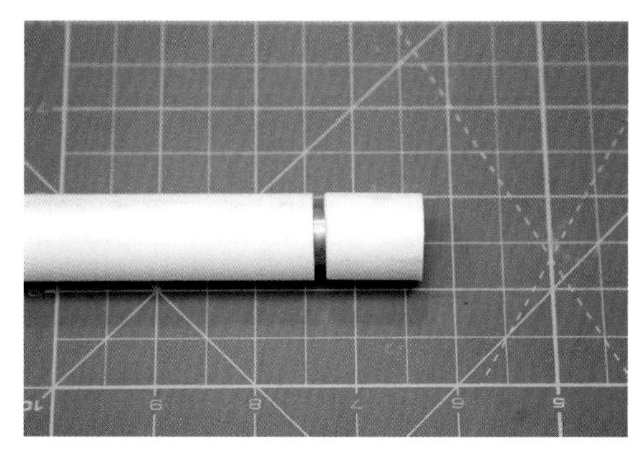

Figure 11-9. *Epoxy the PVC parts to the brass tube so the gap is exactly the width of the O-ring.*

Once the glue dries, slide the O-ring into place. This will be tough! That O-ring doesn't really want to stretch wide enough to fit over the PVC pipe, but it will make it.

Figure 11-10. *Install the O-ring after the glue dries completely.*

Assemble the remainder of the launch tube from a slip coupling, the 4" length of pipe you cut earlier, and the adapter.

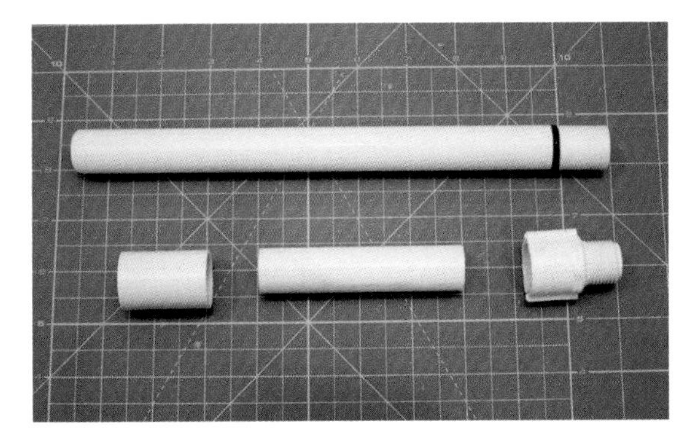

Figure 11-11. *The remaining parts for the launch tube are the 4" PVC pipe, a slip coupler, and the male adapter.*

Glue these parts using PVC cement. See "Building the Base" on page 67 in Chapter 4 if you need a refresher on gluing PVC.

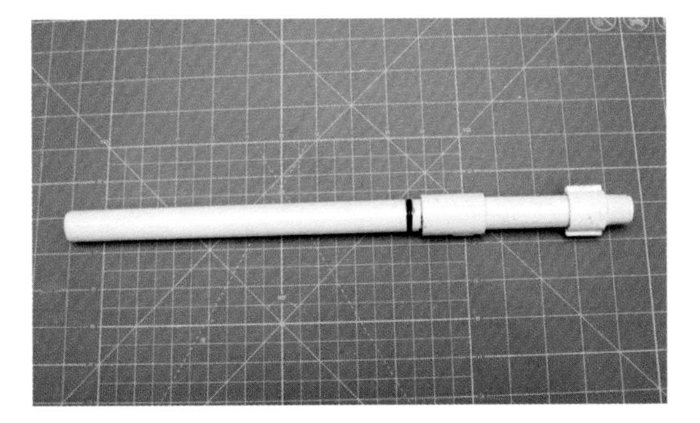

Figure 11-12. *Glue the parts together with PVC cement to complete the launch tube.*

Building the Lower Pressure Tube

We will pressurize the launcher using a 1/4" hose barb mounted in an end cap. Start by drilling a 1/2" hole in the center of the end cap. Thread the hose barb into the hole. You're cutting threads as you screw the hose barb into the end cap, so the first time is particularly important. Be sure the hose barb is perfectly straight as you thread it into the hole. You will probably need pliers and a wrench to cut the threads.

Remove the hose barb and inspect the end cap carefully for cracks. PVC is ductile enough that there shouldn't be any, but if the parts were particularly cold, or if you screwed the hose barb in at a slight angle, the end cap might crack. If it's cracked, throw the part away and try again.

Once you are satisfied with the piece, use Teflon tape and screw the hose barb into place one last time. See "Building an Air Rocket Launcher" on page 132 for tips on working with Teflon tape.

Figure 11-13. *Drill a 1/2" hole into the end cap and screw the hose barb in place. Remove the hose barb and check for cracks, then use Teflon tape for a pressure-resistant seal when you screw the hose barb into place for the final time.*

The lower pressure tube consists of the 1" and 9" PVC tubes cut earlier, along with a female-threaded coupler, a 90° elbow, and the end cap with the hose barb installed.

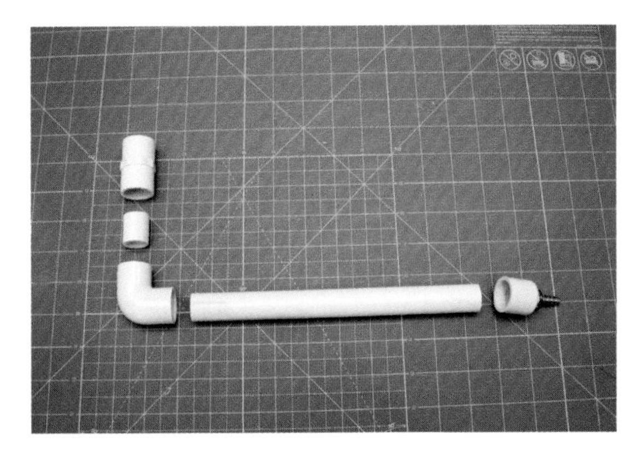

Figure 11-14. *The parts for the lower pressure tube.*

Glue these parts in place using PVC cement.

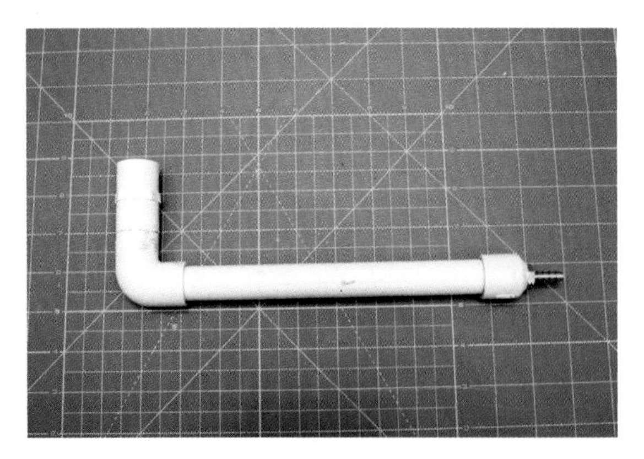

Figure 11-15. *The completed lower pressure tube.*

It's fair to ask why the pressure tube is made in two parts. After all, doesn't that result in another place for a leak? And it definitely adds a dollar or so to the cost of the launcher, and makes it more complex to build. What gives?

If you recall, soda bottles are not all the same size. You may want to build some water rockets from a different brand of bottle with a slightly different neck design. Or perhaps you will want to build some water rockets from smaller 1-liter or 20-ounce soda bottles. All of these will need a different launch tube. With this design, you can make several different launch tubes and change them in the field as your bottles change.

Of course, it's fine to leave out that threaded connection between the launch tube and lower pressure tube if you're building the launcher for a school program or for some other situation where you are sure all of the bottles will be the same size.

Building the Launcher Base

Cut two 12"-long pieces of wood for the base of the launcher. The photos show a nice grade of 1x3 lumber. You don't need to get quite that fancy, though. I wanted a nice-looking launcher that would last a while, but some launchers just use a couple of pieces of 2x4 lumber with no paint at all.

Test fit the T-plate, tube straps, lower pressure tube, and two screw eyes in the locations shown in Figure 11-17. Mark the locations for the various screws, and drill pilot holes in the wood, as shown in Figure 11-16.

Figure 11-16. *Drill pilot holes for the T-plate, tube straps. and screw eyes.*

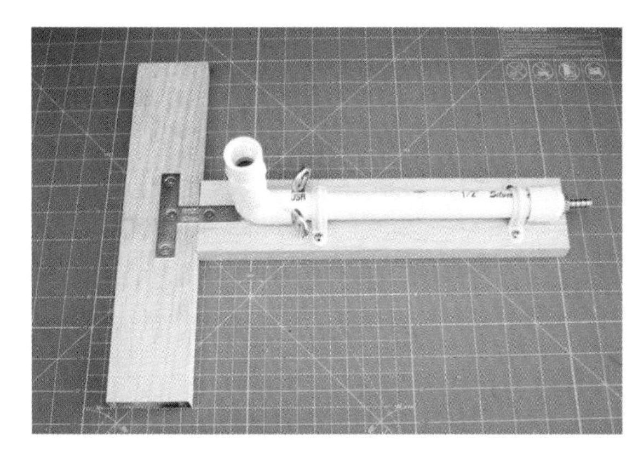

Figure 11-17. *Configuration of the T-plate, lower pressure tube, tube straps, and screw eyes.*

I used some pretty large screw eyes. The ones shown are 1" in diameter. The screw eyes create a guide path for the string we will use to launch the rocket, but they do double-duty as a way to lock the lower pressure tube in place. The hose clamps do that, too, but the screw eyes can be turned to tighten the tube in place. While I like this arrangement, smaller screw eyes will work, too.

You might be tempted to glue the lower launch tube in place. Leaving it loose gives you a way to tip the water rocket so it can be launched at a slight angle. Make sure the tube can rotate back and forth, but also make sure it is pretty hard to rotate it. The launcher base needs to be able to stay upright when there is a heavy rocket filled with water perched on the upper pressure tube.

Add tape inside the tube straps to tighten them up if it is too easy to rotate the lower pressure tube.

Building the Launch Release Mechanism

One of the hardest things to do well on a water rocket launcher is to keep the rocket on the launch pad until you're ready to let it fly. There's a lot of pressure trying to push the water rocket off of the launcher. Ian Clark in Australia developed a system in 1998 that has pretty much become the standard. There are a number of variations, but all of the zip tie–based systems trace back to his original invention.

Like all of the variants, our clamp uses the heads of several zip ties to hold the bottle in place as the pressure builds. The zip ties ring the adapter that sits just under the bottle, held in place by duck tape and a hose clamp.

Place a rubber band around the top of the slip adapter and slide the bottle down until it rests on the top of the adapter. Shove the zip ties under the rubber band, as shown in Figure 11-18. Leave a little space between adjacent zip ties.

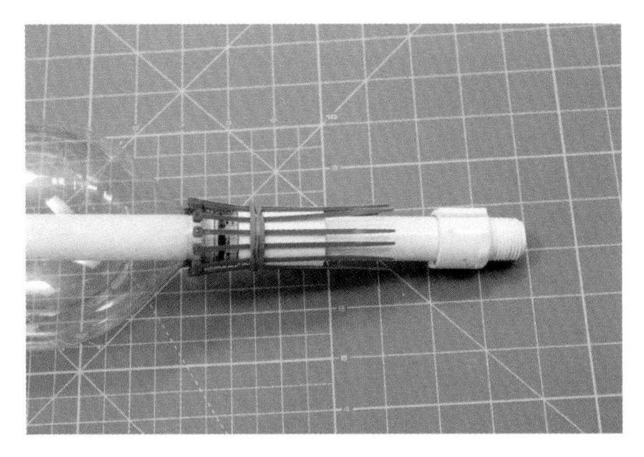

Figure 11-18. *Use a rubber band to position the zip ties around the adapter that sits right below the neck of the soda bottle.*

Once the zip ties are in place, secure them with a strip of duck tape and remove the rubber band. We'll add a clamp in a moment.

Figure 11-19. *Tape the zip ties in place so they clamp the bottle against the PVC coupler.*

It's still pretty easy to pull the bottle off of the launch tube. The zip ties pop open, allowing it to slide off. Once the bottle is under pressure, the zip ties are held in place until launch time by a collar that slips up over the heads of the zip ties. Pulling this collar down releases the zip ties, launching the rocket.

Unfortunately, the necks on soda bottles vary enough that a single size collar won't work for all available bottles. While a 1 1/2″ PVC pipe worked well for the bottles I found, you might have to go up or down a bit in size to fit the specific bottle you selected. The rest of the instructions assume you are using a 1 1/2″ PVC pipe. Change the size if you need to, though.

Cut a 1 1/2″ piece of 1 1/2"-diameter PVC pipe. Actually, any stiff, waterproof tube will do. Pick a tube that will just barely slide over the heads of the zip ties. Drill two 3/16″ holes at the base of the tube. Add duck tape inside until the tube stays in place when it is slipped over the ring of zip ties.

Figure 11-20. *Cut a collar from 1 1/2" PVC that is 1 1/2" long. Drill two 3/16" holes on opposite sides near one end of the collar. Add duck tape until it stays in place when slipped over the ring of zip ties.*

You don't need much of the 1 1/2" PVC, which is usually sold in 10-foot lengths. Ask at your local hardware store; I found a 2-foot length of scrap that worked just fine and kept me from lugging 10 feet of the stuff home when I only needed a short piece.

There are alternatives to this friction fit system. Some people use additional zip ties, or a section of water bottle, to form a spring that holds the collar in place until you tug the launch string. I like the friction fit system a little better because it's so simple to set up and use. Search the Internet for alternatives if you don't like this system.

Use a hose clamp to securely fasten the zip ties near the bottom of the PVC adapter. Position the hose clamp so it keeps the PVC collar from sliding all the way down the launch tube, but gives plenty of clearance for the collar to slide down far enough to expose the heads of the zip ties to release the rocket.

You can trim off the extra parts of the zip ties that stick out at the bottom, but I thought they looked kind of cool, so I left them.

Figure 11-21. *Use a hose clamp to secure the zip ties.*

Adding the Pressurization Tube

We need a way to pressurize the water rocket, and we need to be able to do that from a distance to stay safe. We'll use a Schrader valve, which is the same valve you see on car tires and large bicycle tires. You can find these at auto supply stores and some hardware stores. They generally have a large, bulbous rubber coating to make them easy to seat in a tire. Carve that away carefully with a hobby knife until you expose the inner brass lining.

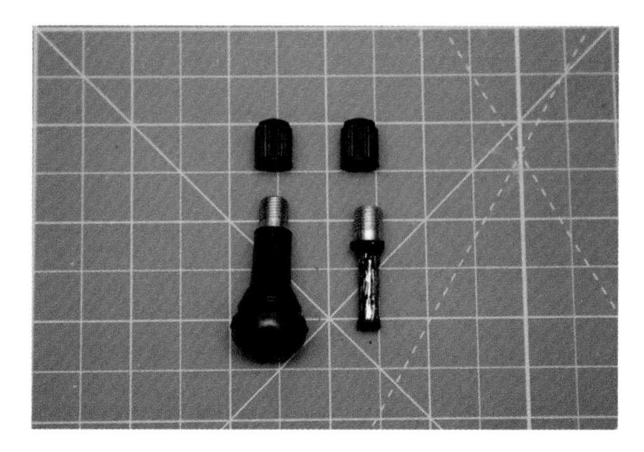

Figure 11-22. *Carve the excess rubber from a tire valve to create the valve for your launcher.*

We also want to be able to release the pressure if something goes wrong and we need to approach the rocket. There are several strategies for doing this. One is to add a special pressure release valve. An even simpler solution is to remove the pin from the center of the Schrader valve so it lets air flow both ways. Disconnect the pump to release all of the air in the system.

The pin in the middle is threaded in the brass tube. There is a special tool for unscrewing it, but unless you're mass-producing water rocket launchers, there isn't much point in buying the tool. Sometimes you can manage to unscrew the pin and its accompanying valve by grabbing it with very thin pliers and twisting. You can also remove the pin by grabbing it with a normal pair of needle-nose pliers and pulling very hard. If the pin is really stubborn, you can also drill it out from the bottom using a 3/32″ drill bit.

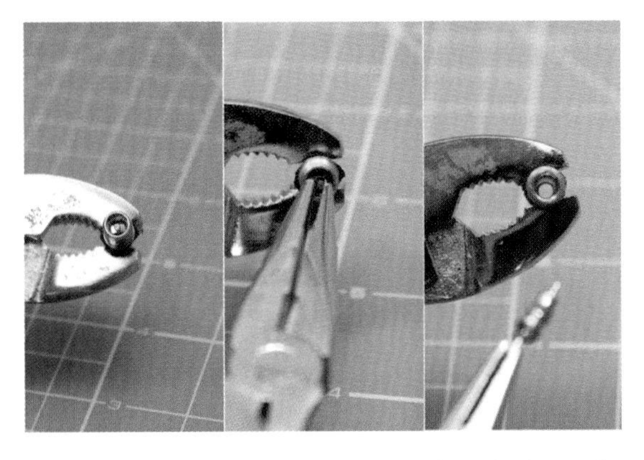

Figure 11-23. *Remove the pin in the Schrader valve by yanking it out with needle-nose pliers or drilling it out from the back with a 3/32″ drill bit.*

Connect the Schrader valve to the hose barb using 5/16″ clear vinyl tubing. You have a choice on the length of the tubing:

10 feet
: Use 10 feet of tubing if you intend to launch rockets using 60 psi or less of pressure. The person launching the rocket and anyone else less than 20 feet from the rocket should wear safety glasses.

20 feet
: Use 20 feet of tubing to launch rockets with more than 60 psi of pressure. Anyone less than 40 feet from the rocket should be wearing safety glasses. You can also launch rockets using 60 psi of pressure or less without safety glasses.

40 feet
: Use 40 feet of tubing to launch rockets using more than 60 psi of pressure without safety glasses.

Attach the tubing to the hose barb and Schrader valve using hose clamps. It may be tough to push the tubing over the hose barb. If you can't muscle it over the barb, try heating the tubing by holding it under hot running water or by dipping it in boiling water to soften it.

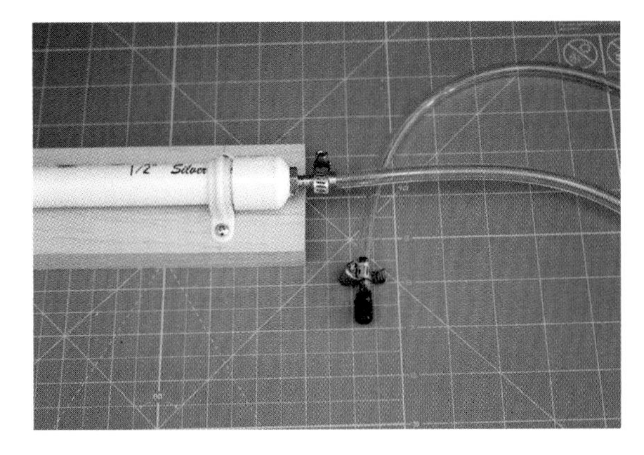

Figure 11-24. *Connect the vinyl hose to the Schrader valve and hose barb using hose clamps.*

Adding the Launch Cord

You'll launch the rocket by pulling on a long launch string that pulls the PVC launch collar down, allowing the zip ties to spring apart. While you can use any kind of string, it does need to be strong. Hardware stores carry #18 mason line in a wide variety of bright colors. It's cheap and makes a great launch cord. Mason line is available in twisted line and braided line. The braided line works a little better. Pick the brightest color you can. The line will be lying across the launch area. A bright color makes it easier to see, and that makes it less likely someone will trip over your line.

Screw the launch tube into the lower pressure tube. Form a loop of string that ties onto one of the holes in the PVC collar, threads through one screw eye, loops back through the second screw eye, and finally ties off on the other hole in the PVC collar. Tie another piece of string to the center of the loop. Cut the string so it is a bit longer than the vinyl tubing used to pressurize the rocket.

At some point you'll probably add a second launch tube and want a fast way to swap between the tubes. Untying and retying the launch string will take too much time. Add snap swivels—slightly larger versions of the ones you used for parachutes on solid propellant rockets—so you can easily detach the launch cord from one launch tube and reattach it to the next one.

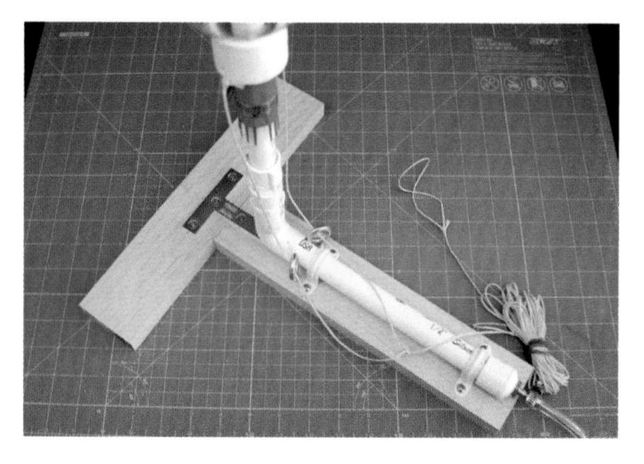

Figure 11-25. *Add the launch string to complete the launcher.*

Now that the launcher is all together, remove all of the screws and paint the wooden parts with a waterproof paint such as a wood sealer, acrylic paint, or enamel paint. You can paint the lower pressure tube, too, but leave the launch tube unpainted so the rocket bottle slides on and off easily. Reassemble the finished launcher once the paint dries.

Themis: A Ballistic Water Rocket

Simple water rockets are extremely fast and easy to build. Don't let the simplicity of this first water rocket lull you into thinking they are always this simple, though. Some advanced water rockets use multiple bottles, multiple stages, and advanced electronics for staging and parachute deployment. In fact, we'll add a parachute mechanism in Chapter 12. Like their solid propellant cousins, water rockets can also carry payloads, like cameras. The current world record altitude for a water rocket flight is 2,044 feet, held by the US Water Rockets team. That water rocket was a tad more advanced than the one we're about to build.

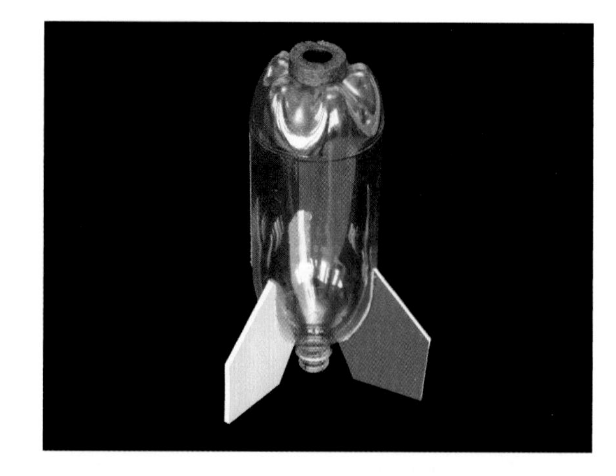

Figure 11-26. *Simple water rockets are fast and cheap to build.*

Themis

Themis is a pretty big asteroid, measuring nearly 200 kilometers in diameter. It was discovered on April 5, 1853 by Annibale de Gasparis, an Italian astronomer. In 2009, NASA's Infrared Telescope Facility found ice on Themis, making it the first asteroid with a confirmed water source. It seems a fitting choice as the namesake for our first water rocket.

Like many asteroids, Themis is named after a Greek goddess: the goddess of natural law and divine order.

Table 11-3 lists the parts you'll need to build our rocket, called Themis.

Table 11-3. Parts list

Part	Description
Two-liter soda bottle	Not all soda bottles are made the same way, so pick one that will fit the launcher. See the instructions for building the launcher for details.
Foam core	A sheet of foam core board from the hobby store makes great fins. Corrugated cardboard works OK, but tends to get soaked. You might also find corrugated plastic—that works really well, too.
Pipe insulation	While not absolutely necessary, it's nice to put some sort of foam padding on the top of the rocket to cushion the impact on landing. This can save a set of fins if the rocket lands on something hard. Any squishy foam will do; the plans show a short piece cut from pipe insulation.

You will also need some glue to attach the fins to the soda bottle. Polyurethane glue, such as Titebond or Gorilla Glue, works well. Wood glue does not do well here.

The body of the rocket is made from a two-liter soda bottle. While two-liter bottles are readily available, there is nothing magic about this size. I've also built water rockets from one-liter bottles, and they can even be made from 20-ounce soda bottles. Be sure to use soda bottles, though, not water bottles. Soda bottles are made to hold in the pressure of carbonated soda, so they'll be better able to withstand the pressure of a water rocket. Plastic bottles used to hold water will not stand up to too much pressure; they don't have to, since the water they hold is not carbonated.

The rocket will fly just fine with the original label on the bottle. The label will add a small amount of drag, but not enough to matter. Still, it's natural to peel it off. In fact, you might have done that already. Usually, a few stubborn spots are left where the glue firmly attaches a few scraps of label to the bottle. An amazingly effective way to remove the rest of the label and glue is to soak a paper towel in WD-40 and lay it over the areas with glue. Let it sit for an hour or two. The glue and remnants of the label will literally wipe away.

Remove any WD-40 from the bottle using rubbing alcohol before you apply your own glue.

Figure 11-27. *After removing the loose parts of the label by hand, soak the remaining scraps and glue on the outside of the bottle in WD-40 for a couple of hours. The scraps of label and glue will wipe away.*

You can make fins from almost any lightweight material, but keep in mind that this is a water rocket, and the fins are going to get wet. While corrugated cardboard is a good choice structurally, it's not going to hold up after getting soaked. You can buy a sheet of foam core from a craft store for a few dollars. It will make dozens of rocket fins. Corrugated plastic works well, too.

This water rocket is pretty light compared to its size, so it's not going to do any damage when it lands. It's more likely that the ground will damage the rocket, especially if it is concrete or some other hard material. Keep the rocket intact by putting some padding on the nose of the rocket to cushion the impact of landing. Figure 11-28 shows a short piece cut from foam pipe insulation. The drab gray is admittedly a little boring. Another great source of foam for the top of the rocket is a short piece from a swim noodle. They come in a lot of fun colors. You can really use any squishy foam that's available, though.

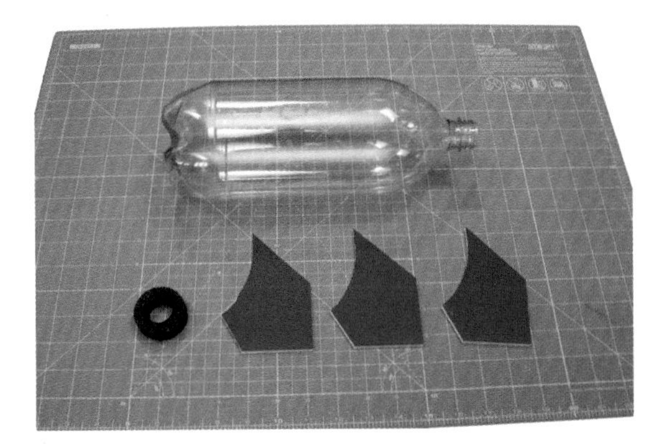

Figure 11-28. *The parts for Themis include a two-liter bottle, three fins cut from foam core, and a nose pad cut from pipe insulation.*

Using the fin guide in Figure 11-29 as a guide, cut three fins with a hobby knife. Attach the three fins with polyurethane glue, such as Titebond or Gorilla Glue. I'm a big fan of hot glue guns for attaching most parts on a water rocket, but this is one place to avoid a hot glue gun. The heat can change the molecular structure of the bottle, weakening it slightly. CA glues can do the same.

Add a glue fillet along the joints between the fins and the bottle to add strength.

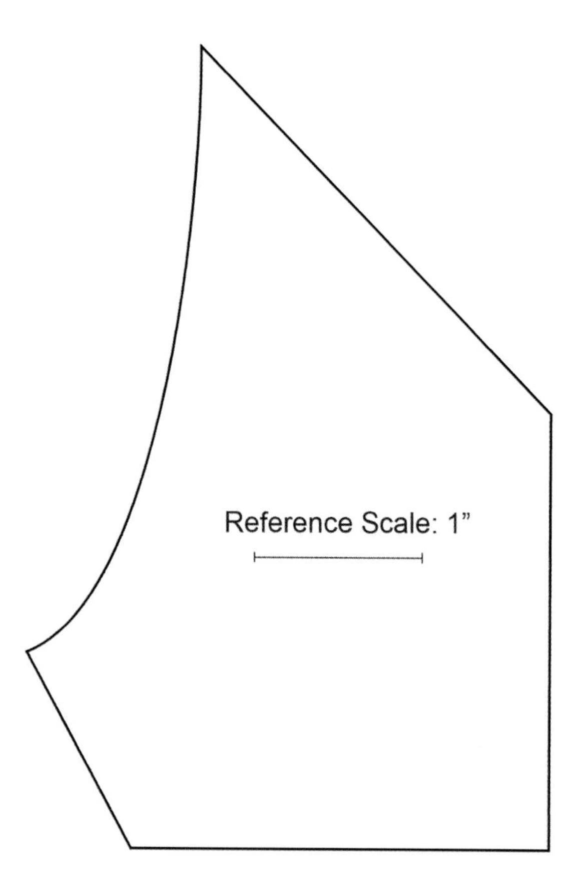

Figure 11-29. *Fin pattern for Themis. You may need to change the curved part slightly to fit the specific soda bottle you are using. A full-size template is available on the author's website (http://bit.ly/byteworks-make-rockets).*

I used blue foam core for the fins, so they did not need to be painted. You can paint the fins if you like.

Flying Water Rockets

Too many people believe that water rockets are not dangerous, so they don't have to follow safety rules. Hogwash. That's not to say that water rockets are particularly dangerous, or that they are any more dangerous than solid propellant rockets, but they do deserve some respect. At least two people have broken their hands because they violated a simple and obvious safety rule by working near a water rocket when it was under pressure. The water rocket is heavy when it lifts off, carrying over a pound of water, and it hits speeds of around 100 mph pretty quickly. You do not want to be in the way when it launches!

The NAR water rocket safety code is reprinted here. Most of the rules have obvious parallels to the solid propellant rocket code we saw in Chapter 5, but a couple of rules are there because of the properties of PVC pipe under pressure. We discussed that back in the launcher construction section.

Check your rockets before you head out to the launch area. Plastic bottles will eventually wear out. If the rocket is showing signs of stress, like discolored plastic, deep folds or gouges in the plastic, or bulges from stress, it's time to retire the bottle and build a new rocket.

About the NAR Water Rocket Safety Code

NAR knows more about model rocket safety than any organization, but it does concentrate on solid propellant rockets. NAR has created two safety codes for water rockets in the past. The one shown here is one of those safety codes. It's considered obsolete, but it's still the best safety code available for water rockets.

NAR is currently revising the water rocket safety code, and I had a chance to work with Steve Lubliner from the NAR safety committee as he drafted the new standard. It's not clear yet whether this new safety code will be fully adopted by NAR.

The launcher, water rockets, and launch procedures in this book were developed to satisfy both the old standard and the draft I have seen of the new standard. Be sure to watch the NAR website for the new water rocket safety code. If it is adopted, I will review the standard and post any required updates to the launcher design, rocket design, and launch procedures. Check the errata page for the book on the author's website (*http://bit.ly/byteworks-make-rockets*) and the publisher's website (*http://bit.ly/make-rockets*) occasionally to look for these updates. Any changes needed will be mentioned in both locations.

Pick a launch area large enough for the pressure you intend to use. There is obviously no fire danger; in fact, you're going to water the launch area, so put the launcher in a grassy area if one is available.

Place the launcher either in the middle of the launch area or toward the direction of any wind, so the wind will carry the rocket into the middle of your launch area. Connect whatever you will use to pump up the launcher. You can use a bicycle pump, but that's a lot of work. I like to use the Black & Decker Portable Power Station with its built-in tire inflator.

Fill the rocket about 1/3 full of water. Turn the launcher over and insert the launch rod into the rocket, then invert the entire assembly gently, trying not to get any water down the launch tube.

NAR Water Rocket Safety Code

Revision J - 16 December 2002 *updated 15-AUG-2007*

1. Definitions. For the purposes of this safety code, a 'Water Rocket' is defined as any rocket whose thrust is generated by expansion of a compressed, non-combustible gas. An inert fluid such as water may be used for thrust augmentation.

2. Scope. This code applies to water rockets having a pressure chamber volume greater than 1200ml or a launch pressure exceeding 35 psi.

3. Materials. The pressure chamber of the rocket shall be constructed of thin, ductile plastic. Only lightweight, non-metal parts shall be used for the nose, body, and fins.

4. Compressed Gas Safety. A safe distance shall be maintained at all times between persons and pressurized water rockets or launchers. The recommended safe distance is as follows:

Launch pressure	With eye protection	Without eye protection
Up to 60 psi	10′	20′
Above 60 psi	20′	40′

The new draft standard specifies 15 feet for pressures up to 80 psi, and 30 feet for pressures above 80 psi. There is no difference with or without eye protection.

5. Pressurization System. Compressed air tanks and gas cylinders shall be stored and transported in accordance with all applicable safety codes. Line fittings near the operator shall be rated by the manufacturer for use with compressed gas at the intended pressure.

6. Launcher. The launcher shall hold the rocket to within 30 degrees of vertical to ensure that it flies nearly straight up. It shall provide a stable support against wind and any triggering forces, and allow the rocket to be pressurized and depressurized from a safe distance. Launchers shall be constructed from materials rated for at least 3 times the intended launch pressure.

7. Launch Safety. I will use a countdown prior to launch to ensure that spectators are paying attention and are a safe distance away. If my rocket does not launch when triggered, I will not allow anyone to approach it until it has been depressurized.

8. Size. A water rocket whose mass (excluding water) exceeds 453 grams (1 lb) shall be considered a "Large Model Rocket" for the purpose of compliance with Federal Aviation Administration regulations.

9. Flight Safety. Water rockets shall not be directed at targets, into clouds, or near airplanes. Flammable or explosive payloads shall not be carried.

10. Launch Site. Water rockets shall be launched outdoors, in an open area at least 100 feet on a side (for rockets using a launch pressure of 60 psi or less), or 500 feet on a side (for rockets using higher pressure).

11. Recovery System. A recovery system such as a streamer, parachute, or tumble recovery shall be used, with the intent to return it safely to earth without damage.

12. Recovery Safety. Recovery shall not be attempted from power lines, tall trees, or other dangerous places.

How Much Water Should You Use?

Themis is propelled by pressurized air forcing water out of the nozzle at high speed, followed by the escape of the air itself. If there is too much water, the rocket gets too heavy and won't fly as high. Typically, about 2/3 of the thrust comes from expelling the water, and 1/3 of the thrust comes from expelling the air. In fact, you can launch Themis without any water at all!

Choosing the perfect amount of water is tricky. It varies depending on the weight of the rocket, the size and shape of the nozzle used, and the pressure in the rocket bottle. Filling the bottle about 1/3 full seems to work well with Themis, but it's certainly an area worthy of experimentation.

Figure 11-30. *The launcher is small enough to turn it over. Insert the launcher into the rocket, then flip them over for launch.*

Wedge the crossbar of the launcher against something firm. You might bring along a couple of tent stakes to drive into the ground. This lets you tug on the launch cord without pulling the launcher across the grass, possibly tipping it over.

Push the rocket down far enough on the launch tube that the zip ties are over the ring on the neck of the bottle. Slide the launch collar up over the edge of the zip ties. Give the rocket one final check to make sure it has not been damaged.

Figure 11-31. *The rocket is ready to pressurize. Note the tent stakes. They hold the rocket launcher in place while the launch string is pulled.*

Move all spectators back behind the pump and start pressurizing the rocket. The rocket will take off pretty fast, so the spectators may want to back up even farther than required so they can see the flight better.

Pressurize the system to between 60 psi and 75 psi, then turn off the compressor. Make sure all spectators are at least as far away as the pump and are aware a launch is about to take place. Check for low-flying aircraft.

How Much Pressure Should You Use?

Soda bottles will burst at about 110 psi. This isn't particularly dangerous if you are following all of the other rules listed here, but it is clearly not something you want to happen, either. Leaving aside the obvious safety issues, you are going to ruin a perfectly good rocket.

Seventy-five psi is plenty of pressure for a soda bottle water rocket like Themis. You might go to a higher pressure if you graduate to bigger and stronger water rockets, but that would mean learning a lot more about the safety of materials under pressure, not to mention building a different launcher and rocket. For now, keep it to 75 psi at most.

Figure 11-32. *Some low-flying geese invaded our flight area the day we were photographing Themis for the book. We had to hold the launch while they vacated the airspace!*

In the unlikely event that something goes wrong after pressurizing the rocket, be sure to release the pressure before approaching the launcher. Assuming you built the launcher as directed, all you have to do is disconnect the air pump. The air will rush out of the launcher. If you did not pull the pin from the Schrader valve, push the button in the middle of the valve and hold it down while the air rushes out. This should remind you to pull that pin out before the next launch.

Once you are sure everything is ready, give a countdown from five and tug on the launch cord to release the rocket. It will launch very quickly, reaching speeds of around 100 mph in a fraction of a second. Be ready to get wet. The spray from the water will frequently give you a quick shower if the wind is blowing toward you!

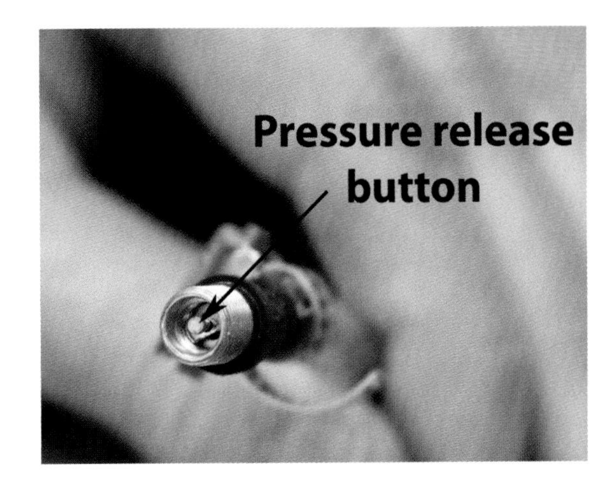

Figure 11-33. *If pulling the launch string does not release the pressure, release the pressure by disconnecting the pump. If the pin on the Schrader valve is still in place, push down on the pressure release button in the center of the valve that is connected to the launcher. This will release all of the pressure.*

Themis is a fairly small water rocket, launched using an open nozzle. It's about as simple as water rockets get. Still, I've tracked it to 165 feet and 91 mph using an altimeter. I've seen it go higher on other flights. Themis is a cheap, easy rocket to build and fly. I'm sure you'll have a lot of fun with it.

Here's a checklist that summarizes the flight preparation.

Flight Checklist

1. Check the water bottle for damage. Don't launch with a damaged bottle.
2. Clear spectators from the launch zone.
3. Fill the rocket about 1/3 full of water.
4. Slide the rocket onto the launch tube.
5. Pressurize the launcher to 60–75 psi.
6. Check for clear skies—no aircraft (or geese!).
7. Count down from five and launch.

Parachute Recovery of a Water Rocket

12

Figure 12-1. *Themis with a parachute deployment system.*

Themis, the simple water rocket from Chapter 11, used featherweight recovery, where a rocket is light enough that it doesn't need a parachute or streamer. While the rocket wasn't particularly light, we could get away with featherweight recovery because it was really large for its weight, which increased the rocket's drag.

As you move forward with water rockets, though, you may want to add an altimeter, a camera, or simply more bottles to get more power and altitude from your rocket. This makes the rocket heavier, making it more likely you will damage the rocket or something it hits if you rely on featherweight recovery. You should use a parachute for anything bigger than a single two-liter bottle. You should also use a parachute with a rocket based on a single two-liter bottle if it has a payload that weighs more than a half ounce or so.

There are a lot of ways to attach a parachute to a water rocket. I've tried several myself, and I've seen others struggle with several more. As a result, I've tried and seen a lot of unworkable or unreliable ways to try to add a parachute to a water rocket! This chapter shows a system that can be built for around $20. It's reliable, effective, and fun to build. It will also fit right on top of Themis or any other water rocket based on a two-liter bottle.

In water rocketry, there are two common ways to deploy the parachute. All of the solid propellant rockets in this book use what's called *axial deployment*, where the parachute fires out of the top of the motor chamber, moving along the main axis of the rocket. This works really well if your propulsion system can provide an ejection charge.

You can create axial deployment systems with water rockets too, even though pressurized soda bottles don't naturally come with an ejection charge. These generally use a spring-loaded system of some sort to stay away from pyrotechnic charges, which may be illegal or have age restrictions in some areas. The system we'll build in this chapter uses a different method that looks a little odd at first, but actually works better for water rockets. It's *side deployment*, in which a small door on the side of the rocket pops open at the right moment to deploy the parachute. There is a bit more drag with the side deployment method, but it tends to be a little simpler and lighter than axial systems. By all means, check the Internet for some axial deployment systems to see if they appeal to you. In this chapter, however, we'll stick with side deployment.

We need a way to eject the parachute. Lots of ways have been developed and tried, from black powder charges to metal springs. We'll use a clever mechanism that turns the side of a soda bottle into a powerful spring that will pop the parachute out of the rocket. The main problem with this method is that the spring is so powerful, we need an equally powerful mechanism to hold it in place until it is time to release the spring on command. A cheap servo will do nicely.

There are two basic ways to time the deployment of a parachute. One is to time the release of the parachute so it occurs roughly when the rocket reaches apogee. That's how all of the solid propellant rockets in this book deploy parachutes and streamers. The ejection charge timer is essentially a timer built from a smoke delay.

The other way to decide when to deploy the parachute is to use an onboard altimeter that deploys the parachute at a precise altitude, usually apogee. We're going to build an electronic deployment system that will deploy the parachute using either method. Deploying the parachute with an altimeter requires adding an altimeter, though, and they cost anywhere from $80 to $350, so we'll stick with a timed mechanism here. Still, the system is expandable. This system will grow with you if you eventually get to the point where you are flying multibottle or multistage rockets.

Building the Water Rocket Parachute System

Table 12-1 lists the parts required to build the parachute recovery system for Themis. Table 12-2 lists some additional supplies you'll need for this project.

Table 12-1. Parts list

Part	Description	Part number
Two-liter soda bottle (2)	These will need to have flat sides. The photos show parts from a clear bottle and a green 7 Up bottle, but that was just to make the door easier to see.	
Foam core or corrugated plastic	This is used for the structural parts of the system. The leftovers from cutting the fins for Themis will work just fine.	
Ping-Pong ball	While you can use just about anything, including the original bottle cap, a Ping-Pong ball makes a great nose cone tip for a water rocket.	
#18 mason line	The same braided nylon cord used to make the launch string for the water rocket launcher is perfect.	Home Depot 65405
Wire ties or twister seals	You will need four sturdy fasteners. The instructions show small nylon wire ties, and they work great. Twister seals like the ones used to tie off grocery bags work well, too.	
MSP430 LaunchPad	This is a small microcontroller from Texas Instruments. This is the computational heart of the parachute deployment system. It's also pretty cheap—generally about $11.	Mouser Electronics 595-MSP-EXP430G2
Sub-micro servo	The parachute's spring-loaded door is released by this servo. While the one shown is from SparkFun, pretty much any small servo will work. The smaller the better, though.	SparkFun ROB-09065
Battery holder	You have a lot of choice here. See the text for thoughts.	Radio Shack 270-411
AAA batteries (4)	These fit the battery holder listed, but again, you have choices. See the text.	

Part	Description	Part number
Male-to-female jumper wires	These are used to make the various electrical connections. The text shows these jumper wires, but there are lots of choices.	Amazon 50 PCS Jumper Wires Premium 200mm M/F Male-to-Female, sold by Mysmile
Male headers	These are used as power connectors on the MSP430 LaunchPad. They are not essential, but are shown in the description that follows.	SparkFun PRT-00116 Break Away Headers - Straight

Table 12-2. Additional tools and supplies

Part	Description	Part number
Duck tape or electrical tape	You will use electrical tape or heat shrink tubing to protect the wiring, and electrical tape or duck tape to fasten the parachute bottle to the rocket bottle. The tape shown in the text is 3/4"-wide yellow electrical tape.	
Clear shipping tape	Wide clear tape is used as a hinge for the parachute door. Duck tape will work, too.	
Cellophane tape	Normal clear tape like the kind used to wrap presents is used in a couple of places to hold parts down. Lots of substitutions will work.	
Mounting tape	Thick mounting tape is used to hold the battery pack and electronics in place.	Scotch Permanent Clear Mounting Tape or similar.
Glue	A hot glue gun works well for assembling the foam core parts, but the polyurethane glue used to attach the fins works well, too.	
Windows-based computer	You will need access to a Windows-based computer to load the software onto the microcontroller. You only need to do this once, but plan ahead if you are a Mac or Linux user.	
Scissors	A pair of office scissors is a great way to cut water bottles.	
Hobby knife	Used for small cuts in the water bottle.	
Hand drill	You will be drilling several holes in the plastic.	
Soldering iron	Several wires need to be soldered together. Depending on the specific model of the LaunchPad microcontroller, you might also be soldering wires or headers onto the microcontroller.	

Installing the ServoChron Software on the LaunchPad Microcontroller

We'll use a small microcontroller from Texas Instruments called, appropriately enough, the LaunchPad to time the parachute release. The folks at US Water Rockets developed software for the device so it works as a parachute deployment system. You will need to download their software and install it on the LaunchPad microcontroller using a free installer program.

The installer only runs on Windows. This is a bit inconvenient for Macintosh and Linux users, but you only have to do this once. Plan ahead to make sure you have access to a Windows machine when you install the software.

Start by downloading the software for the LaunchPad microcontroller. It's called *ServoChron*, and you can download it directly from the US Water Rockets site (*http://bit.ly/uswr-servochron-firmware*). This software is free, and there are no restrictions on downloading or using it. The download is a zipped file; extract everything from the zipped file and save it in a convenient spot.

While you are on the US Water Rockets site, check out the manual for the ServoChron software (*http://bit.ly/uswr-servochron*). We're going to set up the controller to deploy a parachute using a single servo based on a timer. However, the software and hardware can support up to two servos and can use an external altimeter. You can find details about these additional features in the manual.

Loading software into the nonvolatile memory of a microcontroller is called *flashing*. It's done with a specialized piece of software. The software needed to flash the LaunchPad is called FET-Pro430 Lite. It's a free piece of software produced by Elprotronic, Inc. that you can download from Elprotronic.com (*http://www.elprotronic.com/download.html*). It's only available on Windows, so plan ahead if you normally use Macintosh or Linux computers. Follow the installation instructions to install the software on your computer.

Connect the LaunchPad to the computer using the black USB cable that comes with the LaunchPad, and start FET-Pro430. A pair of LEDs, one red and one green, will blink near the bottom of the LaunchPad, while an LED near the USB cable will glow solid green.

Warning: Disconnect the Battery

Do not connect a battery or other power supply to the LaunchPad while using the USB cable. The USB cable supplies all of the power needed during flashing. Additional power can harm the LaunchPad or the USB port on the computer, or cause the battery to overheat.

The initial display you see when you run the software might seem a little intimidating at first, but you really don't need to deal with all of the features in this program. Go to the File menu and select Open Project. Navigate to the place where you extracted the ServoChron software and open the *ServoChron_V2.10.FET430.prj* file.

Start by erasing any old programs on the device by clicking the ERASE FLASH button. You may get a scary-looking dialog like the one in Figure 12-3. Just ignore the warning by pressing Yes.

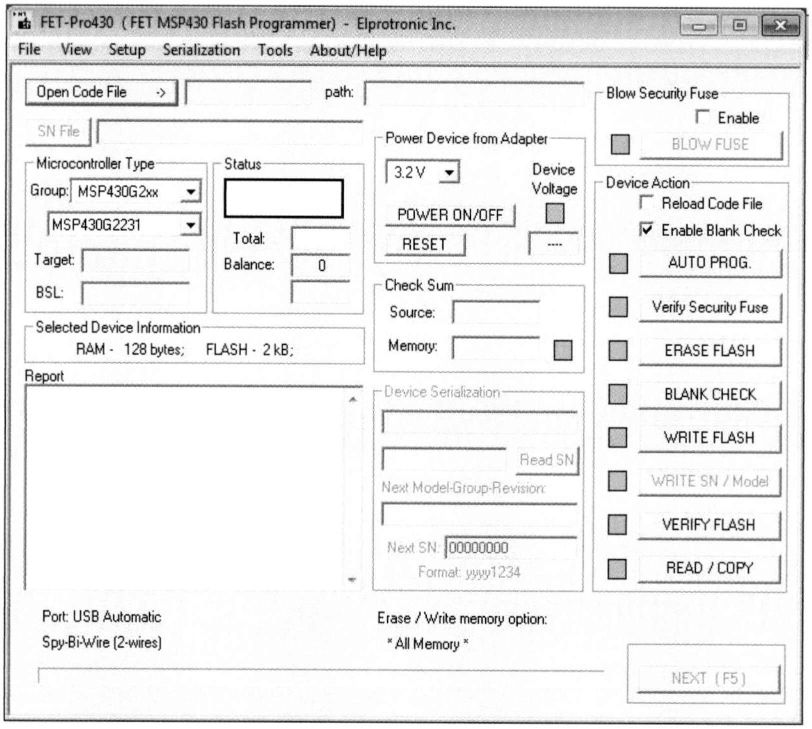

Figure 12-2. *The initial screen on the FET-Pro430 software. Load the ServoChron software using the Open Project command, found in the File menu.*

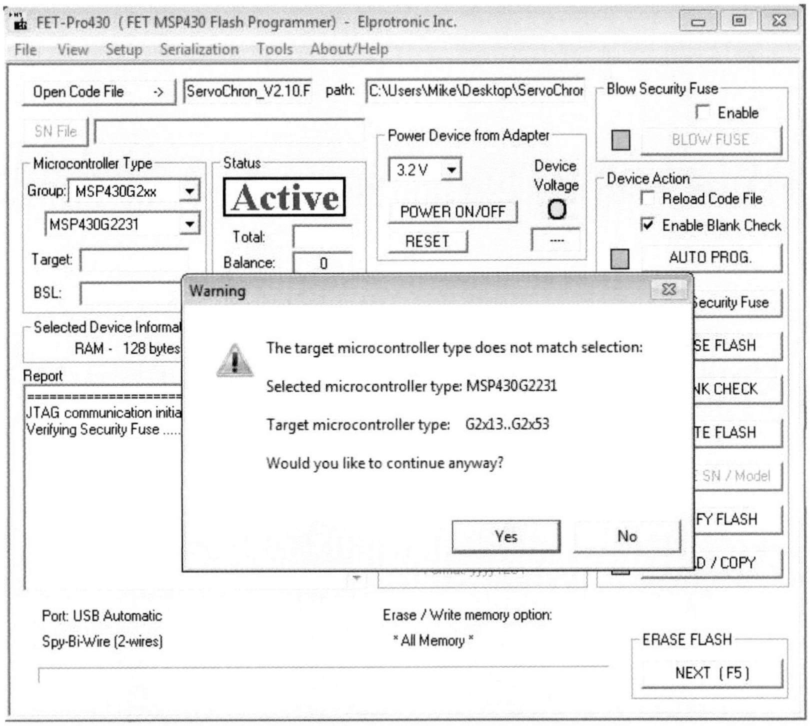

Figure 12-3. *Depending on the specific versions of the software and LaunchPad, you may see a warning. Ignore the warning, and click Yes.*

Once the memory on the LaunchPad is erased, the Status field will change to a big green Pass, and the LEDs at the bottom of the device will stop blinking. Flash the ServoChron software by clicking the WRITE FLASH button. You will probably get the same warning message; ignore it again. After a moment, the Status will again change to Pass. The LaunchPad microcontroller is now a dedicated device designed to deploy water rocket parachutes. Disconnect it from the computer.

Building the Electronic Timer

I've seen photos of several different models of the LaunchPad microcontroller. Some have the male header pins you see in Figure 12-4, while others have holes along the sides. I actually prefer the version with holes, since it makes it easier to solder the connections so there is no possibility they will come loose during flight or landing, but you can work with either.

We'll look at just one way to handle the wiring. Don't be afraid to adapt the wiring to fit variations in the board or the wires and tools you have available. The big advantage to the method shown is that nothing is soldered to the LaunchPad or servo except some connection headers, so both can be recycled for other projects. The disadvantage is the wires can come loose during a rough landing, so always do a visual check of the wiring before a flight.

The various parts are shown in Figure 12-4. The LaunchPad microcontroller is at the bottom left.

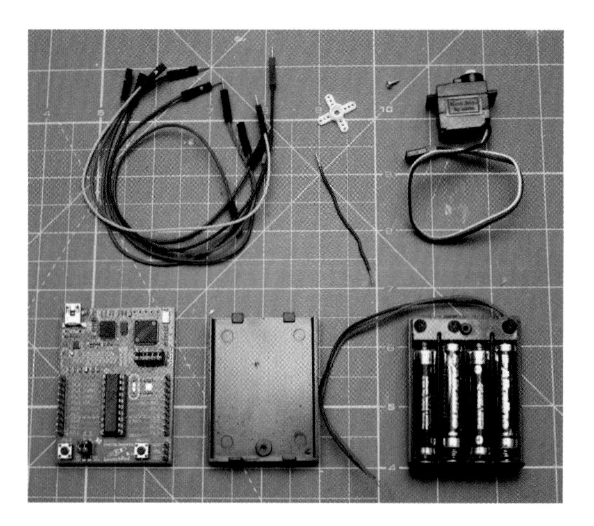

Figure 12-4. *The parts needed for the electronics.*

You have a lot of choices when it comes to power. The servo and microcontroller both work well at 6 V, so I selected a battery pack from Radio Shack that holds four AAA batteries. It's enclosed, so the batteries don't pop loose during a hard landing. This is a good choice if you want a rugged power source that will last for many flights, but it is heavier than some of the alternatives. Check around for other batteries if you are willing to sacrifice the convenience and long battery life of this package for a lighter, perhaps rechargeable alternative.

Pretty much any servo will do. Since weight is a factor, get a small one, like the sub-micro servo shown. There are a variety of control horns—that white cross in the center—that will come with the servo. We'll use the control horn to hook a string to the spring-loaded parachute door. It needs to hold that string to keep the door from opening, and then release it when the servo turns. This cross shape is a great choice.

Finally, you will need some connecting wire. I chose to work with some male-to-female jumper wires because they could connect to the servo and LaunchPad without soldering. However, you can solder 22-gauge wire directly to both devices if you prefer. The small scrap of 22-gauge connecting wire will be used for the launch detection switch. You can also use scraps of wire from other projects or old, recycled cables. Use whatever you have that is convenient.

We're going to build three wiring harnesses to make the connections shown in Figure 12-5, and use one additional male-to-female jumper.

The first harness will connect the ground pin on the LaunchPad, the black wire on the battery, and the black wire on the servo. It will also have a female jumper that will serve as one half of the launch detection switch.

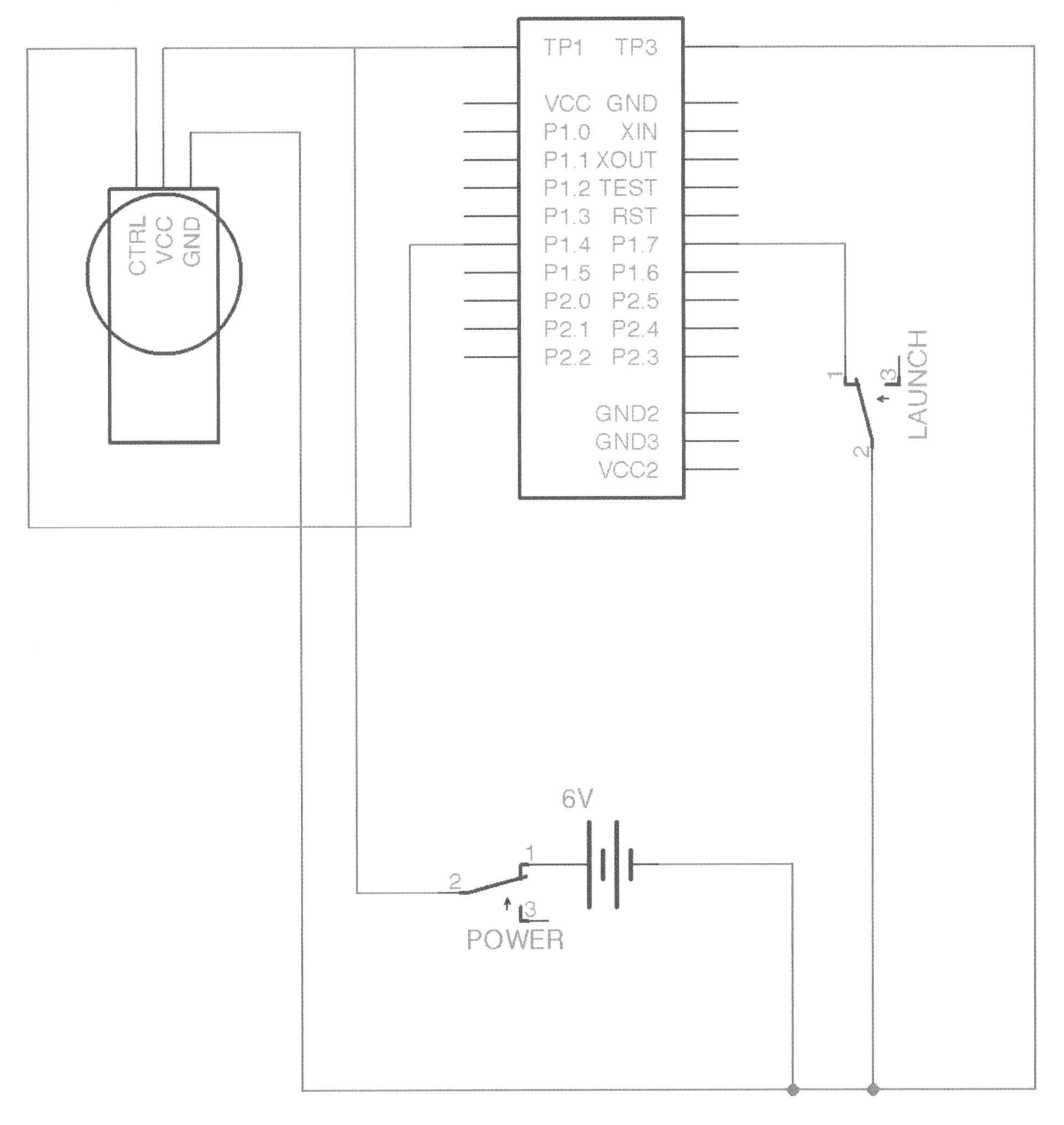

Figure 12-5. *Circuit diagram for the launch controller. On the servo, CTRL is the white wire, positive power is the red wire, and negative power is the black wire.*

Figure 12-6. *The wiring harness for the ground (negative) connection.*

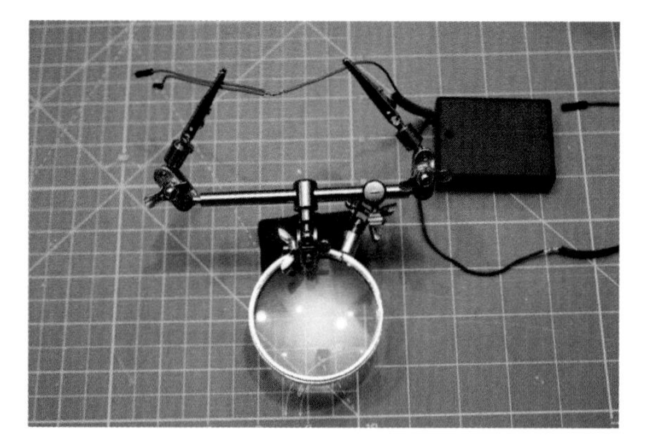

Figure 12-7. *The power harness carries +6 V from the battery to the servo and LaunchPad.*

Cut a black male-to-female jumper in half. The male half will eventually connect to the servo, while the female half will connect to the TP3 pin on the LaunchPad.

Cut the male pin off of another jumper, leaving the female connector on the long part of the wire. This will go to the launch detection switch on the outside of the rocket.

Strip about 1/2" of insulation from all three wires, as well as from the black wire on the battery holder.

Twist all three wires together with the black wire from the battery holder and solder the wires. Cover the connection with heat shrink tubing or electrical tape. These wires will get bounced around in flight, and you don't want them brushing against an exposed connector on the LaunchPad, making an unwanted short circuit.

The second harness will carry +6 volts to the LaunchPad and servo.

Cut a red male-to-female jumper in half. As with the ground harness, the male connector connects to the servo, while the female connector connects to the TP1 pin on the LaunchPad.

Strip about 1/2" of insulation from the two jumper wires and from the red wire on the battery holder. Twist the three wires together and solder them. Protect the exposed wire with heat shrink tubing or electrical tape.

The last wiring harness is really just a jumper wire with a female connection on both ends. If you happen to have a female-to-female jumper, you're done. If you bought a package of male-to-female jumpers for this project, cut the male connections from two of them, strip the wires, and solder them together. Add heat shrink tubing or electrical tape to protect the exposed wire.

This jumper will connect pin P1.7 on the LaunchPad to the other end of the launch detection switch.

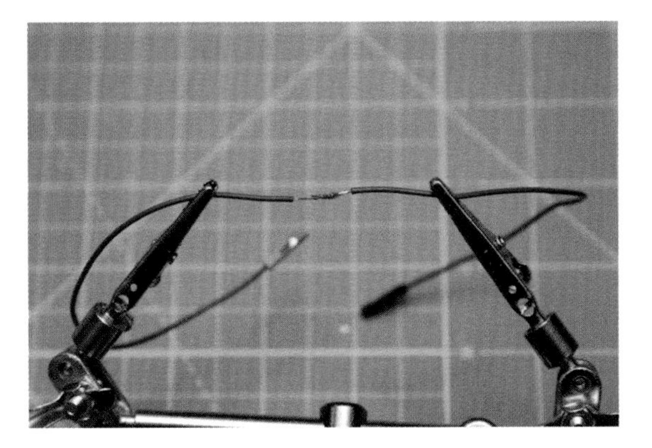

Figure 12-8. *Create a female-to-female jumper for the second connection on the launch detection switch.*

Finally, we need to connect power to the LaunchPad itself. The LaunchPad has power connectors labeled VCC and GND, but those are used for conditioned 3.3 V power supplied directly to the microcontroller. The battery pack is an unconditioned 6 V power supply. There are two power connections on the test portion of the LaunchPad, labeled TP1 and TP3; these are located right next to the USB connector at the top-left edge of the circuit board. They go through the power conditioning on the boards, so we can provide 6 V to those connections to power the microcontroller. Unfortunately, these are bare holes.

There are several ways to connect power to these locations. One is to solder the power wires directly into the holes. The technique shown here makes use of the power harness by providing two male headers, allowing you to disconnect the LaunchPad completely from the parachute ejection system and reuse it in other projects.

Solder two male headers into the holes for TP1 and TP3, as shown in Figure 12-9. The pin names TP1 and TP3 are not terribly descriptive. I expect I'll forget which is the positive connector and which is the negative connector before long, so I also painted the base of the male header for TP1 red. Red is almost universally used for the positive connection, so that reminds me to connect the red power wire to TP1 and the black ground wire to TP3.

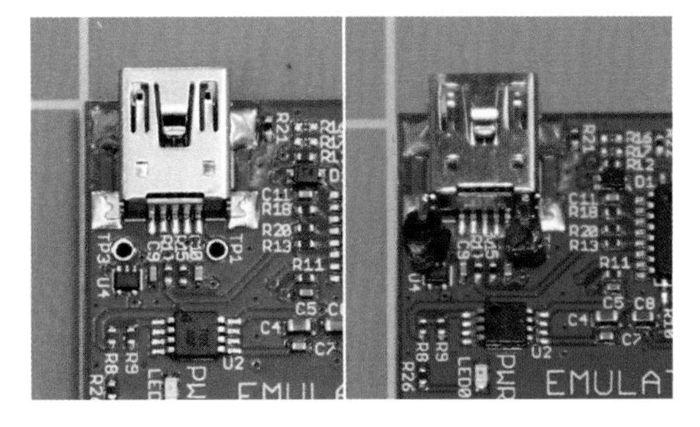

Figure 12-9. *Solder two male headers into the holes labeled TP1 and TP3. The image on the left shows the locations of the holes, right next to the USB connector. The image on the right shows the headers installed. Note that the right header is painted red to remind us that TP1 is the positive power connection, while TP3 is the negative connection.*

You can skip ahead a bit of you want to test the electronics now. We'll get to it eventually, but it is pretty fun to play with the electronics even before the parachute bay is complete.

Building the Framework for the Parachute Chamber

The parachute chamber is the part of the device that holds the servo, electronics, and parachute spring. It will slide into the parachute capsule, described in the next section. You will need some foam core or corrugated plastic, a soda bottle, and some zip ties or twist ties. The mechanical parts are shown in Figure 12-10, after cutting the parts from the foam core and soda bottle.

Figure 12-10. *Parts for the parachute chamber. The green spring is cut from a 7 Up soda bottle.*

Start by cutting the spring. This is a section cut from the side of a two-liter bottle. The one shown was cut from a green 7 Up bottle rather than a clear soda bottle to make it a little easier to see in photographs.

It's pretty easy to make large, straight cuts on soda bottles using office scissors. Sewing scissors are a poor choice. It's not that they won't cut the plastic; they cut the plastic really well. The problem is the consequences when the owner of the sewing scissors finds out you used the scissors for anything but fabric. Trust me, it's not something you want to experience. A hobby knife works well for small, detailed cuts, or for making slots to initially insert the scissors.

Using scissors or a hobby knife, cut the bottom from the soda bottle. Make another cut to form a 3 1/4" ring, then cut the ring to form an 8"-long section that is 3 1/4" tall. The door in Figure 12-10 shows the direction of the curl. It is 3 1/4" tall and 8" wide, although the curve of the bottle makes it look less than 8" wide in the photo.

The two circles and the rectangular piece are cut from foam core, available at almost any craft store or office supply store. Another great choice is corrugated plastic. It looks like corrugated cardboard used for shipping boxes, but it's made from thin plastic instead of brown paper. You can use other materials, too, such as 1/8"-thick balsa wood.

Use the leftover part of the bottle as a template to draw two circles on the foam core. Cut these circles so they just barely fit inside the bottle. They should slip in and out, but be fairly snug.

From the foam core, cut another piece that measures 3 1/2" by 7". Cut it down the middle to form two 3 1/2"-square panels, but only cut the foam core halfway through. Later, you will bend the foam core to create the walls of the parachute chamber.

Use the pattern in Figure 12-11 to drill eight 1/8" holes in the rectangular piece of foam core. Drill eight matching holes in the spring following the pattern in Figure 12-12. Use a piece of scrap wood behind the plastic or foam core to get a nice, clean hole, as shown in Figure 12-13. These holes will be used later to fasten the spring to the foam core.

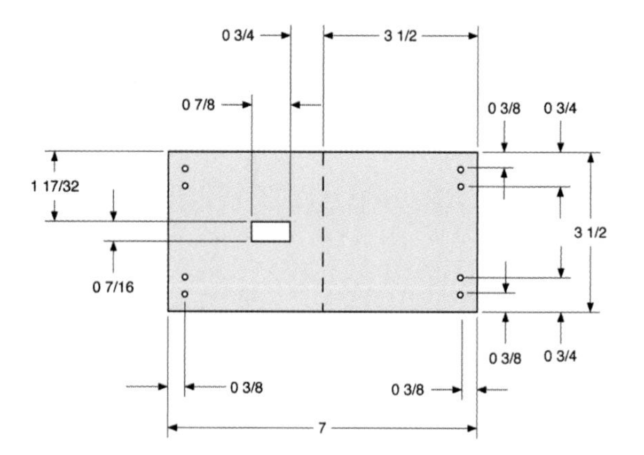

Figure 12-11. *Pattern for the parachute capsule walls. Cut from foam core or similar material. All round holes are 1/8". The rectangular cut should fit the particular servo you use. Cut halfway through on the dashed line and fold to form a hinge.*

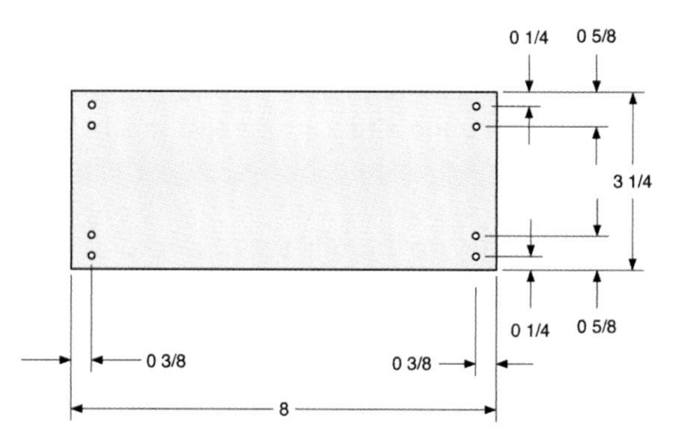

Figure 12-12. *Cut the parachute capsule spring from a two-liter soda bottle. All holes are 1/8" in diameter.*

Figure 12-13. *Use a piece of scrap wood under the foam core or soda bottle when drilling. This gives a stiffer back while drilling, resulting in a cleaner hole.*

Drill a 1/4" hole 1" from the edge on one of the disks. You will use this hole to feed the wire from the servo up from the parachute chamber to the top platform where the electronics will sit.

Use a hobby knife to cut a rectangular chunk from the side of the foam core. This hole should be the same size as the base of the servo. The pattern in Figure 12-11 shows a hole that is the perfect size for the SparkFun servo in the parts list. Adjust the size to fit the particular servo you have, but check the clearance on the servo to make sure you will be able to fold the hinge to about a 60° angle without hitting the servo.

I like to use a hot glue gun for assembling foam core. It's fast and easy, and forms good joints. Almost any glue will work if you're willing to wait for it to dry, though. Whatever you pick, glue the rectangular piece to the disk without the hole as shown in Figure 12-14. The edges of the walls should just touch the outside edges of the disk.

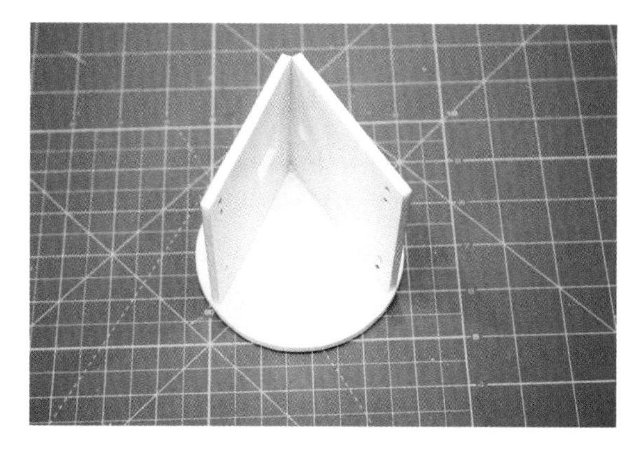

Figure 12-14. *Glue the walls to the bottom disk so the edges of the walls just touch the edges of the disk at the ends and the hinge.*

Glue the top platform in place with the hole centered near the fold in the walls.

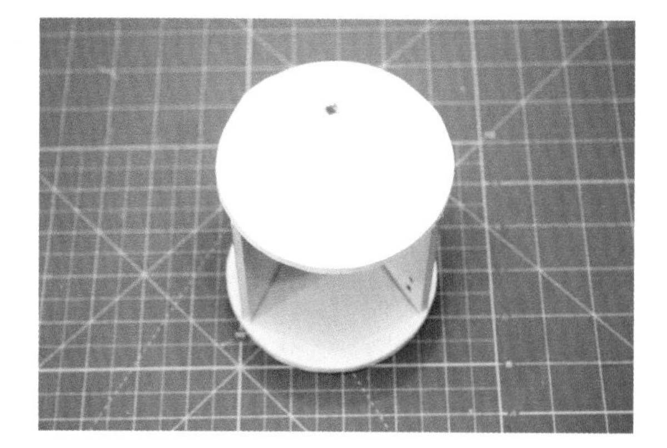

Figure 12-15. *Glue the top platform in place so the hole is near the hinge in the walls.*

Building the Parachute Bay

The parachute bay is the outer enclosure that holds the parachute chamber. It forms the nose of the rocket. A door on the side will pop open to release the parachute. We'll eventually cut some access holes so the interior components can be manipulated without taking the rocket apart. You will need a soda bottle with a flat side, a Ping-Pong ball, and some tape to build the parachute bay. You can get by without the Ping-Pong ball.

Use a hobby knife to cut the top portion off of the soda bottle. Use scissors or a hobby knife to cut off the base. Cut the base off just above the curved part, leaving as much of the flat side of the bottle as possible.

Cut the Ping-Pong ball in half and glue it in the forward portion of the bottle to form a nose cone. While this is a nice touch, it's not essential. You could just leave the original top on the bottle and screw the lid back on.

Figure 12-16. *Cut the top and base from a soda bottle. Cut a Ping-Pong ball in half and glue it inside the nose to form a nose cone.*

The next step is to cut the door. Since bottle sizes and shapes vary a bit, it's tough to give precise measurements for the position of the door. We'll use the partially completed parachute chamber as a guide.

Slide the parachute chamber into place. Push it as far forward as possible. Once it is in place, use a hobby knife to cut the door. Cut the top and bottom sides first, right up against the bottom edge of the top disk and the top edge of the bottom disk. Cut the sides of the door right up against the outside of the opening formed by the walls. Figure 12-17 shows a close-up of the finished door, showing both the size and the way it will look when completed.

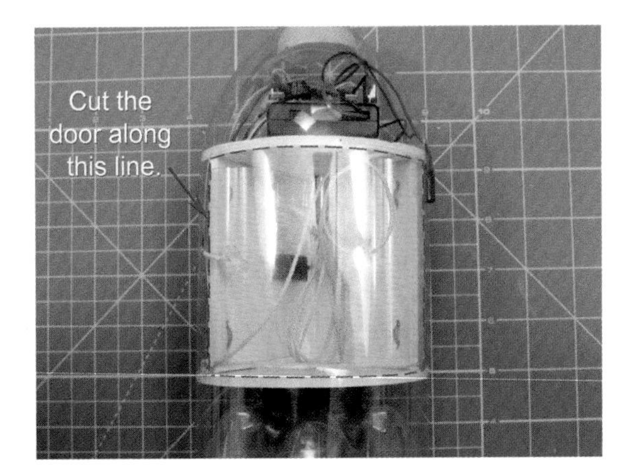

Figure 12-17. *Looking ahead to the finished module helps visualize the door. The top cut is along the bottom of the top disk, the bottom cut is along the top of the bottom disk, and the left and right cuts are on the outside edges of the walls.*

Push the parachute chamber out of the parachute bay.

Drill a 1/8" hole in the left side of the door, centered vertically and about 3/8" from the left edge. This is for the string that holds the door closed.

Use a piece of wide tape to tape the right side of the door back in place, forming a hinge for the door. Add an extra layer of wide tape to the inside for a more durable hinge.

Set the parachute bay aside while we finish the parachute chamber.

Adding the Spring and Electronics to the Parachute Chamber

Push the servo into place and secure it with the hot glue gun. While this will attach it securely, the hot glue can be peeled off with a fingernail if you ever tear the parachute chamber down and repurpose the servo. Some people advocate using bolts to mount the servo, but I think foam core is a bit too soft to hold bolts well. It probably won't matter if the fit is snug, though.

Feed the wire for the servo through the hole in the top platform.

Attach the spring to the walls using wire ties as shown. Twist ties like the kind used on many garbage bags work well, too.

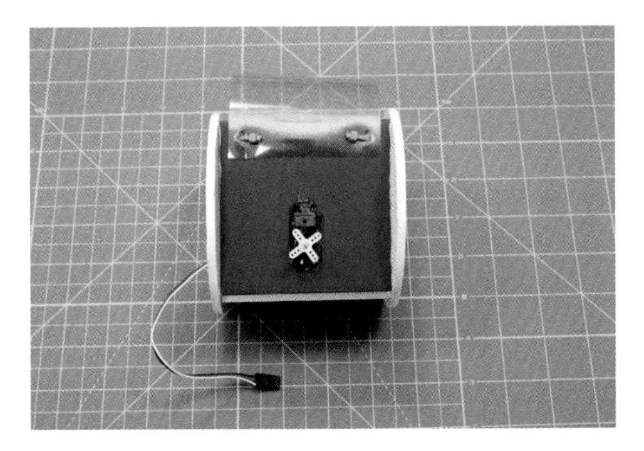

Figure 12-18. *Glue the servo in place and thread the connector through the hole in the top disk. Attach the spring using wire ties or twist ties.*

Insert the batteries into the battery holder and attach the battery holder to the top platform using mounting tape. Mounting tape is thick, double-sided tape designed for mounting pictures or fastening small objects to the wall.

The battery holder has a screw that will fasten it closed. Leave it out, but add a piece of electrical tape to hold the battery holder closed. This leaves a chance that the battery holder will pop apart on a hard landing, but if you leave the screw out, you won't have to remove the battery holder from the top platform to change the batteries. It's a lot easier to pull the parachute chamber from the parachute bay to reattach the battery holder than it is to peel the battery holder from the foam core once it is attached with mounting tape. You also risk tearing the foam core if you have to remove the battery pack, which might force you to rebuild the whole thing.

Use another piece of mounting tape to fasten the LaunchPad microcontroller to the battery holder. Be sure to leave the on/off switch on the battery holder exposed.

Figure 12-19. *Mount the battery holder and microcontroller to the top platform with mounting tape.*

Connect the battery holder to the microcontroller using the black and red female jumpers. Connect the red jumper to the pin labeled TP3; this is the positive power input, located just below the USB connector on the right. Connect the black wire to the ground pin, labeled TP1. This is just to the left of TP3.

Figure 12-20. *Connect the red wire to TP3 and the black wire to TP1.*

Connect the servo to the wiring harness. This time there are three wires to connect. Most servos have a black, a red, and a white wire. The black wire is the ground, or negative power wire. The red wire is the positive power wire. The white wire carries the digital signal that tells the servo which position to turn to. If your servo has wires with different colors, check your data sheet to see which wires carry negative power, positive power, and the signal, and connect it accordingly.

The black and red wires from the servo are the power wires, which get connected to the black and red wiring harnesses using the male connectors. The white wire carries the digital signal telling the servo how far to turn. Connect this wire to pin P1.4 on the microcontroller using a male-to-female jumper wire. The connection of the signal wire to pin P1.4 isn't done with any of the wiring harnesses; this connection is made with an unmodified male-to-female jumper wire. The wire shown for this connection is yellow, since the package of jumper wires I bought did not have a white jumper wire. Use a small piece of mounting tape or cellophane tape to fasten the servo and male jumper connections to the top disk so it doesn't bounce around during launch. We don't want those connections to break!

Figure 12-21. *Connect the servo to its power source by connecting the black and red pins from the wiring harness to the black and red wires of the servo. Add another jumper to connect the white servo wire to pin P1.4 on the microcontroller.*

Connect one end of the female-to-female jumper to pin P1.7 on the microcontroller. Leave the other end loose for now.

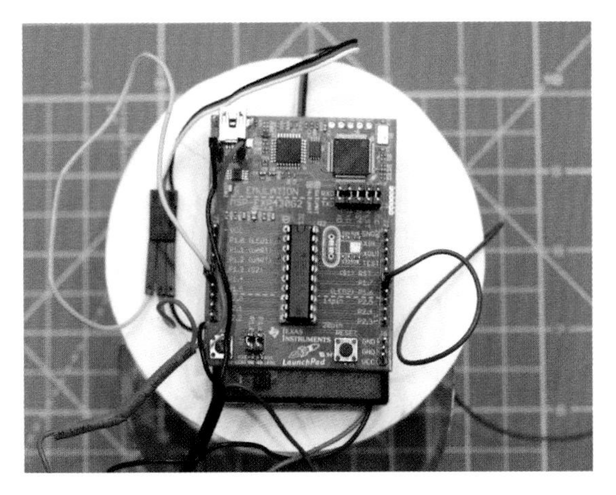

Figure 12-22. *Connect one end of the female-to-female jumper to pin P1.7 of the microcontroller.*

The wiring is complete, but there are still two loose female connections. These will form the launch detection switch. Stretch the two unconnected female jumpers to the side, pull the remaining wires on top of the microcontroller, and use a piece of cellophane tape to tack the loose wires down. This makes it less likely they will bounce around during flight or landing.

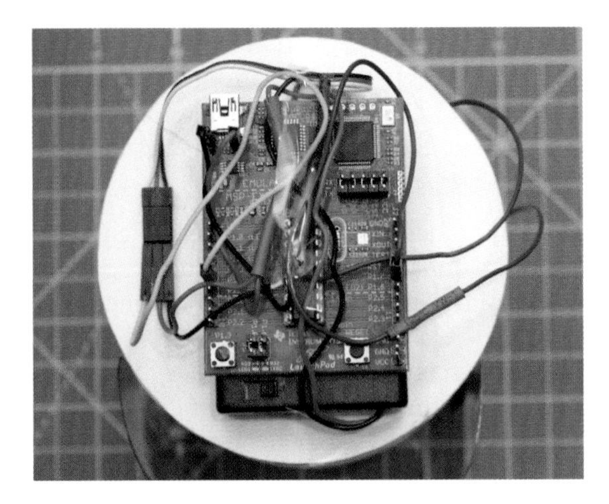

Figure 12-23. *Tape the loose wires to the microcontroller.*

We'll need some way to connect the parachute. Cut a 5-foot-long piece of string from a roll of #18 mason line. This is the same string as the launch string for the water rocket launcher.

You might notice there is no hole for mounting the shock cord. Foam core and thin plastic are so weak that the string might tear through the material. Instead, wrap the string completely around the bottom of the walls, as shown in Figure 12-24. Form a loop at the other end. Use a small dab of glue on each knot to keep them from coming undone. An alternative many people use is to wrap the shock cord around the outside of the rocket at the center of gravity, then tape it in place. That's a little easier to build, but adds a little drag. As it turns out, the center of gravity on this rocket is at the bottom of the parachute door, so it works just as well to wrap the shock cord securely around the foam core walls.

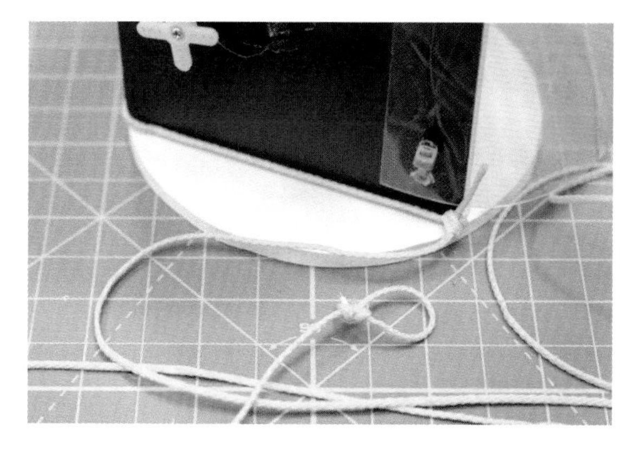

Figure 12-24. *Tie a 5-foot piece of #18 mason line around the bottom of the parachute chamber. Form a loop at the other end. This is the shock cord.*

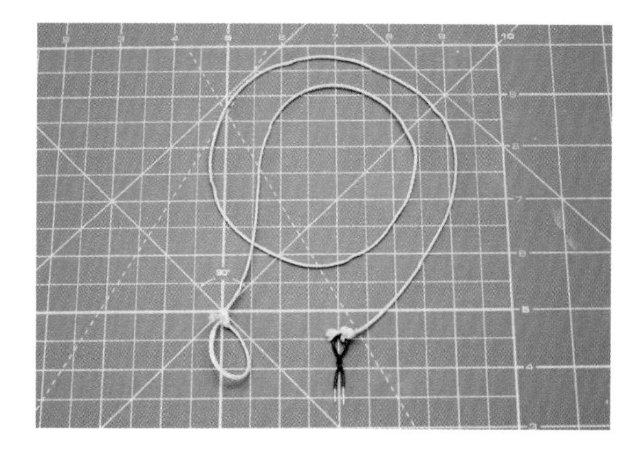

Figure 12-25. *The launch control switch connects the two remaining female jumpers. It is tied to the launcher so the wire pulls out, breaking the connection when the rocket lifts off.*

While the glue and string are out, let's make the other half of the launch detection switch. This is basically a wire that connects the two female jumpers. The wire is tied to the launcher so it will be pulled out when the rocket lifts off. The microcontroller will see an open circuit and start the timer. Once the preprogrammed time elapses, the microcontroller will tell the servo to rotate, releasing the parachute.

Cut a piece of 20- or 22-gauge wire about 2" long and strip the ends to about 1/4". Twist the wire to form an eye at one end with two parallel connectors about 1/10" apart, as seen in Figure 12-25.

Cut a 2-foot length of #18 mason line. Tie one end to the loop in the wire, and form a loop in the other end about 1" in diameter. Apply a spot of glue to both knots to keep them from coming loose.

Programming the Servo Position and Delay Time

The ServoChron software lets you select parachute deployment times for two servos. We're only using one servo, but you still have to program reasonable times for both. The software will not work properly if you don't put in a time for the second servo, or if the time you do put in makes no sense. Keeping that in mind as you work through the programming will save you a lot of confusion.

The software is programmed with the two buttons at the bottom of the microcontroller. There are two LEDs, one red and one green, that give you an indication of what is happening inside the microcontroller. Locate these in Figure 12-26 and on the LaunchPad microcontroller you have installed on the parachute chamber.

There is a third LED near the USB connector. This is a power indicator. It isn't used for programming.

Figure 12-26. *The two buttons at the bottom of the microcontroller are used to program the delay times. The two LEDs show you what is happening.*

There are two steps to programming the microcontroller. The first step defines where the servo should be at launch time and how far it should rotate to release the parachute. The second step sets the delay time. Read through the directions for each step carefully, making sure you remember them before starting. You will want to be looking at the LaunchPad microcontroller during the programming process, not looking back and forth between the book and the microcontroller. It also works well if you work with a second person. One of you can read the instructions out loud while the other programs the microcontroller.

We'll begin by programming the servo position. Hold down the mode button while you turn on the power. The red LED will light, showing that you are setting the position of the first servo. This is the servo we're using. Tap the mode button to change the position of the servo. It will rotate in 30° increments. Once the servo gets to the end of its range, it will start rotating back in the opposite direction. We're going to want the servo to rotate in a clockwise direction to release the parachute spring. Tap the mode button to go through the range of motion until you are sure which position is the farthest in the counter-clockwise direction. Once you get there, press and hold the mode button for at least one second, then release it. This sets the launch position for the servo. The red LED will go dark at this point.

Continue tapping the mode button until the servo is rotated all the way in the clockwise position. Press and hold the mode button to select this position as the release position for the parachute.

The green LED will light, telling you it is time to set the position for the second servo. Press and hold the mode button for at lest one second to set the launch position for the second servo. Tap the mode button once so the release position is not the same as the launch position, then press and hold the mode button for longer than one second to finish the programming process.

Once you successfully set the positions for the second servo, the servo you're actually using will spin back to the launch position.

ServoChron Position Programming Checklist

1. Press the mode button while you power on. The red LED will light.
2. Tap the mode button until the launch position is reached.
3. Press and hold the mode button for 1 second to set the launch position. The red LED will dim.
4. Tap the mode button until the release position is reached.
5. Press and hold the mode button for 1 second to set the release position. The green LED will light.
6. Repeat the process for the second servo, which is unused in this project.

The next step is to set the ejection delay. Turn on the power if it is not already on. The LEDs will blink once per second. Feel the rhythm. It's how we will set the delay time.

Push and hold the mode button. While you keep it pressed down, tap the reset button. This starts the timer for the first servo. Themis needs about a two-second delay time. Count off two flashes, and then release the mode button. Count off a couple more flashes, and then tap the mode button again. This sets the delay time for the second servo.

The LEDs will start to blink. This tells you the delay times. The red LED will blink until the deploy time is reached, and then it will stay lit while the green LED continues to count off the delay for the second servo. Once the second timer elapses, the LEDs go dark for a moment, and then the process repeats. It's a handy way to check your work, but it's also nice when you are attaching the parachute module to a rocket. You can visually double-check the delay time before launch.

Check the delay time now to be sure it's two seconds. If not, repeat the process until you get a two-second delay time on the first servo.

ServoChron Timer Programming Checklist

1. Press and hold the mode button.
2. Tap the reset button to start the timer for the first servo. The red LED will flash once per second.
3. Release the mode button to set the timer for the first servo.
4. Tap the mode button to set the release time for the second servo.

Once you finish programming the servo, turn on the power so the servo rotates to the launch position. Remove the control horn. Put it back on the servo so the topmost bar is rotated to the left, forming a slightly lopsided X, as seen in Figure 12-27. The string that will eventually hold the parachute door closed will sit in the back corner of the X during launch. When the timer fires, the servo will rotate clockwise, releasing the string so the door can pop open.

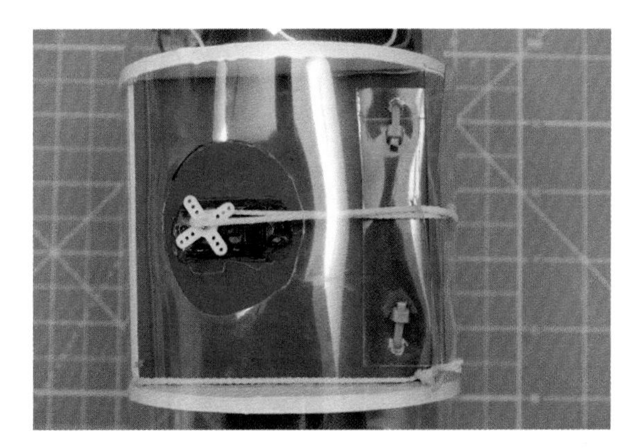

Figure 12-27. *With the servo in the launch position, remove the horn and reattach it in the orientation shown. The V on the left holds the launch bay door latch string securely until it is time to deploy the parachute. The photo shows the string in place so you understand why the servo should be positioned as it is, but you have not installed the string yet.*

Finishing Touches

We still need to attach the string that will run between the parachute bay door and the servo, cut a couple of access holes, and fasten the female headers that form the release switch.

Slide the parachute chamber into the parachute bay. You will need to press the spring into the chamber to get the assembly into position. Once it reaches the right position, the spring will pop out, forcing the parachute bay door open.

Mark off a large access hole around the servo horn. The one shown in Figure 12-27 is about 1 3/4" in diameter. Give yourself room to work. If you like, you can save the piece you cut out and tape it in place after getting the door ready for launch. I just leave the access port open.

Cut an 8"-long piece of mason line to serve as the door latch. Feed it through the hole in the door and tie a knot so the string is just long enough to reach the V in the control horn. Test the fit to make sure the string is the perfect length and the access hole is large enough to make it easy to attach the string.

Use a hobby knife to cut a hole in the nose just above the on/off switch on the battery holder. Make the hole just big enough to shove a pencil through. This is how you will turn the power on before the launch.

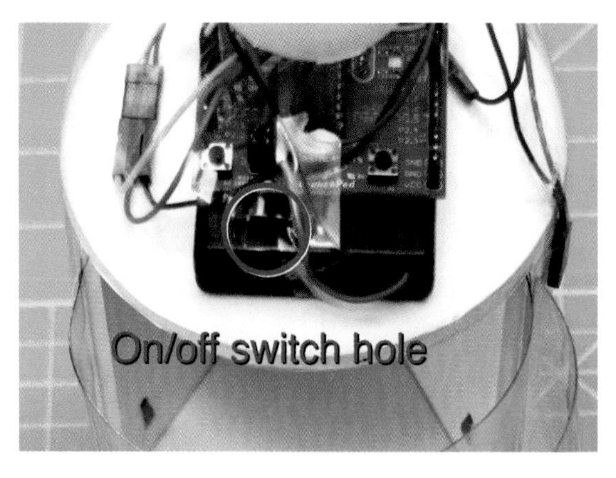

Figure 12-28. *Cut an access hole just above the on/off switch.*

Cut another hole just above the top of the platform and to the right of the door. Pull the parachute chamber down a bit to give yourself room to work, then feed the two female launch detection switch wires through the hole. Slide the parachute chamber back into place and tape the connectors to the side of the rocket, pointing the open end of the female connectors down.

Figure 12-29. *Cut an access hole to the right of the door. Feed the launch detection connectors through and tape them in place.*

The completed parachute module is ready to install on Themis, or any other water rocket based on a two-liter bottle. Use 3/4"- to 1"-wide electrical tape or duck tape to attach the parachute bay to the rocket. Both are available in a variety of fun colors.

Figure 12-30. *The completed parachute bay, ready for installation.*

Flying Themis with a Parachute

Stick a pencil through the access point over the on/off switch and power on the system. This will rotate the servo to the launch position. Pack a 24" parachute, powdering it just as you did for Juno (see "Packing the Parachute" on page 123). Attach the parachute to the loop on the shock cord. Shove it into place, pushing the spring into the parachute chamber and wrapping the door latch string around the servo. If the fit is a little tight, stick the pencil back into the access hole and push the mode button. This will rotate the servo to the release position. Slip the string over the control horn and release the mode button. Releasing the mode button rotates the servo back to the launch position.

Prepare the booster just as you did back in Chapter 11 when flying Themis without a parachute.

Connect the Launch Detection Switch Before Turning On the Power

The ServoChron software starts the timer when the launch detection switch changes either from open to closed or closed to open, so follow the directions in the next few steps carefully. If you forget to turn the power off before connecting the switch, the timer will count down and fire, dumping your carefully packed parachute out onto the launch pad!

Attach the string for the launch detection switch to the base of the launcher. Stick the loop through the screw eye on the launcher base, then feed the other end through the loop to form a secure attachment.

Figure 12-31. *Stick the loop of string through the screw eye on the launcher, then feed the loose end through the loop. Draw it tight for a secure attachment to the launcher.*

Turn the power off on the LaunchPad. Plug the two wires from the launch detection switch into the female headers taped to the parachute bay, then stick a pencil through the access hole and turn the power back on. The red and green LEDs will start to blink out the delay times. Make sure the delay time for the red LED is two seconds.

Figure 12-32. *Turn the power off, insert the launch detection switch, and then turn the power back on.*

Clear the launch area and pressurize the rocket.

You're ready for flight! Check the area, check the airspace, and count down to launch.

Here's a checklist that summarizes the flight preparation.

Flight Checklist

1. Clear spectators from the launch zone.
2. Pack the parachute.
3. Fill the rocket about 1/3 full of water.
4. Slide the rocket onto the launch tube.
5. Arm the parachute:
    - Turn the power off.
    - Attach the launch detection switch.
    - Turn the power back on.
6. Pressurize the launcher to 60–75 psi.
7. Check for clear skies—no aircraft (or geese!).
8. Count down from five and launch.

Subsonic Aerodynamics

13

A lot of rocketry is what we call *sport rocketry*. We're doing it for fun. At some point, though, you will want to do better. Maybe there is a local club competition coming up, a race for the Highest Altitude or Longest Duration Aloft Under a Parachute, and you want to compete and do well. Maybe you're getting ready for a more serious competition, like the annual high school TARC competition, or NARAM (the National Association of Rocketry Annual Meet). Maybe you just want to see how much you can coax from your rockets. Whatever the reason, you want your rocket to fly higher or faster. That's when you need to get a basic understanding of aerodynamics.

This chapter is a brief introduction to aerodynamics as it applies to high-performance model rockets. For the most part, we'll look at drag—the force that opposes the motion of the rocket as it moves through the air. Minimizing drag is one key aspect of improving the performance of a rocket. In the next chapter, we'll put this to use to see how to predict the altitude of a model rocket.

You can get a PhD concentrating on aspects of aerodynamics that we'll gloss over in a sentence or two, so don't expect this to be a complete picture of how the atmosphere works. Think of this chapter as a good start. There are some references at the end of the chapter that expand on some of the topics covered here, so if you are one of those folks for whom aerodynamics is a gas, not a drag, be sure to look some of them up.

Feel free to skip this chapter if you are new to rocketry, or just don't find this topic interesting yet. You can always come back to it later. Missing the information in this chapter won't keep you from building and flying any of the rockets in this book. It won't even keep you from designing your own rockets. The information in this chapter will help you get better performance out of the rockets you build, though.

Reynolds Number and Viscosity

Let's start off by getting an understanding of the different kinds of fluid flow, and exactly where model rocketry fits into the bigger picture. In fact, let's start off by talking about why we're even discussing fluid flow. After all, our rockets travel through air, not water, and air is a gas, not a fluid, right?

Well, yes and no. It turns out that liquids and gases work pretty much the same way when we're talking about aerodynamics. Many of the ideas we'll discuss here apply whether we're designing a submarine or a rocket. Aerodynamics is all about fluid flow, and gas is just one of the fluids. As you look around for references about drag on a rocket, you're looking for articles and books about fluid flow.

The real difference in fluid flow isn't whether the fluid is a liquid or gas. The difference is something called the *Reynolds number*. The Reynolds number describes the relative importance of *inertial forces* and *viscous forces* within a fluid. Inertial forces occur because the molecules of the fluid hit and bounce off of the object moving through the fluid. At the microscopic level, it's like billions of tiny marbles smacking into the object, then bouncing off it, and also bouncing into each other. Viscous forces occur because the air molecules—the "marbles" in our metaphor—aren't really smooth, frictionless spheres. Think of them as covered with double-sided sticky tape. Sometimes they stick to each other, or to the object they impact with. That stickiness causes the object to trail along more molecules than it normally would.

Mathematically, the Reynolds number is:

$$R_e = \frac{inertial\ force}{viscous\ force} = \frac{\rho\, v\, L}{\mu}$$

Let's pull this apart to see how it works. As we do this, think about lots of tiny, sticky marbles hitting the rocket as it moves, and imagine what they do to slow the rocket down.

The first term in the inertial force is the density of the fluid, ρ. This term says it takes more force to move an object through the fluid if the fluid is denser. In our case, it will take more force to move a rocket through the air at sea level than in the upper atmosphere, where the density of the air is thinner.

Next is the velocity, v, of the object as it moves through the fluid. This term tells us that the faster we go, the more force we will need.

Finally we have L, which is a characteristic dimension of the body as it moves through the fluid. This tells us that bigger objects need more force to move through a fluid than smaller ones.

The viscous force, μ, is in the denominator. This accounts for the stickiness of the fluid particles. Air isn't very sticky compared to water, which will cling to most surfaces just a bit. Water isn't as sticky as honey, which will stick so well you can use an open stack of disks to pull it from a jar.

Figure 13-1. *Honey is so viscous that a wooden stack of disks can pull it from a jar. That certainly won't work well with water, which is much less viscous than honey!*

Viscosity can be measured. The units are *pascal-seconds*, abbreviated as Pa•s. In more familiar units, this is kg/(m•s). Table 13-1 shows the viscosity for some common fluids at about 20°C.

Table 13-1. Viscosity of fluids at 20°C

Fluid	Viscosity
Air	1.83×10^{-5}
Water	1×10^{-3}
Honey	1,000

For rocketry, we're interested in fluid flow for high Reynolds numbers, where the inertial forces are far, far more important than the viscous forces. In fact, they are so much more important that for the most part we just ignore viscous forces.

Mach Number

The *Mach number* is the speed measured in units of the speed of sound. An object traveling at Mach 1 is traveling at the speed of sound. Mach 0.5 is half of the speed of sound, while Mach 2 is twice the speed of sound. Let's see why this is important.

If you are new to aerodynamics, you probably think the best way to reduce drag is to use pointy objects that will slip through the air easily. That's backed up by looking at the shape of things that move really, really fast, like the X-15 high-altitude rocket plane, which still holds the speed record for a manned aircraft (4,510 mph, set back in 1967!). Pointy shapes poke through the air better than blunt ones, right?

Figure 13-2. *Neil Armstrong poses in front of an X-15. Notice that the leading edges of the flying surfaces are sharp, which is common of high-performance supersonic aircraft.*

Well, yes and no. It turns out that air behaves very differently depending on how fast you are going. If you are moving a lot faster than the speed of sound, which is about 340 m/s in air at sea level, pointy shapes work a lot better than blunt ones. If you are traveling at less than the speed of sound, though, an airfoil shape works better. And then there's the transonic region, from about Mach 0.8 to Mach 1.2. Things are just plain weird at those speeds, as the turbulence of subsonic speed starts to mix with the shock wave of supersonic travel. You can usually tell when a model rocket passes through the transonic region. The smoke trail will suddenly wiggle, as does the rocket. The buffeting is severe enough that most rockets built to travel above Mach 0.8 are built from fiberglass or carbon fiber. These rockets are high-power rockets or very high-performance mid-power rockets, though, not the rockets covered in this book.

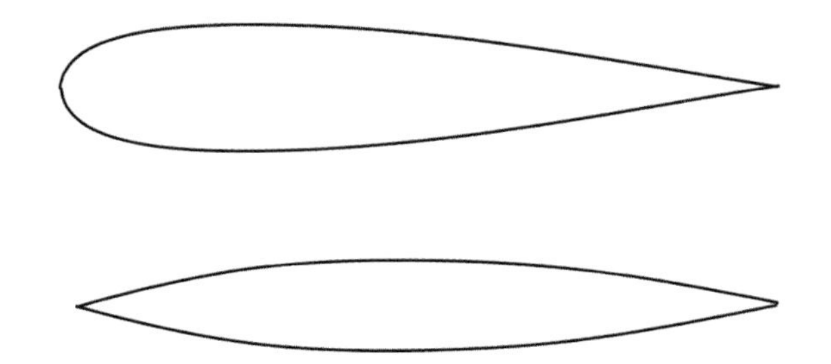

Figure 13-3. *At subsonic speeds, airfoil shapes like the top fin cross section shown here slip through the air with less drag than pointy ones. The bottom airfoil is appropriate for supersonic speeds. Thicknesses are exaggerated for clarity.*

The Speed of Sound

The speed of sound isn't a constant. It varies depending on what the sound is traveling through. In air, the speed of sound also depends on the temperature. Sound travels faster in hot air, where the molecules move more quickly than in cold air.

At 15°C, or 59°F, the speed of sound is about 340 m/s, 1,116 ft/s, or 761 mph.

High-power rockets can definitely break the sound barrier. With currently available motors, though, model rockets with an E or smaller motor will not break the sound barrier. That's why this chapter will ignore supersonic and transonic flight.

To summarize, then, the information in this chapter deals with high Reynolds number, subsonic aerodynamics. Low Reynolds number aerodynamics and supersonic flight are interesting topics you can explore (see the references for some pointers), but they are topics we don't need to deal with for this class of model rockets.

High Reynolds Number Drag

The force of drag at high Reynolds number subsonic flow is:

$$D = \frac{1}{2} C_d \rho v^2 A$$

Carpe Fudge

C_d is a fudge factor because it's accounting for a lot of real effects that are generally considered too complex to reduce to an exact form. For example, Newtonian physics with relativistic corrections (at least in a drag-free environment like space) is absolutely perfect as far as can be measured. Thermodynamics—the study of massive numbers of particles—is literally a statistical average. The drag equation is even further from precise reality than thermodynamics. We know it's wrong. It's just the best we have, and it is useful. C_d is the fudge factor that hides all of the effects that are too small to show up as other terms in the equation.

The C_d term is called the *coefficient of drag*. This is a fudge factor that accounts for the shape of the object. The drag on a flat plate is different from the drag on a streamlined airfoil, for example. We'll dig deeper into the coefficient of drag in a moment.

The density of the air is ρ. Since you lose about 20% of the air density by going from sea level to 5,000 feet, this tells us drag is about 20% higher at sea level than it is where I live, in Albuquerque. This makes some sense. More air means more drag.

The v^2 term tells us that drag goes up as the square of velocity. That's really important, since our rockets tend to go fast. It tells us that, all other things being equal, we need to increase the thrust by about four times if we want to double the speed.

Finally, the force of drag goes up with the area presented by the rocket. The A term is the *frontal area* of the rocket. For most rockets, that's the area of the cross section of the largest body tube plus the area of the fins and launch lug. This term tells us to keep the rocket skinny if we want to reduce drag.

Kind of a Drag: The Three Types of Drag

A lot of information is wrapped up in the drag equation and coefficient of drag. Before moving on to look at drag on individual rocket components, let's take a moment to understand where the drag comes from.

The Cross Section of Two Rockets

Juno is the rocket we built in Chapter 3. Hebe is a high-performance rocket we'll build in Chapter 15. The drag equation already tells us the source of the biggest difference in performance between these rockets.

Juno is built around a BT-50 body tube, which has a diameter of 0.976". Using the equation for the area of a circle gives the characteristic area, A, from the drag equation:

$$A = \pi \frac{d^2}{4} = 0.748 \text{ in}^2$$

To cut drag, Hebe will be built around the smallest body tube that will house a standard motor, the BT-20. The diameter of the BT-20 is 0.736". This gives an area of 0.425 square inches. That's just 57% of the frontal area of Juno, cutting the drag almost in half!

Pressure drag, also called *form drag* or *profile drag*, is the kind of drag you would expect from the "billions of marbles" model we started with at the beginning of the chapter. The air molecules hitting the rocket and bouncing off cause pressure drag. If the rocket is sitting still, about the same number of molecules are hitting it from every direction—top, bottom, side, front, back—so the pressure drag is zero. As soon as the rocket moves, though, the pressure becomes imbalanced. The molecules hitting the front edges of the rocket pick up a little of the rocket's momentum as they bounce off, causing the rocket to slow down. There is also a negative pressure drag that pulls on the back of the rocket. As the rocket picks up speed, it outruns more and more air molecules, and the ones that do hit the rear of the rocket hit at a lower relative speed. The imbalance between the molecules smacking into the front of the rocket, but molecules not being able to smack as hard into the back, causes a sucking motion that literally pulls the rocket backward.

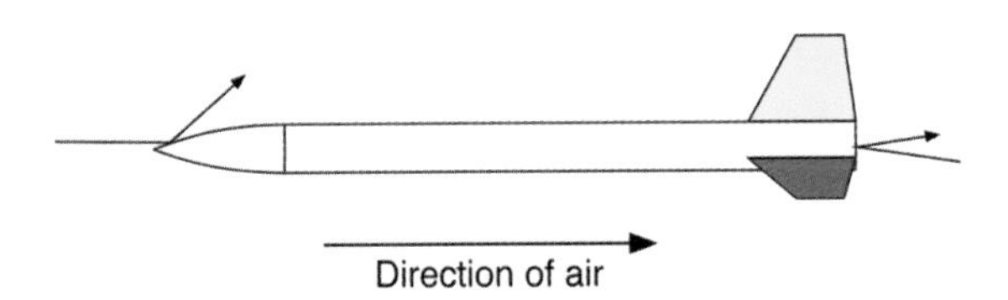

Figure 13-4. *Molecules hitting the front of the rocket deflect off, pushing the rocket back a bit. This is the source of pressure drag. At the rear of the rocket, the molecules may still reach the rocket, but they are traveling slower compared to the speed of the rocket, and don't impart as much energy. This creates a partial vacuum that literally sucks back on the rocket.*

The air flowing parallel to the rocket, such as along the body tube or the flat surface of a fin, causes *friction drag*. Even though viscosity is relatively unimportant for a rocket traveling in air, it does still have some effect. The molecules of air right next to the surface of the rocket actually stop moving. The molecules next to them tend to slow down a bit, aided by the slight stickiness caused by the electromagnetic properties of individual air molecules. This forms a thin region of air around the body tube called the *boundary layer*, where the air molecules are moving more slowly than the surrounding air. The air in the boundary layer can even swirl around a bit. Smooth-flowing air in the boundary layer is called *laminar flow*, while turbulent air is called *turbulent flow*. The boundary layer is the source of friction drag, which is greater if the boundary layer is turbulent—although there are some odd cases where turbulent flow can actually reduce the overall drag on an object.

If the rocket tips ever so slightly, the fins work to straighten it out. The pressure on one side of the fin is higher than the other, pushing the rocket back toward straight flight. Right at the edge of the fin, though, some of the high-pressure air leaks around the edge, starting a twirling motion of air that creates a vortex, like a little tornado. The energy to set up that motion comes from somewhere, and of course causes drag. This is called *induced drag*. It won't happen at all on a rocket with perfectly straight fins, which flies perfectly straight for the entire flight. Since there's no such rocket or trajectory in real life, this means that induced drag happens all the time. While we don't usually include induced drag in our drag calculations, since we're assuming straight flight, it's still an important aspect of drag that impacts the design of fins.

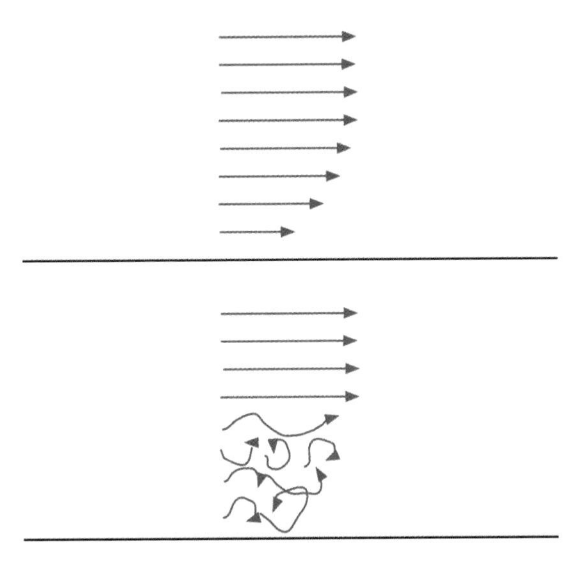

Figure 13-5. *Air flow slows down and finally stops right next to the body tube in what is called the "boundary layer." If the rocket is smooth enough, the flow is laminar. Rougher surfaces or higher Reynolds numbers change the boundary layer to turbulent, with higher drag.*

Figure 13-6. *The vortex caused by induced drag is striking in this photo of a crop duster. The same thing happens on a rocket fin when it creates lift to straighten a rocket in flight.*

If you dig around a bit, you will find that different authors divide drag in different ways. This is usually because they are concerned with drag on a particular kind of object. Ships, for example, deal with wave drag. Airplanes and rockets traveling near the speed of sound discuss interference drag. The three kinds of drag listed here are convenient for discussing subsonic drag on rockets, but don't be surprised when you dig into the literature if you find other terms being thrown about. Take the time to make sure you understand what each author means by each term.

The Importance of Streamlining

So far, the discussion of drag has been pretty academic. We've spent our time discussing equations and what they mean and defining various kinds of drag. Now that we have the vocabulary to talk about aerodynamic drag, let's stop and take a look at what it can mean for various shapes. Figure 13-7 shows some common shapes and their coefficients of drag. If the frontal area, density, and velocity are the same, the coefficient of drag is proportional to the overall drag.

Let's concentrate on just three of these shapes to get an idea of how streamlining can affect our rockets. The short cylinder is essentially a short body tube with a flat nose cone. It has the highest coefficient of drag of any shape in the figure. It makes sense that something with a flat front would be the worst shape in terms of minimizing drag.

Rounding the front and back to form a sphere reduces the coefficient of drag from 1.15 to 0.47. We've reduced the force of drag by a factor of 2.45! That seems pretty good. But compare that to a streamlined body, with a coefficient of drag of 0.04. By properly shaping the rocket, we can cut the drag from 1.15 to 0.04, reducing the force of drag by a factor of 28.75. That's an amazing savings in terms of the energy needed to push our rocket through the air.

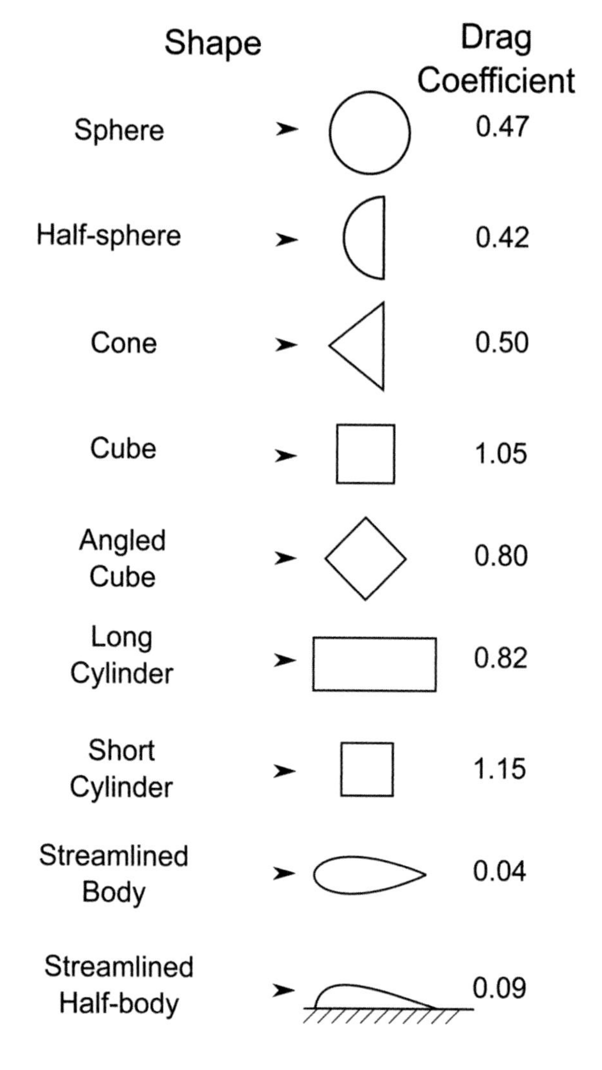

Figure 13-7. *Coefficients of drag for some common shapes.*

Understanding and Reducing Drag on a Model Rocket

There are several ways to find the coefficient of drag on a model rocket. By far the best is to put the rocket in a wind tunnel and measure the drag directly. There are all sorts of things that impact drag that are very difficult to estimate. What is the coefficient of drag of the finish on the rocket? How smooth are the fin fillets? Are the airfoils on the fins perfectly shaped? And so forth. The wind tunnel takes all of that into account.

The second-best way is to use a rocket simulator. Analyzing the drag on a rocket is a labor-intensive process, with lots of places to make mistakes. It's worth doing at least once to see how each component contributes to the overall drag, but if you had to calculate the coefficient of drag for each rocket by hand, you might not bother.

There's also a little detail most books don't talk about. While the coefficient of drag is roughly constant from about Mach 0.3 to Mach 0.8, it actually varies with speed. That's something hand analysis rarely takes into account, and we won't, either. Rocket simulators do take this into account, though. They may not do it perfectly, but by varying the coefficient of drag with speed as best they can, rocket simulators do a better job of estimating the drag during the flight of the rocket than we will try to do with direct mathematical calculations.

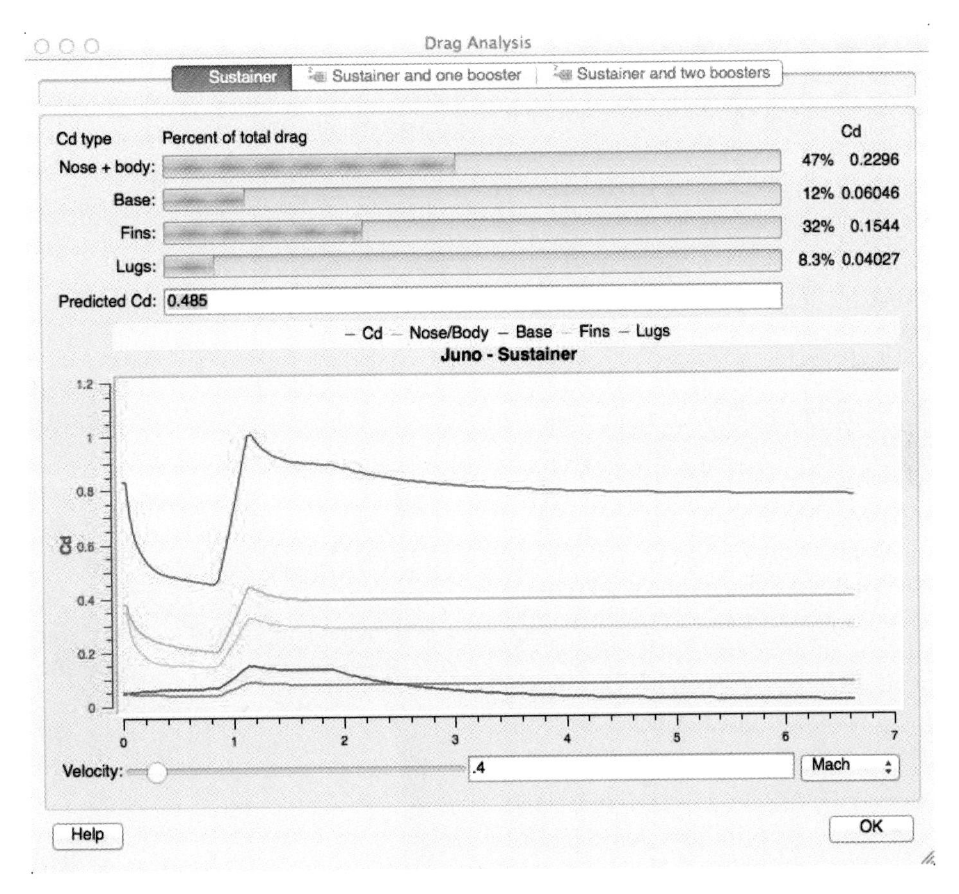

Figure 13-8. *This is the C_d analysis for Juno, generated by RockSim. It shows how the coefficient of drag varies with speed.*

The next-best way to calculate the coefficient of drag is to break the rocket into component parts and analyze the drag on each component. This is actually what the rocket simulators are doing; they just do it a lot faster than we can. This is also a great way to understand the sources of drag, and how to reduce the drag on your rocket.

We'll break our rocket up into the components shown in Figure 13-9. After finding the coefficient of drag on each part, we will combine the individual drag elements to get an overall coefficient of drag for the rocket.

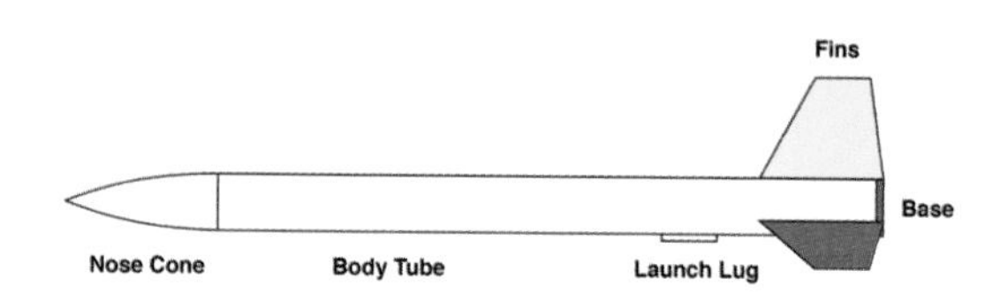

Figure 13-9. *For drag analysis, we'll split the rocket up as shown. The base is the blunt rear of the rocket, not a separate part.*

Nose Cones

Nose cones contribute to drag in two different ways. The first is pressure drag, which we'll deal with here. Nose cones also have friction drag, but we'll lump that in with the body tube friction drag calculation.

You might think that rockets have been around long enough and been studied carefully enough that the pressure coefficient of drag for common nose cone shapes would be very easy to look up from established sources. You would be wrong. The information used for most rocket simulations is drawn from a number of sources, and there is a good amount of estimation involved. That's especially true for rockets moving at subsonic speeds. We'll look at the coefficient of drag as it is calculated in OpenRocket for this section. It's a pretty good estimate for simulation, but this is certainly an area that could stand more research and wind tunnel testing.

The pressure drag varies enormously for different shapes of nose cones. Figure 13-7 tells us the coefficient of drag for a blunt tip is about 0.82. Some of the nose cones we'll look at have a coefficient of drag that is so low for subsonic velocities that OpenRocket simply ignores the pressure drag.

Conical nose cone

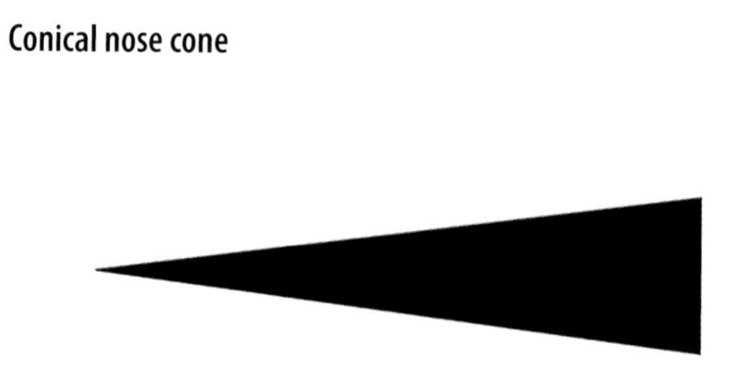

Figure 13-10. *Conical nose cone with a fineness ratio of 4.*

Let's start with a conical nose cone. It's easy to make, and appears on a lot of rockets. The cone can be long and slender or short and fat. The *fineness ratio* is the ratio of the length of the nose cone to the body tube diameter, and of course that has an effect on the coefficient of drag. For Juno, with a body tube diameter of 0.976" and a nose cone length of 2.75", the fineness ratio is 2.81. That's pretty typical; most nose cones have a fineness ratio between 2 and 4.

Figure 13-11 shows the approximate pressure coefficient of drag for conical nose cones with fineness ratios of 2, 3, and 4.

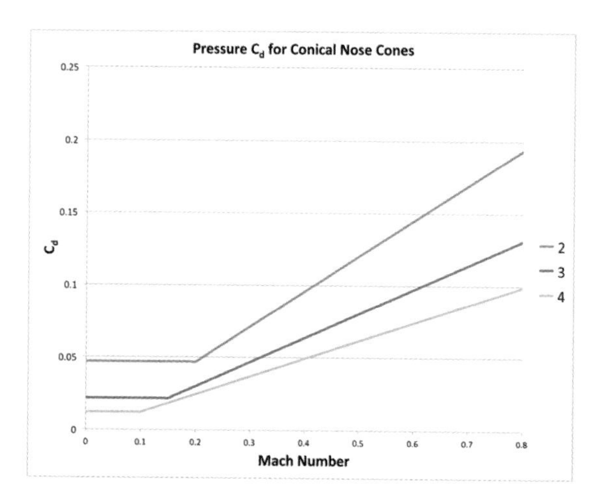

Figure 13-11. *Coefficients of drag for conical nose cones with fineness ratios of 2, 3, and 4, as used in OpenRocket.*

Ogive nose cone

The ogive is a very common nose cone shape because it looks nice and is easy to manufacture.

Figure 13-12. *Tangent ogive nose cone with a fineness ratio of 4.*

The shape is part of the edge of a circle. The tangent ogive is a special subset of the general shape, in which the base of the nose cone is tangent to the body tube. This is the form seen most often in model rocketry. The equation for the surface of rotation for a tangent ogive is:

$$y(x) = \sqrt{\rho^2 - (L - x)^2} - \sqrt{\rho^2 - L^2}$$

$$\rho^2 = \frac{(L^2 + R^2)^2}{4R^2}$$

where L is the length of the nose cone and R is the radius of the base of the nose cone. Ogive noses are considerably better than cones for subsonic rocket flight. The coefficient of drag for an ogive nose cone is about one-tenth the coefficient of drag for a conical nose cone.

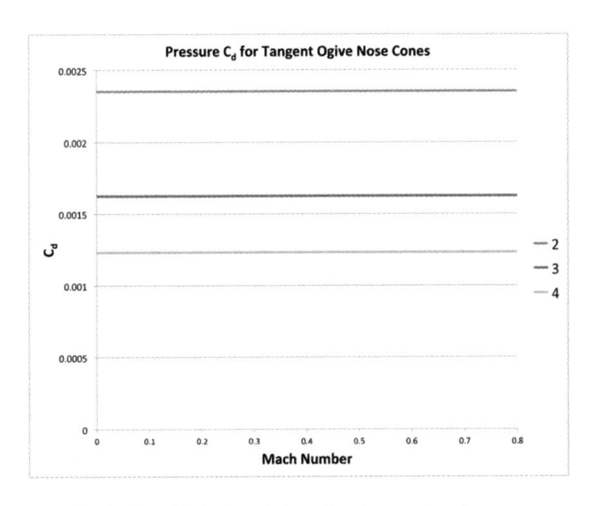

Figure 13-13. *Coefficients of drag for tangent ogive nose cones with fineness ratios of 2, 3, and 4.*

Elliptical nose cone

Figure 13-14. *Elliptical nose cone with a fineness ratio of 4.*

Different authors give different definitions for the elliptical nose cone. The one used in OpenRocket is the most common. The shape of the nose cone is a half-ellipse, given by:

$$y(x) = R\sqrt{1 - \left(1 - \frac{x}{L}\right)^2}$$

According to OpenRocket, elliptical nose cones are better than conical nose cones, but not as good as tangent ogives.

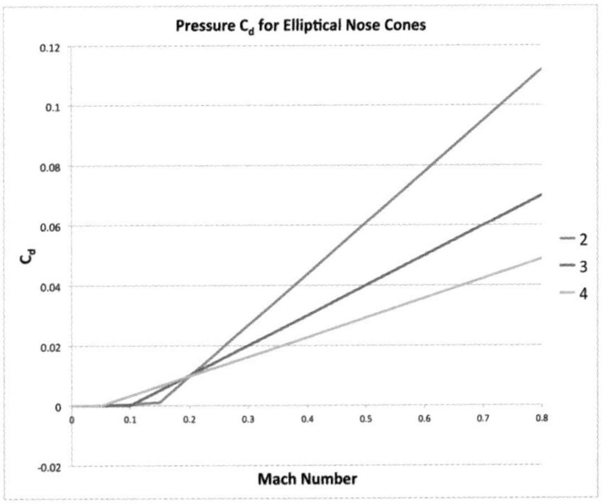

Figure 13-15. *Coefficients of drag for elliptical nose cones with fineness ratios of 2, 3, and 4.*

Figure 13-16 gives you a good idea of how conical, elliptical, and tangent ogive nose cones compare. It shows the coefficients of drag for these shapes with a fineness ratio of 3.

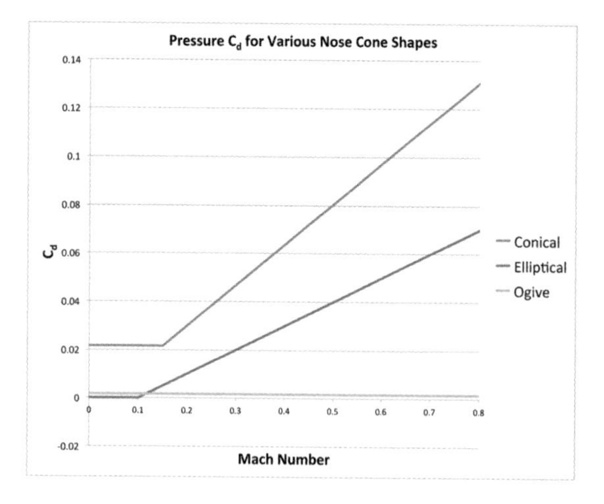

Figure 13-16. *Subsonic pressure coefficients of drag for several nose cone shapes compared. In all cases, the fineness ratio is 3.*

Parabolic series nose cone

Figure 13-17. *Parabolic nose cone with a fineness ratio of 4.*

At first glance, the parabolic series nose cone might look like a tangent ogive, but there are important differences in the shape. The actual equation for a parabolic series nose cone is:

$$y(x) = R\frac{x}{L}\left(\frac{2 - K\frac{x}{L}}{2 - K}\right)$$

The term *K* controls the shape. With *K* set to 0, the shape degenerates into a cone. The nose cone shown has a *K* of 1.0. The performance is good enough that OpenRocket uses a coefficient of drag of 0 for subsonic speeds.

Power series nose cone

The power series nose cone has a rather odd shape, but it's another great performer.

Figure 13-18. *Power series nose cone with a fineness ratio of 4.*

The equation for the shape is:

$$y(x) = R\left(\frac{x}{L}\right)^K$$

Once again, the term *K* controls the shape. With *K* set to 1, the shape degenerates into a cone. The shape shown is for a *K* of 0.5. As with the parabolic series nose cone, OpenRocket uses a coefficient of drag of 0 for this nose cone at subsonic speeds.

This shape is often modified slightly near the base so the joint with the body tube is smoothed.

Haack series nose cone

Figure 13-19. *Von Kárman nose cone with a fineness ratio of 4.*

The Haack series nose cone is based on the following equation:

$$y(x) = R\sqrt{\frac{\theta - \frac{1}{2}\sin(2\theta) + K\sin^3\theta}{\pi}}$$

$$\theta = \arccos\left(1 - \frac{2x}{L}\right)$$

The term K controls the shape. A K of 0 gives a shape called the *Von Kárman nose cone*, which is the one shown in Figure 13-19. This is another excellent performer, which OpenRocket simulates with a coefficient of drag of 0 for subsonic flight. This is also regarded as one of the best nose cones for supersonic flight, so it's a great choice for high-power rockets, too.

Selecting a high-performance nose cone

The shape of a nose cone obviously has a big impact on its coefficient of drag. For high-performance rockets, pick one of the last three shapes if you can. While I have not found solid data to back it up, I suspect the power series nose cone is best for subsonic flight. That is based solely on the fact that it looks a lot like the leading edge of a high-performance airfoil.

Unfortunately, you may not have much choice. The last three nose cones are less common than the first group.

Rocket simulators don't do a particularly good job with nose cone drag. The pressure coefficient of drag is set to zero for some shapes, presumably because they perform well, but also because good data from wind tunnel tests is hard to come by. My own experiments with RockSim indicates rocket simulators are not very sensitive to the length of the nose cone, either, even though nose cone length adds to friction drag.

This is clearly an area where some experimenting could pay off in a competition. Start with nose cones with a fineness ratio of three to four and one of the last three shapes, but be sure to experiment.

Body Tube

Body tube drag at a zero angle of attack comes entirely from friction drag. From Estes TR-11, the coefficient for friction drag on a body tube is:

$$C_{d_f} = 1.02\,C_f\left(1 + \frac{1.5}{\left(\frac{L}{d}\right)^{1.5}}\right)\frac{S_w}{S_{BT}}$$

There are a lot of new ideas here, so let's pull this equation apart and see what it means.

The body tube itself has a length L and diameter d. Those are two of the three things we can control when designing and building our rocket; we'll talk about the third, C_f, in a moment. Looking at the equation for the coefficient of drag, it looks like we want a body tube that is as long as possible compared to the diameter, since that will make the L/d term large. Making L/d large forces the fraction

$$\frac{1.5}{\left(\frac{L}{d}\right)^{1.5}}$$

to be small, reducing the drag. But there is a catch, of course.

The S_w term is the *wetted surface* of the body tube. That's a fancy aeronautical term that basically means the area that is in contact with the air as the rocket flies through the atmosphere. You can find that from the length and diameter of the body tube:

$$S_w = \pi L\ d$$

This time the L term is in the numerator, so making the body tube longer doesn't look like such a good idea anymore.

The term S_{BT} is just the area of the body tube:

$$S_{BT} = \pi\frac{d^2}{4}$$

The last term, C_f, is the *skin friction drag coefficient*. It depends on many things, including how the air flows over the surface, the size of the surface, the altitude, and the temperature. If you look around a bit, you will find a tremendous amount of information for calculating C_f. For our brief discussion, we'll assume turbulent flow, which is pretty likely for a model rocket body tube. A decent approximation of C_f for turbulent flow is:

$$C_f = 0.0576\, R_e^{-\frac{1}{5}}$$

R_e is the Reynolds number, which we discussed briefly earlier. For air, a workable approximation of the Reynolds number is:

$$R_e = \frac{V\,L}{v}$$

Here, V is the speed of the rocket, L is the length from the tip of the nose to the base of the body tube, and v is the kinematic viscosity of the air. The viscosity depends on a number of things, such as the temperature, humidity, and altitude. Rocket simulators go into a great deal of detail to get this term right. For a quick calculation, though, we can use a viscosity of 1.525×10^{-5}. The units are m^2/s, so be sure to use meters for the length and m/s for the velocity of the rocket when calculating the Reynolds number with this viscosity value. See the references for more detailed ways of calculating this value.

There are a couple of ways we can make use of this information. In a moment, we'll calculate the C_d for Juno. Clearly, we'll need the equations to do that. The other thing we can do is to try to understand what the equations are telling us about our choice of body tube length. After all, in some parts of the coefficient of drag equation making the body tube longer increases the coefficient of drag, while in others it makes the coefficient of drag smaller. Why is this happening, and which is the dominant process?

Figure 13-20 shows a plot of the coefficient of friction drag for a BT-50 body tube like the one used on Juno. This particular plot was created for sea-level air at 20°C with the rocket traveling at 100 m/s, although the general shape of the curve and what it tells us doesn't change significantly as the conditions change.

You can clearly see that there is an optimal body tube length, and it's about one inch! This shouldn't be much of a surprise, since the drag contribution from the body tube is entirely friction drag. This tells us that the body tube should be as short as possible to reduce drag (within the limits that are possible based on the motor length and nose cone shoulder, that is). There are other considerations, of course. A shorter body tube cuts the distance to the nose cone, which pulls the center of gravity back, making the rocket less stable. If the body is too short, we will need to add nose weight to make the rocket stable, and the weight will have a negative impact on the performance of the rocket, too. It's all a very careful balancing act, both literally and figuratively.

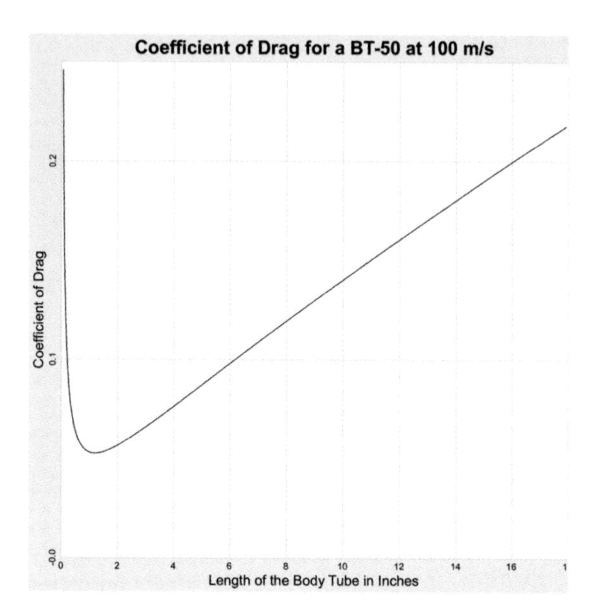

Figure 13-20. *Coefficient of friction drag for a BT-50 traveling at 100 m/s.*

Base

Base drag occurs because the base of the rocket is a flat shape. There are many equivalent but slightly different ways to look at where the base drag comes from. They all boil down to the same thing. The air flowing past the bottom of the body tube can't make a sharp turn to fill in the gap at the base of the rocket. This causes a partial vacuum at the base of the rocket, which pulls backward on the rocket.

Calculating the base drag can be very tricky. After a lot of experimentation, a good approximation has been found. It is:

$$C_{d_B} = \frac{0.029}{\sqrt{C_{d_N} + C_{d_f}}}$$

This uses the coefficients of drag for the nose cone and body tube, which we've already found.

You can reduce the base drag a great deal by using a *boat tail*. A boat tail is like a nose cone at the back of the rocket. It serves the same purpose as the tailing edge of an airfoil, gently returning the air from the widest point along the body of the rocket to a narrower point. We can't bring the back of the rocket to a nice point because of the rocket motor, which leaves a blunt back end that a boat tail can't fix. In fact, on the rockets where we really care about maximum performance, we usually use a minimum-diameter rocket, which means we can't use a boat tail at all.

The effect of the boat tail depends a lot on the slope angle between the rocket and the sides of the boat tail. Figure 13-21 shows Juno with a gentle boat tail, where there is a 5° angle between the side of the body tube and the side of the boat tail. This is a pretty good value to shoot for when designing a boat tail. For this slope, the boat tail reduces the base drag by:

$$r_{bt} = \left(\frac{d_b}{d}\right)^3$$

where d_b is the diameter of the boat tail and d is the diameter of the body tube. We need to leave d_b at 0.7 to leave room for the motor, and we remember that d is 0.98 for the BT-50 body tube used for Juno, so the reduction in drag is:

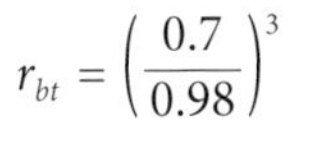

$$r_{bt} = \left(\frac{0.7}{0.98}\right)^3$$

$$r_{bt} = 0.36$$

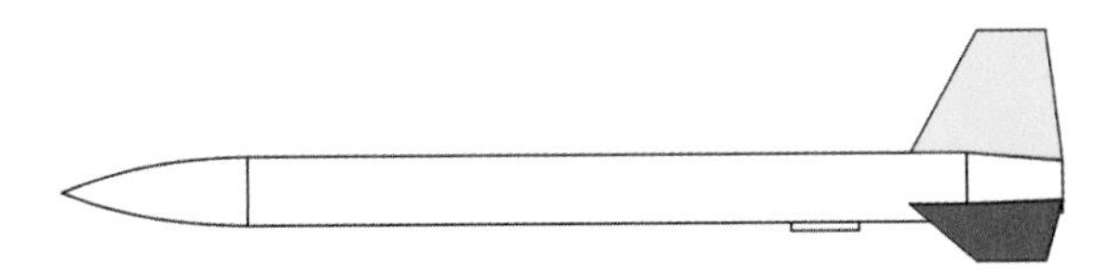

Figure 13-21. *Juno with a 5° boat tail. This reduces the base drag by 74%.*

That's a pretty substantial decrease in drag! The decrease is even better with an ogive-shaped boat tail, just as the drag of the nose cone is lower for an ogive than for a cone.

So, with a conical boat tail with a slope of 5°, the base drag becomes:

$$C_{d_B} = \frac{0.029}{\sqrt{C_{d_N} + C_{d_f}}}\left(\frac{d_b}{d}\right)^3$$

This same idea applies to a tube transition used to transition from a wide payload bay to a smaller body tube, as you might see in an egg lofter. Most tube transitions are really too short to get the kind of reduction in drag shown here, but if you build your own with a 5° slope, you can get a dramatic reduction in drag.

Fins

Fins are the second-biggest source of drag on a typical model rocket, so it makes sense to pay attention to the fins when designing a high-performance rocket. The drag on fins is caused in three different ways, and each needs to be looked at separately to really understand what is happening.

Let's start with pressure drag. We saw that for nose cones, pressure drag comes from the shape of the nose, and fairly small changes in shape can result in dramatic differences in the pressure drag. For fins, the shape of a fin's cross section is the source of pressure drag. There are three shapes that are used most often for subsonic fins. These are shown in Figure 13-22, along with typical drag coefficients.

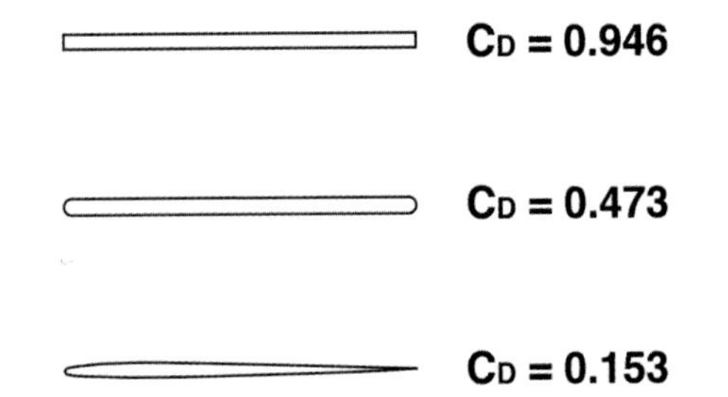

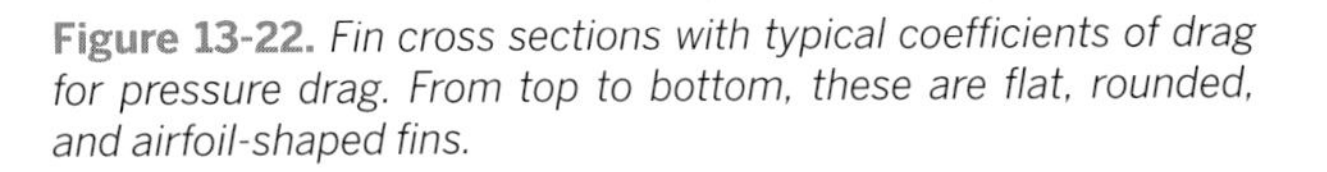

Figure 13-22. *Fin cross sections with typical coefficients of drag for pressure drag. From top to bottom, these are flat, rounded, and airfoil-shaped fins.*

Clearly, flat fins are not especially good. They are also more prone to damage than rounded fins, since the sharp edges that form the corners are more easily dinged than rounded edges. The combination of cutting the drag in half and making the fin less prone to damage is why the instructions for almost every rocket in this book tell you to round the edges of the fins.

But the airfoil shape is better still, since it has about one-third the drag of rounded fin edges. Why not use airfoil-shaped fins on all of our rockets? There are two reasons. First, airfoils are devilishly hard to create accurately. (There are some good techniques for doing so, and you'll see them when building Hebe in Chapter 15 and Icarus in Chapter 20.) Remember the discussion about nose cone shapes? It's tough to tell a Von Kárman from a tangent ogive at a casual glance, but the difference in drag is substantial. The same is true for airfoils. A clumsily carved airfoil might be hard to distinguish from an accurately made one, but will be far less effective. Still, a clumsily carved airfoil is probably better than rounded fins, so don't feel like you shouldn't try. Just do your best to find a way to get the airfoil you really intend to use.

The other reason we don't use airfoil-shaped fins all the time is that they are weak. Those thin trailing edges are very easy to damage on landing, and they can even make the fin so thin it will shake apart from the turbulence of flight.

For the most part, then, reserve airfoil-shaped fins for ultra-high-performance contest rockets, where every little advantage counts. Rounded edges are much sturdier and easier to create for sport rockets and most payload rockets.

If the exact shape of the fin is so important, what shape should you use? It turns out that the answer depends a lot on the rocket. One factor is the Reynolds number the rocket operates at. Different airfoils perform better with different Reynolds numbers. Thickness is also a factor. A very thin fin would be great from a drag perspective, but it would be too weak to fly. Some airfoils work better with thicker fins, and some work better with thin fins. When you get to the point where you really need to answer this question, do an online search for "airfoil." Be prepared to spend a lot of time digging through the material. You will find all sorts of information, including huge databases showing specific airfoils that are good in different situations. The techniques shown later in the book will help you create a fairly accurate rendition of the airfoil you choose.

The next aspect of fin drag is the friction drag. It's computed just like the friction drag for the body tube, but in this case the wetted area is the surface area of the fins. Remember to include the area from both sides of the fin, since the air flows across both sides!

The last kind of drag on the fin is induced drag. This is the drag that occurs when the fin is not flying perfectly straight, and the fins are doing their work to straighten it. As shown back in Figure 13-6, induced drag occurs when the high pressure from one side of a fin spills over to the low-pressure side, causing a vortex that robs the rocket of energy in the form of drag. This can't be avoided, but it can be minimized.

Minimizing induced drag depends on the shape of the fin and the shape of the tip of the fin. Let's consider the four common fin shapes shown in Figure 13-22.

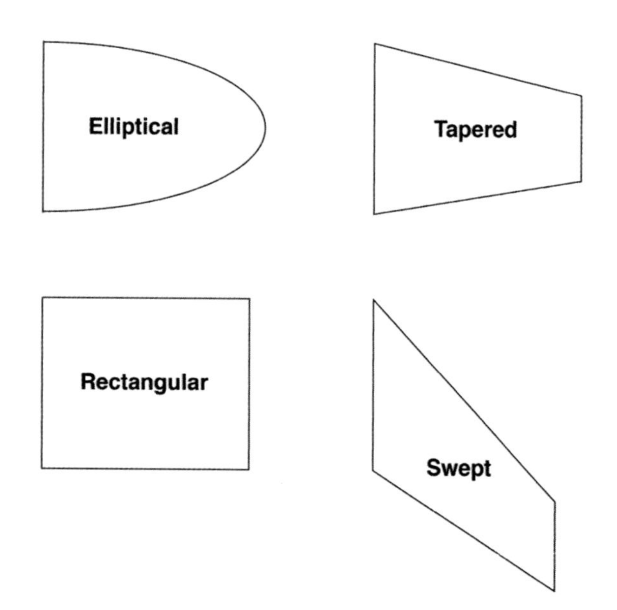

Figure 13-23. *Typical fin planforms. In terms of induced drag, from lowest drag to highest, they are elliptical, tapered, rectangular, and swept.*

These shapes have been studied extensively for subsonic flight because most airplanes fly at subsonic speeds. A fin, after all, is just half of a wing, so all of the information gathered for subsonic airplanes works just fine for rocket fins. The general equation developed for induced drag is:

$$C_{d_i} = \frac{C_L{}^2}{\pi\, A.R.\, e_w}$$

The C_L term is the coefficient of lift. For a symmetric airfoil like a rocket fin, lift occurs when the edge of the fin is not facing straight into the wind. This can happen due to less than perfect construction, a motor with a slightly off-center thrust, or a change in the wind as the rocket ascends. When the fin isn't flying straight into the wind, it produces a lift force that pushes the rocket back into alignment. The lift is a good thing, but some drag comes along with that lift.

The *A.R.* term is the aspect ratio. For a rectangular fin, this is the width of the fin from one fin tip to the tip of the opposite fin, called the *span*, divided by its length. This assumes there is an opposite fin, of course. If not, an equivalent way to think of the span is twice the distance from the fin tip to the center of the body tube. The length of the fin is generally called the *chord*. For more complicated shapes, the aspect ratio is the square of the span divided by the area of the fin. From this we can see that the induced drag is lower for a long, thin fin than a short, stubby one. That's why sailplanes have very long, thin wings. Unfortunately, that also makes the wing weaker, which is one of the reasons why high-speed fighters that need lots of strength have short, stubby wings. Most fins end up with an aspect ratio of 2 or less because they, too, need strength to withstand the vibration from high-speed flight.

The last term, e_w, is a fudge factor that accounts for the shape of the fin, also called the *planform*. The very best fin shape for reducing induced drag is the elliptical fin, which has an e_w of 1 by definition. The exact value for e_w depends on a lot of factors, but we can make some general observations.

The elliptical planform is best for reducing induced drag. Nothing has been found that is any better. A tapered fin with a 0° sweep angle and a tip that is about 40% of the thickness of the root is second best. The difference in performance compared to the elliptic planform is about 4%, and the tapered fins are a lot easier to cut, so that's what is used most often. Next in line is the rectangular fin. It's really not all that bad. The worst of the lot is the swept fin.

So, if the swept fin is so bad, why do we see swept fins on many rockets and aircraft? The answer is the same as for nose cones and airfoils: swept fins work better at supersonic speeds, and many of the rockets and aircraft we see are designed for those speeds. Model rockets of the size shown in this book are subsonic, and a swept fin doesn't work well at subsonic speeds. Swept fins are pretty, so a lot of sport rockets have swept fins, but they are not the correct choice for high-performance subsonic rockets.

For fins with a straight tip, which includes all of the ones we've been discussing except the elliptical fin, the shape of the tip is also important. The tip should be left flat, not rounded or brought to a point. This gives the fin another small advantage by reducing the tendency to form a vortex.

Finally, since the shape of the airfoil is so important, what happens if we use an elliptical or tapered fin as the chord changes? Near the fin tip, the fin is thinner, changing the shape of the airfoil. Is that important? Well, yes. That's why the fin thickness should ideally be proportional to the chord. If the fin is 1/8" thick at the root, it should be 1/16" when the chord is one-half of the root chord. Of course, that makes the fin weaker. Changing the shape with the chord can be done fairly accurately for tapered fins, as you will see later when you build Icarus. It's very hard to do well for an elliptical fin.

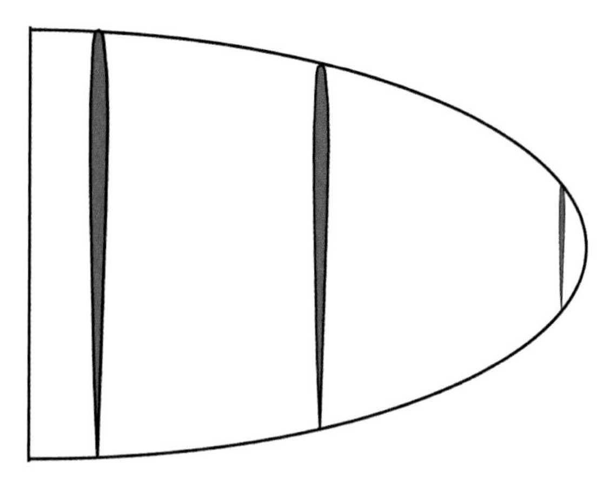

Figure 13-24. *For lowest drag, the thickness of the fin should be proportional to the chord. Note how the fin cross section, shown shaded, gets thinner as the chord shrinks.*

You've seen a lot of factors that go into choosing a fin shape. For sport rockets and rugged payload rockets, which include most of the rockets in this book, the fin design is usually selected to minimize fin breakage or to look really cool. For a high-performance rocket, you want a thin fin with an elliptical planform and a large aspect ratio, sanded to a perfect airfoil. Such a fin will probably break, either from the aerodynamic stress of flight or from impact on landing. It is also very hard to create an accurate airfoil. Some designers go to the other extreme, picking the rectangular planform so they can create an accurate airfoil without worrying about varying the fin thickness.

This is part of what makes rocketry both fun and challenging. The right answer is part science, part craftsmanship, and part engineering. Each designer picks from a selection of factors, trying to balance drag, ease of construction, and ruggedness to get the very best from a particular fin and rocket.

Putting all of this together, the total drag on the fins for a zero angle of attack is:

$$C_{d_F} = C_{d_{FF}} + C_{d_{FP}}$$

The fin friction term is just the surface area of the fins. It's calculated the same way as the friction drag on a body tube:

$$C_{d_{FF}} = 1.02\, C_f \left(1 + \frac{1.5}{\left(\frac{L}{d}\right)^{1.5}}\right)\frac{S_w}{S_{BT}}$$

The wetted area is calculated by using the surface area of the fins. Remember, both sides of the fin count, since the air flows over both.

The second term is the fin pressure drag. A good place to start is the drag coefficients from Figure 13-22, but these need to be scaled to the reference area for the total drag coefficient, which is the body tube. The equation is:

$$C_{d_{FP}} = C_f \frac{S_f}{S_{BT}}$$

Here, C_f is the coefficient of drag from Figure 13-22, while S_f is the frontal surface area of the fin. For a fin of uniform thickness, this is:

$$S_f = t\, l\, n$$

where *t* is the thickness of the fins, *l* is the length of the fin, and *n* is the number of fins.

Launch Lugs

There is a lot of conflicting information about launch lugs. Let's start with a great engineering estimate from Estes Technical Report 11, "Aerodynamic Drag of Model Rockets" (*http://sargrocket.org/Documents/Estes/TR-11.pdf*), where Dr. Gerald Gregorek develops a classic engineering estimate for the drag on a launch lug by looking at the bounding case. He calculates the drag for a solid rod with a flat front for the pressure drag, and then adds a term for the friction drag. He arrives at this equation for launch lug drag:

$$C_{d_{LL}} = \frac{1.2\, S_{LL} + 0.0045\, S_{LLW}}{S_{BT}}$$

Here, S_{LL} is the frontal area of the launch lug and S_{LLW} is the wetted surface area of the launch lug. Don't forget that there is an inside and an outside component for the wetted surface area. Here's how it is calculated:

$$S_{LLW} = \pi l (d_{in} + d_{out})$$

where l is the length of the launch lug, d_{in} is the inside diameter of the launch lug, and d_{out} is the outside diameter of the lunch lug.

It's interesting to compare the theoretical calculation and the estimate used by RockSim with some actual wind tunnel tests. John DeMar did a nice wind tunnel study as an R&D project for NARAM-37 in 1995. According to his wind tunnel tests, a 2"-long launch lug accounted for about 23% of the coefficient of drag for the rocket. This experiment used a rocket based around a BT-20 body tube, so the relative contribution from drag from the launch lug was higher than for Juno, which is built using a larger body tube. Scaling the coefficient of drag for Juno puts the contribution of the launch lug at 19%. This is considerably higher than the percentage suggested by our theoretical calculations, and the 8.3% that RockSim uses. Which is right? It's possible all of them are, depending on where the launch lug is placed, the way it is glued to the body tube, and the finish used. It's also possible that one or more of these estimates is flawed. While we can use the theory as a starting place, this is clearly an area that is ripe for further investigation.

For a high-performance rocket, though, the answer is deceptively simple. Even if the lowest estimate is correct, adding 8.3% to the drag hurts performance. The solution is to get rid of the launch lug entirely, using a tower launcher that supports the rocket with three adjustable rods. That eliminates the launch lug and any drag it might contribute. For anything other than a high-performance contest rocket, however, the launch lug is just too convenient to do without.

Putting It All Together

The final drag coefficient for the rocket is the sum of all of the individual drag coefficients. The reason this works is because all of the individual components were scaled to the same cross section area, which is the frontal area of the body tube. This gives:

$$C_d = C_N + C_{d_F} + C_{d_B} + C_{d_F} + C_{d_{LL}}$$

Example: The Coefficient of Drag for Juno

Let's put all of this together to calculate the coefficient of drag for Juno. Some of the values are speed dependent. For our example, we'll use a speed of Mach 0.1, about 34 m/s. We will stick with the metric system for all measurements, for consistency. (Some of the constants were developed under the metric system, and converting a few lengths to metric lengths will be easier than reworking all of the constants.)

Let's start with the nose cone. This appears to be a tangent ogive. It's 2.75" long and just under 1" wide, giving a fineness ratio of 2.75. (Since both measurements were in inches, the dimensions go away, and we don't have to convert to metric.) Refer back to Figure 13-11 for the coefficient of drag. Juno flies fairly slowly on most motors, so we'll pick a value from the flat part of the plot, a bit lower than the value for a fineness ratio of 3. This gives:

$$C_N = 0.03$$

Next is the friction drag. The basic equation is:

$$C_{d_f} = 1.02\, C_f \left(1 + \frac{1.5}{\left(\frac{L}{d}\right)^{1.5}}\right) \frac{S_w}{S_{BT}}$$

C_f is:

$$C_f = 0.0576 \left(\frac{V\,L}{\nu}\right)^{-\frac{1}{5}}$$

We've picked a speed of 34 m/s. Juno's length from the nose tip to the bottom of the body tube is 14.25", or 0.362 m. We'll use our earlier approximation for the viscosity of air, 1.525x10^{-5}. This gives:

$$C_f = 0.0576 \left(\frac{34 \times 0.362}{1.525 \times 10^{-5}}\right)^{-\frac{1}{5}} = 0.00379$$

The outside diameter of a BT-50 body tube is 0.0248 m, so that's the value we will use for *d*. We can also get the frontal area of the body tube from the diameter. The frontal area will appear in several places:

$$S_{BT} = \pi \frac{d^2}{4} = 4.83 \times 10^{-4}\,\text{m}^2$$

The wetted area is the outside area of the body tube and nose cone. We could calculate the precise area of the nose cone, but treating it as an extension of the body tube is close enough for our purpose here. The total length of the body tube and nose cone is 0.362 m. That gives a wetted area of:

$$S_W = \pi\, d\, L = 0.0282\,\text{m}^2$$

Putting all of this together, the friction drag from the body tube and nose cone is:

$$C_{d_f} = 1.02 \times 0.00379 \left(1 + \frac{1.5}{\left(\frac{0.362}{0.0248}\right)^{1.5}}\right) \frac{0.0282}{4.83 \times 10^{-4}}$$

$$C_{d_f} = 0.232$$

The base drag is calculated from the drag on the nose cone and body tube:

$$C_{d_B} = \frac{0.029}{\sqrt{C_{d_N} + C_{d_f}}} = \frac{0.029}{\sqrt{0.03 + 0.232}} = 0.0567$$

We're assuming the rocket is at a zero angle of attack, so there is no induced drag. The fin drag, then, is:

$$C_{d_F} = C_{d_{FF}} + C_{d_{FP}}$$

Let's start with the friction drag. Most of the equation for friction drag is identical to the one for friction drag on the body tube; only the wetted area changes. The wetted area for one side of one fin is just the area of the trapezoid that forms the fins. Figure 13-25 shows a fin from Juno with the dimensions in meters.

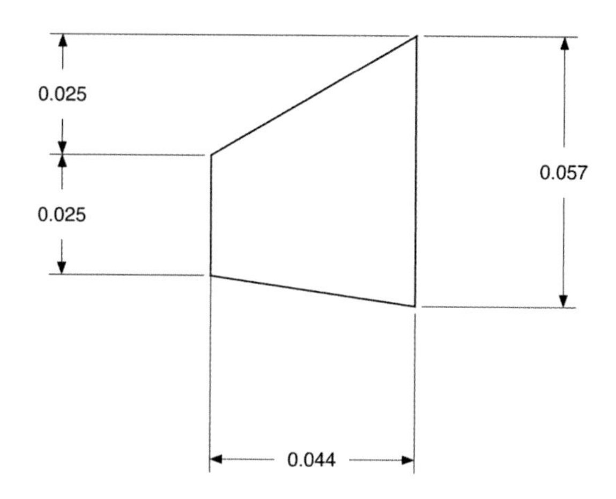

Figure 13-25. *One fin from Juno, with the dimensions shown in meters.*

We can find the surface area for the fin from the equation for the surface area for a trapezoid:

$$A = \frac{r+t}{2} l = \frac{0.057 + 0.025}{2} 0.044 = 0.001804\,\text{m}^2$$

There are three fins with two surfaces each, so the wetted area is six times this value. Plugging this back into the equation for friction drag gives:

$$C_{d_{FF}} = 1.02 \times 0.00379 \left(1 + \frac{1.5}{\left(\frac{0.362}{0.0248}\right)^{1.5}}\right) \frac{6 \times 0.001804}{4.83 \times 10^{-4}}$$

$$C_{d_{FF}} = 0.0890$$

Juno uses rounded fins, so the pressure drag coefficient is 0.473. The fins are 3/32" thick, or 0.00238 m, and 0.044 m wide. This gives a frontal area of:

$$S_f = 0.00238 \times 0.044 = 1.05 \times 10^{-4}\,\text{m}^2$$

Scaling this for the dimensions of the rocket gives:

$$C_{d_{FP}} = C_f \frac{S_f}{S_{BT}} = 0.473 \frac{1.05 \times 10^{-4}}{4.83 \times 10^{-4}} = 0.103$$

There are three fins, so the total contribution to the coefficient of drag from fins, then, is:

$$C_{d_F} = 0.0890 + 3 \times 0.103 = 0.398$$

The last term is the drag on the launch lug:

$$C_{d_{LL}} = \frac{1.2\, S_{LL} + 0.0045\, S_{LLW}}{S_{BT}}$$

S_{LL} is the frontal area of the launch lug. The launch lug has an outside diameter of 4.41 mm and an inside diameter of 3.99 mm. This gives a frontal area of:

$$S_{LL} = \pi \frac{d_{out}^2}{4} - \pi \frac{d_{in}^2}{4}$$

$$S_{LL} = \frac{\pi}{4}(0.00441^2 - 0.00399^2) = 2.77 \times 10^{-6}\, \text{m}^2$$

The launch lug is 1″ long, which, by definition, is 2.54 cm, so the wetted area of the launch lug is:

$$S_{LLW} = (\pi\, d_{in} + \pi\, d_{out})\, L$$

$$S_{LLW} = \pi\, (0.00441 + 0.00399)\, 0.0254$$

$$S_{LLW} = 6.70 \times 10^{-4}$$

The drag from the launch lug, then, is:

$$C_{d_{LL}} = \frac{1.2 \times 2.77 \times 10^{-6} + 0.0045 \times 6.70 \times 10^{-4}}{4.83 \times 10^{-4}}$$

$$C_{d_{LL}} = 0.0131$$

This gives us the final coefficient of drag for Juno as:

$$C_d = C_N + C_{d_f} + C_{d_B} + C_{d_F} + C_{d_{LL}}$$

$$C_d = 0.03 + 0.232 + 0.0567 + 0.398 + 0.0131$$

$$C_d = 0.730$$

This is in pretty good agreement with the values shown for the coefficient of drag in Figure 13-8, which showed the C_d from RockSim. At Mach 0.1, RockSim reports 0.3041 for the nose and body—a bit higher than the combined 0.262 from our calculation, but still in the right range. The value for base drag of 0.05259 is very close to ours, as is the launch lug coefficient of drag of 0.01502. The fin drag is again a bit lower, at 0.239. The total drag coefficient from RockSim at Mach 0.1 is 0.611, which shows that our estimate is pretty good—easily good enough for understanding where drag comes from on our rocket.

Table 13-2. Sources of drag

	RockSim	Calculated
Nose and body	0.3041	0.262
Base	0.05259	0.0567
Launch lug	0.01502	0.0131
Fins	0.239	0.398
Total drag	0.611	0.730

So where does it come from?

We lumped the nose cone in with the body tube for friction drag, but if you look at the contribution from pressure drag on the nose cone, its coefficient of drag of 0.03 accounts for just 4% of the overall drag. Only the launch lug is lower, contributing 1.8%. Base drag is a bit higher, but it is still just 7.8% of the total.

Most of the drag—54.5% of the total—comes from the fins. The friction drag on the body tube and nose cone is a close runner-up, accounting for the remaining 35.9% of the drag. This tells us that when trying to reduce drag, we should concentrate on the fins and body tube. Here are some tips that will help:

1. Keep the body tube as short as possible to reduce the friction drag.
2. Keep the fins as small as possible to reduce friction drag and pressure drag.
3. The fin aspect ratio is a trade-off. A higher aspect ratio (long, thin fins) reduces induced drag, but a small aspect ratio (short, stubby fins) reduces pressure drag. Experiment with both simulators and real flights to find a good trade-off. Most rockets end up with an aspect ratio in the 2 to 4 range.
4. Use an accurate airfoil to reduce pressure drag.

5. Put a very good finish on the rocket. This makes the air flow more smoothly over the surface of the rocket for as long as possible. Eventually, when the speed is high enough, the flow of air will become turbulent, which is the drag condition we assumed. A smooth finish delays the transition to turbulent flow.
6. Use fin fillets. This reduces the interference drag between the fins and the body tube. We didn't discuss interference drag—this introduction was long enough already!—but smooth fin fillets will help reduce it.

Further Reading

There are lots of refinements that can be made to our calculation of the coefficient of drag, including looking at the effects of different angles of attack, examining the interaction between the fins and the body tube, and looking more closely at turbulent and laminar flow and how they vary with the Reynolds number. This simplified discussion is good enough to help understand where drag comes from and how to modify our rockets to reduce it, but, for example, we would need to go into a lot more detail to create a really good rocket simulator. If you would like to dig deeper, these references are a great place to start:

- OpenRocket technical documentation (*http://openrocket.sourceforge.net/techdoc.pdf*)
- Estes TR-11, "Aerodynamic Drag of Model Rockets" (*http://bit.ly/TR-11pdf*)
- "Model Rocket Drag Analysis using a Computerized Wind Tunnel," NARAM-37 R&D Project (*http://www.interactiveinstruments.com/pdfs/28.pdf*)

14 How High Will It Go?

Rockets obey the same basic laws of physics as anything else, and we can use those laws to figure out what will happen to a rocket during flight. We can predict (roughly) how high the rocket will go, what its maximum speed will be, and what the acceleration force will be on the rocket and anything it carries. Being able to predict the altitude is especially useful, since, as we saw in Chapter 5, that tells us how large a flying field we need. It might also clue us in on whether we even want to fly a rocket that day. For example, if I expect a rocket to break 1,000 feet in altitude, I generally won't fly it unless the wind is pretty calm. Part of the reason is that I don't want to lose the rocket. Part of the reason is I don't want to chase the rocket. Yes, I'm lazy that way.

There are three common ways to predict the altitude of a model rocket. The first is to look at the altitude listed for a rocket kit or plan. For example, Table 3-1 in Chapter 3 showed the predicted altitudes for Juno when using various motors. Table 14-1 shows a few of those values again.

Table 14-1. Juno's predicted altitude with various motors

Motor	Altitude
A8-3	200 ft
B6-4	550 ft
C6-7	1,280 ft

Chances are, even if you are not flying one of the rockets used in this book, the rocket you are flying will be pretty close in specifications. If a rocket is built around a BT-50 body tube, is about a foot long, and doesn't have any crazy protuberances or heavy payloads, the Juno altitude prediction table will be fairly close—close enough to pick a flying field and decide if it is too windy, anyway.

Another way to predict the altitude of a model rocket is from basic physics. Of course, depending on how far you have gone in your study of physics, the formulas and calculations we use may not seem basic at all. Still, if you are at all mathematically inclined, calculating your rocket's estimated height is a lot of fun. After examining the equations, we'll use a very simple computer program to predict the altitude for Juno.

The best way to routinely predict the altitude and other performance characteristics of a rocket, though, is to use a rocket simulator. Rocket simulators are not perfect, but they are very, very good. Compared to measured values, the altitude predicted by rocket simulators is generally accurate to within 25%, although occasionally it may be outside that range. "Predicting Altitude with a Rocket Simulator" on page 310 explores how to use a rocket simulator to predict the performance characteristics of a rocket.

Predicting Altitude Using Physics

The most fun way to predict the altitude of a model rocket is using physics, because—tautology alert—physics is phun. That said, if you don't share my love of everything mathematical, you can skip this section. You will miss gaining a deeper understanding of how rocket simulators work and what really makes your rocket fly, but you can build and fly every rocket in this book, and predict their altitude perfectly well, without understanding the physics.

The problem of predicting the altitude of a model rocket can be as complicated as you want it to be. For example, Sampo Niskanan developed the OpenRocket simulator as a thesis while working on his MS in Physics from the Helsinki University of Technology. You can find his thesis online at *http://openrocket.sourceforge.net/thesis.pdf*. We're going to simplify things a lot compared to what he did!

For our simulation, we're going to track the rocket until it reaches apogee. We will assume the rocket travels vertically, ignoring the launch rod angle and any contributions from the wind. We will assume the coefficient of drag, C_d, does not vary with speed. With these assumptions, there are three forces working on the rocket, as shown in Figure 14-1.

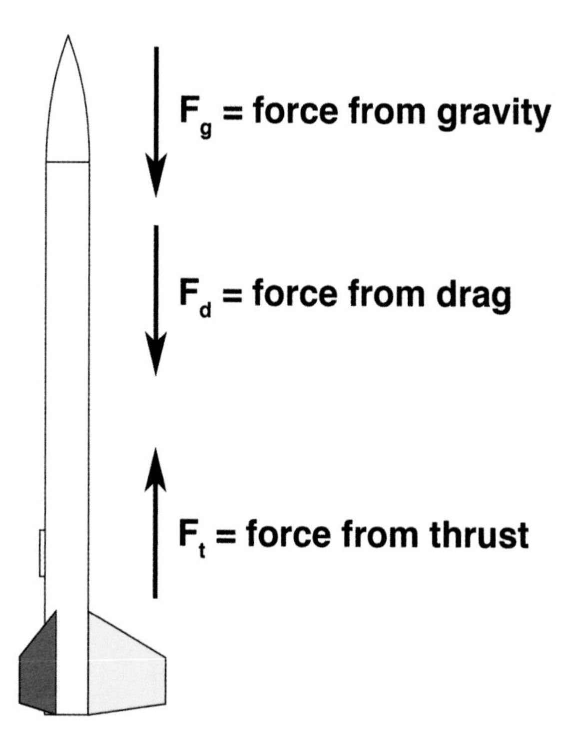

Figure 14-1. *The force on the rocket includes the force of gravity and the force of drag pushing down and the force from the motor thrust pushing up.*

Tracking the flight of the rocket uses three basic equations from the first few weeks of high school physics, combined with the forces you just saw. The first of the equations is Newton's Second Law:

$$F = m\,a$$

This equation relates the force needed to accelerate an object of a given mass. F is the force, usually measured in newtons or pounds; m is the mass, measured in kilograms or slugs; and a is the acceleration, measured in meters/second^2 or feet/second^2. We know the forces acting on our rocket: they are a mix of the force of gravity, the thrust of the motor, and the force of atmospheric drag. We also know the initial mass of the rocket, because we've weighed it. The mass will change as the propellant burns during flight, and there are good sources of information, which we'll look at in a moment, to track this change. From these, we can find the acceleration of the rocket.

The second equation is the equation of motion:

$$d = d_0 + v_0 t + \frac{1}{2} a t^2$$

This tells us the position of the rocket, d, relative to a starting position, d_0; a starting velocity, v_0; the acceleration, a; and the elapsed time, t. We already have an idea of how to find the acceleration. For just a moment, let's assume we have a rocket lifting off of the launch pad under a constant acceleration that is about eight times the force of gravity, or 78.4 m/s^2. (Acceleration does not stay constant during the flight of a rocket, really, but we'll deal with that when the time comes.) The term d_0 is the starting distance: in our case, that's the launch pad, and a rocket on the pad has traveled zero distance. Similarly, the v_0 term is the starting velocity, which again starts at zero. Let's apply the acceleration for one-tenth of one second and see how far the rocket goes:

$$d = 0 + 0 \times 0.1 + \frac{1}{2} 78.4 \times 0.1^2 = 0.392\ \text{m}$$

In that first tenth of a second, the rocket travels 0.342 meters, or about 1.1 feet. It's still on the launch rod. The last equation we will need tells us how fast it is going:

$$v = v_0 + a\,t$$

$$v = 0 + 78.4 \times 0.1 = 7.84\,\mathrm{m/s}$$

This gives the new velocity, v, after accelerating for t seconds at a meters/second^2. That's about 17.5 mph, already over half of the 30 mph recommended speed before a rocket leaves the launch rod.

In the next tenth of a second, the rocket already has some altitude and speed, and these add to the new altitude and speed. The previous value for d becomes the new value for d_0, and the previous value for v becomes the new value for v_0. The math looks like this:

$$d = 0.392 + 7.84 \times 0.1 + \frac{1}{2}78.4 \times 0.1^2 = 1.568\,\mathrm{m}$$

$$v = 7.84 + 78.4 \times 0.1 = 15.68\,\mathrm{m/s}$$

The rocket continues to accelerate (while we continue to calculate its distance and velocity) until the motor burns out, at which point the acceleration becomes negative. The thrust from the motor drops to zero, but the forces of gravity and drag are both acting to slow the rocket. Once the velocity drops to zero, the simulated rocket has reached its maximum altitude.

This is all complicated somewhat by the fact that the mass and acceleration change with time, too. Let's take a look at why this happens, and how we can account for it in our calculation.

First of all, the motor itself does not generate a constant amount of force. Figure 14-2 shows the thrust curve for the Estes C6 motors.

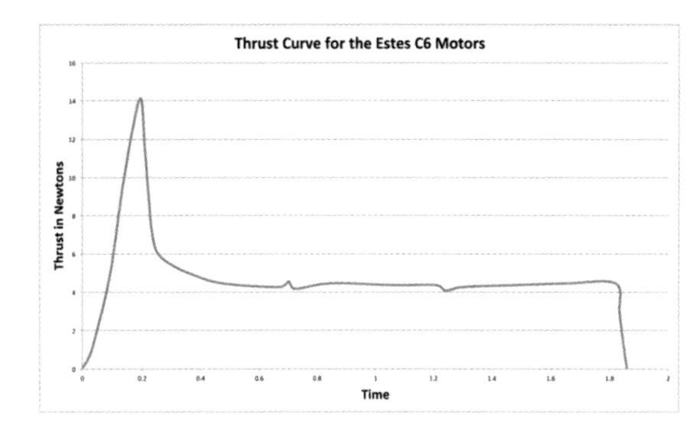

Figure 14-2. *Thrust curve for an Estes C6 motor.*

We can look at this plot and estimate the thrust at any particular time. Thrust curves are often included in the motor's documentation, and also available digitally—John Coker has set up and maintains a wonderful resource at *http://www.thrustcurve.org*. I've been able to find thrust curves for any motor I've ever looked for, in two of the common digital file formats: RockSim and RASP. Let's take a look at the RASP (Rocket Altitude Simulation Program) file format, originally developed by Harry Stine for the RASP simulator, since it is a little easier to read. Here's the file for the Estes C6 motor (the semicolons indicate comment lines; the comments have been reformatted slightly from the original file so they will fit on the page better):

```
;
;Estes C6 RASP.ENG file made from
;NAR published data
;File produced October 3, 2000
;The total impulse, peak thrust,
;average thrust and burn time are
;the same as the averaged static test
;data on the NAR website in the
;certification file. The curve drawn with
;these data points is as close to the
;certification curve as can be with such
;a limited number of points (32) allowed
;with wRASP up to v1.6.
C6 18 70 0-3-5-7 0.0108 0.0231 Estes
0.031 0.946
0.092 4.826
0.139 9.936
0.192 14.09
0.209 11.446
0.231 7.381
0.248 6.151
0.292 5.489
0.37 4.921
0.475 4.448
```

```
0.671 4.258
0.702 4.542
0.723 4.164
0.85 4.448
1.063 4.353
1.211 4.353
1.242 4.069
1.303 4.258
1.468 4.353
1.656 4.448
1.821 4.448
1.834 2.933
1.847 1.325
1.86 0
```

According to the comments, this file was created in 2000. While high-power motors seem to come and go, many low-power motors have been around for decades with practically no change. The C6-5 of today is basically the same as the motors I flew back around 1970.

From left to right, the values on the first line that is not a comment are the motor designation (C6), the diameter in millimeters, the length in millimeters, the available delay times separated by dashes, the propellant mass in kilograms, the total motor mass in kilograms, and the manufacturer.

The rest of the lines have two values. The first is a time since ignition in seconds, and the second is the thrust in newtons at that time.

This data is everything we need to figure out the force on the rocket from the thrust at any given time. If the flight time is after the last time in the table, the thrust is zero. For any other time, we can get the thrust by interpolating between the two adjacent times.

If we allow ourselves to make one simple assumption about the relationship between thrust and fuel mass, then this table also gives us the change in total mass of the motor, which in turn gives the change in the force of gravity. If we work through the values, we can find the total impulse for the motor up to any given point in time. That gives us the total energy expended by the motor. It is very reasonable to assume that the energy expended is equal to the loss in fuel weight. The fraction of total impulse up to any time, divided by the total impulse, is proportional to the fuel mass loss at that time.

Let's work through a quick example to see how this works. According to the first line of data in the table, during the first 0.031 seconds the thrust goes from 0 to 0.946 newtons. We don't know the exact shape of the thrust curve during this time, but we can make the assumption that it was linear, so the average thrust was 0.473 newtons. We can then calculate the total impulse up to this point in time as follows:

$$I_t = I \times t = 0.473 \times 0.031 = 0.0147 \text{ N - s}$$

So, the total impulse delivered in the first 0.031 seconds is 0.0147 N-s. With some tedious calculations, we can work through the rest of the table to find the total impulse of the C6 motor. Using a spreadsheet to do the calculations, I found the total impulse is 8.817 N-s for this thrust curve.

The motor data tells us the propellant weight for the motor is 0.0108 kg. Making the assumption that the total impulse is proportional to the mass of the fuel burned, the lost mass for the first 0.031 seconds is:

$$m = \frac{0.0147}{8.817} 0.0108 = 0.000018 \text{ kg}$$

We can continue this calculation for the rest of the burn time for the motor to find the mass of the rocket at any given time, and use that information, along with the forces on the rocket, to find the acceleration at any time.

The force on the rocket also changes due to changes in the force of drag, which occur as the speed of the rocket increases. We saw how speed affects the force of drag in Chapter 13, where we looked in detail at the equation that gives the force of drag:

$$D = \frac{1}{2} C_d \rho v^2 A$$

Add all of the forces and divide by the current mass of the rocket, and you have the acceleration at any particular time.

Of course, this is a very simple look at how to calculate the altitude for a model rocket. There are all sorts of other factors we could add for a more accurate simulation, like the variation of the coefficient of drag with speed, the deflection from a crosswind, or the effect of tipping the launch rod away from the vertical.

Back when we first started looking at the forces on the rocket, we used a time of 0.1 seconds to see what happened in the first tenth of a second of flight. We could use the information about mass loss and drag and continue stepping through the flight of the rocket in 0.1-second increments to get an approximation of how high and how fast the rocket will fly.

The biggest problem with doing this is that this method is painfully tedious. And that's what computers are for. The equations in this chapter have been encoded in a very simple BASIC program called Juno Simulation, which appears in Appendix A. This program is a text-based program with no inputs. You can change the coefficient of drag, the mass of the rocket without a motor, the motor, and other characteristics near the start of the program. Running this program with a time step of 0.1 seconds predicts Juno will fly to an altitude of 265 meters (870 feet) on a C6-7 motor. We'll look at how that compares to the other simulators and to an actual flight of Juno at the end of the chapter. Here is the output for the program for Juno on a C6 motor:

```
Maximum altitude = 265.127533 meters.
Maximum velocity = 96.716843 m/s.
Maximum acceleration = 230.18779 m/s^2.
Ideal ejection delay = 4.700055 seconds.
```

Of course, using time steps of 0.1 seconds introduces error. Reducing the time step helps. If you have an understanding of differential equations and how to solve them numerically, you also know there are great ways to reduce the error without decreasing the time step, such as the Runge-Kutta method. That's beyond what we will get to in this book, but if your education has taken you to the point where you understand mathematical tools like Runge-Kutta, this is a great time to put them to use.

The text program works, but a user interface would be even better. User interfaces depend a lot on the device you are running on. Still, there is a second program in Appendix A that has the simple user interface shown in Figure 14-3. This one lets you change the coefficient of drag, rocket mass, frontal area of the rocket, air density, time step, and motor. It calculates the altitude, maximum speed, maximum acceleration, and ideal ejection delay.

Figure 14-3. *This iPhone program simulates a one-dimensional rocket flight. The source code for this program and a more portable text-based program can be found in Appendix A.*

This particular simulator is written in techBASIC, the technical computing language for iOS. The simulator is also available as a free standalone app on the iPhone, iPod touch, and iPad—search the App Store for SRockS, the Simple Rocket Simulator. The source code is available in the app, so you don't need to type it in; look in the *Maker Books* folder.

Predicting Altitude with a Rocket Simulator

There are a number of great model rocket simulators available. Back in Chapter 7 we took a look at OpenRocket, which we used to determine if a rocket would be stable. Let's use it again to predict what will happen when our rocket flies.

We already covered how to start OpenRocket, how to load the Juno simulation file, and how to load a motor —refer back to "Checking Stability with a Rocket Simulator" on page 158 if you need a refresher. Following those instructions, run OpenRocket and set Juno up for a simulation with an Estes C6-5 motor. Go ahead and click the "Run simulation" button. Once you do, your screen will look like Figure 14-4.

We get a lot of useful information right away. The first value is the velocity at the moment the rocket lifts off of the launch rod. From Chapter 7, you know that we want a velocity of at least 13.4 m/s at that moment. OpenRocket is telling us the liftoff speed is well above the minimum recommended value.

Next is the apogee, which is the rocket's maximum altitude. The value of 350 meters may seem a bit high if you worked through the last section, where our simulation showed the rocket would only reach 265 meters, but we'll get to at least part of the reason in a moment.

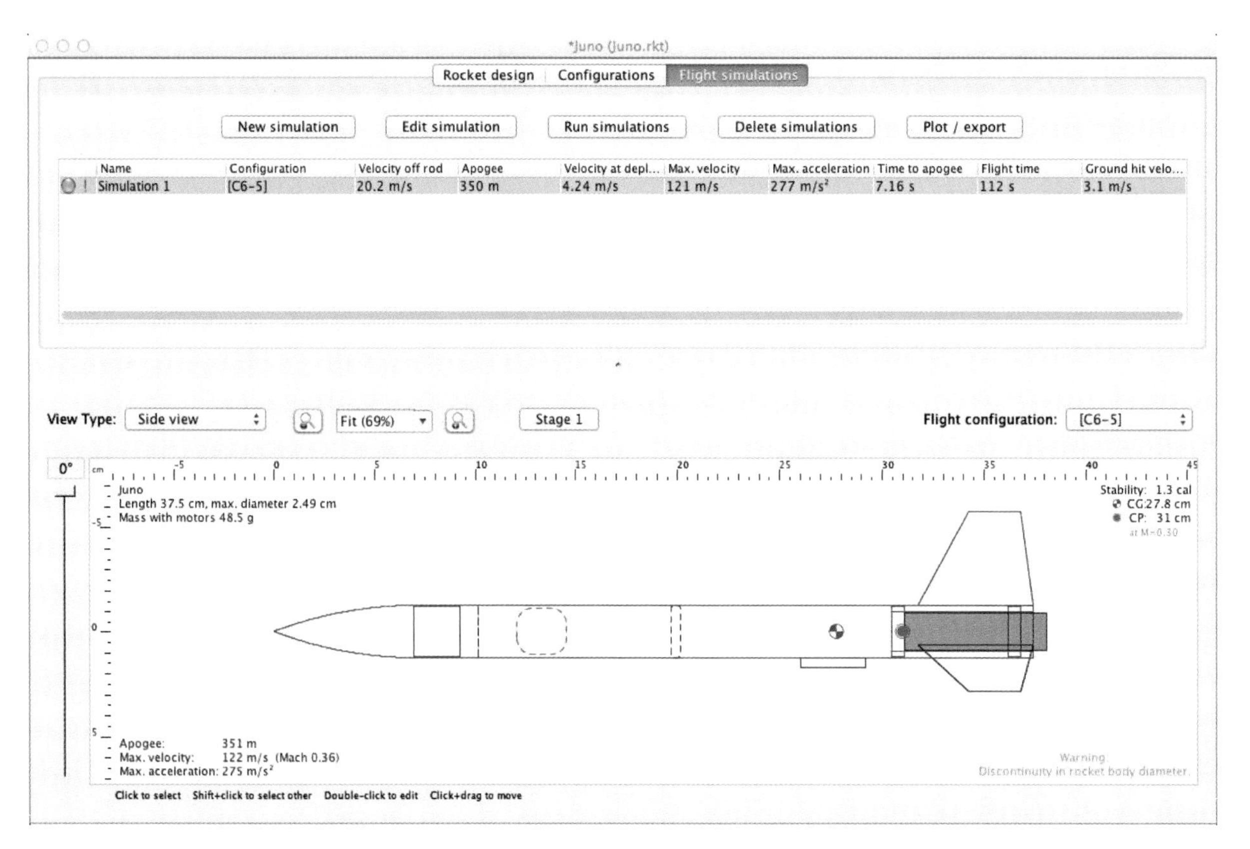

Figure 14-4. *OpenRocket after running the Juno simulation using the default configurations.*

The next number is velocity at deployment. This is the speed the rocket is moving at when the ejection charge fires. Very bad things can happen if this speed is too high: the parachute can rip, or the shock cord can break. If the shock cord is stronger than the paper body tube, the shock cord can literally tear apart the body tube! OpenRocket is telling us the speed of Juno at deployment is about 4 m/s, which is fine. If the speed gets above 10 m/s or so, try looking for a motor with an ejection charge that will fire closer to apogee.

Units

This chapter has a lot of calculations, and they were all done using the metric system. OpenRocket defaults to the metric system, so I'm leaving it that way for this chapter to match the other calculations. If you prefer other units, change the units in the program's preferences.

The next four values are interesting, but generally not something we need. They are the rocket's maximum velocity, maximum acceleration, time to apogee, and total flight time. The last value tells you the speed the rocket will be traveling at when it touches down. The predicted value of 3.1 m/s is slightly slower than the optimal recovery speed of 3.5 to 4.5 m/s cited in Chapter 10, but it is still a good value.

Of course, a lot of things can change the performance of the rocket. Let's see the way to determine how those are set for this flight and change them.

Rocket Finish

We saw in Chapter 13 that the quality of the finish on a rocket can change the speed at which the boundary layer changes from laminar to turbulent flow, and that in turn changes the amount of friction drag on the rocket. Double-click on the nose cone, body tube, or fins, and you will see a dialog like the one in Figure 14-5.

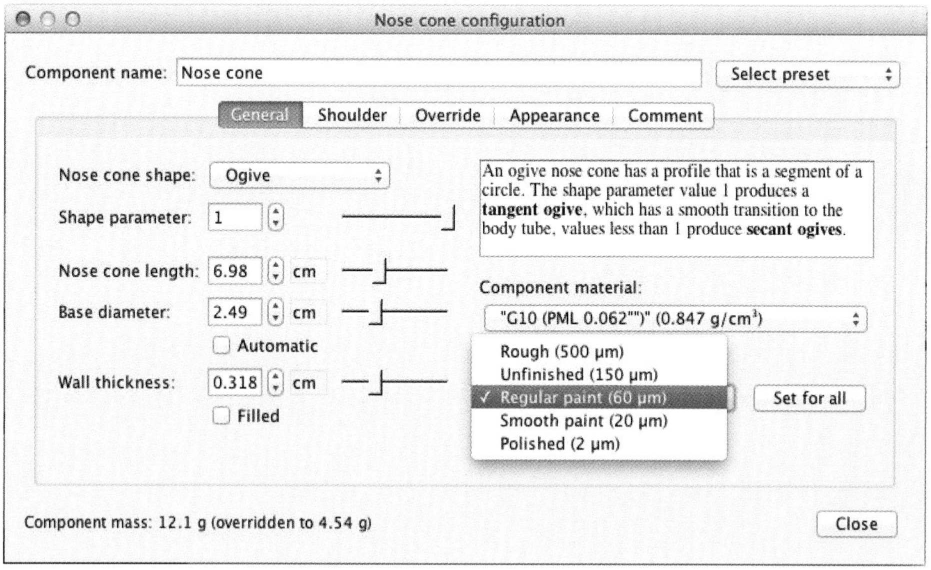

Figure 14-5. *All components have a pop up that shows the selected finish. Make sure your finish matches what is on the rocket as closely as possible to get the most accurate values for friction drag.*

One of the values you can change is the finish on the rocket. Most of the time, rockets are painted, and that's it. The difference between the finish on a rocket and the finish on, say, a car is the way the finish is sanded and polished. It's worth putting in some extra time and effort for a competition rocket, but probably not for most sport rockets.

Launch Conditions

Back in the main window, select the simulation and click on the "Edit simulation" button. You will see a dialog like the one in Figure 14-6.

This is where you set the conditions for your flight.

The wind conditions can have an effect on the flight trajectory by causing the rocket to tilt into the wind, especially when it is traveling slowly. For most purposes, you can just leave these values alone, but if you are really trying to predict the altitude for a specific flight as closely as possible, set these values as close to what you expect as you can.

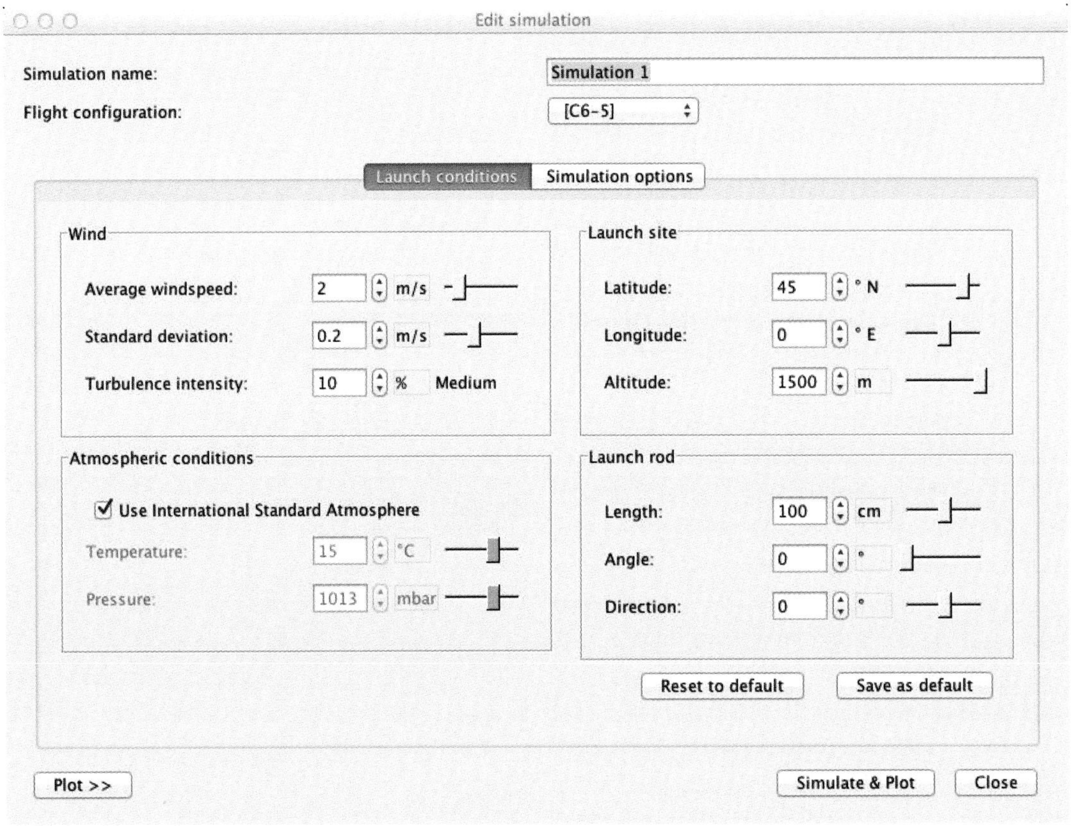

Figure 14-6. *Be sure to check the launch conditions for your launch before trusting the simulation.*

The latitude and longitude of the launch site are not that important for the low-power rockets in this book, but they become important when you start to fly high-powered rockets that can crack 100,000 feet in altitude. That high up, the shape of the Earth and the way it spins start to seriously affect the calculations.

The launch altitude is important for those of us who live significantly above sea level. You can see that I've changed the value to 1,500 meters (5,000 feet), which is about the altitude of my favorite launch sites here in Albuquerque. (One of the launch sites I fly at is over 7,000 feet!) As we saw in Chapter 13, the drag on a rocket depends on the density of air, and goes down as the rocket flies at or to higher altitudes.

The barometric pressure is also important. The local density of the atmosphere changes with the weather, and just like altitude, those density changes affect the drag on the rocket.

You don't need to change the pressure setting on the atmosphere model to match the altitude. That's because OpenRocket, like weather stations, calculates the barometric pressure after adjusting for altitude. You can change it to match the local weather conditions, though, such as a high- or low-pressure zone moving through, as reported by the local weather stations.

The speed of sound, instead of being a hard and fast number, actually varies with temperature; higher temperatures mean a faster speed of sound, and vice versa. Temperature is therefore very important when simulating a rocket flight, because the speed of sound has a big impact on how rockets fly if they travel at an appreciable fraction of the speed of sound. This effect has more impact on high-power rockets, which often hit the transonic region above Mach 0.8, or even the supersonic region. Refer back to Chapter 13 for details.

The launch rod section lets you change the characteristics of your launcher. The 100 cm launch rod length, for example, is a bit too long for most US launchers, which use a 36-inch launch rod—about 91 cm. You can also set the launch rod angle from the vertical and its direction relative to the wind.

The "Simulation options" tab lets you change the way the simulation is performed. Leave those factors alone until you really know what you are doing, or are experimenting to learn more *so* you really know what you are doing.

Rocket Mass

The last variable is the rocket mass. OpenRocket can estimate the mass of the rocket from the components you add, but, as always, using a measured value is better. For example, I used very thick primer and paint for my build of Juno so it would photograph well; as a result, my rocket weighs 36 grams. Based just on the parts I used to build the rocket, OpenRocket calculated that Juno should weigh 25.4 grams. Almost 30% of my rocket's mass is therefore cosmetic. Getting the weight correct is very important for payload rockets, so try to weigh your rocket if you can.

One way to account for the weight of payloads, paint, and such is to add what is called a *mass object* to the simulation. This can be thought of as adding a new part to the rocket that has weight, but no other purpose as far as the simulation is concerned.

Select the "Rocket design" tab at the top of the window, then click on "Body tube" in the list at the top left of the window. The "Add new component" section shows all of the components you can add to the rocket. Click on the "Mass component" button. You may need to scroll down to find it.

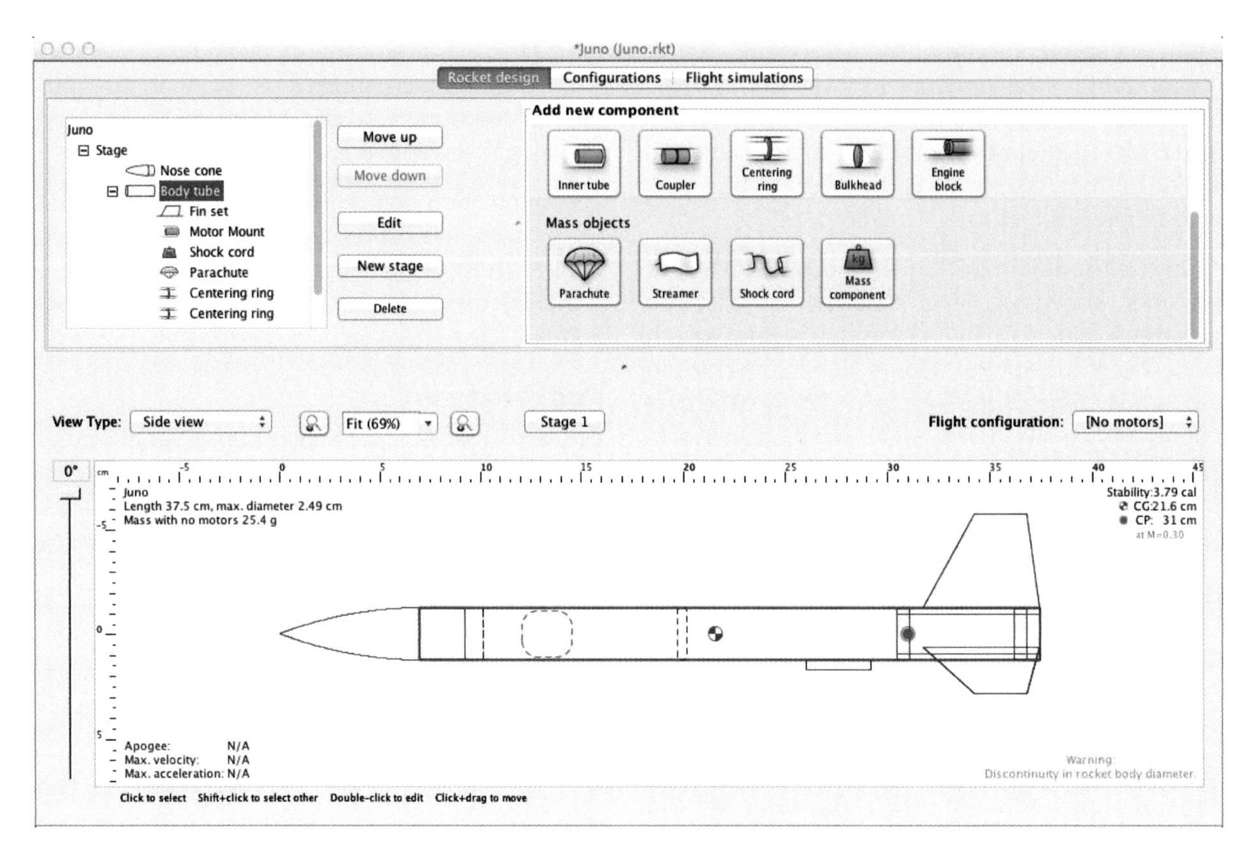

Figure 14-7. *Add a mass component to account for the weight of payloads, paint, and so forth. Select the component where the mass will be, then click on the "Mass component" button.*

A dialog like the one in Figure 14-8 will appear. Add the mass, in this case 10.6 grams. Change the location of the mass component using the "plus" value; for added weight from paint and such, it should be somewhere near the center of gravity of the rocket so it does not change the performance much.

All of these values have fine-tuned the simulation. With these changes in place, the simulator recalculated the predicted altitude from 350 m to 360 m. While the added weight had an effect, the lower drag due to the altitude in Albuquerque more than compensated.

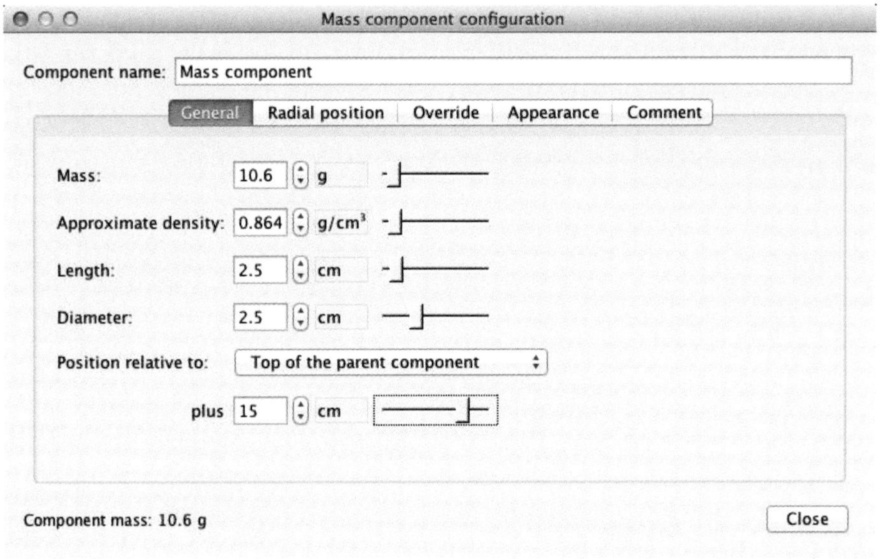

Figure 14-8. *Set the mass and location of the mass using the dialog that appears when you select the "Mass component" button.*

Comparing Simulators to Flight Data

So far in this chapter we've calculated the predicted altitude for Juno using two very different methods. Let's test these methods by measuring the actual altitude with a commercial altimeter. I inserted a MicroPeak altimeter into Juno's hollow nose cone and tried an actual flight with a C6-5 motor.

The original calculation, based on a one-dimensional simulation that duplicated the physics from "Predicting Altitude Using Physics" on page 305, predicted the rocket would reach 265 meters, but that was at sea level and without the altimeter. After correcting for the altitude in Albuquerque and the actual weight of the rocket with the altimeter, the simple simulator predicted an altitude of 290 meters. OpenRocket predicted 350 meters, or 360 meters after correcting for the actual weight of the rocket and the altitude in Albuquerque. That seems like a big difference: for the same flight conditions, the two simulations are 70 meters different (more than 20%!).

Another simulator, RockSim, predicted Juno would fly to 407 meters.

The MicroPeak altimeter said the actual altitude achieved on my flight was 307 meters.

Table 14-2 summarizes these results. It also shows the altitudes in feet, and the percentages by which each simulation varied from the measured altitude.

Table 14-2. Juno's predicted and measured altitude on a C6-5 motor

Method	Meters	Feet	% from measured
1D simulation	290	951	–6%
RockSim	407	1335	32%
OpenRocket	360	1181	17%
Altimeter	307	1007	

The point here isn't to criticize one simulation compared to another. With a different rocket, different flight conditions, or a different person making different assumptions while setting up the simulation, the results might also be very different. The point is to get some idea of how much trust we can place in simulations. Yes, they help. Like the spelling and grammar checkers I relied on while writing this book, though, they are not perfect. In the end, it's up to us, the rocket scientists, to understand the limitations of the tools we use and make reasonable judgments based on our knowledge and experience. Rocket simulators are great tools, but they are not exact. Checking the simulation with experimentation is always important.

And keep in mind that the altimeter is not exact, either! Due to the way altimeters work, it's probably closer than any of the simulators, but any measurement has error. The documentation for the MicroPeak altimeter claims it has an accuracy of 0.2 meters, but that is based on the accuracy of the barometer and ignores a lot of sources of error. Do you really think the altitude reached was 307 ± 0.2 meters? I don't. But then, I can't prove it either way. Gee, this sounds like a science fair project…

Minimum-Diameter Rockets

15

You saw in Chapter 14 that drag is one of the three forces that control how high and fast a stable rocket will travel; the others are the force of gravity acting on the mass of the rocket, and the thrust of the motor. You also saw that most of the drag on a rocket is proportional to the area the rocket presents in the direction of travel. That means one of the easiest ways to make a rocket go faster and higher is to slim it down. And that's what's done with Hebe.

Hebe: Skinny Rockets Fly Faster and Higher

Hebe's body tube is the same diameter as the motor, keeping the frontal area presented by the rocket as small as possible. That's what a minimum-diameter rocket really is; it uses the smallest possible body tube for a given rocket motor. This also helps keep the rocket light. This combination means Hebe will fly much, much higher on a given class of motor than Juno. Table 15-1 shows the expected altitudes for both rockets on equivalent motors. Juno doesn't use exactly the same motors, though. Because Hebe flies so much higher and faster on, say, a B6 motor, it needs a longer ejection charge delay. The ejection times on the motors shown are appropriate for Hebe, but not Juno.

Hebe

Hebe (pronounced "HEE-bee") is an asteroid you may have seen—or, at least, you may have seen a piece of it. Long ago, an impact on Hebe knocked off some chunks that became near-Earth asteroids, and these in turn got hammered into smaller pieces. By some estimates, 40% of all meteorites you see in museums on Earth originated from Hebe! If you've seen a few meteorites, there is a good chance you've seen a piece of Hebe.

Karl Hencke, a German astronomer, discovered the asteroid on July 1, 1847. With an average diameter of nearly 200 km, Hebe is a pretty big asteroid, even with all the pieces missing.

Hebe is named for the Greek goddess of youth.

Table 15-1. Recommended motors for Hebe and altitude comparison with Juno

Motor	Hebe altitude	Juno altitude
A8-5	530 ft	340 ft
B6-6	1,150 ft	760 ft
C6-7	2,230 ft	1,600 ft

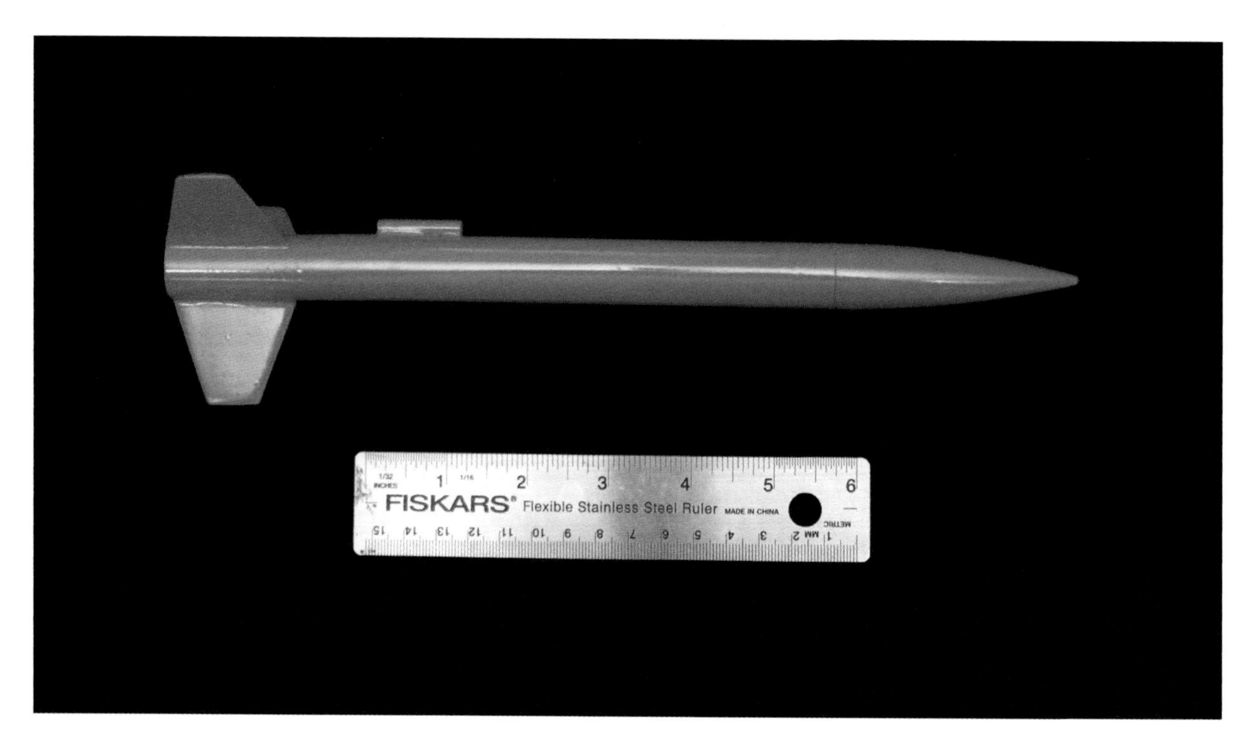

Figure 15-1. *Hebe is a minimum-diameter rocket, built for high-altitude flight.*

In fact, the ejection delays are still too short for the B6 and C6 motors, because the rocket is still going up when the ejection charge fires. Simulation shows Hebe is pretty close to apogee, however, so adding another second to the B6 and two seconds to the C6 so the rocket has time to reach apogee would only add 10–20 feet to the altitude. These also happen to be the longest ejection charge delays available with the motors listed.

Table 15-2 lists the parts you'll need to build Hebe. Additional items you may want to have on hand for this project are listed in Table 15-3.

Table 15-2. Parts list

Part	Description
BT-20 body tube, 7 3/4″	The main body tube for the rocket. This is a magical length for this rocket, not because of aerodynamics, but because there is a precut length of white BT-20 in the Estes Designer's Special! I've made many rockets similar to this one with 9″ lengths of BT-20, though, so if you buy an 18″ stock BT-20 body tube, just cut it in half.
BT-20 size nose cone, 2 3/4″	The plastic nose cone from the Estes Designer's Special or a similar plastic nose cone is a good choice for Hebe. We're going to modify the nose cone slightly to hold a MicroPeak altimeter.
1/16″ balsa fin stock; about 2 1/2″ x 6″	The balsa wood will be used for the fins.
1/8″ launch lug	Use about 1″, cutting it from a longer piece if needed.
Engine block	This is usually a short, dark gray tube about 2/10″ long that slides into the BT-20. Anything of similar size that will slide into the BT-20 will work.
1/8″ shock cord	You will need 14″ of shock cord.
Streamer, approximately 1 1/5″ x 32″	This can be made from just about any thin, strong material, and the specific dimensions are not critical. A two-to-three-foot piece of wide ribbon, Mylar from an old helium balloon, crêpe paper, or plastic from a large bag will work fine. The streamer shown is from the Estes Designer's Special.
String	Used to attach the streamer.
Tape	Used to attach the streamer.

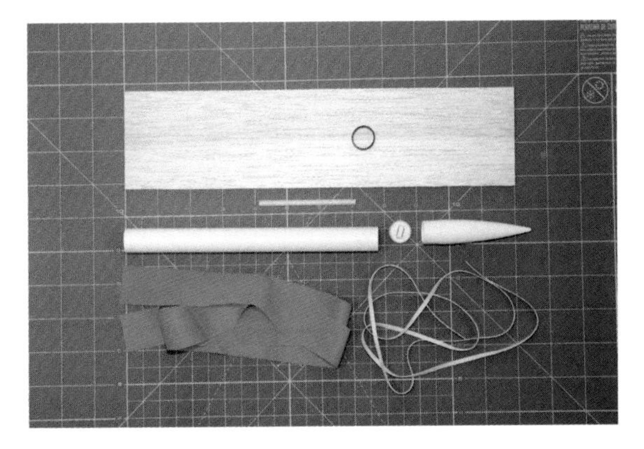

Figure 15-2. *Visual parts list.*

Table 15-3. Optional parts list

Part	Description
Snap swivel	You will need one or two snap swivels for attaching the streamer and shock cord.
MicroPeak altimeter	This tiny altimeter will fit in Hebe's nose cone.
Fine sandpaper	Use this if you want a very fine finish. You will need 800, 1000, 1200, and 1500 grit. This is available from auto paint supply stores, and can be found online at stores like Amazon.
Polish	Used to remove sanding imperfections. Get something like Turtle Wax T-415 Grade Rubbing Compound.
Wax	Any good grade of car wax will work.
Tower launcher	A tower launcher lets you fly without a launch lug. Your local club may have one.

Construction

Refer back to Chapter 3 for basic construction techniques, tools, and supplies.

The Estes Designer's Special comes with a precut length of BT-20 body tube that is 7 3/4″ long. That works great for this rocket. If you are working from a longer piece, start by cutting a 7 3/4″ to 9″ piece of body tube from the longer tube.

Most people make a very pragmatic choice when shaping fins for model rockets. You know from Chapter 13 that rounding the leading and trailing edges of fins cuts the drag on the fins in half compared to a flat leading and trailing edge. You also know that using an airfoil shape reduces it even more, cutting drag to about 16% compared to flat leading and trailing surfaces. So why don't we sand all of our fins into a nice airfoil shape?

Well, there are three good reasons. The first is that it's a lot harder to sand a fin into a good airfoil than to just round the edges a bit. It also reduces the strength of the fins, especially on the trailing edge that usually hits first when the rocket lands. Fins with a rounded edge are a lot stronger. The final reason is that drag is generally not that much of a concern. We are rarely trying to coax the most possible speed or altitude from a rocket, so why bother with the extra effort, especially since it weakens the fins? That's why the instructions for every other rocket in this book suggest rounding all of the edges of the fins except the root, which is glued to the body tube.

Still, you can always use an airfoil shape if you like. In fact, for this rocket, I suggest you give it a try. It's labor-intensive, but I know you're up for the challenge.

This book shows two good ways to sand an airfoil in a fin. The most accurate way involves building hardwood tips to aid in accurate sanding, but this is very difficult to do well with a thin fin. We will see this technique later, when we carve the airfoil for the wings of Icarus, a rocket glider described in Chapter 20. The method we'll use here is not as accurate, but it's pretty easy to do even on thin fins.

First, cut five fins, not three. Sanding an airfoil is a bit of an art. If there is too much balsa, you can always sand off more, but if you sand off too much, the fin is ruined. Unless you're a lot better at sanding than I am, the fins also won't be exactly the same shape. The more time you take, the closer you are likely to get them, but not all of the fins will be perfect. Once you finish sanding, take a close look at your work with a very critical eye and select the three best fins.

Figure 15-3. *Cut five fins from 1/16" balsa using the fin pattern in Figure 15-4, then copy and cut out the airfoil templates (Figure 15-5).*

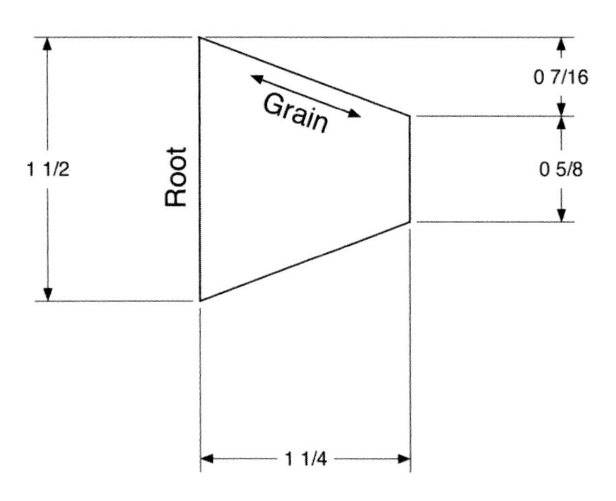

Figure 15-4. *Fin pattern for Hebe.*

As for the actual shape, the NACA006 airfoil shown in Figure 15-5 is a great choice. Like most low-drag subsonic airfoils, this one is symmetrical, with the thickest portion 1/4 of the way back from the front of the fin. The front is rounded, while the back tapers to a point. It's been adjusted to account for the swept leading edge on the fin, so the shape of the fin at the tip is proportionally a little thicker compared to the length than the shape at the root of the fin. These shapes were created from a graph at Wikimedia Commons (*http://bit.ly/UcqDCW*). I used Photoshop to adjust the size to 1 1/2" for the root, and then shrank the airfoil horizontally to 1" and 5/8" to match the length of the fin roughly at the center and tip. Cut the templates following the blue line and the top of the airfoil. This gives a template that will sit flat on the table top. As you sand, check one side of the fin while it is lying on a flat surface, then flip it over and check the other side.

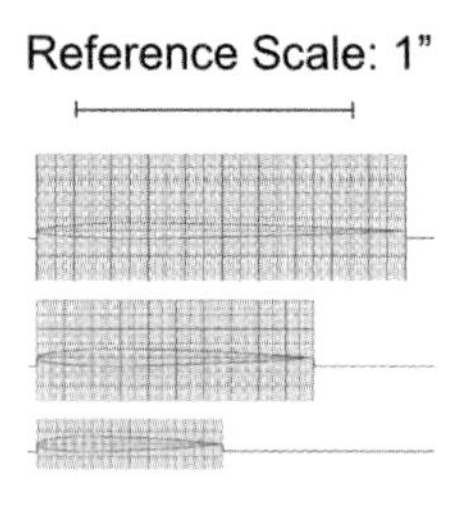

Figure 15-5. *Templates for checking the shape of the airfoil. Full-size patterns are available on the author's website (http://bit.ly/byteworks-make-rockets).*

You might wonder why I changed the shape of the airfoil as the cord of the fin changed, rather than leaving the shape the same and making the fin get gradually thinner at the tip. If so, give yourself an A+ for aerodynamic intuition. You know from Chapter 13 that the airfoil shape should stay the same, so the fin should be thinner at the tip than at the root. There are two reasons why I didn't suggest doing this, though. The first is that it's really hard, especially on a thin fin. The second reason is that it weakens the tip of the fin.

Sand the fins using a sanding block, starting with the block touching the thickest point on the airfoil, as shown in Figure 15-6. Gradually move the sanding block toward the edge of this fin as you sand back and forth along the direction indicated. This sands the leading and trailing edges more than the center, since the sandpaper is in contact with the edge the entire time you sand, but only touches the thick portion of the airfoil for a short time. Use the patterns to check the shape regularly, and adjust the amount of time you spend on various parts of the fin as needed to get the right shape.

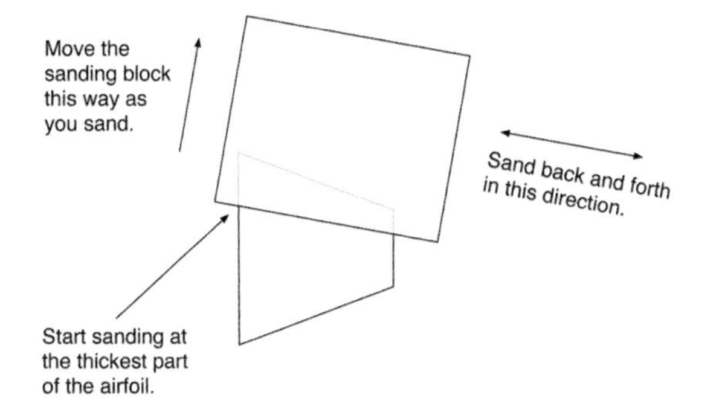

Figure 15-6. *Here's how to sand the leading edge. The trailing edge is similar.*

Don't try to bring the trailing edge of the fin to a perfect point. The balsa isn't strong enough to stand up to flight, let alone landing, if it gets too thin. You get almost all of the benefit of the airfoil shape as long as the trailing edge is no more than 8% as thick as the thickest part of the airfoil. Even if you don't sand them to a sharp edge, you will notice the fins are getting pretty weak along the trailing edge. We'll strengthen them later.

Figure 15-7 shows the airfoils as they near completion. Note that the fins are not perfect. The human brain is very good at comparative pattern matching. Stacking the fins this way for an occasional check points out the problems.

Leave the tips of the fins flat. This reduces induced drag.

Figure 15-7. *Comparing the fins to each other from time to time helps find flaws.*

When you're done with the fins, cut a 1" piece from a longer launch lug. In Figure 15-8, you see a 1/8" x 3" launch lug cut into three pieces.

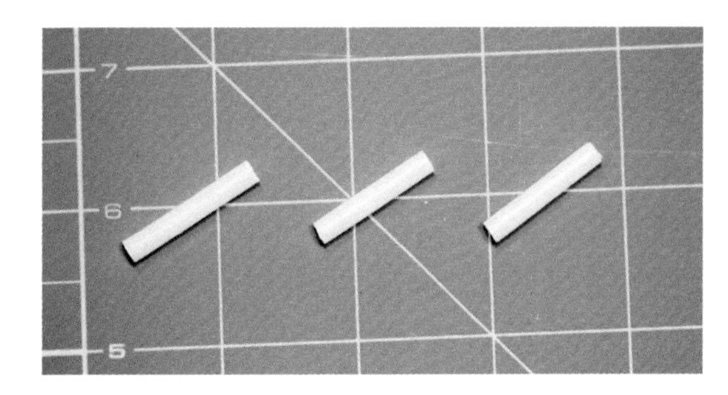

Figure 15-8. *Cut a 1" piece of 1/8" launch lug.*

Before you decide to glue on the launch lug, though, it's worth discussing an alternative. Various sources put the contribution of drag from the launch lug between 8% and 23%. One option is to leave off the launch lug, but if you do that you will have to use a very special launcher called a *tower launcher*. Instead of a single rod, tower launchers have three rails. These are usually adjustable so you can change the size of the gap between them to accommodate different sizes of body tubes. With this arrangement, the rocket can be launched without a launch lug at all, which reduces the overall drag on the rocket.

Building or buying a completely separate launcher for minimum-diameter rockets is quite a commitment, and one you probably don't want to make for just this rocket. An alternative is to join your local rocket club and use theirs. Figure 15-9 is a photo of the rail launcher from the Albuquerque Rocket Society.

With all of the pieces cut and sanded, it's time to move on to assembly. Start by marking the body tube for three fins and a launch lug (if you're using one) using the fin guide from Chapter 3 (Figure 3-24).

Figure 15-9. *A rail launcher is used for rockets that do not use a launch lug. The rails form a cage around the rocket to guide it for the first few feet of flight. This launcher is a bit large for Hebe, but it is large enough to see the tower itself. It's holding a high-power rocket flown by Todd Kerns, standing with his rocket as he gets it ready for launch. Photo by Jim Jewell.*

The motor mount in a minimum-diameter rocket like Hebe is very different from the one you saw on Juno. There is no engine hook, and there is no separate body tube for the motor. Instead, you'll glue an engine block into the body tube to prevent the motor from sliding forward into the rocket. Of course, the motor will push back when the ejection charge fires. With no engine hook to hold it in, we need a way to make sure the streamer pops out, not the motor. We'll discuss specifics in the next section, where we describe flight preparation, but basically we'll wrap tape around the motor until it is snug in the body tube. This is called a *friction fit*.

Use a stick to apply a liberal amount of wood glue about 2 1/4" into the body tube. Put the engine block into the tube and, using your motor, push the block forward into the glue so the motor sticks out 1/4" from the back of the rocket. Pull the motor out right away.

Figure 15-10. *Glue the engine block 2 1/2" into the body tube. Use the motor as a ruler. It will stick out 1/4" when the engine block is pushed in far enough.*

Glue the fins and launch lug into place. The fins go flush with the back of the body tube, which is the same side where the engine block was mounted. The launch lug should be glued 2 1/2" from the base of the rocket. Set the assembly aside to dry.

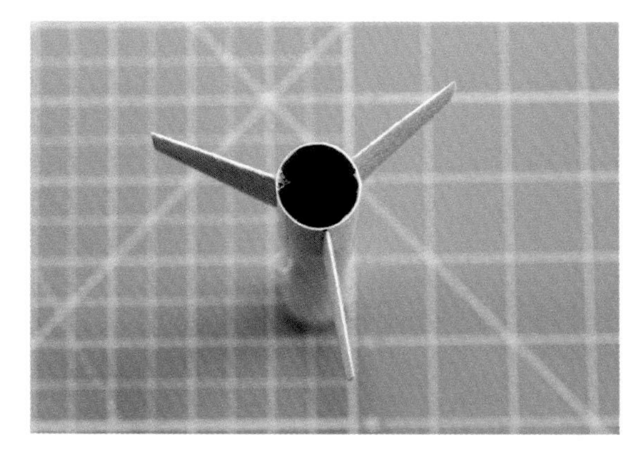

Figure 15-11. *Glue the fins and launch lug in place. They are on the same end of the rocket as the engine block.*

Let's work on the nose of the rocket while the fins dry. The instructions that follow show how to mount a MicroPeak altimeter in the nose cone. If you don't plan to use the MicroPeak altimeter, just glue the two pieces of the nose cone together.

The altimeter cannot be exposed to the ejection charge. Take a look at the base of the nose cone—there is a gap that will allow the gases to pressurize the forward part of the nose cone. If we install the altimeter in the nose cone, but leave this gap open, the gases from the ejection charge will leak into the nose cone itself when the ejection charge fires. We need to block that hole so the altimeter does not feel the pressure from the ejection charge when the streamer is ejected.

Cut a piece of paper that just fits into the base of the bottom of the nose cone. Plain writing or printing paper will work fine. The paper won't block the ejection charge, of course, but it's really just a support. Mix enough epoxy glue to generously coat both sides of the paper, then insert the coated paper into the bottom of the nose cone base, as seen in Figure 15-12. Use tape, not glue, to attach the two pieces of the nose cone, since you need to get it apart to insert and remove the altimeter.

The altimeter itself is a pretty snug fit. Stuff some cotton into the end of the nose cone as padding, then insert the altimeter into the top of the nose cone.

Add a ring of tape around the base of the nose cone to attach it to the top of the nose cone. The nose cone should be snug enough that it won't fall out of the body tube if you turn the rocket over and shake it, but not much tighter. If the tape makes the nose cone too tight in the body tube, sand the nose cone a bit to reduce its diameter. If it's too loose, add more tape. You will want to check (and often replace) this tape before each flight, so be sure you have tape in your range box.

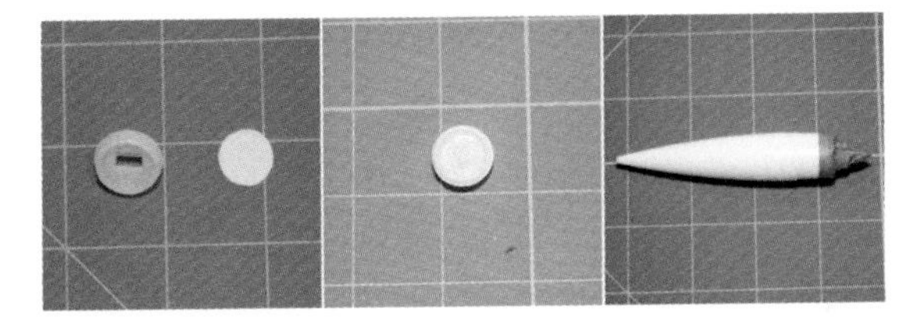

Figure 15-12. *Use epoxy to glue a paper disk in the base of the nose cone to keep the ejection charge gases from pushing into the nose cone. Use tape to hold the two parts of the nose cone together.*

Moving back to the base of the rocket, we see that the fins, thinned by sanding, are fairly weak. Reinforce them with thin CA glue. Dribble generous amounts on each surface of the fins. The CA glue will soak into the wood and strengthen it. Don't worry if a little glue gets on the body tube, too. It will also strengthen the body tube.

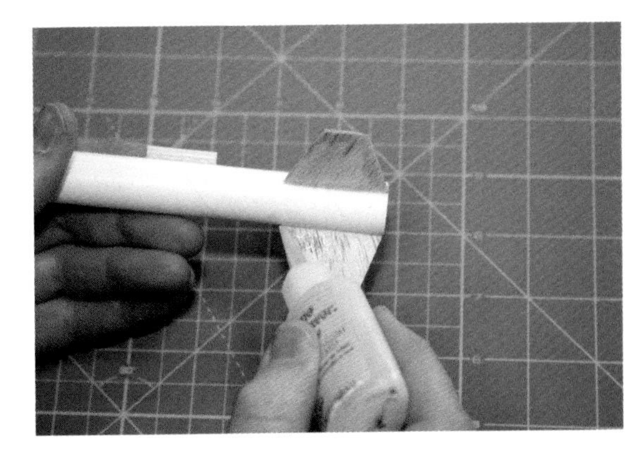

Figure 15-13. *Soak the fins in thin CA glue to strengthen them.*

Once the CA glue dries, apply the standard wood glue fillets where the fins join the body tube.

When everything is dry, install the shock cord. Attach one end to the inside of the body tube, just like you did for Juno. Tie the other end of the shock cord to the base of the nose cone using a buntline knot (see Figure 3-56), and add a small dab of glue to keep the knot from coming undone.

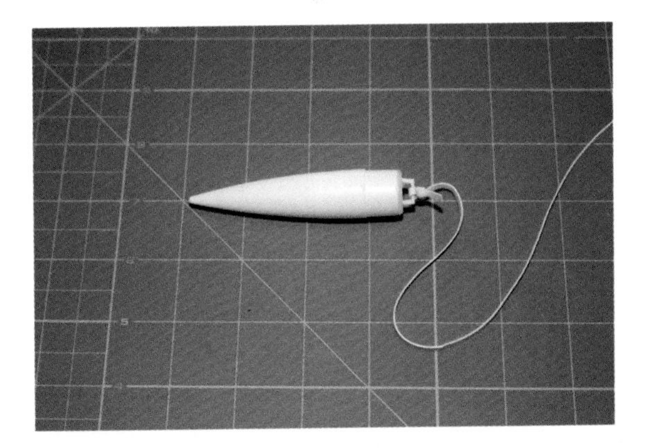

Figure 15-14. *Tie the shock cord to the nose cone using a buntline knot.*

An altimeter uses a barometer to sense air pressure. Insert the nose cone in the rocket and drill a 1/16" hole 3/8" below the joint between the nose cone and body tube. This vent hole will let outside air into and out of the nose cone so the altimeter's barometer will work. Ideally this hole would be one body tube diameter behind the seam or curve from the nose cone, but this nose cone is not long enough to make that possible. This placement of the vent hole can cause noise in the altimeter readings due to irregular air flow past the hole, but in tests, the results are good.

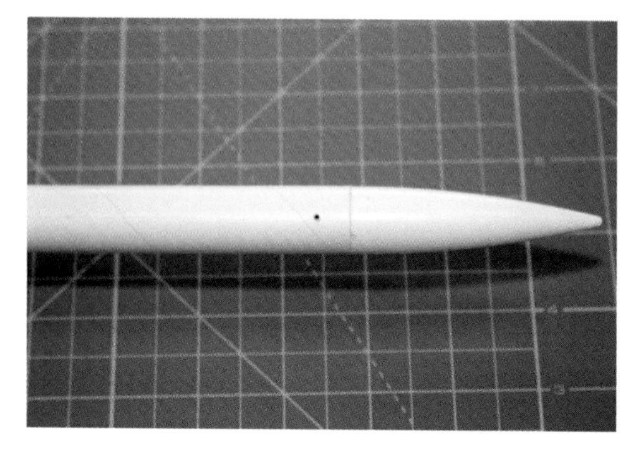

Figure 15-15. *If you are installing the optional altimeter, drill a 1/16" hole 3/8" from the seam between the nose cone and body tube to provide a vent hole for the barometer on the altimeter.*

This completes the rocket itself. You can finish it the same way as you did Juno, but this is a high-performance rocket, so perhaps it deserves a high-performance paint job. We'll look at some finishing tips in a moment.

Figure 15-16. *Once everything dries, the rocket is ready for primer and paint.*

Recovery

Hebe is a very light rocket. While a parachute would work for recovery, it would also carry the rocket down very, very slowly, and the rocket is going to start out very, very high. That's a good way to lose a rocket. We will use a streamer instead of a parachute.

A streamer is basically just a long ribbon, usually plastic, that flaps in the wind as the rocket falls. That's all you really need for a light rocket like this one.

There is a 32" streamer in the Estes Designer's Special; that's a great length for the streamer. Use a small piece of tape to fasten about 4" of string to the streamer. Put a loop in the string under the tape for added strength. Tie the other end to the base of the nose cone, or better yet, to a snap swivel so you can switch the streamer from rocket to rocket.

Crêpe paper also makes a great streamer. See "Streamer Design" on page 234 in Chapter 10 for details.

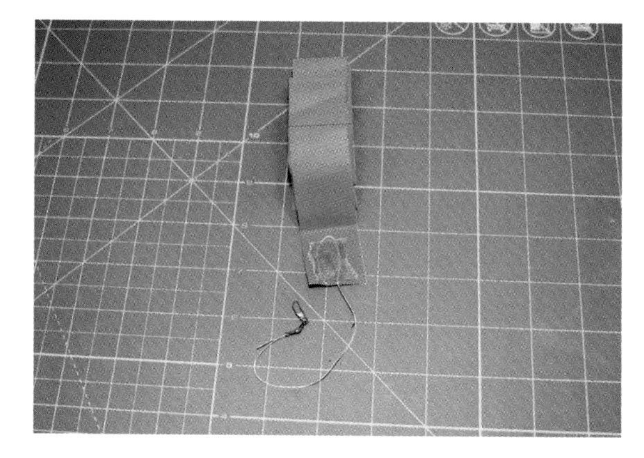

Figure 15-17. *Build the streamer using a 32" piece of plastic ribbon (or something similar). Attach a short piece of string using tape, then tie it off to a snap swivel or the nose cone of the rocket.*

High-Performance Finishes

You probably take pride in your work, and put some effort into painting Juno. If you used spray-on enamel, you were using the same type of paint used to paint a car. Still, the finish on Juno won't have been anywhere near as smooth as the finish on a car. That roughness isn't just cosmetic. As we saw in Chapter 13, that minor roughness in the paint also increases the friction drag. Based on RockSim simulations, using a polished finish like you find on a car increases the altitude of a flight with Hebe using a C6-7 motor from 1,979 feet to 2,252 feet when compared to a matte finish like the one on Juno. That's a 14% increase in altitude!

The first few steps in a high-performance finish are pretty much the same as those you took with Juno (see Chapter 3). Begin by applying a good-quality sandable primer. Once it is dry, sand it until you see the original material showing through. Repeat until the rocket is perfectly smooth.

The joint where multiple colors meet forms a small bump, so apply a single color of paint to the entire rocket. Use several light coats of enamel paint.

At this point you have a finish like the one on Juno. It's good. If you want to take it to the next level, you need clear paint, some very fine grit sandpaper, polish, and wax. You will want 800 grit, 1000 grit, 1200 grit, and 1500 grit sandpaper. You can buy fine grit sandpaper at auto paint centers or order it online through retailers like Amazon. You will also need a polish, such as Turtle Wax T-415 Grade Rubbing Compound. Car wax is perfect for the last step.

Make sure the paint is absolutely dry. This doesn't mean dry to the touch; it means *really* dry, as in waiting at least a day, and preferably two or three days. Dampen the 800 grit sandpaper and gently sand the entire rocket in a circular motion. Wipe it down with a microfiber cloth to remove the paint and sandpaper particles.

Add two or three light coats of clear enamel paint. Again, let this dry completely. Wet sand it with the various grits of sandpaper, working your way from 800 grit to 1000 grit to 1200 grit, and finally to 1500 grit. The finish should be very smooth to the touch at this point, but there will still be some very fine sanding marks in the finish.

Buff the entire rocket with polish. A Dremel tool with a fiber head works great for this step. You are finished when there are no visible marks or swirls from the sandpaper.

Finally, apply a light coat of car wax.

Is it worth it? That depends. A fine paint job takes a lot of time and effort. I would not bother if this is your first minimum-diameter rocket. It's going to fly really well even with a standard paint job. That extra 14% is pretty nice for a contest, though.

Flying Hebe

Start with the A8-5 motor for the first launch. This gives you a pretty good chance of not losing the rocket.

While Hebe uses a streamer and Juno uses a parachute, the big difference between flying the two rockets is the way the motor is inserted. Hebe does not have an engine hook to hold the motor in place. Its engine block keeps the motor from pushing forward into the rocket, but when the ejection charge fires, it is going to pop out the back of the rocket unless the motor is held in tighter than the streamer and nose cone.

Test fit the motor. It will almost certainly slip easily into and out of the body tube. Wrap some tape around the base of the motor and try again. Keep adding tape until the motor fits tightly. It should be tight enough to make it difficult to get the motor out, but not so tight you risk damaging the rocket sliding it in or out.

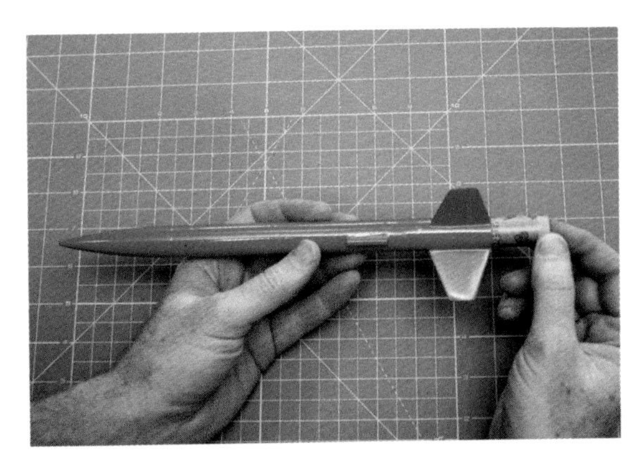

Figure 15-18. *Wrap tape around the motor for a snug fit, then push it into the body tube.*

With the motor in place, stuff about 1 1/2" of recovery wadding into the body tube. Push as much of the shock cord as possible into the body tube. Roll the streamer into a tight bundle, then wrap the bundle loosely with the string to keep it rolled. Stuff the streamer into the body tube, but make sure it is not a tight fit.

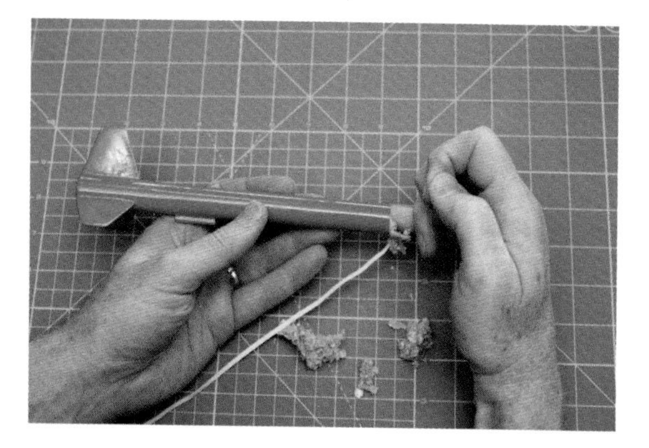

Figure 15-19. *Put about 1 1/2" of recovery wadding into the body tube.*

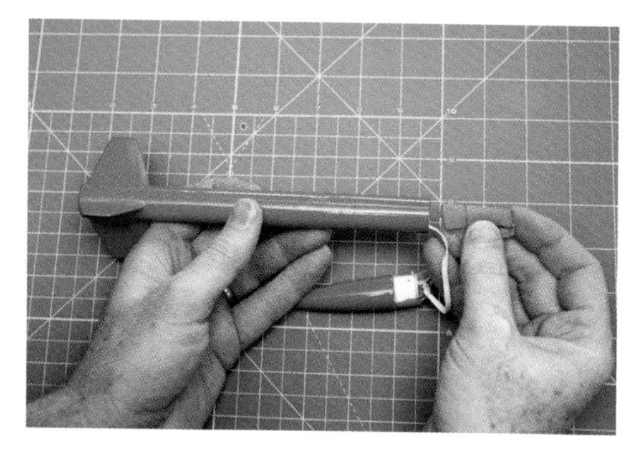

Figure 15-20. *Roll up the streamer, wrapping it in string, then put it into the body tube.*

If you are flying with a MicroPeak altimeter, pull the nose cone apart, start the altimeter, and then insert the altimeter in the nose cone. Tape the base of the nose cone securely to the top. Check the fit—the nose cone should not fall out when the rocket is held upside down, but it should not be tight, either. Adjust the amount of tape or sand the nose cone to get the proper fit.

Your rocket is almost ready to fly. Install the igniter, place the rocket on the pad, and go through your launch procedures, just like you did with Juno. Use the complete flight preparation and launch checklists for the flight.

Hebe should go noticeably higher than Juno on the same size motor!

Hebe Launch Preparation Checklist

1. Insert the motor, using tape for a snug fit.
2. Insert recovery wadding.
3. Attach the streamer.
4. Insert the streamer and shock cord; make sure they are not too tight.

Hebe Flight Checklist

1. Clear spectators from the launch zone.
2. Install the igniter.
3. Remove the launch key from the launch rail, slide the rocket onto the launch rail, and replace the launch key on the launch rail.
4. Attach the igniter clips.
5. Get the launch key from the pad.
6. Insert the launch key. Make sure the continuity light is on.
7. Check for clear skies—no aircraft or low clouds.
8. Count down from five and launch.
9. Remove the launch key and place it on the launch rod.
10. Recover the rocket.

16 Mini-Rockets

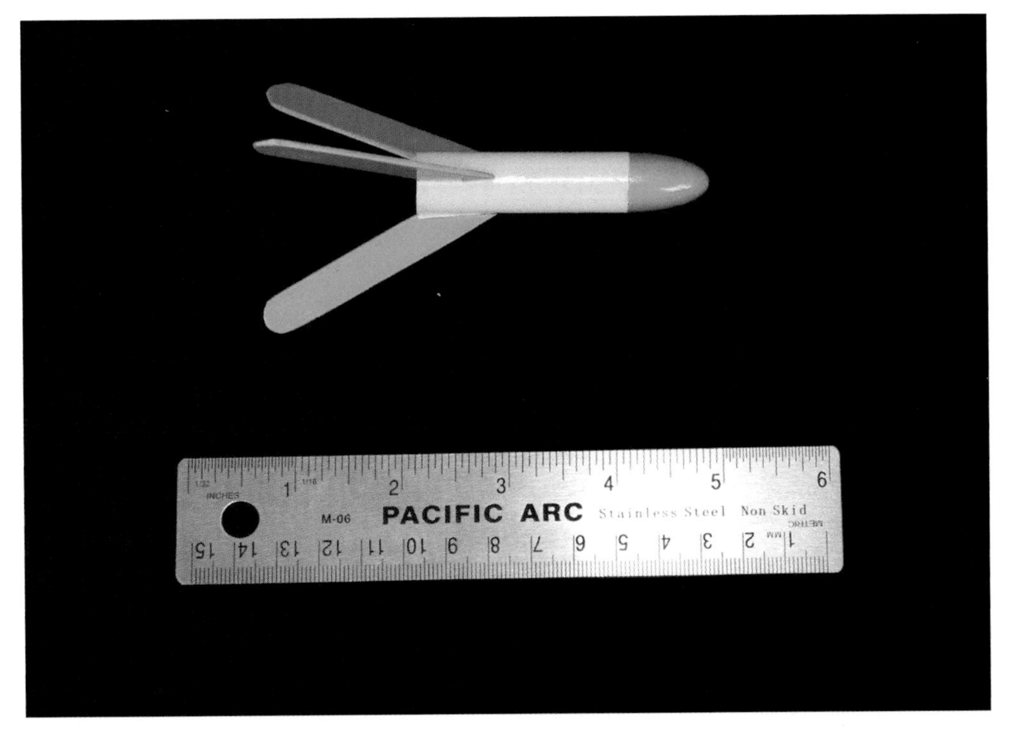

Figure 16-1. *Toutatis, a featherweight recovery rocket that is slightly over 4" tall.*

Model rockets come in all sizes, from the common 12–24"-tall rockets to impressive 6-foot Goliaths. Some of my favorites go in the other direction, though. We're going to take a look at two mini-rockets in this chapter. These rockets use the undersized T motors, which measure only 1 3/4" long and 1/2" in diameter. They are fast, easy builds, and look like they won't do much—but don't let the size fool you. These rockets can easily exceed altitudes of 1,000 feet!

Along the way, we'll also explore two new recovery techniques: featherweight recovery and nose-blow recovery.

Toutatis: The Powerful Midget Rocket

We'll start with the 4" Toutatis. Don't let its size fool you! This tiny titan is so small and light that the T-series motors can loft it quite high (see Table 16-1 for a list of recommended motors and expected altitudes). In fact, I don't recommend using the larger motors except in wide-open fields and perfect flying conditions. The rocket is small and hard to track visually; without a parachute to make it more visible during descent, it's very easy to lose.

Table 16-1. Recommended motors

Motor	Approximate altitude
1/4A3-3T	460 ft
1/2A3-4T	800 ft
A3-4T	1,370 ft
A10-3T	1,000 ft

Table 16-2 lists the parts required to build our first mini-rocket.

Table 16-2. Parts list

Part	Description
BT-5 body tube, 2"	The main body tube for the rocket. This will be cut from a longer tube. There is a 4" long BT-5 tube in the Estes Designer's Special, or you can order an 18" tube.
BT-5 size nose cone, 3/4"	Any BT-5 nose cone of balsa or plastic will do. The nose cone shown is from the Estes Designer's Special; similar nose cones are available from many sources. A blunt nose cone is best; see the flying instructions for an explanation.
1/16" balsa fin stock; about 1 1/3" x 2 1/2"	This is for the fins.
1/8" launch lug	Use about 1", cutting it from a longer piece if needed.

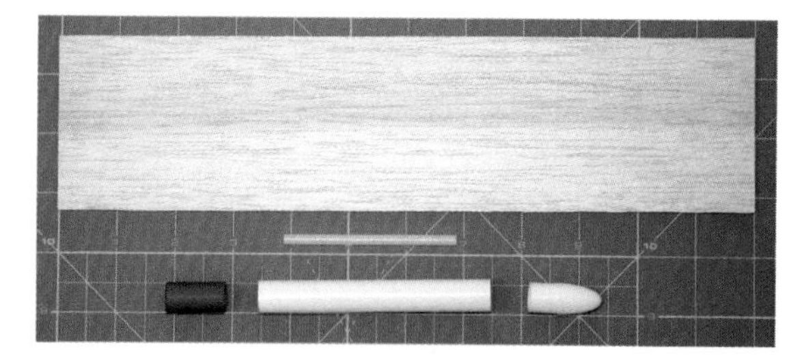

Figure 16-2. *Visual parts list. The red tube coupler is actually a tool, not part of the finished rocket. They don't get any simpler than this one!*

Toutatis is an apt namesake for the smallest rocket in this book, since it is one of the smallest named asteroids. Discovered in 1989 by the French astronomer Christian Pollas, Toutatis is a potato-shaped asteroid that is only 4.5 km across at its widest point.

It's also one of the family of near-Earth objects, which are asteroids and comets that, according to NASA, come near "the Earth's neighborhood." On December 12, 2012, it passed less than 4.5 million miles from the Earth. The Chinese lunar probe Chang'e 2 got some photos during this near encounter.

Toutatis was named for a Celtic god, generally thought to be a tribal protector.

Construction

Refer back to Chapter 3 for basic construction techniques, tools, and supplies.

The exact length of the body tube will depend a bit on the nose cone you select. That's because the nose cone also functions as the engine block in this rocket. We want the motor to stick out the back by about 1/4". Most BT-5 nose cones have a shoulder length (the length of the nose cone that sticks into the rocket) of 1/2", and the motors are 1 3/4" long. A little quick arithmetic or laying the parts out next to a ruler will show that you want a 2" body tube. If you end up with a nose cone with a 3/4" shoulder, increase the body tube length to 2 1/4", and so on.

The Estes Designer's Special comes with two 4"-long BT-5 body tubes. These are the white tubes that are about 1/2" in diameter. If you are buying parts *à la carte*, the BT-5 body tube will probably be 18" long. Either way, you will need to cut the body tube to size. Save the extra piece—we'll use it to build Eros later in this chapter.

Figure 16-3. *Cut a 2" length of BT-5 body tube from a longer piece. As shown in Chapter 3, use a support when cutting the tube. Here, a tube coupler is used. An old motor works well, too.*

Launch lugs come in a variety of sizes. The ones in the Estes Designer's Special are 3" long. Cut a 1" piece from a longer length of launch lug.

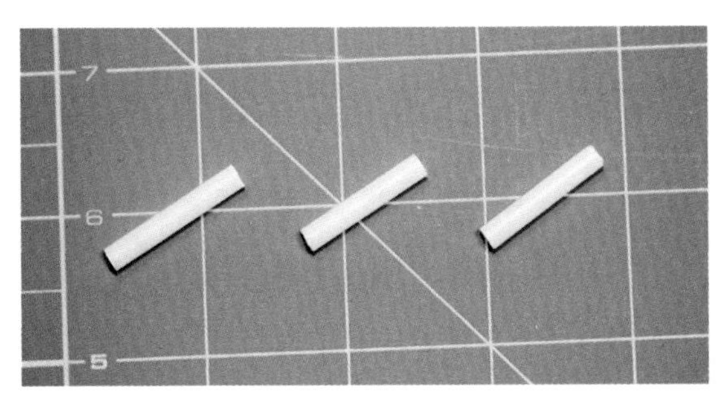

Figure 16-4. *Cut a 1" length of 1/8"-diameter launch lug.*

Cut three fins from 1/16"-thick balsa using the fin pattern in Figure 16-5. The grain direction is critical. The wood grain must run along the length of the fin, parallel to the leading edge.

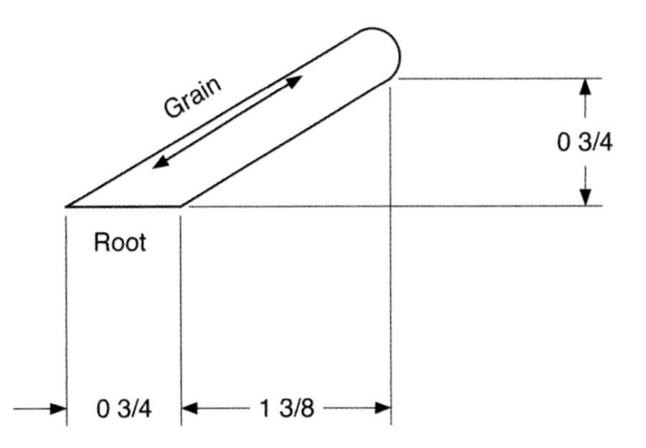

Figure 16-5. *Cut three fins from this pattern for Toutatis.*

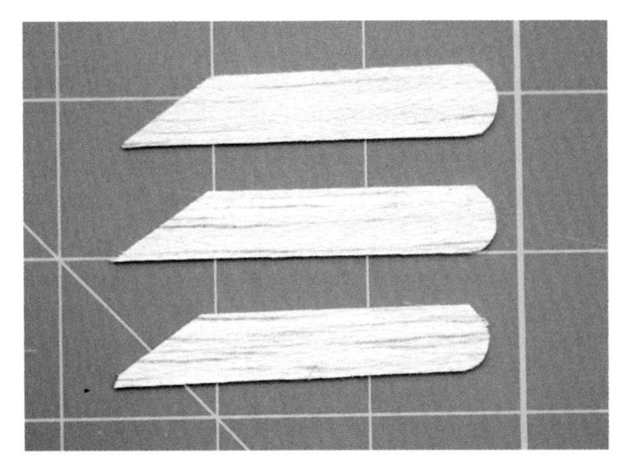

Figure 16-6. *Cut three fins from 1/16" balsa.*

Round all edges except the root edge, which will be glued to the body tube. Sand the fins smooth using medium grit sandpaper.

Use the fin marking guide from Figure 3-24 in Chapter 3 to mark the locations for the three fins and the launch lug. Glue on the fins and launch lug, checking the alignment from above to make sure they are straight.

The fins are at the base of the body tube. Position the launch lug 1/4" from the base of the body tube.

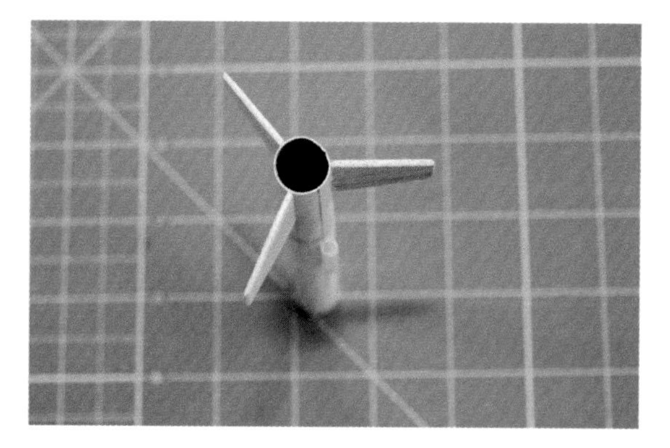

Figure 16-7. *Attach the three fins and launch lug.*

Once the fins are dry, glue the nose cone in the top of the body tube. The nose cone comes as two pieces. The base of the nose cone provides the attachment point for the parachute and shock cord. We don't need it in this rocket, so discard that piece, leaving just the top portion shown in Figure 16-2.

Apply fillets to the fins for extra strength. You can see one of the fins starting to sag in Figure 16-8; a quick swipe with a carefully designed fillet tool (my finger) fixed that issue.

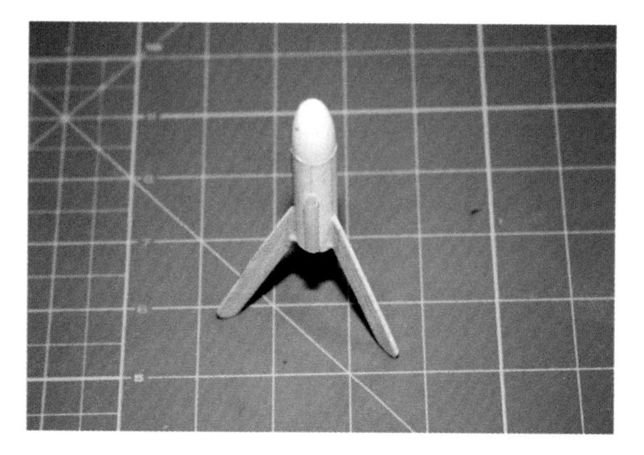

Figure 16-8. *Glue on the nose cone and apply fillets to the fins.*

After the glue dries, the rocket is ready for priming and final painting.

Flying Toutatis

This rocket is as easy to fly as it was to build. It uses a recovery technique called *featherweight recovery*. Basically, that means the rocket is so light for its size that you can let it fall without any recovery system. Let's see how that works.

Insert the motor. Like with Hebe in Chapter 15, this is a minimum-diameter rocket, so we've wrapped tape around the back of the motor to make sure it fits snugly in the body tube. Unlike with Hebe, though, the motor won't stay in the rocket. This time, we actually want to motor to pop out and fall separately to the ground when the ejection charge fires. Apply enough tape to make sure the motor doesn't fall out of the rocket before it fires, but don't go for a really tight fit.

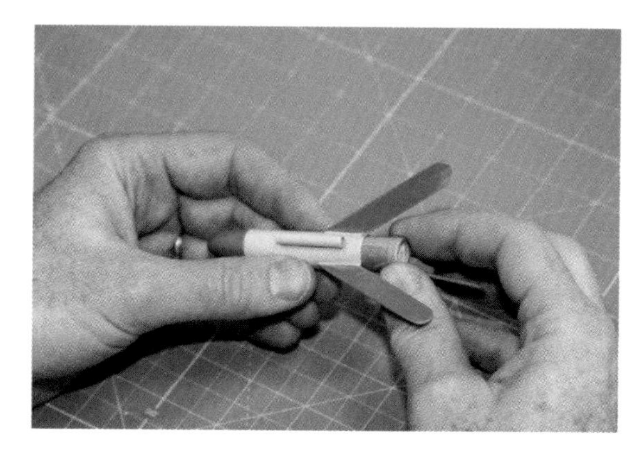

Figure 16-9. *The motor should fit snugly enough to stay in place until it fires, but not too tightly. It is supposed to pop out when the ejection charge fires.*

There is no parachute or other recovery mechanism to prepare. Install the igniter, slip the rocket onto the launch rod, and launch in the normal way. Toutatis will launch and fly very quickly, easily flying out of sight (even for someone with 20/20 vision!) on an A motor. The ejection charge will pop the motor out of the rocket, and both will fall separately to earth. They are light enough compared to their size that they will not do any damage.

Now, being lightweight does not mean the rocket is unstable. In fact, it's still very stable after the motor is ejected. The rocket will continue in a ballistic trajectory, rising to its apogee, then turning over and coming right back down. That's why the parts list recommends a blunt nose cone. That's also why the motor must have an ejection charge—an A3-0T, for example, is not suitable. The extra weight of the motor on a very stable ballistic rocket makes the combination heavy enough to do a little damage if it hits something delicate (you, for example).

All rockets should be flown on a small engine for their maiden flight. That's especially true for Toutatis. By now, you've built and flown a few rockets, and you know it's the parachute you usually see as the rocket descends. Toutatis doesn't have a parachute, and is very difficult to spot. Use smaller motors until you are used to spotting the rocket—or searching for it on the ground!

So what about the motor? It too will fall to the ground, but lacking fins, it is unstable. It will tumble, and probably take a little longer to get to the ground than the rocket. You may not find the motor, but if you do, be sure to collect it and place it in the garbage.

Toutatis Launch Preparation Checklist

1. Insert the motor, using tape for a snug fit.

Toutatis Flight Checklist

1. Clear spectators from the launch zone.
2. Install the igniter.
3. Remove the launch key from the launch rail, slide the rocket onto the launch rail, and replace the launch key on the launch rail.
4. Attach the igniter clips.
5. Get the launch key from the pad.
6. Insert the launch key. Make sure the continuity light is on.
7. Check for clear skies—no aircraft or low clouds.
8. Count down from five and launch.
9. Remove the launch key and place it on the launch rod.
10. Recover the rocket.

Eros: A Small Rocket That Flies High

Our second mini-rocket is called Eros. Table 16-3 lists the recommended motors for this rocket, and the expected altitudes.

Table 16-3. Recommended motors

Motor	Approximate altitude
1/4A3-3T	250 ft
1/2A3-4T	580 ft
A3-4T	1,190 ft
A10-3T	1,090 ft

Eros uses a recovery system that is very similar to the featherweight recovery used by Toutatis, but there is a key difference. While there is no streamer or parachute, the nose cone does pop off when the ejection charge fires. This makes the rocket unstable, so it will tumble down more slowly than if it remained streamlined. Preparation and launch of Eros is almost as easy as with Toutatis, yet this rocket flies just like the larger cousins you've built up to this point.

Table 16-4 lists the parts you'll need to build Eros.

Table 16-4. Parts list

Part	Description
BT-5 body tube, 4″ and 2″	The main body tube for the rocket. The instructions here use the 4″ BT-5 body tubes from the Estes Designer's Special, but it works even better to cut a single 6″ piece from a stock 18″ BT-5 body tube.
BT-5 tube coupler	This is used as an engine block and to join the two smaller tubes together.
BT-5 size nose cone, 1 1/4″	Any BT-5 nose cone of balsa or plastic will do. The nose cone shown is from the Estes Designer's Special; similar nose cones are available from many sources.
1/16″ balsa fin stock; about 1.3″ x 3″	This will be used for the fins.
1/8″ launch lug	Use about 1″, cutting it from a longer piece if needed.
1/8″ shock cord	You'll need 12″ of shock cord, which you can cut from a longer piece. Elastic from the fabric store works fine, too.

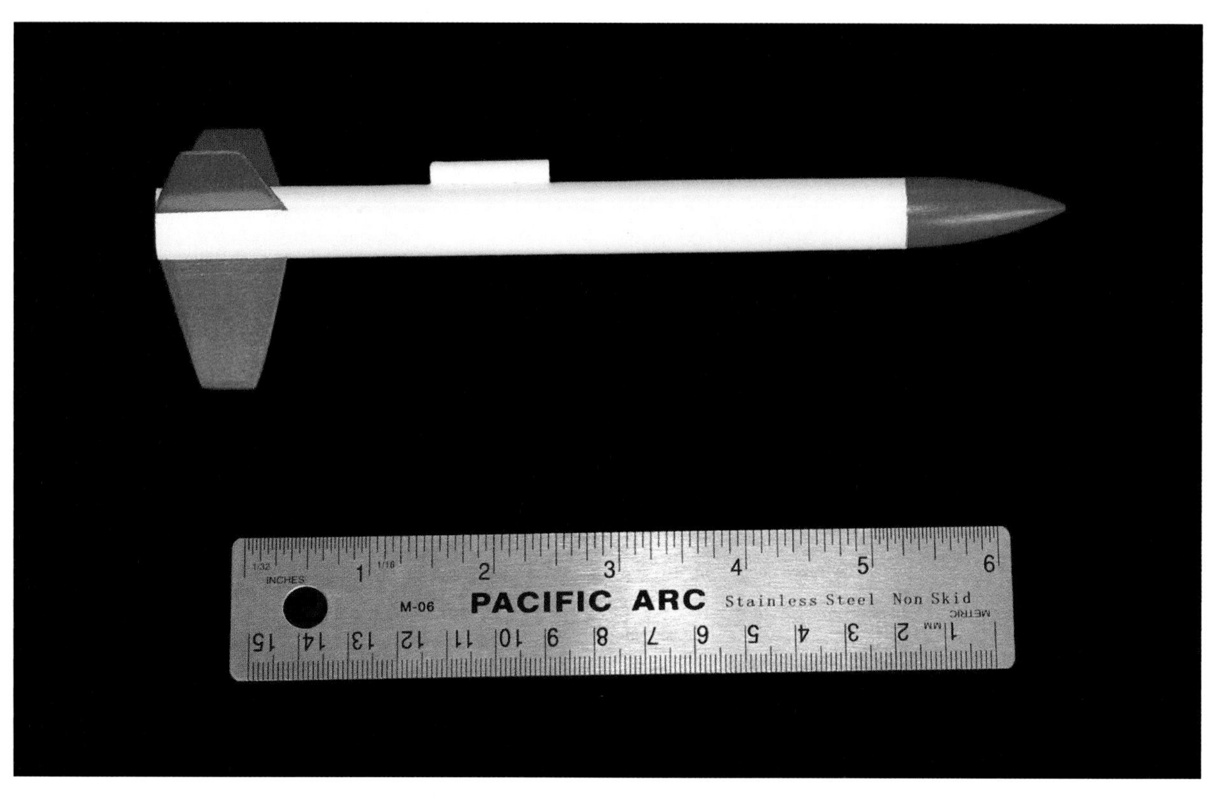

Figure 16-10. *Eros is a simple-to-build, high-flying rocket that can be flown with or without a streamer.*

Eros

Eros was the first asteroid known to orbit the sun partially inside the orbit of Mars. It was first seen by the German astronomer Carl Witt, on August 13, 1898.

Eros is another near-Earth asteroid. Due to interactions with the inner planets, Eros's orbit is expected to change over the next few hundred thousand years, eventually dropping inside the orbit of Earth. It is possible the asteroid will collide with Earth, and it's large enough to cause another extinction event like the one that wiped out the dinosaurs 65 million years ago.

Eros holds the distinction of being the first asteroid ever to have a spacecraft land softly on its surface. NASA's NEAR Shoemaker probe landed on February 12, 2001.

Eros was named for the Greek god of love.

Figure 16-11. *NASA created this image of Eros from a 3D model built from high-resolution measurements from the NEAR Shoemaker probe.*

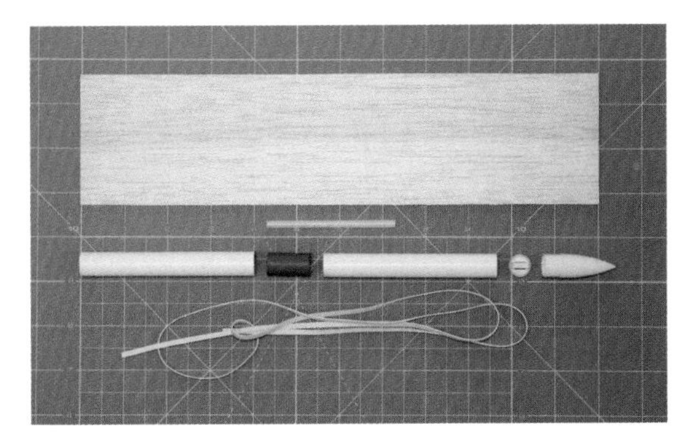

Figure 16-12. *Visual parts list.*

Construction

Refer back to Chapter 3 for basic construction techniques, tools, and supplies.

Eros is built around a 6″ BT-5 body tube. The Estes Designer's Special doesn't actually come with a 6″ or longer BT-5, so we'll need to build one up from shorter pieces. These instructions assume you will use the 2″ section of body tube left over from building Toutatis, and the remaining 4″ tube. Even better, of course, is to cut a single 6″ tube from an 18″ tube purchased from a hobby store or online store.

The red tube coupler will do double-duty. Not only does it hold the two shorter pieces of body tube together, but it also serves as an engine block, just like the one we used on Hebe in Chapter 15. That means the position of the tube coupler is rather important. As with all of our rockets, we want the motor to stick out of the back of the rocket by 1/4″ so the exhaust doesn't scorch the rocket. The tube coupler should sit exactly at the end of the motor.

As shown in Figure 16-13, use a motor to help position the tube coupler. Slide the motor in until 1/4″ is left sticking out the back. Test fit the tube coupler. It's 1″ long, the motor is 1 3/4″ long, and the body tube section is 2″ long. If 1/4″ of the motor sticks out one end of the tube, 1/2″ of the tube coupler should stick out the other end when the motor and tube coupler sit against each other. Use a ruler during the test fit to make sure the parts fit together as expected.

Use wood glue to glue the tube coupler into the body tube. You can leave the motor in while the tube coupler is positioned in the tube, but pull the motor out as soon as you are sure the tube coupler is in the right spot. You don't want to glue the motor permanently into the body tube!

Figure 16-13. *Using a motor to check positioning, glue the tube coupler into the 2″ section of body tube.*

You can wait until the glue dries from the previous step, or take extra care to make sure you don't move the tube coupler while performing the next step. Apply wood glue to the inside of the 4″ section of body tube and insert the tube coupler to permanently join the two tubes into a single 6″ body tube. If no glue squirted out through the crack between the body tubes, apply a small fillet of glue to fill in the crack. Wipe away any excess glue. If you do this before letting the glue dry on the 2″ section, be sure to slide the motor in and right back out to check the fit.

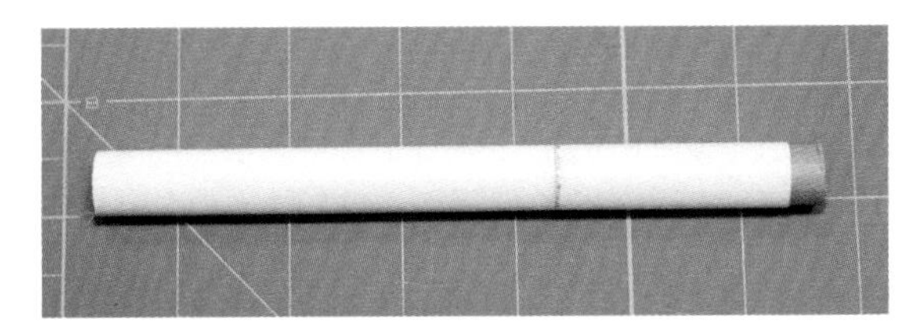

Figure 16-14. *Glue the 4″ section of tube to the 2″ section. Use a motor to check the spacing to make sure the tube coupler did not shift position.*

Cut three fins from 1/16″ balsa using the fin pattern in Figure 16-15. Follow the techniques introduced in Figure 3-38 to make sure they are all the same size. Round the forward, trailing, and outer edges of the fins, but not the fin root. Sand the fins smooth with medium grit sandpaper.

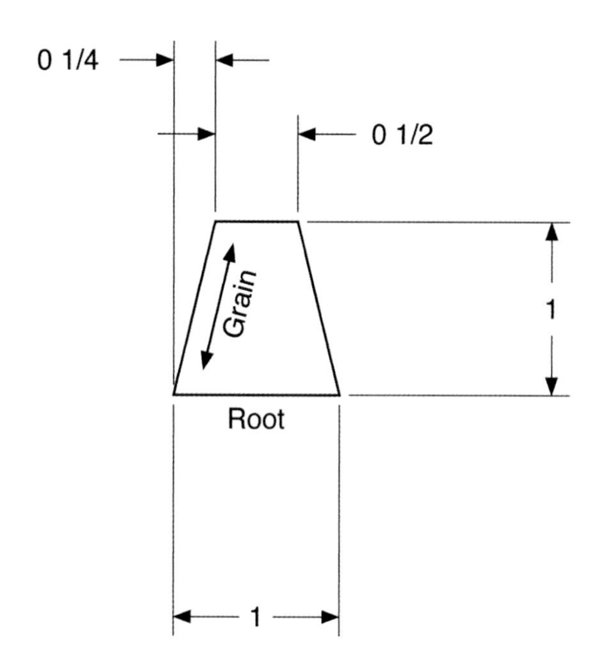

Figure 16-15. *Fin pattern for Eros.*

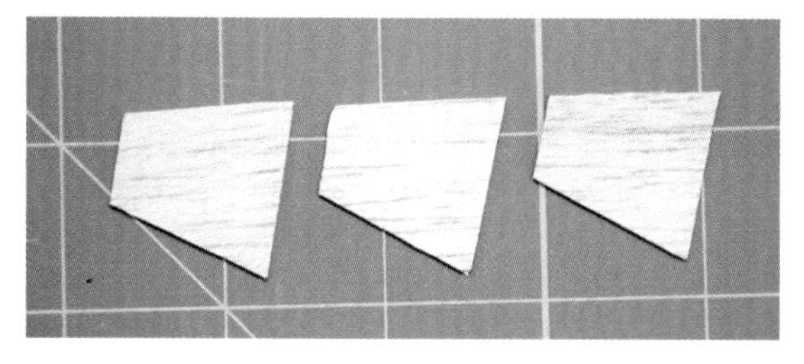

Figure 16-16. *Cut and shape three fins from 1/16" balsa.*

Use the fin guide from Figure 3-24 in Chapter 3 to locate the correct positions for the three fins and the launch lug, and then glue them in place. The fins should be right against the trailing edge of the body tube, while the launch lug should be 2 1/4" from the bottom of the rocket.

Remember, the 2" section of body tube goes at the bottom of the rocket so the motor can rest against it during powered flight!

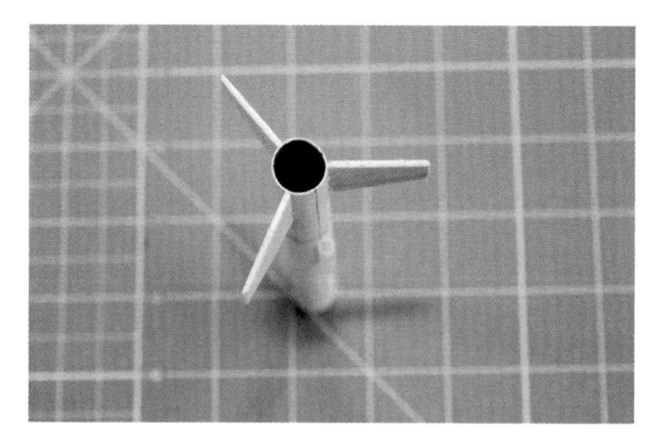

Figure 16-17. *Glue the fins and launch lug in place.*

Plastic nose cones come in two pieces. Use plastic model cement or some other glue that works on plastic to attach the base to the nose cone. You won't need to glue parts together if you are using a balsa nose cone, but you will need to add a screw eye to serve as an attachment point.

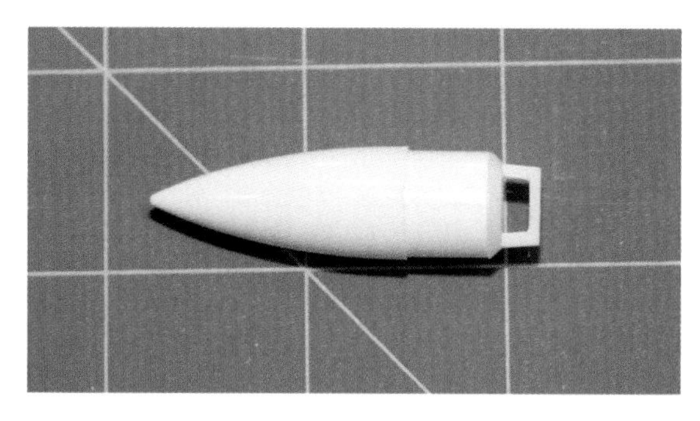

Figure 16-18. *Glue the two halves of the nose cone together.*

Shock cord installation is pretty routine, except for one minor point. I don't know about you, but my fingers are too big to position the shock cord in a 1/2"-diameter body tube. I used a pencil to do the job.

Tie the other end of the shock cord to the nose cone using a buntline knot. Apply a small dab of glue to keep the knot from coming undone.

Once all of the glue dries, Eros is ready for primer and paint.

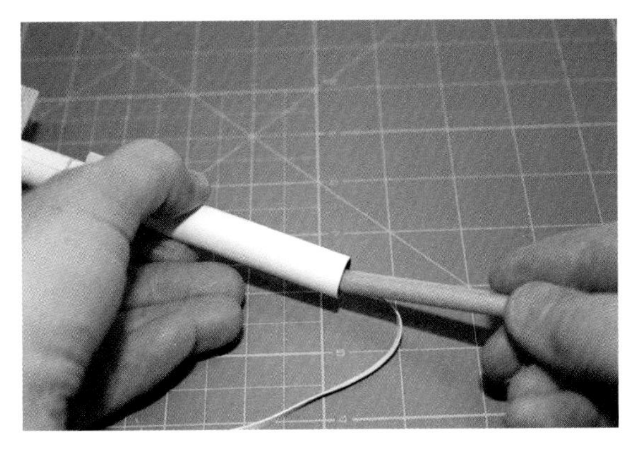

Figure 16-19. *Use a shock cord mount to secure the shock cord in the body tube. Tie the other end to the nose cone.*

Figure 16-20. *The completed Eros, ready for paint.*

Flying Eros

Flying Eros is very similar to flying Hebe from Chapter 15. The only difference is the streamer—there isn't one. Eros is so light that the nose cone and body act as a streamer, as long as the rocket parts are flopping around and not forming a streamlined body. This recovery method is called *nose-blow recovery*, since the nose cone is blown off by the ejection charge. Nose-blow recovery is a kind of *tumble recovery*, where something changes the geometry of the rocket so it tumbles rather than returning to earth ballistically.

Start by packing 1" of recovery wadding into the body tube. There is no streamer to protect from hot gases, but the shock cord still needs some protection. Pile the shock cord on top, and then insert the nose cone.

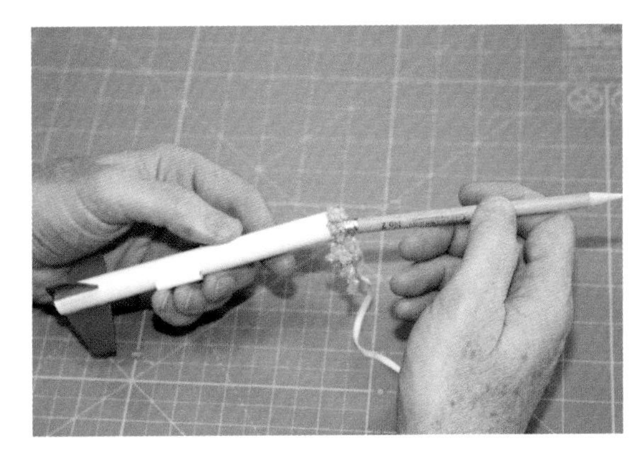

Figure 16-21. *Insert enough recovery wadding to fill about 1" of the tube, then stuff in the shock cord and insert the nose cone.*

The motor is friction mounted, just like the motors for Hebe and Toutatis. Wrap tape around the motor to get a nice, snug fit. Unlike with Toutatis, we don't want the motor to pop out when the ejection charge fires, so the fit does need to be tight—significantly tighter than the nose cone. The nose cone should be snug enough to not fall out if you turn the rocket over and shake it, but not much tighter. Use sandpaper to reduce the size of the nose cone if it is too tight, or wrap it in tape if it is too loose.

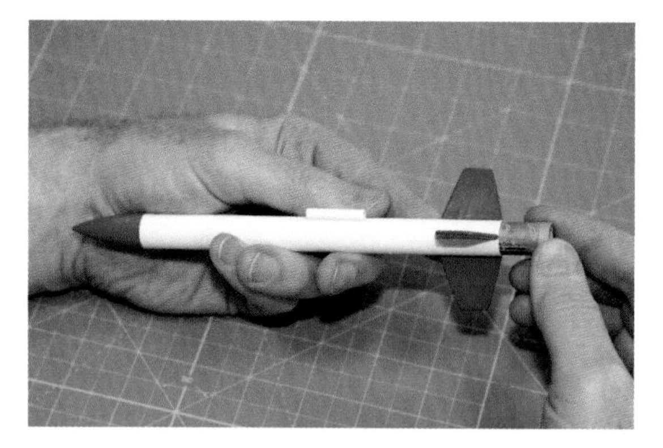

Figure 16-22. *The motor should fit snugly—far more snugly than the nose cone. It should be difficult to remove. Don't make it so tight that you damage the rocket sliding it in or out, though!*

With the motor in place, install the igniter, slide the rocket onto the launch rail, and lift off!

You might be wondering about the lack of a streamer or parachute. In very light rockets like Eros, you can get away with leaving them off. With the nose cone flopping around in the wind, the rocket is unstable. It will fall fairly slowly.

On the other hand, there is nothing wrong with using a streamer with Eros. It will work quite well, and will make the rocket a little easier to see as it descends. If you used a snap swivel on the streamer for Hebe, you can swap it out and give it a try with Eros.

Eros Launch Preparation Checklist

1. Insert the motor, using tape for a snug fit.
2. Insert recovery wadding.
3. Attach a streamer for added visibility (optional).
4. Insert the streamer (if including one) and shock cord; make sure they are not too tight.

Eros Flight Checklist

1. Clear spectators from the launch zone.
2. Install igniter.
3. Remove the launch key from the launch rail, slide the rocket onto the launch rail, and replace the launch key on the launch rail.
4. Attach igniter clips.
5. Get the launch key from the pad.
6. Insert the launch key. Make sure the continuity light is on.
7. Check for clear skies—no aircraft or low clouds.
8. Count down from five and launch.
9. Remove the launch key and place it on the launch rod.
10. Recover the rocket.

17 Multistage Rockets

Staging is a common concept in rocketry. Using one powerful motor to launch a large rocket has mass penalties: the rocket tube must be thick enough (and therefore heavy enough) to withstand the thrust of the large motor. By stacking multiple rockets on top of each other in stages, you can pack more power into a thinner (and lighter) tube.

The concept of staging is very simple: a multistage rocket is launched using only the motor in the first stage. When the first motor burns out, the rocket drops the weight of the entire first stage—motor, fins, and tube—leaving a smaller rocket to continue the journey. The smaller rocket fires, taking advantage of the speed and altitude provided by the first stage to fly higher and faster than it could have flown on its own. You can do this more than one time. In model rocketry, two-stage rockets are common, but three-stage rockets are not unusual, and you can use even more stages.

This chapter shows how it's done. You will build Romulus, a two-stage rocket that can fly to altitudes of 1,460 feet, even though it is a big, impressive rocket.

The Two-Stage Romulus

You can fly Romulus with any combination of an A8-0, B6-0, or C6-0 motor in the first stage and an A8-5, B6-6, or C6-7 motor in the second stage. Table 17-1 shows some of the more common combinations and the calculated maximum altitudes. Table 17-2 lists the parts you'll need to build Romulus.

Table 17-1. Recommended motors

First stage	Second stage	Approximate altitude
A8-0	A8-5	270 ft
A8-0	B6-6	500 ft
B6-0	B6-6	680 ft
C6-0	A8-5	800 ft
C6-0	B6-6	1,040 ft
C6-0	C6-7	1,460 ft

Romulus

There are two solar system bodies named Romulus. One is a main-belt asteroid about 21 kilometers in diameter orbiting the sun between Mars and Jupiter. The other Romulus is one of the moons of the asteroid Sylvia. Students of history won't be too surprised to hear that the second moon is named Remus; in Roman mythology, Romulus and Remus were the twin sons of princess Rhea Silvia. Standing on Sylvia and looking up at Romulus, you would see an irregularly shaped asteroid. The shorter dimension would appear to be about 10 times the size of our moon from Earth, while the longer axis would look 32 times larger than the moon. That's a sky-full.

The twins of legend were raised by a wolf, and went on to found Rome. Both as one of a pair of moons and as one of a pair of twins, Romulus is an apt name for a two-stage rocket.

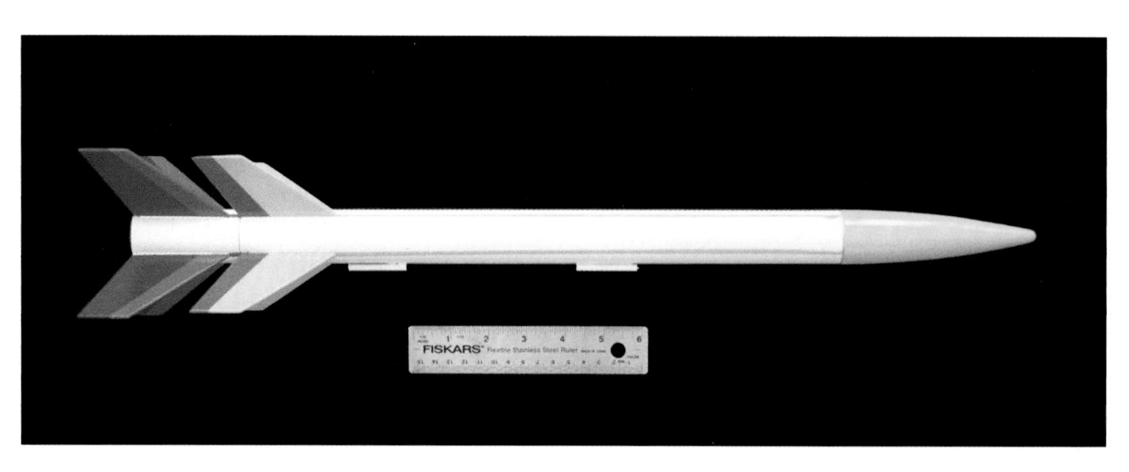

Figure 17-1. *The 24" Romulus is a sleek, impressive flyer that uses two stages to reach altitudes over 1/4 mile.*

Table 17-2. Parts list

Part	Description
BT-55 body tube, 18"	The main body tube for the rocket.
BT-20 body tube, 2 3/4" (2)	You will need two motor mounts. The Estes Designer's Special has precut white ones; it's also fine to cut them from a longer length of BT-20 body tube.
BT-55 size nose cone, 5"	Any BT-55 nose cone of balsa or plastic will do. The nose cone shown is from the Estes Designer's Special; similar nose cones are available from many sources.
3/32" balsa fin stock; about 4" x 12"	The balsa wood will be used for the fins.
1/8" launch lug	You will use two pieces. Any length from 1" to 1 1/2" is fine; the photos show a 3" launch lug cut in half.
BT-55 tube coupler	This is used to join the two stages. It needs to be 1" to 1 1/4" long.
Engine block	This is usually a short, dark gray tube that slides into the BT-20, about 1/5" to 1/4" long. Anything of similar size that will slide into the BT-20 will work.
BT-55 to BT-20 centering ring (4)	These are used to mount the BT-20 motor mount tubes in the BT-55 body tube.
1/8" shock cord	You will need about 30" of shock cord.
15" parachute	There is a nice 15" parachute in the Estes Designer's Special, or you can make one following the instructions from Chapter 3.
Snap swivel (2), optional	For attaching the parachute and shock cord.

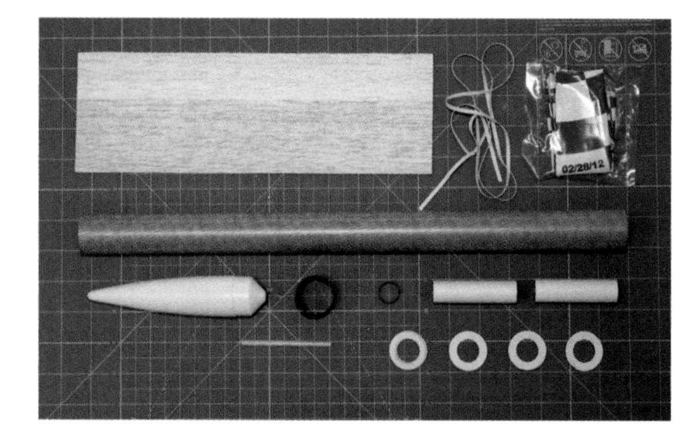

Figure 17-2. *Visual parts list.*

Figure 17-3. *Use a hobby knife to cut the tabs holding the four centering rings in the sheets of centering rings in the Estes Designer's Special. There are two BT-55 to BT-20 centering rings on each sheet.*

Equivalent Kits

The Estes Mongoose is about the same length as Romulus, but uses a slimmer BT-50 body tube. The molded plastic fins make it a very easy build. The Estes Loadstar II adds a clear payload bay. (You can also easily add a payload to Romulus using the techniques from Chapter 9.)

The Quest Navajo AGM is a nice two-stage rocket that features the use of engine hooks to hold the motors in place. While I prefer friction fit for staging, due to the higher reliability of taped motors, the decision to use hooks or friction is, admittedly, a personal preference. You can, if you want to, also build the Navajo AGM with friction fit motors, like Romulus. It's an exceptionally cool looking rocket.

It's truly sad to watch your rocket plunge back to certain destruction because the second stage failed to ignite. That has not happened to me often, but even once is enough to instill some extra caution.

How Staging Works

The rocket motors you have used up to this point all work essentially the same way. There is a propellant charge that provides the power to push the rocket into the air. This is followed by a smoke delay that gives the rocket time to slow down. Finally, the ejection charge fires to deploy the recovery system.

The first stage is different in a two-stage rocket. It doesn't have a smoke delay or ejection charge. Instead, just before the propellant burns out, it shoots burning particles of propellant into the empty space above the motor. This burning propellant ignites the second-stage motor. The hot gases shooting out of the base of the second-stage motor push the first stage away from the rocket, and the second stage continues on. The first stage has literally given the rocket a boost, lifting it to a higher altitude and giving it a high initial speed, so the second stage travels higher and faster than it would have if it were launched on a single motor.

The second stage of a two-stage rocket works just the same as a normal rocket, although the smoke delay is generally a bit longer because the rocket is expected to be traveling faster than usual.

Romulus is a classic two-stage rocket that uses a tube coupler to hold the stages together, so the second stage slides smoothly onto the first stage.

The thrust of the first stage is enough to keep the pieces together until the propellant burns through. That's probably all it will take for enough burning black powder to ignite the second-stage motor, but it's still a good idea to tape the two motors together to increase the chance that the first stage will ignite the second, and not just blow it off. A ring of tape completely around the motors does the job nicely. You will see details on how to apply the tape in the launch instructions at the end of the chapter.

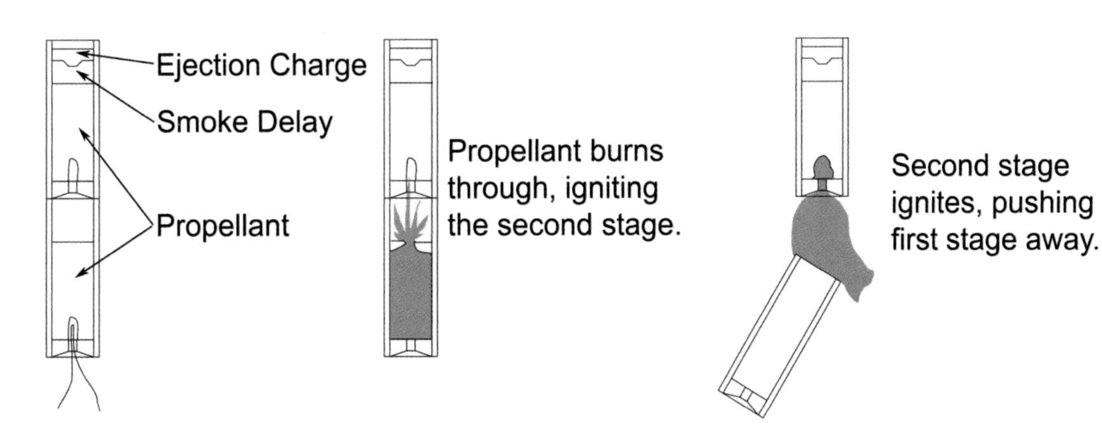

Figure 17-4. *Two-stage ignition sequence.*

Once the two stages separate, the first stage is unstable. It tumbles, which messes up the smooth air flow past the fins. That's a good thing, as it means the first stage slows down very quickly instead of coasting to a high altitude. The first stage continues to tumble as it descends. This is a classic example of *tumble recovery*, where the rocket's configuration changes so it is unstable, giving a gentle recovery without a parachute or streamer.

What about three-stage rockets? They're not as common as two-stage rockets, but I've built and flown a few. There is no theoretical reason why you can't just keep adding stages. At some point, though, the weight of the rocket or the amount of propellant will get too large. Also, bear in mind that a three-stage rocket flown with C motors is an E-class rocket, with 30 N-s of total impulse, so it needs to be launched from a distance of 30 feet, not 15 feet.

Alternate Two-Stage Strategies

This is not the only way to make a two-stage rocket. It's possible, for example, to build a minimum-diameter two-stage rocket. You could add a booster to Hebe using a 2 3/4"-long piece of BT-20 body tube, and additional fins that are about 1/4" larger than those on the original rocket. Using C6-0 and C6-7 motors, the resulting rocket will scream to an altitude of about 3,400 feet, reaching over 500 miles per hour—about 2/3 the speed of sound! You'll be lucky to find it again, but what a flight!

In this case, the upper-stage motor will protrude into the lower stage by 1/4". Both the upper-stage and lower-stage motors are friction fit, and there is a layer of tape holding the two motors together. When the lower-stage motor burns through, the upper-stage motor ignites and pushes the lower-stage motor away, easily breaking the tape and popping the lower stage off. Figure 17-5 shows the construction of the staging mechanism for both Romulus and a two-stage Hebe, with fins removed for clarity.

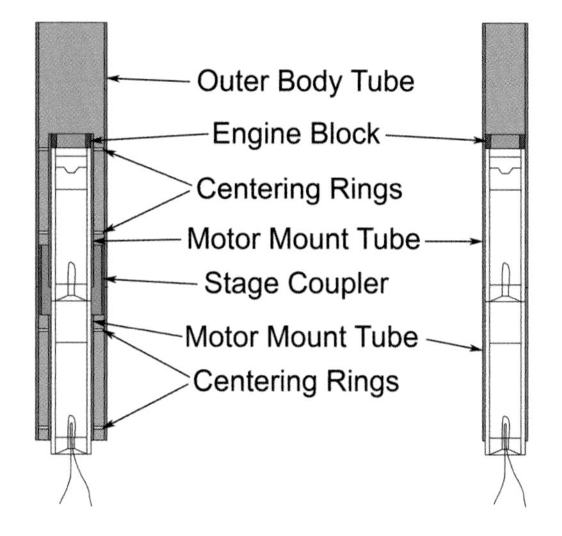

Figure 17-5. *The staging mechanisms for Romulus (left) and a two-stage version of Hebe (right) seem different, but they accomplish the same goals.*

Construction

Refer back to Chapter 3 for basic construction techniques, tools, and supplies.

The first stage of Romulus is formed from a 2 3/4" length of BT-55 body tube, while the second stage is 15 1/4" long. Mark the 18" body tube 2 3/4" from one end and, using the tube coupler for support, cut the tube.

Figure 17-6. *Cut a 2 3/4" piece from the 18" BT-55 body tube. The shorter piece will be the first stage, while the longer piece will be the second stage.*

While it is not absolutely essential, it helps a lot when prepping the rocket for flight if the motor mount on the lower stage doesn't rest right up against the motor mount for the second stage. That leaves room for the tape that holds the motors together until stage separation, without the need to maneuver the motor mounts over the tape.

Assuming you are using the 2 3/4″ motor mount tubes from the Estes Designer's Special or from a motor mount kit, cut one of the motor mounts to make room for the tape. Measure 1/2″ from the end of one of the motor mounts. Using the engine block as a support, cut the tube.

If you are making your own motor mounts from a longer piece of BT-20 body tube, cut one motor mount 2 3/4″ long and the other 2 1/4″ long.

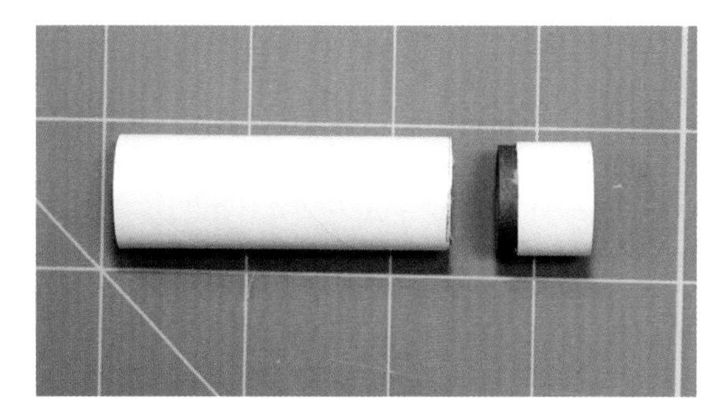

Figure 17-7. *Cut 1/2″ from one of the two motor mount tubes in the Estes Designer's Special.*

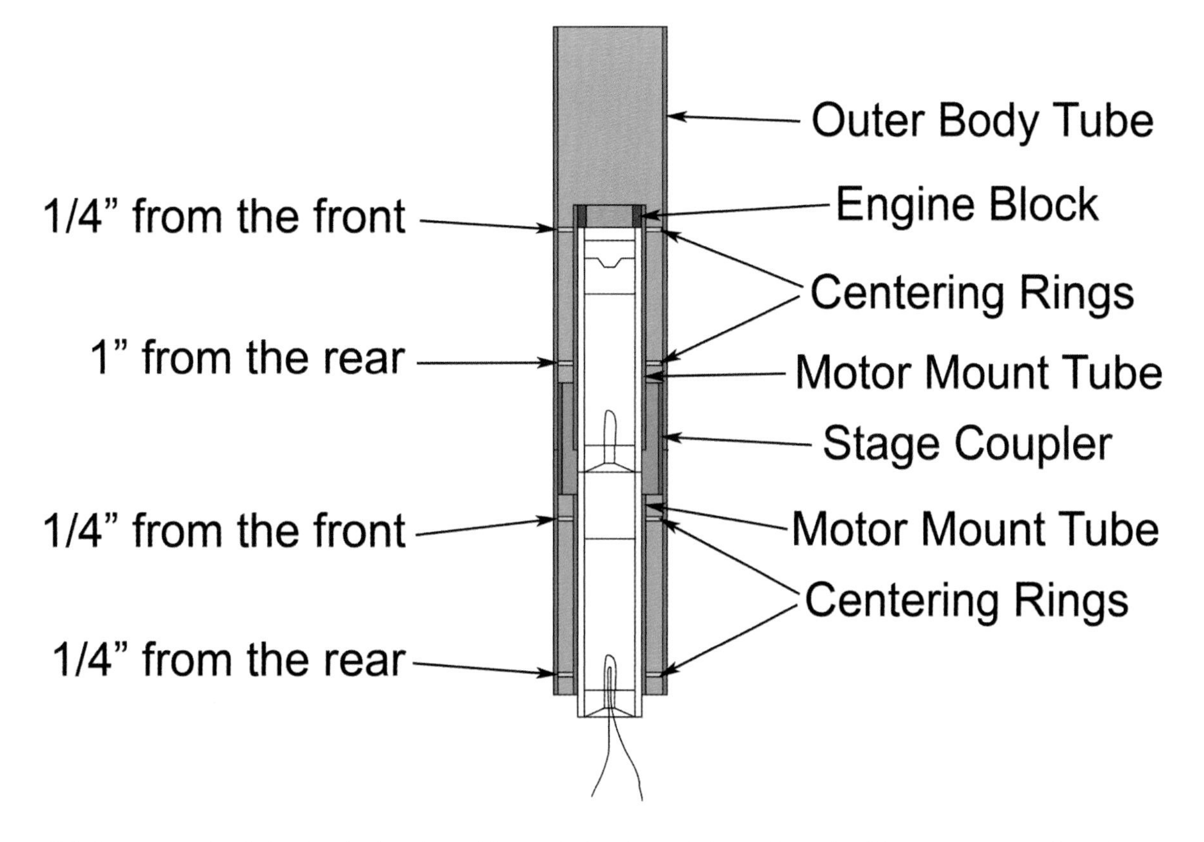

Figure 17-8. *Stage area detail showing the locations of the centering rings relative to the ends of the motor mounts. From this view, it is clear why one motor mount is shorter, and why the centering rings are positioned where they are.*

Referring back to the visual parts list in Figure 17-2, you see four thick paper centering rings. These are used to center the two motor mount tubes in the first- and second-stage body tubes. Figure 17-3 showed the two sheets of centering rings that come with the Estes Designer's Special. If you bought individual centering rings, they will generally be two to a sheet, but the idea is similar. Detach the centering rings, using a hobby knife to trim any tabs holding them to the sheet.

Glue the centering rings to the motor mounts. The first two rings should go 1/4" from the ends on the shorter motor mount, which will go in the first stage of the rocket. The longer motor mount goes in the second stage. A tube coupler mounted in the first stage will hold the stages together until separation. We need plenty of room for the tube coupler to slide into the second stage, so the centering ring that will be closest to the bottom of the second stage needs to be 1" from the end of the longer motor mount tube. The centering ring at the other end should go 1/4" from the end.

Glue the engine block in the end of the second-stage motor mount tube. It should rest right up against the end of the tube, and be on the side where the centering ring is closest to the end of the tube.

Apply a generous fillet of glue where the centering rings meet the motor mount tube, but only on one side. Set the motor mounts aside to dry. There's a lot of glue there, and drips are possible, so be sure to put wax paper or some other material that will peel away from glue easily under the motor mounts until the glue is firmly set.

Once the glue is hard enough that there is no danger of drips, turn the motor mounts over and put fillets on the other side.

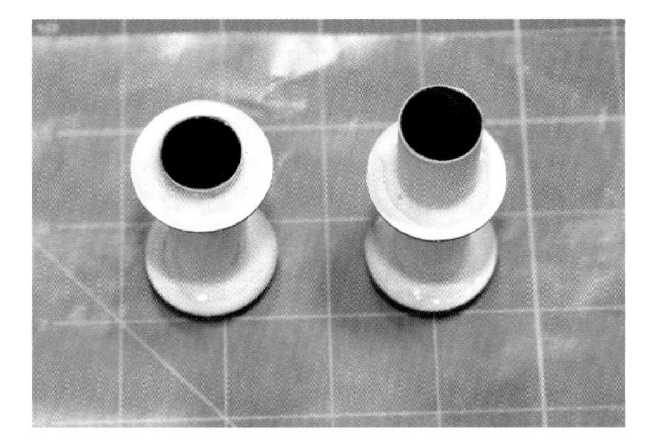

Figure 17-9. *Glue two centering rings 1/4" from one end of each tube. Glue the second centering ring 1/4" from the other end of the shorter tube, and 1" from the other end of the longer tube. Glue the engine block in the longer tube at the end where the centering ring is 1/4" from the end.*

Each stage has four fins. As you know, three fins will cause less drag, but there is a lot of weight near the back of this rocket. Rather than using three very large fins, the rocket uses four slightly smaller ones. This does add some drag, but the fins would not fit well on the first stage if they were much larger.

Use the fin guides from Figures 17-10 and 17-11 to cut the fins from 3/32"-thick balsa. Round the leading, trailing, and outer edges of the fins, but not the root edge. Use a sanding block and medium grit sandpaper to sand the fins so they are smooth to the touch.

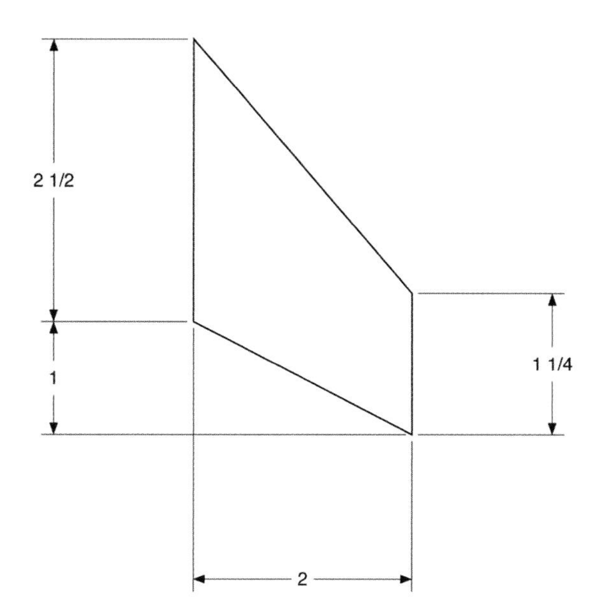

Figure 17-10. *Fin guide for second-stage fins. Full-size templates are available from the author's website (http://bit.ly/byteworks-make-rockets).*

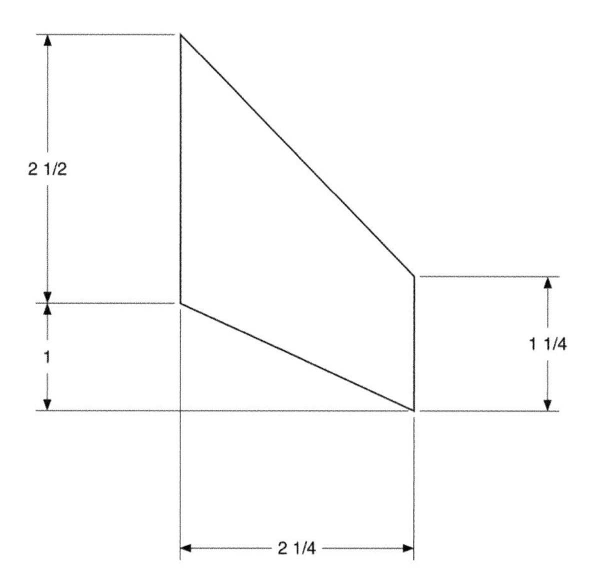

Figure 17-11. *Fin guide for first-stage fins.*

Figure 17-12. *Cut four fins for each stage from 3/32" balsa. Sand until smooth, and round all edges except the root edge of the fins.*

The Estes Designer's Special comes with 1/8"-diameter launch lugs that are 3" long. Cut one in half; we will use both pieces on Romulus, which is a fairly long rocket. There is nothing magical about the 1 1/2" length. If you have two 1 1/4" launch lugs or a 2"-long piece that you would like to cut in half, go ahead. Using 1" or 1 1/4" launch lugs will work fine.

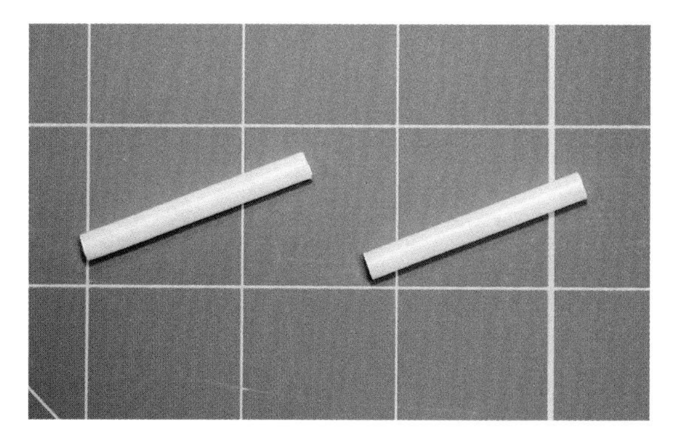

Figure 17-13. *Cut two launch lugs that are 1" to 1 1/2" long each. The photos show 1 1/2" launch lugs.*

Using the tube marking guide from Figure 3-25 in Chapter 3, mark the locations for the four fins and the launch lug on the second-stage body tube. Use the tube coupler to temporarily hold the tubes together while you use a door jamb to extend the marks to lines that run from the base of the first stage to about 4" above the base of the second stage. Marking the tubes together this way will help you get the fins positioned in exactly the same spot on the two stages. If you're off slightly on one of the fins, both stages will be off by the same amount.

Extend the line for the launch lug all the way up the second-stage tube.

Glue the four fins in position. Look down from above to check the alignment of the fins. Glue one launch lug so it is 2 3/4" from the bottom of the second stage, and the other so it is 5" from the top of the second stage. This puts the launch lugs on either side of the center of gravity, holding the rocket straight as it slides up the launch rail.

Check the fins occasionally as they set. After they are dry enough to handle, but before the glue is completely dry, pull the stages apart. There are two reasons for doing this. The first is that it lets you turn the stages over and sit them gingerly on a flat surface to make sure the backs are aligned. It's not too late to reset a fin if they are really out of whack. The other reason to pull the stages apart is to make sure no glue dripped into the crack between the stages. You don't want to accidentally glue the stages together!

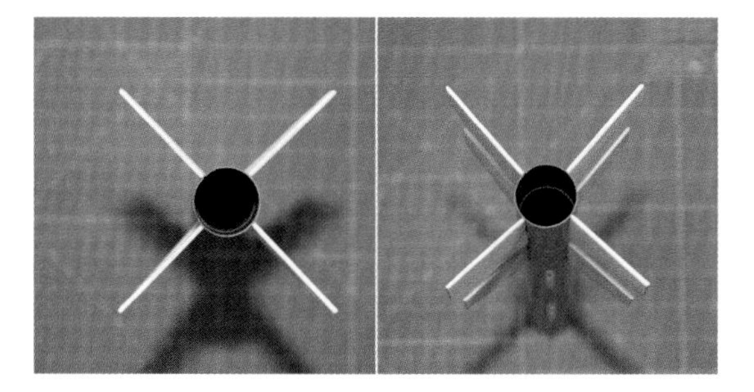

Figure 17-14. *A view of the fins from above and from a slight angle as the glue sets. Check the fins visually a couple of times as the glue dries. Pull the stages apart once the glue is firm, but before it dries completely.*

That's a lot of glue. Rather than sitting around watching it dry, let's do something useful.

Grab the nose cone and check it for flash (the extra plastic that squeezes into the cracks between molded parts). Remove any extra plastic from the center of the eyelet and scrape any flash from the nose cone. Refer back to Chapter 3 for details.

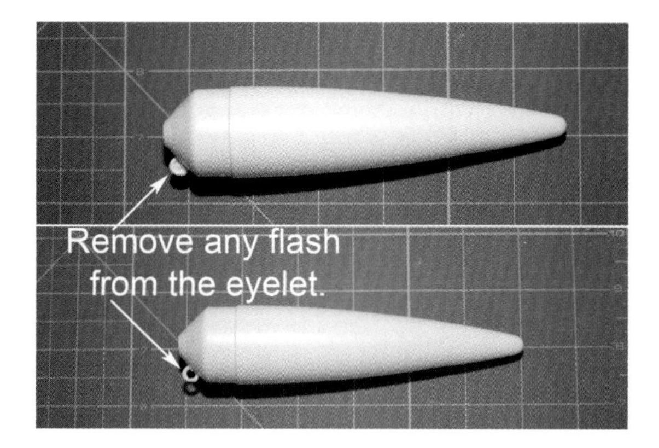

Figure 17-15. *Remove any plastic blocking the eyelet, and scrape any excess plastic forming a line down the length of the nose cone.*

Fasten snap swivels to one end of the shock cord and to a 15" parachute. If you're making a parachute from scratch, now is a good time to make it.

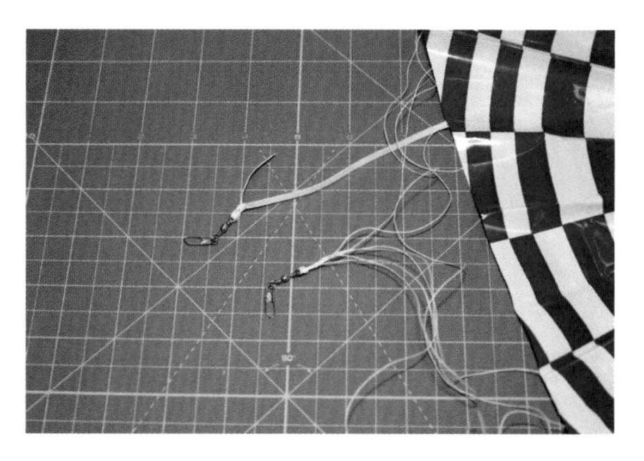

Figure 17-16. *Fasten snap swivels to one end of the shock cord and to a 15" parachute.*

Once the glue dries, mount the motor mounts in the body tubes. Start with a dry test fit to make sure that all of the pieces will slide in easily and that everything fits. Be sure and check the clearance on the tube coupler—that's the red tube in Figure 17-17. The shorter motor mount goes in the first stage. It should be flush with the back end of the body tube. The longer motor mount goes in the second stage. Be sure the end with the engine block goes in first, and the end with 1" between the end of the tube and the centering ring is on the bottom. Both stages should fit together with the centering ring and motor mounts in place.

Once you are satisfied with the fit, apply glue inside the body tube and push the motor mounts into place. Apply glue to the top of the first stage and slide the tube coupler into place. It should slide in halfway, so 3/4" of the tube coupler is glued in the first stage, and 3/4" sticks up to join to the second stage. Check the fit one last time by sliding the first stage into the second stage, but pull the stages apart immediately to prevent the two stages from getting stuck together.

Once the glue sets, apply generous fillets to the exposed joints between the centering rings and body tubes. Apply fillets to the joints between the fins and body tubes, too.

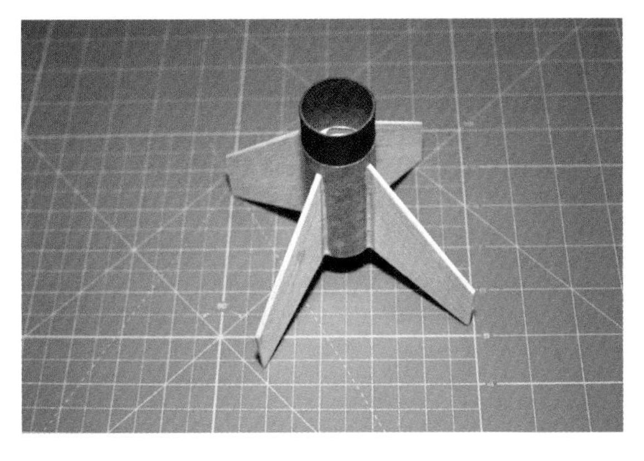

Figure 17-17. *The completed first stage sitting aside to dry.*

Once the fillets dry, glue in the shock cord. The rocket is now ready for final finishing.

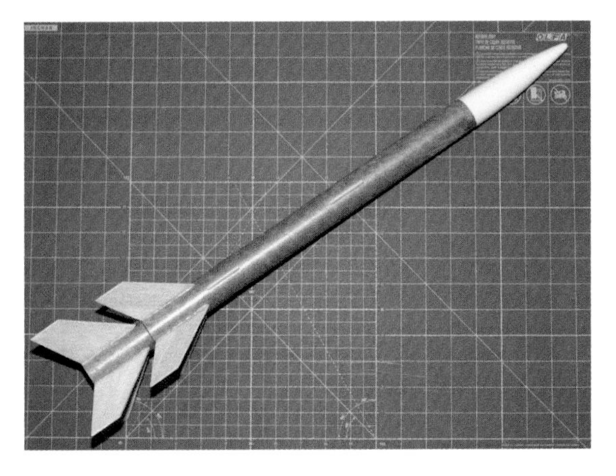

Figure 17-18. *The fully assembled but unpainted rocket. It stands an impressive 24" tall.*

Flying Romulus

While there are a few extra steps, prepping Romulus for flight uses the same techniques you have already practiced on Juno and other rockets.

Begin by stuffing enough recovery wadding into the second stage to fill about 2 1/2" of the tube with wadding.

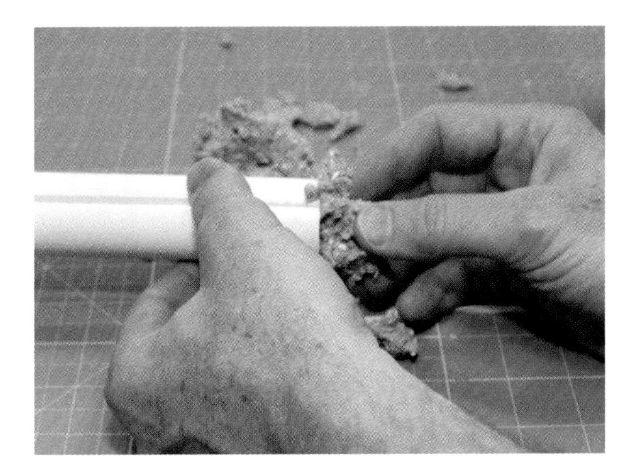

Figure 17-19. *Insert 2 1/2" of recovery wadding into the second stage.*

Make sure the snap swivels for the shock cord and parachute are both attached to the eyelet on the nose cone. Be sure both snap swivels are fully closed.

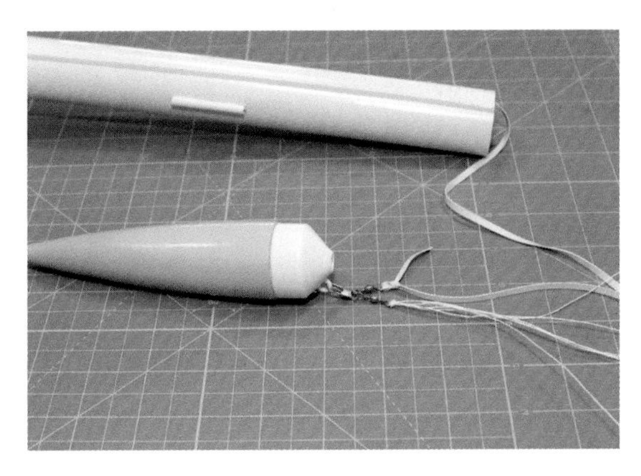

Figure 17-20. *Attach the parachute and shock cord to the nose cone.*

Pack the parachute. Insert the shock cord into the body tube, then the parachute. Push the nose cone into place.

The parachute should slide in easily. The nose cone should be tight enough so it does not fall out if the rocket is turned over and shaken gently, but it should not be any tighter. Sand the nose cone or apply tape if you need to adjust the fit.

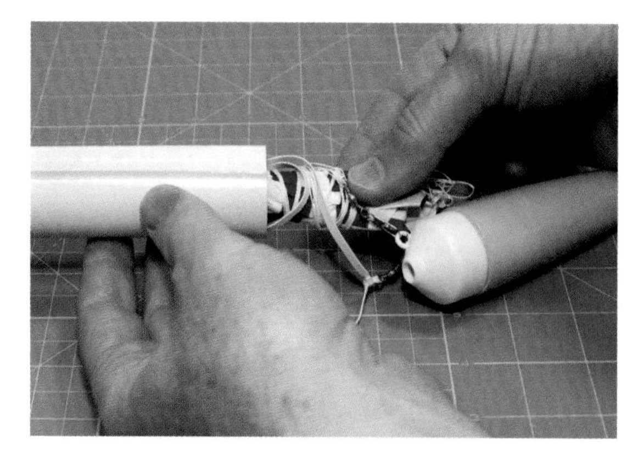

Figure 17-21. *Pack the parachute and shock cord in the rocket.*

Line the motors up so the nozzles point the same way. The top of the first-stage motor—the A6-0, B6-0, or C6-0—should face the nozzle on the second-stage motor.

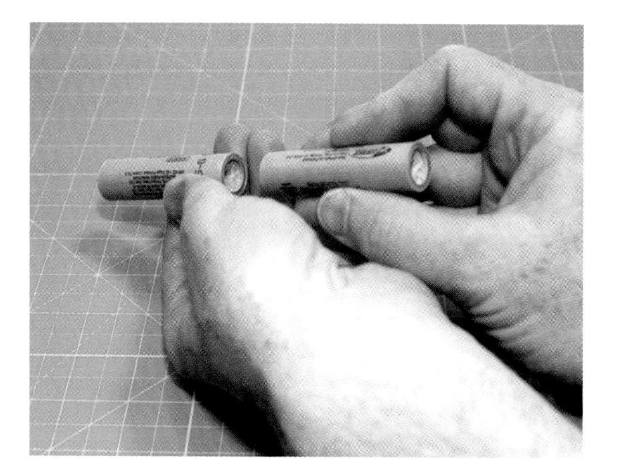

Figure 17-22. *Line the motors up so they point the same way, nozzles down. The nozzle of the second-stage motor will rest on the top of the first-stage motor.*

Apply one layer of tape completely around the motors to hold them together.

If you need to, apply more tape to the second-stage motor. Mount it in the second stage of the rocket so the motor rests against the engine block. It should be quite snug, so the motor stays put when the ejection charge fires. We want the parachute to pop out, not the motor!

Figure 17-23. *Tape the motors together.*

If the tape holding the two motors together makes the motor too tight to fit into the rocket, tear it back so only 1/4″ of tape is holding the second-stage motor to the first-stage motor. A total of 1/2″ of tape—1/4″ on each motor—is plenty to hold them together until the stages separate.

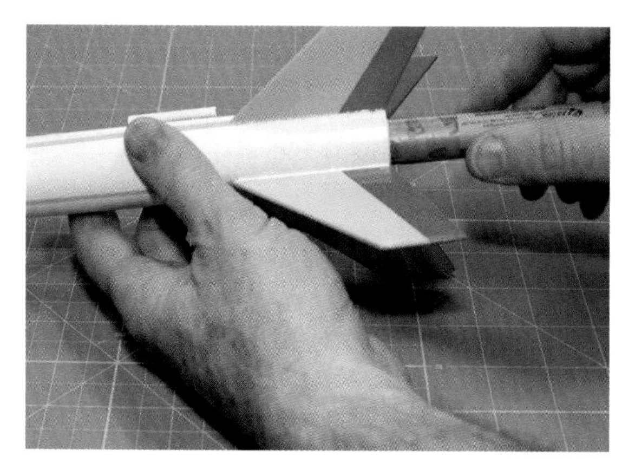

Figure 17-24. *Apply enough additional tape for a snug fit.*

Repeat the process, applying tape if needed to hold the first-stage motor in place. Slide the first stage into the second stage until the body tubes meet. While you will probably display the rockets with the fins on the two stages neatly lined up, it's better to fly it with the fins rotated 30° or so. Lining the fins up causes turbulence from the second-stage fins to hit the first-stage fins, causing a rough flight and reducing the effectiveness of the first-stage fins. Rotating the fins relative to each other fixes this issue.

The first stage should still be snug, but it can be a little looser than the second stage. There won't be an ejection charge trying to push the first-stage motor out of the rocket.

Figure 17-25. *Apply tape as needed to hold the first stage in place as it slides over the assembled motors.*

Multistage model rockets have a lot of fin area, and tend to be a bit overstable. As you know, that can cause weathercocking in high winds. In general, you should fly Romulus and other multistage rockets on calm days. That helps prevent weathercocking, and also makes it more likely that you will find these high-flying rockets, since a calm wind won't carry them as far as a stiff breeze!

The rocket is ready for an igniter and launch. Before you push that button, though, think about what is going to happen. Your rocket is going to shoot into the air, the stages will separate, and the second stage will blast a considerable height into the sky—an additional 250 to 1,500 feet, depending on the motors used. Everyone will be watching with excitement as the rocket climbs higher and higher until the parachute deploys.

Now quick—where's the first stage?

As I write this, the first stage was lost on the two most recent two-stage flights at the Albuquerque Rocket Society. One of them was mine. It may not be a glamorous job, but be sure to assign someone to track the first stage. You need to recover both parts!

Romulus Launch Preparation Checklist

1. Insert recovery wadding.
2. Attach the parachute and shock cord.
3. Insert the parachute and shock cord; make sure they are not too tight.
4. Tape the motors together.
5. Double-check—is the ejection charge visible from the exposed end of the second stage?
6. Insert the motors, using tape for a snug fit.
7. Make sure the two stages fit completely together.

Romulus Flight Checklist

1. Clear spectators from the launch zone.
2. Install the igniter.
3. Remove the launch key from the launch rail, slide the rocket onto the launch rail, and replace the launch key on the launch rail.
4. Attach the igniter clips.
5. Assign at least one person to watch the first stage.
6. Get the launch key from the pad.
7. Insert the launch key. Make sure the continuity light is on.
8. Check for clear skies—no aircraft or low clouds.
9. Count down from five and launch.
10. Remove the launch key and place it on the launch rod.
11. Recover both stages of the rocket.

Cluster Rockets

18

Figure 18-1. *The Saturn 1B booster is a Tinkertoy cluster rocket. It's made up of eight tanks from Redstone rockets holding fuel and liquid oxygen, a liquid oxygen tank from a Jupiter rocket, and eight H1 rocket motors.*

In a cluster rocket, two or more motors are used to propel the rocket with the force of multiple motors. It's a popular technique in both model rocketry and full-scale rocketry because it's an easy way to use relatively inexpensive, lower-power motors you already have on hand to do the job of a much larger and often more expensive motor.

Things to Consider When Clustering Rockets

We're going to look at two clustered rocket designs later in this chapter, but first, let's take a look at some of the things you need to consider when designing a cluster rocket.

Why Cluster Rockets?

Back in Chapter 7, you saw that a rocket needs to reach a speed of about 44 ft/s, or 13.4 m/s, to be stable. A quick-and-dirty calculation to tell if a rocket will make it is to divide the maximum thrust of the rocket by the weight. If the result is over 11, the rocket will be stable when it leaves the launch pad. Of course, that's not the only factor, but it is one of the important ones.

Let's take a look at what that means if we're trying to launch a heavy rocket. The C6-5 motor, a mainstay of model rocketry, has a maximum thrust of 14 newtons. It can safely lift a rocket with a weight of about 1.3 newtons, or 4.6 ounces. That's a fairly heavy rocket, but nowhere near the 53-ounce maximum for a model rocket.

How about the Estes D12 series? That has more than twice the average thrust of a C6 series motor, thrusting in at 30 newtons. It can lift 9.8 ounces.

And the Estes E9? With twice the total impulse, it must be able to lift more, right? Actually, no—it lifts less. It's the maximum thrust that matters, or more specifically, the thrust while the rocket is on the launch rail, which is fortunately when most motors deliver their maximum thrust. The total impulse and average thrust really aren't important for launch weight. The maximum thrust of the E9 series motors is 25 newtons, so it can only safely lift 8.1 ounces.

A single motor, then, is limited to lifting about 10 ounces. If you want to lift a heavier payload, you have to make the same choice NASA did: use more motors. That's what clustering is all about.

Of course, there is another reason to build a cluster rocket. It's fun!

Figure 18-2. *Bill Beggs's 13-motor cluster rocket being prepared for flight. With C6 motors, this rocket tops in at 140 grams of propellant and a total impulse that makes it a G class rocket. That means it must be flown from a distance of 100–200 feet and requires an FAA clearance.*

Picking the Right Number of Motors

There are several factors to keep in mind when picking the number of motors for a cluster rocket. Some are legal, some are technical, and some are practical.

From a practical standpoint, cluster rockets cost more to launch. Cerberus, the first rocket we'll build in this chapter, uses three motors on each launch, so it will cost about three times as much per flight as a single-motor rocket.

From a legal and safety standpoint, there are limits on how much propellant we can use in a model rocket. If you go over the limit for propellant—125 grams, or about 4.4 ounces—you hit the limit allowed by the FAA in the United States. That means you've entered the domain of high-power rocketry. While that's fine if you're a member of a rocket club with FAA clearances, it does present a bit of a problem for those of us who want to take a rocket to the local park for a flight.

Table 18-1 shows the maximum number of motors for common Estes motors based on this limit of 125 grams of propellant. As you can see, this generally isn't much of an issue.

Table 18-1. Maximum number of motors you can cluster without FAA clearance for the flight

Motor class	Single motor propellant weight (grams)	Maximum motors without FAA clearance
C6	10.8	11
D12	21.1	5
E12	35.9	3

The second legal and safety limit is the minimum launch distance. Under NAR rules, anything with less than 20 N-s of total impulse—a D rocket—can be launched from a distance of 15 feet. Cerberus uses three motors, though, so if you are flying it with C motors, the minimum allowed distance is 30 feet. That's fine if you built the mono launcher in Chapter 4 with the recommended 35-foot cord, but most commercial launchers only have a 17-foot cord.

Another factor is the number of igniters. Looking back at Table 4-4 from Chapter 4, you'll see that the mono launcher can handle a cluster rocket with four motors using eight AA batteries, but only if you are using the Quest Q2G2 igniter. Six-volt launchers like those in most starter kits won't do the job at all—but then, they also don't have a long enough launch cord to safely fly a cluster of three C motors. Switch to a 12-volt lead-acid battery like a car battery, though, and the mono launcher can handle a cluster of 4 motors using Estes igniters, or 13 motors using Quest Q2G2 igniters.

Putting all of this together, here's a quick checklist to run through when designing cluster rockets or getting ready for a cluster rocket flight.

Cluster Checklist

1. Does the propellant weight exceed the maximum allowed propellant weight? (Check Table 18-1.)
2. If the rocket has more than 20 N-s of total impulse (more than one D motor, two C motors, four B motors, or eight A motors), is the launch cord at least 30 feet long?
3. Will the launcher ignite all of the igniters? (Check Table 4-4 in Chapter 4.)
4. Is the thrust sufficient for the weight of the rocket?
5. Is the rocket stable? Multiple motors add a lot of weight at the rear of the rocket. Be sure to run a simulator or use a proven design to ensure stability.

Configuration of Clustered Motors

Clustered motors should be close together. There are several reasons for this, but the most important is stability. If the rocket motors are too far apart, say in pods at the ends of the fins, the rocket could easily veer off course. One reason for this is that the most common launch failure is an igniter failure—some of the motors may not light! If the motors are close together, the rocket will not veer off course as dramatically. Also, just like anything humans build, all rocket motors are not created exactly the same. There can be as much as a 10% difference in thrust from one motor to the next. Once again, if the motors are physically close together, this effect is hidden somewhat, but if they are spread apart, the rocket could veer dramatically off course.

With that in mind, Figure 18-3 shows some common cluster configurations.

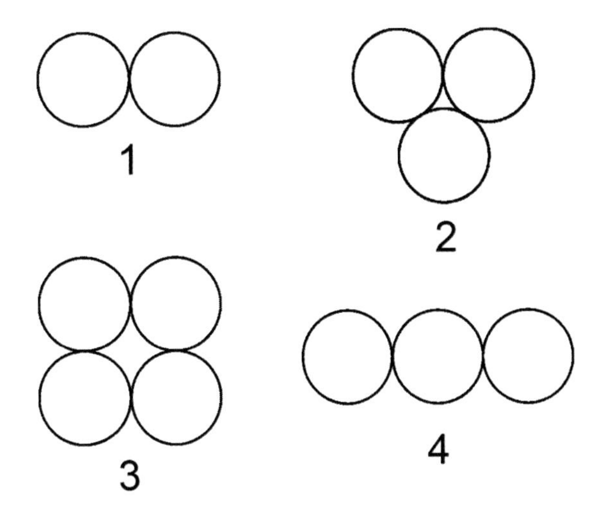

Figure 18-3. *Common cluster configurations. The configurations are numbered for reference.*

The two rockets in this chapter use the triple configuration labeled 2 in Figure 18-3. One reason is because it makes a very pretty rocket, and the other is because it's a practical way to arrange three C6 motors to get more liftoff thrust than you would with a D12 or E12 motor. One of the things that make this configuration so attractive is that three BT-20 motor mount tubes slide perfectly into a BT-60 body tube.

One disadvantage of this configuration is that all of the motors need to be the same. If any are different, the rocket could veer off course due to uneven thrust.

Staging Cluster Rockets

Another reason to switch from the motor configuration used on the rockets in this chapter to one of the others is staging. It's perfectly reasonable to add a second stage to a cluster rocket. There are ways to electrically ignite a second-stage motor, which also allows the second stage to be clustered, but it is simpler to use the staging technique from Chapter 17. That means we can't cluster the second stage, since the motors on the first stage are unlikely to burn out at exactly the same time. It also means we cannot use the motor configuration shown in this chapter, since the second-stage motor would not sit right on top of any of the first-stage motors. The three inline motors (configuration 4) would work fine, though.

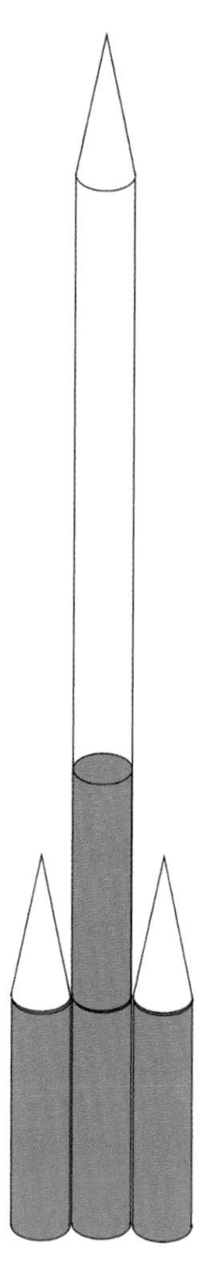

Figure 18-4. *Layout of a two-stage rocket with a cluster of three motors on the first stage and one motor for the second stage. Motors are shown in gray, and fins are omitted for clarity.*

The Three-Motor Cerberus

The three-motor Cerberus is a large, attractive rocket that is fun to fly. You can fly Cerberus on several motor combinations. Note that the motors listed in Table 18-2 are usually reserved for the top stage of multistage rockets. The longer delay times are a must with the extra thrust from three motors.

Cerberus

Most asteroids orbit between Mars and Jupiter. They might have formed another rocky planet past Mars if not for the heavy pull of Jupiter. Cerberus is a Mars-crosser asteroid, which means its orbit crosses inside the orbit of Mars. Maybe it got knocked out of the normal orbit for asteroids by collisions, or the gravitational interaction with the planets and other asteroids may have disturbed the orbit.

For whatever reason, the orbit is very odd. At aphelion, it's 1.58 AU from the Sun. (An AU, or astronomical unit, is the average distance from the Earth to the Sun.) Mars orbits at 1.52 AU, so Cerberus gets just a bit farther from the Sun than Mars. At perihelion, Cerberus is just 0.58 AU from the Sun—well inside the orbit of Venus, which is at 0.72 AU—so Cerberus crosses the orbits of three planets!

Cerberus is pretty small, with a diameter of just 1.2 km. Discovered in 1971 by Luboš Kohoutek of Czechoslovakia, it's named for the three-headed guard dog of the Greek mythological underworld. One head (and one planet crossing!) for each motor makes the name a great fit for our rocket.

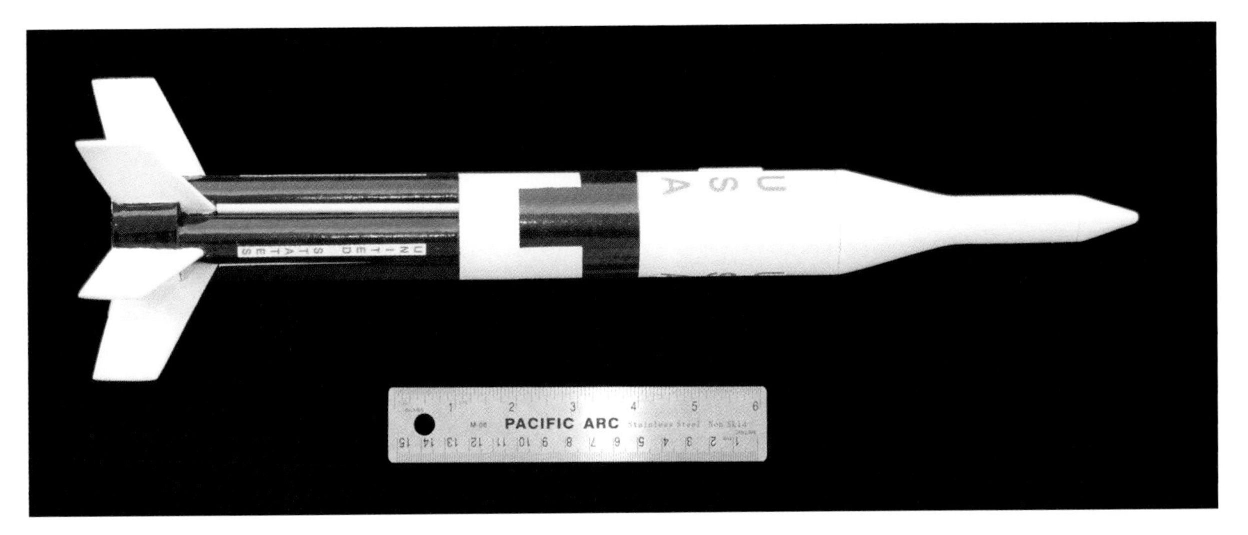

Figure 18-5. *The 24" Cerberus is a cluster rocket sure to get attention at any launch.*

Table 18-2. Recommended motors

Motors	Approximate altitude
A8-5	520 ft
B6-6	1,130 ft
C6-7	2,210 ft

Table 18-3 shows the parts needed for Cerberus, all of which are in the Estes Designer's Special.

Table 18-3. Parts list

Part	Description
BT-60 body tube	You will cut two pieces from a longer tube. The rocket uses a 5"-long piece and a 1"-long piece.
BT-20 body tube (2)	One BT-20 is cut into three 6" pieces to form the triple motor tubes. A 3 3/4" piece is cut from the other tube to form the payload bay.
BT-20 nose cone, 1"	While any BT-20 nose cone will do, the small, 1"-long conical nose cone in the Estes Designer's Special is reminiscent of the Apollo capsule carried on some Saturn 1B rockets.
1/8" balsa fin stock	The 1/8" balsa wood will be used for the fins and for tube coupler bulkheads.
1/16" balsa fin stock	The 1/16" balsa wood will be used for the motor mount bulkheads.
3/16" launch lug	You will use two pieces, each 1" long.
BT-60 tube coupler	This is used as the base of the nose section. You will need a 1"-long piece. It can be cut from a longer piece in the Estes Designer's Special.
Engine block (3)	You will need three engine blocks, one for each motor tube.
Tube adapter	The tapered section connecting the BT-60 to the BT-20 is made from paper. The Estes Designer's Special comes with a template. Figure 18-9 shows a pattern you can use to cut your own from card stock.
1/4" shock cord	You will need 24-30" of the thicker 1/4" shock cord.
18" parachute	There is a nice 18" parachute in the Estes Designer's Special, or you can make one following the instructions from Chapter 3.
Screw eye	You will need a small screw eye for the base of the nose section. These are available at hardware stores.
1/8" dowel	Used for the white accent tubes that show between the black tubes. These are available at hardware stores and hobby stores.
Clear label paper	Used to create the USA decals that appear on the white background. You will also need access to a laser printer. Use Avery Clear Full Sheet Labels, part 18665, or equivalent. This is available at most office supply stores.
White label paper	Used to create the white UNITED STATES decals that appear on the black tubes. Use Avery White Shipping Labels, part 8165, or equivalent. This is available at most office supply stores.
Snap swivel (2), optional	For attaching the parachute and shock cord. These are available at sporting goods stores, in the fishing section.
Putty	You will need model putty or wood putty to smooth the joints on the tube adapter.

Equivalent Kit

The Cerberus from *http://jonrocket.com* is a nice three-motor cluster rocket similar to the three-motor payload rocket. It also shares a name with ours!

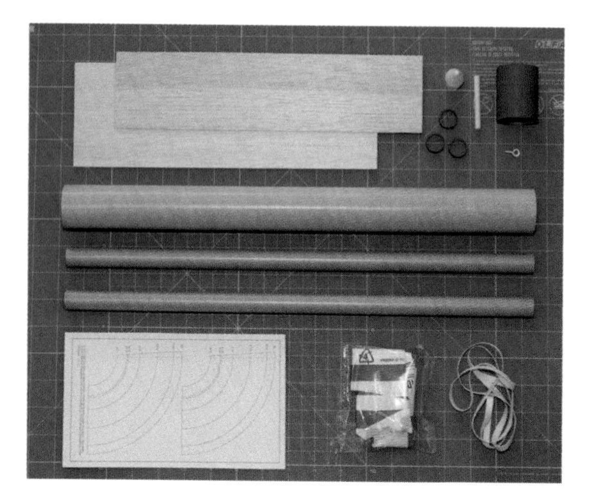

Figure 18-6. *Visual parts list. The sheets of labels and dowel (which starts out 3 feet long) are not shown.*

Construction

Refer back to Chapter 3 for basic construction techniques, tools, and supplies.

Cerberus is easily the most complicated rocket in this book. It's not that any individual step is that hard, but there are a lot of them. Plan to spend some time on construction and finishing, which will stretch over several days as you wait for various parts to dry. If you rush this rocket, it will definitely show—and perhaps won't fly.

Cutting the parts

Begin by cutting the various tubes.

The large outer white tube and the tail ring on Cerberus are made from BT-60 body tubes. Start by cutting a 1″ piece for the tail ring and a 6″ piece for the main body tube.

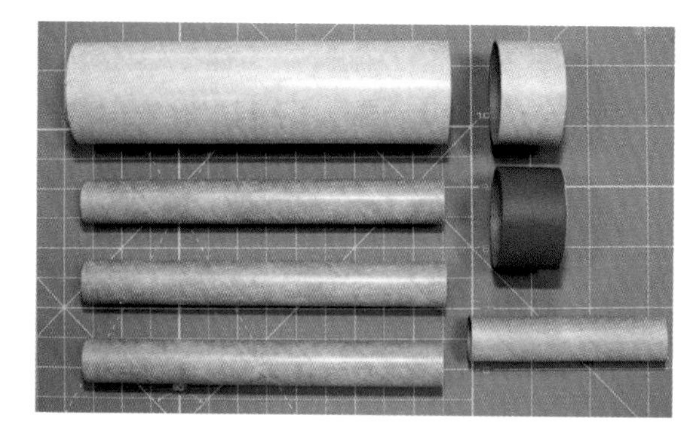

Figure 18-7. *Cut 1″ and 6″ pieces from a BT-60. Cut three 6″ pieces from an 18″ BT-20, and another 3 1/4″ piece from another BT-20. Cut a 1″ piece from a BT-60 tube coupler—the tube coupler in this photo is red, but many are gray.*

The three black motor tubes are cut from a single 18"-long BT-20 body tube. Each is 6″ long, so you're cutting the tube in thirds. Very small differences in tube length won't show up in the final rocket; they will be hidden inside the 6″ BT-60 body tube.

The skinny top section of the rocket is another BT-20 body tube. This one extends through the conical tube adapter, which will be made from paper. It is 3 3/4″ long. Cut it from another BT-20 body tube.

The conical tube adapter and skinny top section form the nose of the rocket. We'll eventually build a shoulder for this nose section from a 1″ section of tube coupler. The Estes Designer's Special comes with a 3"-long section of tube coupler; cut 1″ from this longer piece. Other tube couplers may be different lengths. You might be tempted not to bother cutting a quarter inch from a 1 1/4″ tube coupler if that's what you buy, but this rocket chooses form over function in a few places, and the upper section that holds the parachute is one of them. You really have to trim the tube coupler back to 1″, even if you are only cutting off a small piece, to leave room for the parachute.

Next, we'll cut the paper parts.

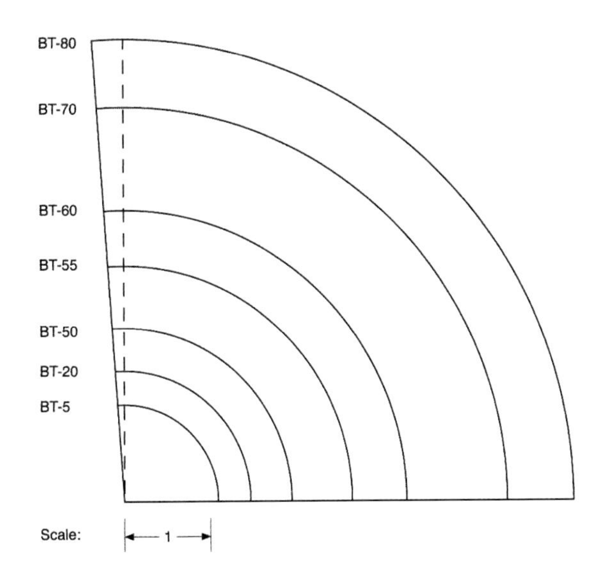

Figure 18-8. *Cut two 1" pieces from a 3/16" launch lug. Cut the tube coupler from the tube coupler sheet in the Estes Designer's Special.*

Cut two 1" pieces of 3/16" launch lug. These are the fatter launch lugs; this is a heavy rocket that needs a thicker launch rod.

There are a couple of ways to create the paper tube coupler that will smoothly join the BT-60 to the BT-20. The Estes Designer's Special comes with two sheets of tube couplers. Locate the arcs for BT-60 and BT-20 and use a hobby knife to cut out the tube coupler. Take your time and cut accurately along the lines. Accuracy is critical in this step. Even a small gouge or bulge will make it very difficult to get a good seal and finish later.

If you did not get the Estes Designer's Special, don't rush out and pay money for a tube coupler. You can make your own from card stock and a printer or copier. Figure 18-9 shows a template for cutting paper tube couplers with about a 15° angle to the body tube. "Custom tube transitions" on page 357 also shows you how to create tube transitions from scratch.

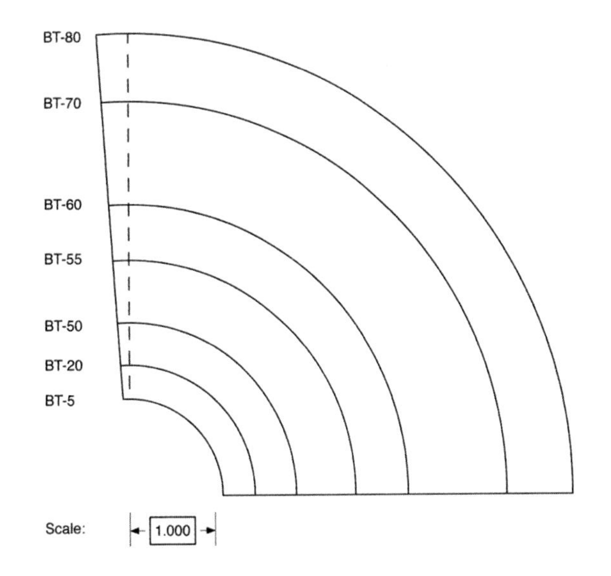

Figure 18-9. *Tube transition template for common body tube sizes. The angle between the body tube and the wall of the transition is about 15°. See the author's website (http://bit.ly/byteworks-make-rockets) for a full-size template.*

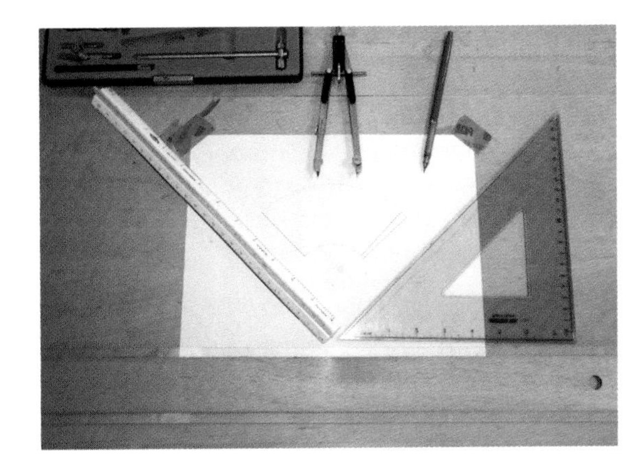

Figure 18-10. *A protractor, a compass, and a ruler are all you need to create custom tube transition templates.*

The remaining parts are cut from 1/16" and 1/8" balsa sheets.

Figure 18-11. *Cut the bulkheads and fins from 1/8" balsa, and the oddly shaped motor mounts from 1/16" balsa.*

Working from the top left, the three large bulkheads appear to be the same size in Figure 18-11, but actually are not. Two fit inside the tube coupler to form a bulkhead. The third is the same size as the outside of the BT-60 body tube. The smaller disk fits inside the BT-20 body tube. Use the body tubes and tube coupler as templates to cut these parts. Cut all four of the bulkhead disks from 1/8" balsa.

The fins are also cut from 1/8" balsa. The extra thickness is needed if the rocket is loaded with C6 motors.

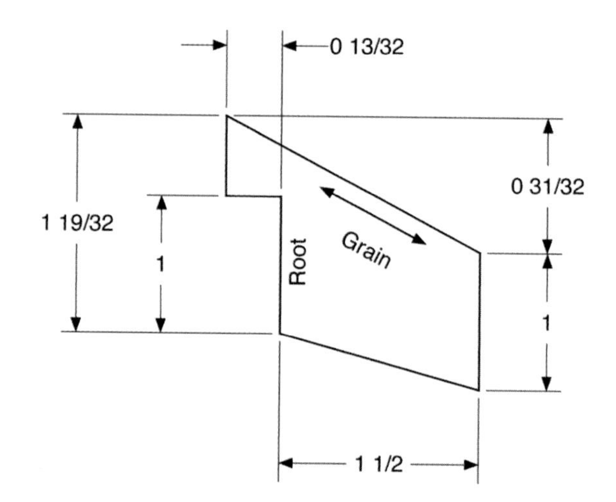

Figure 18-12. *Cut three fins with a long tab from 1/8" balsa. These fins are mounted over the groove between the BT-20 body tubes. The tabs are a bit long; this allows you to sand a point on the tab so the fin fits snugly into the groove between the BT-20 body tubes.*

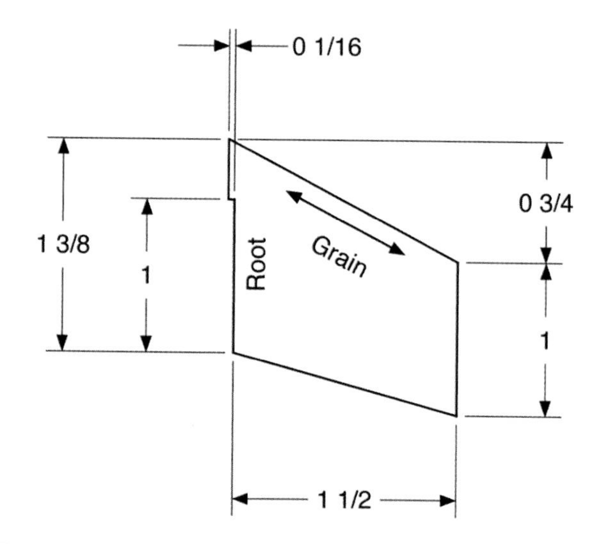

Figure 18-13. *Cut three fins with a short tab from 1/8" balsa. These fins are mounted where the BT-20 motor tubes touch the BT-60 outer body tube.*

Take a look at Figure 18-14, and you can see why the odd pieces at the bottom are shaped as they are. The two tiny triangles fit between the three motor tubes that are painted black in the final rocket. The 12 curved pieces hold the tubes in place. All of these are cut from 1/16" balsa. They don't need to be made from thicker balsa, since there will be four layers of these bulkheads in the finished rocket. And, don't forget, the motor tubes are also glued directly to the BT-60 main tube, so with all this bracing it's not likely they're going to be moving around.

Use a BT-20 body tube and the outside of the BT-60 tube coupler to draw an actual-size version of Figure 18-14. Use this as a template to cut the various parts shown in Figure 18-11.

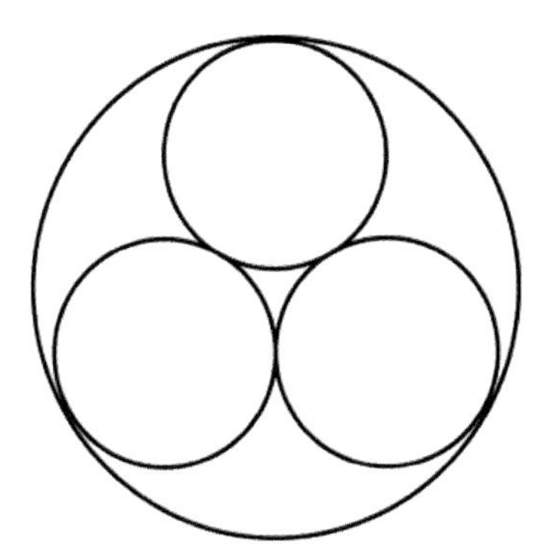

Figure 18-14. Three BT-20 body tubes fit perfectly inside a BT-60 body tube. Use a BT-20 body tube and the BT-60 tube coupler to create an actual-size version of this drawing. Use it as a template to cut the parts.

Custom tube transitions

What if you have an odd tube size, or you want a specific angle between the body tube and transition for a scale model or a competition rocket? I hope you paid attention in trig class!

The first thing we need to find is the distance along the transition itself from one body tube to the next. We either know or can measure the diameters of the body tubes, so d_1 is just one half of the difference in the body tube diameters.

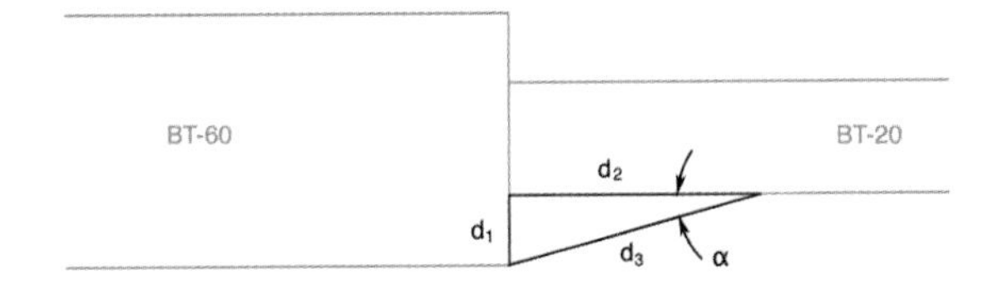

Figure 18-15. Use basic geometry and trigonometry to find d3, the length of the transition cone.

Depending on the source of our information, we may know either the length of the tube coupler in the direction of the body tubes, d_2, or the angle between the tube coupler and body tubes, α. If we know d_2, the length along the transition tube, d_3, is easy to find using the Pythagorean theorem:

$$d_3 = \sqrt{d_1^2 + d_2^2}$$

If we know the angle, the length of the hypotenuse is:

$$d_3 = \frac{d_1}{\sin \alpha}$$

Either way, we have the length along the side of the transition cone, which is what we need for the next step.

Measuring the angle in radians, the length along part of a circle is:

$$l_1 = r_1 \theta$$

l_1 is the diameter of the smaller body tube, while l_2 is the diameter of the larger body tube. We need to find the two radii, r_1 and r_2, and the angle θ to draw a template for the tube adapter.

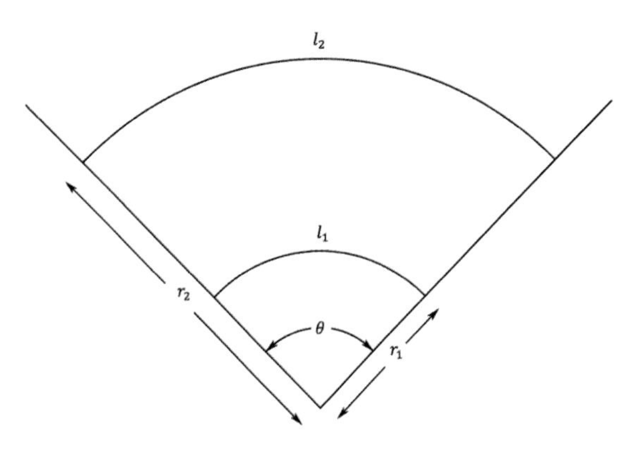

Figure 18-16. Location of radii, arc lengths, and angle used to calculate the size for a transition template.

Fortunately, we just calculated d_3, which is also the difference between r_2 and r_1. This gives us two equations and two unknowns:

$$l_1 = r_1 \theta$$

$$l_2 = r_2 \theta = (r_1 + d_3)\, \theta$$

Solving for θ gives:

$$\theta = \frac{l_2 - l_1}{d_3}$$

For example, our rocket will use a BT-20 and a BT-60, with outer diameters of 1.637" and 0.736". This makes $d_1 = 0.45"$.

The length of the transition is 1.75". Using the Pythagorean theorem, the transition tube length, d_3, is 1.81".

Plugging the circumferences of the body tubes, l_2 and l_1, and the transition tube length into our equation, we find that θ is 1.56. Converting the angle from radians to degrees, we get an angle of 90°.

With the angle in hand, we can go back and find r_1 and r_2, which are 1.47" and 3.27".

Now, with a little work with a protractor, a compass, and a ruler, we can draw our transition tube directly onto card stock. Add an extra 4–5° for a glue flap.

Motor compartment assembly

With all of the parts cut, it's time to start assembly. Begin by gluing the three 6" BT-20 body tubes together, making sure the ends match precisely on one end, even if the ends of the tubes are a bit off at the other end. The end where the tubes match precisely will be the bottom of the rocket.

Glue the 1" BT-60 tail ring to the motor mount cluster on the end where the motor mount tubes match precisely.

Figure 18-17. *Glue the three motor mount tubes together, then glue them into the 1" section of BT-60 body tube.*

Once the glue on the body tube assembly is dry enough to handle, glue the two triangular pieces of the bulkhead in the gap between the three BT-20 body tubes, one on each end of the cluster. Fill any gaps with glue; it is important that you make a good seal so the ejection charge gases push the parachute out and are not vented through the gap in the middle of the three motor mount tubes.

Glue the 12 arced pieces of the bulkhead in place. Three form the base of the rocket, and are glued flush with the bottom of the rocket, at the base of the BT-60 tail ring. Three are placed at the top of the tail ring, forming a bulkhead flush with the top of the BT-60 tube. You can see a tiny sliver of these in Figure 18-18.

Another three arced pieces form the bulkhead at the top of the assembly. Glue the final three pieces 1/2" from the end of the tube. Make sure these line up perfectly with each other; these three pieces will form another bulkhead once the remaining BT-60 body tube is glued in place.

Figure 18-18. *Glue the two triangular pieces at either end of the assembly to block ejection gas from passing through the gap between the motor mount tubes. Glue the 12 arced pieces in place to form the four bulkheads. Two of the bulkheads are still exposed on the right side of this photo. A third is just barely visible where the three tubes go into the tail ring. The fourth bulkhead is at the base of the tail ring.*

Once the glue sets, slide the 6″ BT-60 body tube over the end of the assembly, checking the fit carefully. Sand the balsa pieces as needed to get a snug, but not tight, fit. Once you are satisfied, put a generous amount of glue about 1/4″ inside one end of the BT-60 body tube and slide the assembly into place. Twist the assembly to spread the glue evenly.

Check the alignment of the tube assembly by rolling it across a flat surface. Is there any wobble? If so, straighten the tubes so the assembly rolls smoothly.

The Tubes Must Align Perfectly!

Check and recheck the alignment of the tubes using this technique of rolling the assembly across a table. The glue will set quickly, but you have time to make adjustments. The rocket may not fly straight if the tubes are misaligned, so it is very important to do this properly.

Check inside the BT-60. If sliding the tube into place did not form a generous fillet between the body tube and balsa parts, apply more glue.

Set the entire assembly aside to dry. Prop it up vertically so the glue fillet at the top of the rocket doesn't drip down the tube. It's also a good idea to place the rocket on wax paper or plastic wrap so any glue at the base doesn't stick to the cutting board.

Nose assembly

Gather the parts for the nose section. You will need the 3 3/4″ section of BT-20 body tube, the 1″ piece of BT-60 tube coupler, the tube transition, the four circular bulkhead parts, and the screw eye.

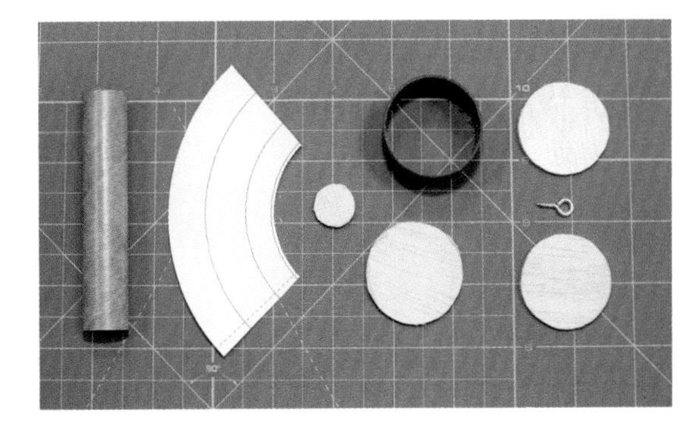

Figure 18-19. *The parts for the nose section.*

Test fit the two slightly smaller circular bulkheads in the tube coupler. When you are satisfied with the fit, glue these together to form a 1/4"-thick bulkhead. Glue them so the grain on the two parts is perpendicular; this adds strength to the overall part.

Test fit the smallest disk in the BT-20 body tube. It should be snug enough to keep the body tube in place, but not tight. The remaining large bulkhead disk should be the same size as the outside of a BT-60 body tube; it will overhang the tube coupler, which should slide into the BT-60. Sand the tube coupler if it is too tight.

Glue the small bulkhead in the exact center of the large bulkhead.

The Center Disk Location Is Critical

Check and recheck the position of the small disk with a ruler. It is extremely important that it be properly centered. If it is not, the tube transition may not fit, or the rocket may not fly straight.

Figure 18-20. *Glue two bulkheads in the base of the tube coupler. Glue the BT-20 bulkhead in the exact center of the largest bulkhead, which should be the same size as the outside of a BT-60.*

Glue the large bulkhead to the top of the tube coupler with the small BT-20 bulkhead on the outside. Use a 90° angle to check the alignment, and glue the BT-20 over the smaller bulkhead, which holds it centered on the assembly.

Figure 18-21 shows a drafting tool being used to check the angle. The specific tool is not critical; the corner of a piece of card stock or an uncut corner of a sheet of balsa wood will work well, too, as will any 90° angle. Whatever tool you use, check the angle on several sides to make sure the BT-20 is perfectly perpendicular to the top of the assembly.

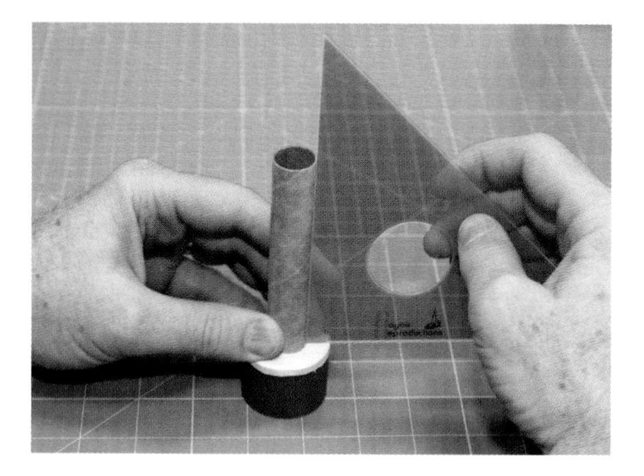

Figure 18-21. *Glue the bulkhead and BT-20 in place, checking carefully to make sure the BT-20 is perfectly aligned.*

Set this assembly aside until the glue sets thoroughly. In the meantime, you can begin shaping the tube coupler.

Use a pencil or thick dowel to shape the tube coupler. Work slowly, curling the stiff paper tube coupler around the pencil gradually. You will end up with large wrinkles in the paper if you work too quickly. Keep at it until the tube coupler wraps around on itself, and has to be enlarged to fit over the body tubes.

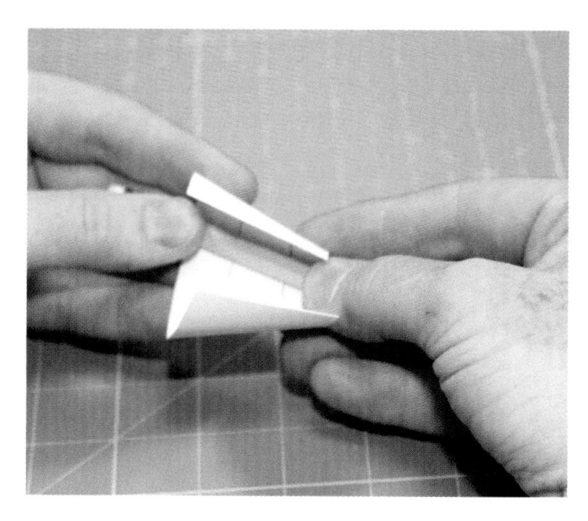

Figure 18-22. *Use a pencil to gradually curl the tube coupler until it needs to be expanded to fit over the body tubes.*

Once the glue has dried on the tube assembly, use a sanding block to shape the edge of the bulkhead so the tube coupler sits flush against the balsa. It's not much of an angle—just 18°—so don't get carried away and sand off too much. Test fit the tube coupler often while you do this, recurling it on the pencil as needed.

You can slip the nose cone into place at this point, but don't glue it. The BT-20 body tube forms a small payload section. Add tape to the nose cone if it is too loose in the body tube.

Figure 18-23. *Bevel the edge of the bulkhead so the tube coupler sits flush against the bulkhead.*

Glue the tube coupler in place. Make sure it overlaps the bulkhead, and that there is plenty of glue holding it in place. Apply tape to keep the tube coupler in exactly the right place while the glue dries.

Figure 18-24. *Tape the tube transition in place while the glue dries.*

Set the nose assembly aside to dry overnight.

Fin assembly

Cerberus has six fins, not three or four, so the fin alignment guides from Chapter 3 won't do much good. Figure 18-25 will work for six fins, though.

The fins are designed to fit in specific locations. While all six are glued to the tail ring, a small tab at the top of each fin will also get glued to the motor mount tubes. Three of the fins get glued to the outer edges of the tubes, while three are glued into the crack formed where the motor mount tubes join.

Line the body tubes up carefully so the outer edges of the motor tubes also align with three of the fin locations. These are the locations where the fins with the smaller tab will be placed. Mark the locations of the six fins and the launch lug.

Using a doorjamb, extend the fin lines along the length of the tail ring. Extend three of the lines up the edge of the motor mount tubes. Draw the launch lug line all the way up the side of the rocket, since one of the launch lugs will be placed high on the 6" section of BT-60 body tube.

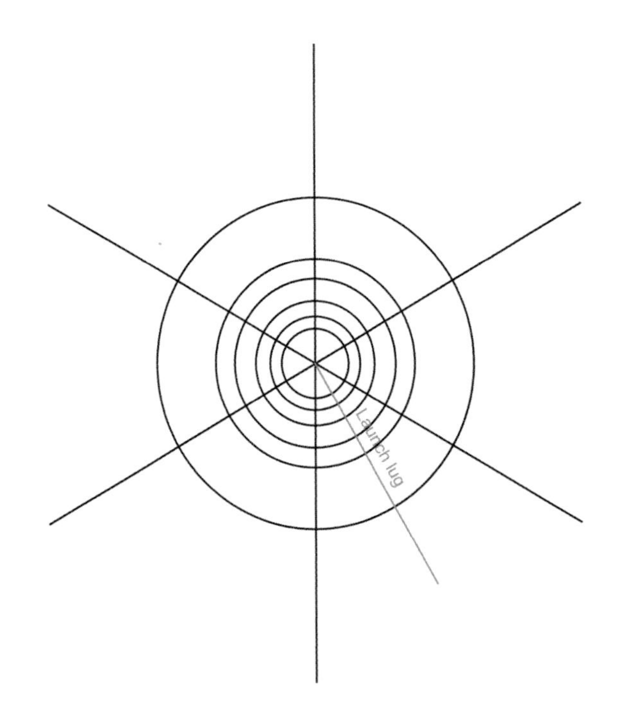

Figure 18-25. *Fin marking guide for six-finned rockets.*

Round the outer edges of all six fins, and test fit the fins in place. The tabs on the three fins that fit into the cracks formed by the motor mount tubes are a bit long; sand them to a point so the fins fit snugly in the cracks formed by the body tubes.

You have a choice on the launch lugs. The rocket is designed with the launch lugs glued against the body tube. But this rocket isn't just a pretty face. It is capable of lifting more weight than any other rocket in the book, since the three C6 motors deliver more initial thrust than a D12 motor. Remember how the rockets in Chapter 9 had standoff tabs for the launch lugs so the rockets could lift oversize payload bays like the egg capsule and camera capsule? You can add those same standoff tabs to this rocket and use it to loft those oversize payloads, too.

So, make a decision based on what you will do with this rocket, then glue on the launch lugs. Use standoff tabs on the launch lugs if you want to be able to use it with some of the larger payloads from Chapter 9. One launch lug is mounted on the tail ring, and the other 1″ from the top end of the 6″ BT-60 body tube.

Figure 18-26. *Glue the fins and launch lugs in place.*

Final assembly

There are a few details left before the rocket is completely assembled.

First, glue the three engine blocks in place. You thought I forgot, right? No, but leaving the engine blocks until now reduces that chance of a silly mistake. It's very unlikely you will mount the engine block on the wrong end of the engine tube now that the rocket is mostly assembled!

Use a scrap of wood or your finger to rub a generous amount of glue about 2 1/4″ into the back of one of the motor mount tubes, then use a rocket motor to push the motor mount 2.5″ into the motor tube. The rocket motor should stick 1/4″ out of the back of the rocket. Repeat the process with the other two motor mount tubes.

The shock cord for this rocket is one of the thicker ones. It should be 1/4″ wide, not the narrower 1/8″ version, and should be at least 24″ long. The shock cords in the Estes Designer's Special are about 30″ long; go ahead and use the whole thing. A longer shock cord won't do any harm, and what are you going to do with a few inches of leftover shock cord, anyway? Glue it in place in the normal way.

Mount the screw eye in the bulkhead of the motor assembly. As usual, screw it in, remove it, insert some glue in the hole, then screw it back in. This creates a nice, strong joint.

Finally, apply model putty or wood putty to the joint on the tube coupler and to the uneven area where the tube coupler and bulkhead join together. Let it dry completely, then carefully sand the putty until you get a smooth joint.

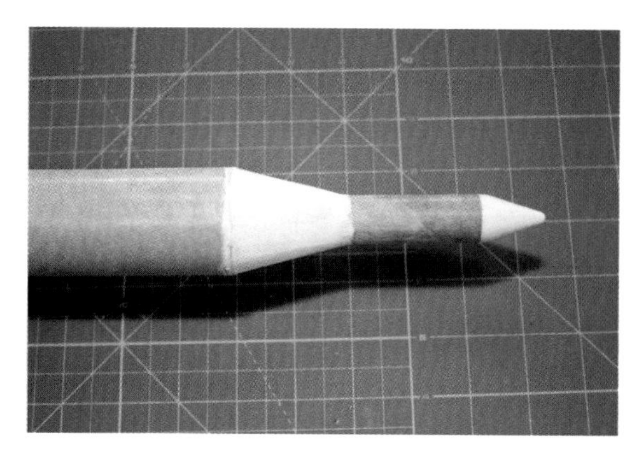

Figure 18-27. *Use putty to smooth out the paper joint and to fill imperfections.*

If you like, attach snap swivels to the shock cord and an 18″ parachute. If you don't use snap swivels, use a buntline hitch to attach the shock cord to the screw eye. Either way, add a small dab of glue to keep the knot from working loose.

You might have noticed that we still have not used the 1/8″ dowels. It actually works better to glue those in place as part of the finishing step, so we're leaving that for later.

Finishing

Cerberus is a bit involved to build, and it's also a bit involved to finish. Since it's a little beyond priming, sanding, and painting, let's go through the finishing steps in detail.

Of course, the first step is still priming, sanding, and painting. Pay particular attention to all those corners and crevices where the fins attach to the motor mount tubes. Yes, they are hard to reach. Be patient and work the sandpaper into those hard-to-reach areas until they feel silky smooth.

Once the rocket feels smooth, spray the entire rocket gloss white. You will end up painting the motor mount tubes, too, but that's not important. They will eventually be black.

While you are working on the rocket, get the dowel ready, too. Prime and sand at least 12" of the dowel, then paint it white. This will be cut up to provide decorative trim after the rocket is painted.

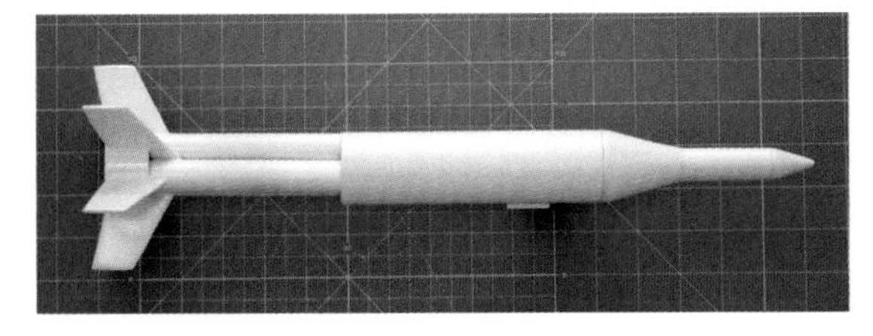

Figure 18-28. *Prime and sand the rocket carefully, then paint the entire rocket white.*

The top section of the body will be painted with black rectangles that extend down from a black stripe. These cover 1/6 of the body tube. Cut a piece of paper and wrap it around the body tube, marking the location where it overlaps.

From the end of the paper up to the mark you just made, carefully fold the paper in thirds. Then fold the result in half. This gives you a very precise tube marking guide. The distance between folds is exactly 1/6 of the diameter of the body tube.

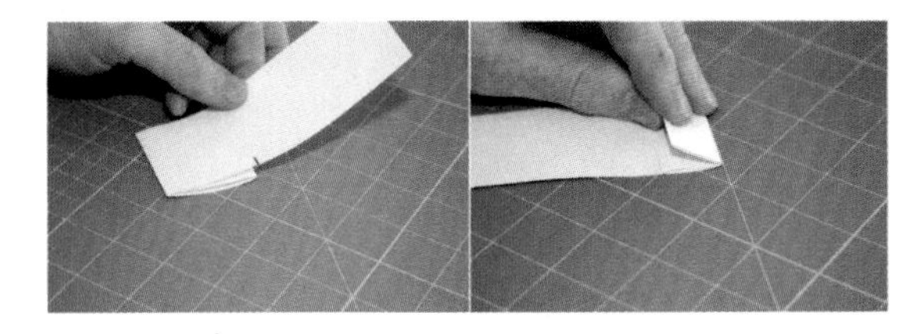

Figure 18-29. *Create a tube marking template by marking a piece of paper with the diameter of the body tube, then folding it into sixths.*

Wrap a line of tape around the top section of BT-60, so one edge is 1" from the base of the tube. This should reach right to the edge of the body tube. If the tape is too narrow, use another piece to reach the edge of the body tube. If it is too wide, trim the tape to a 1" width using a hobby knife.

Wind a second piece of tape around the body tube so the bottom edge of the tape is 3" above the bottom of the body tube, forming a 2" gap between the first piece of tape and the second.

With the tape in place, use a ruler to extend the line from one fin to the white gap between the tape stripes, and mark this location with a pencil. Use the tube marking guide to mark the other five locations around the tube.

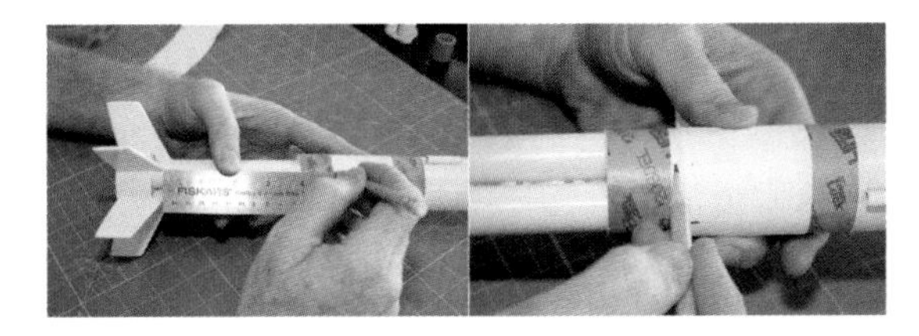

Figure 18-30. *Add a band of tape that covers the lower 1" section of the BT-60, and another that leaves a 2" gap between the bands. Use the tube marking guide to mark six evenly spaced locations over the fins.*

Use the tube marking guide again to mark the width of one rectangular section on a piece of tape. Use a hobby knife to trim the tape to width, and to cut a nice, even end on one end of the strip of tape.

Figure 18-31. *Use the tube marking guide to make a piece of tape to tape off the white part of one rectangular band.*

Place the tape between two of the pencil marks, extending up 1" from the bottom band of tape. This forms the white part between the black rectangles that you will eventually paint on the rocket. Repeat the process two more times to cover three rectangular areas around the rocket.

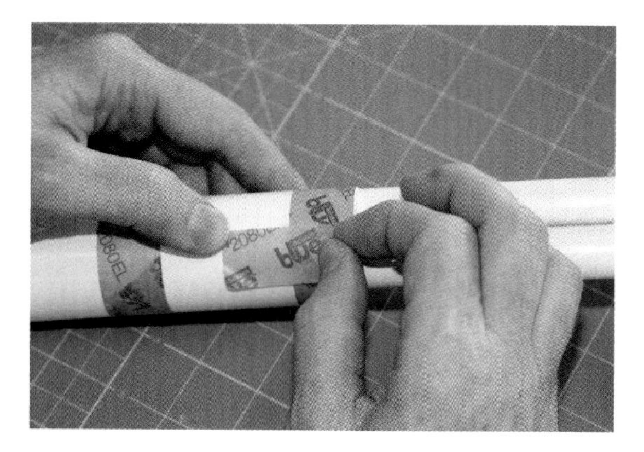

Figure 18-32. *Use the trimmed tape to mask off three rectangles at the bottom of the 2" gap.*

Finish taping the rocket. The fins remain white, so they are completely covered in tape. The tail ring is striped, though. The sections below the taped areas in the 2" gap on the main body tube are also taped so they stay white; the areas below the untaped part of the ring are also untaped, so they are painted black. You can see one of the untaped sections in Figure 18-33.

It's pretty tough to accurately tape the area covered by the balsa motor mounts, so I left those open.

The top of the rocket is covered in paper and tape. Seal the edges down carefully before you apply the black paint. There are a lot of edges, and any that are not sealed perfectly will allow paint to seep through. Once all of the tape is secure, apply black paint and allow it to dry.

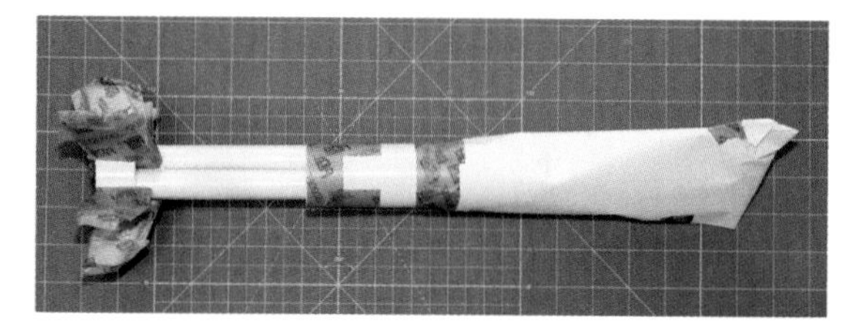

Figure 18-33. *After final taping, paint the exposed parts of the rocket black.*

Perhaps you're better at sealing the tape than I am. I had a few leaks. Look back at "Painting the Rocket" on page 56 in Chapter 3 for tips on touching up the mistakes.

Once the paint is dry, there are just a couple more steps before the rocket is complete.

Start by trimming three sections of the dowel to length, just under 4", and glue these in place in the cracks formed by the motor tubes. Most of the time we're pretty generous with glue when building a rocket; this is one time to be stingy. Use just enough to tack the dowels in place, and not enough to show in the cracks between the dowels and the motor tubes. I usually use wood glue for practically every part of the rocket, but this is one place where I used some thick CA glue—although wood glue would have worked, too.

Download the decal masters from the author's website (*http://bit.ly/byteworks-make-rockets*) and print them on white and clear label paper. Cut the vertical UNITED STATES decals from the white label sheet and attach them to the motor mount tubes. Cut the USA decals from the clear label sheet and fasten them above the black stripe on the main body tube.

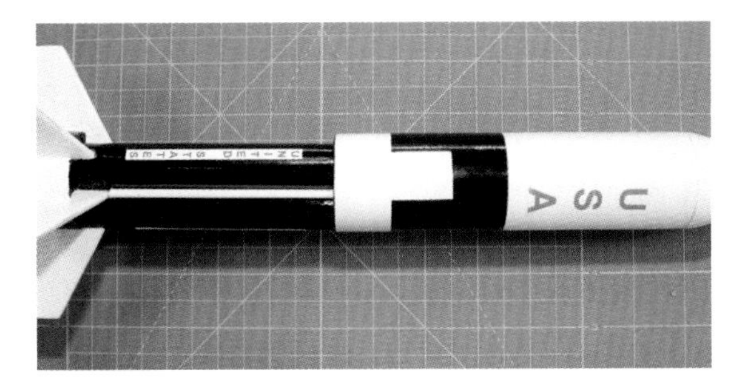

Figure 18-34. *Cut sections of the painted 1/8" dowel to fit and glue them in place in the cracks between the motor tubes. Apply the decals as shown.*

While this rocket is patterned to look like the Saturn 1B, it's clearly not a scale model. You can easily change those decals to fit your locale. Perhaps it is time to start the Roswell Space Agency, or the Canadian Mounted Rocket Police. Have some fun and make your trim unique!

Flying Cerberus

The recovery section works pretty much like all of our other parachute recovery rockets, although it is a bit of a tight fit. Begin by stuffing enough recovery wadding into the parachute bay to fill about 2 1/2" of the tube.

Figure 18-35. *Insert 2 1/2" of recovery wadding into the parachute bay.*

Make sure the snap swivels for the shock cord and parachute are both attached to the eyelet on the nose section. Be sure both snap swivels are fully closed.

Apply talcum powder to the parachute to make opening go smoothly. Fold the parachute into an appropriate-sized bundle and wrap it with the parachute shroud lines to keep the bundle tight (see "Packing the Parachute" on page 123 if you need a refresher). Stuff the shock cord, and then the parachute, into the parachute bay, then slip the nose section into place. While there is no room to spare, there should be plenty of room for the recovery wadding, shock cord, and parachute if the parachute is folded well.

The parachute should slide in easily. The nose cone should be tight enough so it does not fall out if the rocket is turned over and shaken gently, but it should not be too tight. Sand the nose cone or apply tape if the fit needs to be adjusted.

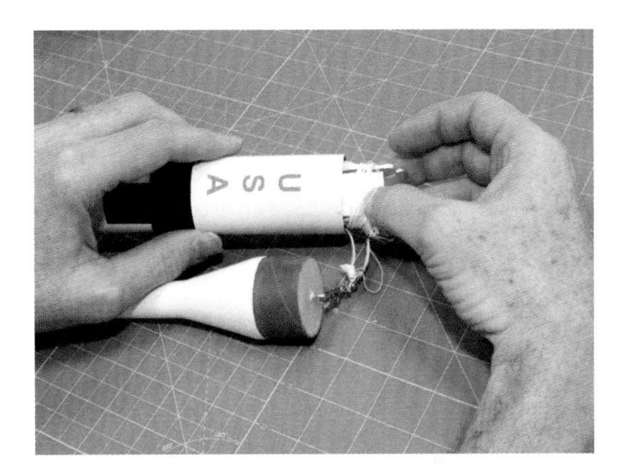

Figure 18-36. *Attach the parachute and shock cord to the nose section, then fold and insert the shock cord and parachute.*

Make sure all of the motors are the same kind. I'm going to say that again: make sure all of the motors are the same make, and have the same rating. The first flight should be made with A8-5 motors. The motors should fit tightly in the motor tubes so the ejection charge blows the parachute out, not the motors. The fit must be considerably tighter than for the parachute and nose section.

Figure 18-37. *Insert the three motors, using tape to ensure a tight fit.*

The igniters must be connected in parallel, not series.

Install them so one wire from each igniter is in the middle of the rocket, and twist all three of these wires together.

Use a paper clip or scrap of wire to form a loop around the outside of the rocket, connecting the other wire from each igniter to this loop.

Connect one igniter clip to the central bundle of wires, and the other to the outer ring of wires.

Figure 18-38. *Insert the igniters and connect them in parallel.*

Cerberus is ready for flight.

Run through the checklists to make sure everything is ready, and then prepare to be amazed. A cluster launch is truly a wonderful sight.

Launch Preparation Checklist

1. Insert recovery wadding.
2. Attach the parachute and shock cord.
3. Insert the parachute and shock cord; make sure they are not too tight.
4. Insert the three motors, making sure they are all the same kind of motor. Use tape for a tight fit.
5. Install the igniters in parallel.
6. If you are using Estes igniters, make sure you are using a lead-acid battery (e.g., a car battery) for the power source. You can use AA batteries with Quest Q2G2 igniters.
7. If you are using C motors, make sure the launch cord is at least 30 feet long.

Flight Checklist

1. Clear spectators from the launch zone.
2. Remove the launch key from the launch rail, slide the rocket onto the launch rail, and replace the launch key on the launch rail.
3. Attach the igniter clips.
4. Get the launch key from the pad.
5. Insert the launch key. Make sure the continuity light is on.
6. Check for clear skies—no aircraft or low clouds.
7. Count down from five and launch.
8. Remove the launch key and place it on the launch rod.
9. Recover the rocket.

The Ceres C Booster

Chapter 9 showed how to build two boosters for payload rockets, one for D and E motors, and another for A, B, and C motors. That chapter also mentioned that there was a third possibility—a clustered booster that could lift a heavier payload than any of the other rockets.

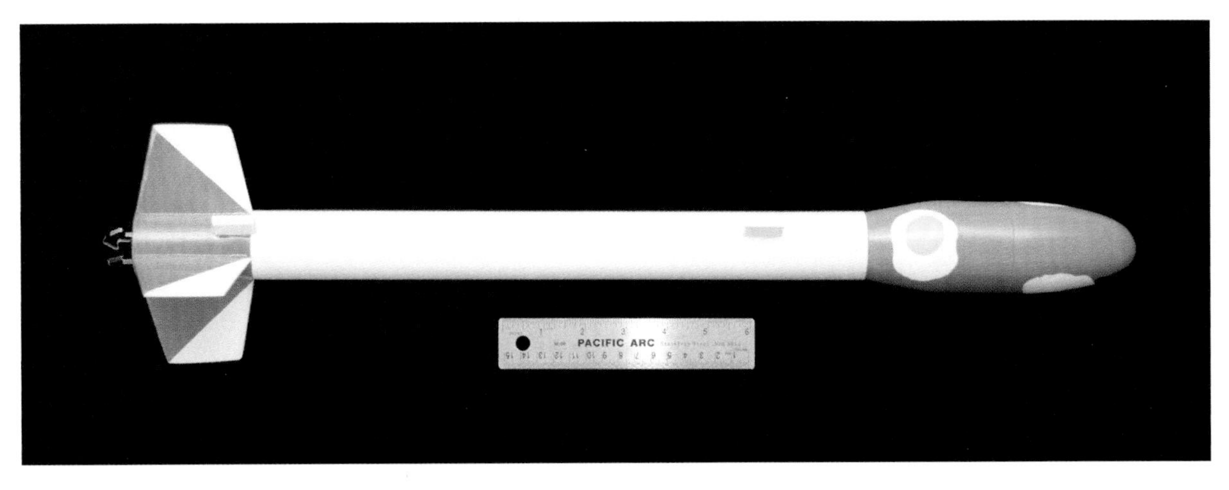

Figure 18-39. *The 24" Ceres C is a cluster rocket designed for lifting heavy payloads.*

Here it is. The Ceres C is the third variation on that basic booster. It can lift a payload that weighs up to 5 ounces with three C6 motors.

You can use any of the payload modules from Chapter 9. This section describes assembly of the booster and launch procedures with the Thin Man payload bay.

It's worth pointing out that the 5-ounce payload puts the rocket a bit below the recommended liftoff speed. It's still within the error limit applied to many commercial rockets, but the liftoff will be pretty slow. A launch with a 5-ounce payload should only be attempted with very calm winds.

Table 18-4 through Table 18-8 list the recommended motors to use with the Ceres C for each of the five payloads from Chapter 9.

Table 18-4. Ceres C/Thin Man recommended motors

Motors	Payload weight	Liftoff weight	Approximate altitude
A8-3	0 oz	5.8 oz	240 ft
	1 oz	6.8 oz	180 ft
	2 oz	7.8 oz	135 ft
B6-6	0 oz	6.0 oz	650 ft
	1 oz	7.0 oz	530 ft
B6-4	2 oz	8.0 oz	430 ft
	3 oz	9.0 oz	350 ft
C6-7	0 oz	6.5 oz	1,520 ft
	1 oz	7.5 oz	1,360 ft
	2 oz	8.5 oz	1,170 ft
	3 oz	9.5 oz	1,030 ft
	4 oz	10.5 oz	900 ft
C6-5	5 oz	11.5 oz	750 ft

Table 18-5. Ceres C/Fat Man recommended motors

Motors	Payload weight	Liftoff weight	Approximate altitude
A8-3	0 oz	6.7 oz	180 ft
	1 oz	7.7 oz	130 ft
B6-6	0 oz	6.9 oz	470 ft

Motors	Payload weight	Liftoff weight	Approximate altitude
B6-4	1 oz	7.9 oz	370 ft
	2 oz	8.9 oz	340 ft
	3 oz	9.9 oz	280 ft
C6-7	0 oz	7.4 oz	1,200 ft
	1 oz	8.4 oz	1,080 ft
	2 oz	9.4 oz	950 ft
C6-5	3 oz	10.4 oz	840 ft
	4 oz	11.4 oz	740 ft

Table 18-6. Ceres C/ICU recommended motors

Motors	Payload weight	Liftoff weight	Approximate altitude
A8-3	0 oz	6.2 oz	210 ft
	1 oz	7.2 oz	160 ft
	2 oz	8.2 oz	120 ft
B6-6	0 oz	6.4 oz	570 ft
	1 oz	7.4 oz	470 ft
B6-4	2 oz	8.4 oz	390 ft
	3 oz	9.4 oz	320 ft
C6-7	0 oz	6.9 oz	1,360 ft
	1 oz	8.4 oz	1,220 ft
	2 oz	9.4 oz	1,090 ft
	3 oz	10.4 oz	950 ft
C6-5	4 oz	11.4 oz	720 ft

Table 18-7. Ceres C/ICU2 recommended motors

Motors	Payload weight	Liftoff weight	Approximate altitude
A8-3	0.6 oz	7.0 oz	170 ft
B6-4	0.6 oz	7.1 oz	480 ft
C6-7	0.6 oz	7.7 oz	1,200 ft

Table 18-8. Ceres C/Over Easy recommended motors

Motors	Payload weight	Liftoff weight	Approximate altitude
A8-3	2 oz	8.1 oz	120 ft
B6-4	2 oz	8.3 oz	390 ft
C6-7	2 oz	8.8 oz	1,080 ft

Table 18-9 lists the parts you will need to build the Ceres C booster.

Table 18-9. Parts list (booster only)

Part	Description
BT-60 body tube	You will need an 18"-long tube. Some companies sell 34"-long body tubes. Cutting one in half and using 17" works fine, too.
BT-20 motor mount (3)	You can use three precut 2 3/4" motor mount tubes or cut your own from a longer piece of BT-20 body tube.
1/8" balsa fin stock	The balsa wood will be used for the fins and for motor mount bulkheads.
3/16" launch lug	You will use two pieces, each 1" long.
Engine hooks (3)	Use three of the 2 3/4"-long engine hooks.
Engine sleeves (3)	You can use tape if none are available.
1/4" shock cord	Use about 30" of thick shock cord.

The Estes Designer's Special has enough parts for one of the payload boosters. Chapter 9 described two of those boosters, the Ceres A and Ceres B. The Ceres A uses D or E motors, while the Ceres B uses A, B, or C motors. If you are working with the parts in the Estes Designer's Special, plan on building just one of these boosters, or place a separate order for an extra BT-60 body tube and some extra engine hooks, depending on what other rockets you are building. You may also need a sheet of 1/8" balsa from the local hobby store or hardware store.

Other than the motor mount, the parts and construction of this booster are identical to the Ceres A and Ceres B. The body tube, fins, and shock cord are all the same. Figure 18-40 shows the parts needed for the motor mount.

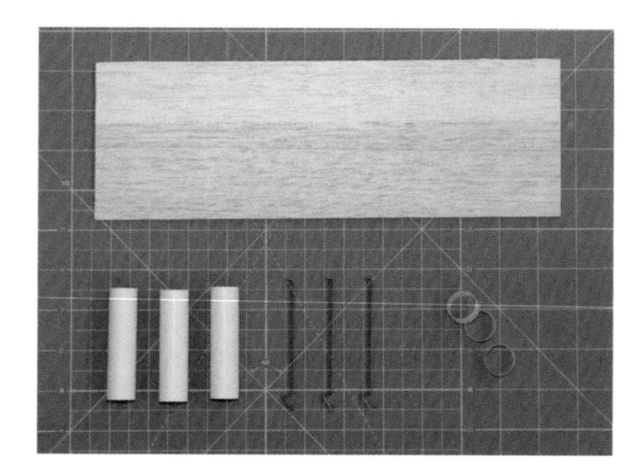

Figure 18-40. *Parts for the motor mount on the Ceres C.*

Construction

Refer back to Chapter 3 for basic construction techniques, and to Chapter 9 for payload bay construction.

There are no body tubes to cut if you are starting with precut 2 3/4" motor tubes. If you are starting with a BT-20 body tube, cut three 2 3/4" pieces from the tube for the motor tubes.

The main body tube is an uncut 18" BT-60 body tube.

There are also several balsa parts to cut.

You will need four fins cut from 1/8" balsa and rounded on all edges except the root edge. See Figure 9-14 in Chapter 9 for the fin pattern; these are the same fins used on the Ceres A.

You will also need six motor mount pieces and two small triangular pieces. These are cut from the same 1/8" balsa sheet used for the fins. See Figure 18-14 for a method to cut the pieces.

With all of the pieces cut, you're ready to start assembly of the motor mount. Figure 18-41 shows the parts you will need.

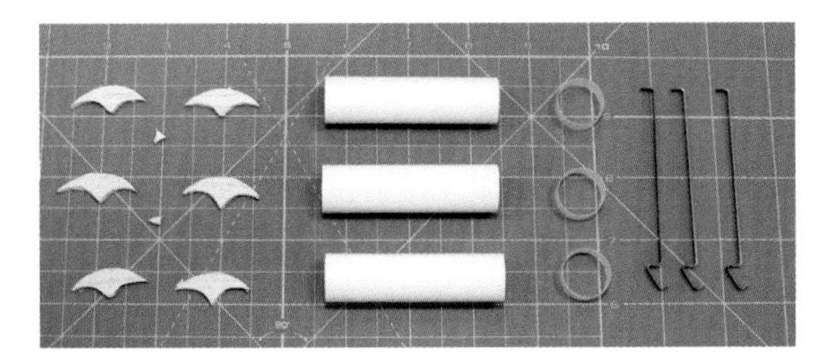

Figure 18-41. *Ceres C motor mount parts, ready for assembly. Cut the balsa pieces from 1/8" balsa, not the 1/16" balsa used for Cerberus.*

Looking back at the construction details for Cerberus, you know that this rocket could be built with engine blocks rather than engine hooks. It's an interesting design decision. For Cerberus, I recommended engine blocks because the rocket was all about appearance. With the engine blocks, there are no ungainly engine hooks hanging out the back of the rocket. The Ceres C is meant to be a workhorse rocket, though; you want to be able to quickly reload it and get in an additional flight. It is easier to install a motor in a rocket with engine hooks, since there is no tape to deal with to get a tight fit. Engine hooks are also more reliable, since there is no chance of putting on too little tape and having the motor pop out when the ejection charge fires. You can certainly decide to use engine blocks instead of engine hooks, though.

Assuming you are not changing the design, begin by mounting the 2 3/4"-long engine hooks on each of the motor tubes. Cut a small slot 1/4" from the end of each tube and poke the end of one of the engine hooks through that slot. Place some glue under the engine hook, from the end that's inserted in the motor mount tube to about halfway down the tube. Glue an engine sleeve 3/4" from the bottom end of the engine mount. When you are finished, you will have three engine tubes that look like the ones in Figure 18-42.

Remember, tape supported by a layer of glue works just fine if you run out of motor sleeves. Just be sure to leave 3/4" between the edge of the tape or engine sleeve and the lower end of the motor tube.

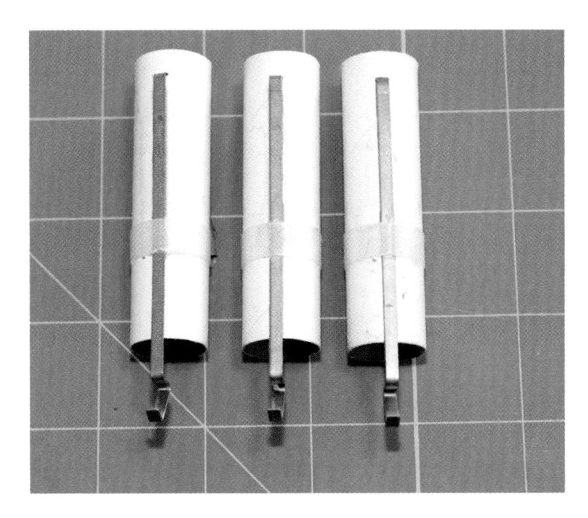

Figure 18-42. *Three assembled motor tubes.*

Once the glue sets, test fit the motor assemblies in the BT-60 body tube. While three BT-20 motor tubes fit perfectly into the BT-60 body tube, adding the engine sleeves causes enough of a bulge to distort the body tube slightly. That's another design decision. The plans shown here opt for ruggedness and simplicity. Another way to build the assembly is to leave off the motor sleeve and use epoxy glue to glue the engine hook to the engine tube, extending the glue to 3/4" from the bottom of the tube. If you use epoxy, be sure to apply the epoxy between the engine hook and motor tube, then coat the top of the engine hook to form a solid mound of epoxy resin over the hook that extends onto the tube. This creates a nice, almost indestructible seal.

Twist the motor mounts around. What's the best orientation for the engine hooks? It would be nice to put them all in the center, but if you do that, there isn't enough room to open the hooks to insert a motor. Putting them all on the outside edge won't work, either: it adds to the bulge in the body tube, and you'll tear the BT-60 when you bend the hooks back to insert a motor. The best configuration is to mount the hooks in a spiral fashion, as shown in Figure 18-43.

Figure 18-43. *Position the motor tubes inside the body tube with the engine hooks pointing toward the large gaps between the BT-20 motor tubes and the BT-60 body tube.*

Once you are comfortable with how the engine hooks are oriented, glue the three engine tubes together. Once the glue dries, add the balsa motor mounts 1/4" from the front edge of the assembly and 3/4" from the rear end, just below the engine sleeves. Glue the two small triangular pieces on either end of the assembly (in the gap between the three BT-20 motor tubes) to prevent ejection gases from exiting between the tubes.

Once the glue dries thoroughly, test fit the motor assembly in the BT-60 body tube. Sand the balsa parts so the fit is snug but not tight. You can also sand the motor sleeves to reduce the bulge in the BT-60 body tube a bit, but don't sand all the way through the sleeves.

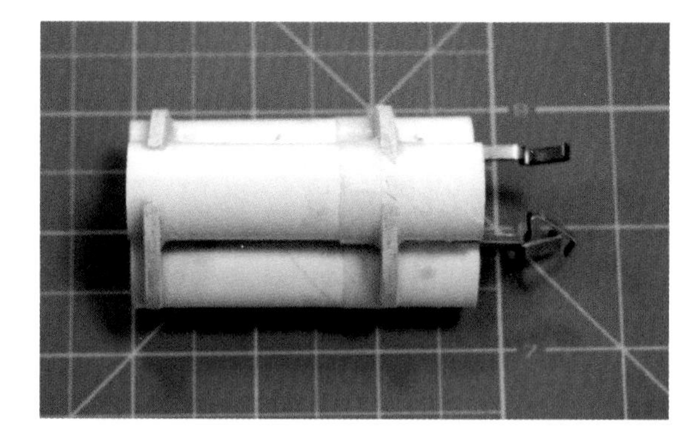

Figure 18-44. *The completed motor assembly.*

Use the fin guide in Figure 3-25 from Chapter 3 to mark the BT-60 body tube for four fins and a launch lug, but don't glue the fins on just yet. Most of the time it doesn't really matter whether you glue the fins on first or glue the engine mount in place first. I generally glue the fins on first because it is convenient to stand the rocket up on both the top and the bottom while checking fin alignment and when allowing the glue to dry. In this particular rocket, though, the slight bulge caused by the motor mount assembly can push the fins slightly out of alignment, so it's important to install the motor mount first.

After a test fit to make sure there will be no issues sliding the motor assembly into place, apply a generous amount of glue approximately 2 1/2" into the lower end of the BT-60 body tube using your finger or a piece of scrap wood. Apply another layer of glue 3/4" in. Slide the motor assembly into place, twisting slightly to spread the glue. The bottom of the motor assembly should be flush with the bottom of the BT-60 body tube, just like they are in Figure 18-43.

Now that the motor assembly is in place, you can glue the fins and launch lugs into place. Glue the four fins even with the base of the rocket, checking the alignment to make sure the fins are perfectly straight and in line with one another. Glue the 1"-long 3/16"-diameter launch lugs 2" from each end of the body tube on 1/2" standoffs so oversize payload bays can be used with the rocket. Refer back to "Construction" on page 217 in Chapter 9 for details.

You don't want the glue fillet formed by pushing the motor mount through the glue to drip, so the rocket needs to be supported in a vertical position while the glue on the motor mount and the fins sets. One way to do that is to tape the rocket to a launch rod and stand it up in the launcher.

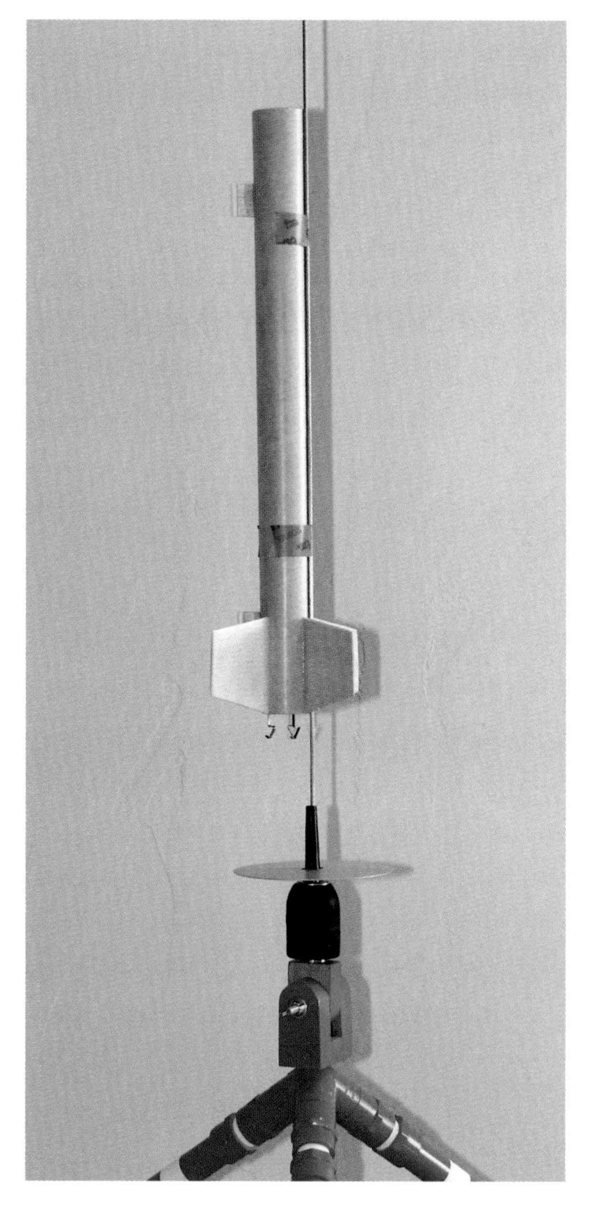

Figure 18-45. *Tape the rocket to a launch rod to hold it vertically while the glue at the front of the motor assembly dries.*

Once the glue sets, turn the rocket over and apply a generous fillet to the bottom motor mount. Cover the entire balsa motor mount with glue, especially where the balsa meets the BT-60 body tube or one of the motor tubes. Try to leave a small gap where the engine hook pokes through the motor mount, though. Wait for the glue to set, then apply fillets to the fins and launch lugs, turning the rocket occasionally to prevent drips in the fin fillets.

The completed rocket, ready for priming and painting, is shown in Figure 18-46, with the optional clear payload bay. The green cutting board used for most of the photos in this book was too small for this impressive rocket, so I've shown it on a blue cutting board, temporarily swiped from my wife's quilting closet. (Naturally enough, I'm not allowed to build rockets on this board.) The lines are 1" apart; the entire rocket is 29 3/4" long, not counting the engine hooks sticking out the back.

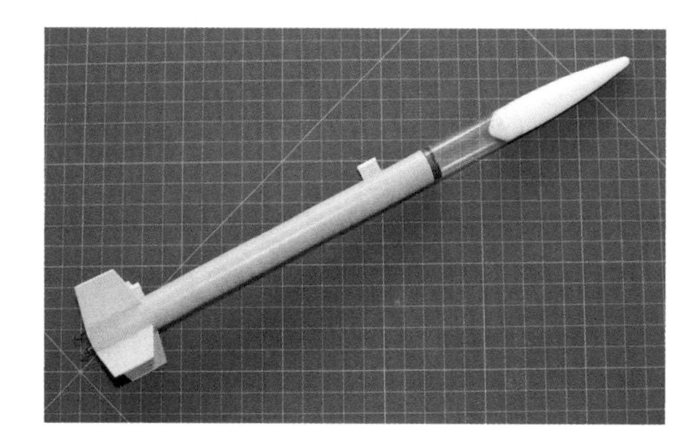

Figure 18-46. *The completed Ceres C booster, shown with the clear payload bay from Chapter 9. It's ready for priming and painting.*

Flying the Ceres C

The Ceres C can be flown with any of the payload bays from Chapter 9. That's one of the attractions of the Tinkertoy nature of the payload rockets from that chapter.

Begin your flight preparations by stuffing enough recovery wadding into the parachute bay to fill about 2" to 3" of the tube with wadding.

Figure 18-47. *Insert 2 1/2" of recovery wadding into the body tube.*

Make sure the snap swivels for the shock cord and parachute are both attached to the eyelet on the payload section. Be sure both snap swivels are fully closed.

Apply talcum powder to the parachute to make opening go smoothly. Fold the parachute into thirds, wrapping it with the parachute shroud lines to keep the bundle tight. Stuff the shock cord and then the parachute into the booster, then slip the payload bay into place.

The parachute should slide in easily. The nose cone should be tight enough that it does not fall out if the rocket is turned over and shaken gently, but it should not be too tight. Sand the nose cone or apply tape if the fit needs to be adjusted.

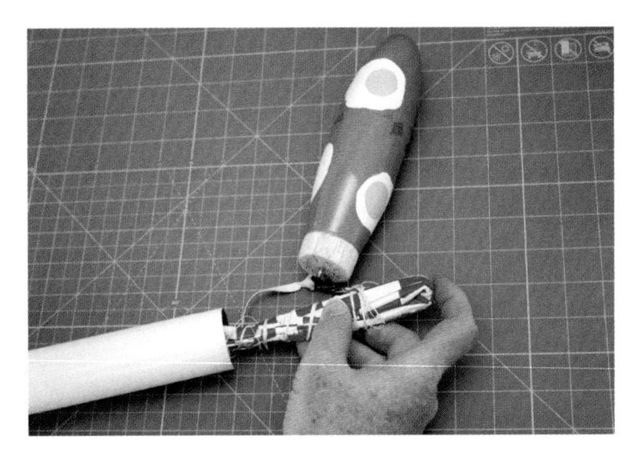

Figure 18-48. *Attach the parachute and shock cord to the nose section, then fold and insert the shock cord and parachute.*

Make sure all of the motors are the same kind. Use Table 18-4 to pick motors that are appropriate for the weight of the rocket.

As with Cerberus, the igniters must be connected in parallel, not series. Install them so one wire from each igniter is in the middle of the rocket, and twist all three of these wires together.

Use a paper clip or scrap of wire to form a loop around the outside of the rocket, connecting the other wire from each igniter to this loop.

Connect one igniter clip to the central bundle of wires, and the other to the outer ring of wires.

Figure 18-49. *Insert the igniters and connect them in parallel.*

The Ceres C is designed to carry a payload. Now is the time to get it ready. Maybe it's breakfast—pack that egg carefully! Or you can start the camera, or press the pairing button on your Bluetooth low-energy sensor, or do whatever other preparation is needed.

The Ceres C is ready for flight. Run through the checklists to make sure everything is ready, and then launch the rocket.

Launch Preparation Checklist

1. Insert recovery wadding.
2. Attach the parachute and shock cord.
3. Insert the parachute and shock cord; make sure they are not too tight.
4. Insert the three motors, making sure they are all the same kind of motor.
5. Install the igniters in parallel.
6. If you are using Estes igniters, make sure you are using a lead-acid battery (e.g., a car battery) for the power source. You can use AA batteries with Quest Q2G2 igniters.
7. If you are using C motors, make sure the launch cord is at least 30 feet long.
8. Install the payload (if any).

Flight Checklist

1. Clear spectators from the launch zone.
2. Remove the launch key from the launch rail, slide the rocket onto the launch rail, and replace the launch key on the launch rail.
3. Attach the igniter clips.
4. If needed, start the payload (for cameras or other electronics).
5. Get the launch key from the pad.
6. Insert the launch key. Make sure the continuity light is on.
7. Check for clear skies—no aircraft or low clouds.
8. Count down from five and launch.
9. Remove the launch key and place it on the launch rod.
10. Recover the rocket.

19 Helicopter Recovery

Have you ever watched a maple seed drop from a tree and twist slowly to the ground? It's nature's version of helicopter recovery. By deflecting the air to create spin, the seed slows its fall. In the case of the maple tree, this keeps the seed in the air longer so the wind can carry it farther from the parent tree. In the case of a rocket, helicopter recovery gives us a showy way to slow the descent of a rocket. Helicopter recovery is a constant draw at any launch.

Figure 19-1. *Nature's expert at helicopter recovery, the maple tree spreads its seeds by keeping them in the air longer so the wind can move them farther away.*

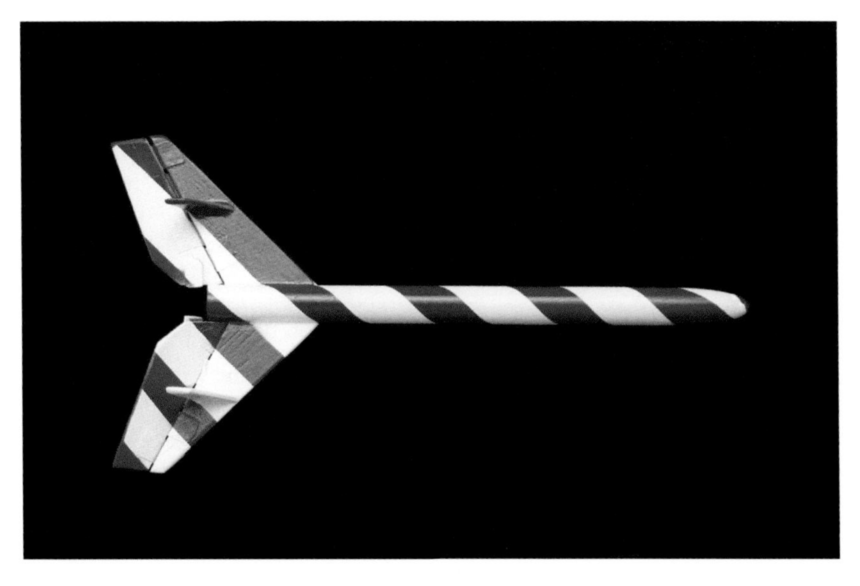

Figure 19-2. *The Nicomachus helicopter recovery rocket uses the ejection charge to flip control surfaces over when the ejection charge fires.*

There are several ways to build a rocket with helicopter recovery. Some rockets have helicopter blades that fold along the outside of the body tube and spring out when the ejection charge fires. Some have helicopter blades attached to the nose cone, keeping the blades folded inside the body tube during liftoff. We're going to build a rocket with movable control surfaces on the fins, sort of like ailerons on a delta wing airplane, and snap them over to spin the rocket on descent.

Nicomachus

Nicomachus is a somewhat odd-looking rocket. Figure 19-3 shows the overall layout, which is rather different from a typical three- or four-finned rocket like most of the ones we've seen so far in this book. That's what makes Nicomachus so much fun—it's truly different from a typical rocket.

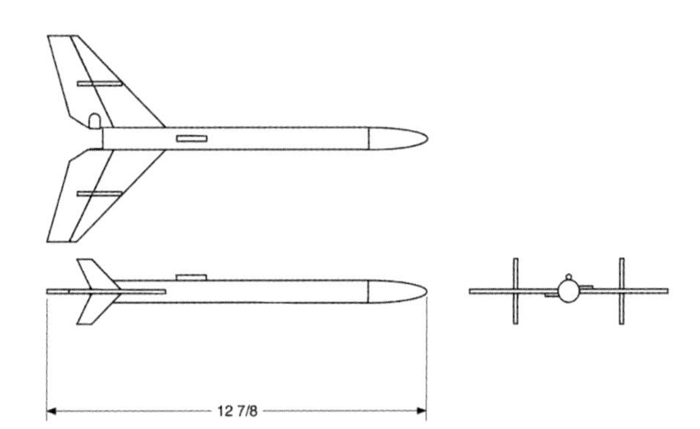

Figure 19-3. *Nicomachus plan.*

Table 19-1 shows the recommended motors. This is a rocket I like to fly on the smaller motors, though. I almost always fly it on an A8-3 motor, and only occasionally on a B6-4 motor. Other than for test flights, I just don't use a C6-5. I want to see the flight!

Table 19-1. Recommended motors

Motors	Approximate altitude
A8-3	380 ft
B6-4	830 ft
C6-5	1,300 ft

The parts for Nicomachus are shown in Table 19-2. If you are building the rockets in this book using the Estes Designer's Special, you will run short of a few parts. Plan ahead a bit to make sure you have the correct parts.

Table 19-2. Parts list

Part	Description
BT-20 body tube	Use a 9″ piece cut from a longer tube.
BT-20 nose cone	You will run short of BT-20 nose cones if you build all of the rockets in this book using the Estes Designer's Special. You can order additional ones from many sources. I like balsa nose cones; these can be ordered from JonRocket (*http://www.jonrocket.com*) and Balsa Machining (*http://www.balsamachining.com*).
1/8″ balsa plank	You may be running short of 1/8″ balsa, so check your supplies before starting the rocket. This is used for the rather large fins.
Engine block	Use an engine block rather than an engine hook. We want the motor to slide out when the ejection charge fires.
1/8″ launch lug	You will need a 1″ piece of launch lug.
Weight	You will need a nose weight. See the text for details.
Elastic thread	You can buy elastic thread at fabric supply stores. Get the 1/16″ round thread.
Popsicle stick	You will need a small wooden stick about 3/8″ x 1/16″. A popsicle stick works well. If you're not hungry, you can find them in sacks at most craft stores.
1/2"-wide ribbon	You will need 8″ of 1/2"-wide ribbon. Use something thin and strong. Ribbon is available from fabric stores. If you have a sewer in the house, you might be able to requisition the material from her stock.

Nicomachus

8128 Nicomachus isn't notable for much, but it is a very fast-rotating asteroid, with a rotational period of just 4.67 hours. That's not a record, but it's pretty quick for an asteroid. The speed record is currently held by 2008 HJ, a tiny asteroid that rotates once every 43 seconds. 8128 Nicomachus was selected as the namesake for our helicopter rocket, though, because 2008 HJ just doesn't roll off the tongue.

8128 Nicomachus was discovered in 1967 by C. U. Cesco and A. R. Klemola at Argentina's El Leoncito Astronomical Complex. The asteroid is named for the Greek mathematician who, among other things, was noted for his work on perfect numbers. Perfect numbers are numbers that are the sum of all of their divisors. The first two perfect numbers are 6 (1 + 2 + 3) and 28 (1 + 2 + 4 + 7 + 14). Can you find the next two without resorting to the Internet?

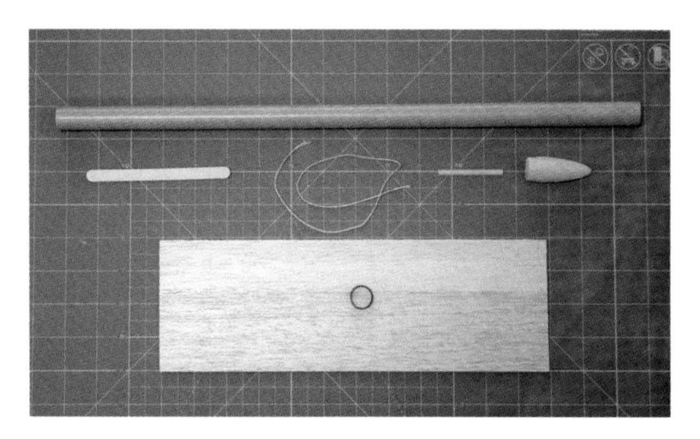

Figure 19-4. *Visual parts list for Nicomachus.*

Equivalent Kit

The Texas Twister from Apogee Components is a bit different from Nicomachus, but uses the same basic idea of a variable-geometry fin.

The mechanism for deploying long blades is challenging to build unless you have good machine tools. A kit with premolded plastic parts is a good alternative.

The Estes Sky Twister and HeliCat both feature helicopter fins that attach to the nose cone, which separates from the rocket at apogee.

The Heli-Roc from Apogee Components has large helicopter blades that fold along the outside of the rocket. The ejection charge moves a release mechanism, allowing rubber bands to deploy the blades.

Construction

Refer back to Chapter 3 for basic construction techniques, tools, and supplies.

The fins (or wings, or whatever you want to call them) on Nicomachus are larger than normal. You could get by with 3/32" balsa, but the rocket will be a lot more durable if you use 1/8" balsa, so that's what the plans call for. Begin by cutting two fins, two flaps, and four rudders from 1/8" balsa, using the plans in Figures 19-5, 19-6, and 19-7 as a guide.

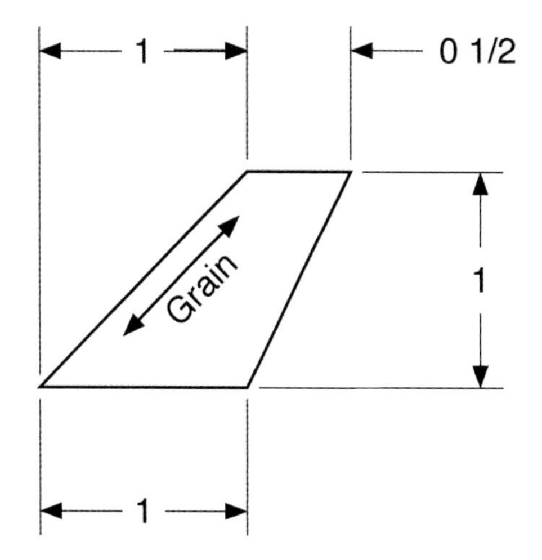

Figure 19-5. *Rudder plan. Cut four from 1/8" balsa.*

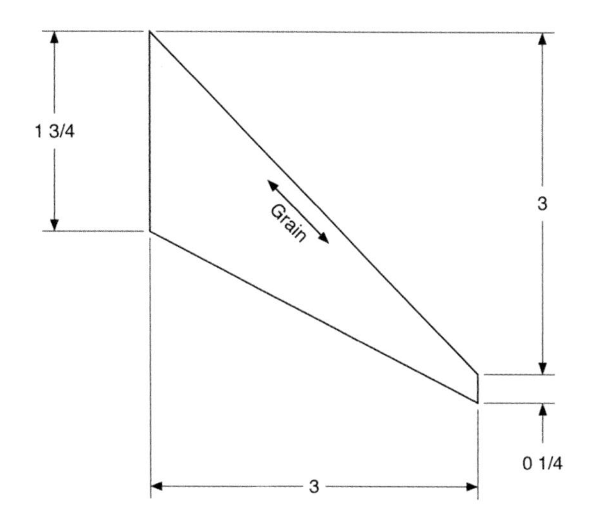

Figure 19-6. *Wing plan. Cut two from 1/8" balsa.*

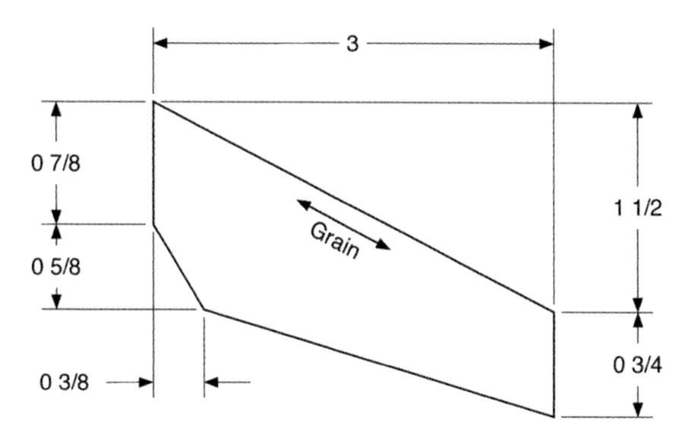

Figure 19-7. *Aileron plan. Cut two from 1/8" balsa.*

Round all edges on the rudders except the root edge.

Round all edges on the fins and flaps except the root edge and the edge where the two join.

Cut two 1/2" sections from the ends of a popsicle stick. Sand the flat end so it is at a 45° angle.

Cut 9" from a BT-20 body tube, and 1" from a 1/8" launch lug.

Figure 19-8. *The finished cut pieces. The launch lug is not shown.*

The rudders will be glued to the fins. Mark each fin 1 1/2" from the root edge with a line parallel to the root edge. This is the line you will use to glue the rudders to the fins, just like you use guide lines drawn on the body tube to glue fins to the body tube. Don't glue the rudders on yet, though. It's easier to put the guide lines on the fins now, but there is a lot to do before the rudders get glued to the fins.

Cut eight 1" lengths from a strip of 1/2"-wide ribbon. Fabric stores sell ribbon of all sorts of descriptions. Shop around for something fairly thin and strong that is premade in a 1/2" width.

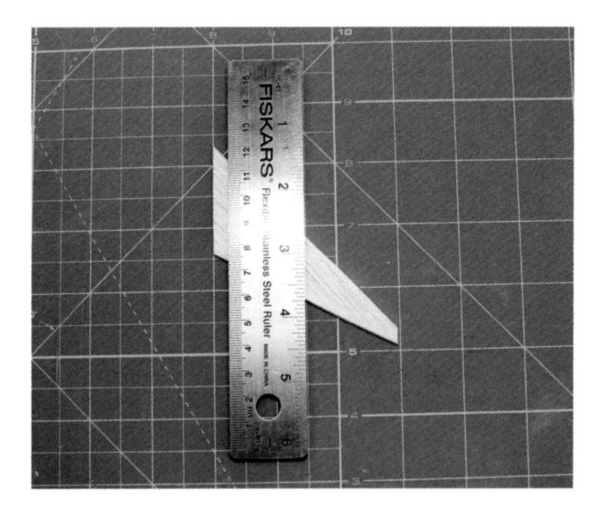

Figure 19-9. *Mark each fin on the top and bottom with a guide line. This will be used later to glue the rudders to the fins. The guide line is 1 1/2" from the root of the fin.*

Glue the ribbon to the top and bottom of the fin and flap. The leftmost ribbon in Figure 19-10 is glued to the top of the fin and the bottom side of the flap. The ribbon next to it is glued to the bottom of the fin and the top of the flap. Alternating the ribbons in pairs forms a nice, tight hinge. This type of hinge has been used on small model airplanes for years, and works really well. Just be sure to put the glue on the top and bottom part of the hinge, and not in the joint between the flap and fin. Keep all glue about 1/8" away from the joint between the flap and fin.

Repeat this process for both fins.

Take Your Time with the Hinges

Don't try to glue the entire hinge assembly in one step. Glue the four hinges to the fin, sandwich it between two pieces of wax paper, and lay a book on the top piece of wax paper to hold the hinges in place while the glue dries.

Once the glue is dry, thread the hinges over and under the flap and apply glue, then again sandwich the assembly between pieces of wax paper and place a book on top while it dries.

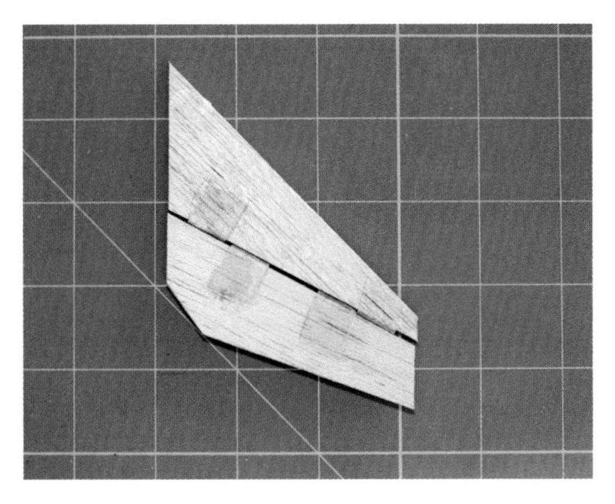

Figure 19-10. *Form hinges with 1" strips of 1/2"-wide ribbon. Keep glue away from the joint between the flap and fin.*

Mark the body tube using the four-fin guide in Figure 3-25 from Chapter 3. You will only use three of the lines, of course—two for the fins and one for the launch lug. The lines on the opposite sides of the body tube are used for the fins, and one of the remaining lines is used for the launch lug.

Glue the fins, but not the flaps, to the body tube. The front of each fin should be 2 1/8" from the base of the body tube.

Save Time by Taping the Flaps in Place

Having the flaps flapping around is a real problem, especially when gluing the fins to the body tube. Use painter's tape to hold them flat with the fin during most construction. While the tape was removed for photos so you could see the structure, the flaps and fins were firmly taped while this rocket was built.

Glue the four rudders in place. Looking down from the top, the opposing rudders should form a perfectly straight line. Make sure the fins and rudders are pointed straight into the air flow when the rocket is in flight. The base of each rudder should be right at the base of the fin. If the flap is moved up, the base of the rudder should stop it, lying perfectly along the edge of the flap.

Add the launch lug. Place the base of the launch lug 2 1/2" from the base of the body tube.

As always, once the initial glue is dry, add a nice glue fillet to all joints—well, all of them except the joints between the flaps and fins, of course. Add extra-thick fillets between the fins and body tube.

Strengthening the Fins

The large fins on Nicomachus can break on landing if the landing site is hard. If possible, fly Nicomachus where it will land in grass or on very soft dirt.

If the fins break, they tend to break by tearing the paper body tube. If you expect to fly in areas with harder surfaces like packed dirt, reinforce the fin joint by dribbling a thin line of CA glue on the inside of the body tube where the fins are attached. This strengthens the body tube, making it less likely the fins will pop off on a hard landing.

Figure 19-11. *Glue the fins, rudders, and launch lug in place. Use tape to hold the flaps flat relative to the fins. The tape is removed here so you can see the construction, but you can still see remnants of the tape next to the hinge on the right flap.*

All of our rockets up to this point were designed so the motor stuck 1/4" out the back of the body tube. This was done to reduce the scorching of the back of the rocket by the hot gases from the motor. This time, though, we will be using the motor to keep the flaps flat while the rocket lifts off. Put a mark on an expended motor casing 1/2" from the end, and use this as a guide for positioning the engine block in the body tube. Make sure the motor sticks out by a full 1/2".

The flaps should be able to move up and down freely with the motor in place. You will probably find that there is a bit of contact between the flap and the body tube or motor. Wedge a piece of sandpaper between them and sand the edge of the flap until it moves freely.

Figure 19-12. *Install the engine block so the motor will stick out 1/2" from the base of the body tube.*

Once the engine block and all other glue joints are thoroughly dry, glue the two flap blocks you made from the popsicle stick in place. The 45° edge you sanded on the end of each piece helps it rest perfectly against the back end of the motor. With the motor in place and the flap blocks resting against the motor, the flaps should be perfectly aligned with the fins.

Glue one fin stop to each flap. They should be on opposite sides of the flaps, so the flap stop that is not shown will be on the other side of the rocket.

You might find there is some play in the fins. If so, be absolutely sure the flap stops are positioned so the flaps will not be able to move past the motor. The flaps will eventually be pulled over by a piece of elastic thread, but that must not happen until the motor is ejected by the ejection charge.

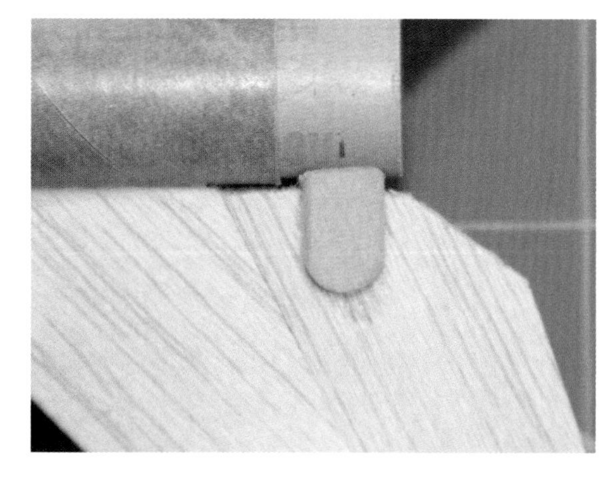

Figure 19-13. *Glue the flap stops to the flaps. With the motor in place, the flap stops should rest against the motor when the flap and fin are perfectly aligned.*

Nicomachus will be unstable without some nose weight. You will need about 4 grams of weight. I used a small chunk of lead designed for trimming model airplanes. A small fishing weight, curtain weight, or buckshot will all work well, too. Whatever weight you use, put enough in the base of the nose cone so the rocket balances right at the front edge where the fin meets the body tube, 2 1/8" from the base of the body tube. Check the balance with a C6-5 motor in place. With no motor, the balance point is 3 3/4" from the base of the body tube, well in front of the launch lug.

Glue the weight firmly in place. Once the rocket is properly balanced, glue the nose cone into the body tube.

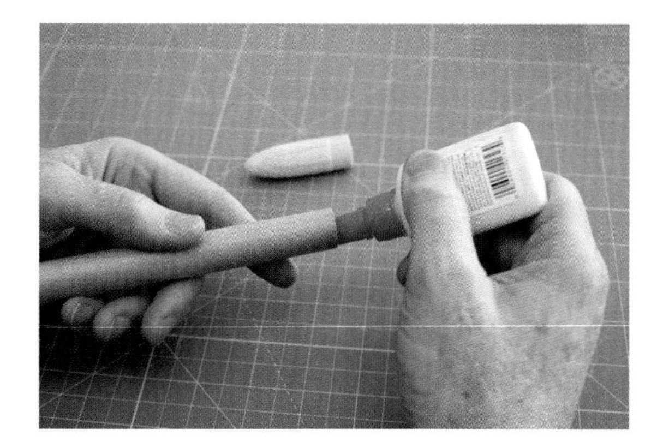

Figure 19-14. *Add weight to the nose cone until the rocket balances where the leading edge of the fin meets the body tube, then glue the nose cone in place.*

The rocket isn't quite finished yet. We still need to add the elastic thread that will pop the flaps up when the motor ejects. We don't want to get paint all over the thread, though, so we'll stop now and paint the rocket.

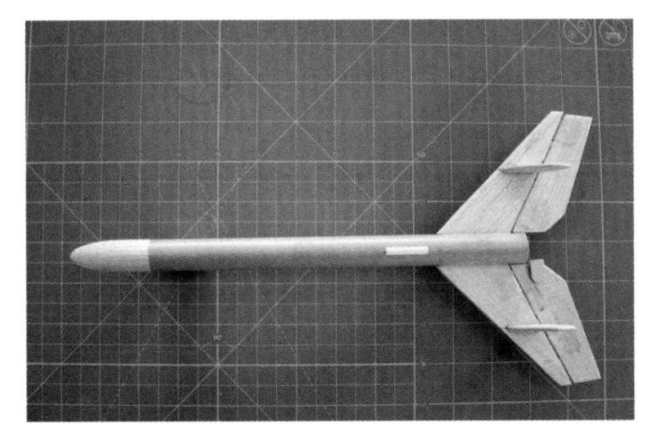

Figure 19-15. *Most of the construction is complete, but we need to paint the rocket before installing the elastic thread that will pull the flaps into position when the motor ejects.*

Painting Nicomachus

The paint job you see was managed with white and red spray paint and 1"-wide painter's tape.

Start by painting the entire rocket white.

Once the white paint dries, place a piece of tape along the front of one fin. When the tape hits the body tube, just keep going—it forms a perfect barber's pole stripe up the body tube of the rocket.

When the tape gets to the base of the nose cone, cut it gradually narrower with scissors so it comes to a point just as it gets to the tip of the nose cone.

With the main strip in place, go back to the fins and apply additional stripes of tape parallel to the existing stripe on the body tube.

Make sure all of the edges are sealed tight, and then paint the entire rocket red. When the paint dries and the tape is removed, you'll have a rocket that looks like it ought to twist—and it will!

Once the rocket is painted, drill four small holes for the elastic thread. I used a 1/16" drill bit, but the balsa is so soft and the working space is so tight that I just drilled the holes by twisting the bit with my fingers.

Drill one hole in two of the rudders, 1/2" along the leading edge from the root and 1/8" back from the leading edge of the rudder. Which two? Look at how the flaps will be pulled up when the ejection charge fires. You will drill the holes in the two rudders that are on the opposite sides of the flaps from the flap stops.

Pull the first flap up against the rudder with a hole in it. Drill a hole in the flap that is 1/8" from the trailing edge of the flap and 1/8" from the edge of the point where the flap meets the rudder. The hole should be on the same side of the rudder as the body tube. Repeat this process for the second flap.

Figure 19-16. *Drill holes in the two rudders on the opposite sides of the flaps from the flap stops. Drill two more holes in the flaps.*

Thread some 1/16"-thick elastic thread through the hole in one of the rudders. A large sewing needle is a great help, but patience and a toothpick will work, too. Once the thread is in place, tie a knot on the side farthest from the body tube. Make the knot large enough that the elastic thread will not pull through the hole. Don't tug too hard, though—the balsa wood is soft. We'll reinforce the knot in a moment.

Pull the elastic thread through the hole in the flap. Tie another knot on the side of the flap with the flap stop. The elastic thread should be just long enough so it is limp right when the flap is in contact with the rudder.

Figure 19-17. *Use a large needle or toothpick to install the elastic thread that pulls the flap into place.*

Add glue to the knots to keep them from pulling through the weak balsa wood.

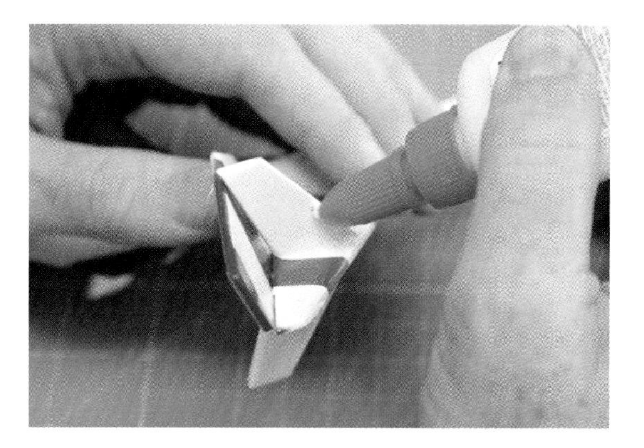

Figure 19-18. *After knotting the elastic thread, add a dab of glue to reinforce the knot and the hole where the thread goes through the balsa.*

Test the joints thoroughly, making adjustments if needed. Pull the flaps straight and insert a motor. The flap stops should hold the flaps flat compared to the fins. Pull out the motor. The elastic thread should pop the flaps up against the rudders, one up and one down, forming the helicopter blades that will return Nicomachus gently to the ground.

Check the balance to make sure the paint and elastic thread did not shift the center of gravity too far back. If the center of gravity with a C6-5 motor installed is behind the front of the fins, add more weight to the nose. You can do that by drilling a small hole in the nose cone, adding the weight, then using putty to smooth the hole.

Flying Nicomachus

While Nicomachus is fairly challenging to build, it's a very easy and fun rocket to fly—you just pull the flaps straight and insert a motor. If the motor is so loose it falls out when the rocket is upright, add a little tape to the outside of the motor to tighten it a bit. The motor is *supposed* to pop out when the ejection charge fires, though, so don't add too much tape.

I like to use a smaller motor with this rocket so I can watch the whole flight, but it flies well with larger motors, too.

Launch Preparation Checklist

1. Install the motor.
2. Insert the igniter.

Flight Checklist

1. Clear spectators from the launch zone.
2. Remove the launch key from the launch rail, slide the rocket onto the launch rail, and replace the launch key on the launch rail.
3. Attach the igniter clips.
4. Get the launch key from the pad.
5. Insert the launch key. Make sure the continuity light is on.
6. Check for clear skies—no aircraft or low clouds.
7. Count down from five and launch.
8. Remove the launch key and place it on the launch rod.
9. Recover the rocket and, if you can find it, the spent motor.

Rocket and Boost Gliders 20

If you've ever built a model airplane, you've probably already thought a bit about how to combine rockets and airplanes. That's what we'll do in this chapter. We'll start out by exploring what's possible and what's not with rocket-powered airplanes, then build a classic pop-pod glider. In the next chapter we'll build a variable-geometry air rocket glider.

Rocket-Powered Aircraft

The obvious approach to using a model rocket motor to power an airplane is to simply replace the airplane motor with a rocket motor. There are several reasons why that's a bad idea. The first is simply practical. Most model airplanes are designed to fly with a relatively low-powered motor that delivers power for the entire flight. Model rocket motors are designed to deliver a lot of power for a very short period of time. A model rocket motor would rip most model airplanes apart, and would not power them long enough for a useful flight.

Another problem with this approach is safety. Model rockets, even those attached to airplanes, must be launched no more than 30° from vertical. It's just not safe to point anything powered by a model rocket motor in a nearly horizontal direction and ignite it.

There is one class of model airplane that is ideally suited to model rocket motors, though, and that's the glider. Think about how a glider works. It needs some form of propulsion to lift it off the ground so it can glide back to earth. Model glider builders usually use hand launching, giant rubber bands, tow lines, or small motors to do the job. We're going to use a rocket motor and launch the glider vertically.

If this still seems like an odd approach, stop and consider the Space Shuttle—it took off vertically under rocket power, then glided back to Earth after its mission was complete.

Figure 20-1. *Like the Space Shuttle, our rocket plane will take off vertically, returning to earth after jettisoning its solid rocket booster.*

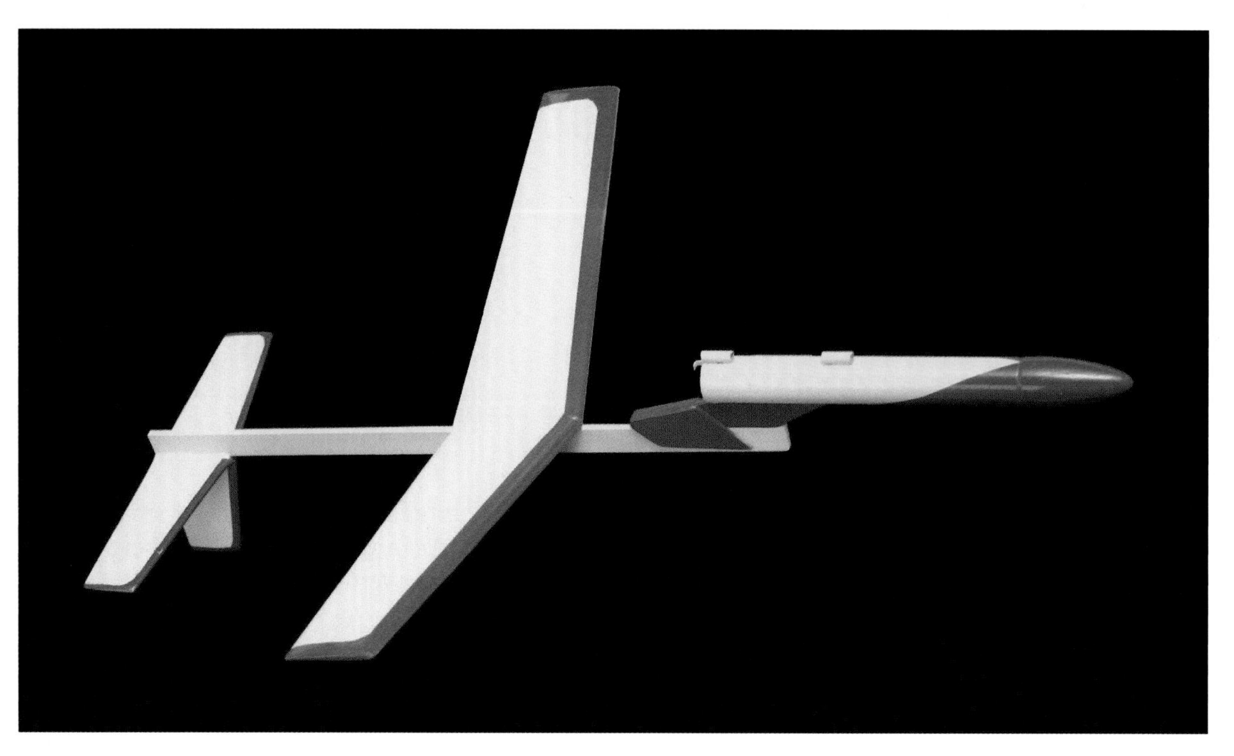

Figure 20-2. *The Icarus boost glider.*

Icarus

Icarus is a boost glider that uses a *pop pod*. The pop pod holds the rocket motor and attaches to the airplane. The combined pop pod and airplane slides onto a normal launch rail and is launched vertically. When the ejection charge fires, the pop pod separates from the glider, and the glider circles slowly back to earth. The pop pod has a streamer, and falls fairly quickly back to the ground with its long streamer flapping in the air.

Table 20-1 lists some recommended motors for this project, and the approximate altitudes Icarus will reach with them.

Table 20-1. Recommended motors

Motors	Approximate altitude
A8-3	200 ft
B4-2	500 ft
C6-3	1,000 ft

Table 20-2 lists the parts you'll need to build the Icarus boost glider. If you are building the rockets in this book using the Estes Designer's Special, you will run short of a few parts. Icarus also uses some wood parts not found in the Estes Designer's Special. Plan ahead a bit to make sure you have the correct parts.

Table 20-2. Parts list

Part	Description
BT-20 body tube	Use a 5 1/4″ piece cut from a longer tube.
BT-20 nose cone	You will run short of BT-20 nose cones if you build all of the rockets in this book using the Estes Designer's Special. You can order additional ones from many sources. I like balsa nose cones; these can be ordered from JonRocket (*http://www.jonrocket.com*) and Balsa Machining (*http://www.balsamachining.com*).
1/16″ balsa plank	The thinner wood is used for the side panels of the pop-pod base. You won't need much.
1/8″ balsa plank	Used for the small strut on the pop pod and the rudder and stabilizer on the glider.
1/4″ balsa plank	Used for the glider wings.

Part	Description
1/8″ x 1/4″ hardwood strip	This is used for the body of the glider. Basswood, spruce, or poplar will work well, and are available at hobby stores, hardware stores, or online. You will need a total of 16″. 12″ is used for the fuselage, and the rest for airfoil templates. Other thin hardwood, such as aircraft plywood, can be used for the templates.
Screw eye	A small screw eye is used to attach the shock cord to the nose cone. You won't need this with plastic nose cones.
Engine hook	You will run short of these in the Estes Designer's Special. You can bend and cut one of the longer ones, buy an additional engine hook from an online source, or switch to an engine block.
Engine sleeve	Use one of the Mylar sleeves in the Estes Designer's Special, or make your own from paper or tape.
1x2 scrap wood	You're going to need a very long sanding block. A 10″ to 12″ piece cut from a 1″ x 2″ strip of pine works very well.
Streamer	You can use the same streamer that is used for Hebe.
1/8″ shock cord	You will need 12″ of shock cord.
1/8″ launch lug	You will need two 1/2″ pieces of launch lug.
Clay	A small amount of modeling clay is used to weight the nose of the glider. There is some clay in the Estes Designer's Special.

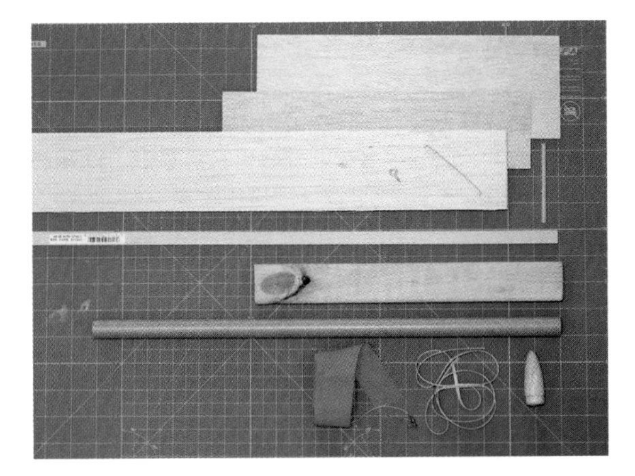

Figure 20-3. *Visual parts list for the Icarus boost glider.*

Icarus

1566 Icarus is one of the Apollo asteroids that cross the orbit of the Earth. Unlike most Apollo asteroids, though, Icarus has an orbit that carries it far into the inner solar system. It gets about 0.18 AU from the Sun, half the distance of the planet Mercury! Discovered in 1949 by the German astronomer Walter Baade, it was another 51 years before another asteroid was found that got closer to the Sun.

1566 Icarus is named for an aviator from Greek mythology. Icarus the flyer crafted wings from feathers and wax, and took to the skies. Unfortunately, he got too close to the Sun, and the wax melted, causing him to plummet to his death. I thought the Greeks had more pluck than that. Since those wings worked well enough to get him so close to the Sun that the wax melted, why didn't they start an immediate search for better glue?

For years, the myth of Icarus was treated as a lesson against letting one's ambition get too big, against "flying too high." But lately that seems to be changing; as film director Stanley Kubrick said in 1998, it might just as well have been "do a better job on the wings!" We're going to do a very good job indeed on Icarus's wings.

While it gets pretty close to the Sun, 1566 Icarus also gets close to the Earth—close enough that in 1967 it inspired Professor Paul Sandorff from the Massachusetts Institute of Technology to task his students with finding a way to destroy the asteroid so it would not impact the Earth. Their report was the basis for the 1979 movie *Meteor*.

Equivalent Kit

The overall design of the Estes Tercel is very similar to that of Icarus.

Construction

Refer back to Chapter 3 for basic construction techniques, tools, and supplies. A few specialized tools are included in the parts list in Table 20-2.

Cutting the parts

There are two ways to get templates for the various glider parts. The first is to grab them from the author's website (*http://bit.ly/byteworks-make-rockets*). These plans are full-size drawings in PDF and JPG formats that you can print on your own printer. The other way to create templates is to draw them yourself, using the dimensions given in the various drawings in the chapter.

Start by cutting the glider wing, stabilizer, and rudder, working either from the downloaded plans or from templates you create yourself from the design shown in Figure 20-4.

Cut the vertical stabilizer (rudder) and horizontal stabilizer from 1/8" balsa with the grain running parallel to the trailing edge. Note that the base of the rudder is not at a perfect 90° angle. The rudder will be mounted on the horizontal stabilizer, which is mounted at a slight angle on the fuselage.

Cut two wings from 1/4" balsa, each with the grain running parallel to the trailing edge.

Cut the fuselage from 1/8" x 1/2" hardwood. The best choice for wood is a light, strong wood like poplar, spruce, or basswood. Strips of wood precut to 1/8" by 1/2" are available from many hobby stores and hardware stores. If you can't find them locally, check online sources like *http://nationalbalsa.com*.

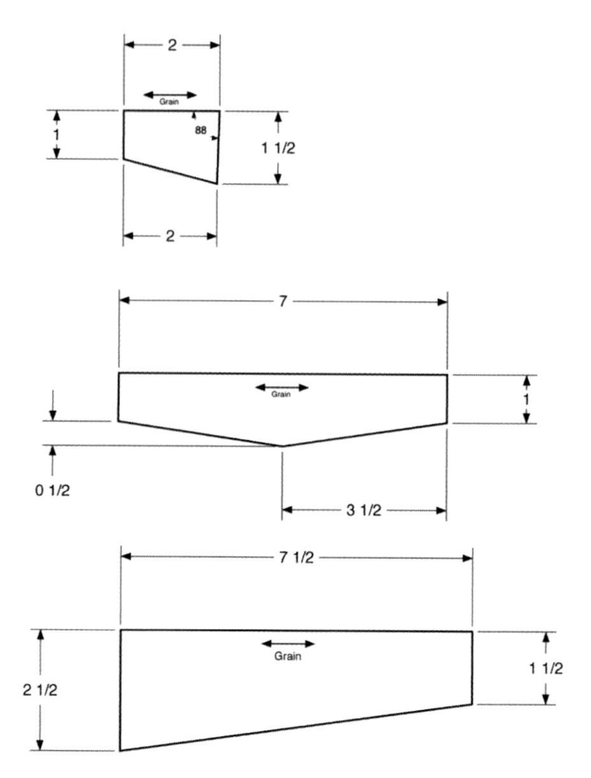

Figure 20-4. *Plans for the glider control surfaces. Cut the rudder and horizontal stabilizer (top pieces) from 1/8" balsa. Cut two wings (bottom piece) from 1/4" balsa.*

Hardwoods are (obviously!) much harder to cut than balsa, but the cuts are small and can still be made with a hobby knife. Work slowly, cutting with a sharp knife. Balsa usually doesn't cut all the way through with a single cut, and of course the hardwood won't either. It will take a lot of cuts, but eventually you will cut all the way through.

Save the piece cut from the notch at the front of the fuselage. This piece will be used for the pop pod.

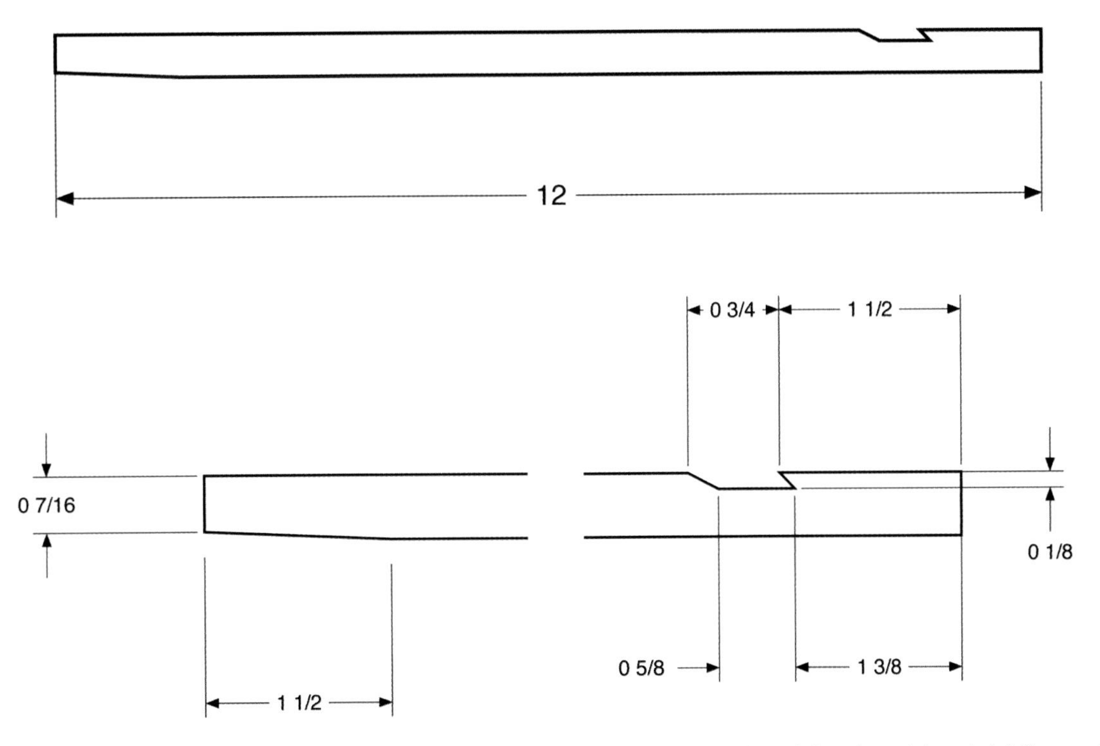

Figure 20-5. *Cut the fuselage from 1/8" x 1/2" basswood, spruce, or some other light hardwood. Cut the notch and stabilizer seat using a hobby knife.*

Figure 20-6 shows the layout for the pop pod, while Figure 20-7 shows the pieces needed for assembling the pylon. The small poplar piece is the chunk cut out of the fuselage on the previous step, although one end is trimmed a bit. Cut the main pylon from 1/8" balsa and the two side supports from 1/16" balsa.

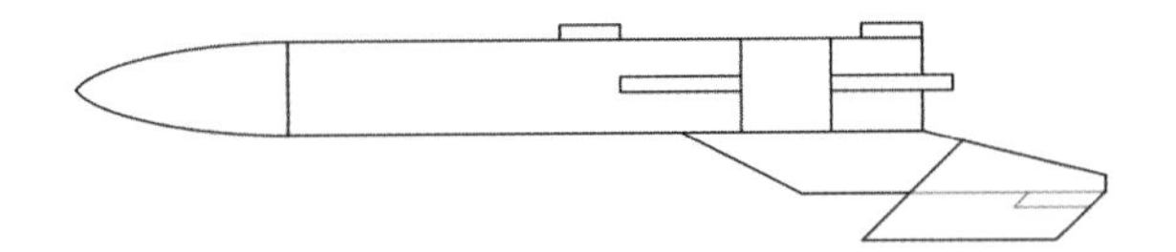

Figure 20-6. *Pop pod plan.*

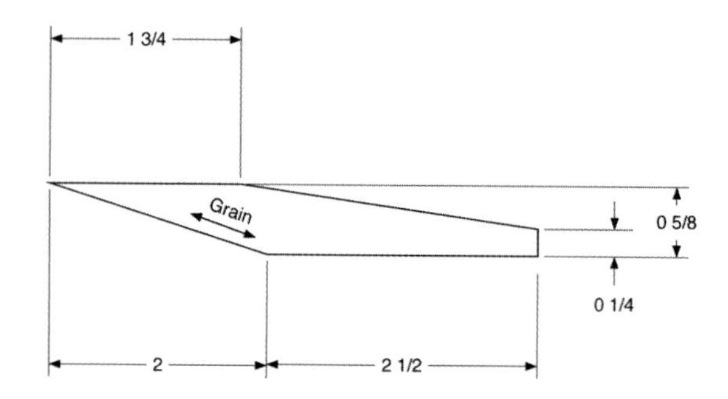

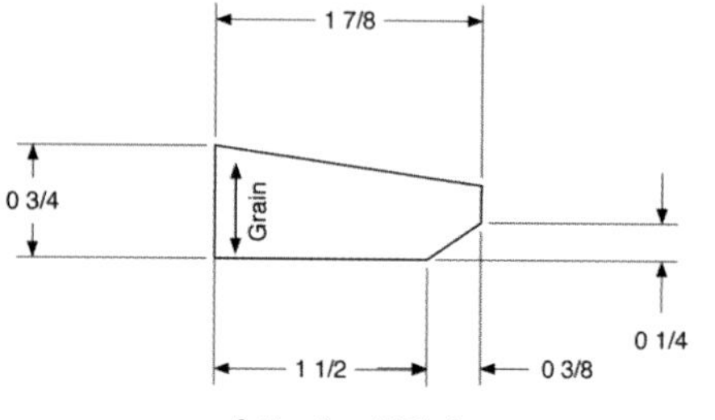

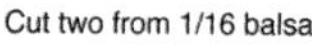

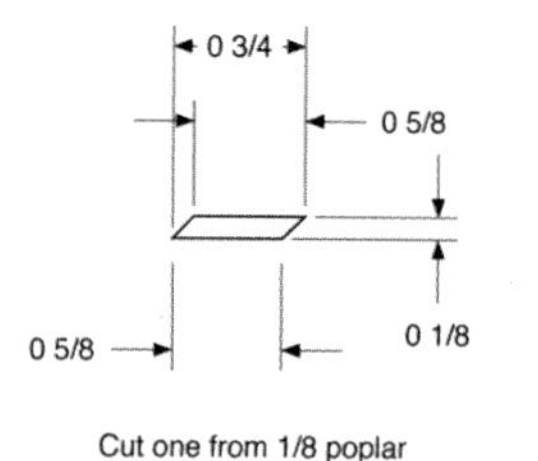

Figure 20-7. *Pop pod part details.*

Cut two 1/2″ pieces from a 1/8″ launch lug and trim 5 1/4″ from a BT-20 body tube. You could make the pop pod a bit longer if you buy extra body tubes. A 6″ pop pod would be better, but the 5 1/4″ length leaves enough BT-20 body tube for the other projects in the book.

With all of the pieces cut, you will have the assortment of cut parts shown in Figure 20-8.

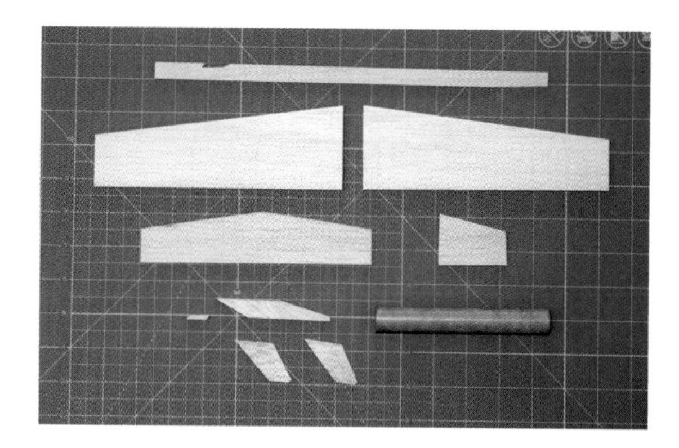

Figure 20-8. *You should end up with this assortment of wood and body tube parts after finishing the basic cuts.*

Building the glider

A carefully cut airfoil is a superb way to reduce drag. Back in Chapter 13 you found out that an airfoil can cut the force of drag by about ten times compared to a rod of the same thickness. Airfoils that are used in airplane wings also serve another purpose. In addition to reducing drag, a properly designed airfoil generates lift. The 1/4"-thick wings on the glider are needed for strength, but they are also needed to give us something thick enough to work with to form a good airfoil.

Choosing the proper shape for an airfoil is one of those scientific arts. Yes, there are good scientific reasons for choosing one specific airfoil shape over another, but there are also trade-offs. Picking the theoretically "best" airfoil for your task may not work out so well if the airfoil is too difficult for you to make with your available tools and skills.

We're also picking an airfoil for a wing that will be subjected to two very different kinds of flying. At liftoff, the airplane will function as the fins for the rocket, so we want a high-speed, low-drag airfoil with no lift. But once the pop pod separates, the glider will return to earth as a sailplane, so we want a low-speed, high-lift airfoil for the best glide characteristics. Assuming we want the longest-duration glide possible, we could choose to optimize the launch so the glider starts very high, or optimize the glide so it flies well, even if it doesn't start as high. These are the sorts of choices engineers make on every design project.

There are a huge number of resources available on the Internet for wing airfoils. I encourage you to poke around a bit to see what is available. You can certainly try a different airfoil than the one I'll present in a moment. Keep an eye out for airfoils called *low Reynolds number airfoils*. Recall from Chapter 13 that the Reynolds number tells you how a fluid interacts with an object moving through it. For low Reynolds numbers, fluids behave like syrup, sticking to the surface and preventing the object from moving easily. For high Reynolds numbers, the fluid behaves more like air, with a tendency to generate turbulent eddies as the object moves through it. It might seem odd that you should look for a low Reynolds number airfoil—wouldn't a glider moving through air require a high Reynolds number? You're right, but in this case, "low" is a relative term. Model airplanes are small enough compared to large jetliners that the nature of the way air flows over the airfoil changes a bit. While the flow of air across a model airplane wing is still considered a high Reynolds number flow, the smaller size of the model airplane wing makes the Reynolds number a lot lower than for a large airplane, so the airfoils you want are called "low Reynolds number airfoils."

The particular airfoil I've selected has the classy name S7055. It's a flat-bottomed airfoil, which will make it a lot easier to carve from the plank of balsa we've cut for the wing. The airfoil is shown in two sizes in Figure 20-9.

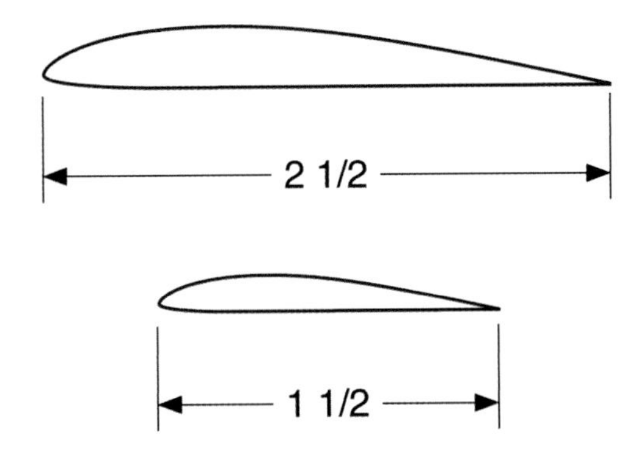

Figure 20-9. *Full-size wing airfoil patterns.*

The shape of the airfoil is very important. The wings we've cut are wider near the fuselage and narrower at the tip. Ideally, that means the thickness of the wing should change, too, so the shape of the airfoil stays constant. Back in Chapter 15 you saw one way to carve airfoils, which is to use multiple templates and check the shape of the airfoil as you sand it. This time, we'll use a different technique that gives a consistent cross-section shape, narrowing the wing thickness near the wing tip. Be sure to read through the description to understand what you will be doing before you start.

Begin by cutting two full-size templates from Figure 20-9. You can make the templates by photocopying the page from the book, or download full-size templates from the author's website (*http://bit.ly/byteworks-make-rockets*). You should cut these from a hard wood. One option is using some of the leftover poplar from the fuselage. Make the templates as close a representation of the originals as you possibly can. This is a situation where you should take your time and be very precise. If you mess up, start over with another piece of wood.

With the two templates in hand, use thick CA glue or wood glue to glue the templates to the ends of the wing, as shown in Figure 20-10. Putting some wax paper underneath is a great way to keep the wing from getting glued to the cutting board.

Figure 20-10. *Glue the templates to the ends of the wing.*

Create a very long sanding block by taping or gluing a piece of medium grit sandpaper—about 100 grit—to a 10"- to 12"-long piece of 1" x 2" wood. Make sure the wood is very flat by sighting along its length. Use the sanding block to sand the wing into shape. You need to be careful as the wing approaches its final shape: the balsa wing is much softer than the poplar airfoil templates glued to the ends of the wing, but the sandpaper can certainly sand the poplar, too. Use the poplar templates as a guide; don't depend on them to stop the sandpaper.

Figure 20-11. *Carefully sand the wing until the sanding block just touches the templates on both ends of the wing. Don't sand so aggressively that you sand the templates, too.*

Figure 20-12. *Here you see the front edge nearly complete (top), and the completed airfoil from one wing tip (bottom right).*

Once the wing is perfectly sanded, cut the templates off of the ends with a hobby knife and glue the clean edges of the templates to the other wing. Make sure you are making one left wing and one right wing! Sand the other wing's airfoil, then cut the templates from that wing's tips, too.

The wings sit at a *dihedral*, a fancy name for the angle of the wings where they meet the fuselage. This helps the plane to stay stable as it flies. If the craft gets tipped to one side by a gust of wind, the wing that is more nearly horizontal generates lift in a vertical direction, while the tipped wing generates lift to the side. This levels the airplane out, and also turns it toward the more horizontal wing.

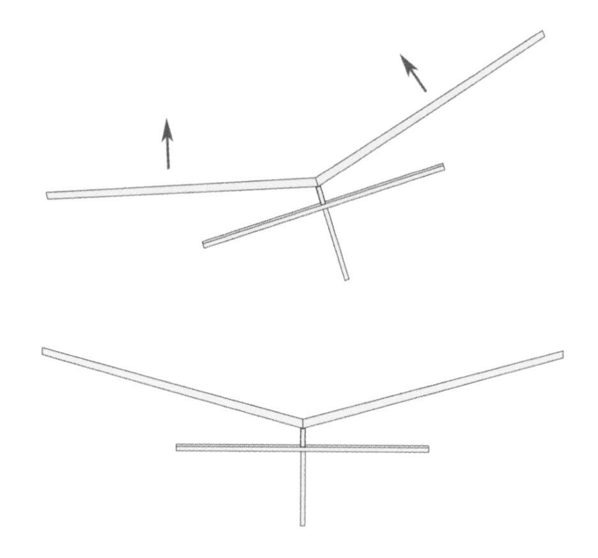

Figure 20-13. *A tipped airplane with a dihedral will right itself due to the uneven lift on the wings. Seen from the front, the glider will roll clockwise and turn as it levels out.*

The dihedral angle is 15° from the horizontal. To get the proper angle, lay one wing flat on some wax paper covering the cutting board. Cut a piece of scrap wood that is exactly 3 3/4" long. Use this to prop up the edge of one wing.

The wings will not fit together properly—at least, not at first. Take a look at how they do fit, and sand the root edges of both wings until they fit perfectly at the correct angle. Glue the wings together and let them dry.

Figure 20-14. *Sand the root edges of the wings until they fit together perfectly at the dihedral angle.*

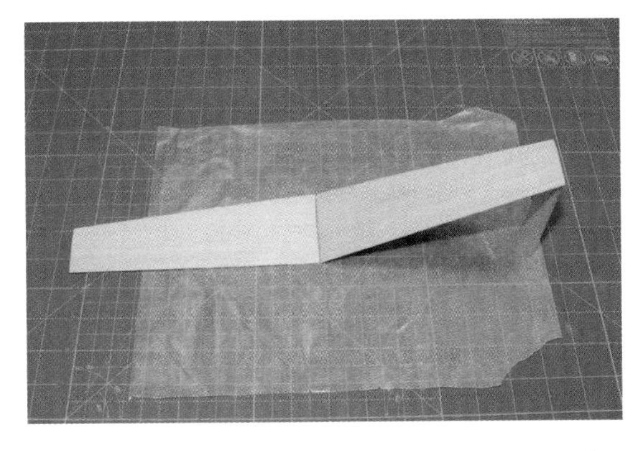

Figure 20-15. *Check the dihedral angle with a piece of scrap wood 3 3/4" long. When they fit perfectly, glue the wings together.*

Sand the vertical and horizontal stabilizers while the glue dries on the wings. You could just sand a rounded edge on each of them, like a fin, but they are thick enough to sand an airfoil. The portion of the horizontal stabilizer that will eventually be glued to the fuselage needs to stay flat on both the top and bottom, though. Figure 20-16 shows a piece of tape across the central band, providing guidance and a little protection as the stabilizer is sanded.

Figure 20-16. *Sand the horizontal and vertical stabilizers with a neutral airfoil, like a fin. Keep the part of the horizontal stabilizer that will be glued to the fuselage and vertical stabilizer flat, protecting it with a thin strip of tape. Here you see two vertical stabilizers, one with an airfoil (right) and the other still flat (left).*

Once the glue dries on the wings, sand the bottom joint flat to give the wings a flat surface that will be glued to the fuselage. Use a sanding block and check your work often. When you finish, the wings will ideally sit flat on this surface, with both wing tips an equal height off of the cutting board. The closer you are to this ideal, the easier it will be to glue the wings to the fuselage.

Figure 20-17. *Sand a flat area on the bottom of the wing joint. This is where the wings will be glued to the fuselage, so take your time.*

It's time for final assembly of the glider. Figure 20-18 shows the overall plan.

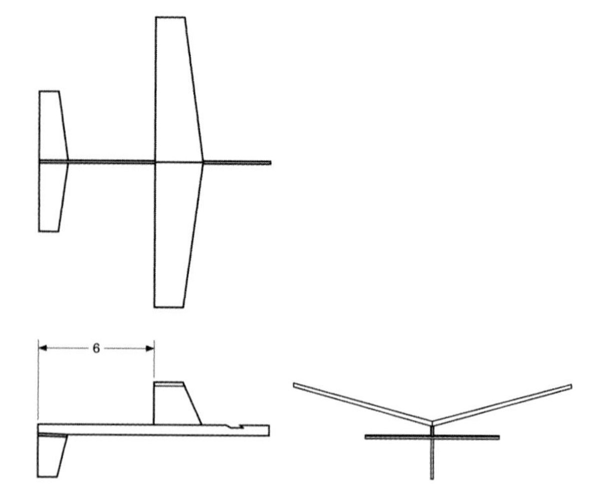

Figure 20-18. *Overall layout of the glider.*

There is one pretty obvious difference between this glider and what you might think of as a normal airplane. The vertical stabilizer, or rudder, is pointed down instead of up. This might seem a bit odd, but as it turns out, this is the best way to point a rudder. A rudder that is pointed down helps the airplane to stay level in flight, since the weight tends to tip the airplane level rather than turn it over. The reason most full-size airplanes are not built this way is that the landing gear would have to be exceptionally long to accommodate the rudder. We don't have landing gear, so we can build the rudder the best way for aerodynamic stability.

Figure 20-19. *While most full-size airplanes point the rudder up so they can have short landing gear, a few opt for better aerodynamics and really long landing gear.*

What Makes a Glider Fly Level?

There is one other design aspect worth a closer look. Have you ever noticed that a hand-launched glider pitches up when it's going very fast, and starts to go nose down when it's going very slow? That's a deliberate design decision. The idea is to build the glider so it will go nose down if it is going too slow, which will cause it to pick up speed. If it is going too fast, the nose will pitch up, slowing it down. Combined with the dihedral, the natural tendency to fly level at the correct speed helps the glider perform well. Many full-size airplanes are designed the same way, so the pilot really doesn't have to "fly" the airplane so much as steer it.

We make this happen by balancing two different aerodynamic properties. Let's start with the case of a glider that is sitting still and is simply dropped. The center of pressure for the glider needs to be a bit behind the center of gravity —just as it is for a rocket. This will cause the glider to pitch forward, pointing its nose down. With a nose-down, streamlined shape, it will start to pick up speed quickly.

As the glider picks up speed, we want the wings and stabilizer to start pitching the nose up. The faster it goes, the bigger the force rotating the nose upward should be. There are several ways to accomplish this. The main one used by our glider is to angle the horizontal stabilizer slightly so it pushes the tail down, and thus the nose up, as air flows over its surface. That's why there is a small angle cut in the fuselage where the horizontal stabilizer is mounted.

Glue the parts together as shown using wood glue. Be sure all of the parts line up perfectly. Look at the airplane from the front and back to make sure the wings, vertical stabilizer, and horizontal stabilizer are aligned and level.

Once everything dries, add some thin CA glue to the center of the leading edges of the wings, right near the root. The pop pod can separate with considerable force, and often flips back, striking the leading edge of the wings. If you don't strengthen the wings right where the pop pod will hit, the pop pod will eventually dig two grooves in the front of the wings. Strengthening the leading edge of each wing right at the root prevents this.

Building the pop pod

The pop pod is a pretty standard small streamer rocket, but instead of fins, it has a mechanism to hook it to the glider. You may already have a 5 1/4″ length of BT-20 for the pop pod, but if not, cut one now.

Start by marking the body tube for fins using the four-fin guide in Figure 3-25 from Chapter 3, but only mark three of the four fins. The pop pod mechanism and launch lugs will be on opposite sides of the body tube. Use the third fin line to line up the engine hook.

Mount the engine hook first, placing it so the hook allows the rocket motor to stick 1/4″ out of the back of the body tube. Use tape, glue-soaked paper, or a Mylar engine sleeve to secure the engine hook. The main 1/8"-thick portion of the pylon is glued over the ring of tape or Mylar.

Glue the two 1/2"-long launch lugs to the body tube, opposite the pylon. One is glued flush with the base of the body tube, and the other 2″ from the base of the body tube.

Glue the screw eye into the nose cone. Check the fit of the nose cone, sanding the shoulder of the nose cone if it fits too tightly in the body tube.

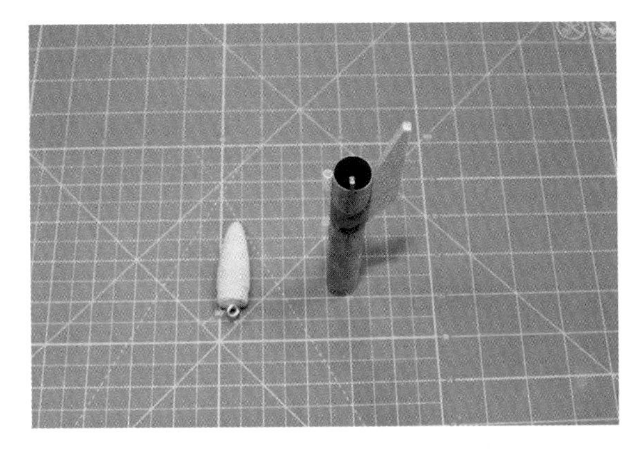

Figure 20-20. *Glue the engine hook, main pylon, launch lugs, and screw eye in place.*

Glue one of the two thin pylon side walls to the main pylon. The side walls will sandwich the fuselage between them, clamping it firmly in place and preventing the glider from wobbling from side to side as the rocket launches.

The pylon is designed so the small piece of poplar hooks into the notch in the glider's fuselage. We don't want the glider to wobble up and down on the pop pod as the rocket motor fires, so the poplar hook needs to be mounted midway between the ends of the flat surface of the main pylon piece. That will hold the glider firmly against the flat bottom of the pylon. Place the poplar hook 1/2″ from the rear edge of the pylon.

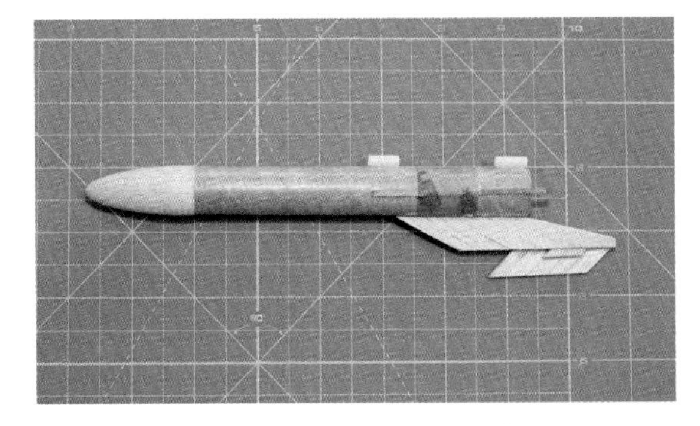

Figure 20-21. *Add the first side wall and the poplar hook.*

Add the remaining side wall, checking the fit on the glider before committing glue. The glider should fit snugly into the slot formed by the two pylon sides, and not wobble from side to side or up and down with gentle shaking. Jerking the pop pod up and down quickly should dislodge the glider.

If the glider is too tight, sand the edges of the fuselage a bit. If it is too loose, sand the side of the main pylon. Once the fit is just right, glue the second side wall in place.

Add some extra glue around the front part of the engine hook and the tape or Mylar ring holding it in place. Wood glue will shrink a bit as it dries, and light sanding will create a good surface for painting.

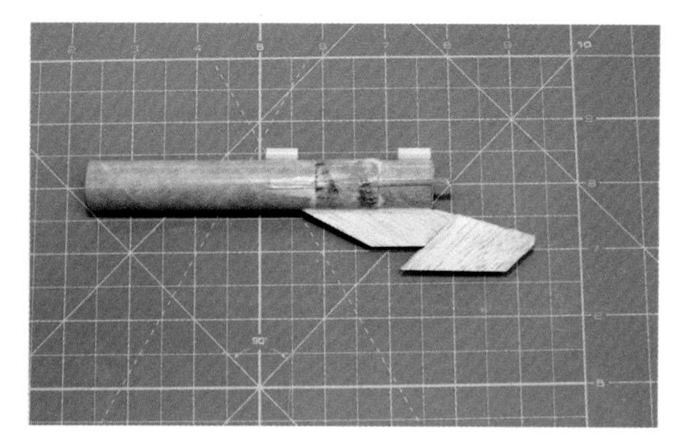

Figure 20-22. *Test the fit, making sure the glider does not wobble but can shake loose easily, then glue the final side wall in place.*

Add a 12"-long piece of 1/8"-wide shock cord to complete the pop pod assembly.

Figure 20-23. *The completed pop pod.*

Once all of the glue dries, check the fit and alignment of all parts. The glider is now ready for paint—or is it? Read the next section before deciding.

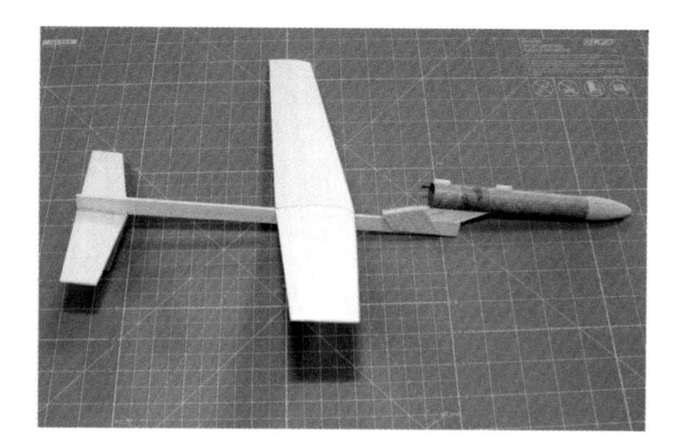

Figure 20-24. *With construction complete, Icarus is ready for paint.*

Finishing

If you've read the book in order and haven't skipped around, you already know my preference for making rockets and launchers look nice. You may not agree with my definition of "nice," but the effort does pay off in rockets that are easier to see in flight and easier to find on the ground, as well as rockets that last longer when they land in damp grass or get caught in a sudden rain shower. The paint even reduces drag and makes fins a bit stronger.

There is another school of thought for gliders, though. Paint adds a lot of weight to a glider, and that can cause it to fly like a brick. The smaller the glider, the greater the weight of the paint compared to the weight of the glider. Many people feel a glider flies a lot better if it is not painted. In my experience, it depends on the size of the glider. For Icarus, I tried it both ways. Icarus will fly fairly well with or without paint. It flies better without, but it isn't as strong and won't stand up to as much abuse. This is another of those engineering choices you have to make for yourself. Are you looking for an attractive glider that will stand up to some rocky landings, or do you want a glider that will fly longer, even if it has to be replaced or repaired more often?

Why Does Paint Matter More on Small Gliders?

A little geometry can help us understand a lot about the world around us. This is one of those cases. The weight of the parts of a glider goes up as a volume, increasing with the cube of the overall size. The weight of paint goes up as an area, increasing as the square of the overall length. Let's simplify to see how this works.

Start with a cube of balsa 1″ on a side. It will weigh about 0.074 oz, perhaps a bit more or less depending on the chunk of wood you cut it from. The paint will cover six square inches—one square inch per side—and will weigh about 0.06 oz. The paint is 19% of the total weight.

Now build the same block 12 times bigger, so each side is a foot long. The volume of the balsa wood goes up to 1,728 cubic inches, and the weight goes up by the same factor, increasing to 127.872 oz, or almost 8 pounds. The paint stays the same thickness, so its weight goes up with the surface area. Each of the six faces is now 144 square inches, for a total of 864 square inches. The paint will weigh 8.64 oz, which is just 7% of the total weight.

Different shapes have a different amount of surface area compared to their volume, so the specifics vary a bit. A sphere has the lowest surface area compared to volume of any shape, while the glider has a pretty high surface area compared to its volume. That's why the paint is 50% of the weight of the painted glider. The principle is the same, though. Double the size of the glider, and the weight of the wood will go up by a factor of eight, but the weight of the paint only goes up by a factor of four. For a glider twice the size of Icarus, the paint would only be 1/3 of the total weight. I'll leave it to you to work out the math to see why, and to figure out the percentage of the weight of the paint if Icarus were 232 feet long, the size of a Boeing 747.

There is another factor—a glider can be *too* light. A large glider that weighs too little will fly like a feather, fluttering around in the slightest gust of wind, rather than slicing through the air. That's not the case with Icarus, though.

It's fair to ask how much weight a coat of paint really adds. After all, it doesn't seem like much. The amount will vary a lot depending on how you finish the model. As shown on the right in Figure 20-25, Icarus is painted with two layers of sanding filler followed by enamel spray paint. The unpainted version of Icarus on the left weighs 15 grams (about 0.5 oz), while the painted version weighs 30 grams (about 1 oz).

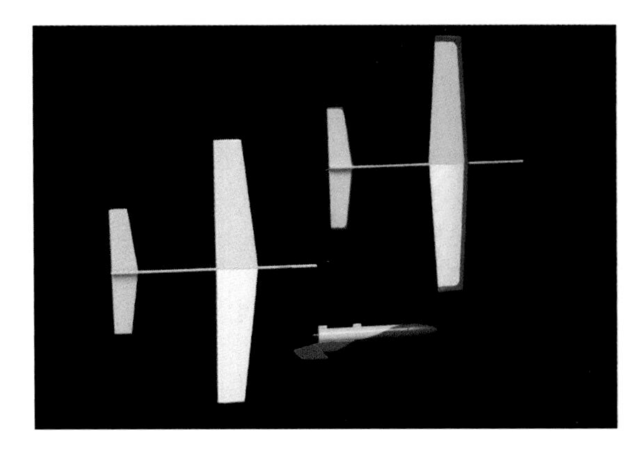

Figure 20-25. *You can paint the glider or not, depending on whether you want an attractive, rugged glider or one that will fly better but not last as long.*

One of the other things paint does is make the wood flame-resistant. A totally unpainted Icarus tends to get scorched near the trailing edge of the wing, right where it joins the fuselage. Even if you decide not to paint Icarus, add about a one-inch-wide strip of clear model airplane paint along the top root of each wing.

Flying Icarus

With the finishing done, you're now ready for your first flight.

Balancing the glider

Icarus is a bit tail-heavy to fly well. We'll need to add a small amount of clay to the nose to balance it. Start by adding enough clay so the glider balances about 5/8″ in front of the trailing edge of the wings. Icarus is now ready for its first flight!

No, put away the pop pod. The first few flights will be hand-launched flights so you can get the balance perfect. Pick a place with no obstructions and a soft area to land, like a grassy lawn. Give the glider a firm toss horizontally. It may pitch up or drop a bit at first, but then should settle down into level flight.

If Icarus pitched up, stopped, and then nosed over, add a little weight to the nose. If it nosed over and dove into the ground, take some weight off the nose. Keep trimming the airplane until it flies really well with a nice horizontal toss.

One of the problems with a perfectly balanced glider is that it might fly a bit *too* well, and end up sailing far, far away. We can counteract this with a tiny bit of clay under one wing tip. This will cause that wing to tend to dip, which in turn causes the glider to circle. It will still fly well, but it will spiral down in a circle, and not head off to who-knows-where in a straight line. Add a small piece of clay under the wing tip and try a few more tosses. Icarus should circle in a very gentle turn. If it was high enough to make a complete circle, the circle should be about 30–50 feet in diameter. Icarus won't fly that far tossed from your hand, but you can look at the curve and estimate how big a complete circle would be.

Should You Turn Right or Left?

Some authors claim that, in the Northern Hemisphere, you should add the clay to the left wing tip so the glider circles in the opposite direction of thermals, which supposedly all rotate clockwise. This action would tend to get the glider out of the thermal, so it doesn't get carried up too far and lost. The rule would be reversed in the Southern Hemisphere, where thermals spin in the opposite direction.

Hang glider pilots have busted this myth. While a thermal might circle to the right if there were no other factors involved, there are a lot of other factors in real, moving air. Thermals can turn in either direction. That means it doesn't matter which direction you make your glider turn, as long as you make it turn.

So why did I add the clay to the left wing tip? Tradition, I guess. Truthfully, I didn't notice which wing tip the clay was on until I was writing this section of the book.

Now you're ready for the first rocket-powered flight!

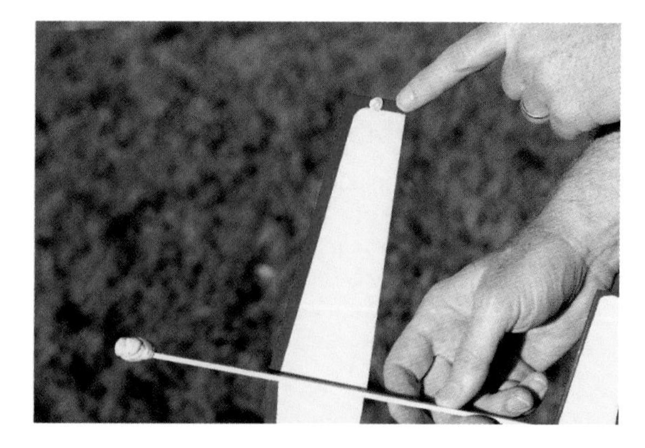

Figure 20-26. *Add clay to the nose to get a good glide path, then add a tiny bit to the wing tip so Icarus flies in a circle to the left.*

Flight preparation

Flight preparation proceeds like for most streamer rockets, up to a point. Insert about 1 1/2" of recovery wadding. Roll up the streamer and insert it and the shock cord, followed by the nose cone. Make sure the nose cone is not too tight, but that it is tight enough to not fall out if shaken upside down gently. Install the motor and igniter. Finally, slide the glider into the slot in the pylon, hooking it on the poplar support.

Most of your rockets slide almost to the bottom of the launch rail, resting against a small standoff or a clothespin near the bottom of the rail. Icarus is different. You must use something to keep the pop pod from sliding down too far and dislodging the glider. You also need to support the launch wires, which are heavy enough to pull the igniter out of the motor. The blast deflector plate usually does that job, but with Icarus, it is too far away from the motor.

Check out Figure 20-27, which shows one of many methods to secure the pop pod and launch wires.

Here, we use a clothespin to stop the pop pod from sliding all the way down the launch rod. The paper supports found on many launch wires (or the support tags you make from tape) also rest on the clothespin. This separates the leads so they don't create a short circuit.

Figure 20-27. *Use a clothespin or some other clamp to support the pop pod and igniter.*

Just be sure the clothespin is angled so it doesn't block the wings of the glider—and don't check by leaning all the way over the launch rod! As you know from Chapter 5, we *never* lean over a launch rod, especially with a live rocket on the pad.

With Icarus on the pad, the launch proceeds as usual. While Icarus is designed to fly with C6-3 motors, start with a small A8-3 motor.

Does the rocket go up in a perfect, straight line? Probably not. If it arced over backward, add a small shim at the rear of the pylon so the motor points a bit downward on the nose. If it arced forward, add a shim under the front part of the pod. Increase the motor size when the rocket is flying fairly straight.

Launch Preparation Checklist

1. Insert recovery wadding.
2. Attach the streamer and shock cord.
3. Insert the streamer and shock cord; make sure they are not too tight.
4. Install the motor.
5. Insert the igniter.

Flight Checklist

1. Clear spectators from the launch zone.
2. Remove the launch key from the launch rail, slide the rocket onto the launch rail so the pop pod is supported by a clothespin or other stop, and replace the launch key on the launch rail.
3. Attach the igniter clips so the launch wires are supported by a clothespin or other stop.
4. Make sure the glider will clear the clothespin as it slides up the rail.
5. Assign one person to watch the pop pod and another to watch the glider. Both parts need to be found!
6. Get the launch key from the pad.
7. Insert the launch key. Make sure the continuity light is on.
8. Check for clear skies—no aircraft or low clouds.
9. Count down from five and launch.
10. Remove the launch key and place it on the launch rod.
11. Recover the pop pod and glider.

More About Rocket Gliders

Rocket-powered airplanes are a huge topic because they combine the challenges of rocketry and the challenges of airplane design and construction. Icarus shows one way to build a rocket-powered airplane, but there are many, many more.

Some gliders stay attached to the rocket pod, expelling the motor to change the center of gravity so the rocket shifts from vertical, powered flight to horizontal, airplane-style flight. The CiCi Glider from Edmonds Aerospace is an example of this kind of rocket.

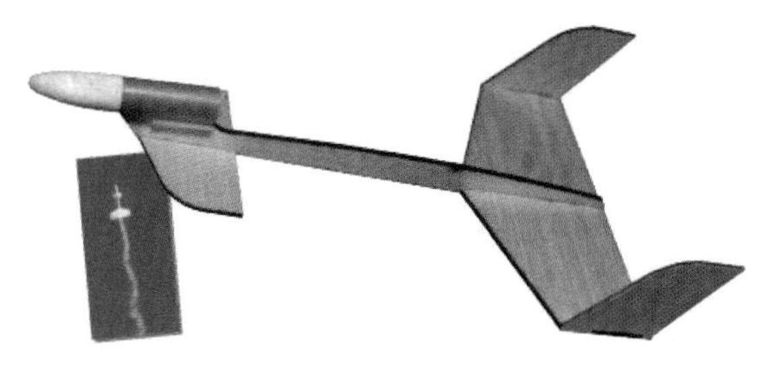

Figure 20-28. *CiCi from Edmonds Aerospace goes up and comes down in one piece.*

The fundamental design problem faced by all rocket gliders is that the way you balance a rocket for flight is very different from the way you balance a glider for flight. Icarus solves this problem by ejecting the pop pod, and CiCi does something similar by ejecting the motor. Both techniques shift the center of gravity dramatically, helping the glider make a transition from vertical to horizontal flight. A more advanced technique is to change the shape of the glider. In some cases this means using the ejection charge of the motor to flip up a control surface, to move wings from a rearward fin-like position into a more forward position, or even to unfold wings that were tucked against the sides of the rocket. That's the approach used by the air rocket glider in Chapter 21. For a challenge, you could mount that glider on a pop pod or try flying Icarus from an air rocket launcher.

Perhaps the biggest challenge (and the most fun), though, are radio-controlled rocket gliders. This is an advanced, complex form of rocketry all its own. There is even a separate safety code (*http://www.nar.org/NARrcrbgsc.html*) for flying radio-controlled rocket gliders. You can find out more on the NAR website.

Figure 20-29. *Some rockets, like this air rocket glider from Chapter 21, change their shape in flight.*

Further Reading

If Icarus piqued your interest, you may want to explore the topic of rocket gliders further. Here are some places to start:

Handbook of Model Rocketry, Seventh Edition, Chapter 13 (http://www.amazon.com/Handbook-Model-Rocketry-Edition-Official/dp/0471472425)

This classic book by G. Harry Stine and Bill Stine has a great chapter on rocket gliders, including lots of history and extensive descriptions of the various kinds of rocket gliders. Icarus is designed using the overall principles laid out in this chapter.

TR-4 Model Rocket Technical Report: Boost Gliders (http://www2.estesrockets.com/pdf/2266_TR-4_Boost_Gliders.pdf)

Estes Industries has a great series of technical reports on various aspects of model rocketry. TR-4 gives a good overview of boost gliders.

NAR Rocket Glider Plans (http://www.nar.org/competition/plans/rocketglide.html)

The National Association of Rocketry has a nice collection of rocket glider plans on its website. This is a good place to start if you want to explore some of the many ways of building rocket gliders before launching off on your own.

Air Rocket Glider

21

When an airplane goes into a dive, the lift of the wing or the tilt of the horizontal stabilizer causes the plane to pitch up at high speed, so it recovers. That doesn't work well for a rocket, though, which would generate enough speed to spiral around in loops! Icarus, from Chapter 20, solves this problem by moving the motor well ahead of the wings and tail surface. Here, we'll look at another approach.

Daedalus

There is another way to prevent the rocket glider from looping as soon as it leaves the pad, and that's to change the shape of the aircraft. That's the approach Rick Schertle used in a rubber band–launched glider he presented in *MAKE* volume 31 (also available online (*http://bit.ly/fold-wing-glider-MAKE*)). The glider uses a really cool mechanism patented in 1939 by Jim Walker. When it is launched, the wings are folded next to the body of the rocket. Air pressure holds them in place. Once the aircraft slows down, the wings flip out and the aircraft glides to the ground. Later, Rick teamed up with Keith Violette, who built the air rocket launcher from Chapter 6. Their modified glider uses a plastic tube for the fuselage that allows it to be used with the air rocket launcher, and requires only a slight modification to the launcher to hold the wings back until launch. They are currently creating a kit based on this concept that uses plastic parts. We'll take a look at how to build a similar air rocket glider here, using some of the techniques from Rick's original rubber band–launched rocket and the concepts Rick and Keith developed to turn it into an air rocket glider.

While launching the glider from the air rocket launcher is loads of fun, you might also consider that the basic design would work for any sort of vertical launch. It's pretty easy to see how this could be modified to use a solid propellant rocket motor.

Figure 21-1. *Daedalus takes off vertically from an air rocket launcher with its wings folded. At apogee, the wings deploy, and the aircraft returns as a glider.*

Daedalus

Like 1566 Icarus, 1864 Daedalus is one of the Apollo asteroids that cross the orbit of the Earth. It was discovered on March 24, 1971 by Tom Gehreis at the Palomar Observatory. In most respects, this is a relatively unremarkable asteroid. I chose it as the namesake for the air rocket glider because of the historical tale of Daedalus.

You see, in Greek mythology, Daedalus (who was Icarus's father) was the *other* guy who strapped wings to his back and flew. Father and son used their wings in an attempt to escape from Crete, where they were being held prisoner. Daedalus warned Icarus not to fly too close to the Sun, but Icarus ignored the warning. The wax holding his wings together melted, and he plunged to his death. Daedalus is less well known in mythology, but he made it to freedom on his wings by staying low—perhaps the first example of flying under the radar!

The parts list for the air rocket version is shown in Table 21-1, and a list of tools and supplies you'll need for this project is found in Table 21-2. The major parts are shown visually in Figure 21-2.

Table 21-1. Parts list

Part	Description
BT-50 body tube	Use a 12″ piece cut from a longer tube.
BT-50 nose cone	The nose cone shown is the BNC50Y from Balsa Machining (*http://www.balsamachining.com*). Other nose cones will work fine.
BT-50 nose block	The nose block shown is a solid balsa plug fit to a BT-50 that is 1″ long. Most of the time it's fine to substitute a nose block made from a tube coupler capped on each end with a sheet of balsa, but this one is used as a structural component of the rocket. It doesn't have to be a precise fit, so you can carve one from a block of balsa if you like.
BT-50 tube coupler	This is used to tighten the fit so the body tube fits more snugly on the launcher. You can substitute tape if you don't have a handy tube coupler.

Part	Description
3/32" balsa plank	Used for the wings, stabilizer, and rudder.
Screw eye	A small screw eye is used as an attachment point for the rubber band that moves the wings.
Rubber band	Small rubber band, about 2 1/2" long and 1/16" thick. A #16 rubber band from an office supply store works well.
Staples (2)	Use two staples for the attachment points for the rubber bands. You won't need the stapler, just the staples.
Music wire (thin)	A piece of thin, rigid music wire about 9" long. The diameter should be 1/16" or less, but not so thin it's too flexible. I used 3/64" (0.047") wire.
Music wire (thicker)	A piece of thin, rigid music wire about 24" long. The diameter should be 1/16" or more, but not so thick it is difficult to bend. I used 1/16" wire.
Aluminum soda can	We'll cut the metal wing supports from an aluminum soda can.
5/16" brass tube	This is used for the outside of the wing pivot.
9/32" brass tube	This is used for the inside of the wing pivot. The exact sizes of the two brass tubes are not important, as long as one slides smoothly into the other.
Clay	A small amount of modeling clay is used to weight the nose of the glider. There is some clay in the Estes Designer's Special.

Table 21-2. Additional tools and supplies

Tool	Description
Dremel tool or similar with a cutting disk	This will be used to cut the metal parts. Be sure you have safety glasses, of course.
Hobby knife	Used to cut the balsa and paper parts.
Needle-nose pliers	Used to bend the wire.
Drill	You will need a drill with a 1/16" bit and a 5/16" bit.
Wood glue	Used to glue the rocket together.
Epoxy glue	Used to glue the metal wing root supports to the root of the wings.
Thick CA glue	Used to glue the staples to the balsa wings. Epoxy can be used instead.
Light machine oil	A light oil, like WD-40, to lubricate the metal parts.
Paint	Paint and brushes for finishing.

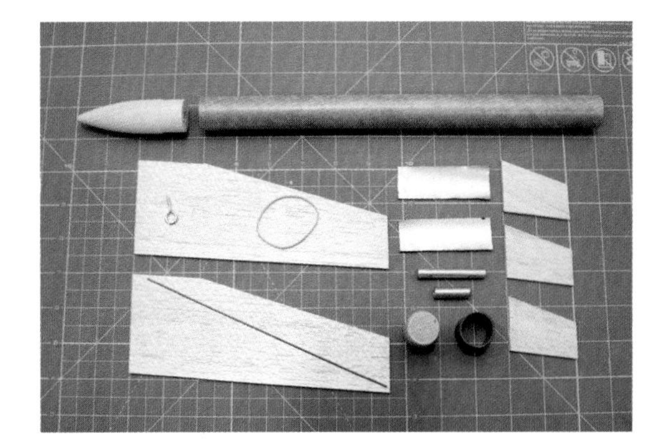

Figure 21-2. *Visual parts list for Daedalus.*

Equivalent Kit

I built and examined prototypes of the upcoming air rocket glider kit from Keith Violette and Rick Schertle while designing and building Daedalus. The premade parts for the wing rotation mechanism make their kit considerably easier to build than the scratch-built version from this chapter. On the other hand, Daedalus is a bit lighter in weight than the kit, and can be built from parts you may already have on hand after building other rockets in this book.

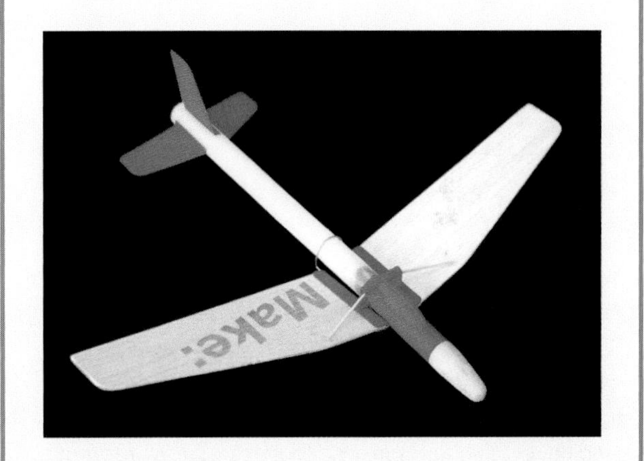

Figure 21-3. *One of the early prototypes of Keith and Rick's glider. The red parts are printed with a 3D printer. The nose is molded from soft rubber.*

Construction

In many ways, the air rocket glider is built like most of the solid propellant rockets in this book. The nose cone, body tube, and fins could easily be used in a standard model rocket. Refer to Chapter 3 if you need more details on basic construction techniques, tools, and supplies. The only real difference between building Daedalus and building a standard solid propellant rocket is the wing rotation mechanism, so I'll spend a lot of time describing how it is built.

Cutting the parts

The visual parts list in Figure 21-2 shows all of the balsa and aluminum parts precut. Let's start by getting our parts to that state.

Cut a 12" section from a BT-50 body tube. This will be the fuselage of the aircraft.

Use the patterns in Figures 21-4 and 21-5 to cut two wings and three pieces for the tail section from 3/32" balsa wood. The rudder and stabilizer are all the same size. Round all of the edges on the wings and tail surfaces except the root edges.

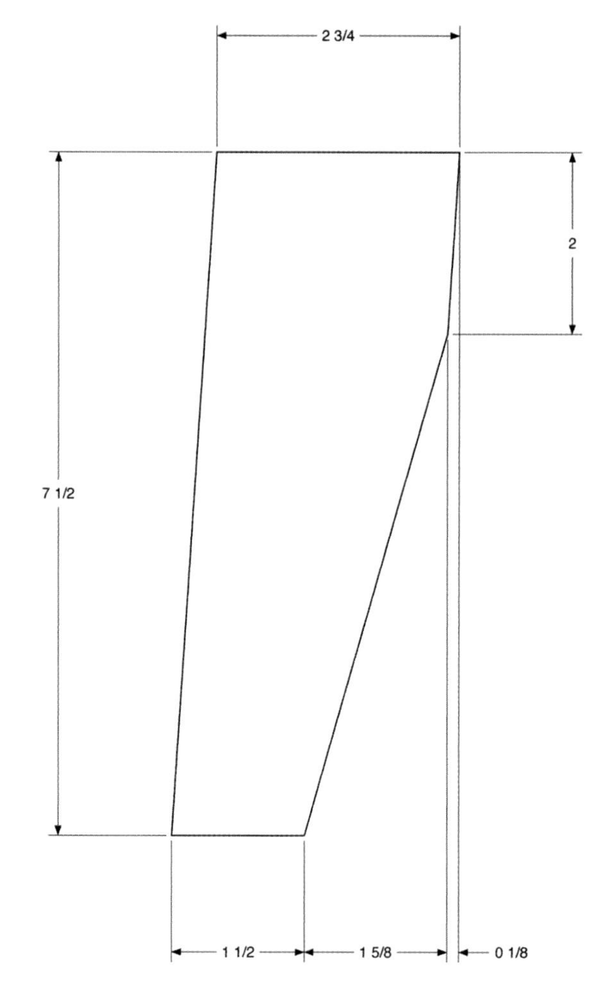

Figure 21-4. *Plans for the wing. Cut two from 3/32" balsa with the grain parallel to the trailing edge.*

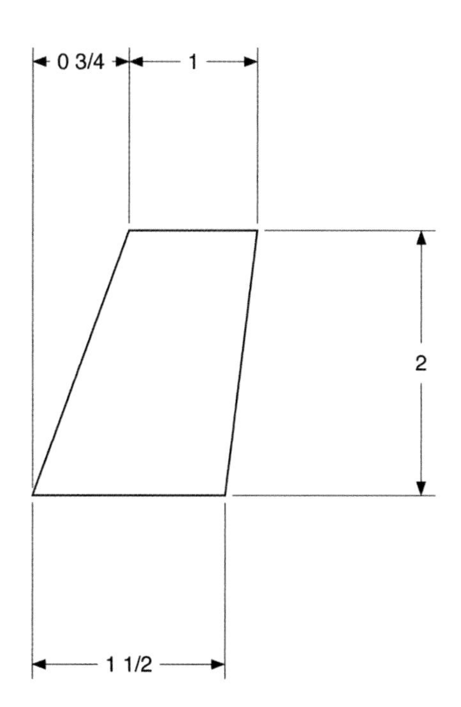

Figure 21-5. *Plans for the rudder and horizontal stabilizer. Cut three from 3/32" balsa.*

We'll use a solid balsa nose block for strength where the wing attaches to the fuselage. These are available from several sources, including Balsa Machining (*http://www.balsamachining.com*) and JonRocket (*http://www.jonrocket.com*). These generally come in long lengths; you will need to cut off a piece about 1" long. Some of them have a hole in the middle used to mount a hardwood dowel. If yours does, fill in the hole with the dowel that comes with the nose cone. We need a solid block of wood for the wing support.

This piece is going to be glued deep in the body tube, where minor mistakes are hidden from prying eyes, and where a dab of extra glue can fix small imperfections. In short, it doesn't have to be perfect. If you don't have a nose block, you can carve one yourself from a block of balsa. Whatever method you choose, you will need a 1"-long plug that fits reasonably snugly in the BT-50 body tube.

The wing needs to rotate in two directions to move from the launch configuration to the glide configuration. Take a look at Figure 21-6. It shows the detail view of the wing attachment. There is a wire running through the aluminum wing root support. This wire runs through a brass tube that slips through a second, slightly larger brass tube. The pair of brass tubes let the wing rotate from a position where the wing root is perpendicular to the body tube to a position where the wing root is parallel to the body tube. The wire lets the wing rotate from a folded position to an extended position. You can see the motion of the wings in Figure 21-7. They had to be held in place in the first three shots, since the rubber band really wants to snap the wings into the flight configuration!

Figure 21-6. *The wings need to rotate in two directions. The pair of brass tubes that run through the body allow the wings to rotate so the wing root is parallel to the body, while the wire allows the wings to rotate from a folded position to the extended position.*

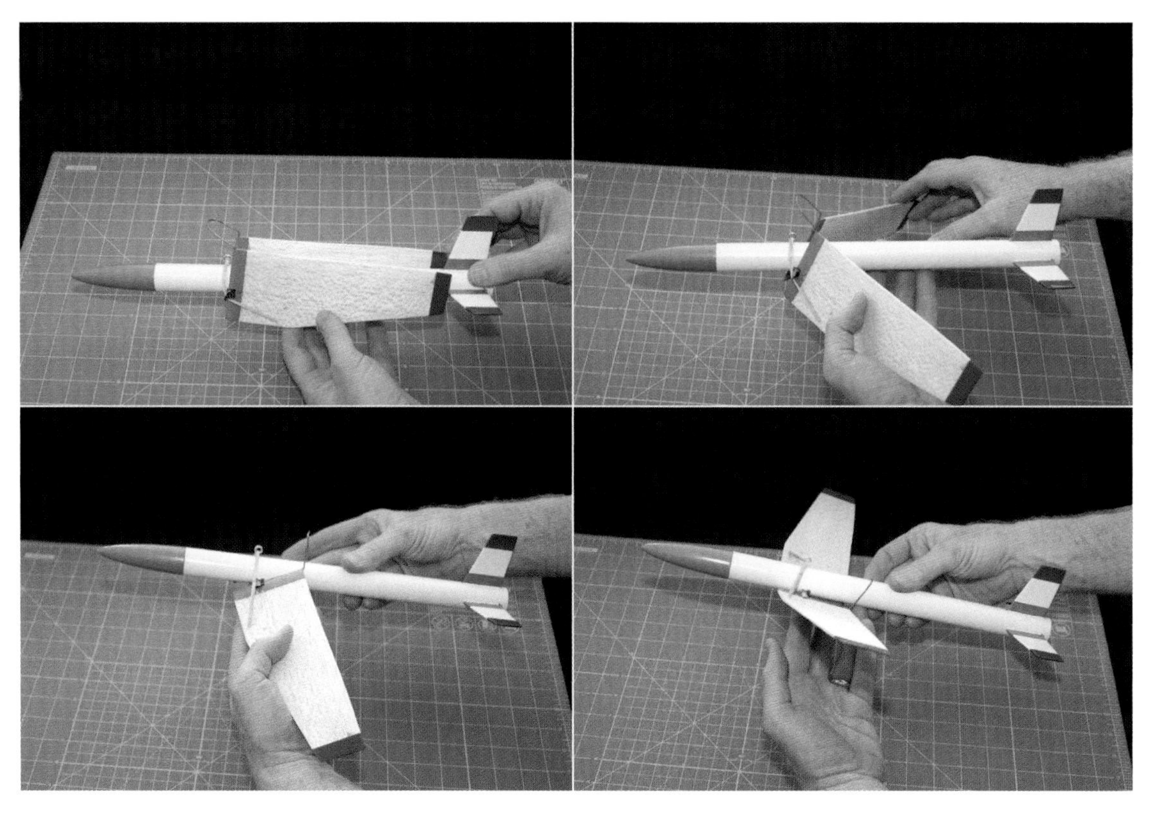

Figure 21-7. *Starting from a folded position, the wings will rotate forward and out to reach the flying position.*

Start the construction of the wing root by cutting the two metal supports from an empty soda can. Even if you've built a lot of rockets, you may not have worked much with metal. It might seem a bit intimidating at first, but it's really pretty easy. We will be cutting pieces of metal from a soda can and from brass tubes. A Dremel tool with a cutting wheel works very well for this. Frankly, it's easier to cut the metal parts than the balsa parts.

Wear Eye Protection

Be sure to wear eye protection while cutting the metal parts. The cutting process will throw off small bits of metal. They are usually not dangerous, and won't hurt your skin, but you don't want to get hit in the eye with one.

Mark the soda can near the base using a felt-tip marker and a piece of paper, just as you would mark a body tube before cutting it. Mark the can again 2 3/4" away from the first mark. Cut the can in both places using a cutting wheel, forming a 2 3/4"-long cylinder of aluminum.

Figure 21-8. *Use a cutting wheel on a Dremel tool to cut a 2 3/4"-long cylinder from a soda can. This will be used to form the supports for the wing root.*

Cut the cylinder lengthwise, then cut two 1"-wide pieces. This gives you two pieces, each 1" by 2 3/4".

Figure 21-9. *Cut two 1" pieces from the cylinder to form the two 1" by 2 3/4" wing root supports.*

The wing pivot is cut from two brass tubes, the smaller of which should slide into the larger. While the parts list calls for 5/16"- and 9/32"-diameter tubes, any tubes that are about this size will work fine as long as one tube slides easily into the other. You can find these at well-stocked hobby stores, where you can test the fit of the tubes before buying them. Of course, if you do substitute different-sized tubes, you will need to adjust the various holes and patterns accordingly.

Cut a 1"-long piece from the larger brass tube. Clean off the cut edge, both inside and out, so it is free of burrs or rough edges. You can clean up the edges with sandpaper, a grinding wheel attached to a Dremel tool, or both. When you are finished, the smaller tube should still slide easily through the larger one.

The smaller tube forms the base for the wings when they are deployed. The positioning of the holes for the wire that will eventually hold the wings in place and the angle of the cut where the wings will rest are both critical. The easiest way to get these right is to use a template like the one in Figure 21-10 and trace the edges on the brass tube with a permanent felt-tip marker. Use the pattern for drilling and cutting the part. The figure has a length marked that should be exactly 1" long in the final pattern. This lets you check for scaling problems caused by copying or printing. Be sure the pattern is exactly the right size. You can also get PDFs of perfectly sized patterns at the author's website (*http://bit.ly/byteworks-make-rockets*).

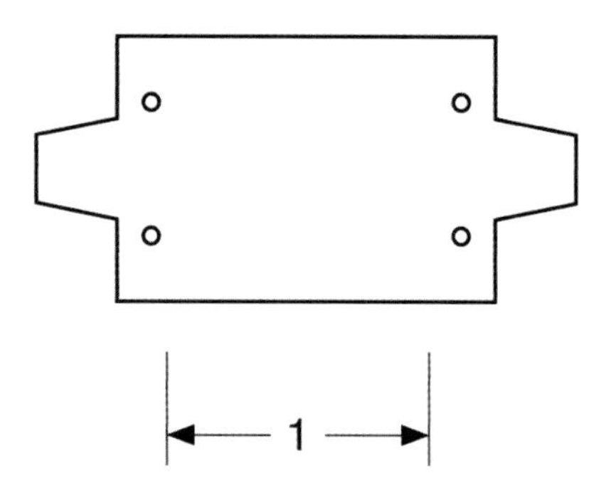

Figure 21-10. *Template for the thinner brass tube. Print and cut this template, making sure the indicated dimension is exactly 1". Use the template to mark the thinner brass tube with a permanent marker, then cut and drill as indicated.*

Drill the four holes for the wire using a 1/16" drill bit. Bits that small are flexible, and you're drilling into a curved metal surface. Choke the bit up into the chuck until only a small part sticks out. This will keep it rigid enough to go through the brass tube where you want to drill.

Don't try going all the way through and drilling the top and bottom holes with one pass. It is way too easy to drill the bottom hole in the wrong place that way. Drill the hole on one side, rotate the piece, and drill the hole on the other side. Once all four holes are drilled, check the piece carefully to make sure the holes are aligned with each other.

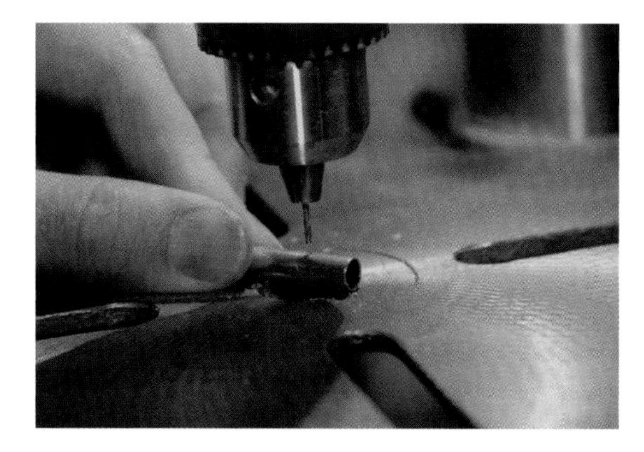

Figure 21-11. *Insert the drill bit far back into the chuck so only a small piece sticks out. This makes it rigid enough to drill the holes in the brass tube without the thin bit deflecting.*

Finish this piece by grinding or sanding off any burrs or rough edges. Make sure the piece slides easily in and out of the larger brass tube.

Building the wing root

The balsa nose block provides strength for the wing root attachment. Glue the nose block firmly in place 2 1/4" from one end of the body tube. Use the same technique you would use to install a motor mount: first, apply the glue inside the body tube, and then use an old D motor or stick to quickly shove the nose block into place. Wood glue doesn't dry fast, but it will lock the porous balsa nose plug inside the paper body tube with unexpected quickness, so don't dawdle once you start this process. Dry fit all of the parts first, then apply the glue and quickly get the nose block shoved into place.

Once the glue sets, drill a 5/16" hole 2 3/4" from the end of the body tube and as close to the edge of the body tube as you can get it without drilling through the body tube. Figure 21-12 shows what the final result should look like.The hole should pass right through the nose block that you glued in place in the last step. Insert the 1"-long piece of 5/16"-diameter brass tube into the hole and glue it in place. It should be exactly centered in the hole, and absolutely must be perpendicular to the body tube.

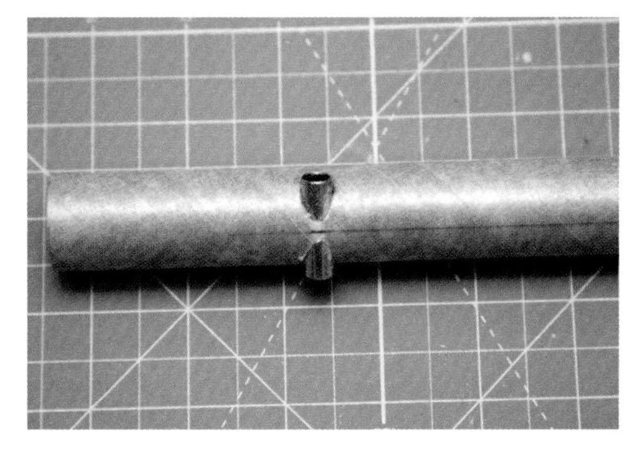

Figure 21-12. *Drill a 5/16" hole 2 3/4" from the end of the body tube. Glue the larger brass tube in this hole.*

Mark the body tube for four fins using the guide from Figure 3-25 in Chapter 3. The bottom fin mark should run right through the middle of the wing root tube, as shown in Figure 21-12. The fin guide on the top side will be used for the rudder and the screw eye that holds the rubber band in place. The two side fin marks will be used for the horizontal stabilizers.

Form the two wing root supports from the two pieces of aluminum cut from the soda can. Start by bending them in half to form a piece that is 2 3/4" long and 1/2" wide. Use a ruler or other thin, rigid tool to help you bend the aluminum. When you are finished, the aluminum pieces should have a gap between the top and bottom that is roughly 3/32" deep.

Figure 21-13. *Bend the aluminum wing root supports around a metal ruler or other support.*

Test fit the wing root supports. They should form a slot along the root of the wing that is just large enough for 1/16"-diameter music wire to slide easily through the opening. Glue the pieces in place using epoxy glue.

Figure 21-14. *Glue the wing root supports in place, forming a channel large enough for 1/16"-diameter music wire.*

Clamp the wing root supports in place while the glue sets. You can do this with small modeling clamps like the ones shown in Figure 21-15, or you can lay a heavy book across the parts while they dry. Be sure and put some wax paper over and under the wing if you use a book—you don't want to squeeze any glue out and glue the wing to the book or table!

Figure 21-15. *Use small clamps or a book to hold the wing root supports firmly in place while the glue sets.*

While the glue sets, begin forming the wing root wire. This piece is bent from an 8 1/2"-long piece of music wire. This wire should be 1/16" in diameter or thinner, but not too thin. The wire will be much too hard to bend if it is thicker than 1/16", and too flexible if it is very thin. I used 3/64"-diameter wire, which was a great compromise.

Using needle-nose pliers, slowly bend the wire using the pattern in Figure 21-16. It will be a lot easier to slide the wire into place if one end is about 1/8" longer than the other. That will let you concentrate on one side of the wing at a time.

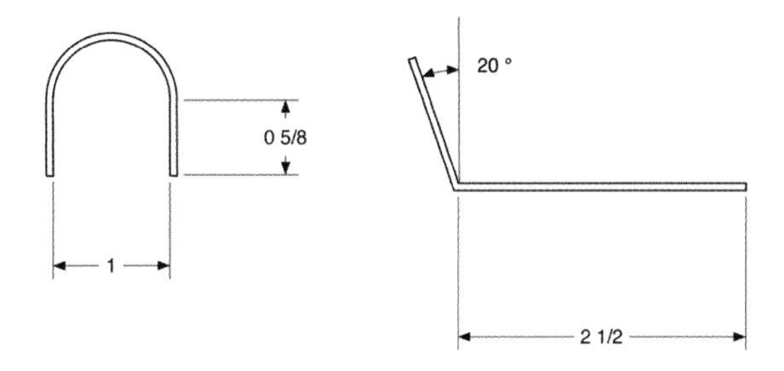

Figure 21-16. *Bend the middle of the wire into a curve, as shown on the left. Then, 5/8" from the end of the curve on each side, bend the straight section of wire to an angle about 20° from the vertical, as shown in the side view on the right.*

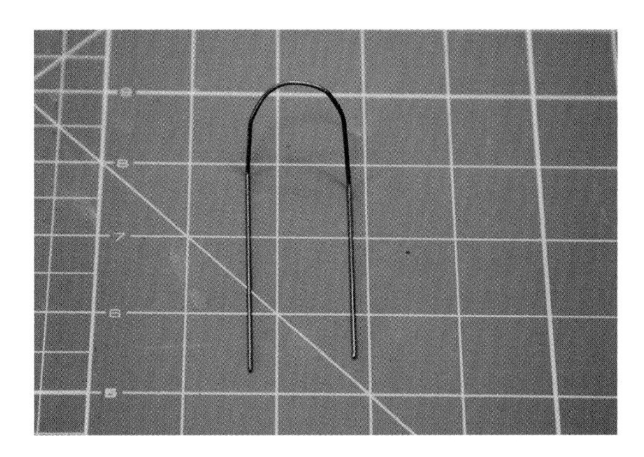

Figure 21-17. *Finished wing root wire. Note that the lefthand side is a bit longer. This helps a lot when threading the wire through the wing roots and the holes drilled in the brass tube.*

Once the glue on the wings dries, use the cutting wheel to carve out a 1/4"-deep, 5/16"-long slot in each wing root, as seen in Figure 21-18. Insert a staple from the lower side of each wing to the top. Figure 21-18 shows a pattern for the slot and the exact location of the staples.

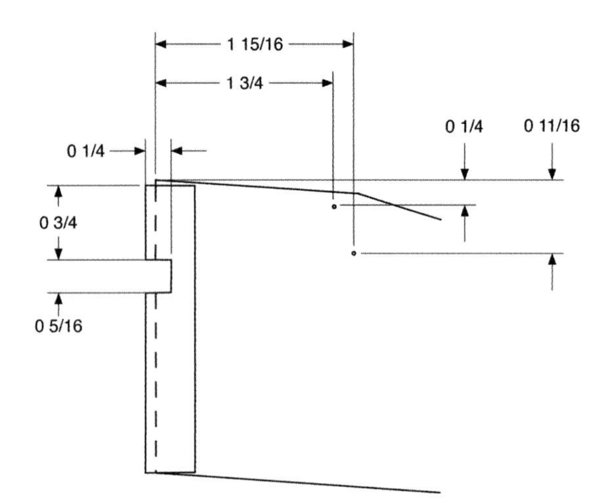

Figure 21-18. *Use a cutting wheel to cut a 1/4" by 5/16" notch in the wing root where the wing root wire will slide through the holes in the brass tube. The locations of the staple holes are shown at the top right of the drawing.*

Working from the top side of the wing, bend the side of the staple farthest from the leading edge of the wing so that is is flush with the top of the wood. Bend the other end to form a small hook with the opening in the direction shown in Figure 21-19. This hook will be used to secure the rubber band that pops the wing into place. Glue the staples in place with CA glue or epoxy.

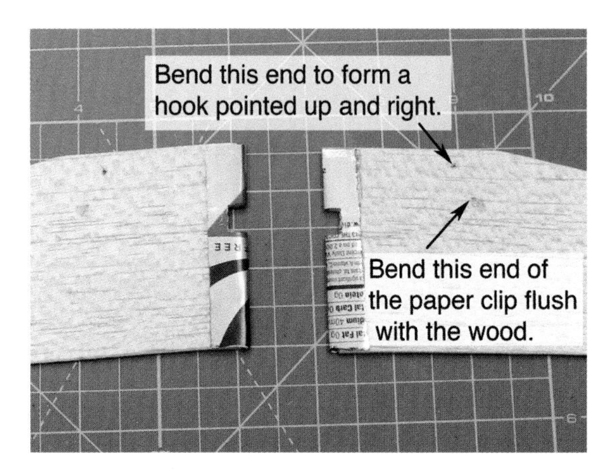

Figure 21-19. *Cut slots in the wing supports as shown. Install the hooks for the rubber bands using staples.*

Insert a screw eye 2 1/2" from the end of the body tube and exactly opposite the wing root support. Slide the wire into place so it goes through the channels in the center of the wing root tube and the holes drilled in the smaller brass tube, which slips into place in the wing root tube. Run a rubber band through the screw eye, hooking the ends over the staples.

Picking a Rubber Band

The parts list calls for a #16 rubber band. You can find these at an office supply store. They are 2 1/2" long and 1/16" wide.

Don't get too hung up on the size, though. Pretty much any rubber band about this size will work. For example, I also use #32 rubber bands. These are 3" long and 1/8" wide. This is a common size for newspapers, so you might get them delivered every morning for free. Yes, there are advantages to real, paper newspapers!

Unhook the rubber band when storing the glider. It will last longer that way. It's still going to wear out, though, so keep a few in your field box to replace it if it breaks or starts to weaken.

You can slip the nose cone into place, too, but don't glue it! The nose block prevents any compressed air from getting to the front of the glider. Later you can drill a hole in the nose cone and insert any weight needed to balance the glider, hiding the weight inside the rocket rather than attaching clay to the outside. Leaving the nose cone unglued allows you to remove it later to adjust the weight.

Figure 21-20. *Glue a screw eye in place as shown. Slide the smaller brass tube into the wing root tube, then use the wire and rubber band to attach the wings. Do not glue the nose cone in place.*

Test the wing action by folding the wings down, then rotating them to the rear of the rocket. When you release the wings, they should pop into place.

Building the tail section

Once you are satisfied with the wing action, glue a tube coupler into the back end of the body tube. The BT-50 is just a bit too large for the PVC pipe used on the air rocket launcher. The tube coupler gives a snug fit so the expanding air shoots the rocket off the pad, rather than escaping through a gap between the body tube and launcher tube.

Glue the three tail pieces in place just as you would the fins for a rocket. These form the horizontal stabilizer and rudder for the airplane.

Finishing

You can paint Daedalus, but I do recommend leaving most or all of the wings unpainted. The paint can add quite a bit of weight, as we discussed when we were constructing Icarus.

Once Daedalus is painted, check the alignment of the wings carefully. From the front, the wings should form a dihedral angle, which helps keep the plane level in flight. While this is usually specified as an angle, the easiest way to get it right is to lay one wing flat on a table and measure the height of the other. For Daedalus, one wing tip should be about 3 1/2" to 4" off the table when the other is lying flat. Make very small adjustments to the brass tube if they are not close to this range.

Figure 21-21. *When deployed, the wings should bend up. When one wing is flat against the table, the other should be 3 1/2" to 4" off the table. Adjust or replace the thinner brass tube if the dihedral is off.*

From the side, the wings should tip up about 3–5°. This forms an angle between the wing and tail that causes the plane to pitch up if it is going too fast. The combination of this angle and the dihedral is what causes the plane to settle into a steady glide. You can change this angle by bending the wing root wire. Bend the wire so the angle is closer to 0° from the vertical to increase the pitch angle; bend it so it is flatter to decrease the angle. During test flights, increase the angle if the glider tends to go into a nosedive and not pull out, even at high speed. Decrease the angle if the plane pitches up even at fairly low speeds, so it stalls and falls forward again.

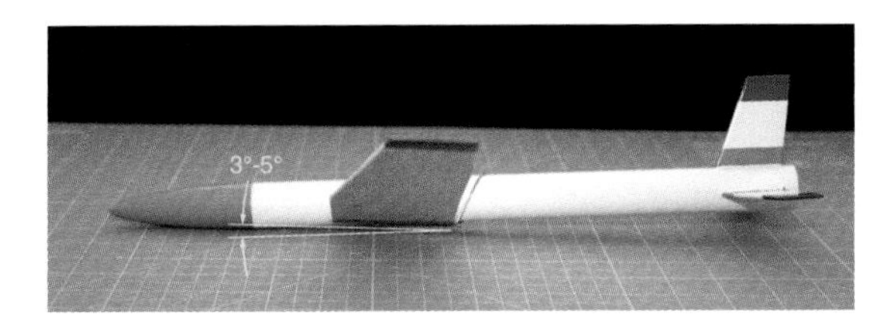

Figure 21-22. *The wing should form an angle of about 3° to 5° relative to the horizontal stabilizer.*

Finally, add weight to the nose until the center of gravity is 1/2" from the rear of the wing root. This may not be perfect, but it will be close enough to start fine adjustments to the trim.

The wing holder

You know by now that those wings simply do not want to stay put when you fold them back along the body tube. It doesn't take a lot of force to hold them in place —simply blowing on the wing will keep it folded—but the wings will pop into place as soon as all of the forces holding them back are removed.

We need some way to keep them in place until the rocket picks up speed and clears the launcher. Cut a 24" piece of 1/16"-diameter music wire. Use needle-nose pliers to bend it 1 1/2 times around the air rocket launch tube. Bend the wire perpendicular to this ring, adding small loops at the end of the wire to make it easier to slide the wing past the top of the wire. You can slip this cage onto the launcher for a glider flight, and then remove it for regular air rocket flights.

Figure 21-23. *Use 24" of 1/16"-diameter wire to form a cage that pins the wings in place until the rocket launches.*

Flying Daedalus

Balancing the glider

Just like Icarus, Daedalus must be trimmed before its first flight. Follow the tips in "Balancing the glider" on page 395 from Chapter 20 to trim the aircraft, making sure it flies without tipping up and stalling or going into a nosedive. Add a small amount of clay to one wing tip so it will glide in a wide circle.

Now you're ready for the first powered flight!

Powered flight

You saw how the wing cage holds Daedalus in place until launch time. Once the rocket picks up enough speed to clear the launcher, air pressure will hold the wings in place. As the rocket slows at apogee, the air pressure will no longer hold the wings in place, and they will pop out, turning the rocket into a glider.

Daedalus is a bit heavier than Icarus, so it will descend a bit quicker. It will also behave better in the wind, though.

Air rockets can be launched at angles up to 30° from the vertical. Daedalus cannot. If you launch Daedalus at an angle, it might travel in an arc, never slowing enough for the wings to deploy! It is very important to launch Daedalus vertically so it slows to a stop at apogee, allowing the wings to deploy.

Try to fly Daedalus in a large, flat grassy area. Keep in mind that you can't predict which way it will glide, so you should be in a fairly large open area.

This is a very fun aircraft to fly. Since it's an air rocket, there isn't much time between flights. It may not fly as high as Icarus, but over an hour of flying, it just might be in the air longer because it takes so little time to get it ready for another flight.

Flight Checklist

1. Clear spectators from the launch zone.
2. Slide the rocket onto the launch tube, making sure the wing cage holds the wings in place.
3. Pressurize the launcher to 40–75 psi.
4. Check for clear skies—no airplanes.
5. Count down from five and launch.

Rocket Clubs and Contests

22

Model rocketry is a surprisingly social activity. Whether it's a couple of friends headed down to the park to fly rockets together, a school class, or a huge national gathering like NARAM, rocket scientists of all ages gather together all the time to share ideas, to share equipment, to share the joy of success, and to learn from each other's failures. Here are a few of the organizations and larger meets you can consider getting involved with as you explore rocketry.

National Association of Rocketry (NAR)

Figure 22-1. *Logo for the National Association of Rocketry.*

The oldest and largest organization of model rocket scientists is the National Association of Rocketry, better known as NAR. Founded in 1957, NAR has promoted model rocketry pretty much since its inception. This was the same year Orville Carlisle sent a sample of his prepackaged model rocket motor to Harry Stine, a safety officer at the White Sands Missile Range who was looking for a safe way to introduce people to rocketry. Orville's patent on the model rocket would not be granted until the following year. Harry Stine went on to write *The Handbook of Model Rocketry*; he was a cofounder of NAR and one of the founders of the first model rocket company.

Since that time, NAR has sponsored over 50 national meets, established accepted rules for model rocket contests, and promoted safe rocketry through the NAR Safety Codes (*http://www.nar.org/safety.html*).

NAR membership includes a subscription to *Sport Rocketry* magazine, a print publication with six issues each year; a book showing the basics of model rocketry; access to NAR technical reports; and $2,000,000 rocket flight liability insurance. Personally, I think anyone involved in rocketry should be a member of NAR, TRA, or both. You can join NAR via the website (*http://www.nar.org/NARjoin.html*).

The Pink Book

One of the cool things NAR has done is to standardize the rules for rocket contests. These rules are used for setting national records for all sorts of events, as well as to guide NAR clubs in setting up their own contests. The rules are in the Pink Book, which you can find at *http://www.nar.org/pinkbook*.

Rocket contests give you a way to test your own skills and knowledge of rocket science. You are pitting yourself and your rockets against the skills and rockets of others. This can take the form of anything from a friendly meeting in the park to see who can keep their rocket in the air the longest, to an attempt to set a new record for the highest altitude on a C motor.

Here is a sample of some of the categories covered in the 2014 edition of the Pink Book. This gives you an idea of the huge variety of events you might find at a contest:

Altitude
: The goal of the altitude contest is to gain the highest altitude possible on a given size motor. There are 12 different altitude events for motors with a total impulse ranging from 1/8A to G. Other categories exist for cluster rockets, for achieving a specific altitude, or for combinations of altitude and rocket length.

Payload
: The payload competition is similar to the altitude competition, but this time the rocket must carry a payload as high into the air as possible. Once again, there are several subcategories based on motor size. There are also three different payloads. One payload is a cylinder of sand about 19 millimeters in diameter and 60–80 mm long, weighing no less than 28 grams. This represents carrying the approximate dimensions of an instrument package. My favorite, though, is the egg loft competition, where each rocket must carry a raw grade A egg, weighing 60±3 grams and no larger than 45 mm in diameter. There is also a variant where the rocket carries *two* eggs.

Duration
: There are several events where the goal is to keep a rocket in the air as long as possible. These are divided up by recovery method, including parachute, streamer, helicopter, and three kinds of gliders. Other categories include keeping an egg (or two) aloft, or keeping a rocket aloft for a specific amount of time.

Craftsmanship
: Building a model rocket is an exacting hobby. Building a model rocket that looks like a real rocket, like the Saturn V that carried American astronauts to the Moon, or the Shenzhou 5 that carried the first Chinese astronaut into space, is even more exacting. Or imagine building a flying replica of the Millennium Falcon from *Star Wars*. These are the sorts of rockets built for the various craftsmanship competitions. Some of the subcategories are scale, sport scale (where the rocket looks like the original, but is not measured for precision), plastic model conversion, and concept scale (a model of a rocket that never really flew; either fictional, like the Millennium Falcon, or real, like the Boeing X-20 Dyna Soar).

Miscellaneous
: There are four other competition categories: spot landing, the drag race, radio-controlled gliders, and research & development. Past entries to the research & development category have appeared in this book, both for calculating the center of pressure and for calculating drag.

National Association of Rocketry Annual Meet (NARAM)

NARAM is an annual weeklong gathering of rocket scientists sponsored by NAR. It's held in a different location each year. There isn't enough time for all of the competitions in the Pink Book, even in a whole week, but you'll find several of them. There are lots of rocket vendors there, too. Many people go each year to visit their rocketing friends at the event.

To give you an idea of the scale of NARAM, the 56th annual event, scheduled for July 2014 in Pueblo, Colorado, has 7,000 acres of flying space with two separate flying fields, one for sport launches and one for competitions!

Find out more about NARAM at *http://naram.org*.

National Sport Launch

NAR sponsors a national sport launch each year, and sponsors a number of smaller sport launches through local member clubs. These sport launches are a great place to meet people with an interest in rocketry. You will be amazed how far they push the hobby!

Visit the NAR website (*http://www.nar.org/NARlaunches.html*) for a list of upcoming launches, including the National Sport Launch and gatherings by NAR member clubs.

World Space Modelling Championships (WSMC)

Competitions and records go beyond the borders of the United States. These competitions are currently held in even years, usually in Europe. NAR organizes the United States team; you can find out more at *http://www.nar.org/internats*.

Tripoli Rocketry Association (TRA)

Figure 22-2. *Logo for the Tripoli Rocketry Association.*

TRA has been around since 1964, but really came into its own only recently. While both NAR and TRA cover all sizes of model rockets, from those flying with tiny 1/8 A motors all the way up to massive O motors, they do specialize a bit. NAR does a lot more with low-power rockets, like the ones in this book. TRA, on the other hand, concentrates on high-power rocketry. TRA was also instrumental in getting high-power rocketry approved in the United States.

As an illustration of this division, the national records kept by NAR stretch back for years, but stop with the G motor. TRA, on the other hand, maintains a list of TRA records, but the lowest total impulse currently covered is the F motor. TRA also changes its rules as technology changes. The oldest record following the rules that are current as of 2014 was set in 2004.

You can join TRA by filling out the membership form (*http://bit.ly/joinTRA*) on the organization's website. Membership includes access to TRA's private web forum, and $1,000,000 insurance at any TRA-sanctioned launch—essentially, any launch that follows the TRA safety code. See the TRA website for details.

Another difference between NAR and TRA is that TRA sponsors research launches. These launches recognize that in the field of high-power rocketry, there are many advanced rocket scientists who have the skills and facilities to go beyond commercially tested devices. Research launches allow people to build and test their own motors. This does not mean anything goes, though—there are still strict safety requirements. Flying at research launches is also limited to TRA members with a high-power level 2 certification. You can find more information about research launches at the TRA website (*http://bit.ly/TRA-research-launches*).

You might see references to commercial launches on the TRA website as well. These aren't paid launches, just non-research launches (i.e., launches conducted with approved commercial motors and launch techniques). This term is simply a way of distinguishing research launches from non-research launches.

It's fair to ask which organization you should belong to, if either. You should definitely join one or the other. While you will find people who will strongly encourage you to join whichever organization is their favorite, I would recommend joining NAR if your primary interest is low-power rocketry, and TRA if you are reading this book as a first step toward high-power rocketry.

As for me, I'm a member of both. My high-power certifications are through TRA, but the organizations have a reciprocity agreement, so those certifications can be transferred to NAR.

Large and Dangerous Rocket Ships (LDRS)

The big annual meeting for TRA is LDRS. The location changes from year to year, and the website changes for each yearly event. Find out about the next LDRS by doing a web search for "TRA LDRS."

This is a very well known event that has been featured on the Science Channel. One of their better-known categories is called Odd Rockets, where people try to make odd things fly. Past entries have included flying outhouses and flying pigs! Search the Web for "LDRS Science Channel" for some fun video clips.

LDRS has some days for commercial launches and some for research launches. Check the schedule in advance if you plan to attend and want to fly.

Badd Ass Load Lifting Suckers (BALLS)

TRA has a second regularly scheduled national event called BALLS.

BALLS is always a research launch. It's held every year in Black Rock, Nevada. This is where the really big rockets fly! All high-power launches require an FAA clearance. Launches at the local club I belong to have clearances to 9,000 feet, 15,000 feet, or 25,000 feet, depending on which launch site we are using and how high people want to fly. We're pretty lucky to get clearances like that. The clearance for BALLS 23 in September 2014 is 491,000 feet! That's literally flying into space, which is usually defined as 100 km high, or about 328,084 feet.

It's fair to ask why BALLS has such a high altitude limit. After all, the current TRA altitude record for an O motor, the largest allowed high-power rocket motor, is 30,168 feet—a record set by Gerald Meux Jr. at BALLS 22 in 2013. Some rockets combine several large motors to achieve a total impulse greater than O, though. There are also a few individuals and groups who have the knowledge and facilities to build and test motors. Since BALLS is a research launch, it gives them a venue to test their new designs.

You must be a TRA member with at least a level 2 certification to fly at BALLS, and your rocket must use a K or larger motor. Find out more by searching the Web for "BALLS TRA."

Black Rock, Nevada is a favorite launch site for high-altitude launches year-round. It's where the Civilian Space Exploration Team (CSXT), led by Ky Michaelson, launched the first amateur rocket into space. Their Go-Fast rocket was built by about 25 people. Parts arrived from six different states, and were assembled at the launch site. At 11:12 A.M. on May 17, 2004, the rocket started its flight to 72 miles, or 116 kilometers high.

Local Clubs

Local clubs range from informal gatherings to clubs that are sponsored by TRA or NAR. The club I currently belong to is the Albuquerque Rocket Society (ARS) (*http://www.arsabq.org*). I'll tell you a little about our club to give you an idea of what you can expect from a well-run local rocket club.

Figure 22-3. *ARS members getting ready for a launch. Photo by Jim Jewell.*

ARS has about 50 members, and one way or another we meet about two dozen times a year. The big events are the launches. Once a month we meet in an open area, a few miles from any structure or regularly traveled road, for a launch. Non-club members are welcome at these launches, although they are limited to flying low-power and mid-power rockets. The club handles obtaining FAA clearances and provides launch facilities. The club owns a trailer packed full of equipment for these launches, including low-power and high-power launchers, a PA system, and a weather station. We typically set up eight low-power pads and three high-power pads. Two of the high-power pads are set up 100 feet from the LCO table, so they can handle rockets with a total impulse up to 1,280 N-s (J class). The third pad is set up at 200 feet for K motors and complex launches. On occasion, you might see a pad set up at 300 feet or 500 feet for even larger rockets, but since we only get FAA clearance to 9,000 or 15,000 feet, depending on need, it's unusual to see those pads set up for the regular launches.

The club will always have a designated LCO and RSO at these launches. There is almost always someone there who can handle certification launches for both NAR and TRA.

I've seen all sorts of rockets flown at this event, from stomp rockets to L motors. Most months you'll see a couple of nervous people working on a level 1 or level 2 high-power certification, or occasionally a level 3 certification. We've hosted Cub Scouts working on merit badges, curious locals, and high-school rocket scientists working on the TARC competition (*http://www.rocketcontest.org/*). You will see everything from A to K motors on a typical day in the summer.

The club also meets most months for a business meeting. Well, really it's to get together and eat, share photos, bring in our latest project, ask questions of more experienced rocket scientists, see demonstrations of new construction techniques, and take level 2 exams… you get the idea. But we do conduct some business, too.

Several times a year we collect at a ranch near the Very Large Array, west of Socorro, New Mexico. This is a fairly isolated spot, so we're able to get a clearance to 25,000 feet. While you will almost always find a low-power pad or two set up, these ranch launches tend to concentrate on the larger motors. This is where you will find K and L motors, as well as a few composite motors.

These launches are occasionally followed by a second day designated as a TRA research launch.

Your local club will be different, of course. It might be a smaller club just getting started, or a huge organization. However it is set up, though, your local club is a fantastic place to meet other model rocket scientists. You'll learn quite a lot, and hopefully mentor a few newer rocketeers, too. I highly recommend joining your local rocket club.

Canadian Association of Rocketry (CAR)

Don't feel left out if you live a bit north of the United States! The Canadian Association of Rocketry is an organization of Canadian model rocket scientists. Like NAR and TRA, CAR has a safety code and certification procedures for high-power rocketry, maintains Canadian records, publishes a magazine, and organizes national launches. See the CAR website (*http://www.canadianrocketry.org*) to learn more.

Figure 22-4. *Logo for the Canadian Association of Rocketry.*

Team America Rocketry Challenge (TARC)

Figure 22-5. *President Obama meets a TARC team from Presidio, Texas in 2012.*

While there are lots of model rocket competitions, one deserves special mention. The Aerospace Industries Association (AIA) has sponsored an annual contest for high-school students for over 10 years. The Team America Rocketry Challenge uses model rocketry to promote science, technology, engineering, and math education, better known these days by the acronym *STEM*. The rules change a bit from year to year to keep things fresh.

The rules for the 2013 contest were pretty typical. The goal was to lift two raw eggs to a height of 825 feet, with a total flight time of between 48 and 50 seconds. There were various restrictions on motor size, parachute configuration, and the overall weight of the rocket.

Each rocketry team was made up of 3 to 10 students in grades 7 to 12. Teams could be sponsored and financed by an organization. Each team had an adult sponsor and mentors, often from local rocket clubs. The club I belong to has provided mentors for TARC many times.

The competition starts in September. By the end of March, all qualifying flights must be completed, and the results submitted to the AIA.

The top 100 qualifiers are invited to the final fly-off, held in May near Washington, DC. The top 10 finishers split prizes that include $60,000 in scholarships.

One of the neat things about TARC is that it's not just about building the rocket. The way the contest is set up encourages the teams to experience engineering the way it's done in the real world. Sure, this includes a lot of science, math, and engineering, as you would expect, but participants also learn proposal writing (to get grants from sponsors), budgeting, project planning, and team management skills.

See the TARC website (*http://www.rocketcontest.org*) for complete rules and entry information.

Table 22-1. Past winners at TARC

School year	Winner
2013–14	Creekview High School (Team 1), Canton, GA
2012–13	Georgetown 4H, Georgetown, TX
2011–12	Madison West High School (Team 1), Madison, WI
2010–11	Rockwall-Heath High School (Team 1), Heath, TX
2009–10	Penn Manor High School (Team 1), Millersville, PA
2008–09	Madison West High School (Team 3), Madison, WI
2007–08	Enloe High School (Team 2), Raleigh, NC
2006–07	Newark Memorial High School, Newark, CA
2005–06	Statesville Christian School, Statesville, NC
2004–05	Dakota County 4-H Federation, Farmington, MN
2003–04	Penn Manor High School, Millersville, PA
2002–03	Boonsboro High School, Boonsboro, MD

Onward and Upward

23

This book has led you through all of the major kinds of rockets in the fields of water rocketry, air rocketry, and low-power solid-propellant rocketry. You've learned how to build and fly rockets using all of the major recovery techniques, and you've seen propulsion systems like clustering and multistage rockets.

This chapter closes with a bridge to mid-power rocketry, a challenge for your new skills, and, just for fun, some really odd rockets.

Most mid-power rockets and nearly all high-power rockets use reloadable motors. While these are uncommon in low-power rocketry, there are a few reloadable motors you can use with the rockets in this book. Let's start by taking a look at one of them.

Reloadable Motors

All of the motors used so far in this book have been single-use black powder motors. There is another kind of motor that is less expensive per flight, and often smaller as well. *Reloadable motors* use the same aluminum perchlorate solid propellant rocket fuel that was used in the solid propellant strap-on boosters for the Space Shuttle. For model rockets, chemicals are sometimes added to produce different-colored flames or more smoke. Like the Space Shuttle motors, reloadable model rocket motors are reusable. The motor case is made from aluminum, while the *forward closure* and *aft closure* are made from brass or aluminum. We'll see what these parts are in a moment. The disposable parts include propellant, a smoke charge, an ejection charge, a nozzle, and various O-rings and protective liners.

> **Build as Designed**
>
> A lot of engineering goes into these motors. It is very important that you build them as designed. Leaving out a part could lead to a motor that burns through, creating a safety hazard and destroying your rocket.
>
> Don't try to intentionally modify the motor or the propellant load in any way, either. It probably isn't safe, and even if it is, you are violating the safety codes that keep our hobby safe—and keep your insurance valid!

The first flight with a reloadable motor is more expensive than a single flight with a single-use motor, because you need to buy both the motor and the reload materials. After a few flights, though, the cost evens out. In the long run, you can save quite a lot of money, since each individual reload costs far less than a single-use motor.

Due in large part to the fact that the fuel used in reloadable motors has a higher *specific impulse*, which means more power for a given weight, reloadable motors are about the same weight or perhaps even a bit lighter than their single-use counterparts. The motor we will look at in a moment can be loaded with premade fuel grains that generate between 20 N-s and 57 N-s of total impulse, making it a D, E, or F motor, depending on the reload used. When loaded with a D fuel grain, the motor weighs slightly more than an Estes single-use D motor; when loaded with an E fuel grain, it weighs less than an Estes E motor.

There are two drawbacks to these motors. The first is that you must be 18 to buy reloadable motors. Unlike the restrictions on black powder motors, this is not a restriction imposed by a few states; it is a federal law. The second disadvantage is that aluminum perchlorate motors need different igniters to ignite the fuel. The igniters require a 12-volt ignition system. These igniters will work fine with the launchers from Chapter 4 and some high-end launchers, but they will not work with the 6-volt or 9-volt launchers in starter sets.

There are several companies that make reloadable motors, but most of them cater strictly to the high-power rocketry crowd, creating F and above motors. The particular motor we will look at here is from Aerotech. It is 24 mm in diameter, and will work in most model rockets built for Estes D and E motors. It works quite well in the Ceres B from Chapter 9.

Building the Aerotech RMS 24-40 Motor

Figure 23-1 shows what you will need to buy to use reloadable motors. The package contains three reloads. In this case they are D15-7T motors. The reusable part of the motor is at the top right. You will also need some kind of grease. Different model rocket scientists get pretty passionate about their choice of grease, but in truth, almost any synthetic or petroleum-based grease will work. Based on the instructions from the manufacturer and personal experience, ordinary Vaseline will work fine with this motor. Finally, you will need some tape and a hobby knife.

Figure 23-2 shows the various parts after removing them from their packaging. It's a good idea to lay them out on a clean surface as you build the motor. Naturally, you should do this well away from any source of flame or high heat. While people do reload motors at the launch site, and I've done it myself, my preference is to build the motor the evening before a launch in an indoor environment.

Reloadable Motor Designations

Most companies add an extra letter to the end of the motor designation for motor reloads. This letter tells you a bit more about the characteristics of the rocket fuel used, which can vary.

While reloadable motors use aluminum perchlorate as the fuel, other chemicals are often added to get a different effect. The T in the designation of the D15-7T motor is for Blue Thunder motors, which have a bright blue flame during the propellant burn. There are two other fuel types used in reloads for the RMS 24-40 motor. They are W, for White Lightning; and J, for Black Jack. White Lightning fuel gives the same brilliant white flame seen in most full-size solid propellant rockets, while Black Jack gives off a dense black smoke trail.

Blue Thunder fuel gives the highest thrust of the three, burning very quickly. White Lightning is next, with Black Jack giving lower thrust but a longer burn.

Most people make the choice of fuel for the visual effects, but you can also tweak the performance of the rocket by carefully selecting the fuel. Shorter, high-thrust burns from a Blue Thunder motor are great for lifting heavy payloads. A slower burn from Black Jack fuel maintains thrust, allowing for a gentler lift at lower speeds, keeping drag down. White Lightning lies between the two.

Aerotech and other companies produce many other fuel types, but they are not available for the motor we're describing here. See the manufacturers' websites for the fuel types available for other motors.

Figure 23-1. *You will need a motor, a reload, and some grease, in addition to the tape and hobby knife used throughout the book.*

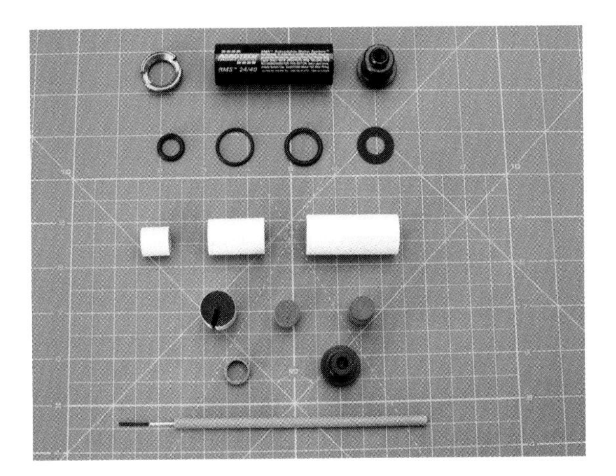

Figure 23-2. *The parts for the reloadable motor.*

Follow the Directions Carefully

These instructions show the various techniques used to assemble a reloadable motor, using a specific motor as an example. Different motors, even from the same manufacturer, will have slightly different components and assembly techniques. Be sure to read and follow the instructions for the specific motor you are assembling.

The reusable part of the motor is along the top row in this photo. From left to right, the parts are the aft closure, the motor tube, and the forward closure.

All of the rest of the parts are disposable. The O-rings and forward seal disk are in the second row. These prevent hot gases from the motor from coming in contact with the motor tube. Next are three paper tubes. The smaller one will hold the delay charge. The middle tube is a spacer that fits inside the right tube, which is a liner that prevents the rocket fuel from burning the aluminum motor tube. The spacer is shorter for E and F reloads.

The next row has the pyrotechnics. The rocket fuel is on the left. It's a premade pellet called a *grain*. This one has a slot to expose more surface area when the motor starts to burn. Next is the delay charge. The red cylinder is the ejection charge, which is loose black powder shipped inside two nested plastic caps that will be used to seal the forward and aft ends of the motor.

The next-to-last row has another spacer. This one will be used when building the ejection system, which will fit in the forward closure. The nozzle is on the right.

Finally, the cardboard tube at the bottom is a protective tube that contains the igniter.

Begin assembly by applying a thin layer of grease to all three O-rings, but not to the forward closure disk. You should use enough grease so the O-ring is shiny, but not so much that you can see or feel deposits of grease on the O-ring. Apply a little grease to your fingertip and run the O-ring through your fingers, being sure to apply the grease around the entire O-ring. Make sure you don't get any grit or sand on the O-rings. If you do, wipe them off and apply more grease if needed.

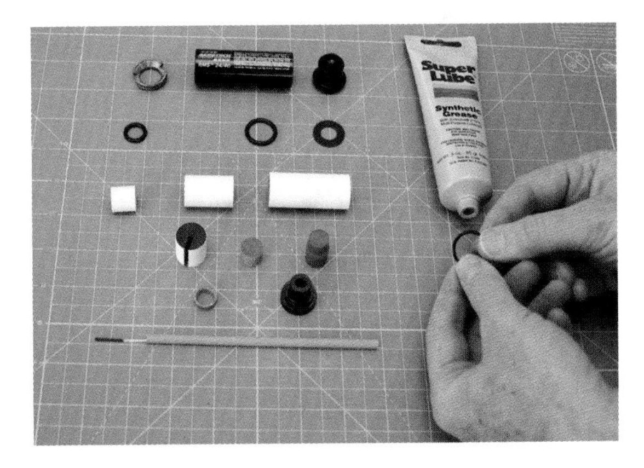

Figure 23-3. *Apply grease to the three O-rings.*

You will be inserting tight-fitting components into the small tube, which holds the delay charge, and the large tube, which holds the rocket fuel. The tubes usually have a slight indentation on the inside edges of the ends from the cutting process. That makes it hard to slide in the components. Run your thumbnail around the inside of the ends to smooth out this burr.

Figure 23-4. *Chamfer the edges of the small and large tubes.*

Slide the delay charge into the smaller of the tubes. It should be a snug fit. Don't be afraid to use a bit of force to push the delay charge into the tube.

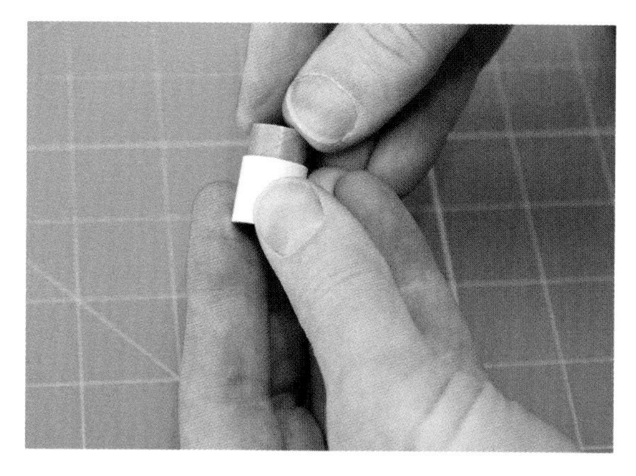

Figure 23-5. *Install the delay charge.*

Slide the smallest spacer tube into the other end of the delay charge housing. The delay charge and spacer seen here are for a 7-second delay. Longer and shorter delay charges will be different lengths, but the charge and spacer will still fit together so they fill the tube.

Figure 23-6. *Install the spacer in the delay charge tube.*

The small, fat O-ring prevents hot gases from reaching the side of the forward closure and traveling down the side of the tube. Install it in the forward closure.

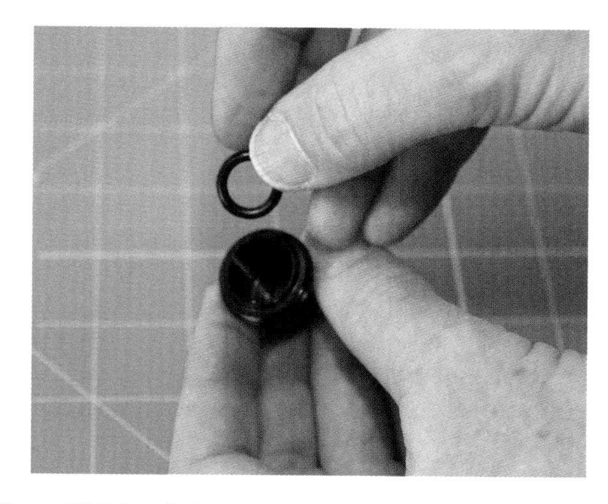

Figure 23-7. *Install the small, fat O-ring in the forward closure.*

Slide the delay charge into the forward closure with the delay charge going in first, and the spacer on the exposed end. The delay charge needs to rest right up against the forward O-ring.

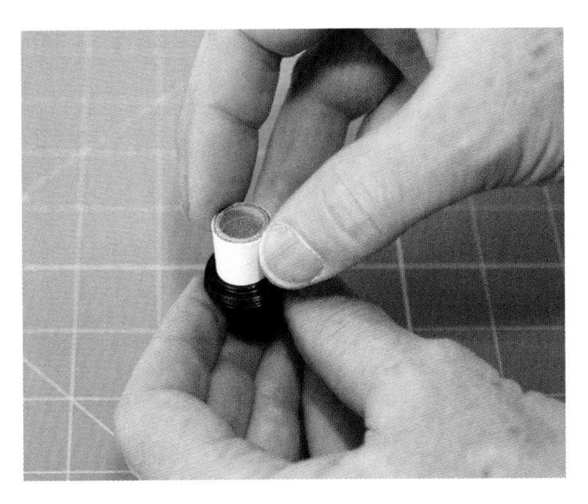

Figure 23-8. *Install the delay charge in the forward closure.*

Set the forward closure aside. The next step is to assemble the rocket fuel and place it in the motor tube.

Begin by placing a small piece of tape on one end of the fuel grain. This should completely cover the slot in the fuel grain. It's used to block the igniter so you don't push it too far into the motor.

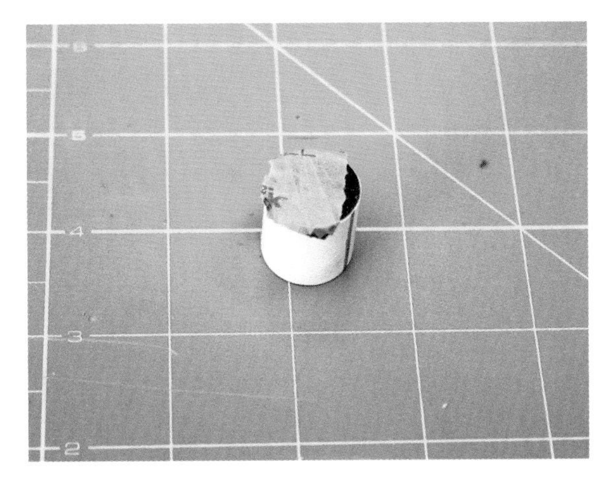

Figure 23-9. *Use tape to cover the slot in one end of the fuel grain.*

Slide the fuel grain into the larger of the remaining tubes until the end of the grain is flush with the end of the tube. The taped end should be on the inside of the tube.

Figure 23-10. *Insert the fuel grain with the taped end going in first.*

Slide the remaining tube into the other end of the motor liner. The fuel grain will be longer for E and F reloads, and the spacer will be shorter. The spacer is cut so it locks the fuel grain in place. The fuel grain and spacer should fill the motor liner tube.

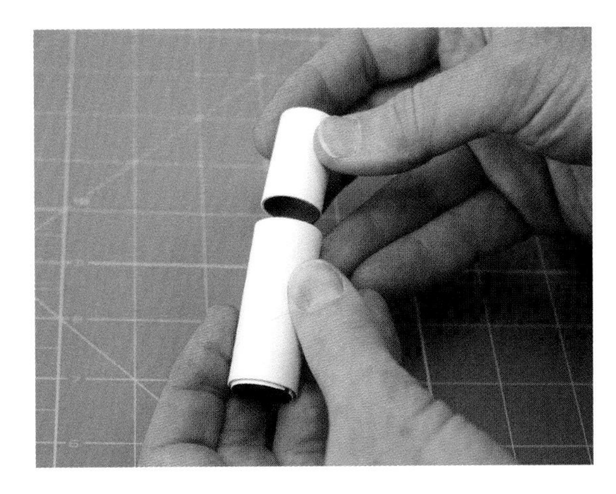

Figure 23-11. *Install the spacer.*

The next step is optional, but I recommend it. Coating the outside of the motor liner with grease makes it a little easier to install, and also makes cleanup easier after the flight. Spread the grease across the entire outside surface of the motor liner, but do not get any grease on the end of the fuel grain or inside the spacer tube. If you accidentally get grease on the fuel, wipe it off immediately. Be sure to clean your fingers carefully before the next step.

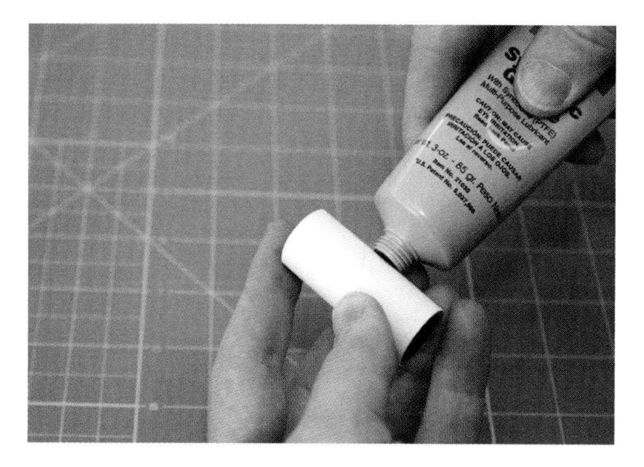

Figure 23-12. *Grease the outside of the motor liner tube.*

Slide the fuel assembly into the motor tube. This might be a tight fit, but stop if there is extremely stiff resistance. Remember, you're going to have to get this tube out again, too!

I've occasionally found that the paper tube was just a bit too big to slide into the aluminum motor tube. It's OK if you have to push a bit, but don't force the fuel assembly into the motor tube. Tear off the outer layer of paper on the motor assembly tube if the tube is too tight, then reapply grease and try again.

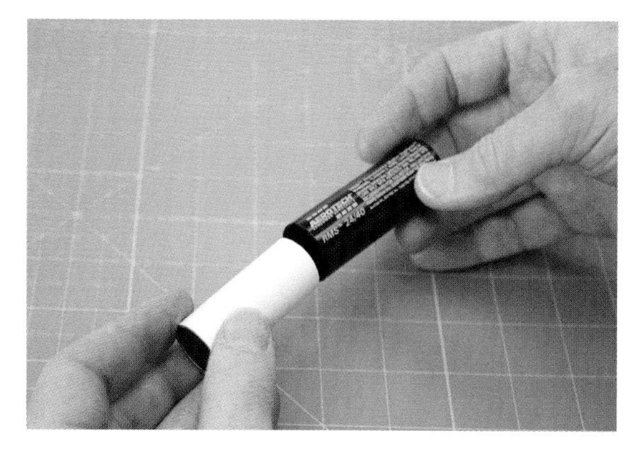

Figure 23-13. *Insert the fuel assembly into the motor tube.*

Place the forward seal disk in the motor tube, over the end with the spacer.

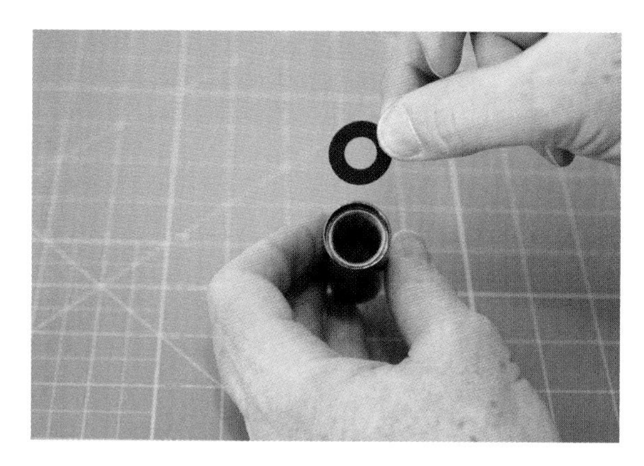

Figure 23-14. *Insert the forward seal disk.*

Follow the forward seal disk with the skinny O-ring. Press the O-ring up against the forward seal disk.

Figure 23-15. *Insert the thin O-ring.*

If the motor is new, apply grease to the outside threads of the forward closure. If it's not new, there will be some grease there from the cleaning procedure we'll go through in a moment.

Screw the forward closure in place. It goes on the end with the spacer, the forward seal disk, and the thin O-ring. It should screw all the way in.

Figure 23-16. *Install the forward closure.*

The next step, as recommended by the manufacturer, is to install the igniter. This is done now to make it easier to get the igniter into the slot in the fuel grain. High-power motors are never built this way, though. For high-power motors, the igniter is always left out of the motor until the motor is on the launch pad. Most RSOs have no problem with preinstalling the igniter in a low-power motor, but if yours does, skip this step. You can install the igniter at the pad with a lot of care and probing, or you can partially disassemble the motor at the pad and repeat the following steps to complete the assembly of the aft end of the motor.

The smaller Aerotech motors ship with the igniter shown, which is a Copperhead igniter. These are pretty temperamental, to say the least. I've had a lot of misfires with Copperhead igniters, and so have other members of my club. By now, you are probably in the habit of taking extra igniters to the launch site in case you have a misfire. I recommend against buying more Copperhead igniters as spares. Get the First Fire Jr. or the Estes Pro Series 2 igniters, instead. Do not get the larger First Fire igniters—they are far too large for this motor.

Take a close look at the igniter before installing it. There appears to only be one wire running out of the black pyrogen end. Actually, it's two flat copper plates separated by an insulator. You will need to attach one igniter lead to one side of this strip, and the second lead to the other side. We'll discuss how that's done in a moment. First, though, take a very close look at the end of the igniter—the one on the far end from the pyrogen charge. Make sure there are no burrs or imperfections that will short the igniter. If there are, trim them away.

Slide the igniter into the slot in the fuel grain. Keep it as close to the center of the fuel grain as possible, right next to the edge of the slot. Slide the igniter in until it touches the tape at the end of the fuel grain.

Figure 23-17. *Install the igniter.*

The next step is to install the nozzle. This can be a little surprising if you are not expecting it, but there may not be a hole in the nozzle—it may be blocked by a thin piece of plastic left over from the molding process. If there is a hole, it may have rough edges. Use a pencil to poke a hole through the plastic if necessary, and/or to smooth any rough edges.

Figure 23-18. *Use a pencil to smooth out any imperfections in the nozzle.*

Thread the igniter through the hole in the nozzle and slide the nozzle into the motor, letting it rest against the bottom of the fuel grain. The big end goes in first.

Figure 23-19. *Install the nozzle.*

Slide the last O-ring over the igniter so it rests against the bottom of the nozzle.

Figure 23-20. *Install the last O-ring.*

Put a little grease on the threads of the aft closure if it is new, and then thread it into the base of the motor. Be careful, because the igniter can still fall out. You may need to apply a bit of force to tighten the closure all the way, but it should meet the motor tube. Don't use tools to tighten it, though. A paper towel or rag will help; this keeps the sharp edges from the aft closure from pushing into your hand.

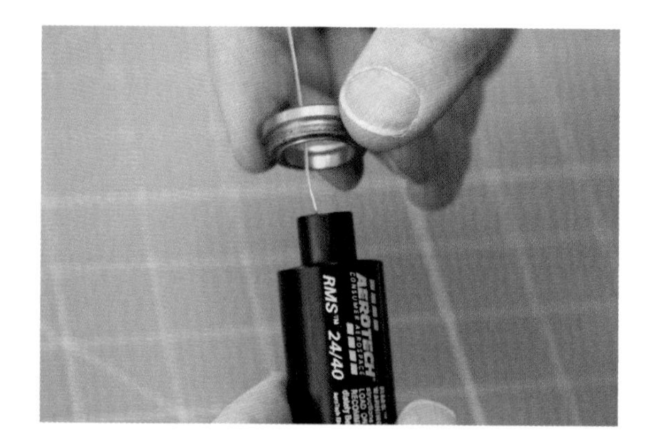

Figure 23-21. *Install the aft closure.*

The black powder ejection charge is snuggled safely in the two overlapping red plastic caps. Hold the caps and carefully remove the larger one, leaving the black powder in the smaller cap. As you do this, take care not to spill any of the black powder.

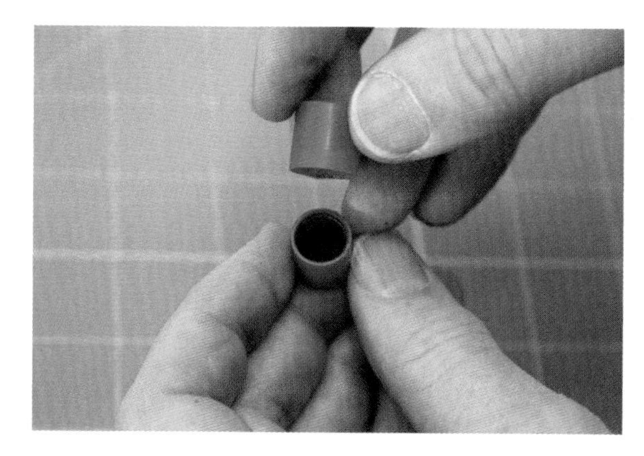

Figure 23-22. *Remove the larger plastic cap, leaving the ejection charge in the smaller cap.*

Place the ejection charge cap onto a flat surface and press the forward closure into the cap. It will fit snugly, creating a seal that will hold the black powder in place until it fires. Turn the motor over and tap gently to get some of the black powder into the forward closure, resting right on top of the delay charge.

Figure 23-23. *Insert the motor into the ejection charge cap.*

Use a hobby knife to cut a small slot in the closed end of the larger red cap. The hole should be large enough for the igniter.

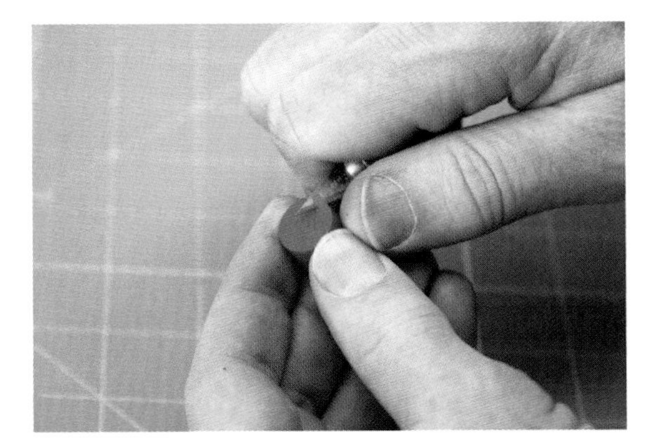

Figure 23-24. *Cut a small slot, about 1/16" long, into the large red cap.*

Directions from the manufacturer show the Copperhead igniter bent over, with the cap holding the igniter in place on the side of the nozzle. I've tried this, and judging from discussions in rocketry forums, others have, too. This seems to lead to misfires. I don't recommend installing a Copperhead igniter that way. Thread it through the hole you just cut, instead.

Slide the cap down and over the end of the nozzle. Depending on the size of the hole, the igniter may be loose enough to fall out. If it is, use a piece of tape to hold it in place.

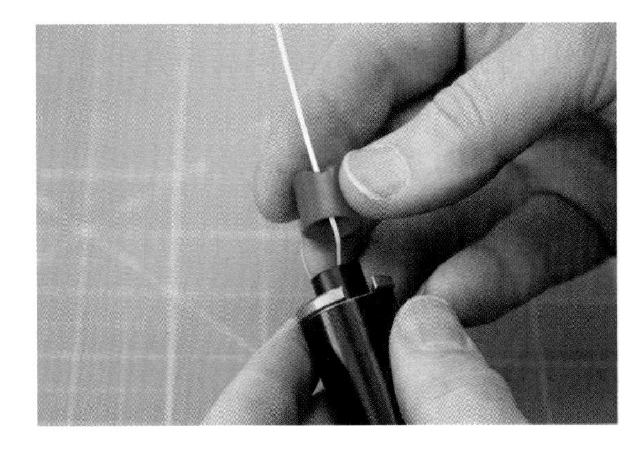

Figure 23-25. *Install the igniter cap.*

This completes the motor assembly. As with any complicated task, it's easy to miss something. Here's a checklist for this specific motor that will help you assemble it correctly each time.

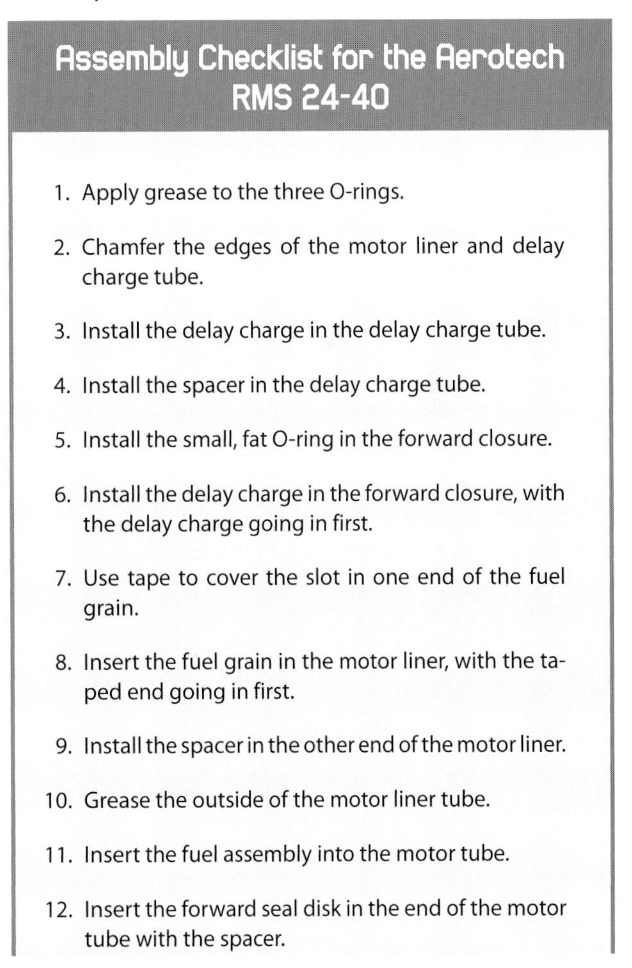

Assembly Checklist for the Aerotech RMS 24-40

1. Apply grease to the three O-rings.
2. Chamfer the edges of the motor liner and delay charge tube.
3. Install the delay charge in the delay charge tube.
4. Install the spacer in the delay charge tube.
5. Install the small, fat O-ring in the forward closure.
6. Install the delay charge in the forward closure, with the delay charge going in first.
7. Use tape to cover the slot in one end of the fuel grain.
8. Insert the fuel grain in the motor liner, with the taped end going in first.
9. Install the spacer in the other end of the motor liner.
10. Grease the outside of the motor liner tube.
11. Insert the fuel assembly into the motor tube.
12. Insert the forward seal disk in the end of the motor tube with the spacer.

13. Place the thin O-ring over the forward seal disk.
14. Thread the forward closure into the end with the forward seal disk and thin O-ring.
15. Install the igniter.
16. Use a pencil to smooth out any imperfections in the hole in the nozzle.
17. Install the nozzle, large end first.
18. Place the last O-ring over the base of the nozzle.
19. Install the aft closure.
20. Remove the larger plastic cap from the ejection charge, leaving the ejection charge in the smaller cap.
21. Insert the motor into the ejection charge cap.
22. Cut a small slot, about 1/16" long, into the closed end of the large red cap.
23. Install the igniter cap, threading the igniter through the slot cut in step 22.

Attaching Igniter Clips to a Copperhead Igniter

We've assembled the motor, but there is still the issue of connecting the igniter clips to the igniter. There are two ways to do this. The first is to apply tape to one end of the igniter clip so the clip only makes contact with one side of the copper strip. While this works, it's kind of obnoxious: the next person to use the igniter clip will need to remove your tape and clean your excess glue from the igniter clip.

The second way is shown in Figure 23-26. Put a piece of tape on each side of the copper lead, leaving some of the lead exposed on the back side of each piece of tape. Clip the leads to the igniter so one side of the clip touches the tape and the other touches one side of the igniter.

If you look at the clip, you'll see that one side of the clip is soldered to the launch wire, while the other side is fastened to the clip with a metal bar and spring. Attach the clip so the side of the clip that is soldered to the wire touches the copper igniter lead, and the other side touches the tape. This helps a little by cutting the resistance to the lead slightly.

Attach the clips as close to the rocket nozzle as practical to reduce the overall resistance.

Figure 23-26. *Attach the leads using tape so each clip only touches one side of the copper igniter wire.*

Cleaning the Motor

Single-use motors have a paper case that is a pretty good insulator. While they may be warm or even hot, it's easy enough to pull them out of the rocket right after a flight. That's not true with a reloadable motor. The metal case will be quite hot after the flight. Wait a few minutes before trying to remove the motor.

Once the motor cools enough to handle, it's time to clean it. You might leave this job until you get back from the flying field, but don't leave it for the next day. The longer you leave the motor before cleaning it, the harder it will be to clean. It's just like washing the dishes—the longer you let the mess sit, the harder it gets. The burned fuel and ejection charge are also mildly corrosive, and can damage the motor if left too long.

I find it very strange that rocket supplies are frequently mislabeled and shelved in the oddest locations in stores. My personal favorite tool for cleaning reloadable motors is called "baby wipes," and they are found in the grocery store where baby supplies are sold. They are perfectly designed for rocketeers who need a convenient package that will travel well, be easy to open, and seal easily to keep the content moist. I suppose they are convenient for wiping babies, too. An alternative to baby wipes is rubbing alcohol and paper towels. A pencil or dowel is handy for pushing out the motor liner, and needle-nose pliers can be useful for pulling out stubborn delay charge liners.

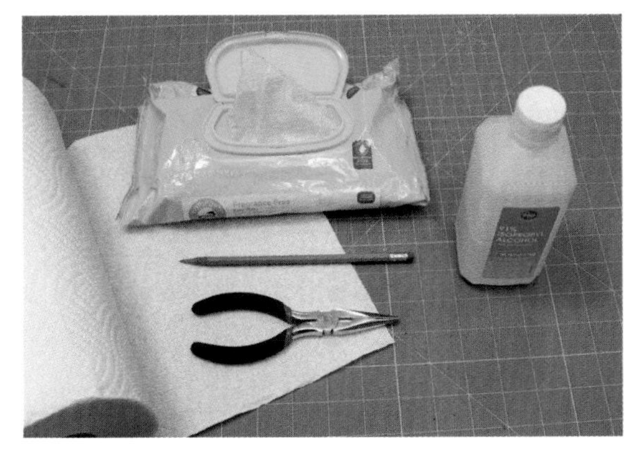

Figure 23-27. *Supplies and tools for cleaning reloadable motors.*

Begin by removing the forward and aft closures. Pull out the nozzle and dispose of it. Use a pencil, a dowel, or your finger to push the motor liner and any remaining O-rings out of the motor case. Dispose of the O-rings, motor liner, and any ashes from the spent fuel. These are safe to throw out with the normal garbage.

Figure 23-28. *After removing the forward closure, aft closure, and nozzle, use a pencil, a dowel, or your finger to push out the motor liner, O-rings, and ashes.*

Remove the delay charge liner and O-ring from the forward closure. This is usually pretty easy to do, but if it is stuck, try gripping the end of the liner with needle-nose pliers. You may also need to pry out the O-ring that was in front of the delay charge liner. Use something soft, like a toothpick. Never use sharp implements like a hobby knife, which might slip and score the metal.

Dispose of everything you remove.

Figure 23-29. *If the delay charge liner doesn't slide out easily, use needle-nose pliers to pull it out. If the O-ring stays behind, pry it out with something softer than metal, such as a toothpick.*

Use a baby wipe or a paper towel soaked in rubbing alcohol to clean the inside and outside of the motor tube, forward closure, and aft closure. Take your time, wiping all surfaces until they are perfectly clean. Be sure to clean the threads. One way to do this is to fold the wipe over your fingernail and scrape the inside of the threads with your nail. When you finish, all surfaces should be as clean as new.

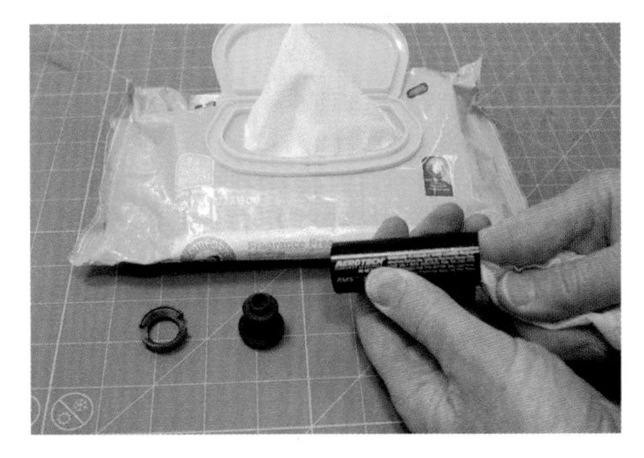

Figure 23-30. *Wipe the inside of the motor tube, especially the threads, until there is no more residue.*

Figure 23-31. *Use your fingernail to push the wipe into the thread grooves to clean them. Turn the closure as if you were screwing it in to move the wipe along the thread groove.*

Once all parts are clean, wipe some grease on the threads and replace the forward and aft closures. The motor is ready for storage and for reloading just before the next flight.

You might want to use the same motor several times in a day. While I prefer to build motors and clean them at home, it's perfectly reasonable to clean and rebuild a motor at the flying field, too. Just be sure to bring along a trash bag and all of the cleaning supplies.

Designing Your Own Rockets

This book gives you all the basic tools you need to design and build model rockets that use 40 N-s of total impulse or less—an E motor or smaller, or a combination of motors with a combined total impulse below that level. You've seen how to build rockets of all sizes and descriptions, from tiny sport rockets like Eros, which flies on an Estes T motor, to big, impressive rockets using multiple motors, either for staging or for clustering. You learned all sorts of recovery methods, from parachute to glider. You also learned a lot about how to select parachute sizes, how to design fins, how to reduce drag, and how to select an appropriate motor for a rocket. That's a lot of information. Some of it is critical, and some of it is just there to give you a deeper understanding of rocketry.

The only way to make sure you've really grasped all of the important information is to put it to use. That's what we're about to do. The last project in the book is to design and build your own rocket. We'll use this project to organize the information in the book, making sure you can put all that theory and construction experience to good use by building your own rockets.

The project I'm going to suggest is a rugged and powerful sport rocket, capable of using either disposable D and E motors like those sold by Estes, or reloadable D and E motors like those sold by Aerotech. It will essentially be a scaled-up version of Juno that can take the larger motors. As with Juno, you'll be able to add a payload bay later, as long as the body tube size is not larger than the size you select for the rocket.

Of course, you don't have to build the rocket I'll describe. Maybe you're on a TARC team, and you've been reading this book to figure out how to build a winning rocket. Or perhaps you have a small camera you would like to fly to photograph your school or house. Whatever project you choose, you can follow the steps we'll go through to design and build a successful model rocket.

Collecting the Requirements

Every successful engineering project begins with a requirements phase. This can be very formal or very simple, depending on how many people are involved. For a model rocket, this means deciding what the rocket will do, what size motor it will use, and what recovery system it will use. If the goal is to break your club's altitude record for a C motor, you'll be building a minimum-diameter rocket with the thinnest possible fins. You will need to decide on the shape and size of the fins, as well as whether they will be sanded to an airfoil. You will need to pick the shortest body tube that will give you a stable design, trading off between fin size and body tube length to minimize drag while maintaining stability.

If there is more than one person involved, it's also convenient to pick a name early in the design phase so you have something to call the project. All of the other solid propellant rockets in this book have been named for asteroids. This one is beyond the scope of the predesigned rockets in the book, so let's reach a bit further and call this one Eris, after the largest known trans-Neptunian minor planet.

Eris

Eris is a big part of the reason that Pluto was demoted from its status as the ninth planet.

The team of Mike Brown, Chad Trujillo, and David Rabinowitz announced the discovery of Eris on January 5, 2005. Eris has a moon named Dysnomia, which gives a good way to calculate the mass of Eris. Based on those calculations, the mass of Eris was estimated to be about 27% larger than the mass of Pluto, making Eris the largest trans-Neptunian object known. Later observations put its size close to that of Pluto, so either it is denser than Pluto, or there is some error in the observations. Still, for a time, NASA and its discoverers called Eris the tenth planet.

Discovery of two other large objects was announced at about the same time. They were smaller than Pluto, but still big objects. Other large objects are probably out there, too, in the asteroid belt beyond Neptune. With these new discoveries, the outer solar system was getting crowded. Were these new bodies really planets? This led to an agonizing reappraisal by the International Astronomical Union, centered around the existential question, "What *is* a planet?" The debate raged for a while, and in the end, all of the objects in the trans-Neptunian asteroid belt were designated as minor planets, thereby stripping Pluto of its title as a planet.

The conflict over the reclassification of Pluto makes it particularly fitting that Eris is the Greek goddess of strife and discord.

Eris will be a sport rocket that will use D or E motors, which are 24 mm in diameter. These fit nicely in a BT-50 body tube. Since this is a rugged rocket, we'll use a BT-55 body tube for the airframe. The rocket will use parachute recovery. Eris is a bigger version of Juno.

One nice thing about the choice of a BT-55 body tube (for those who have been following along, building the rockets in this book from the Estes Designer's Special) is that you have a leftover BT-55 body tube and nose cone. Reusing available parts is a great way to keep a project's budget under control. It's a technique NASA and other large players use. NASA's Mercury-Redstone, Mercury-Atlas, and Gemini-Titan all used recycled military missiles for their boosters. Even the Saturn IB used recycled motors and tanks from earlier rockets.

Initial Design

With the requirements in place, it's time to lay out an initial design for the rocket. This means picking a body tube length and nose cone; deciding on the number, size, and shape of the fins; and selecting an initial parachute size. The best way to do this is to create the design in a rocket simulator. Both RockSim and OpenRocket have design tools that help you quickly create a rocket design. You'll need to simulate the design to make sure it is stable anyway, so why not set up the design in the simulator right away?

You've seen rocket simulators before, in Chapter 7 and Chapter 14. You should find it pretty easy to crank up OpenRocket or another simulator and put the rocket together. I want to make sure you know how to design the rocket, but I don't want to design it for you. Instead, I'll show you how to use OpenRocket to design a rocket like Eris from scratch.

Sizes in OpenRocket are given in centimeters by default. You might want to switch the units to inches in the preferences dialog. That will also make it easier to follow along with this example, which shows all measurements in inches.

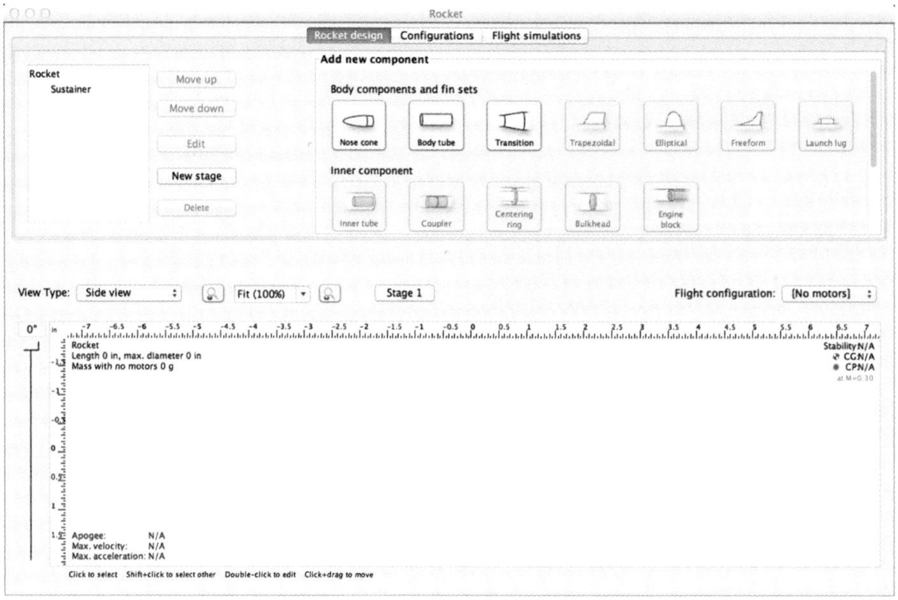

Figure 23-32. *OpenRocket starts with a window for a new rocket. You can also create a new rocket with the New menu command.*

OpenRocket starts with a window for a new rocket. The first step is to add the various components that make up the airframe. Working from left to right in the components area, we start with a nose cone. Clicking the nose cone icon opens a dialog that shows the design parameters for this component (Figure 23-33).

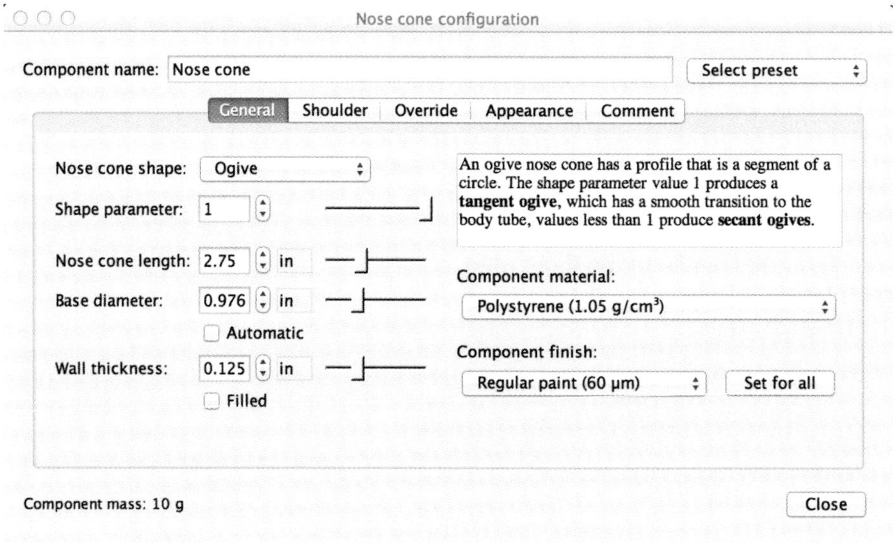

Figure 23-33. *Fill in the values for the nose cone you have selected. The values shown are for Juno.*

Fill in the values for the nose cone you have selected. There are two parts to a nose cone, the tip and shoulder. The General tab we started on only includes the tip. Click on the Shoulder tab, shown in Figure 23-34, and fill in the appropriate values for the shoulder. So what are the appropriate values? One way to find out is to measure the nose cone. You might also find dimensions on the manufacturer's website.

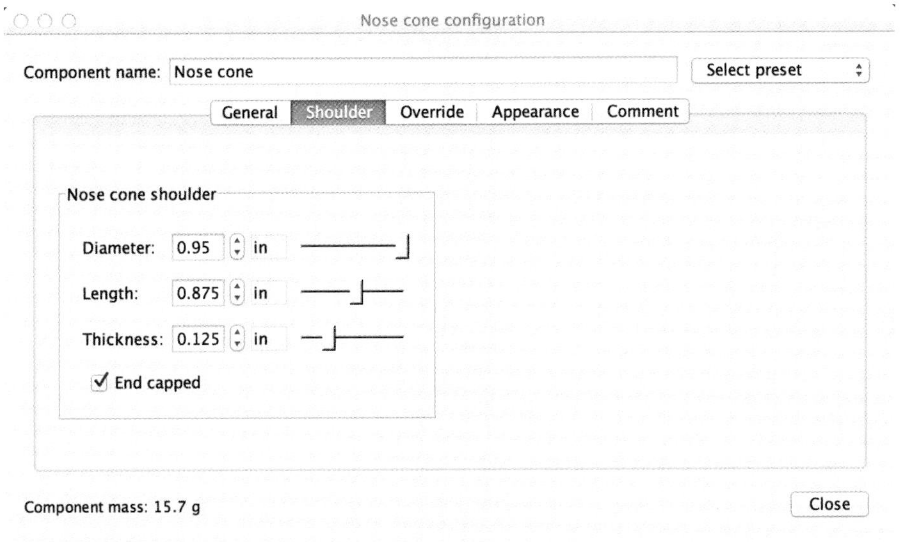

Figure 23-34. *The part of the nose cone that fits inside the rocket's body tube is the shoulder. Click on the Shoulder tab to fill in the proper values for your nose cone.*

Close the nose cone dialog and click on the body tube icon to add a body tube to the rocket. This opens a dialog where you can enter the size of the body tube (Figure 23-35). You can find common body tube sizes back in Table 3-4 from Chapter 3. Manufacturers will also provide details on the sizes and materials of their body tubes, so you can check the manufacturer's website for less common body tube sizes. Fill in the values for your rocket and close the dialog.

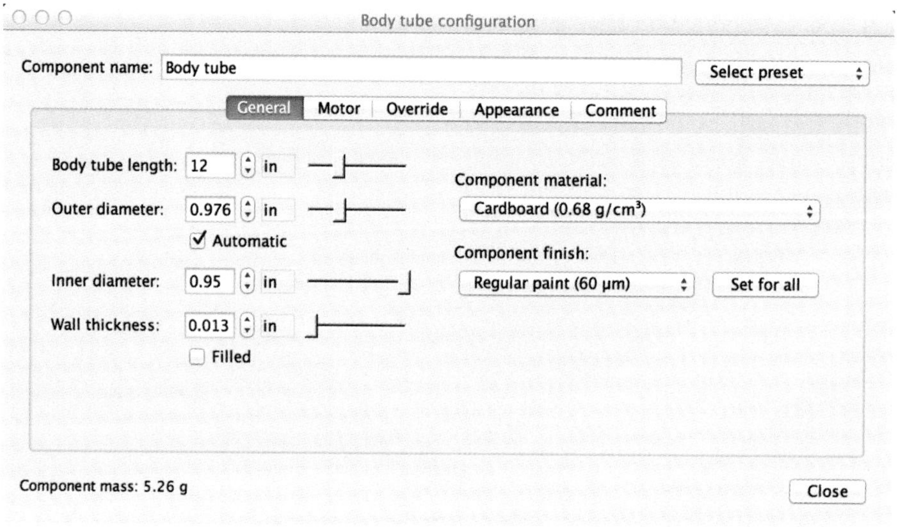

Figure 23-35. *Click on the body tube icon and fill in the appropriate values for your rocket's body tube.*

Fins are next. You want to add the fins to the body tube, not the nose cone, so be sure and select the body tube by clicking on it in either the list of components or the image. Then click on the fin icon that matches the shape you've selected and fill in the values for your rocket in the resulting dialog (Figure 23-36). Don't forget the values on the right. You need to select the proper fin material, fin cross section, thickness, and finish, too.

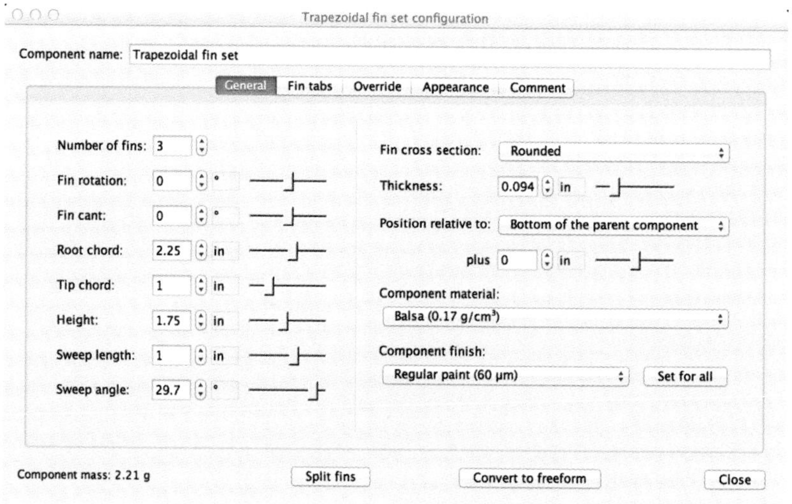

Figure 23-36. *There are several basic fin shapes, each with its own icon and fin dialog. This one is for a trapezoidal fin, the most common shape used in this book and the shape of Juno's fins. Other fin shapes will have different dialogs, but a little exploration should be all you need to create appropriate fins.*

Follow the same pattern to fill in the details for the launch lug (Figure 23-37). Again, the launch lug is attached to the body tube, so be sure to select the body tube by clicking on it before clicking on the icon for the launch lug.

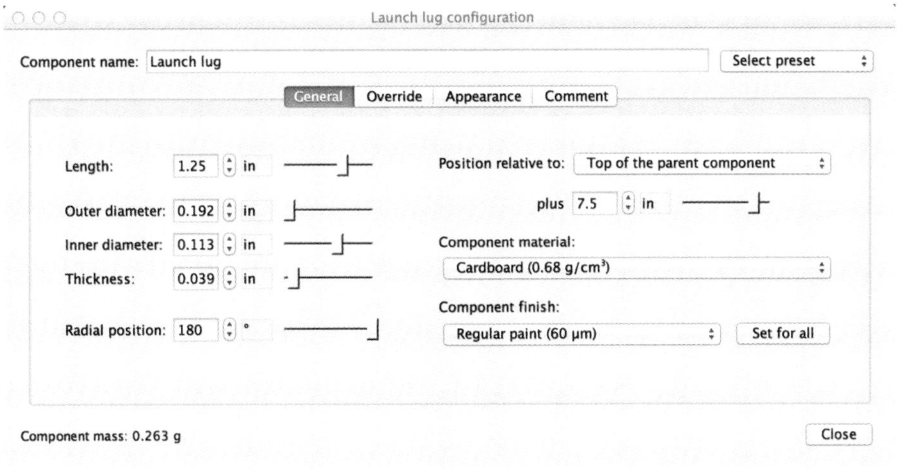

Figure 23-37. *Add the launch lug to the body tube. The values shown are for a 1/8" launch lug. Is that what you will use? If not, select an appropriate size.*

With the outside of the rocket complete, it's time to turn our attention to the interior. Select the body tube, and then click on the inner tube icon.

There are two tabs where values need to be set, and a few others you may want to look through for other options. The first is the General tab, where you set the tube size and length (Figure 23-38).

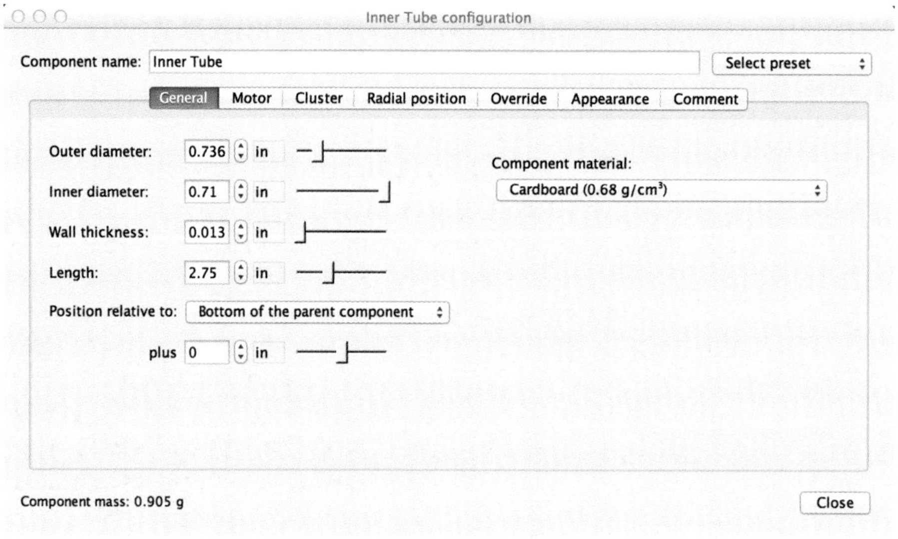

Inner Tube configuration

Component name: Inner Tube | Select preset

General | Motor | Cluster | Radial position | Override | Appearance | Comment

Outer diameter: 0.736 in

Inner diameter: 0.71 in

Wall thickness: 0.013 in

Length: 2.75 in

Position relative to: Bottom of the parent component

plus 0 in

Component material:

Cardboard (0.68 g/cm³)

Component mass: 0.905 g

Close

Figure 23-38. *Create an inner tube that will be used for the motor mount.*

This tube will be used as a motor mount, so click on the Motor tab (Figure 23-39) and select the checkbox indicating that this is a motor mount. The motor on Juno is also designed to hang 1/4" out of the rear of the rocket, so the motor overhang value needs to be set to 0.25".

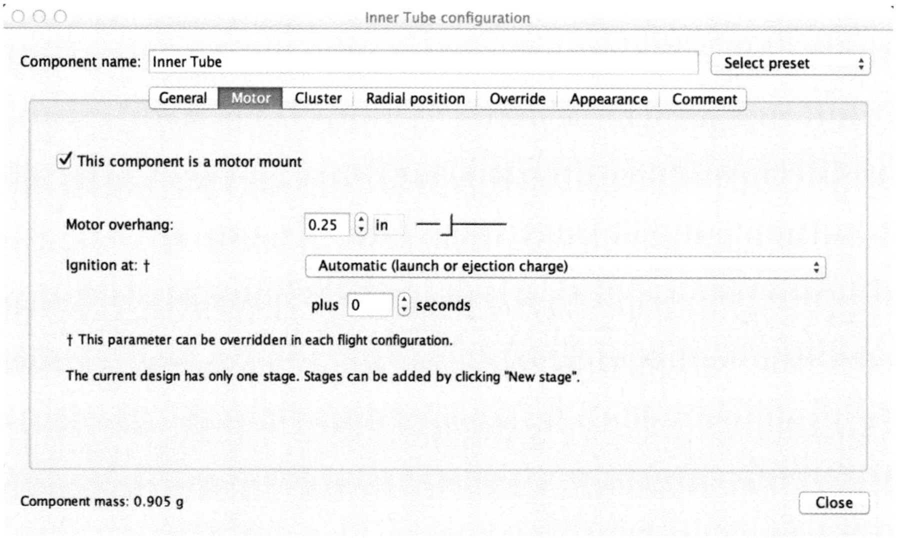

Inner Tube configuration

Component name: Inner Tube | Select preset

General | Motor | Cluster | Radial position | Override | Appearance | Comment

✓ This component is a motor mount

Motor overhang: 0.25 in

Ignition at: † Automatic (launch or ejection charge)

plus 0 seconds

† This parameter can be overridden in each flight configuration.

The current design has only one stage. Stages can be added by clicking "New stage".

Component mass: 0.905 g

Close

Figure 23-39. *Select the motor mount checkbox.*

The motor mount is held in place with centering rings. Close the motor mount dialog, select the body tube again, and then add a centering ring. Fill in the values as shown in Figure 23-40. The outer and inner diameter will be filled in automatically when the centering ring is positioned properly. Add a second centering ring near the other end of the motor mount.

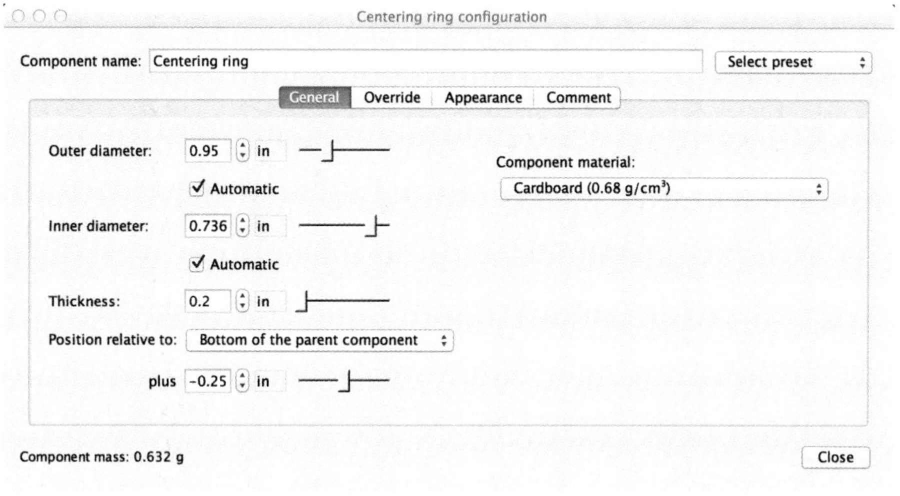

Figure 23-40. *Add two centering rings to the body tube.*

Sometimes your rocket will have a component that isn't on the list of components. These components still matter, since their mass affects both the center of gravity and the overall mass of the rocket. The engine hook is a good example. We add these components as mass objects—things that have a general shape and mass, but otherwise have no impact on the rocket simulation.

Figure 23-41 shows how to configure a mass component for the engine hook. The specific size is not all that important, as long as the overall mass and the location of the center of mass are reasonably accurate. It might be tough to find the mass of a component, though, especially a light one. You may find the mass on a manufacturer's website, but how can you accurately measure the mass of something like an engine hook that is very light, when typical scales may only measure to the closest gram? One way is to weigh 10 or so of the objects at once and divide the overall weight by the number of objects.

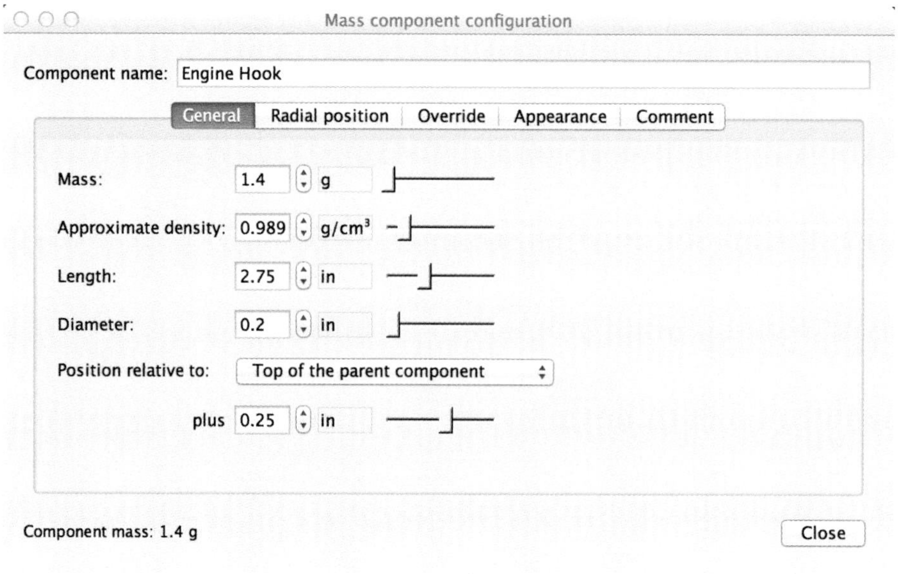

Figure 23-41. *Add an engine hook as a mass object.*

Follow the same pattern to add a parachute and shock cord to the body tube, as seen in Figures 23-42 and 23-43. While they tend to be near the nose cone when packed, I have always suspected they slide down toward the motor when the motor fires. Putting the mass of the parachute and shock cord lower in the rocket is the conservative choice, so that's where I put them.

If the rocket is stable with the parachute near the motor, it will still be stable if the parachute actually stays near the nose cone, since shifting the center of gravity forward makes the rocket more stable.

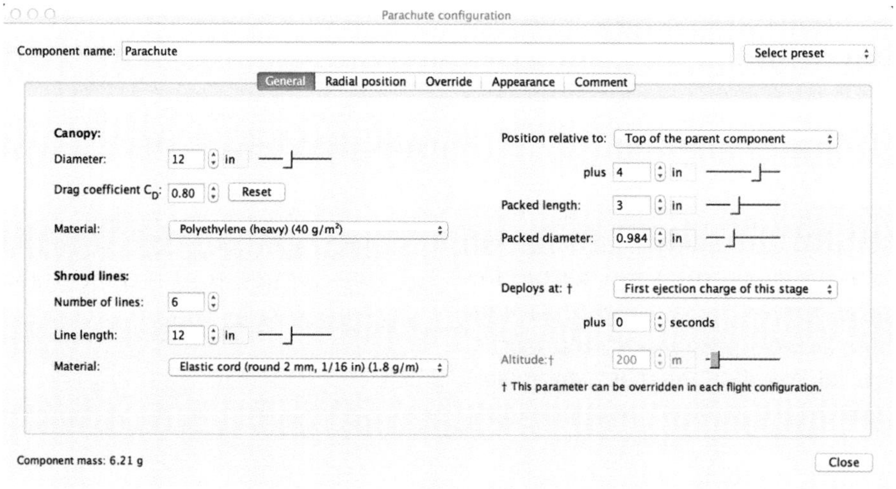

Figure 23-42. *Add a parachute.*

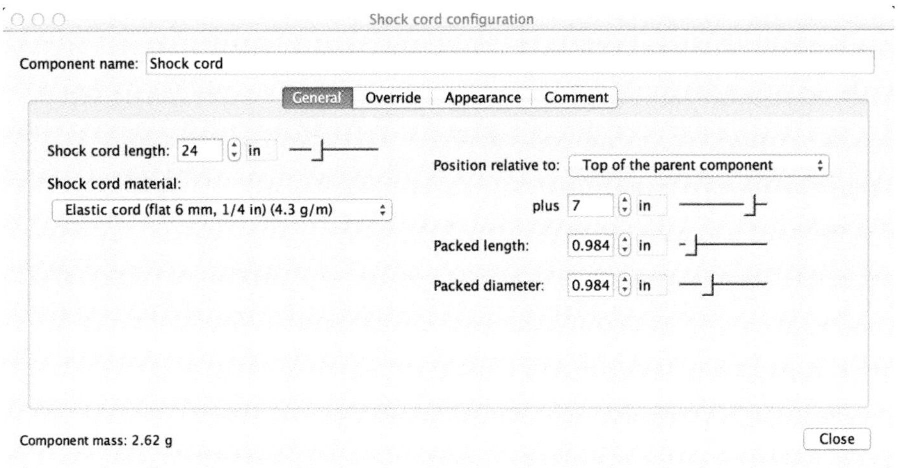

Figure 23-43. *Add a shock cord.*

Simulating the Rocket

The reason for designing the rocket in OpenRocket was to make it easier to run the simulation. At this point you should check the performance of the rocket. Of course, you need to select a motor, and the weight and power of the motor could affect whether the rocket is stable. What if you want to use A motors, but they are not powerful enough to lift the rocket off the pad quickly enough? Or what if you want to use a C motor, but the weight of the C motor makes the rocket unstable? I've seen both of these situations with rockets. A common engineering practice in this situation is to bracket the problem. The simulation needs to be done with two motors: the least powerful you intend to use and the heaviest you intend to use. For Eris, this means Estes D and E motors.

We covered how to check the center of gravity back in Chapter 7. Once the motor is loaded into the simulation, it's right there on the main screen, above and to the right of the rocket, as seen in Figure 23-44. The rocket will be stable as long as the Stability factor is at least 1. A value of 1 to 2 is ideal. While values over 2 are acceptable, they indicate an overstable rocket that might have trouble in high wind. Check for stability with the heaviest motor. If that one works, lighter motors will, too.

You will need to make design changes if the rocket is not stable. You can make the fins larger, add more fins, or move them further toward the rear of the rocket, perhaps by using swept fins. You can also make the body tube longer, moving the center of gravity toward the nose of the rocket. Another alternative is to add weight to the nose.

The ground hit velocity tells you how fast the rocket will be traveling when it hits the ground, which tells you if the parachute size is appropriate. When you bracket engines, this value won't change much unless the motor sizes are dramatically different, such as changing from a 2 3/4"-long D motor to a 3 3/4"-long E motor, or unless the payload weight changes. The landing speed should be between 3.5 m/s and 4.5 m/s. Try a larger or smaller parachute if the speed is well outside this range. Select the parachute that gives a landing speed closest to this range.

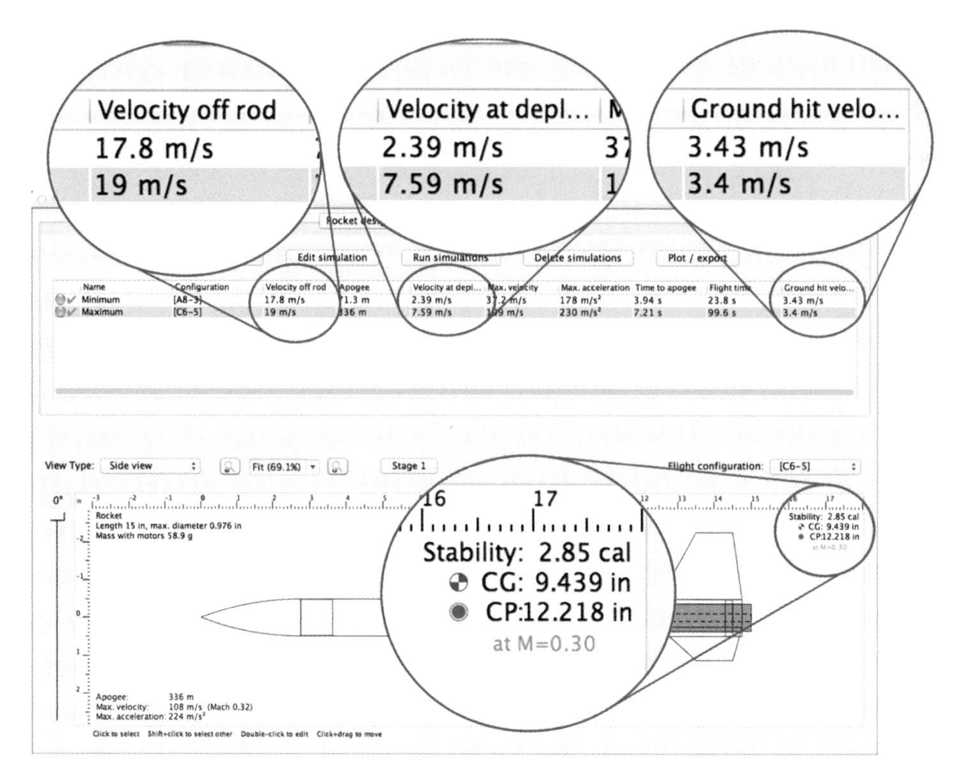

Figure 23-44. *After loading a motor and simulating a flight, check the stability, velocity off of the launch rod, velocity at parachute deployment, and ground hit velocity.*

The remaining two values change a lot from motor to motor, so you need to check them for each motor you plan to use in the rocket. The first is velocity off of the rod. As you know from Chapter 7, the rocket should be traveling at 30 mph, or about 13.4 m/s, when it leaves the rod. Rockets sometimes fly successfully coming off the rod a bit slower, but it is best to stick with this speed or a higher speed whenever possible. If the rocket is not leaving the launch rod at an appropriate speed, you will need to either reduce the weight of the rocket, lengthen the launch rod, or increase the initial thrust of the rocket.

The last thing to check is the velocity at deployment. This is the speed the rocket is traveling when the parachute opens. The best way to check this value is to try all of the available delay times for a particular motor. Choose the one with the lowest speed at ejection. If two ejection charges have values for speed at deployment that are close, it is likely that the rocket is traveling up when the ejection charge fires for the shorter delay, and down for the longer delay. In that case, you will generally pick the longer delay time to allow the rocket to reach the maximum possible altitude.

Not all simulators do this, but OpenRocket has another nice feature. There is a colored dot to the left of each simulation. This indicates if OpenRocket found any issues. It will, for example, turn red if the rocket is not stable. As engineers, we should always check the values ourselves, but it's nice to have the software double-check our results.

Construction

Once the design is complete, it's time to build the rocket. You're well prepared to build it after reading this book and building a few of the rockets.

Once you have assembled the rocket, double-check the simulation results for center of gravity. If you can, check the overall mass, too. You can check the mass on a kitchen or school laboratory scale. It's easy to check the location of the center of gravity by installing the parachute and motor, then balancing the rocket on your finger. You can change the simulation if the measured values are significantly different from the simulated values, and then rerun the simulations to make sure the rocket will fly safely.

Flight Testing

Simulations have their place, but you should always test the rocket in flight. The general practice is to fly the rocket on the smallest possible motor for the first flight. That way, if there are any mistakes in the design or the build, they're not amplified by the power of a large motor! It's also important to test the rocket with the heaviest motor to check its stability.

Test flights should be done with the smallest possible crowd. You should make sure all of the spectators are aware the launch is a test flight, and make sure they are all standing, facing the rocket, and paying attention when it is launched. At our club launches, the LCO yells "Heads up!" for any test launch so people are aware they should pay closer than normal attention.

If you've done everything carefully up to this point, pushing the launch button should just confirm your engineering skills with a flawless flight. On the occasions when that doesn't happen, take the time to identify the cause of the failure. Decide right away what you will do differently in the future to prevent the same failure. Everyone makes mistakes, and everyone has failures. The mark of a good engineer is learning from those mistakes so they don't happen again. The mark of a great engineer is carefully following accepted practice and watching other people's mistakes so very few mistakes are made in the first place.

Odd Rockets

Up to this point, all of the rockets in this book have been relatively normal, and each has had a purpose. They were carefully chosen to teach some aspect of rocketry, demonstrating unique recovery methods, different kinds of payloads, or how small or large model rockets can be.

Enough with normal—let's have some fun!

Odd rockets, sometimes called *oddrocs*, are rockets that don't look like rockets at all. You can make anything fly if you have enough power, and if the center of gravity is far enough ahead of the center of pressure. One example is a Styrofoam Easter egg. Drill it out for a motor and launch lug and add some 18"-long dowels, and it flies quite nicely. The rocket is very light once the motor pops out, so featherweight recovery works well. You can come up with lots of holiday or other unusually themed rockets.

Figure 23-45. *This rocket uses a foam Easter egg from a craft store and 18"-long dowels for fins. A little paint changed it into a nice alien.*

Fins come in all sorts of odd configurations. A favorite at many launches is called a *tube rocket*, where a cluster of body tubes replaces the traditional fin. Be a bit careful when designing these rockets, though. The tubes add a significant amount of weight to the rear of the rocket. You don't want to make the fin tubes very long, as that would move the center of gravity back far enough to make the rocket unstable.

Figure 23-46. *Tube rockets use body tubes in place of fins. They are fun rockets at any launch.*

When I said you could make anything fly with enough power and an appropriate choice of centers of gravity and pressure, I meant that literally. Figure 23-47 shows a great example—it's a flying Tardis, paying homage to the *Dr. Who* TV series.

Figure 23-47. *Anything will fly if you use a big enough motor and make sure the center of gravity is far enough ahead of the center of pressure. Rocket and photo by Marc Bonem.*

Some model rockets aren't that odd, but they are very cool. Scale rockets fall into that category. Researching a rocket to build a good scale kit is a great way to learn about the history of space flight, as well as technical aspects of aerodynamics and astrodynamics. Figure 23-48 shows a rocket that is both an odd rocket and a scale model—it's a scale model of the very first liquid-propelled rocket ever built. Robert Goddard flew the original on March 16, 1926. This rocket is a kit produced by FlisKits (*http://www.fliskits.com/products/rocketkits/kit_detail/nell.htm*).

Figure 23-48. *The FlisKits scale model of the first liquid-fueled rocket. Model built and flown by Tom Tweit. Photo by Roger Smith of JonRocket, provided courtesy of JonRocket.com.*

Finally, let's take a look at an odd use for an otherwise normal rocket. Figure 23-49 shows the garter toss at the wedding of Steve and Susan Foy. The rocket is painted like a tuxedo, and carried the garter! A second rocket, white with silver lace, carried a part of the bouquet for the bouquet toss. It was quite a sight seeing guys in tuxedos and women in dresses vaulting fences to chase down the rockets!

Figure 23-49. *The garter toss at the wedding of Steve and Susan Foy.*

The Outer Limits

Rocketry can take you in many different directions. One person in our club is really good at photography from rockets. Another is an expert on the V2. Another has published papers on supersonic flight based on high-power rockets that fly at around Mach 3.5.

Whether rocketry takes you in one of these directions or somewhere totally different, I wish you the best in your experiences.

Programs from the Book

A

This appendix contains the source code for the programs in the book. You can also find the source code online at the author's website (*http://bit.ly/byteworks-make-rockets*), as well as built into techBASIC and techBASIC Sampler. Look in the *Maker Books* folder in the techBASIC *Programs* directory.

techBASIC, QBASIC, and Other BASICs

techBASIC is a technical programming environment that uses the familiar BASIC programming language. The original version of BASIC and ANSI Standard BASIC both include extensive mathematical features, such as arrays as built-in data types and mathematical operations like + and * that work on matrices. Microsoft removed these features when it created small BASIC interpreters for some of the original microcomputers, like the Apple][and IBM PC. techBASIC is a technical programing language, so it puts them back in.

The programs in this appendix all run under techBASIC. One program uses a graphical user interface to show how a simple rocket simulator can easily be ported to a modern platform. That sample, called SRockS (Simple Rocket Simulator), needed to be developed for a specific implementation of BASIC, since a graphical user interface is generally tied to a specific platform. All of the other programs, including a text-based version of SRockS, also run in QBASIC, and should run with little or no modification in pretty much any modern BASIC that supports subroutines.

Find out more about techBASIC at *http://www.byteworks.us/Byte_Works/techBASIC.html*. You can find out more about QB64, a free implementation of QBASIC, at *http://www.qb64.net*.

Juno Simulation

This text-based BASIC program appears in Chapter 14. It is a one-dimensional simulation of the flight of Juno. The program will run in almost any implementation of BASIC with little or no change.

You can easily change the program to simulate other rocket flights. The lines to edit appear near the top of the listing.

`rocketMass` is the mass of the rocket in kilograms. The best way to get this value is to weigh the rocket on a kitchen scale. The weight includes the weight of the parachute and recovery wadding, but not the weight of the motor:

```
REM Rocket mass in kg, w/o the motor.
rocketMass = 0.036
```

You can calculate the coefficient of drag using the techniques in Chapter 13. You can also estimate the C_d. Use a value of 0.4 to 0.6 for most rockets, with 0.4 reserved for polished rockets with airfoil fins. Go higher if the rocket has lots of decorations hanging off it:

```
Cd = 0.524: REM Coefficient of drag.
```

The frontal area of the rocket is the area at the base of the nose cone. Given the diameter of the base of the nose cone, the area is:

$$A_r = \pi \frac{d^2}{4}$$

The area is in square meters:

```
REM Frontal area of the rocket.
Ar = 0.000483
```

ρ (rho) is the density of the atmosphere in kilograms per cubic meter. The default value is the air density at sea level:

```
rho = 1.225: REM Density of the air.
```

Subtract about 4% per 1,000 feet if your launch site is significantly above sea level.

dt is the simulation time step. Lower numbers are generally more accurate, but make the program take longer to run:

```
dt = 0.01: REM Simulation time step.
```

motorName$ is the filename of a motor file in RASP format. You can find these files at *http://www.thrustcurve.org*. There are three preloaded in techBASIC (the other two are *Estes_A8.eng* and *Estes_B6.eng*):

```
motorName$ = "Estes_C6.eng"
```

You can install other motor files using iTunes; see the techBASIC Quick Start Guide for details. You can find the techBASIC Quick Start Guide at the techBASIC documentation page (*http://bit.ly/techBASIC-doc*).

Example A-1 lists the complete source for the Juno Simulation program.

Example A-1. Text-based rocket simulator

```
REM Simple Text-Based Rocket Simulator.
REM
REM This simulation is set up to simulate a sea-level flight of Juno,
REM a rocket in the book "Make: Rockets, Down to Earth Rocket Science."
REM Change the variables in the next section to simulate a different rocket
REM rocket flight. See the book for details on how this simulator works.

DIM SHARED thrust(50), time(50), mass(50)
DIM SHARED propellantWeight, motorWeight
DIM SHARED bounds AS INTEGER
DIM SHARED tokens(20) AS STRING, tokenCount AS INTEGER

debug = 0

REM Initialize the rocket data. Change these values for different rockets,
REM rockets, motors, or flight conditions.
DIM rocketMass, d, d0, v0, a, fg, fd, ft, t, f, dt, Ar, rho, Cd
REM Rocket mass in kg, w/o the motor.
rocketMass = 0.036
Cd = 0.524: REM Coefficient of drag.
REM The Frontal area of the rocket.
Ar = 0.000483
rho = 1.225: REM Density of the air.
dt = 0.01: REM Simulation time step.

motorName$ = "Estes_C6.eng"
```

```
REM Load the motor data.
CALL LoadMotor (motorName$)
IF debug THEN
  PRINT propellantWeight, motorWeight
  FOR i = 1 TO bounds
    PRINT time(i), thrust(i), mass(i)
  NEXT
END IF

REM Run the simulation.
MaxA = 0.0
MaxV = 0.0
fg = (rocketMass + motorWeight)*9.81
DO
  t = t + dt
  fd = Cd*rho*v0*v0*Ar
  ft = Interpolate(t, time(), thrust())
  f = ft - fd - fg
  a = f/(rocketMass + Interpolate(t, time(), mass()))
  IF a > MaxA THEN MaxA = a
  d = d0 + v0*dt + a*t0*t0/2
  d0 = d
  v0 = v0 + a*dt
  fg = (rocketMass + Interpolate(t, time(), mass()))*9.81
  IF v0 > MaxV THEN MaxV = v0
  IF debug THEN PRINT t, d, v0, a
LOOP UNTIL (t > 1) AND (v0 < 0)

REM Print the results.
PRINT "Maximum altitude = "; d; " meters."
PRINT "Maximum velocity = "; MaxV; " m/s."
PRINT "Maximum acceleration = "; MaxA; " m/s^2."
PRINT "Ideal ejection delay = "; t - time(bounds); " seconds."
END


REM Use a specific time to interpolate in a table of time-based values.
REM
REM Inputs:
REM    t - The desired time.
REM    time - An array of times, from lowest to highest.
REM    value - A table of values, one for each elemnt element in the time array.
REM
REM Returns: A linearly interpolated value from the value array. If t is
REM    less than the first time value, the first value in the value array
REM    is returned. If t is greater that than the last time, the last value in
REM    the value array is returned.

FUNCTION Interpolate (t, time(), value())
IF t < time(1) THEN
  Interpolate = value(1)*t/time(1)
ELSE
  IF t < time(bounds) THEN
    DIM index AS INTEGER
    index = 1
    WHILE time(index) < t
      index = index + 1
```

```
    WEND
    Interpolate = value(index - 1) + _
      (t - time(index - 1))*(value(index) - value(index - 1))_
      /(time(index) - time(index - 1))
  ELSE
    Interpolate = value(bounds)
  END IF
END IF
END FUNCTION


REM Load a motor file.
REM
REM Parameters:
REM    fileName - The name of the motor file.
REM
REM Global values set:
REM    propellantWeight - The total weight of the propellant in kg.
REM    motorWeight - The total weight of the motor, with propellant
REM    in kg.
REM    time - The times for which thrust data is available.
REM    thrust - The thrust at each time, in newtons.
REM    mass - The motor mass values at each time step.

SUB LoadMotor (fileName AS STRING)
DIM aLine AS STRING, count AS INTEGER

REM Scan the file, reading the data.
OPEN fileName FOR INPUT AS #1
bounds = -1
WHILE NOT EOF(1)
  LINE INPUT #1, aLine
  IF LEN(aLine) > 0 AND LEFT$(aLine, 1) <> ";" THEN
    CALL Parse(aLine)
    bounds = bounds + 1
    IF bounds = 0 THEN
      propellantWeight = VAL(tokens(5))
      motorWeight = VAL(tokens(6))
    ELSE
      time(bounds) = VAL(tokens(1))
      thrust(bounds) = VAL(tokens(2))
    END IF
  END IF
WEND
CLOSE #1

REM Find the total impulse.
DIM impulse
impulse = TotalImpulse(time(), thrust(), bounds)
IF debug THEN PRINT "Total Impulse = "; impulse

REM Assume the mass loss is proportional to the impulse. Find
REM the motor mass as a function of time.
DIM currentMass, t0, f0, di, i AS INTEGER
currentMass = motorWeight
FOR i = 1 TO bounds
  di = (f0 + thrust(i))*(time(i) - t0)/2
```

```
  currentMass = currentMass - propellantWeight*di/impulse
  mass(i) = currentMass
  f0 = thrust(i)
  t0 = time(i)
NEXT
END SUB


REM A utility subroutine for LoadMotor that breaks a line into space-
REM delimited strings. The tokens are placed in the global array tokens.
REM
REM Parameters:
REM    aLine - The line to parse.

SUB Parse (aLine AS STRING)
REM Declare and initialize our local variables.
DIM index AS INTEGER
tokenCount = 0

REM Strip any leading space from the line.
WHILE LEFT$(aLine, 1) = " "
  aLine = RIGHT$(aLine, LEN(aLine) - 1)
WEND

REM Collect the tokens.
WHILE LEN(aLine) > 0 AND tokenCount < 10
  tokenCount = tokenCount + 1
  index = SpaceIndex(aLine)
  IF index = 0 THEN
    tokens(tokenCount) = aLine
    aLine = ""
  ELSE
    tokens(tokenCount) = LEFT$(aLine, index - 1)
    aLine = RIGHT$(aLine, LEN(aLine) - index)
    WHILE LEFT$(aLine, 1) = " "
      aLine = RIGHT$(aLine, LEN(aLine) - 1)
    WEND
  END IF
WEND
END SUB


REM A utility function for Parse that finds the index of the next space
REM character.
REM
REM Parameters:
REM    sval - The string.
REM
REM Returns: The index of the next space, or 0 if there are none.

FUNCTION SpaceIndex (sval AS STRING)
DIM index AS INTEGER, fund AS INTEGER
found = 0
index = 1
WHILE (index < LEN(sval)) AND (found = 0)
  IF MID$(sval, index, 1) = " " THEN
    found = 1
```

```
  ELSE
    index = index + 1
  END IF
WEND
IF found = 0 THEN
  index = 0
END IF
SpaceIndex = index
END FUNCTION


REM Find the total impulse of the motor.
REM
REM Parameters:
REM     time - The time steps where motor thrust is available.
REM     thrust - The thrust at each time step.
REM     bounds - The number of data points in time() and thrust().
REM
REM Returns: The total impuls impulse in newton-seconds.

FUNCTION TotalImpulse (time(), thrust(), bounds AS INTEGER)
DIM t0, f0, impulse, i AS INTEGER
FOR i = 1 TO bounds
  impulse = impulse + (f0 + thrust(i))*(time(i) - t0)/2
  t0 = time(i)
  f0 = thrust(i)
NEXT
TotalImpulse = impulse
END FUNCTION
```

SRockS

The Simple Rocket Simulator (SRockS) is essentially the Juno Simulation from the previous section with a user interface for the iPhone and iPod touch. It will also run on the iPad, of course. It is preinstalled in techBASIC and techBASIC Sampler in the *Maker Books* folder.

The rocket mass is given in kilograms. The best way to get this value is to weigh the rocket on a kitchen scale. As before, the weight includes the weight of the parachute and recovery wadding, but not the weight of the motor.

You can calculate the coefficient of drag using the techniques in Chapter 13. You can also estimate the C_d. Use a value of 0.4 to 0.6 for most rockets, with 0.4 reserved for polished rockets with airfoil fins. Go higher if the rocket has lots of decorations hanging off it.

Figure A-1. SRockS is a one-dimensional rocket simulator available for the iPhone, iPod touch, and iPad.

The area of the rocket is the frontal area, at the base of the nose cone. Tap on the value to select a new value. A list of common body tube sizes appears. Select the appropriate body tube size and the frontal area will be filled in automatically. The area is in square meters.

Atm. density is the density of the atmosphere in kilograms per cubic meter. Subtract about 4% per 1,000 feet if your launch site is significantly above sea level.

The time step is the simulation time step. Lower numbers are generally more accurate, but make the program take longer to run.

Tap the motor value to select a different motor. A picker will appear that will show all of the installed motor files. Select the new motor and tap the Done button. You can find additional motor files at *http://www.thrustcurve.org*. Follow the instructions in the techBASIC Quick Start Guide to install new motor files using iTunes. You can find the techBASIC Quick Start Guide at the techBASIC documentation page (*http://bit.ly/techBASIC-doc*).

Example A-2 shows the complete source for the SRockS program.

Example A-2. One-dimensional rocket flight simulator

```
REM Simple Rocket Simulator.
REM
REM One-dimensional rocket flight simulator.

REM Motor variables.
DIM propellantWeight, motorWeight, motorName AS STRING
DIM thrust(1), time(1), mass(1)
DIM motorNames(1) AS STRING, motorFileNames(1) AS STRING

REM GUI elements.
DIM titleLabel AS Label
DIM quitButton AS Button, simulateButton AS Button
DIM massTextField as TextField, massLabel as Label
DIM cdTextField as TextField, cdLabel as Label
DIM arValueLabel as Label, arPicker AS Picker, arLabel as Label
DIM rhoTextField as TextField, rhoLabel as Label
DIM dtTextField as TextField, dtLabel as Label
DIM motorValueLabel as Label, motorPicker AS Picker, motorLabel as Label
```

```
DIM altitudeValueLabel as Label, altitudeLabel as Label
DIM speedValueLabel as Label, speedLabel as Label
DIM accelerationValueLabel as Label, accelerationLabel as Label
DIM delayValueLabel as Label, delayLabel as Label
DIM pickerBackground AS Label, pickerDoneButton AS Button

REM Initialize the rocket data. Change these values for different rockets,
REM motors, or flight conditions.
DIM rocketMass, d, d0, v0, a, fg, fd, ft, t, f, dt, Ar, rho, Cd
rocketMass = 0.036 : REM Rocket mass in kilograms, without the motor.
Cd = 0.524 : REM Coefficient of drag.
Ar = 0.000483 : REM Frontal area of the rocket.
rho = 1.225 : REM Density of the air.
dt = 0.01 : REM Simulation time step.

REM Set up the program.
loadMotors
motorName$ = motorNames(1)
motorFileName$ = motorFileNames(1)
initGUI
simulate
showAbout
END


REM Create a label and an initialized TextField.
REM
REM Parameters:
REM    x, y - The top-left location for the label.
REM    title - The title for the label.
REM    labelCtrl - The label control.
REM    textFieldCtrl - The text field control.
REM    value - The initial value.
REM    scientific - Format the value using scientific notation?

SUB createEditable (x, _
                    y, _
                    title AS STRING, _
                    BYREF labelCtrl AS Label, _
                    BYREF textFieldCtrl AS TextField, _
                    value, _
                    scientific AS INTEGER)
labelCtrl = Graphics.newLabel(x, y, 140)
labelCtrl.setText(title)
labelCtrl.setAlignment(3)
textFieldCtrl = Graphics.newTextField(x + 150, y, 140)
textFieldCtrl.setText(format(value, scientific))
END SUB


REM Create a label, button, and picker. The button displays a value.
REM That value can be changed using a picker, which is displayed when
REM the user taps on the button and hidden when the user picks a value
REM or taps on a Done button, which is shared among all picker controls.
REM
REM Parameters:
REM    x, y - The top-left location for the label.
```

```
REM    title - The title for the label.
REM    labelCtrl - The label control.
REM    valueCtrl - The label for displaying a value.
REM    pickerCtrl - The picker control.
REM    values - The values to display in the picker control.
REM    value - The initial value.
REM    scientific - Format the value using scientific notation?

SUB createPicker (x, _
                  y, _
                  title AS STRING, _
                  BYREF labelCtrl AS Label, _
                  BYREF valueCtrl AS Label, _
                  value, _
                  scientific AS INTEGER)
labelCtrl = Graphics.newLabel(x, y, 140)
labelCtrl.setText(title)
labelCtrl.setAlignment(3)

valueCtrl = Graphics.newLabel(x + 150, y, 140)
valueCtrl.setText(format(value, scientific))
END SUB


REM Create a label and an initialized TextField.
REM
REM Parameters:
REM    x, y - The top-left location for the label.
REM    title - The title for the label.
REM    labelCtrl - The label control.
REM    valueCtrl - The label for displaying a value.
REM    value - The initial value.
REM    scientific - Format the value using scientific notation?

SUB createUneditable (x, _
                      y, _
                      title AS STRING, _
                      BYREF labelCtrl AS Label, _
                      BYREF valueCtrl AS Label, _
                      value, _
                      scientific AS INTEGER)
labelCtrl = Graphics.newLabel(x, y, 140)
labelCtrl.setText(title)
labelCtrl.setAlignment(3)
valueCtrl = Graphics.newLabel(x + 150, y, 140)
valueCtrl.setText(format(value, scientific))
END SUB


REM Format a floating-point value for display. This is used for all
REM fields to give a consistent, readable view of values.
REM
REM Parameters:
REM    value - The value to format.
REM    scientific - Format the value using scientific notation?
REM
REM Returns: The formatted value.
```

```
FUNCTION format (value, scientific AS INTEGER) AS STRING
DIM s$
IF scientific THEN
  PRINT $ s$ USING "#######.###^^^^^^"; value;
ELSE
  PRINT $ s$ USING "#######.###"; value;
END IF
s$ = LTRIM(s$)
format = s$
END FUNCTION


REM Set up the user interface.

SUB initGUI
System.showGraphics(1)
Graphics.setToolsHidden(1)

width = Graphics.width
height = Graphics.height

REM Add a title.
titleLabel = Graphics.newLabel(20, 20, width - 40, 40)
titleLabel.setText("Simple Rocket Simulator")
titleLabel.setAlignment(2)
titleLabel.setFont("Sans-Serif", 22, 1)

REM Create labels and initialized text boxes for the editable rocket
REM components.
DIM y, dy
y = 70
dy = 28
createEditable(20, y, "Rocket Mass:", massLabel, massTextField, _
  rocketMass, 0)
y = y + dy
createEditable(20, y, "Cd:", cdLabel, cdTextField, Cd, 0)
y = y + dy
createPicker(20, y, "Area:", arLabel, arValueLabel, Ar, 1)
y = y + dy
createEditable(20, y, "Atm. Density:", rhoLabel, rhoTextField, rho, 0)
y = y + dy
createEditable(20, y, "Time Step:", dtLabel, dtTextField, dt, 0)
y = y + dy
createPicker(20, y, "Motor:", motorLabel, motorValueLabel, 0, 0)
motorValueLabel.setText(motorName$)
y = y + dy

REM Create the buttons.
quitButton = newButton(width - 92, height - 57, -1, "Quit")
simulateButton = newButton(20, y, 280, "Run Simulation")

REM Create the labels and uneditable text fields for the results.
y = y + 47
createUneditable(20, y, "Altitude:", altitudeLabel, _
  altitudeValueLabel, 0, 0)
y = y + dy
```

```
createUneditable(20, y, "Max Speed:", speedLabel, speedValueLabel, 0, 0)
y = y + dy
createUneditable(20, y, "Max Accel.:", accelerationLabel, _
  accelerationValueLabel, 0, 0)
y = y + dy
createUneditable(20, y, "Best Delay:", delayLabel, delayValueLabel, 0, _
  0)

REM Create the pickers and related objects. These are created last so they
REM will show up on top of other controls.
pickerBackground = Graphics.newLabel(0, 70, width, height - 70)
pickerBackground.setHidden(1)

pickerDoneButton = newButton((width - 72)/2, 320, -1, "Done")
pickerDoneButton.setHidden(1)

DIM values(7) AS STRING
values(1) = "BT-5"
values(2) = "BT-20"
values(3) = "BT-50"
values(4) = "BT-55"
values(5) = "BT-60"
values(6) = "BT-70"
values(7) = "BT-80"
arPicker = Graphics.newPicker(0, 80)
arPicker.setHidden(1)
arPicker.insertRows(values, 1)
arPicker.selectRow(3)

motorPicker = Graphics.newPicker(0, 80)
motorPicker.setHIdden(1)
motorPicker.insertRows(motorNames, 1)
motorPicker.selectRow(1)
END SUB

REM Use a specific time to interpolate in a table of time-based values.
REM
REM Inputs:
REM     t - The desired time.
REM     time - An array of times, from lowest to highest.
REM     value - A table of values, one for each elemnt in the time array.
REM
REM Returns: A linearly interpolated value from the value array. If t is
REM     less than the first time value, the first value in the value array
REM     is returned. If t is greater than the last time, the last value in
REM     the value array is returned.

FUNCTION interpolate (t, time(), value())
IF t < time(1) THEN
  interpolate = value(1)*t/time(1)
ELSE IF t < time(UBOUND(time, 1)) THEN
  DIM index AS INTEGER
  index = 1
  WHILE time(index) < t
    index = index + 1
  WEND
  interpolate = value(index - 1) + _
```

```
    (t - time(index - 1))*(value(index) - value(index - 1)) _
    /(time(index) - time(index - 1))
ELSE
  interpolate = value(UBOUND(value, 1))
END IF
END FUNCTION

REM Load a motor file.
REM
REM The values for the motor file are placed in global variables. The
REM variables set are:
REM
REM    propellantWeight - The total weight of the propellant in kg.
REM    motorWeight - The total weight of the motor, with propellant, in kg.
REM    motorName - The name for this motor.
REM    time - The times for which thrust data is available.
REM    thrust - The thrust at each time, in newtons.
REM    mass - The motor mass values at each time step.
REM
REM Parameters:
REM    fileName - The name of the motor file.

SUB loadMotor (fileName AS String)
DIM aLine AS STRING, tokens(1) AS STRING, count AS INTEGER

REM Scan the file, counting the non-comment lines. This gives
REM the number of data points.
OPEN fileName FOR INPUT AS #1
WHILE NOT EOF(1)
  LINE INPUT #1, aLine
  IF LEN(aLine) > 0 AND LEFT(aLine, 1) <> ";" THEN
    count = count + 1
  END IF
WEND
CLOSE #1

REM Dimension arrays to hold the time, thrust, and weight.
DIM tm(count - 1), tr(count - 1), ma(count - 1)

REM Rescan the file, reading the data.
OPEN fileName FOR INPUT AS #1
count = 0
WHILE NOT EOF(1)
  LINE INPUT #1, aLine
  IF LEN(aLine) > 0 AND LEFT(aLine, 1) <> ";" THEN
    tokens = parse(aLine)
    IF count = 0 THEN
      motorName = tokens(1)
      propellantWeight = VAL(tokens(5))
      motorWeight = VAL(tokens(6))
    ELSE
      tm(count) = VAL(tokens(1))
      tr(count) = VAL(tokens(2))
    END IF
    count = count + 1
  END IF
WEND
```

```
CLOSE #1

REM Find the total impulse.
DIM impulse
impulse = totalImpulse(tm, tr)
IF debug THEN PRINT "Total Impulse = "; impulse

REM Assume the mass loss is proportional to the impulse. Find
REM the motor mass as a function of time.
DIM currentMass, t0, f0, di, i AS INTEGER
currentMass = motorWeight
FOR i = 1 TO UBOUND(tm, 1)
  di = (f0 + tr(i))*(tm(i) - t0)/2
  currentMass = currentMass - propellantWeight*di/impulse
  ma(i) = currentMass
  f0 = tr(i)
  t0 = tm(i)
NEXT

REM Copy the values to the global arrays.
time = tm
thrust = tr
mass = ma
END SUB

REM Scans the document directory for motor files. Any that are found are
REM placed in the two arrays of motors. The arrays are:
REM
REM    motorNames - The display names for the motors.
REM    motorFileNames - The filenames for each motor file.

SUB loadMotors
DIM fileName AS STRING, count AS INTEGER, i AS INTEGER
count = 0
fileName = DIR("*")
WHILE fileName <> ""
  IF LCASE(RIGHT(fileName, 4)) = ".eng" THEN
    count = count + 1
    loadMotor(fileName)
    DIM names(count) AS STRING, fileNames(COUNT) AS STRING
    FOR i = 1 TO count - 1
      names(i) = motorNames(i)
      fileNames(i) = motorFileNames(i)
    NEXT
    names(count) = motorName
    fileNames(count) = fileName
    motorNames = names
    motorFileNames = fileNames
  END IF
  fileName = DIR
WEND
print motorNames
print motorFileNames
END SUB

REM Creates a new button with a gradient fill.
REM
```

```
REM Parameters:
REM    x - Horizontal location.
REM    y - Vertical location.
REM    width - Width of the button, or -1 for the default width.
REM    title - Name of the button.
REM
REM Returns: The new button.

FUNCTION newButton (x, y, width, title AS STRING) AS Button
DIM b AS Button
IF width = -1 THEN width = 72
b = Graphics.newButton(x, y, width)
b.setTitle(title)
b.setBackgroundColor(1, 1, 1)
b.setGradientColor(0.6, 0.6, 0.6)
newButton = b
END FUNCTION

REM A utility subroutine for loadMotor that breaks a line into space-
REM delimited strings.
REM
REM Parameters:
REM    aLine - The line to parse.
REM
REM Returns: An array of strings.

FUNCTION parse (aLine AS STRING) (1) AS STRING
REM Declare and initialize our local variables.
DIM tokens(10) AS STRING, count AS INTEGER, index AS INTEGER
count = 0

REM Strip any leading space from the line.
WHILE LEFT(aLine, 1) = " "
  aLine = RIGHT(aLine, LEN(aLine) - 1)
WEND

REM Collect the tokens.
WHILE LEN(aLine) > 0 AND count < 10
  count = count + 1
  index = POS(aLine, " ")
  IF index = 0 THEN
    tokens(count) = aLine
    aLine = ""
  ELSE
    tokens(count) = LEFT(aLine, index)
    aLine = RIGHT(aLine, LEN(aLine) - index)
    WHILE LEFT(aLine, 1) = " "
      aLine = RIGHT(aLine, LEN(aLine) - 1)
    WEND
  END IF
WEND

REM Create an array for the final tokens and return them.
DIM finalTokens(count) AS STRING
WHILE count > 0
  finalTokens(count) = tokens(count)
  count = count - 1
```

```
WEND

parse = finalTokens
END FUNCTION

REM Shows the About alert when the program starts.

SUB showAbout
about$ = "One-dimensional simulation of a model rocket flight."

about$ = about$ & CHR(10) & CHR(10) & "See the book, Make: Rockets, " _
  & "for a complete description of this app."

about$ = about$ & CHR(10) & CHR(10) & "This app takes basic " _
  & "information about a model rocket and simulates its flight to " _
  & "find the maximum altitude, acceleration, and speed and the " _
  & "ideal ejection charge."

about$ = about$ & CHR(10) & CHR(10) & "All units are in the meters-" _
  & "kilograms-seconds (MKS) system. Cd is the coefficient of drag, " _
  & "generally 0.4 to 0.7 for most model rockets. Area is the " _
  & "frontal area of the body tube. Atm. density is the atmospheric " _
  & "density, preset for sea level. Time step is the simulation " _
  & "time step."

about$ = about$ & CHR(10) & CHR(10) & "Tap a value to change it. Tap " _
  & "Run Simulation to run the simulation with the new values."

i = Graphics.showAlert("About This Sample", about$)
END SUB

REM Simulate a flight with the current flight characteristics.

SUB simulate
REM Load the motor data.
loadMotor(motorFileName$)
print motorFileName$
IF debug THEN
  PRINT propellantWeight, motorWeight
  FOR i = 1 TO UBOUND(time, 1)
    PRINT time(i), thrust(i), mass(i)
  NEXT
END IF

REM Get the latest values for the simulation.
rocketMass = VAL(massTextField.getText)
Cd = VAL(cdTextField.getText)
rho = VAL(rhoTextField.getText)
dt = VAL(dtTextField.getText)

REM Run the simulation.
DIM MaxA, MaxV, fg, t, fd, ft, f, a, d, d0, v0
MaxA = 0.0
MaxV = 0.0
fg = (rocketMass + motorWeight)*9.81
DO
  t = t + dt
```

```
  fd = Cd*rho*v0*v0*Ar
  ft = interpolate(t, time, thrust)
  f = ft - fd - fg
  a = f/(rocketMass + interpolate(t, time, mass))
  IF a > MaxA THEN MaxA = a
  d = d0 + v0*dt + a*t0*t0/2
  d0 = d
  v0 = v0 + a*dt
  fg = (rocketMass + interpolate(t, time, mass))*9.81
  IF v0 > MaxV THEN MaxV = v0
  IF DEBUG THEN PRINT t, d, v0, a
LOOP UNTIL t > 1 AND v0 < 0

REM Show the results.
altitudeValueLabel.setText(format(d, 0))
speedValueLabel.setText(format(MaxV, 0))
accelerationValueLabel.setText(format(MaxA, 0))
delayValueLabel.setText(format(t - time(UBOUND(time, 1)), 0))
END SUB

REM Find the total impulse of the motor.
REM
REM Parameters:
REM    time - The time steps where motor thrust is available.
REM    thrust - The thrust at each time step.
REM
REM Returns: The total impulse in newton-seconds.

FUNCTION totalImpulse (time(), thrust())
DIM t0, f0, impulse, i AS INTEGER
FOR i = 1 TO UBOUND(time, 1)
  impulse = impulse + (f0 + thrust(i))*(time(i) - t0)/2
  t0 = time(i)
  f0 = thrust(i)
NEXT
totalImpulse = impulse
END FUNCTION

REM Called when a touch ends, this subroutine checks to see if the
REM touch was inside one of the labels used to trigger the appearance
REM of a picker and, if so, displays the picker.
REM
REM Parameters:
REM    e - The touch event that triggered this call.

SUB touchesEnded (e AS EVENT)
DIM where(1, 2), x, y
where = e.where
x = where(1, 1)
y = where(1, 2)
IF x >= arValueLabel.x _
   AND x <= arValueLabel.x + arValueLabel.width _
   AND y >= arValueLabel.y _
   AND y <= arValueLabel.y + arValueLabel.height THEN
  pickerBackground.setHidden(0)
  arPicker.setHidden(0)
  pickerDoneButton.setHidden(0)
```

```
ELSE IF x >= motorValueLabel.x _
   AND x <= motorValueLabel.x + motorValueLabel.width _
   AND y >= motorValueLabel.y _
   AND y <= motorValueLabel.y + motorValueLabel.height THEN
  pickerBackground.setHidden(0)
  motorPicker.setHidden(0)
  pickerDoneButton.setHidden(0)
END IF
END SUB

REM Handle a tap on one of the buttons.
REM
REM Parameters:
REM    ctrl - The button that was tapped.
REM    time - The time when the event occurred.

SUB touchUpInside (ctrl AS Button, time AS DOUBLE)
IF ctrl = quitButton THEN
  STOP
ELSE IF ctrl = pickerDoneButton THEN
  arPicker.setHidden(1)
  motorPicker.setHidden(1)
  pickerDoneButton.setHidden(1)
  pickerBackground.setHidden(1)

  SELECT CASE arPicker.selection
    CASE 1: Ar = 0.000150
    CASE 2: Ar = 0.000275
    CASE 3: Ar = 0.000483
    CASE 4: Ar = 0.000892
    CASE 5: Ar = 0.001359
    CASE 6: Ar = 0.002463
    CASE 7: Ar = 0.003390
  END SELECT
  arValueLabel.setText(format(Ar, 1))

  motorName$ = motorNames(motorPicker.selection)
  motorValueLabel.setText(motorName$)
  motorFileName$ = motorFileNames(motorPicker.selection)
  loadMotor(motorFileName$)
ELSE IF ctrl = simulateButton THEN
  simulate
END IF
END SUB
```

Theodolite

Theodolite is a simple BASIC program that should run in just about any modern BASIC compiler. It takes the baseline and angles from two theodolites and calculates the altitude of a rocket. See Chapter 8 for a detailed description of how to build and use theodolites to track rockets.

Example A-3. Altitude calculator

```
REM Calculate the altitude of a model rocket using angles from
REM two theodolites.

REM Get the baseline and angles.
INPUT "Baseline length: "; B

REM Find the conversion from degrees to radians.
DTR = 3.1415926535/180

REM Get the angles for the first tracker.
INPUT "Tracker 1 azimuth: "; alpha1
INPUT "Tracker 1 elevation: "; epsilon1
alpha1 = alpha1*DTR
epsilon1 = epsilon1*DTR

REM Get the angles for the second tracker.
INPUT "Tracker 2 azimuth: "; alpha2
INPUT "Tracker 2 elevation: "; epsilon2
alpha2 = alpha2*DTR
epsilon2 = epsilon2*DTR

REM Calculate the altitude using the vertical midpoint method.
A1 = B*SIN(alpha2)*TAN(epsilon1)/SIN(alpha1 + alpha2)
A2 = B*SIN(alpha1)*TAN(epsilon2)/SIN(alpha1 + alpha2)
A = (A1 + A2)/2
C = ABS((A1 - A2)/(2*A))

PRINT
PRINT "Vertical Midpoint Method:"
CALL Report(A, C)

REM Calculate the altitude using the geodesic method.
se1 = SIN(epsilon1)
se2 = SIN(epsilon2)
ce1 = COS(epsilon1)
ce2 = COS(epsilon2)
sa1 = SIN(alpha1)
sa2 = SIN(alpha2)
ca1 = COS(alpha1)
ca2 = COS(alpha2)
f = se1*se2 - ce1*ce2*(ca1*ca2 - sa1*sa2)
d1 = B*(ce1*ca1 + f*ce2*ca2)/(1-f*f)
d2 = B*(ce2*ca2 + f*ce1*ca1)/(1-f*f)
A = d1*d2*(se1 + se2)/(d1 + d2)
C = B*ABS((ce2*se1*sa2 - ce1*se2*sa1)/(A*SQR(1-f*f)))

PRINT
PRINT "Geodesic Method:"
CALL Report(A, C)
PRINT
END

REM Print the results.
REM
REM Inputs:
REM     A - The altitude.
```

```
REM    C - Track closed value.

SUB Report (A, C)
PRINT "  Altitude: "; A
IF C <= 0.1 THEN
  PRINT "  Closed track; C = "; C
ELSE
  PRINT "  Track not closed; C = "; C
END IF
END SUB
```

Places to Buy Stuff or Find Information

B

This appendix lists sources of materials and information you need to make rockets.

Model Rocket Parts and Supplies

Here are some suppliers you'll find useful.

Aerotech Consumer Aerospace

http://www.aerotech-rocketry.com

Aerotech manufactures rocket motors and reloads. While concentrating on high-power motors, the company also has lines of single-use motors and low-power and mid-power reloadable motors.

Aerotech makes some of its own motors, but focuses more on the reloads for reloadable motors. Many of the motors people refer to as Aerotech reloadable motors are actually built by a partner company called Reloadable Motor Systems, or RMS for short.

Chapter 23 shows how to prepare and fly the RMS 24-40 motor using Aerotech reloads.

Altus Metrum

http://www.altusmetrum.org

Altus Metrum makes a series of altimeters, including one small one called the MicroPeak that will fit in low-power model rockets. Free software and information about the altimeters is available on the website. The altimeters are offered through other retailers, such as Apogee Components.

Apogee Components

http://www.apogeerockets.com

Apogee Components sells both low-power and high-power rocketry supplies, kits, and motors. It's a great resource for parts, supplies, and kits, and also has a fantastic collection of technical reports and how-to guides that it sends out with qualifying orders.

Apogee Components also markets RockSim, the most popular rocket simulator for model rocketry.

Art Applewhite Rockets

http://www.artapplewhite.com

A small company owned by a couple of disabled vets, this is a really fun place to look for kits, especially oddrocs.

Balsa Machining Service

http://www.balsamachining.com

Balsa Machining Service has the widest selection of balsa nose cones and nose blocks of any source I am aware of. I personally like balsa nose cones better than plastic ones; they are softer on impact and easier to modify.

The payload rockets from Chapter 9-- particularly the ICU2 camera payload and the egg lofter payload—are examples of rockets that can easily be built with balsa parts, but are difficult to build from plastic.

Balsa Machining also has a good selection of other hard-to-find parts, like very long body tubes and tube couplers.

Custom Rocket Company

http://www.customrocketcompany.com

This small rocket company makes a number of fun kits. Its Skybird payload rocket is one example of the unusual twist the designers give to many rockets: rather than the conventional clear payload bay, this one sports a transparent red payload tube.

Estes Industries

http://www.estesrockets.com

The 800-pound gorilla of the low-power model rocket world, Estes Industries has been around since the dawn of model rocketry.

The company makes and sells the Estes Designer's Special, as well as various replacement parts, motors, and kits.

You can frequently find Estes parts, especially the Estes Designer's Special, discounted through other retailers like Amazon.

Estes starter kits, rocket kits, and motors are also sold through independent hobby stores and at chain hobby stores like Hobby Lobby.

FlisKits, Inc.

http://fliskits.com

FlisKits is one of the smaller rocket companies. It specializes in low-power rocket kits and has some really fun ones, like the scale model of Nellie, the first liquid-propelled rocket, that was featured in Chapter 23.

Hobby Lobby

http://www.hobbylobby.com

Hobby Lobby is one of the more reliable suppliers of rocketry supplies you are likely to find in your local area. While specialized hobby stores are great, they aren't easy to find in towns and smaller cities, and even in large ones they may be fairly far away.

Hobby Lobby has a nice selection of rocket motors, launchers, igniters, recovery wadding, and rocket kits, as well as balsa wood and other modeling supplies.

Best of all, it frequently offers 40% off coupons for a single item. I recently got an Estes starter kit for a relative for just $20. Do an online search for "hobby lobby 40 off" to find the coupons.

JonRocket.com

http://www.jonrocket.com

JonRocket has a pretty good selection of rocket parts and kits. In addition to body tubes and other supplies, it's also a great source for balsa parts, including balsa nose cones and nose blocks.

Several of the rockets in the book, particularly the egg lofter and ICU2 camera payload, use nose cones or nose blocks from JonRocket.

Maker Shed

http://www.makershed.com

This is the headquarters for the retail arm of Make:, publishers of this book.

It's a great place to find all sorts of kits, including some of the air rocket kits from this book. Watch Maker Shed for other upcoming rocket kits, too.

Micro Center

http://www.microcenter.com

This is a good place to shop for electronic components and rockets.

Rocketarium

http://www.rocketarium.com

Rocketarium has a good selection of low-power rocket kits, including a very fun section devoted to oddrocs.

Quest Aerospace

http://www.questaerospace.com

One of the few competitors for Estes in the low-power model rocket arena, Quest has a nice online store featuring motors, parts, and kits. They are frequently less expensive than the Estes parts, and the quality is just as good. Estes has a better selection for most items, though.

Stomp Rocket

http://www.stomprocket.com

Stomp Rocket® makes a line of air rockets that are launched by stomping on a bulb that delivers a blast of air to the rocket.

Clubs and Information Websites

Here are some websites for various clubs and other resources.

AirRocketWorks

http://airrocketworks.com

This is a great place to explore air rocketry: it's sort of a combination store, forum, and blog site for air rocket enthusiasts, run by the folks who developed the air rocket launcher from Chapter 6 and the air rocket glider from Chapter 21.

Author's Website

http://bit.ly/byteworks-make-rockets

Full-size templates, decal patterns, and rocket simulation files for most of the rockets in this book are available from the author's website. The author also created techBASIC, which includes copies of all of the BASIC programs from this book. You can find out more about techBASIC at *http://www.byteworks.us/Byte_Works/techBASIC.html*.

Elprotronic

http://www.elprotronic.com

These folks make the software used to flash the Texas Instruments MSP430 LaunchPad used in the water rocket parachute deployment system. The software is a free download.

OpenRocket

http://openrocket.sourceforge.net

OpenRocket is a free rocket simulator that runs on Linux, Mac OS X, and Windows. It started life as a thesis project. This website is the place to go for downloads and further information, or, if you are a programmer and inclined to help, to join the project team.

Rocket Reviews

http://www.rocketreviews.com

This site is dedicated to customer reviews of rocketry-related products. It's a good place to browse to find out what is new and what other model rocket scientists think about various products.

The Rocketry Forum

http://www.rocketryforum.com

This forum is a great place for doing research and getting your questions answered by experienced model rocket scientists. It's the next best thing to a 24/7 model rocket club!

The rocket forum is a part of a larger site, Rocketry Online (*http://www.rocketryonline.com*).

ThrustCurve.org

http://www.thrustcurve.org

This site houses a collection of thrust curves for most commercially available motors. You can check plots of the thrust curves right on the site, allowing you to pick appropriate motors for a given liftoff weight. You can also download the motor data in common formats suitable for most rocket simulators. These formats use ASCII files to store the data, so you can open the files in any text editor to look at the raw data yourself.

US Water Rockets

http://www.uswaterrockets.com

This site caters to the water rocket crowd, with great tutorials on various aspects of water rocketry, a list of world records, and occasionally links to places to buy water rocket–related specialty items. This site and the people behind it were huge resources for writing Chapter 11 and Chapter 12. The water rocket launcher and parachute recovery system derive directly from designs on the site.

Electronics Parts and Supplies

You'll need to know where to get specialty electronics parts; here are some sources.

Digi-Key

http://www.digikey.com

One of just a few online stores that sells a wide selection of electronics parts and will still sell a single part, Digi-Key is a great place to look for electronic components like those used in the launchers and the water rocket parachute system.

Jameco Electronics

http://www.jameco.com

Jameco caters to hobbyists and professionals building prototypes. It has a good selection of both parts and tools.

Mouser Electronics

http://www.mouser.com

Mouser is a great source for electronics parts. The biggest problem is that there are so many! The site has built a great filter-based search engine, though, so you can narrow down the parts pretty fast. Mouser is unusual in that it will sell individual parts with no minimum order, but also sells in bulk.

Newark element14

http://www.newark.com

This online store has some very hard-to-find parts, such as the 9-pin connectors used for the quad launch controller in Chapter 4.

Radio Shack

http://www.radioshack.com

Radio Shack is a great place to find components for electronics projects like the launch controllers in Chapter 4.

While it tends to be a little more expensive than some other shops, Radio Shack makes up for that with brick-and-mortar stores in most cities, so you can stop by and grab a part without waiting for shipping. And, after all, shipping adds enough to the price of mail order parts that the overall cost is often comparable.

SparkFun

https://www.sparkfun.com

SparkFun is an electronics supply house that caters to hobbyist builders.

It has a good selection of common parts as well as a nice selection of microcontrollers, sensors, and construction materials.

A nice feature about the site is that many parts and tools have how-to blogs and videos. There is also a part-based forum where you can ask for help or see what others have tried.

General Supplies and Hardware

These companies can supply you with various tools and other items you'll need for the projects in this book.

Ace Hardware

http://www.acehardware.com

Ace Hardware has both an online presence and brick-and-mortar stores in many cities. It's a favorite haunt of mine because of the great selection of nuts, bolts, and other small hardware items, all available in loose bins so you can buy exactly the number of parts you need.

Amazon

http://www.amazon.com

You can get all sorts of tools and supplies from Amazon. It's worth searching the site for pretty much anything, from an Estes Designer's Special to a hobby knife.

Bolt Depot

http://www.boltdepot.com

This online store specializes in bolts and other fasteners. Try it if you can't find a bolt locally.

Home Depot

http://www.homedepot.com

Home Depot is a big-box hardware store that probably has an outlet near you, and also has an online store. It's a good source for general hardware parts like PVC, brass, paint, and other supplies.

Lasco

http://lascofittings.sitewrench.com

Basically a plumbing supply house, Lasco is a great source for hard-to-find parts for water rocket and air rocket launchers.

Lowe's

http://www.lowes.com

Lowe's is a big-box hardware store that has both an online store and brick-and-mortar stores. It's a good source for PVC, brass, paint, and other supplies used for launchers and water rockets.

Jo-Ann Fabric Store

http://www.joann.com

This is a great place to pick up all sorts of supplies for model rocketry, from cutting boards to elastic for shock cords to nylon for heavy-duty parachutes. Jo-Ann's regularly has coupons for 40% off. Check online by searching for "joann coupon" when you plan a trip.

National Balsa

http://www.nationalbalsa.com

Many cities have well-stocked local hobby stores that carry balsa wood and hardwoods for modeling. If you're not that lucky, or if you are looking for something a little unusual, try National Balsa. And remember, this supplier isn't just for balsa wood; they also carry other woods suitable for building model rockets and airplanes and even have a few model airplane kits.

OnlineMetals.com

http://www.onlinemetals.com

Some metal parts, like brass tubes, may be difficult to find locally. This online store has a great selection of brass tubes and other metal parts, and they can be ordered in small quantities.

SIG Mfg. Co.

http://www.sigmfg.com

SIG makes dope, a kind of paint used for model airplanes, and is a good source for sanding sealer, which is a great wood sealer for porous materials like balsa wood.

SIG also has a great selection of parts and supplies for building model airplanes. This is a great place to shop if you are trying to build a really fancy rocket plane.

The Projects in This Book

C

This book shows you how to build a dizzying array of rockets, launchers, and trackers. Each project was selected to teach a specific concept in rocketry, such as a particular construction technique or a special recovery method. This appendix gives you a quick scorecard so you can quickly find a project to explore some aspect of rocketry.

While the rockets in the book cover the basics, there are a lot of variations on the themes they cover. Each chapter discusses alternative methods of doing what you just learned. For example, the helicopter and glider recovery chapters also discuss different methods of implementing helicopter or glider recovery.

The Rockets

Air Rockets

Air rockets launched from a straw are a fun way to explore rocket stability. See Chapter 1.

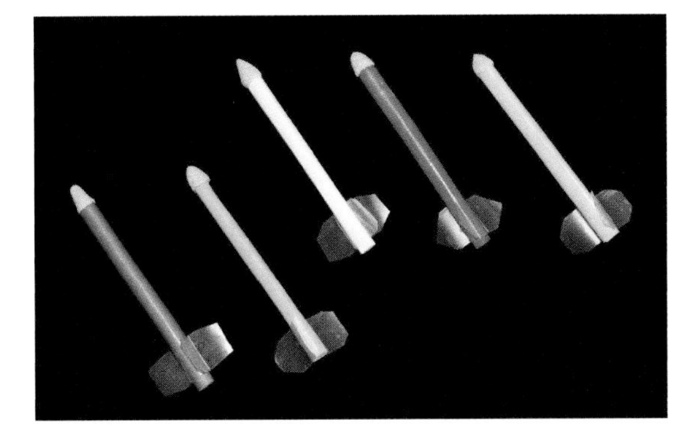

Figure C-1. *Colorful air rockets are a cheap and fun way to explore rocket aerodynamics.*

Balloon Rockets

Unstable rockets are usually dangerous—unless they are large rubber balloons! Chapter 1 introduces rocket stability using balloons with crude paper fins.

Figure C-2. *Balloon rockets are used to explore stability.*

Cerberus

Chapter 18 tells how to build and successfully fly cluster rockets. Cerberus is one of two cluster rockets that appear, with complete construction details. Cerberus uses three motors at once, lifting this fun sport rocket quickly from the pad.

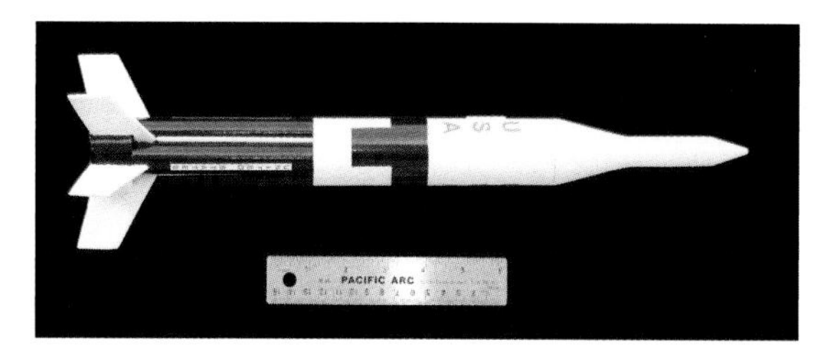

Figure C-3. *Cerberus is both a cluster rocket and a fun-looking sport rocket, reminiscent of the Saturn 1B.*

Ceres

Chapter 9 is all about building rockets to carry payloads. It shows how to build two versions of the Ceres booster, one using a standard rocket motor and the other using powerful D and E motors. Later, in Chapter 18, you build a cluster version of the booster that uses three motors at once!

Figure C-4. *The Ceres booster is designed for lifting payloads.*

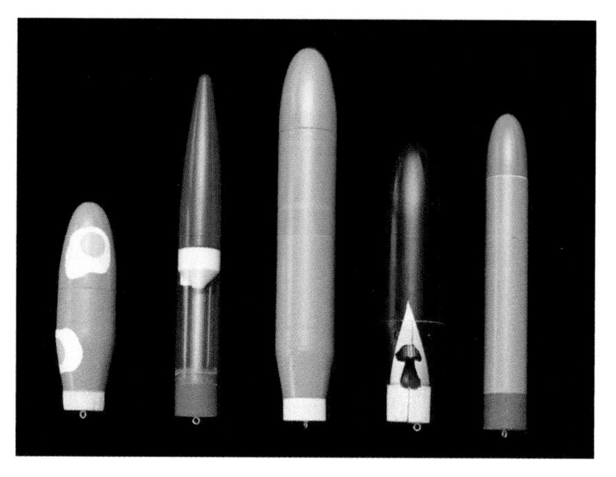

Figure C-5. *Payload bays for Ceres include an egg lofter and a camera bay!*

Any of the boosters can lift a payload designed for a BT-60 body tube. Chapter 9 shows five of them, including an egg lofter for launching a raw egg; a rocket that uses a small, cheap movie camera to take in-flight video; and three general-purpose payload bays that accommodate different-sized payloads.

Compressed Air Rockets

Compressed air rockets are propelled by a blast of air, sending them hundreds of feet into the sky. You can make a compressed air rocket in a few minutes for a few cents' worth of parts. It's a fun way to introduce some basic science at a birthday party or picnic.

Figure C-6. *Compressed air rockets are cheap and fast to build—great for parties!*

Daedalus

Chapter 21 shows how to build a variable-geometry glider. The wings on Daedalus fold flat against the fuselage so the aircraft can launch vertically from the air rocket launcher. At apogee, the wings deploy, and the rocket becomes a glider that sails back to earth.

Figure C-7. *This variable-geometry glider lifts off from the air rocket launcher.*

Eros

Flying on small, inexpensive T motors, Eros (from Chapter 16) introduces nose-blow recovery, a variant on tumble recovery.

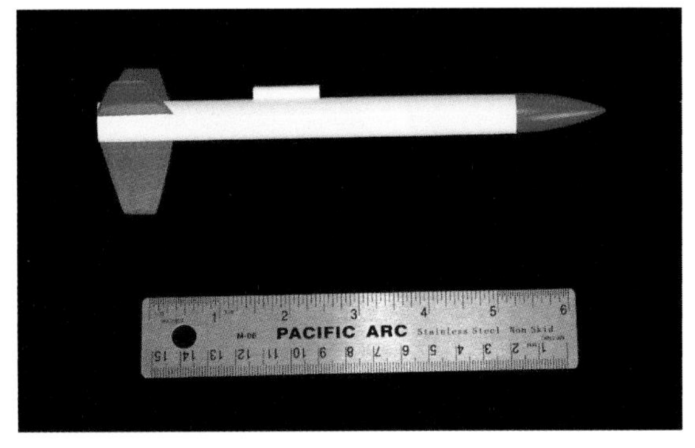

Figure C-8. *Nose-blow recovery with a small, fun rocket.*

Hebe

Chapter 15 shows how to build minimum-diameter rockets that fly high and fast. Hebe is just 3/4" in diameter, but flies higher and faster on a given motor than any other rocket in the book. Chapter 8 shows how to use a small altimeter to accurately track Hebe's flight. On one flight, Hebe topped 500 mph!

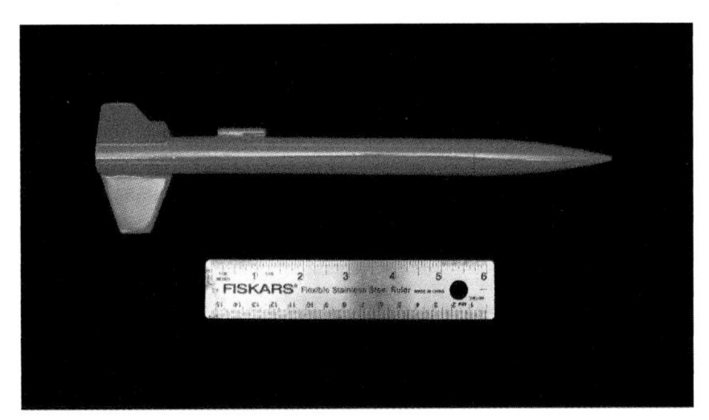

Figure C-9. *Hebe is a fast, high flier!*

Icarus

Learn the basics of glider recovery in Chapter 20, where you will build this classic pop-pod glider. The chapter also shows how to build stable airplanes, discussing dihedrals and how a rudder and wing work together to achieve stable flight.

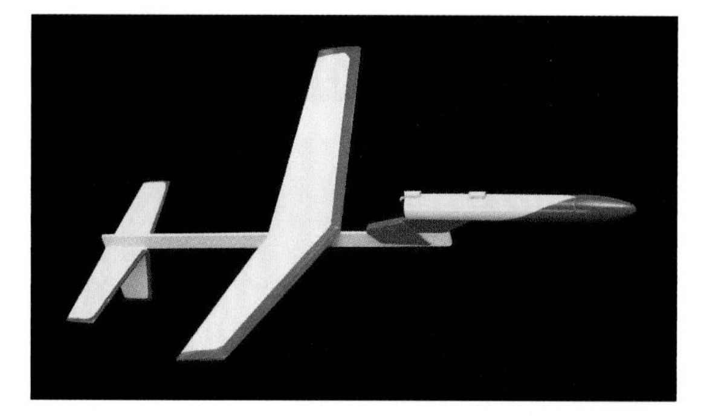

Figure C-10. *A pop-pod boost glider, Icarus ascends as a rocket, then glides back to the ground, just like the Space Shuttle!*

Juno

Juno is a great first rocket. It's used in this book to show basic construction techniques (Chapter 3), as well as how to prepare a rocket for flight and safely fly the rocket (Chapter 5).

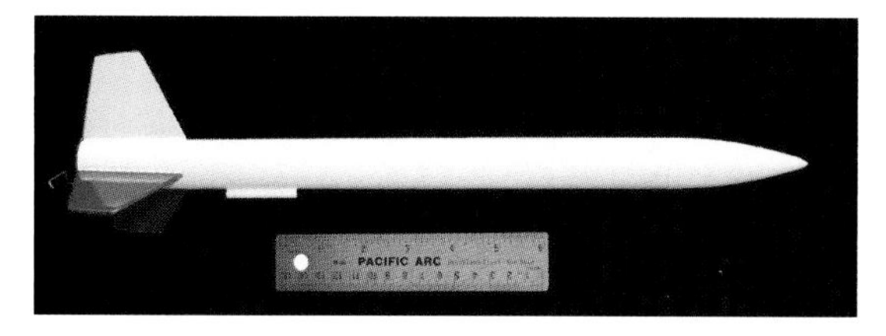

Figure C-11. *Juno is a classic sport rocket, used to show basic construction and flight techniques.*

Chapter 3 also shows two alternate fin designs for Juno.

Juno Payload Conversion

It's easy to add a payload bay to almost any model rocket. Chapter 9 shows how it's done, using Juno as an example.

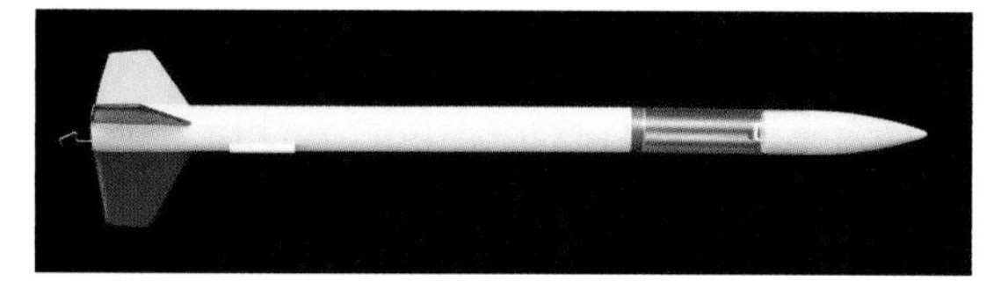

Figure C-12. *The Juno payload conversion shows how to convert almost any rocket to carry a payload.*

Match Head Rockets

Chapter 1 shows how to build a simple solid propellant rocket from a match head, a piece of foil, and a paper clip. This project introduces solid propellant rockets using a cheap, fun experiment suitable for anyone old enough to strike a match.

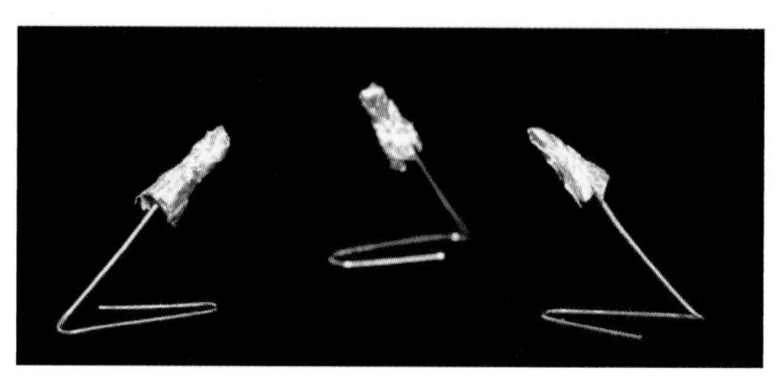

Figure C-13. *Match head rockets are a simple, cheap way to look at the principles of solid propellant rockets.*

Nicomachus

Helicopter recovery is a fun way to safely recover a model rocket. This rocket from Chapter 19 shows one way to use helicopter recovery, flipping two large ailerons to the side when the motor ejects and spinning back to earth.

Figure C-14. *Experiment with helicopter recovery using Nicomachus.*

Romulus

Staging isn't just for the huge NASA rockets! In Chapter 17, you'll learn about multistage rockets by building this gorgeous rocket that's easy to convert for payloads.

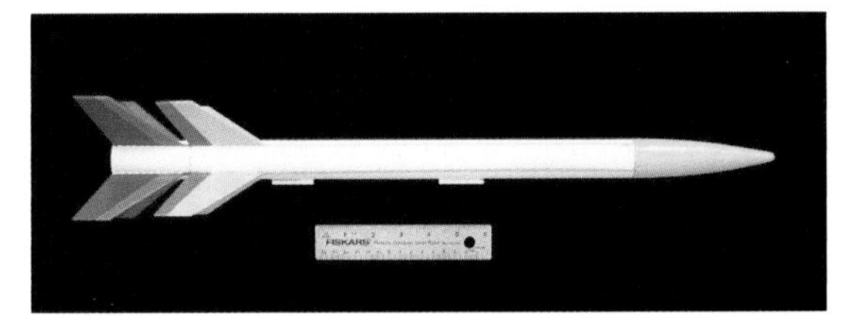

Figure C-15. *Discover staging with the two-stage Romulus.*

The first stage uses tumble recovery, showing another recovery technique.

Themis

Themis is a water rocket built around a two-liter soda bottle. Chapter 11 shows how to build the basic rocket, and also discusses alternatives, like one-liter bottles and 20-oz bottles. Chapter 12 adds the parachute shown here.

Figure C-16. *Themis is a classic water rocket based on a two-liter bottle.*

Water rockets are a cheap way to introduce the science of rocket flight. Parachute recovery is a surprisingly challenging project that can still be handled by a middle-school or high-school student. This version is a great introduction to microcontrollers and servos.

Toutatis

Toutatis is a tiny rocket from Chapter 16 that shows how to use featherweight recovery. It's also a cheap, easy rocket to build and fly.

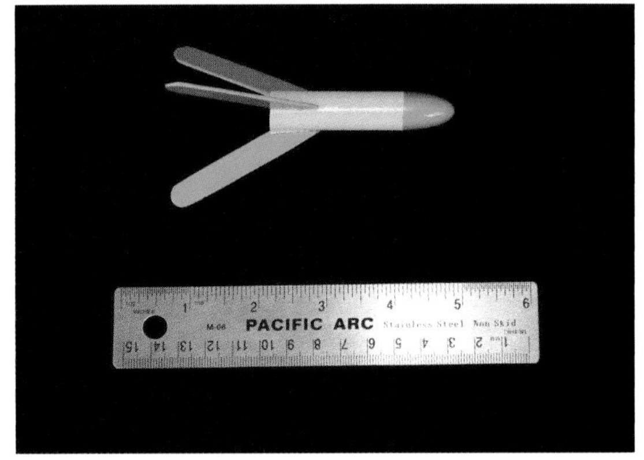

Figure C-17. *This mini-mite is simple to build and fly.*

The Launch Pads

Compressed Air Rocket Launcher

This air rocket launcher is used to fly air rockets and the air rocket glider (Chapter 21). Use a bicycle pump or a portable air source for filling flat car tires to pressurize the launcher. Find out how to build your own in Chapter 6.

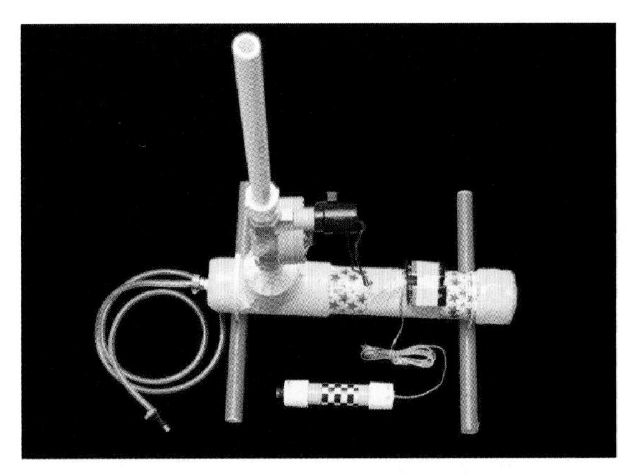

Figure C-18. *The air rocket launcher uses compressed air to launch a rocket hundreds of feet into the air.*

The Mono Launcher

There are a lot of good launchers available from model rocket companies, but almost all of them are starter launchers. The mono launcher is a rugged, portable launcher that can easily handle mid-power rockets far larger than any in this book, yet it stores as a single PVC rod.

Figure C-19. *The mono launcher and mono launch controller are a rugged combination, capable of flying any solid propellant rocket in this book.*

Almost all commercial launch controllers are designed for D or smaller motors. Most can only handle a single igniter, so they cannot launch cluster rockets like the ones from Chapter 18. They also don't have the recommended voltage for lighting igniters for reloadable motors like the one in Chapter 23. The mono launch controller can do it all. Unlike the Estes launch controllers, it is safe to use with both Estes and Quest igniters. It also works well with igniters for reloadable motors and cluster rockets.

You will find detailed assembly directions and all of the theory in Chapter 4.

The Quad Launcher

The quad launch pad and its accompanying controller are designed for schools and clubs that want to make the most of the available launch time. You can set up four rockets at a time and then launch them one at a time, or use a drag race format to launch them all together!

Like the mono launcher, the quad launch pad works well with mid-size rockets. The quad launch controller supports all popular igniters, including igniters for reloadable motors. It also supports cluster rockets.

Figure C-20. *The quad launcher is designed for clubs and schools.*

This launcher is a great project for a school wood shop, and the launch controller is a wonderful way to introduce basic electrical engineering principles. See Chapter 4 for details.

The Water Rocket Launcher

Chapter 11 shows how to build a classic launcher that can send water rockets based on two-liter bottles hundreds of feet into the air. The launcher is powered by a bicycle air pump or any other pressure source designed for car tires, such as a portable air pump for fixing flats.

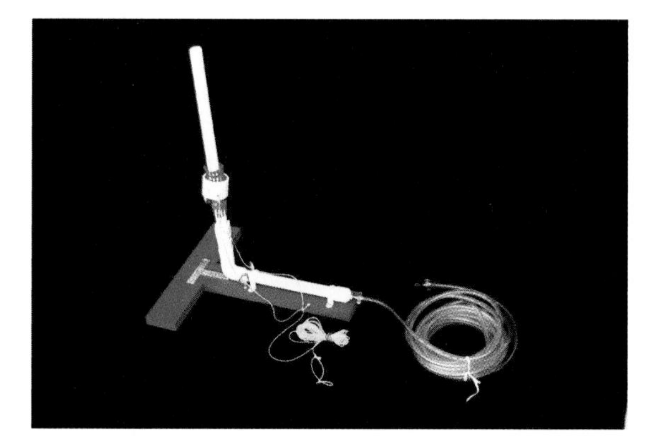

Figure C-21. *Launch water rockets from this simple but effective water rocket launcher.*

Trackers

Single-Axis Tracker

This single-axis tracker from Chapter 8 costs just pennies to make, but gives the altitude of a model rocket. It's also a great project for a beginning algebra or trigonometry class. Chapter 8 introduces the concept of error calculations and measurements, both great skills for young engineers.

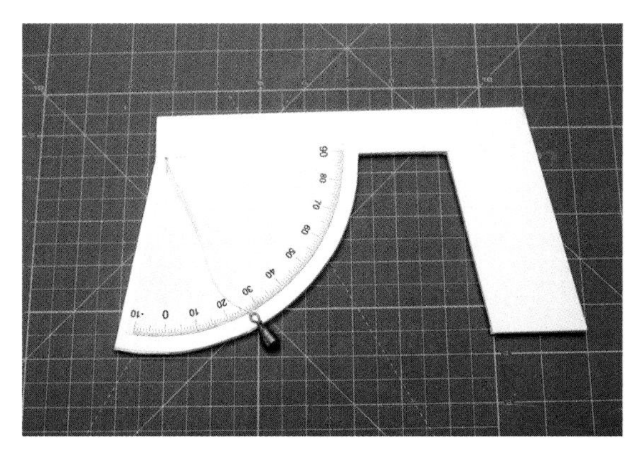

Figure C-22. *This cheap but effective single-axis tracker is an easy way to find out how high a rocket flew.*

Theodolite

The ultimate in optical tracking, a pair of theodolites is an accurate way to determine the altitude of a model rocket—or the height of anything that can't be measured using a simpler method. This project from Chapter 8 is also great in math classes, where it provides practical uses for both trigonometry and vector analysis.

Figure C-23. *Use a pair of theodolites for more accurate altitude determination.*

Science and Math

This is a book about rocket science, after all! While you can skip every bit of the math and science if it suits you, this book gives you a good basic understanding of the science, technology, engineering, and math behind model rocketry.

Aerodynamics

Aerodynamics is central to model rocketry, so the book starts with some simple projects to play with stability. In Chapter 1 you explore stability with balloons and straw rockets.

Chapter 7 dives into the details. You will learn several ways to find the center of pressure and center of gravity for a model rocket, and how to balance the two to create a stable rocket. Like all of the technology sections, you can skip this chapter if you are just building the rockets according to plan.

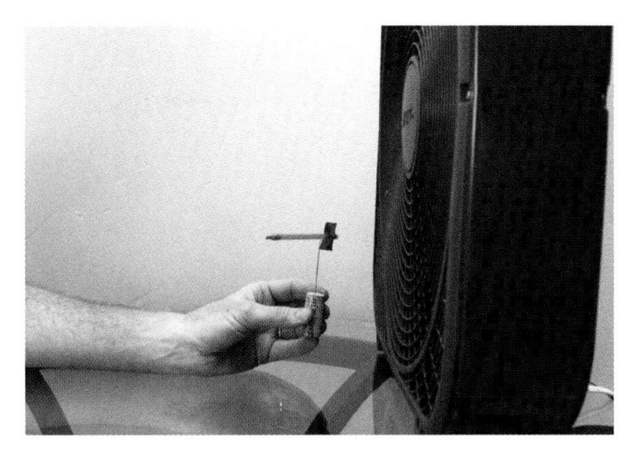

Figure C-24. *You can learn a lot about stability with a fan, a straw, some tape, and a needle.*

If you think fluid dynamics is a gas, not a drag, see Chapter 13. That's where you find out how to make a rocket fly higher and faster by reducing drag. From the Reynolds number to practical tips on selecting nose cones, this chapter helps you pierce the air more efficiently.

We pull this together in Chapter 14, where we figure out how high a model rocket will fly. You'll see all the math behind a simple but practical way to predict the altitude of a model rocket flight, then see a simple computer program that does a pretty good job with the equations. Finally, you'll see how to use a free state-of-the-art rocket simulator to predict what will happen when the rocket takes off.

Computer Science

It's possible to predict the altitude of a model rocket with a pencil and paper, but it's very, very tedious. Chapter 8 shows how to do it with a simple computer program you can enter and modify yourself.

There is a fair amount of trigonometry involved in using a two-axis tracking system. Chapter 8 presents a computer program that does the heavy lifting.

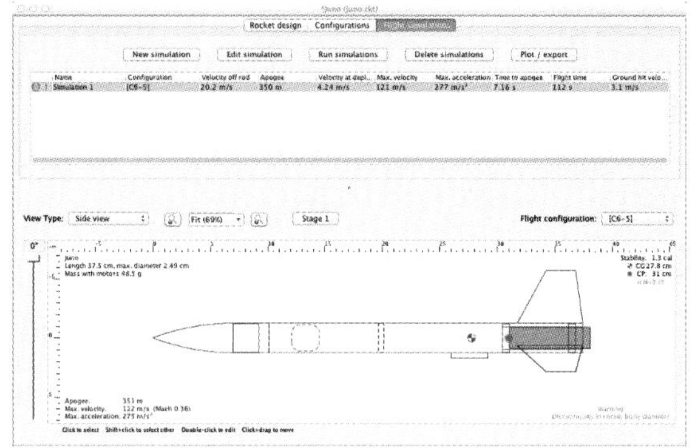

Figure C-25. *Use the free OpenRocket simulator to explore rocket stability and performance.*

We often use software programs to do complex technical tasks. This book uses OpenRocket to find the center of pressure on a rocket (Chapter 7) and predict its altitude (Chapter 8), and even to design a rocket from scratch (Chapter 23).

Electrical Engineering

You find out how to build model rocket launchers in Chapter 4. Along the way, you can explore basic electronics theory to discover why the launchers are built as they are. You can use that theory to adjust the launchers to meet your own specific goals.

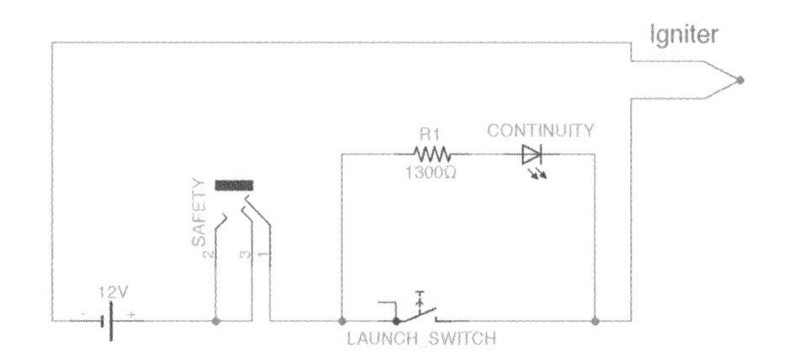

Figure C-26. *Typical launch controller circuit.*

You find out how to use an altimeter in Chapter 8, where you install an altimeter in a rocket to track its altitude and speed. You also use a free computer program to download and analyze the data.

There's a fun little project in Chapter 12 where you build a parachute ejection system for a water rocket based on a servo and an inexpensive microcontroller. It's a great example of a creative use for modern digital electronics.

Mathematics

Mathematics is the language of science and engineering. You can build every rocket and launcher in this book without using much, if any, math, but where's the fun in that?

You will find math scattered throughout the book, used whenever it makes sense. Some of it is pretty simple, like in Chapter 4, where you learn how to calculate the number of igniters a launcher can light at one time. Trigonometry pops up in several places, from a simple calculation in Chapter 8 that finds the altitude of a model rocket using a protractor to a much more involved calculation later in the chapter that shows how to use two theodolites to accurately track a rocket. There's even some calculus thrown in here and there, like in Chapter 7, where we figure out the center of gravity of a nose cone, or Chapter 8, where we do error analysis to determine the accuracy of altitude measurements.

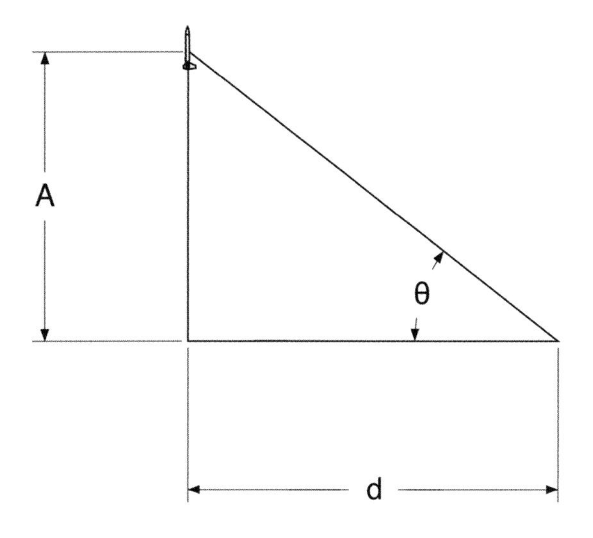

Figure C-27. *Math is all around us, and all through the book. Simple trigonometry is used to find the altitude of a rocket flight.*

Rocket Motors

Explore the basic technology behind rocket motors in Chapter 2 before putting it to use throughout the book. This is where you will learn the difference between an A and a C motor, how to read motor labels, and how motors work. Chapter 23 shows a rocket motor from the inside out, as you learn how to assemble the reusable Aerotech RMS 24/40 motor.

Figure C-28. *Installing the aft closure in an Aerotech 24/40 motor.*

The NAR Model Rocket Safety Code

D

Effective August 2012 (*http://www.nar.org/NARmrsc.html*)

1. **Materials.** I will use only lightweight, non-metal parts for the nose, body, and fins of my rocket.
2. **Motors.** I will use only certified, commercially-made model rocket motors, and will not tamper with these motors or use them for any purposes except those recommended by the manufacturer.
3. **Ignition System.** I will launch my rockets with an electrical launch system and electrical motor igniters. My launch system will have a safety interlock in series with the launch switch, and will use a launch switch that returns to the "off" position when released.
4. **Misfires.** If my rocket does not launch when I press the button of my electrical launch system, I will remove the launcher's safety interlock or disconnect its battery, and will wait 60 seconds after the last launch attempt before allowing anyone to approach the rocket.
5. **Launch Safety.** I will use a countdown before launch, and will ensure that everyone is paying attention and is a safe distance of at least 15 feet away when I launch rockets with D motors or smaller, and 30 feet when I launch larger rockets. If I am uncertain about the safety or stability of an untested rocket, I will check the stability before flight and will fly it only after warning spectators and clearing them away to a safe distance. When conducting a simultaneous launch of more than ten rockets I will observe a safe distance of 1.5 times the maximum expected altitude of any launched rocket.
6. **Launcher.** I will launch my rocket from a launch rod, tower, or rail that is pointed to within 30 degrees of the vertical to ensure that the rocket flies nearly straight up, and I will use a blast deflector to prevent the motor's exhaust from hitting the ground. To prevent accidental eye injury, I will place launchers so that the end of the launch rod is above eye level or will cap the end of the rod when it is not in use.
7. **Size.** My model rocket will not weigh more than 1,500 grams (53 ounces) at liftoff and will not contain more than 125 grams (4.4 ounces) of propellant or 320 N-sec (71.9 pound-seconds) of total impulse.
8. **Flight Safety.** I will not launch my rocket at targets, into clouds, or near airplanes, and will not put any flammable or explosive payload in my rocket.

9. **Launch Site.** I will launch my rocket outdoors, in an open area at least as large as shown in the accompanying table, and in safe weather conditions with wind speeds no greater than 20 miles per hour. I will ensure that there is no dry grass close to the launch pad, and that the launch site does not present risk of grass fires.
10. **Recovery System.** I will use a recovery system such as a streamer or parachute in my rocket so that it returns safely and undamaged and can be flown again, and I will use only flame-resistant or fireproof recovery system wadding in my rocket.
11. **Recovery Safety.** I will not attempt to recover my rocket from power lines, tall trees, or other dangerous places.

Table D-1. Launch site dimensions

Installed Total Impulse (N-sec)	Equivalent Motor Type	Minimum Site Dimensions (ft.)
0.00–1.25	1/4A, 1/2A	50
1.26–2.50	A	100
2.51–5.00	B	200
5.01–10.00	C	400
10.01–20.00	D	500
20.01–40.00	E	1,000
40.01–80.00	F	1,000
80.01–160.00	G	1,000
160.01–320.00	Two Gs	1,500

Glossary

aft closure
: Reloadable motors are typically made from three reusable parts and several disposable parts. The metal case has two threaded parts. The aft closure is the part that screws into the rear of the motor, holding the nozzle in place. See also *forward closure*.

aileron
: The control surface on a wing or wing-like fin.

air rocket
: Air rockets are propelled by a blast of pressurized air from the launcher. They can be pretty simple, like a soda straw rocket, or very complicated, like the air rocket glider.

airfoil
: The shape of the cross section of a wing or other flying surface. In model rocketry, the term airfoil also refers to the shape of a fin that is streamlined to reduce drag.

altimeter
: A small device, usually but not always electronic, for measuring the altitude of a rocket.

ammonium perchlorate
: A rocket fuel used in many high-performance rockets. It was used in the solid boosters for the Space Shuttle. It is also the most common fuel for high-power model rocketry.

amp
: An abbreviation for ampere, a unit of electric power. One ampere is 6.241×10^{18} electrons flowing past a point per second.

angle of attack
: The angle between a flight surface and the direction of flight. For a fin, this is the angle between the flat surface of the fin and the wind direction. For almost any rocket, it's the angle between the body tube and the wind. For a glider, each flight surface is usually looked at separately, although you can also look at the angle between the nominal direction of flight for the glider and the direction of the wind.

aphelion
: The point in an orbit when a body is farthest away from the Sun. Compare this to *apogee*.

apogee
: Technically the point in an orbit when a satellite is farthest from the earth, this term is used in model rocketry to describe the highest point in the rocket's flight.

arming key
: A mechanism to disable the electrical ignition system. This is called the arming key even if it happens to be a plug or switch. Also known as the launch key or safety key.

aspect ratio
: The square of the span of a fin divided by the area of the fin.

astronomical unit
: The astronomical unit, abbreviated AU, is a unit for measuring distances in the solar system. One AU is the average distance from the Sun to the Earth.

AU
: An abbreviation for astronomical unit.

axial deployment
: Describes recovery system deployment along the axis of the rocket, where the parachute or streamer generally pops out of the body tube, pushing the nose cone off in the process. See also *side deployment*.

azimuth angle
: The angle along the horizon from a reference point to an object. See also *theodolite*.

ballistic return
: A rocket falling the same way it flew, nose first and streamlined. Unsafe unless the rocket is very light compared to its size.

BALLS
: The annual research launch for the Tripoli Rocketry Association, Badd Ass Load Lifting Suckers (BALLS) is held each year in Black Rock, Nevada.

balsa
: A tropical tree famous for its lightweight wood. Pound for pound, it's stronger than steel! Balsa wood is available from many hobby and craft stores in blocks and planks. It is commonly used in model rocketry for nose cones, nose blocks, fins, and glider wings.

barometer
: An instrument used to measure air pressure.

base drag
: The drag created by the flat rear end of the rocket.

baseline
: The distance from the rocket launcher to an altitude tracker, or the distance between two altitude trackers.

black powder
: Also called gunpowder, this is a flammable substance used for rocket fuel in most low-power model rockets, and a few mid-power and high-power rockets.

blast deflector plate
: The part of the launcher that deflects the hot exhaust from a rocket motor so it does not scorch the launcher or the ground.

boat tail
: A transition piece that gradually reduces the diameter of a rocket, typically used at the base of a rocket to reduce base drag or below a fat payload bay to reduce drag from the payload bay.

body tube
: The long cylinder that forms the main structure of most rockets. Not all rockets have a body tube, of course.

booster motor
: The first stage of a two-stage rocket, or an intermediate stage of a multistage rocket. Usually designed to ignite the motor for the subsequent stage with no ejection delay or ejection charge. Booster motors have a 0 for the delay time in their designation; a C6-0 motor is a booster motor.

boundary layer
: When an object is moving through a fluid, the fluid right next to the object moves right along with the object. A short distance away, the fluid barely moves at all. The layer of fluid between the skin of the object moving through the fluid and the part of the fluid that doesn't move much is called the boundary layer.

buntline knot
: A good knot for tying shock cords to screw eyes, or any other situation where a string or rope is fastened to a fixed rod or loop.

CA glue
: Cyanoacrylate glue, also known as superglue, is a strong, fast-drying glue. In a thin form, it's used to strengthen porous parts, since it soaks in before drying. Thicker forms are great for gluing porous parts.

caliber
: In model rocketry, the ratio of the distance between the center of pressure and the center of gravity to the diameter of the rocket. If the rocket is 1" in diameter, and the distance from the center of pressure to the center of gravity is 2", the rocket has a caliber of 2.

center of gravity
: The point a body turns around when rotating, given no other forces are acting on it.

center of pressure
: The point where pressure from the wind is the same on either side. An object held in the wind at its center of pressure will not rotate.

centering ring
: A disk or short cylinder used to mount one tube inside another. In model rocketry, centering rings are typically made from paper or wood. They have many uses, the most common of which is to mount a motor tube in a larger body tube.

chamfer
: Technically a beveled edge between two surfaces, in model rocketry chamfer is often used as a verb describing the process of removing a beveled protrusion left when tubes are cut.

chord
: In aerodynamics, the chord is the length of a fin or wing from front to back, measured in the direction of air flow.

closed track
: When tracking with two theodolites, it's possible for the altitudes reported by the two trackers to differ. Under NAR contest rules, if the difference in altitude is less than 10% of the altitude, the altitude measurement is called a closed track, and the result is considered valid. The altitude used is the average of the altitudes reported by the two theodolites. See also *open track*.

cluster rocket
: A rocket with two or more motors on the same stage.

coefficient of drag
: A number that accounts for the shape of an object in the equation for drag.

composite motor
: In model rocketry, generally a motor that uses aluminum perchlorate as the primary fuel, as opposed to black powder.

control horn
: A small projection from a flight surface or a device mounted on a motor or rod to translate movement. For example, a control horn on a servo might be a projection that turns with the motor, allowing the turning motion of the motor to be changed to a push/pull motion. A control horn on an aircraft elevator allows a push/pull motion to rotate the control surface up or down.

cyanoacrylate glue
: See *CA glue*.

Darwin Awards
: Humorous awards given to people who remove themselves from the gene pool in particularly creative ways.

datum
: An arbitrary starting point used, among other things, to calculate the center of gravity of a rocket.

delay charge
: A smoke charge that burns after the rocket propellant is exhausted, providing a delay and usually a smoke trail until the rocket reaches apogee and the ejection charge fires.

delay time
: The time between the end of the thrust phase of a rocket and the firing of the ejection charge. The delay time is part of the motor designation. For example, a B6-4 motor has a 4-second delay time.

density
: The ratio of mass to volume. The density of air is the mass of the air divided by the volume of the air.

dihedral
: In aircraft, the angle between the wing and the horizontal flight plane.

drag
: The force on a model rocket caused by pushing through air. It acts in the opposite direction from the direction of travel of the rocket.

duck tape
: Duck tape is a cloth-based waterproof tape. It was originally invented in World War II, where it was used to tape ammunition boxes shut, giving them a waterproof seal. This earned the tape the affectionate name of duck tape, which stuck. Occasionally, people who are unaware of the origins of the name mislabel it as duct tape.

dynamic stability
: Dynamic stability refers to to how a rocket moves in the air, and how it reacts to changes in the forces pushing against it. This is particularly important in understanding weathercocking, where a rocket might be stable, but could still fly in unpredictable directions because of the combination of its stability, the wind, and the speed of the rocket. See also *static stability*.

ejection charge
: A small black powder charge generally used to eject the parachute. The ejection charge fires after the motor finishes the thrust phase and coasts to apogee.

ejection delay
: The time between the end of thrust from the motor and firing the ejection charge.

elevation angle
: The angle between the horizon and an object. See also *theodolite*.

elliptical nose cone
A nose cone whose cross section is a mathematical curve called an ellipse.

engine block
A device inserted in the motor tube to prevent the motor from moving forward. This is usually a small paper or fiber ring glued firmly into the body tube, but with a hole to allow the ejection charge to pass through.

engine hook
Used to hold a rocket motor firmly in place during flight. Usually made from small, flat pieces of steel, engine hooks hold the motor in place during flight while making it easy to remove and insert motors.

epoxy glue
Epoxy resins are mixed just before use. Rather than drying to form a bond, like wood glue or CA glue, epoxy bonds due to a chemical reaction between the two components that are mixed just before the glue is used. It is very strong and will set and dry even when the parts being glued are not exposed to air.

equilibrium
In terms of stability, equilibrium is when something is not moving away from its current state. For a rocket, this generally means the nose is pointed in the direction of flight, although technically it could be pointed in any direction, as long as it isn't flipping about. See *stable equilibrium* and *unstable equilibrium*.

featherweight recovery
Rockets that are extremely light compared to their size can use featherweight recovery, where the rocket may return ballistically, but is so light that it will not damage itself or anything it hits.

fillet
A narrow strip of glue added to a joint to increase the strength of the joint. These are often applied by running a bead of glue along the joint, then wiping your finger across the bead to make it smooth and regular.

fin
Small projections near the base of the rocket used to make the rocket stable. Feathers on an arrow are fins, for example.

fin root
The part of a fin that attaches to the body tube.

fineness ratio
The ratio of the length of a nose cone to the diameter of the body tube. A 3"-long nose cone on a 1"-diameter body tube has a fineness ratio of 3.

first stage
The initial stage in a multistage rocket.

flash
The excess plastic squeezed into the crack between segments of a multipart mold. In model rocketry, usually found on things like nose cones.

flashing
Loading software into the nonvolatile memory of a microcontroller.

fluid flow
The way objects move through liquids and gases.

form drag
Another name for pressure drag.

forward closure
Reloadable motors are typically made from three reusable parts and several disposable parts. The metal case has two threaded parts. The forward closure is the part that screws into the front of the motor, either to block the front end or to hold the ejection charge. See also *aft closure*.

friction
A force that acts in the opposite direction of motion or in the opposite direction of a propulsive force. Friction is what makes things stop—without friction or some other force, a moving object would keep moving indefinitely.

friction drag
The component of drag caused by the viscous effects of the surrounding fluid.

friction fit
In rocketry, the art of holding a motor in place in a minimum-diameter rocket by wrapping tape around the motor until it fits very tightly. This is only used so the ejection charge ejects the parachute or streamer instead of the motor. An engine block is used to prevent the motor from sliding forward in the rocket.

frontal area
In the drag equation, the frontal area is a representative area used to account for the size of the object. In model rocketry, this is the area of the cross section of the largest body tube plus the fins and launch lug.

fuselage
: The body of an airplane.

grain
: Cylindrical pellets of fuel used in reloadable motors.

Haack series nose cone
: A nose cone with a profile defined by a mathematical shape known as a Haack series. See *Von Kárman nose cone*.

heat shrink tubing
: A thin, insulated tube of plastic that will shrink when exposed to heat, usually supplied from a specialized heat blower. It is used to create an insulated cover on electric wiring.

high-power model rocketry
: Flying a rocket that requires FAA clearance before flights, but that is small enough not to require FAA approval of the individual rocket and flight. While there are a couple of exceptions, this typically includes rockets that use motors in the H to O classes.

hobby knife
: A cutting instrument with a blade that can be easily replaced when it becomes worn. X-ACTO makes the most widely available hobby knives, which can be found in almost any hobby or craft store.

igniter
: A device used to start combustion in a rocket motor, and sometimes to fire pyrotechnic charges to eject parachutes or deploy other recovery devices. All igniters in model rocketry are triggered by electricity.

igniter clips
: Small clips used to connect the launcher's wire to the igniter in the rocket motor.

imperial units
: The system of units commonly used in the United States; units like inches, pounds, and seconds.

impulse
: The force exerted by a rocket motor over a given time, usually measured in newton-seconds, where one newton is one kilogram-meter/second^2. Impulse is also frequently measured in pound-seconds.

induced drag
: When a fin or wing produces lift, either to lift an airplane or to rotate a rocket back to straight flight, some air slips over the tip of the fin or wing. This forms a vortex, spinning air that contributes to the drag on the rocket or airplane. This drag is called induced drag.

inertial forces
: In fluid flow, the forces from the molecules of the fluid bumping into an object. The ratio of inertial force to viscous force is the Reynolds number.

internal resistance
: The resistance exhibited by a battery when used in a circuit.

joule
: A unit of energy, equivalent to the kinetic energy of a one-kilogram mass moving at one meter per second. The units are kilogram-meter^2/second^2.

kinetic energy
: The energy of a body in motion, equal to $1/2\ m\ v^2$.

laminar flow
: The smooth flow of a fluid past an object with very little, if any, turbulence. See *turbulent flow*.

land shark
: A rocket that turns sideways just after launch and flies more or less horizontally. This can be caused by a marginally stable rocket, a rocket that isn't traveling fast enough when it reaches the end of the launch rail, or a structural failure.

launch cable
: The wire that runs from the launch controller to the launch pad.

Launch Control Officer (LCO)
: Directs the setup, organization, and operations at the launch site.

launch controller
: The electrical ignition system used to ignite the rocket motor.

launch key
: A plug or actual key used to enable the launch system. Also known as the arming key or safety key.

launch lug
: Attaches a rocket to a rod or rail to guide it for the first few feet of flight. Typically made from thin tubes for launch rods or button-shaped tabs for launch rails.

launch pad
: The structure that holds the launch rod, providing a stable launch mechanism for the rocket. Sometimes the term is used to include the launch controller, too.

launch rod
: Guides a rocket for the first few feet of flight. Sometimes a launch rail is used instead.

Launch Safety Officer (LSO)
: Another name for the Range Safety Officer. See *Range Safety Officer (RSO)*.

law of sines
: For a triangle, the ratio of the length of the side opposite the angle and the sine of the angle is a constant for all three combinations of angles and sides.

lawn dart
: A rocket that has arced over and returns to earth in a nose-first ballistic trajectory, without a working recovery system.

LCO
: Launch Control Officer.

LDRS
: The main annual TRA meet, Large and Dangerous Rocket Ships (LDRS) includes high-power launches and competitions like the Odd Rocks competition.

leading edge
: The part of the rocket that hits the air first, usually used when referring to a fin or wing. The leading edge of the fin, for example, is the edge closest to the nose cone.

low Reynolds number airfoil
: An airfoil designed for small wings. Under these conditions, the Reynolds number is still high compared to that of, say, a ship going through water, but it is lower than the Reynolds number for most wings.

low-power model rocketry
: For the purposes of this book, low-power rockets are those that use E or smaller motors and weigh one pound or less.

LSO
: Launch Safety Officer.

Mach number
: The speed as a fraction of the speed of sound. Mach 1.0 is the speed of sound, while Mach 2.0 is twice the speed of sound.

mass drive
: A concept proposed as a cheap way to launch material, especially from an airless body like the Moon. The mass drive fires the material from a ground propulsion system. It is then snagged in orbit or continues to propel itself with onboard rockets.

metric units
: Used in virtually every country except the United States, and used extensively in science and engineering in the US. Measurements are in units like meters, kilograms, and seconds. The metric system is technically known as the International System of Units, and is abbreviated as SI for the name of the standard in French, or MKS, which is short for meters-kilograms-seconds.

microcontroller
: A small, self-contained computer with a central processing unit, memory, and I/O of some kind. A microcontroller can be housed in a single chip, or might come in the form of a small circuit board.

mid-power model rocketry
: Flying a rocket too big to be considered a low-power model rocket, but not large enough to need FAA clearance before flights. In general, mid-power model rockets are made from heavy cardboard tubes with aircraft plywood fins, and use F or G motors.

minimum-diameter rocket
: A rocket whose body tube is just big enough to hold the motor.

misfire
: When the launch button is pressed and the rocket does not launch.

motor mount
: The assembly used to hold the rocket motor in the main body tube.

motor retention system
: The mechanism used to hold the motor in place. See *engine hook* and *engine block*.

multistage rocket
: Rockets often use two or more stages. Each stage has a motor, fins, and its own recovery system. After the motor in a lower stage completes its burn, the stage separates, allowing the upper stage to continue its flight without the weight and drag of the lower stage.

NAR
: The National Association of Rocketry.

NAR Pink Book
: The official set of rules for NAR contests. It is available online at the NAR website (*http://www.nar.org/pinkbook/*).

NARAM
: An annual, week-long NAR event featuring sport launches, contests, and vendors.

National Association of Rocketry
: Generally called NAR, this is the oldest and largest organization for model rocketry. NAR covers all forms of rocketry, from tiny 1/8A motors to massive O motors, but concentrates on low-power model rocketry.

newton
: A measure of force. One newton (N) is one kilogram-meter/second^2. It is equal to about 0.2248 pounds of force.

newton-second
: Abbreviated N-s, this corresponds to one newton of force applied for one second. The most common unit of measure for model rocket motor power.

Newton's Third Law of Motion
: For every action, there is an equal and opposite reaction.

nichrome wire
: A nickel and chromium alloy wire that heats when current is applied. Frequently used in igniters.

normal force
: A force at right angles to a direction. In model rockets, the normal force is used extensively in stability analysis, where it refers to a force that is at right angles to the main axis of the rocket.

nose-blow recovery
: A form of tumble recovery used for small, light rockets. The nose cone is ejected, remaining attached to the rocket with a shock cord, but there is no parachute or streamer. The shape of the rocket keeps it from returning to the ground ballistically.

nose cone
: The tip of most rockets, which gradually narrows to a point or small blunt end. In most cases, the nose cone is a separate plastic or balsa wood piece that slips into the top end of the body tube.

nozzle
: The part of the rocket motor that speeds up and directs the flow of ejected material. In the rockets from this book, this is generally some sort of clay or plastic constriction the gas shoots from.

O-ring
: A ring of material, usually something flexible like rubber, used as a pressure seal. O-rings are used in water rockets to seal the launcher and in reloadable rocket motors to prevent hot gases from reaching the metal rocket motor case.

odd rockets
: Also called oddrocs, these are rockets that just don't look like normal rockets. They can be anything from a rocket with strange fins to a flying pig.

ogive nose cone
: A nose cone with a shape that is a segment of a circle.

Ohm's law
: The relation between voltage, resistance, and current in a circuit: $V = IR$.

open track
: When tracking with two theodolites, it's possible for the altitudes reported by the two trackers to differ. Under NAR contest rules, if the difference in altitude is more than 10% of the altitude, the altitude measurement is called an open track and is not considered valid. See also *closed track*.

parabolic nose cone
: A nose cone with a cross section formed from a mathematical shape called a parabolic series.

parachute
: A large plastic or fabric umbrella-like device used to slow the descent of a model rocket.

pascal
: A unit used for pressure, stress, or tensile strength, a pascal is one kilogram/(meter second^2). The pascal-second, abbreviated Pa•s, is a unit used to measure viscosity in fluids.

perihelion
: The point in an orbit when an object is closest to the Sun.

Pink Book
: See *NAR Pink Book*.

planform
: The shape of the fin, such as rectangular or elliptical.

pop pod
: A pop pod attaches to something else, usually a glider. It holds a motor and recovery system, and is designed to pop off of the airplane or mother ship when the ejection charge fires.

power
: See *impulse*.

power series nose cone
: A nose cone with a profile in the shape of a mathematical curve called a power series.

pressure drag
: The component of drag resulting from air molecules bouncing off of the surface of a rocket. It is sometimes called form drag or profile drag.

profile drag
: Another name for pressure drag.

pylon
: An extension on an aircraft or rocket intended for attaching something.

pyrogen
: A flammable plastic that lights at a relatively low temperature but burns at a high temperature. It is typically used as the tip of an igniter, where it starts to burn due to the heat from an electric current flowing through a wire, and in turn ignites the black powder or aluminum perchlorate rocket fuel.

Range Control Officer (RCO)
: Another name for the Launch Control Officer. See *Launch Control Officer (LCO)*.

Range Safety Officer (RSO)
: Responsible for the overall safe operations at the launch site. The RSO has the final say on any safety question.

RCO
: Range Control Officer.

recovery wadding
: Flame-resistant material inserted into the body tube of a rocket to protect the parachute from the heat of the ejection charge.

reloadable motor
: A rocket motor that can be reused after a flight, like the boosters on the Space Shuttle. In model rocketry, most reloadable motors are made with an aluminum tube and brass or aluminum end caps, all of which are reusable. The motor liner, nozzle, various O-rings, and rocket fuel are disposable, and are cleaned from the case after each flight.

resistance
: In electronics, resistance is the opposition to passing electric current. See *Ohm's law*.

Reynolds number
: The ratio of inertial forces to viscous forces, the Reynolds number is a key property of liquids and gases used when computing drag.

rocket simulator
: See *simulator*.

rod whip
: When a long launch rod flexes back and forth as the rocket lifts off. It can be violent enough to whack the rocket at the end of the rod, causing serious damage. In general, 1/8" launch rods should be 3 or, at most, 4 feet long. A 3/16" launch rod is usually acceptable for lengths of up to about 6 feet and will work fine for low-power rockets, but you will want a thicker rod or launch rail to support heavy mid-power and high-power rockets.

root
: See *fin root*.

root chord
: The length of the fin that attaches to the body tube.

RSO
: Range Safety Officer.

rudder
: On an airplane, the rudder is the part that sticks up or down from the fuselage. It is also called the vertical stabilizer.

safety cap
: A blunt cap put on top of a launch rod to reduce the danger of injury if someone trips and falls on the launch rod.

safety interlock
: The part of a launch controller that disables the controller until a launch key is inserted.

sanding block
: A small piece of wood around which sandpaper is wrapped to allow sanding of flat surfaces.

Schrader valve
: The kind of valve commonly used on tires. The Schrader valve fits most common bicycle and car pumps.

screw eye
: A screw with a metal loop on the head, frequently used in model rockets to form an attachment point for parachutes and shock cords at the base of a balsa nose cone.

semichord
: The length from the center of the root of a fin to the center of the tip. This length is important for stability analysis.

servo
: A small package containing a motor and associated electronics that allow the motor to be positioned accurately based on a timed input signal. The kinds

of servos typically used in model rockets are designed for use in model airplanes.

shock cord
: Used to attach the parachute or nose cone to a rocket. Typically made from strips of rubber or elastic, shock cords on larger rockets may be made from nylon cord or rope.

shoulder
: The part of a nose cone or payload bay that slides into the body tube.

shoulder length
: The distance a nose cone or payload bay shoulder slides into the body tube.

shroud lines
: The strings on a parachute that connect the parachute to the rocket.

side deployment
: A parachute ejection method commonly used for water rockets. The parachute deploys through a door in the side of the rocket rather than pushing out of the top of the rocket.

simulator
: A computer program that takes the design of a rocket and predicts whether the rocket will be stable, how high and fast it will fly, what delay time is needed, and so forth.

skin friction drag coefficient
: This specialized drag coefficient is used when finding the friction drag on a rocket. It accounts for the roughness of the finish, which can cause the air flow to switch from laminar flow to turbulent flow, increasing the drag.

slug
: In biology, a small, slimy animal. In physics, a unit of mass in the imperial system. The units are pounds-seconds^2/feet, so one slug is the mass of an object that requires a force of one pound to accelerate it at a rate of one foot/second^2.

solid propellant motor
: Motors powered by chemical combustion of a material that includes both the propellant and an oxidizer. That means they don't need an external supply of air to burn. Typical solid propellants for model rocketry include black powder and ammonium perchlorate.

snap swivel
: Snap swivels are used at the end of shock cords and parachutes to make it easier to change the payload bay or parachute on a rocket. A snap swivel has a clasp at one end that opens and closes for easy removal, and a loop at the other end where the parachute or shock cord is tied. They can be found with fishing supplies.

span
: The distance from one fin tip to the opposite fin tip, assuming there are four fins on the rocket. It is also twice the distance from the fin tip to the center of the body tube. Used when calculating drag. When describing a fin in most other circumstances, the span is the distance from the fin tip to the root chord in a direction perpendicular to the body tube.

specific impulse
: A way to measure the efficiency of rocket fuel. It represents the force generated with respect to the material used in a given amount of time.

spectator
: Anyone at a launch site not specifically involved in the launch.

stability
: The tendency of the rocket to fly straight. We obtain stability by making sure the center of pressure for the rocket is about one to two body tube diameters behind the center of gravity.

stable equilibrium
: When a rocket is properly designed, it will fly in stable equilibrium. This means if something nudges the rocket so it is not pointed in the direction of flight, it will return itself to the proper orientation.

standard deviation
: Used for specifying error in this book, a standard deviation is a value from probability theory. For our purposes, if you measure something—say, the altitude of a model rocket—and give the standard deviation, you are claiming that the actual value has a 68% chance of being within one standard deviation. For example, if you say the altitude is 350±25 feet, you are saying the actual altitude has a 68% chance of being between 325 feet and 375 feet. Extending the concept, the actual altitude has a 95% chance of being within two standard deviations, or in the range 300 feet to 400 feet, and a 99.7% chance of being within three standard deviations, or 275 feet to 425 feet.

static stability

Static stability is determined by finding the center of pressure and center of gravity of a rocket. For a rocket to be stable, the center of pressure should be at least one body tube diameter behind the center of gravity. See also *dynamic stability*.

streamer

A long, thin ribbon of material used instead of a parachute, especially in light rockets that fly very high.

streamlining

In aerodynamics, streamlining refers to selecting a shape to minimize the force of drag.

subsonic region of flight

Literally, flight at less than the speed of sound. But since the transonic region is so important for understanding flight, when discussing aerodynamics, subsonic flight usually means flight below about Mach 0.8.

supersonic region of flight

Literally, flight faster than the speed of sound. In aerodynamics, though, supersonic flight often means flight above Mach 1.2, since the transonic region from Mach 0.8 to Mach 1.2 behaves very differently from speeds above and below this range.

tangent

The length of the opposite side of a triangle divided by the length of the adjacent side, where the opposite and adjacent sides must meet in a right angle.

tangent ogive nose cone

A nose cone with a profile formed from a segment of a circle; the place where the nose cone meets the body is tangent to the circle, forming a smooth joint.

theodolite

A device used to measure angles. It reports the angle from the horizontal to a point above or below the horizon (the elevation angle), and the angle from a reference point to an object along the horizontal plane (the azimuth angle). In model rocketry, two or more theodolites are used to optically track model rockets.

thrust

The force from a rocket motor that propels the rocket. Thrust is usually measured in newtons. A newton is 1 kg m/s^2.

thrust curve

A plot of the thrust from a motor as a function of time.

total impulse

A measure of the overall power of a rocket motor. It is the average thrust delivered by the motor times the number of seconds for which the thrust is delivered. Impulse is usually measured in newton-seconds (N-s), although it is occasionally measured in pound-seconds or other units.

total power

See *total impulse*.

tower launcher

Used for model rockets that do not have a launch lug. It uses three stiff rails or rods that form a cage around the rocket to guide it during the first few feet of flight. These rails or rods are sometimes adjustable for different body tube diameters.

TRA

The Tripoli Rocketry Association.

trailing edge

The part of the rocket—for example, the edge farthest from the nose cone—where the air leaves the rocket. Usually used when referring to a fin or wing.

transonic region of flight

Generally defined as Mach 0.8 to Mach 1.2, the speed at which the effects of subsonic and supersonic flight mix, causing flight behavior that is very different from that of flight well below or well above the speed of sound.

Tripoli Rocketry Association

Often abbreviated TRA or just Tripoli, this organization is dedicated to the advancement of high-power model rocketry. It publishes a set of launch rules adhered to by many ranges in the US.

tube coupler

A small tube that slips inside a body tube, used to join two shorter pieces to make a longer tube. Tube couplers are usually made from a denser, stronger paper material than body tubes. They are also used on payloads (so they slip into a body tube like a nose cone), in some motor mounts, and for all sorts of other creative tasks.

tumble recovery

Some rockets are light enough that they do not need a parachute or streamer, but still too heavy to return ballistically. Tumble recovery, where something changes the geometry of the rocket so it is no longer stable, is often used in such cases. The rocket tumbles as it falls back to earth, slowing it enough for a safe landing.

turbulent flow
: The flow of a fluid past an object with much mixing and swirling. See *laminar flow*.

two-stage rocket
: See *multistage rocket*.

unstable
: A rocket that does not fly properly is unstable. This usually means the center of gravity was not at least one body tube diameter in front of the center of pressure.

unstable equilibrium
: Unstable equilibrium occurs when something is in a particular state, such as a rocket with the nose pointed in the direction of flight, but a slight disturbance will start a process that will push it further and further from the equilibrium state. The rocket will be unstable.

vertical stabilizer
: On an airplane, the vertical stabilizer is the part that sticks up or down from the fuselage. It is also called the rudder.

viscous forces
: The forces from the molecules of the surrounding fluid sticking to an object. The ratio of inertial force to viscous force is the Reynolds number.

voltage
: The electric pressure in a circuit. See *Ohm's law*.

Von Kárman nose cone
: A special case of the Haack series nose cone, the Von Kárman nose cone is formed by setting one of the terms in the Haack series to zero.

vortex
: In aerodynamics, a vortex is a twirling motion of air, like that seen at the tip of a wing or fin that is producing lift.

weathercocking
: An overly stable rocket tends to turn into the wind very quickly. This phenomenon, called weathercocking, is particularly problematic if the rocket is traveling slowly as it leaves the pad and the wind is fairly strong. In severe cases, the rocket can end up in powered flight traveling in a horizontal direction; this is known as a *land shark*.

wetted surface
: The surface area of the rocket that is exposed to the air. A larger wetted surface generally means a large friction drag.

white glue
: White glue, such as Elmer's Glue, is similar to wood glue. It comes in plastic squeeze bottles. While wood glue is a bit better, white glue is a perfectly adequate substitute for wood glue when building model rockets. It's also a common supply item for schools, making it convenient for school rocketry projects.

wood glue
: Wood glue is great for gluing wood and paper. It forms a bond that is generally stronger than the original wood or paper. If you glue a fin to a body tube with wood glue, generally the body tube will tear or the fin will break, but the wood glue will remain intact.

zipper
: One of the failures that can occur when a parachute ejects when the rocket is traveling too fast is a zipper. This is when the parachute and shock cord don't fail, but the speed is high enough for the pull from the shock cord to rip the body tube open. This is rare in low-power model rockets, which tend to use elastic shock cords. It's more common in mid- and high-power rockets that use nylon shock cords.

Index

Symbols

A

B

C

D

E

F

G

H

I

J

K

L

M

N

O

S

T

U

V

W

X

Z

About the Author

Mike Westerfield has a deep interest in science and technology that shows up in his work and hobbies. Educated in physics and a programmer by profession, Mike started his career in the Air Force as the resident physicist on a classified satellite program.

He sold his car to get the money needed to buy an Apple II computer and started writing assemblers and compilers.

He has developed numerous compilers and interpreters, including APW, which Apple Computer shipped to developers for the Apple IIGS. He has worked on plasma physics simulations for Z-pinch machines, disease surveillance programs credited with saving the lives of Hurricane Katrina refugees, and advanced military simulations that protect our nation's most critical assets.

Mike currently runs the Byte Works, an independent software publishing and consulting firm specializing in scientific and technical programming, Bluetooth LE technology, iOS development, and cross-platform Java development. He lives with his wife in Albuquerque, New Mexico, where you will find him programming, flying rockets with the Albuquerque Rocket Society, and teaching scuba diving.

Colophon

The cover and body font is Benton Sans, the heading font is Serifa, and the code font is Bitstream Vera Sans Mono.